Traffic and Highway Engineering

Nicholas J. Garber
Lester A. Hoel
Department of Civil Engineering
University of Virginia

West Publishing Company
St. Paul   New York   Los Angeles   San Francisco

Production: Technical Texts, Inc./Sylvia Dovner

Cover Design: Roslyn Stendahl

Cover Photo: Jeff Hunter/The Image Bank

Printed in the United States of America

96 95   8 7 6 5 4 3

Library of Congress Cataloging-in-Publication Data

Garber, Nicholas J.
  Traffic and highway engineering / Nicholas J. Garber, Lester A. Hoel.
    p.   cm.
  Includes bibliographies and index.
  ISBN 0-314-60176-7
  1. Highway engineering.   2. Traffic engineering.   I. Hoel, Lester A.   II. Title.
TE145.G35   1988
625.7--dc19

                          87-24196
                          CIP

This book is dedicated to our daughters,

Allison, Elaine, and Valerie

and

Julie, Lisa, and Sonja

with the hope that one day they or their children will read this book and by so doing better understand the professional work that was their fathers'.

Contents

Preface

Traffic and Highway Engineering is intended primarily for use as an undergraduate text. It is designed for students in engineering programs where introductory courses in transportation, highway, or traffic engineering are offered. In most cases, these courses are taught in the third or fourth year to students who have little knowledge or understanding of the importance of transportation. Further, they are often unaware of the professional opportunities for engineers in this field.

Since the subject of transportation is a broad one, several approaches can be used to introduce this topic to undergraduates. One approach is to attempt to cover the technical details of all transportation modes—air, highway, pipeline, public, rail, and water—in an overview-type course. This approach assures comprehensive coverage, but tends to be superficial and lacking in depth. Another approach that can be effective is covering the subject of transportation by generic elements, such as the vehicle, guideway, terminal, cost, demand, supply, human factors, administration, finance, operations, and so forth. This approach poses problems for some students, however, because each of the modes does not always fit neatly within generic categories, and there may be fundamental differences between modes even within each generic topic. A third approach is to emphasize one mode, such as highways, airports, or railroads. There is considerable merit in focusing on one mode. While some material may be useful in other contexts, the material is specific and unambiguous, and the subject matter can be used later in practice.

In this textbook, we have followed this third approach and elected to emphasize the area of traffic and highway engineering. This focus represents the major area within the civil engineering field that deals with transportation, and undergraduates can relate directly to problems created by motor vehicle travel. We believe that such an approach is appropriate for an introductory transportation course. It provides an opportunity to present material that is not only useful to engineering students who are pursuing careers in transportation engineering but also interesting and challenging to those who intend to work in other areas. While written at an introductory level, this book can also serve as a reference for practicing transportation engineers and for students in graduate courses. Our overall objective is to provide a way for students to get into an area, develop a feel for what it is about, and thereby experience some of the challenges of the profession.

The introductory chapters present material that will help students understand the basis for transportation, its importance in society, and the extent to which transportation pervades our daily lives. Later chapters present information about the basic areas in which transportation engineers work: traffic operations, planning, design, construction, and maintenance. Thus, the chapters in this book have been categorized into five parts. These are Part 1, Introduction; Part 2, Traffic Operations; Part 3, Planning; Part 4, Location and Geometric Design; and Part 5, Materials and Pavement Design.

The topical division of the book organizes the material so that it may be used in two separate courses. For a course in transportation engineering, which is usually offered in the third year and in which the instructor prefers to primarily emphasize traffic and highway aspects, we recommend that material from Parts 1, 2, and 3 (Chapters 1–12) be covered. For a course in highway engineering, in which the emphasis is on highway location, design, materials, and pavements, we recommend that material from Parts 2, 4, and 5 (Chapters 3 and 13–20) be used.

The book is also appropriate for use in a two-semester sequence in transportation engineering, in which traffic engineering and planning (Chapters 3–12) would be covered in the first course, and highway design (Chapters 13–20) would be covered in the second course.

The authors are indebted to many individuals who assisted in reviewing various chapters and drafts of the manuscript. We especially wish to thank the following for their helpful comments and suggestions: Edward Beimborn, David Boyce, Christian Davis, Michael Demetsky, Richard Gunther, Jerome Hall, Jotin Khisty, Lydia Kostyniak, Michael Kyte, Winston Lung, Kenneth McGee, Carl Monismith, Ken O'Connell, Anthony Saka, Robert Smith, Egons Tons, Joseph Wattleworth, Hugh Woo, and Robert Wortman.

We also appreciate the help of those who typed various chapters of the manuscript and assisted with the graphics. Special thanks to Lucy Bates, Gayle Edwards, Shirley Fuller, Ruth Martin, Doris Roberts and Jeanne Roberts. We also thank our wives, Ada and Unni, for their encouragement and support throughout the years that this book was being written.

Nicholas J. Garber
Lester A. Hoel

PART 1 Introduction

Transportation is essential for a nation's development and growth. In both the public and private sector, opportunities for engineering careers in transportation are exciting and rewarding. Elements are constantly being added to the world's highway, rail, airport, and mass transit systems, and new techniques are being applied for operating and maintaining the systems safely and economically. Many organizations and agencies exist to plan, design, build, operate, and maintain the nation's transportation system.

1. The Profession of Transportation Engineering

Importance of Transportation

Overview of U.S. Transportation History

Employment in Transportation

Summary

Problems

Additional Readings

2. Transportation Systems and Organizations

Developing a Transportation System

Modes of Transportation

Transportation Organizations

Summary

Problems

References

Additional Readings

CHAPTER 1

The Profession of Transportation Engineering

For as long as the human race has been in existence, transportation has consumed a considerable portion of society's time and resources. The primary need for transportation is economic and has involved personal travel in search of food or work, travel for trade or commerce, travel for exploration, conquest, or personal fulfillment, or travel for the improvement of one's status in life. The movement of people and goods, which is what is meant by transportation, takes place to accomplish those basic objectives or tasks that require transfer from one location to another. For example, a farmer must transport produce to market, a doctor must see a patient at home or in the hospital, and a salesman must visit clients located throughout a territory. Every day millions of workers leave their homes and travel to factories or offices.

IMPORTANCE OF TRANSPORTATION

Tapping of natural resources and markets and maintaining a competitive edge over other regions and nations are closely linked to the quality of the transportation system. The speed, cost, and capacity of available transportation have a significant impact on the economic vitality of an area and the ability to make maximum use of its natural resources. Examination of most developed and industrialized societies indicates that they are noted for their high-quality transportation services. Nations with well-developed maritime systems (such as the British Empire in the 1900s) ruled vast colonies located around the globe. In more modern times, countries with advanced transportation systems, such as the United States, Canada, Japan, and Western Europe, are leaders in industry and commerce. Without the ability to transport manufactured goods, raw materials, and technical know-how, a country is simply unable to maximize the comparative advantage it may have in natural or human resources. A country such as Japan, for example, with little in the way of natural resources relies heavily on transportation in order to import raw materials and export manufactured products.

Transportation and Economic Growth

Good transportation in and of itself will not assure success in the marketplace, however, the absence of excellent transportation services will contribute to failure. Thus, if a society wishes to develop and grow, it must have a strong internal transportation system as well as excellent linkages to the rest of the world. Transportation is a derived demand, created by human needs and desires to move themselves or their goods from one place to another. It is a necessary condition for human interaction and economic survival.

The availability of transportation facilities can strongly influence the growth and development of a region or nation. Good transportation permits the specialization of industry or commerce, results in reduced costs for raw materials or manufactured goods, and increases competition between regions, resulting in lower costs and greater choice for the consumer. Transportation is also a necessary element for government services such as delivery of mail, defense of a nation, and retaining control of its territories. Throughout history, transportation systems, such as existed in the Roman Empire and exist now in the United States, were developed and built to assure easy mobilization of armies in the event of a national emergency.

Social Costs and Benefits of Transportation

The improvement of a region's economic position by virtue of improved transportation does not come without costs. Building vast transportation systems requires enormous resources of energy, material, and land. In major cities, transportation can consume as much as one-half of all the land area. An aerial view of any major metropolis will reveal vast acreage used for railroad terminals, airports, parking lots, and freeways. Transportation has other negative effects as well. Travel is not without danger; every mode of transportation brings to mind some major disaster, be it the sinking of the *Titanic*, the disaster of the zeppelin *Hindenburg*, the infrequent but dramatic passenger air crashes, and highway fatalities that each year claim about fifty thousand lives. In addition, transportation creates noise, spoils the natural beauty of an area, changes the environment, pollutes air and water, and consumes energy resources.

Society has indicated a willingness to accept some risks and change of the natural environment to gain the benefits from efficient transportation systems. A major task for the modern transportation engineer is to balance society's need for fast and efficient transportation with the costs involved so that the most efficient and cost-effective system is created. In carrying out this task, the transportation engineer must work closely with the public and with elected officials and must be aware of modern engineering practices to assure that the highest quality transportation systems are built consistent with available funds and accepted social policy.

There are also many social benefits of transportation that society values as well. Bringing medical and other services to rural areas and enabling people to socialize who live some distance apart are but a few of the nontangible benefits that transportation provides.

Transportation in the United States

The importance of transportation in the United States is illustrated by several statistics. Approximately 20 percent of the gross national product of the United States is accounted for by expenses related to transportation. About 90 percent of the energy used to drive transportation comes from petroleum resources and this represents 53 percent of all the petroleum used in the United States. Every U.S. citizen travels an average of 1 hr per day, and over 80 percent of eligible drivers have a driver's license, traveling an average of 12,000 mi per year. Transportation industries in the United States employ over 10 percent of the work force in jobs as diverse as gas station attendants, airline pilots, truck drivers, highway construction workers, barge operators, pipeline welders, railroad trainmen, bus drivers, and highway patrolmen.

OVERVIEW OF U.S. TRANSPORTATION HISTORY

Transportation in the United States could be the topic of a separate book. Transportation history covers a 200 year period and involves the development of many separate modes of transportation. Among the principal topics are travel by foot and horseback, automobile and truck travel, development of roads and highways, building of canals and inland waterways, the expansion of the West and construction of railroads, public transportation such as bus and metro systems in cities, and the development of air transportation, including the aircraft, airports, and air navigation facilities that compose the system.

Population Changes

In its early years, the United States was primarily rural with a population of about four million in the late 1700s. Only about two hundred thousand (5 percent) lived in cities; the remainder inhabited rural areas and small communities. That pattern remained until the early 1900s. During the twentieth century, urban population continued to increase such that at present about 75 percent of the U.S. population lives in urban or suburban areas. Although large cities have been declining in population, increases have occurred in suburban areas. These changes have a significant impact on the need for highway transportation. These trends are illustrated in Figure 1.1.

Early Roadbuilding and Planning

During the eighteenth century, travel was by horseback or in animal-drawn vehicles on dirt roads. As the nation expanded westward, roads were built to accommodate the settlers. In 1794 the first toll road, the Lancaster Turnpike, was built to connect the Pennsylvania cities of Lancaster and Philadelphia.

The nineteenth century brought further expansion of U.S. territorial boundaries and the population increased from 3 million to 76 million. Transportation continued to expand with the nation. In 1808 Secretary of the Treasury Albert Gallatin, who served under President Thomas Jefferson, recommended a

Figure 1.1   U.S. Population, 1790–1980

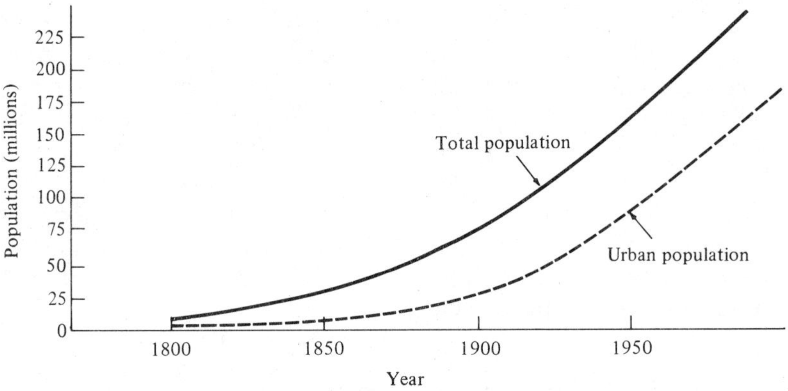

Source: Redrawn from *Statistical Abstract of the United States*, U.S. Bureau of the Census, Washington, D.C., 1985.

national transportation plan. Although his plan was not officially adopted, it did develop a strong case for investing in transportation by the federal government and was the basis for projects such as the Erie Canal that were later constructed. The remainder of the nineteenth century saw considerable activity, particularly in canal and railroad building.

The Short-Lived Canal Boom

An era of canal construction began in the 1820s when the Erie Canal was completed and other inland waterways were constructed. This efficient means of transporting goods was soon replaced by the development of the railroads, which was occurring at the same time. In 1840 the number of miles of canal and railroads were equal (3200 mi), but railroads, which could be constructed almost anywhere in this vast undeveloped land at a much lower cost, superceded the canal as a form of intercity transportation. Thus, after a short-lived period of intense activity, the era of canal construction came to an end.

The Railroad Era

The dominant mode of transportation during the late 1800s was the railroad, and railway lines spanned the entire continent (the first transcontinental railroad was completed in 1869). Railroads dominated intercity passenger and freight transportation from the late 1800s to the early 1920s. Railroad transportation saw a resurgence during World War II but has steadily declined since then due to the competitiveness of the automobile and trucking. Railroad mileage reached its peak of about 265,000 by 1915 but has been reduced to below 200,000 at present.

Transportation in Cities

Each decade has seen continual population growth within cities and with it the demand for improvements in urban transportation systems. City transportation began with horse-drawn carriages on city streets that later traveled on steel tracks. These were followed by cable cars, electric streetcars, underground electrified railroads, and bus transportation. City travel by public transit has been largely replaced by automobiles traveling on urban highways, although new rapid transit systems have been built recently in San Francisco, Washington, D.C., and Atlanta.

The Automobile and Interstate Highways

The invention and development of the automobile created a revolution in transportation in the United States during the twentieth century. No facet of American life has been untouched by this invention, and the automobile, together with the airplane, has changed the way we travel within and between cities. Only 4 automobiles were produced in the year 1895. By 1901, there were 8000 registered vehicles and by 1910 over 450,000 cars and trucks. Between 1900 and 1910, 50,000 mi of surfaced roads were constructed, but major highway building programs did not begin in earnest until the late 1920s. By 1920 more people traveled by private automobile than by rail transportation. By 1930 23 million passenger cars and 3 million trucks were registered and in 1956 Congress authorized a 42,500 mi interstate highway network. Over 95 percent of that system is now in existence.

The Birth of Aviation

Aviation was in its infancy at the beginning of the twentieth century, with the Wright brothers' first flight taking place in 1903. Both World War I and World War II were catalysts in the development of air transportation. The carrying of mail by air provided a reason for government support of this new industry. Commercial airline passenger service was beginning to grow, and by the mid-1930s coast to coast service was available. After World War II, the expansion of air transportation was phenomenal, and the technological breakthroughs developed during the war, coupled with the training of pilots, created a new industry that replaced both ocean-going steamships and passenger railroads. A summary of the historical highlights of transportation development is shown in Table 1.1.

Transportation Facilities Today

The nation's transportation system is now largely in place, having been built over a period of 200 years. However, much of this system, particularly the physical facilities, are becoming old and deteriorated. This is particularly true of the nation's highway system and railroads. In the remainder of the twentieth century, the United States will be involved in some construction of additional segments of existing systems, but considerable effort will be concentrated on rehabilitating and maintaining existing roads, bridges, runways, and terminals.

The United States has constructed one of the most extensive transportation systems in the world, and as new developments occur, they will be reflected in

Table 1.1   Milestones in U.S. Transportation History

1794: First toll road, the Lancaster Turnpike, is completed.

1807: Robert Fulton demonstrates a steamboat on the Hudson River. Within several years steamboats are operating along the East Coast, on the Great Lakes, and on many major rivers.

1808: Secretary of Treasury Albert Gallatin recommends a federal transportation plan to Congress, but it is not adopted.

1825: Erie Canal is completed.

1830: Operations begin on Baltimore and Ohio Railroad, first railroad constructed for general transportation purposes.

1838: Steamship service on the Atlantic Ocean begins.

1857: First passenger elevator in the United States begins operation, presaging high-density urban development.

1865: First successful petroleum pipeline is laid between a producing field and railroad terminal point in western Pennsylvania.

1866: Bicycles are introduced in the United States.

1869: Completion of first transcontinental railroad.

1883: First regular electric street car service.

1887: First daily railroad service from coast to coast.

1903: The Wright brothers fly first airplane 120 ft at Kitty Hawk, North Carolina.

1914: Panama Canal opens for traffic.

1915–18: Inland waters and U.S. Merchant fleet play prominent roles in World War I freight movement.

1916: Interurban electric rail mileage reaches peak of 15,580 mi.

1919: U.S. Navy and Coast Guard crew crosses the Atlantic in a flying boat.

1927: Charles Lindberg flies solo from New York to Paris.

1956: Construction of the 42,500 mi Interstate and Defense Highway System begins.

1959: St. Lawrence Seaway is completed, opening nation's fourth seacoast.

1961: Manned space flight begins.

1967: U.S. Department of Transportation established.

1969: Man lands on moon and returns.

1972: Bay Area Rapid Transit System is completed.

1981: Space Shuttle Columbia orbits and lands safely.

1982–89: Interstate Highway System nears completion.

Source: Adapted from *National Transportation: Trends and Choices (to the Year 2000)*, U.S. Department of Transportation, Washington, D.C., 1978.

the improvement and expansion of this system. However, a major part of the effort will be to keep the existing system in serviceable condition.

EMPLOYMENT IN TRANSPORTATION

Transportation is perhaps one of the broadest fields of employment as it involves many disciplines and many modes. In the United States there are almost 4,000,000 mi of paved roadway, of which 42,000 mi are multilane limited-access freeways connected as the U.S: interstate system. There are approximately 230,000 mi of railroads, over 10,000 airports, 26,000 mi of inland waterways, and 220,000 mi of pipeline. It has been estimated that approximately 20 percent of the country's gross national product is devoted to transportation and that one person in seven has transportation-related employment.

Transportation services are provided within and between cities by rail, air, bus, truck, and private automobile. The automobile is the major mode of travel between cities, and the movement of intercity passenger traffic by public carriers is mainly by air and to a much lesser extent by bus and rail. Within cities, passenger travel is by personal automobile, bus, and rail rapid transit. Freight movement is largely the province of trucking firms within cities, but rail, water, and pipelines provide a substantial portion of the freight movement over the continental United States. The professional skills required to plan, build, and operate this extensive transportation system require a variety of disciplines, including engineering, planning, legal, economic, management, and social sciences.

The physical distribution aspects of transportation, known as business logistics or physical distribution management, is concerned with the movement and storage of freight between the primary source of raw materials and the finished manufactured product. It is an integral part of business administration. The objective of business logistics is to minimize transportation costs so that a company's profits are maximized.

Vehicle design and manufacture is a major industry in the United States and involves the application of mechanical, electrical, and aerospace engineering skills as well as those of technically trained mechanics and other trades people.

The service sector provides jobs for drivers of vehicles, maintenance people, flight attendants, train conductors, bus drivers, and other necessary support personnel. Other professionals such as lawyers, economists, social scientists, and ecologists also work in the fields of transportation when their skills are required to draft legislation, for right-of-way acquisition, and to study and measure the impacts of transportation on the economy, society, and the environment.

Although a transportation system requires many skills and provides a wide variety of job opportunities, we will concentrate on the opportunities for civil engineers. They are primarily responsible for planning, design, construction, operation, and maintenance of the transportation system within the United States. It is the engineer's responsibility to ensure that the system functions efficiently from an economic point of view and meets external requirements concerning energy, air quality, safety, congestion, noise, and land use.

Specialties in Transportation Engineering

Transportation engineers typically are employed by the agency responsible for building and maintaining a transportation system, such as the federal, state, or local government, a railroad, or a transit authority. They also work for consulting firms who carry out the planning and engineering tasks for these organizations. During the past century, transportation engineers have been employed to build the nation's railroads, the interstate highway system, rapid transit systems in major cities, airports, and turnpikes. Each decade has seen a new national need for improved transportation services.

It can be expected that in the late 1980s and 1990s heavy emphasis will be placed on rehabilitation of the highway system, including its surface and bridges, as well as on devising means to ensure improved safety and utilization of the existing system through traffic control and systems management. Highway construction will be required in suburban areas. Building of roads, highways, airports, and transit systems is likely to accelerate in other less-developed countries, and the transportation engineer will be called on to furnish the services necessary to plan, design, build, and operate highway systems throughout the world. Each of the subspecialties within transportation engineering is described in the following paragraphs.

Planning

Transportation planning deals with the selection of projects for design and construction. The transportation planner begins by defining the problem, gathering and analyzing data, and evaluating various alternative solutions. Also involved in the process are forecasts of future traffic, estimates of the impact of the facility on land use, the environment, and the community, and a determination of the benefits and costs that will result if the project is built. The transportation planner investigates the physical feasibility of a project and makes comparisons among various alternatives to determine which one will accomplish the task at lowest cost, consistent with other criteria and constraints.

A transportation planner must be familiar with engineering economics and other means for evaluating alternative systems. The planner must also be knowledgeable in statistics and data-gathering techniques, as well as computer programming for data analysis and travel forecasting. The transportation planner must also be able to communicate with the public and his or her client and have the ability to write and speak well.

Design

Transportation design involves the specification of all features of the transportation system in such a manner that it will function smoothly, efficiently, and according to the laws of nature. The design process results in a set of detailed plans that can be used for estimating the facility costs and for carrying out its construction. For a highway, the design process involves the selection of dimensions for all geometrical features such as the longitudinal profile, vertical curves and elevations, the highway cross section such as pavement widths, shoulders, right-of-way, drainage structures, and fencing. It also involves the design of the pavement itself, including the structural requirement for base and subbase courses and the pavement material such as concrete or asphalt. Also involved in the design are bridges and drainage structures as well as the provision for traffic control devices, roadside rest areas, and landscaping. The highway designer must be proficient in civil engineering subjects such as soil mechanics, hydraulics, land surveying, pavement design, and structural design. The highway design engineer is primarily concerned with the geometric layout of the road, its cross section, paving materials and roadway thickness, and traffic control devices. Special appurtenances, such as highway bridges and drainage structures, are usually designed by specialists in these areas.

The most important aspect of the highway designer's work is to establish the standards that relate the speed of the vehicle with the geometric characteristics of the road. A balanced design is produced in which all elements of the geometrics of the highway—its curve radii, sight distance, superelevation, grade, and vertical curvature—are consistent with a chosen design speed such that if a motorist travels at that given speed, he or she can proceed safely and comfortably throughout the entire highway system.

Construction

Transportation construction is closely related to design and involves all aspects of the building process beginning with clearing of the native soil, preparation of the surface, placing the pavement material, and preparation of the final roadway for use by traffic. Originally, highways were built with manual labor assisted by horse-drawn equipment for grading and moving materials. Today, modern construction equipment is used for clearing the site, grading the surface, compacting the subbase, transporting materials, and placing the final highway pavement. Advances in construction equipment have made possible the rapid building of large highway sections. Nuclear devices for testing compaction on soil and base courses, laser beams for establishing line and grade, and specialized equipment for handling concrete and bridge work are all innovations in the construction industry. Large automatically controlled mix plants have been constructed, and new techniques for improving durability of structures and the substitution of scarce materials have been developed.

Traffic Operations

The operation of the nation's highway system is the responsibility of the traffic engineer. Traffic engineering involves an understanding of vehicle driver and pedestrian characteristics and the utilization of these to improve the capacity of streets and highways. All aspects of the transportation system are included after the street or highway has been constructed and opened for operation. Among the elements of concern are traffic accident analyses, parking, loading, design of terminal facilities, traffic signs, markings, signals, speed regulation, and highway lighting. The traffic engineer works to improve traffic flow and safety, using engineering methods that are reinforced by enforcement and education. Most traffic engineers work directly for municipalities or county governments, however, some are employed as consultants hired to solve specific problems.

Maintenance

Highway maintenance involves all work necessary to ensure that the highway system is kept in proper working order. Maintenance refers to work such as pavement patching, repair, and other actions necessary to maintain the roadway pavement at a desired level of serviceability. Maintenance also involves record keeping and data management of work activities and project needs and analyses of maintenance activities to determine that they are carried out in the most economical manner. Scheduling of work crews, replacement of worn or damaged signs, and repair of damaged roadway sections are important elements of maintenance management. The work of the civil engineer in the area of maintenance involves redesign of existing highway sections, economic evaluation of maintenance programs, testing of new products, and scheduling of manpower to minimize delay and cost. The maintenance engineer must also maintain an inventory of traffic signs and markings and ensure that they are in good condition.

The Challenge of Transportation Engineering

The transportation engineer is a professional who is concerned with the planning, design, construction, operations, and management of a transportation system. (See Figure 1.2.) As a professional, he or she must make critical decisions about the system that will affect the thousands of people who use it. The work depends on the results of experience and research and is challenging and ever changing as new needs emerge and new technologies replace those of the past. The challenge of the transportation profession is to assist society in selecting the appropriate transportation system consistent with its economic development, resources, and goals and to construct and manage the system in a safe and efficient manner.

Figure 1.2   The Profession of Transportation Engineering

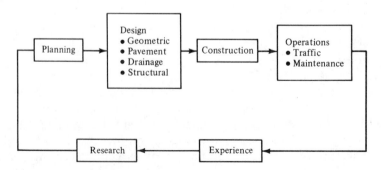

SUMMARY

Transportation is an essential element in the economic development of a society. Without good transportation, a nation or region cannot achieve the maximum use of its natural resources or the productivity of its people. Progress in transportation is not without its costs, both in human lives and environmental damage, and it is the responsibility of the engineer working with the public to develop quality transportation consistent with available funds and social policy and to mitigate damages. Transportation is a significant element in our national life, accounting for one-fifth of the gross national product and employing over 10 percent of the work force.

The history of transportation illustrates that how people move is affected by technology, cost, and demand. The past 200 years have seen the development of several modes of transportation: waterways, railroads, highway, and air. Each mode has been dominant at one point in time, and several have been replaced or have lost market share when a new mode emerged that provided a higher level of service at a competitive price.

The career opportunities in transportation that engineering students have are exciting. In the past, transportation engineers planned and built the nation's railroads, highways, mass transit systems, airports and pipelines. In the coming decades, additional system elements will be required as will efforts toward maintaining and operating the vast system that is in place in a safe and economical manner.

In Chapter 2 we describe in greater detail the elements of transportation systems and how the supply of transportation services interacts with demand to produce the volume of freight and passenger traffic that occurs in various modes. We also discuss how transportation is organized to produce various systems and services.

PROBLEMS

1–1 To illustrate the importance of transportation in our national life, for a one-week period, clip out several transportation-related articles that appear in your local newspaper. Discuss the issue involved and explain why the item was newsworthy.

1–2 Arrange an interview (in person or by telephone) with a transportation professional in your city or state (that is, someone working for a city, county, or state highway, transit, or rail agency). Attempt to learn about the job he (or she) performs, why he entered the profession, and what he sees as future challenges in his field.

1–3 Keep a diary of the trips you make for a period of 3–5 days. Record the purpose of each trip, how you traveled, and the trip time. From this data, try to extrapolate the trip-making characteristics for all students at your university.

1–4 Identify and describe one significant transportation event that occurred in your city or state (for example, the first year that commercial aviation began or the last year that trolley service was offered). Discuss the significance of this event to your city or state.

1–5 For a city in your state, describe how transportation influenced its initial settlement and subsequent development.

1–6 Make an estimate of the number of miles of paved roadway in your city or state. How many miles are interstate highways?

1–7 Estimate the number of motor vehicles in your city or state and the total number of miles driven each year. What is the total revenue raised for each 1 cent/gallon tax?

1–8 Estimate the number of railroad trains that pass through your city each day. What percentage of these are passenger trains?

1–9 Review the classified section of the telephone directory and identify ten different jobs or industries that are related to transportation.

1–10 Estimate the proportion of your monthly budget that is spent on transportation.

ADDITIONAL READINGS

Bowersox, Donald J., Pat J. Calabro, and George D. Wagenheim, *Introduction to Transportation*, Macmillan Publishing Co., New York, 1981.

Haefner, Lonnie E., *Introduction to Transportation Systems*, Holt, Rinehart and Winston, New York, 1986.

Hay, William W., *An Introduction to Transportation Engineering*, John Wiley & Sons, New York, 1977.

Hazard, John L., *Transportation: Management, Economics, Policy*, Cornell Maritime Press, Cambridge, Maryland, 1977.

U.S. Department of Transportation, Federal Highway Administration, *America's Highways 1776–1976*, U.S. Government Printing Office, Washington, D.C., 1976.

U.S. Department of Transportation, Office of Transportation Planning, *National Transportation: Trends and Choices (to the Year 2000)*, U.S. Government Printing Office, Washington, D.C., 1978.

Wright, Paul H., and Radnor J. Paquette, *Highway Engineering*, John Wiley & Sons, New York, 1987.

Yu, Jason, *Transportation Engineering: Introduction to Planning, Design and Operations*, Elsevier, North Holland, 1982.

CHAPTER 2

Transportation Systems and Organizations

The transportation system in a developed nation consists of an aggregation of vehicles, guideways, terminal facilities, and control systems that move freight and passengers. These systems are usually operated according to established procedures and schedules in the air, on land, and on water. The set of physical facilities, control systems, and operating procedures referred to as the nation's transportation system is not a system in the sense that each of its components is part of a grand plan or was developed in a conscious manner to meet a set of specified regional or national goals and objectives. Rather, the system has evolved over a period of time and is the result of many independent actions taken by the private and public sectors who act in their own or the public's interest.

Each day, decisions are made that affect how transportation services are used. The decision of a firm to ship its freight by rail or truck, of an investor to start a new airline, of a consumer to purchase an auto, of a state or municipal government to build a new highway or airport, of Congress to deny support to a new aircraft, and a federal transportation agency to approve truck safety standards are but a few examples of how transportation services evolve and a transportation system takes shape.

DEVELOPING A TRANSPORTATION SYSTEM

Over the course of a nation's history, attempts are made to develop a coherent transportation system, usually with little success. A transportation plan for the United States was proposed by Secretary Gallatin in 1808, but this and similar attempts have had little impact on the overall structure of the U.S. transportation system. Even the interstate highway system, which is national in scope, failed to recognize or account for the impact on other transportation modes or on urbanization. The creation of the U.S. Department of Transportation (DOT) in 1967 had the beneficial effect of focusing national transportation activities and policies within one cabinet-level agency, and in turn, many states followed by forming their own transportation departments.

The Interstate Commerce Commission (ICC), created in 1887 to regulate the railroads, was given additional powers in 1940 to regulate water, highway, and rail modes, preserving the inherent advantages of each and promoting safe, economic, and efficient service. As in the case of the ICC, the intent of Congress was to develop, coordinate, and preserve a national transportation system; however, the inability to implement vague and often contradictory policy guidelines has not helped to achieve desired results. More recently, regulatory reform has been introduced, and transportation carriers are developing new and innovative ways of providing services.

Advantages and Complementarity of Modes

The transportation system that evolves in a developed nation may not be as economically efficient as one that is developed in a more analytical fashion, but it is one in which each of the modes usually complement one another in carrying the nation's freight and passengers. A business trip across country may involve travel by taxi, airplane, and auto, and transportation of freight often requires trucks for pick up and delivery and railroads for long-distance hauling.

Each mode has inherent advantages of cost, travel time, convenience, and flexibility that make it "right for the job" under a certain set of circumstances. The automobile is considered to be a reliable, comfortable, flexible, and ubiquitous form of personal transportation for many people. However, when distances are great and time is at a premium, air transportation will be selected, supplemented by the auto for local travel. If cost is important and time is not at a premium or an auto is not available, then the intercity bus may be used.

Selecting a mode to haul freight follows a similar approach. Trucks have an advantage of flexibility and the ability to provide door-to-door service. They can carry a variety of parcel sizes and usually can pick up and deliver to meet the customer's schedule. Waterways can ship heavy commodities at low cost but at slow speeds and only between points where a river or canal exists. Railroads and airlines can haul an immense variety of commodities between any two points but usually require truck transportation to deliver the goods to a freight terminal or its final destination. In each instance, a shipper must decide if the cost and time advantage is such that the goods should be shipped by truck alone or by a combination of truck and rail or air.

In recent years, many industries have been trying to reduce their parts and supplies inventories, preferring to ship them from the factory when needed rather than stockpiling them in a warehouse. This practice has meant shifting transportation modes from rail to truck. Rail shipments usually are made once or twice a week in carload lots, whereas truck deliveries can be made in smaller amounts and on a daily basis, depending on demand. In this instance, lower rail freight rates do not compete with truck flexibility since the overall result of selecting trucking is a cost reduction for the industry.

Interaction of Supply and Demand

The transportation system that exists at any point in time is the product of several factors that act on each other. These are (1) the state of the economy, which produces the demand for transportation, and (2) the extent and quality of the system that is currently in place, which constitutes the supply of transportation facilities and services. In periods of high unemployment or rising fuel costs, the demand for transportation tends to decrease. On the other hand, if a new transportation mode is introduced that is significantly cheaper to use than those modes that existed in the past, then the demand for the new mode will increase, decreasing demand for the existing mode.

These ideas can be illustrated in graphic terms by considering two curves, one describing the demand for transportation at a particular point in time and the other describing how the available transportation service or supply is affected by the volume of traffic that uses the system.

The curve in Figure 2.1 shows how demand in terms of vehicles/hr could vary with cost. The curve is representative of a given state of the economy and the present population. As is evident, if the transportation cost, C, decreases, then, since more people will use it at a lower cost, the volume, v, will increase. In Figure 2.1, the cost is \$.75/mi and the traffic volume per day is 6000. If cost is decreased to \$.50/mi, then the volume per day increases to 8000. In other words, this curve shows us what the demand will be for our product (transportation) under a given set of economic and social conditions.

Demand can only occur if transportation services are available between the desired points. Suppose the demand we show in Figure 2.1 represents the desire to travel between the mainland of Florida and an island located off the coast and currently inaccessible, as shown in Figure 2.2.

If a bridge is built, people will use it but the amount of traffic will depend on cost. The cost to cross the bridge will depend on the bridge toll and the travel time for cars and trucks. If only a few vehicles cross, then little time is lost waiting at a toll booth or in congested traffic. However, as more and more cars and trucks use the bridge, the time to cross will increase. Lines will be long at the toll booth,

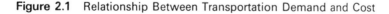

Figure 2.1 Relationship Between Transportation Demand and Cost

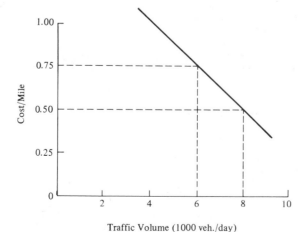

Traffic Volume (1000 veh./day)

Figure 2.2   Location of a New Bridge Between the Mainland and an Island

and there might also be traffic congestion at the other end. The curve in Figure 2.3 illustrates how the cost of using the bridge could increase as the volume of traffic increases, assuming that the toll is $.25/mi. In this figure, if the volume is less than 2000 per day, there is no delay due to traffic congestion. However, as traffic volumes increase beyond 2000 per day, delays occur and the travel time increases. Since "time is money," we have converted the increased time to cost/mile. If 4000 units per day use the bridge, the cost is $.50/mi and at 6000 units/day, the cost is $.75/mi.

Figure 2.3   Relationship Between Transportation Supply and Cost

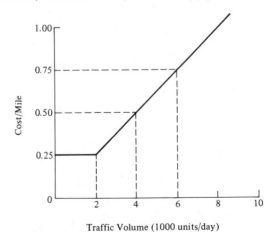

Figure 2.4   Equilibrium Volume for Traffic Crossing Bridge

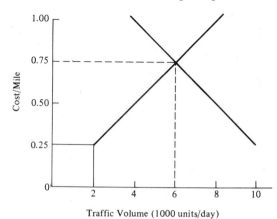

Traffic Volume (1000 units/day)

The curve in Figure 2.3 shows how much transportation volume we can supply at various levels of cost to the traveler.

We can now use the two curves to determine what volume can be expected to use the bridge. This will occur when the demand curve intersects the supply curve, because any other value of volume will create a shift in demand either upward or downward, until the equilibrium point is reached. If the volume increased beyond the equilibrium point, cost would go up and demand would drop. Likewise, if the volume dropped below equilibrium, costs would go down and demand would increase. (See Figure 2.4.) So in both instances we are approaching equilibrium. In this example, the number of units crossing the bridge would be 6000/day. Obviously, we can raise or lower the traffic volume by changing the toll.

Forces That Change the Transportation System

At any point in time the nation's transportation system is in a state of equilibrium as expressed by the traffic carried (or market share) for each mode and the levels of service provided (expressed as travel attributes such as time, cost, frequency, and comfort). This equilibrium is the result of the market forces (state of economy, competition, costs and prices of service), government actions (regulation, subsidy, promotion) and transportation technology (speed, capacity, range, reliability). As these forces shift over time, the transportation system changes as well, creating a new set of market shares (demand) and a revised transportation system. For this reason, the nation's transportation system is in a constant state of flux, causing short-term changes due to immediate revisions in levels of service (such as raising the tolls on a bridge or increasing the gasoline tax) and long-term changes in life styles and land-use patterns (such as moving to the suburbs after a highway is built or converting auto production from large to small cars).

If gasoline prices were to increase significantly, there would be an immediate shift of long-haul freight from truck to rail. In the long run, were petroleum prices to remain high, we might see shifts to coal or electricity or to more fuel-efficient trucks and autos.

Government actions also influence transportation equilibrium to a great extent. The federal government's decision to build the national interstate system affected the truck–rail balance in favor of truck transportation. It also encouraged long-distance travel by auto and was a factor in the decline of intercity bus service to small communities.

Technology has also contributed to substantial shifts in transportation equilibrium. The most dramatic has been the introduction of jet aircraft, which essentially eliminated passenger train travel in the United States and passenger steamship travel between the United States and the rest of the world.

MODES OF TRANSPORTATION

The U.S. transportation system today is a highly developed, complex network of modes and facilities that furnishes shippers and travelers with a wide range of choices in terms of services provided. Each mode offers a unique set of service characteristics in terms of travel time, frequency, comfort, reliability, convenience, and safety. The term *level of service* is used to describe the relative value of these attributes. The traveler or shipper must compare the service offered with the cost in order to make trade-offs and mode selection. Furthermore, a shipper or traveler can decide to use a public carrier or to use private (or personal) transportation. For example, a manufacturer can ship goods through a trucking firm or with company trucks; a homeowner who has been relocated can hire a household moving company or rent a truck, and a commuter can elect to ride the bus to work or drive a car. Each of these decisions involves a complex set of factors that require trade-offs between cost and service.

Freight and Passenger Traffic

The principal modes of intercity freight transportation are highways, railroads, air, water, and pipeline. Traffic carried by each mode, expressed as ton-miles or passenger miles, has varied considerably over the 56 year period between 1930–1986. These changes are illustrated in Figures 2.5 and 2.6.

Railroads, which accounted for 61 percent of freight traffic in 1930, carried only 36 percent by 1986. Pipelines have steadily increased their share of freight traffic from 9 percent in 1940 to 23 percent by 1986. From 1930 to 1980, water transportation has remained relatively constant, although traffic on rivers and canals has increased while Great Lakes traffic has declined. Trucking has steadily increased each year from 10 percent in 1940 to 25 percent by 1986. Air freight is an important carrier for high-value goods but is insignificant on a ton-mile basis, representing only about 0.29 percent of the total.[1]

The four principal carriers for freight movement (rail, truck, pipeline, and water) each account for between 15 percent and 37 percent of the total. The railroad's share is still highest on a ton-mile basis but has been reduced significantly due to competition from truck and pipeline. The railroads have lost traffic

Figure 2.5 Volume of U.S. Domestic Freight Traffic

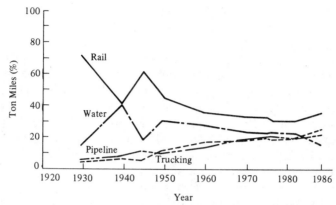

Source: Redrawn from *Transportation in America*, 5th ed., Transportation Policy Associates, Washington, D.C., March 1987.

due to the advances in trucking technology and pipeline distribution. Government policies that supported highway and waterway improvements were also a factor. Subsequent to World War II, long-haul trucking was possible as the U.S. highway system developed. As petroleum became more widely used, construction of a network of pipelines for distribution throughout the nation was carried out by the oil industry.

The distribution of passenger transportation is much different than for freight, with one mode—the automobile—accounting for over 80 percent of all domestic intercity passenger miles carried in the United States. (See Figure 2.6.) With the exception of World War II years, when auto use dropped to 64 percent of passenger miles by 1945 and rail increased to 27 percent, the automobile has

Figure 2.6 Volume of U.S. Intercity Passenger Traffic

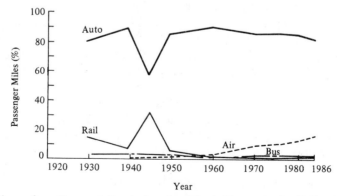

Source: Redrawn from *Transportation in America*, 5th ed., Transportation Policy Associates, Washington, D.C., March 1987.

accounted for between 80 percent and 90 percent of all passenger miles carried. The remaining modes—air, bus, and rail—share a market representing about 19 percent of the total (in 1986) distributed, approximately 17 percent by air, 1.3 percent by intercity bus, and less than 1 percent by rail.[2]

Of the four transportation carriers for intercity passenger movement, two—air and auto—are dominant, representing over 97 percent of all intercity passenger miles. If the public modes (rail, bus, and air) are considered separately from auto, a dramatic shift in passenger demand is evident. The passenger railroad, which accounted for 63 percent of all revenue traffic in 1930, has declined to less than 4 percent in 1985, with most of this traffic occurring in the corridor between Washington, D.C., New York, and Boston. Intercity bus transportation has also declined as a percentage of the total; buses now serve a market consisting primarily of passengers who cannot afford to fly or drive. As air fares have decreased with deregulation, bus travel has declined. The most dramatic increase has occurred in air transportation, which represents almost 90 percent of all intercity passenger miles by public modes. In cities, buses are the major public transit mode, with the exception of those larger urban areas that have rapid rail systems.

Public Transportation

Public transportation is an important element of the total transportation services provided within large and small metropolitan areas. A major advantage of public transportation is that it can provide high-capacity, energy-efficient movement in densely traveled corridors. It also serves medium and low density areas by offering an option for auto owners who do not wish to drive, and an essential service to those without access to an automobile—school children, senior citizens, single auto families, and others who may be economically or physically disadvantaged.

For most of this century, public transportation was provided by the private sector. However, increases in auto ownership, shifts in living patterns to low density suburbs, the relocation of industry and commerce away from the central city, coupled with changes in life styles (which have been occurring since the end of World War II), have resulted in a steady decline in transit ridership. Since the early 1960s, most transit services have been provided by the public sector. Revenues from fares no longer represent the principal source of income, and over a 25–30 year period, the proportion of funds for transit provided by federal, state, and local governments has steadily increased.

Public transportation is a generic term used to describe any and all of the family of transit services available to urban and rural residents. Thus, it is not a single mode but a variety of traditional and innovative services, which should complement each other to provide systemwide mobility. Modes included within the realm of public transportation are

- Mass transit, characterized by fixed routes, published schedules, and vehicles, such as buses and light rail or rapid transit, that travel designated routes with specific stops

- Paratransit, characterized by more flexible and personalized service than conventional fixed-route, fixed-schedule services, available to the public on demand, by subscription or on a shared-ride basis
- Ridesharing, characterized by two or more persons traveling together by prearrangement, such as carpool, vanpool, buspool, or shared ride taxi

A study by the American Association of Highway and Transportation Officials examined the future of public transportation in the United States. Six key factors were identified in establishing public transportation's role in the 1990s: (1) funding alternatives, (2) management alternatives, (3) public–private sector joint efforts, (4) political cooperation, (5) technological advances, and (6) industry involvement.[3]

Funding alternatives include both raising revenue and controlling costs by revising fare and parking policies, increasing productivity levels (controlling labor costs, furnishing realistic levels of service, and fostering support by private carriers), and obtaining funds from a wide range of public and private sources.

Management alternatives are aimed at providing an acceptable level of service at a reasonable cost. These general objectives are value laden with political overtones, nonetheless, management must direct its efforts toward increasing marketing, research, strategic planning, creative financing, and innovative service. Special emphasis will be required to assure greater flexibility, dedicated sources of funding, management training, and private sector involvement.

Public–private sector joint efforts are seen as an innovative approach to developing a healthier transit industry. Privatization can occur in a variety of ways, including joint sharing of transit costs, benefit sharing, private management services, service contracting, deregulation, entrepreneurial activities, and cooperative arrangements such as commuter and mobility clubs or volunteer organizations.

Political cooperation and support are essential if public transportation is to carry out its mission in an effective manner. The commitment of the governor and state legislature to public transportation—and transportation generally—is a vital element for success. Similar cooperative efforts at both the federal and local levels are necessary to assure viable public transportation services in the future.

Technological advances in bus, rail, or ferry modes have not been dramatic in recent years nor are major breakthroughs expected. Rather, innovations are likely to occur in system design and operating procedures. Although only marginal improvements can be expected from advances in vehicle design, payoff should result by making systems more productive and attractive. In this area, wide participation is possible by transit management, the users, and government or private organizations.

Industry involvement in public transportation occurs through several national organizations, and collectively they can help influence key areas of concern, including funding, cost effectiveness and productivity, public–private cooperation, coordination, community relations, and urban planning and development. These organizations (described at the end of this chapter) are the

American Public Transit Association (APTA), the American Association of State Highway and Transportation Officials (AASHTO), the Urban Mass Transportation Association (UMTA), and the Federal Highway Administration (FHWA).

The future of public transportation is expected to include the following elements.

1. As the population ages, the need for public transportation should increase, but mobility will not be as great as desired due to costs of providing service.
2. Less federal funding will be available, placing a greater burden on state, local, and private sources.
3. Increased involvement of the private sector should result in greater management flexibility as well as cost containment.

Little in the way of new technology is expected but system innovations are likely. Increased involvement in public transportation at all levels should result in more effective support by state and local government.

Highway Transportation

Since this textbook emphasizes highway transportation, a brief discussion of this element of the transportation system follows. Further detail is contained in

Figure 2.7 Total Road and Street Mileage in the United States by Surface Type, 1900–1986

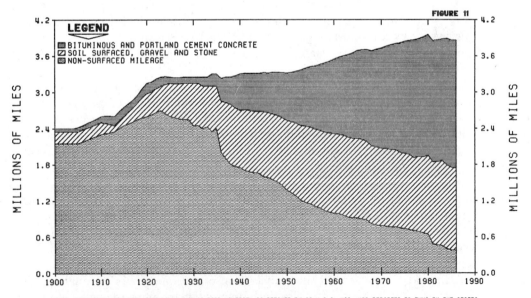

NOTE: BEGINNING WITH 1981 DATA, ONLY PUBLIC ROAD MILEAGE, AS DEFINED BY 23 U.S.C. 402, WAS REPORTED TO FHWA BY THE STATES. PRIMITIVE ROADS AND OTHERS NOT OPEN TO PUBLIC TRAVEL OR MAINTAINED BY PUBLIC AUTHORITY, WHILE INCLUDED IN MILEAGE TOTALS PRIOR TO 1981, ARE NOT REFLECTED IN THE DATA, THEREAFTER. 1986 DATA ARE ESTIMATED.

Source: Reproduced from *Selected Highway Statistics and Charts 1985*, U.S. Department of Transportation, Washington, D.C., 1985.

Chapter 14. Descriptions of other modes can be found in the Additional Readings cited at the end of the chapter. Highway transportation is the dominant mode in passenger travel and one of the principal freight modes. The U.S. highway system comprises approximately 3.9 million miles of highways, ranging from high-capacity, multilane freeways to urban streets to unpaved rural roads. Figure 2.7 illustrates the total highway mileage in the United States as a function of roadway surface. The total number of roadway miles has not increased greatly since 1920; however, the quality of the roadway surface and its capacity has shown a dramatic improvement since 1920.

The Federal Highway System

The federal-aid system, which includes the interstate and other federal-aid routes, consists of a network of roads totaling approximately 930,000 mi. These roads are classified as rural or urban and as arterials or collectors. Urban roads are located in cities of 25,000 or greater population and serve areas of high density land development. Urban roads are primarily used for commuting and shopping trips. Rural roads are located outside of cities and serve as links between population centers.

Arterial roads are intended to serve travel between areas and provide improved mobility. Collector roads and minor arterials serve a dual purpose of mobility and access. Local roads, supported by state and local funds, serve primarily as access and for short trips. Arterial roads have higher speeds and capacities, whereas local roads, which serve abutting land uses, and collector roads are slower and not intended for through traffic. A typical highway trip begins on a local road and continues on to a collector and then a major arterial. This hierarchy of road classification is useful in allocating funds and establishing design standards. Typically, local and collector roads are paid for by property owners and through local taxes, whereas collectors and arterials are paid for jointly by state and federal funds.

The interstate highway system consists of 42,500 mi of limited-access roads. This system, which is almost complete, is illustrated in Figure 2.8. The purposes of the interstate system are to connect major metropolitan areas and industrial centers by direct routes, to connect the U.S. highway system with major roads in Canada and Mexico, and to provide a dependable highway network to serve in national emergencies. Each year about 2000 route miles of the national highway system become 20 years old, which is its design life, and a considerable portion of the system must undergo repair, resurfacing, and bridge replacement. Ninety percent of the funds to build the interstate system are provided by the federal government, whereas 10 percent are provided by states. The completion of the system was originally planned for 1975, but delays caused by controversies, primarily in the urban sections of the system, have caused the completion date to be extended to 1990.

Intercity Bus Transportation

Intercity bus transportation services have benefited from the interstate highway system. Buses provide the greatest nationwide coverage of any public carrier by

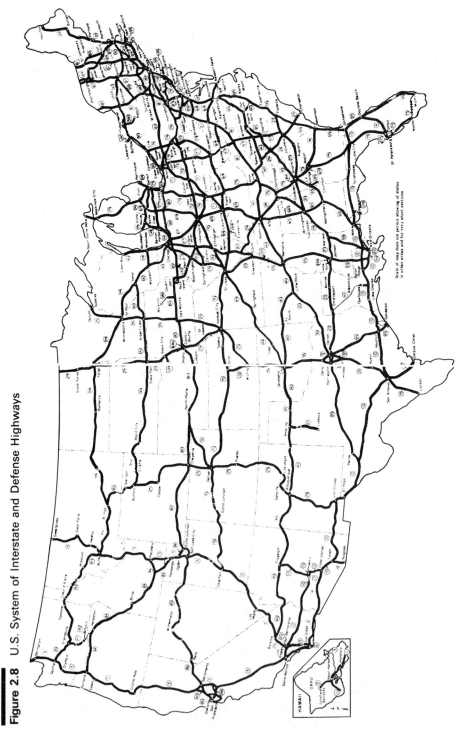

Figure 2.8 U.S. System of Interstate and Defense Highways

Source: Reproduced from *America on The Move!*, U.S. Department of Transportation, Washington, D.C., September 1984.

Table 2.1 Average Energy Efficiency by Mode

	Passenger Miles per Gallon	Seat Miles per Gallon
Auto (avg)	30	100
Intercity bus	110	300
Train	50	210
Airplane	16	52

Source: Adapted from *National Transportation: Trends and Choices (to the Year 2000)*, U.S. Department of Transportation, Washington, D.C., 1978.

connecting about 15,000 cities and towns with bus routes. Intercity buses carry more passengers than any other common carrier (about 60 percent of the total), but since these trips are relatively short, averaging about 110 miles, bus transportation represented only about 11 percent of total passenger miles by public carriers in 1980. Intercity bus service is provided by private companies, the principal ones being Greyhound and Trailways, now merged into one company. They operate at a profit with little or no support from state or federal governments.

Bus transportation is a highly energy-efficient mode. Table 2.1 lists the average number of passenger miles and seat miles per gallon for various transport modes and shows that buses are a superior form of transportation from an energy point of view. Buses are also very safe. Their accident rate of 12 fatalities per 100 billion passenger miles is over 100 times better than automobiles.

In spite of the positive characteristics of safety and energy efficiency, bus travel is generally viewed unfavorably by the riding public. Buses are slower and less convenient than other modes and often terminate in downtown stations that are located in the less attractive parts of the city. Other factors such as lack of through ticketing, comfortable seats, and systemwide information, which the riding public is accustomed to receiving when traveling by air, reinforce the overall negative image of intercity bus transportation.

Truck Transportation

The nation's highway system also carries about 25 percent of all intercity ton-miles of freight, which represents two-thirds of intercity freight revenue. The truck population is very diverse in terms of size, ownership, and use. Figure 2.9 illustrates the distribution of truck sizes and use. Most vehicles registered as trucks are less than 10,000 lbs in gross weight, and over half of these are used for personal transportation. Only about 9 percent are heavy trucks used for over-the-road intercity freight. Many states consider recreational vehicles, mobile homes, and vans as trucks. Approximately 20 percent of trucks are used in agriculture and 10 percent are used by utilities and service industries to carry tools and people rather than freight. Personal truck ownership varies by region and is more prevalent in the mountain and western states than in the middle Atlantic and eastern states.

Figure 2.9   Truck Distribution by Size and Use

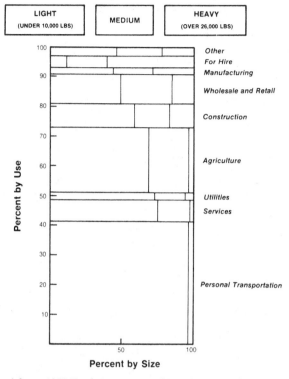

Source: Reproduced from *1972 Truck Inventory and Use Survey*, U.S. Bureau of the Census, Washington, D.C., 1972.

Trucks are manufactured by the major auto companies as well as by specialty companies. The configuration of trucks is diverse and a truck is usually produced to an owner's specification. Trucks are classified by gross vehicle weight. The heaviest trucks, over 26,000 lbs, are widely used in intercity freight, whereas lighter ones transport goods and services for shorter distances. Truck size and weight were regulated by the states with little consistency and uniformity. Trucks were required to conform to standards for height, width, length, gross weight, weight per axle, and number of axles, depending on the state. The absence of national standards for size and weight reflected the variety of truck dimensions that existed. An interstate truck shipment that met requirements in one state may have been overloaded in another along the same route. A trucker would then either carry a minimal load, use a circuitous route, change shipments at state lines, or travel illegally. With the passage of the Surface Transportation Act of 1982, states are required to permit trucks pulling 28 ft trailers on interstate highways and other principal roads. The law also permits the use of 48 ft semitrailers and 102 in. wide trucks, carrying up to 80,000 lbs.[4]

The trucking industry has been called a giant comprised of midgets because, although it is a major force in the transportation picture of the United States, it is very fragmented and diverse. Regulation of the trucking industry has been limited to those carriers who transport under contract (known as *for hire*), with the exception of carriers operating within a single state, or in a specified commercial zone, or those carrying exempt agricultural products. Private carriers, who transport their own company's goods or products, are not economically regulated but must meet federal and state safety requirements. The for-hire portion of the trucking industry accounts for about 45 percent of intercity ton-miles and includes approximately 15,000 separate carriers. With deregulation of various transportation industries, the trucking industry has become more competitive.

The trucking industry has benefited significantly from the improvements in the U.S. highway transportation system. The diversity of equipment and service types is evidence of the flexibility furnished by truck transportation. Continued growth of this industry will depend on how well it integrates with other modes, the future availability of energy, limitations on sizes and weights of trucks, highway speed limits, and safety regulations.

TRANSPORTATION ORGANIZATIONS

The operation of the vast network of transportation services in the United States is carried out by a variety of organizations. Each has a special function to perform and serves to create a network of individuals who, working together, furnish the transportation systems and services that presently exist. The following sections will describe some of the organizations and associations involved in transportation. The list is illustrative only and is intended to show the wide range of organizations active in the transportation field.

Many types of organizations and associations are active in transportation. The following seven categories, described briefly in the following sections, outline the basic purposes and functions that these organizations serve.

- Private companies who are available for hire to transport people and goods
- Regulatory agencies who monitor the behavior of transportation companies in areas such as pricing of services and safety
- Federal agencies such as the U.S. Department of Transportation and the U.S. Department of Commerce who, as part of the executive branch, are responsible for carrying out legislation dealing with transportation at the national level
- State and local agencies and authorities who are responsible for the planning, design, construction, and maintenance of transportation facilities such as roads and airports
- Trade associations, who represent the interests of a particular transportation activity, such as railroads or intercity buses, and who serve these groups by furnishing data and information, by representing them at congressional hearings, and by furnishing a means for discussing mutual concerns

- Professional organizations comprised of individuals who may be employed by any of the transportation organizations but who have a common professional bond and benefit from meeting with colleagues at national conventions or specialized committees to share the results of their work, learn about the experience of others, and advance the profession through specialized committee activities
- Organizations of transportation users who wish to influence the legislative process and furnish its members with useful travel information.

Other means for exchanging information about transportation are through professional and research journals, reports and studies, and university research and training programs.

Private Transportation Companies

Transportation by water, air, rail, highway, or pipeline is either furnished privately or on a for-hire basis. Private transportation, such as automobiles or company-owned trucks, must conform only to safety and traffic regulations. For-hire transportation, regulated until recently by the government, is classified as common carriers (available to any user), contract carriers (available by contract to particular market segments), and exempt (for-hire carriers that are exempt from regulation). Examples of private transportation companies are the Santa Fe Railroad, Greyhound Bus Company, Smith Transfer, United Airlines, and Yellow Taxi, to name but a few.

Regulatory Agencies

Common carriers have been regulated by the government since the late 1800s when abuses by the railroads created a climate of distrust toward these "robber barons" who used their monopoly powers to grant preferential treatment to favored customers or charged high rates to those without alternative routes or services. The Interstate Commerce Commission, a division of the U.S. Department of Commerce, was formed to make certain that the public received dependable service at reasonable rates without discrimination. It was empowered to control the raising and lowering of rates, to require that the carrier had adequate equipment and maintained sufficient routes and schedules, and to certify the entry of a new carrier and control the exit of any certified carrier. Today these concerns are less valid because the shipper has alternatives other than shipping by rail, and the entry and exit restrictions placed on companies no longer tend to favor the status quo or limit innovation and opportunities for new service. For these reasons the Interstate Commerce Commission, which had regulated railroads since 1887 and trucking since 1935, was relieved of these powers; now private carriers may operate in a more independent economic environment that should result in lower costs and improved services. Similarly, the Civil Aviations Board, which regulated the airline industry since 1938, is no longer empowered to certify routes and fares of domestic airline companies; it was phased out in 1985. Other regulatory agencies are the Federal Maritime Commission, which regulates U.S. and foreign vessels operating in international commerce, and the Federal Energy Regulatory Commission, which regulates certain oil and natural gas pipelines.

Federal Agencies

Because transportation pervades our economy, each agency within the executive branch of the federal government is involved in some aspect of transportation. For example, the Department of State develops policy recommendations concerning international aviation and maritime transportation, and the Department of Defense, through the Corps of Engineers, constructs and maintains river and harbor improvements and administers laws protecting navigable waterways. The Department of Transportation is the principal assistant to the president in all matters relative to federal transportation programs. Within the department are seven major administrations that deal with programs for highways, aviation, railroads, urban mass transportation, highway traffic safety, research, and maritime. Other elements of the Department of Transportation are the U.S. Coast Guard, the St. Lawrence Seaway Development Corporation, and the Bureau of Motor Carrier Safety. The department also deals with special problems such as public and consumer affairs, civil rights, and international affairs. An organization chart is shown in Figure 2.10. Table 2.2 furnishes a description of the department's responsibilities.

The U.S. Congress, which represents the people through the Senate and House of Representatives, has jurisdiction over transportation activities through the budget and legislative process. Two committees in the Senate are concerned with transportation: the Commerce, Science, and Transportation Committee and the Committee on Environment and Public Works. The House committees are the Energy and Commerce Committee, Merchant Marine and Fisheries, and Public Works and Transportation.

Figure 2.10 U.S. Department of Transportation Organization and Responsibilities

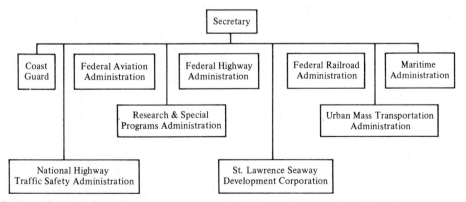

Source: Redrawn from *Transportation in America*, 5th ed., Transportation Policy Associates, Washington, D.C., March 1987.

Table 2.2   Responsibilities of the U.S. Department of Transportation

Secretary of Transportation	Principal assistant to president in all matters relating to federal transportation programs. Major aides include a deputy secretary, general counsel, and assistant secretaries for budget and programs, governmental affairs, administration, and policy and international affairs. The latter formulates transport policy and implementation plans; coordinates U.S. interests in international transport affairs; analyzes social, economic, and energy aspects of transport; assesses performance of transport system; and coordinates environmental and safety programs within DOT. Most of CAB's residual functions were transferred to DOT and placed largely in Office of Assistant Secretary for Policy and International Affairs. They include: economic regulation of international air transport (routes and rates), determination of carrier fitness for domestic air services, protection of consumers (baggage, overbooking, etc.), ensuring adequate service to smaller communities, grant of antitrust immunity, and protection of U.S. international air carriers against discriminatory acts of foreign nations.
Coast Guard	Maintains network of rescue vessels, aircraft and communications facilities to protect lives and property on the high seas and navigable waters of U.S. Enforces federal laws governing navigation, vessel inspection, port safety and security, marine environmental protection, and resource conservation. Sets ship construction and safety standards. Regulates Great Lakes' pilotage. Carries out R&D programs.
Federal Aviation Administration	Promotes civil aviation, including R&D programs; promulgates and enforces safety regulations; develops and operates nation's airways system; administers federal-aid airport programs; and certifies and registers aircraft and commercial/private pilots.
Federal Highway Administration	Administers federal-aid highway program of financial aid to states for highway construction and rehabilitation. Develops and administers highway safety program, including aid to states and local communities. Coordinates and helps fund R&D programs. Its Bureau of Motor Carrier Safety regulates safety performance of interstate commercial motor carriers, including hazardous materials movements and cargo security/noise abatement programs. Makes road checks, prosecutes violators. Develops highway data.
Federal Railroad Administration	Consolidates U.S. support and promotion of railroad transportation; administers and enforces rail safety

Continued

Table 2.2—*Continued*

	regulations; administers rail financial-aid programs; and conducts R&D programs to improve railroad freight and passengers services and safety.
Maritime Administration	Promotes merchant marine; grants ship mortgage insurance; maintains National Defense Reserve Fleet; sponsors R&D for ship design, propulsion and operations. Its Maritime Subsidy Board awards ship operating subsidies to liner carriers (ship construction subsidies have been discontinued) and determines scope of subsidized shipping services and routes—in line with MarAd's determination of ship requirements, services, and routes needed for the U.S. foreign waterborne commerce.
National Highway Traffic Safety Administration	Implements motor vehicle safety programs and issues standards prescribing levels of safety needed; conducts test programs to assure compliance with standards and need for them; helps fund state/local safety programs; maintains national poor-driver register; sets fuel economy standards for autos with EPA and DOE; and assesses penalties for violators of its standards.
Research and Special Programs Administration	Serves as DOT's research, analysis, and technical development arm; and to conduct special programs, with emphasis on: pipeline safety, movements of hazardous materials, cargo security, R&D, university research programs, and commercial air carrier data formerly collected, collated, and published by the CAB.
St. Lawrence Seaway Development Corp.	Administers the operation and maintenance of the U.S. portion of the St. Lawrence Seaway. Works closely with its Canadian counterpart to determine and set toll levels.
Urban Mass Transportation Administration	Develops comprehensive, coordinated mass transport systems for urban areas, including: financial aid for equipment and operations; R&D and demonstration projects; aid for technical studies; planning, engineering, and designing.

Source: Adapted from *Transportation in America*, 5th ed., Transportation Policy Associates, Washington, D.C., March 1987.

State and Local Agencies and Authorities

Each of the 50 states has its own highway or transportation department that is responsible for planning, building, operating, and maintaining its highway system and for administering funds and programs in other modes such as rail, transit, air, and water. These departments may also be responsible for driver licensing, motor vehicle registration, policing, and inspection. The organization and functions of these departments vary considerably from state to state but, since highway programs are a direct state responsibility whereas other modes are either privately owned and operated (such as air and rail) or of greater concern at a local level (such as public transit), highway matters tend to dominate at the state level. However, in many of the larger, more industrial states, the responsibilities of state transportation departments closely parallel those of the U.S. Department of Transportation.

Local agencies are responsible for carrying out specific transportation functions within a prescribed geographic area. Most larger cities have a regional transportation authority that operates the bus and rapid transit lines, and many communities have a separate traffic department responsible for operating the street system, its signing, traffic timing, and parking controls. For example, the Tidewater Transportation District in Virginia operates the bus service and coordinates all car and vanpool programs. On a larger scale, the Washington Metropolitan Area Transportation is planning and building a 100 mile rapid transit system and coordinating this with bus service in an area covering Maryland, Virginia, and the District of Columbia. Local roads are often the responsibility of county or township agencies that vary considerably in the quality of engineering staff and equipment. The jurisdiction and administration of the road in a state can create severe economic and management problems if the responsibility for road improvements is fragmented and politicized.

Trade Associations

Americans are joiners and for each occupation or business involved in transportation there is likely to be an organization that represents its interests. These associations are an attempt to present an industrywide front in matters of common interest, and they promote and develop new procedures and techniques to enhance marketability of their products and provide an opportunity for information exchange. One of the broadest based organizations was the Transportation Association of America (TAA), which represented a wide spectrum of interests in transportation and had a membership of firms representing all modes. The organization, which was dissolved in the early 1980s, published data and reports on transportation and maintained close contact with developments that could affect the transportation industry.

Examples of modally oriented organizations are the Association of American Railroads (AAR), the American Road and Transportation Builders Association (ARTBA), the American Public Transit Association (APTA), and the American Bus Association (ABA). The Highway Users Federation for Safety and Mobility (HUFSAM) is supported by highway-related industries such as oil, trucking, and auto and carries out studies and projects designed to improve the nation's highway system.

Examples of trade associations that improve product performance and marketability are the Asphalt Institute (AI) and the Portland Cement Association (PCA). These groups publish data and technical manuals used by design engineers and maintain a technical staff for special consultations. These and many similar associations are an attempt to keep abreast of changing conditions and provide a communications link for their members and among the many governmental agencies that use or regulate their products.

Professional Societies

Professional societies comprise individuals with a common professional interest in transportation. Their purpose is to exchange ideas, to develop recommendations for design and operating procedures, and to keep their membership informed about new developments in transportation practice. Membership in professional organizations is essential for any professional who wishes to stay current in his or her field.

Examples of professional societies are the Society of Logistical Engineers and the Institute of Transportation Engineers, which represent professionals who work in companies or agencies as transportation managers, planners, or engineers. Members of the American Association of State Highway and Transportation Officials (AASHTO) are representatives of state highway and transportation departments and the Federal Highway Administration. AASHTO produces manuals, specifications standards, and current practices in highway design, which form the basis for practices throughout the country.

The Transportation Research Board (TRB) is a unit of the National Research Council and is responsible for encouraging research in transportation and disseminating the results to the professional community. It operates through a technical committee structure composed of knowledgeable practitioners who assist in defining research needs, review and sponsor technical sessions, and conduct workshops and conferences. The Transportation Research Board is supported by the state transportation departments, federal transportation administrations, trade associations, transportation companies, and individual memberships. Its annual meeting in Washington, D.C. attracts over four thousand people each year.

Consumer Associations

The transportation consumer may have a direct role in the process of effecting change in transportation services and is also represented by several transportation groups or associations. The American Automobile Association (AAA) has a wide membership of highway users and serves its members as a travel advisory service and insurance agency. The American Railway Passenger Association (ARPA) and other similar consumer groups have attempted to influence legislation to improve transportation services (such as requiring nonsmoking sections in airplanes) or maintaining rail service on unprofitable lines. In general, however, the average transportation consumer plays a minor role in the transportation picture and is usually unaware of the way in which the transportation services he or she uses came about or are supported. Quite often the only direct involvement of the consumer is through citizen participation in the selection process for new highways in the community.

SUMMARY

The transportation system in a developed nation consists of a network of modes that have evolved over many years. The system consists of vehicles, guideways, terminal facilities, and control systems, and they operate according to established procedures and schedules in the air, on land, and on water. The system also requires interaction with the user, operator, and environment. The systems that are in place reflect the multitude of decisions made by shippers, carriers, government, individual travelers, and impacted nonusers concerning the investment or use of transportation. The transportation system that has evolved has produced a variety of modes that complement each other. Intercity passenger travel often involves auto and air modes, whereas intercity freight travel combines rail and trucking. Urban passenger travel involves auto or public transit whereas urban freight is primarily by truck.

The nation's transportation system can be considered to be in a state of equilibrium at any given point in time as a result of market forces, government actions, and transportation technology. As these change over time, the transportation system will also be modified. During the past decades, changes in gasoline prices, regulation by government, and new technology have affected the relative importance of each mode. The passenger or shipper thinks of each mode in terms of the level of service provided. Each mode offers a unique set of service characteristics at a given price: travel time, frequency, comfort, convenience, reliability, and safety. The traveler or shipper selects the mode based on how these attributes are valued.

The principal carriers of freight are rail, truck, pipeline, and water. Passenger transportation is by auto, air, rail, and bus. Highway transportation is the dominant mode in passenger travel, accounting for over 80 percent of passenger miles. Trucks carry most freight in urban areas and are a principal mode in intercity travel. The U.S. highway system comprises 3.9 million mi of roadway. The interstate system is 42,500 mi of limited-access roads and represents the backbone of the nation's highway network.

A wide range of organizations and agencies provide the resources to plan, design, build, operate, and maintain the nation's transportation system. These include private companies that furnish transportation, regulatory agencies that monitor the safety and service quality provided, federal, state, and local agencies that provide funds to build roads and airports and carry out legislation dealing with transportation at a national level, trade associations that represent the interests of a particular group of transportation providers or suppliers, professional organizations, and transportation user groups.

In the chapters that follow, the work of the traffic and highway engineer is described. Chapter 3 is concerned with the basic characteristics of the highway environment, which is composed of the driver, the pedestrian, the vehicle, and the roadway. In order to properly design a safe and efficient highway system, the interaction between these four elements must be considered.

PROBLEMS

2–1 Make a comparison of how your typical day might be changed with or without one or more modes of transportation. Consider both your personal transportation as well as goods and services that you rely on.

2–2 What are the most central problems in your state concerning one of the following: (a) air transportation, (b) railroads, (c) water transportation, (d) highways, (e) public transportation. (To answer this question obtain a copy of the governor's plan for transportation in your state or contact a key official in the transportation department.)

2–3 A bridge has been constructed between the mainland and an island. The total cost excluding tolls to travel across the bridge is expressed as $C = 50 + 0.5V$, where V is the number of vehicles per hour and C is the cost/vehicle in cents. The demand for travel across the bridge is $V = 2500 - 10C$.

(a) Determine the volume of traffic across the bridge.

(b) If a toll of 25 cents is added, what is the travel cost per vehicle?

(c) An additional toll booth is to be added, thus changing the travel time to cross the bridge. The new cost function is $C = 50 + 0.2V$. Determine the volume of traffic that would cross the bridge.

(d) Determine the toll to yield the highest revenue.

2–4 A large manufacturer uses two factors to decide whether to use truck or rail for movement of its products to market: cost and total travel time. The manufacturer uses a utility formula that rates each mode. The formula is $U = 5C + 10T$, where C is cost (\$/ton) and T is time (hours). For a given shipment of goods, a trucking firm can deliver in 16 hours and charges \$25/ton, whereas a railroad charges \$17/ton and can deliver in 25 hours.

(a) Which mode should the shipper select?

(b) What other factors should the shipper take into account in making a decision (discuss at least two)?

2–5 Describe the organization and function of your state highway department and/or transportation department.

2–6 List the transportation agencies and organizations located in your town. What services do they provide?

2–7 Obtain from your library a copy of a Transportation Research Record (published by the Transportation Research Board).

(a) Select one article in this publication and write a short summary of its contents.

(b) Describe the general area of transportation covered by this article. Compare topics with other members of the class.

REFERENCES

1. Transportation Policy Associates, *Transportation in America*, 5th ed., Washington, D.C., March 1987.

2. Ibid.

3. American Association of State and Highway Transportation Officials, *A Study on Future Directions of Public Transportation in the United States*, Washington, D.C., 1985.

4. Transportation Research Board, *Twin Trailer Trucks*, Special Report 211, National Academy of Sciences, Washington, D.C., 1986.

ADDITIONAL READINGS

Bowersox, Donald J., Pat J. Calabro, and George D. Wagenheim, *Introduction to Transportation*, Macmillan Publishing Co., New York, 1981.

Gray, George E., and Lester A. Hoel, *Public Transportation: Planning, Operations and Management*, Prentice-Hall, Englewood Cliffs, N.J., 1979.

Harper, Donald V., *Transportation in America*, Prentice-Hall, Englewood Cliffs, N.J., 1978.

Hazard, John L., *Transportation Management-Economics-Policy*, Cornell Maritime Press, Inc., Cambridge, MD., 1977.

Manheim, Marvin L., *Fundamentals of Transportation Systems Analysis*, MIT Press, Cambridge, Mass., 1978.

Morlok, Edward K., *Introduction to Transportation Engineering and Planning*, McGraw Hill Book Co., New York, 1978.

Oglesby, Clarkson H., and R. Gary Hicks, *Highway Engineering*, 4th ed., John Wiley & Sons, New York, 1982.

U.S. Department of Transportation, Federal Highway Administration, *Our Nations Highways: Selected Facts and Figures*, Washington, D.C., 1984.

U.S. Department of Transportation, Office of Transportation Planning, *National Transportation: Trends and Choices (to the Year 2000)*, U.S. Government Printing Office, Washington, D.C., 1977.

Vuchic, Vucan R., *Urban Public Transportation: Systems and Technology*, Prentice-Hall, Englewood Cliffs, N.J., 1981.

Wood, Donald F., and James C. Johnson, *Contemporary Transportation*, Petroleum Publishing Co., Tulsa, Okla., 1980.

PART 2

2

Traffic Operations

The traffic or highway engineer must understand not only the basic characteristics of the driver, vehicle, and roadway but how each interacts with the other. Information obtained through traffic engineering studies serves to identify relevant characteristics and define related problems. Traffic flow is of fundamental importance in developing and designing strategies for intersection control, rural highways, and freeway segments.

Characteristics of the Driver, Pedestrian, Vehicle, and Road

Historically, the three main components of the highway mode of transportation were the driver, the vehicle, and the road. However, the high number of pedestrian accidents over the years has revealed the importance of a fourth component, the pedestrian. To provide efficient and safe highway transportation, a knowledge of the limitations of the characteristics of each of these four components is essential. It is also important to be aware of the interrelationships that exist among these components in order to determine the effect, if any, these characteristics have on each other. These characteristics are also of primary importance when traffic engineering measures such as traffic control devices are to be used in the highway mode. Knowing average limitations may not always be adequate; sometimes it may be necessary to obtain information on the full range of the limitations. Consider, for example, the wide range of drivers' ages in the United States, which is 16 through 80. Sight and hearing vary considerably between the ages of 16 and 80; they can vary even among individuals of the same age group.

Similarly, a wide range of vehicles, from compact cars to articulated trucks, have been designed. The maximum acceleration, turning radii, and ability to climb grades differ considerably among the different vehicles. The road must therefore be designed to accommodate a wide range of vehicle characteristics and at the same time permit its use by drivers and pedestrians with a wide range of physical and psychological characteristics.

This chapter discusses the relevant characteristics of the four components of the highway mode and demonstrates their importance and use in the design and operation of highway facilities.

DRIVER CHARACTERISTICS

One problem that faces traffic and transportation engineers when considering driver characteristics in design is the varying skills and perceptual abilities of drivers on the highway. This is demonstrated by the wide range of abilities possessed by individuals to hear, see, evaluate, and react to information. Studies have shown that these abilities may also vary in an individual under different conditions, such as being under the influence of alcohol, suffering from fatigue, and time of day. Therefore, it is important that criteria used for design purposes be compatible with the capabilities and limitations of most drivers on the highway. The use of an average value such as mean reaction time may not be adequate for a large number of drivers. Both the 85th percentile and 95th percentile have been used to select design criteria, but in general, the higher this percentile, the wider is the range covered.

The Human Response Process

Actions taken by drivers on a road are the result of the evaluation of and reaction to information they obtain through certain stimuli that they see or hear.

Visual Reception

The receipt of stimuli by the eye is the most important method of transmitting information to both the driver and the pedestrian. Some general knowledge of human vision will therefore aid in solving several problems in traffic engineering. The principal characteristics of the eye are visual acuity, peripheral vision, color vision, glare vision and recovery, and depth perception.

Visual Acuity.   Visual acuity is the ability to see fine details of an object. Two types of visual acuity are of importance in traffic and highway emergencies. These are static and dynamic visual acuity.[1] The driver's ability to identify an object when both the object and the driver are stationary depends on his or her static acuity. Factors that affect static acuity include background brightness, contrast, and time. Static acuity will increase with an increase in illumination up to a background brightness of about 10 mililamberts and will then remain constant even with an increase in illumination. When other visual factors are held constant at an acceptable level, the optimal time required for identification of an object with no relative movement is between 0.5 and 1.0 sec.

The driver's ability to detect clearly relatively moving objects, not necessarily in his or her direct line of vision, depends on the driver's dynamic visual acuity. Most persons have clear vision within a conical angle of 3° to 5°, and fairly clear vision within a conical angle of 10° to 12°. Vision beyond this range is usually blurred. This is important when considering the location of traffic information devices. Drivers will clearly see those devices that are within the 12° cone, whereas objects outside this cone will be blurred.

Peripheral Vision.   Peripheral vision concerns the capability of individuals to see objects beyond the cone of clearest vision. Although objects can be seen within this zone, details and color are not clear. Research on the subject has indicated that the cone for peripheral vision could be one subtending up to 160° and that this value is affected by the speed of the vehicle.[2] Age has also been identified as an influential factor of peripheral vision, such that at 60 years, a significant change occurs in an individual's peripheral vision.[3]

Color Vision.   Color vision is the ability to differentiate one color from another, but deficiency in this, usually referred to as color blindness, is not of great significance in highway driving because other means of recognizing traffic information devices may compensate for this. Combinations of black and white and black and yellow have been shown to be those for which the eye is most sensitive.

Glare Vision and Recovery.   There are two types of glare vision: Direct glare occurs when relatively bright light appears in the individual's field of vision and specular glare occurs when the image reflected by the relatively bright light appears in the field of vision. Both types of glare result in a decrease of visibility and cause discomfort to the eyes. It has been shown that age has a significant effect on the sensitivity to glare and that at age 40, a significant change occurs in an individual's sensitivity to glare.[4]

The time required by an individual to recover from the effects of glare after passing the light source is known as glare recovery. Studies have shown that this time is about 3 sec when moving from dark to light and could be 6 or more sec when moving from light to dark.[5] The phenomenon of glare vision is of significant importance during night driving and contributes to the problem of older people having much less vision at night. This phenomenon should be taken into consideration in the design and location of street lighting so that glare effects are reduced to the minimum.

Glare effects can be kept to a minimum by reducing luminaire brightness, increasing mounting height, and increasing the background brightness in a driver's field of view. Specific actions taken to achieve this in lighting design include higher mounting heights, positioning the lighting support further away from the highway, and restricting the light from the luminaire to obtain minimum interference with the visibility of the driver.

Depth Perception.   Depth perception relates to the ability of an individual to estimate speed and distance. It is particularly important on two-lane highways, during passing maneuvers, where head-on accidents may be the result of a lack of proper judgment of speed and distance.

The ability of the human eye to differentiate between objects is fundamental to this phenomenon. It should be noted, however, that the human eye is not very good at estimating absolute values of speed, distance, size, and acceleration. This demonstrates the importance of traffic control devices being standard, which not only aids in distance estimation but also helps the color blind driver to identify signs.

Hearing Perception

The ear receives sound stimuli, which is only important to drivers when warning sounds, usually given out by emergency vehicles, are to be detected. Loss of some hearing ability is, however, not a serious problem since this is normally corrected with the use of a hearing aid.

PEDESTRIAN CHARACTERISTICS

Pedestrian characteristics relevant to traffic and highway engineering practice include those of the driver discussed in the previous section. In addition, other pedestrian characteristics may influence the design and location of pedestrian control devices. Such control devices include special pedestrian signals, safety zones and islands at intersections, pedestrian underpasses, elevated walkways, and crosswalks. Apart from visual and hearing characteristics, walking characteristics play a major part in the design of some of these controls. For example, the design of an all red phase, which permits pedestrians to cross an intersection with heavy traffic, requires a knowledge of the walking speeds of pedestrians. Observations of pedestrian movements have indicated that walking speeds vary between 3.0 and 8.0 ft/sec. Significant differences have also been observed between male and female walking speeds. At intersections, mean male walking speed has been determined to be 4.93 ft/sec and that for females was 4.63 ft/sec. Consideration should also be given to the characteristics of handicapped pedestrians such as the blind. Studies have shown that accidents involving blind pedestrians can be reduced by installing special signals. The blind pedestrian can turn the signal to a red phase by using a special key, which rings a bell, indicating to the pedestrian that it is safe to cross. Ramps are also now being provided at intersection curbs to facilitate the crossing of the intersection by the occupant of a wheel chair.

PERCEPTION AND REACTION PROCESS

The process through which a driver or pedestrian evaluates and reacts to a stimulus received can be divided into four subprocesses.

1. The subprocess of *perception* entails the use of visual reception to see a control device, warning sign, or object on the road.
2. *Identification* is the subprocess through which the driver identifies the object or control device and thus understands the stimulus.
3. During the subprocess of *emotion*, the driver decides what action to take in response to the stimulus; for example, to step on the brake pedal, to pass, to swerve, or to change lanes.
4. *Reaction* or *volition* is the subprocess during which the driver actually executes the action decided on during the emotion subprocess.

Time elapses during each of the above subprocesses. The time that elapses from the start of perception to the end of reaction is the total time required for perception, identification, emotion, and volition, sometimes referred to as PIEV time or more commonly as **perception-reaction time**.

It can easily be seen that perception-reaction time is an important factor in the determination of braking distances, which in turn dictates the minimum sight distance required on a highway and the length of the amber phase at a signalized intersection. Perception-reaction time varies from one individual to another and may, in fact, vary for the same person as the occasion changes. It has been shown that perception-reaction time could vary from 0.5 sec to 7.0 sec. These changes in the value of perception-reaction time depend on how complicated the situation is, the existing environmental conditions, age, whether the person is tired or under the influence of drugs and/or alcohol, and whether the stimulus is expected or unexpected. Results obtained by a study conducted by Johannson and Rumar indicated that reaction times for unexpected stimuli are about 35 percent higher than those for expected stimuli.[6] The reaction time selected for design purposes should, however, be large enough to include reaction times for most drivers using the highways. Recommendations made by the American Association of State Highway and Transportation Officials (AASHTO) stipulate 2.5 sec for stopping sight distances.[7] This time encompasses the decision times for about 90 percent of drivers under most highway conditions. Note, however, that this reaction time of 2.5 sec may not be adequate for some very complex conditions such as at multi-phase at-grade intersections and ramp terminals. Studies have shown that at sites with very complex conditions, perception-reaction times range from about 5 to 10 sec.

Example 3–1 Distance Traveled During Perception-Reaction Time

A driver with a perception-reaction time of 2.5 sec is driving at 55 mph when he observes that an accident has blocked the road ahead. Determine the distance the vehicle would move before the driver could activate the brakes. The vehicle will continue to move at 55 mph during the perception-reaction time of 2.5 sec.

- Convert mph.

$$55 \text{ mph} = \left(55 \times \frac{88}{60}\right) \text{ft/sec}$$

- Find the distance traveled.

$$55 \times \frac{88}{60} \times 2.5 \text{ ft} = 55 \times 1.47 \times 2.5 \text{ ft}$$

$$= 202.1 \text{ ft}$$

VEHICLE CHARACTERISTICS

Criteria for the geometric design of highways are partly based on the static, kinematic, and dynamic characteristics of vehicles. Static characteristics include the weight and size of the vehicle; kinematic characteristics involve the motion of the vehicle, without considering the forces that cause the motion; and dynamic characteristics consider the forces that cause the motion of the vehicle. Since nearly all highways carry both passenger automobile and truck traffic, it is essential that design criteria meet the requirements of the characteristics for different types of vehicles. A thorough knowledge of the characteristics will aid the highway and/or traffic engineer in designing highways and traffic control systems that allow for the safe and smooth operation of a moving vehicle, particularly during the basic maneuvers of passing, stopping, and turning. Therefore, designing a highway involves the selection of a "design vehicle," which has characteristics that will meet the requirements of nearly all vehicles expected to use the highway. The characteristics of the design vehicle are then used to determine criteria for geometric design, intersection design, and sight distance requirements.

Static Characteristics

The size of the design vehicle for a highway is an important input in the determination of design standards for several physical components of the highway; these include lane width, shoulder width, length and width of parking bays, and lengths of vertical curves. The axle weights of the vehicles expected on the highway are important when pavement depths and maximum grades are being determined.

For many years each state prescribed by law the limits of sizes and weights of trucks using its highways, and in some cases local authorities also imposed more severe restrictions on some roads. Table 3.1 shows some features of static characteristics for which limits were prescribed. A range of maximum allowable values is given for each feature.

Since passage of the Surface Transportation Assistance Act of 1982, maximum allowable limits on truck sizes and weights on interstate and other qualifying federal-aid highways are required to be at least

- 80,000 lb gross weight, with axle loads up to 20,000 lb for single axles and 34,000 lb for double axles
- 102 in. in width for all trucks
- 48 ft in length for semitrailers and trailers
- 28 ft in length for each twin trailer

States are no longer allowed to set limits on overall truck length. These provisions will most probably result in wider, longer, and heavier trucks on these highways, which raises the basic safety question of whether there will be an increase in injuries and fatalities resulting from crashes involving larger trucks on these highways. Another impact of these provisions is that reconstruction of the deteriorating highways to meet these new weight and length dimensions will be extremely expensive.[8]

As stated earlier, the static characteristics of vehicles expected to use the highway are factors that influence the selection of design criteria for the highway. It is therefore necessary that all vehicles be classified so that representative static

Table 3.1 Range of State Limits on Vehicle Lengths by Type and Maximum Weight of Vehicle

Type	Allowable Lengths (ft)
Buses	35–60
Single truck	35–60
Trailers, semi/full	35–48
Tractor, semitrailer	55–85
Truck trailer	55–85
Tractor semitrailer trailer	55–85
Truck trailer trailer	65–80
Tractor semitrailer, trailer, trailer	60–105

Type	Allowable Weights (lb)
Single axle	18,000–24,000
Tandem axle	32,000–40,000
State maximum gross vehicle weight	73,280–164,000
Interstate maximum gross vehicle weight	73,280–164,000

Note: Prior to the Surface Transportation Assistant Act of 1982.

Source: Adapted from *State Maximum Sizes and Weights for Motor Vehicles*, Motor Vehicle Manufacturers' Association of the United States, Detroit, Mich., May 1982.

characteristics for all vehicles within a particular class can be provided for design purposes.

AASHTO has provided ten different classifications.[9]

- Passenger car, P
- Single-unit truck, SU
- Single-unit bus, BUS
- Articulated bus, A-BUS
- Intermediate semitrailers, WB-40
- Large semitrailer, WB-50
- Double-bottom semitrailer–full trailer, WB-60
- Motor home, MH
- Car and camper trailer, P/T
- Car and boat trailer, P/B

Table 3.2 shows physical dimensions for the design vehicle for each class. Included in the passenger car (P) class are compacts and subcompacts, all light vehicles, and light delivery trucks (vans and pickups). Single-unit trucks (SU) include all single-unit trucks and small buses. The class single-unit bus (BUS) comprises intercity and transit buses with a wheel base of 25 ft and an overall

Table 3.2 Design Vehicle Dimensions

Design Vehicle Type	Symbol	Dimension (ft)									
		Overall			Overhang		WB_1	WB_2	S	T	WB_3
		Height	Width	Length	Front	Rear					
Passenger car	P	4.25	7	19	3	5	11				
Single-unit truck	SU	13.5	8.5	30	4	6	20				
Single-unit bus	BUS	13.5	8.5	40	7	8	25				
Articulated bus	A-BUS	10.5	8.5	60	8.5	9.5	18		4[a]	20[a]	
Combination trucks											
Intermediate semitrailer	WB-40	13.5	8.5	50	4	6	13	27			
Large semitrailer	WB-50	13.5	8.5	55	3	2	20	30			
Double bottom semitrailer–full trailer	WB-60	13.5	8.5	65	2	3	9.7	20	4[b]	5.4[b]	20.9
Recreation vehicles											
Motor home	MH		8	30	4	6	20				
Car and camper trailer	P/T		8	49	3	10	11	5	18		
Car and boat trailer	P/B		8	42	3	8	11	5	15		

[a]Combined dimension 24, split is estimated.
[b]Combined dimension 9, 4, split is estimated.
WB_1, WB_2, and WB_3 are effective vehicle wheelbases.
S is the distance from the rear effective axle to the hitch point.
T is the distance from the hitch point to the lead effective axle of the following unit.

Source: Adapted from *A Policy on Geometric Design of Highways and Streets*, Washington, D.C.: The American Association of State Highway and Transportation Officials, copyright 1984. Used by permission.

length of 40 ft. The articulated bus (A-BUS) category includes larger than conventional buses that have a permanent hinge near the center, which allows for better maneuverability. Intermediate semitrailers (WB-40) include the majority of medium to large truck tractor-semitrailer combinations, whereas the large semitrailer category (WB-50) includes nearly all of the truck tractor-semitrailer combinations in use. The "double bottom" semitrailer-full trailer category (WB-60) includes nearly all of the tractor-semitrailer–full–trailer combinations. It should be noted that a separate category for triple trailers is not yet included in the

Figure 3.1 Minimum Turning Path for Passenger Design Vehicle

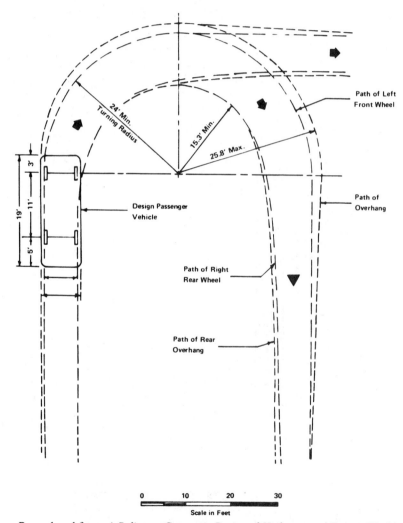

Source: Reproduced from *A Policy on Geometric Design of Highways and Streets*, Washington, D.C.: The American Association of State Highway and Transportation Officials, copyright 1984. Used by permission.

classification system recommended by AASHTO. This may become necessary if the use of these triples becomes widespread.

Minimum turning radii at low speeds (10 mph or less) are mainly dependent on the size of the vehicle. The turning radii requirements for passenger and WB-60 design vehicles are given in Figures 3.1 and 3.2. The turning radii require-

Figure 3.2   Minimum Turning Path for WB-60 Design Vehicle

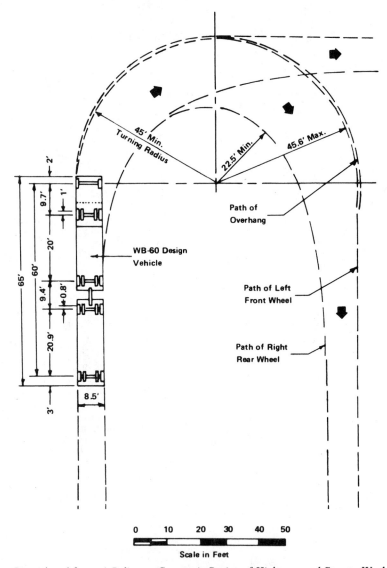

ments for the other vehicles can be found in AASHTO's *A Policy on Geometric Design of Highways and Streets.*[10] These turning paths were selected by carrying out a study of the turning paths of scale models of the representative vehicles of each class. It should be emphasized, however, that the minimum turning radii shown in Figures 3.1 and 3.2 are for turns taken at speeds less than 10 mph. When turns are made at higher speeds, the lengths of the transition curves are increased, and radii greater than the minimum specified are required. These requirements will be described later.

Kinematic Characteristics

The primary element under kinematic characteristics is the acceleration capability of the vehicle. Acceleration capability is important in several traffic operations, such as passing maneuvers and gap acceptance. Also, the dimensioning of highway features such as freeway ramps and passing lanes are often governed by acceleration rates. Acceleration is also important in determining the forces that cause motion. Therefore, a study of the kinematic characteristics of the vehicle primarily involves a study of how acceleration rates influence the elements of motion such as velocity and distance. We therefore review in this section the mathematical relationships that relate acceleration, velocity, distance, and time with each other.

Let us consider a vehicle moving along a straight line from point o to point m, a distance x in a reference plane T. The position vector of the vehicle after time t may be expressed as

$$r_{om} = x\hat{\imath} \tag{3.1}$$

where

r_{om} = position vector for m in T
$\hat{\imath}$ = a unit vector parallel to line om
x = distance along the straight line

The velocity and acceleration for m may be simply expressed as

$$u_m = \dot{r}_{om} = \dot{x}\hat{\imath} \tag{3.2}$$

$$a_m = \ddot{r}_{om} = \ddot{x}\hat{\imath} \tag{3.3}$$

where

u_m = velocity of the vehicle at point m
a_m = acceleration of the vehicle at point m

Two cases are of interest: (1) acceleration is assumed constant, and (2) acceleration is a function of velocity.

Acceleration Assumed Constant

When the acceleration of the vehicle is assumed to be constant, then

$$\ddot{x}\hat{\imath} = a$$

$$\frac{d\dot{x}}{dt} = a \tag{3.4}$$

$$\dot{x} = at + C_1 \tag{3.5}$$

$$x = \tfrac{1}{2}at^2 + C_1 t + C_2 \tag{3.6}$$

C_1 and C_2 are constants that are determined by the initial conditions on velocity and position or by using known positions of the vehicle at two different times.

Acceleration as a Function of Velocity

The assumption of constant acceleration has some limitations as the accelerating capability of a vehicle at any time t is related to the speed of the vehicle at that time. The lower the speed, the higher is the acceleration rate that can be obtained. Figure 3.3 shows acceleration capabilities for different vehicles at different speeds on level roads. One model that is commonly used in this case is

$$\frac{du}{dt} = \alpha - \beta u_t \tag{3.7}$$

where α and β are constants.

Figure 3.3   Acceleration Capabilities for Different Vehicles at Speeds Between 30 and 60 mph on Level Roads

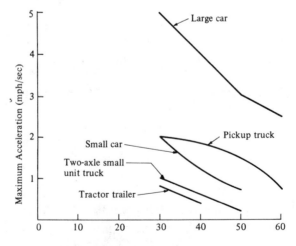

Source: Redrawn from *Transportation and Traffic Engineering Handbook*, 2nd ed., Institute of Transportation Engineers, Washington, D.C., 1982.

In this model the maximum acceleration rate that can be achieved is theoretically α, which means that α has units of acceleration. The term βu_t should also have units of acceleration, which means that β has the inverse of time (for example, sec^{-1}) as its unit.

Integrating Eq. 3.7 gives

$$-\frac{1}{\beta} \ln (\alpha - \beta u_t) = t + C$$

If the velocity is u_o at time zero, then

$$C = -\frac{1}{\beta} \ln (\alpha - \beta u_o)$$

and

$$-\frac{1}{\beta} \ln (\alpha - \beta u_t) = t - \frac{1}{\beta} \ln (\alpha - \beta u_o)$$

$$\ln \frac{(\alpha - \beta u_t)}{\alpha - \beta u_o} = -\beta t$$

$$\alpha - \beta u_t = (\alpha - \beta u_o)e^{-\beta t} \qquad (3.8)$$

$$u_t = \frac{\alpha}{\beta}(1 - e^{-\beta t}) + u_o e^{-\beta t}$$

The distance traveled $x(t)$ at any time may be determined by integrating Eq. 3.8. That is,

$$x = \int_o^t u_t \, dt = \int_o^t \frac{\alpha}{\beta}(1 - e^{-\beta t}) + u_o e^{-\beta t}$$

$$= \left(\frac{\alpha}{\beta}\right)t - \frac{\alpha}{\beta^2}(1 - e^{-\beta t}) + \frac{u_o}{\beta}(1 - e^{-\beta t}) \qquad (3.9)$$

Dynamic Characteristics

Several forces act on a vehicle while it is in motion: air resistance, grade resistance, rolling resistance, curve resistance, and friction resistance. The extent to which these forces affect the operation of the vehicle is discussed in this section.

Air Resistance

A vehicle in motion has to overcome the resistance of the air in front of it as well as the force due to the frictional action of the air around it. The force required to overcome these is known as the *air resistance* and is related to the cross-sectional area of the vehicle in a direction perpendicular to the direction of motion and the square of the speed of the vehicle. It has been shown by Claffey[11] that an estimate of this force can be obtained by

$$F_a = 0.0006 Au^2 \qquad\qquad (3.10)$$

where

    A = cross-sectional area of the vehicle (ft²)
    u = vehicle speed (mph)
    F_a = force (lb)

Grade Resistance

When a vehicle moves up a grade, a component of the weight of the vehicle acts downward along the plane of the highway. This creates a force acting in an opposite direction to that of the motion. This force is the grade resistance. A vehicle traveling up a grade will therefore tend to lose speed unless an accelerating

Figure 3.4   Speed Distance Relationships Observed During Maximum Acceleration Rate

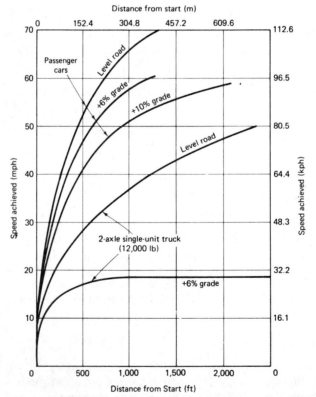

Source: Reproduced from *Transportation and Engineering Handbook*, 2nd ed., Institute of Transportation Engineers, Washington, D.C., 1982.

force is applied. The speed achieved at any point along the grade for a given rate of acceleration will depend on the grade percentage. Figure 3.4 shows the relationships between speed achieved and distance traveled up different grades by different types of vehicles during maximum acceleration.

Rolling Resistance

There are forces within the vehicle itself that offer resistance to motion. These forces are due mainly to frictional effect on moving parts of the vehicle but also include the frictional slip between the pavement surface and the tires. The sum effect of these forces to motion is what is known as rolling resistance. The rolling resistance depends on the speed of the vehicle and the type of pavement. Rolling forces are relatively lower on high-type pavements than on low-type pavements. High-type pavements are those adequately designed for the traffic load they carry with properly constructed bituminous or concrete surfaces. Low-type pavements usually have surfaces that range from untreated base materials to surface-treated earth. A value of 27 lb/ton of weight is normally used for the rolling resistance of modern passenger cars traveling at a speed of 60 mph on high-type pavements. This value increases by 10 percent for each 10 mph increase above 60 mph. The values for low-type pavements vary from 29 lb/ton of weight at 20 mph speed to 51 lb/ton of weight at 50 mph speed. The surface condition of the pavement has a significant effect on the rolling resistance. For example, at a speed of 40 mph on a badly broken and patched asphalt surface, the rolling resistance is 40 lb/ton of weight, whereas at the same speed on a loose sand surface, the rolling resistance is 52 lb/ton of weight.[12]

Curve Resistance

When a vehicle is maneuvered to take a curve, external forces act on the front wheels of the vehicle. These forces have components that have a retarding effect on the forward motion of the vehicle. The sum effect of these components is what constitutes the curve resistance. This resistance depends on the degree of the curve and the velocity at which the vehicle is moving. Table 3.3 gives values for curve resistance for different degrees of curve and speeds.

Table 3.3 Curve Resistance of Passenger Cars on High-Type Surfaces

Degree of Curve	Radius (ft)	Speed (mph)	Resistance (lb)
5	1146	50	40
5	1146	60	80
10	573	30	40
10	573	40	120
10	573	50	240

Source: Adapted from *Transportation and Traffic Engineering Handbook*, 2nd ed., Institute of Traffic Engineers, Washington, D.C., 1982.

Figure 3.5   Forces Acting on a Moving Vehicle

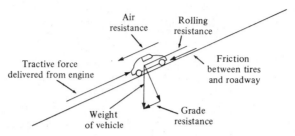

Source: Redrawn from Donald R. Drew, *Traffic Flow Theory and Control*, copyright © 1968, McGraw-Hill Book Company.

Power Requirements

Power is the rate at which work is done and is usually expressed in U.S. units of horsepower, where 1 horsepower is 550 lb ft/sec. The performance capability of a vehicle is measured in terms of the horsepower the engine can produce to overcome air, grade, curve, and friction resistance forces and also put the vehicle in motion. Figure 3.5 shows how these forces are acting on the moving vehicle. The power delivered by the engine is given as

$$P = \frac{Ru}{550} \tag{3.11}$$

where

P = horsepower delivered
R = sum of resistance to motion (lb)
u = speed of vehicle (ft/sec)

The action of the forces (shown in Figure 3.5) on the moving vehicle and the effect of perception-reaction time are used to determine important parameters related to the dynamic characteristics of the vehicles. These include the braking distance of a vehicle and the minimum radius of a circular curve required for a vehicle traveling round a curve with speed u ($u > 10$ mph). Also relationships among elements such as acceleration, coefficient of friction between tire and pavement, the position of the center of gravity of the vehicle above ground, and the track width of the vehicle could be developed by analyzing the action of these forces.

Braking

Consider a vehicle traveling downhill with an initial velocity of u mph as shown in Figure 3.6.

Let

W = weight of the vehicle
f = coefficient of friction between the tires and the road pavement

γ = the angle the grade makes with the horizontal

a = the deceleration of the vehicle when the brakes are applied

D_b = horizontal component of distance traveled during braking (that is, from time brakes are applied to vehicle coming to rest)

Note that the distance referred to as the braking distance is the horizontal distance and not the inclined distance x. The reason is that measurements of distances in surveying are horizontal and, therefore, in highway design are always with respect to the horizontal plane. Since the braking distance is an input in the design of highway curves, the horizontal component of the distance x is used.

$$\text{frictional force on the vehicle} = Wf \cos \gamma$$

The force acting on the vehicle due to deceleration is Wa/g, where g is acceleration due to gravity. The component of the weight of the vehicle is $W \sin \gamma$. Substituting into $\Sigma f = ma$, we obtain

$$W \sin \gamma - Wf \cos \gamma = \frac{Wa}{g} \tag{3.12}$$

The deceleration that brings the vehicle to a stationary position can be found in terms of the initial velocity u as $a = -u^2/2x$ (assuming uniform acceleration), where x is the distance traveled in the plane of the grade during braking. Eq. 3.12 can then be written as

$$W \sin \gamma - Wf \cos \gamma = -\frac{Wu^2}{2gx} \tag{3.13}$$

Figure 3.6 Forces Acting on a Vehicle Braking on a Downgrade

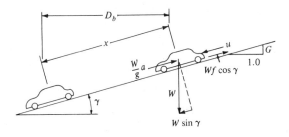

W = weight of vehicle
f = coefficient of friction
g = acceleration of gravity
a = vehicle acceleration
u = speed when brakes applied
D_b = braking distance
γ = angle of incline
G = tan γ (% grade/100)
x = distance traveled by the vehicle along the road during braking

However, $D_b = x \cos \gamma$ and we therefore obtain

$$\frac{Wu^2}{2gD_b} \cos \gamma = Wf \cos \gamma - W \sin \gamma$$

giving

$$\frac{u^2}{2gD_b} = f - \tan \gamma \qquad (3.14)$$

and

$$D_b = \frac{u^2}{2g(f - \tan \gamma)} \qquad (3.15)$$

Note, however, that $\tan \gamma$ is the grade of the incline (that is, percent of grade/100), is G as shown in Figure 3.6.

Eq. 3.15 can therefore be written as

$$D_b = \frac{u^2}{2g(f - G)} \qquad (3.16)$$

If g is taken as 32.2 ft/sec^2 and u is in mph, Eq. 3.16 becomes

$$D_b = \frac{u^2}{30(f - G)} \qquad (3.17)$$

and D_b is given in ft.

A similar equation could be developed for a vehicle traveling uphill, in which case the following equation is obtained:

$$D_b = \frac{u^2}{30(f + G)} \qquad (3.18)$$

A general equation for the braking distance can therefore be written as

$$D_b = \frac{u^2}{30(f \pm G)} \qquad (3.19)$$

where the plus sign is for vehicles traveling uphill and the minus sign is for vehicles traveling downhill.

Similarly, it can be shown that the horizontal distance traveled in reducing the speed of a vehicle from u_1 mph to u_2 mph during a braking maneuver is given by

$$D_b = \frac{u_1^2 - u_2^2}{30(f \pm G)} \qquad (3.20)$$

Table 3.4 Recommended Coefficients of Skidding Friction for Different Running Speeds (Wet Pavements)

Running Speed (mph)	Coefficients of Skidding Friction, f
20	0.40
25	0.38
30	0.35
35	0.34
40	0.32
45	0.31
50	0.30
55	0.30
60	0.29
65	0.29
70	0.28

Source: Adapted from *A Policy on Geometric Design of Highways and Streets*, Washington, D.C.: The American Association of State Highway and Transportation Officials, copyright 1984. Used by permission.

Eq. 3.19 also shows that an important parameter in determining braking distances is the coefficient of friction between the tires and the road pavement. This value will depend on the condition of the vehicle's tires, the type and condition of the pavement, and the weather condition, such as rain or snow. It should also be noted that the coefficient of friction is not actually constant throughout the braking maneuver but varies inversely with speed. In the determination of the braking distance, however, an average value is used. Table 3.4 gives friction values for different running speeds, as recommended by AASHTO for use in design.

The distance traveled by a vehicle between the time the driver observes an object in the vehicle's path and the time the vehicle actually comes to rest is longer than the braking distance, since it includes the distance traveled during perception-reaction time. This distance is referred to in this text as the stopping distance S and is given as

$$S(\text{ft}) = 1.47ut + \frac{u^2}{30(f \pm G)} \tag{3.21}$$

where the first term in Eq. 3.21 is the distance traveled during the perception-reaction time t (sec) and u is the velocity in mph at which the vehicle was traveling when the brakes were applied.

Example 3–2 Determining Braking Distance

A student trying to test the braking ability of his car, determined that he needed 32 ft more to stop his car downhill on a particular road than uphill when driving at 55 mph. Assuming that the coefficient of friction between the tires and the pavement is 0.30, determine the braking distance downhill and the percent grade of the highway at that section of the road.

- Let x ft = downhill braking distance
          $(x - 32)$ ft = uphill braking distance
- Use Eq. 3.17.

$$x = \frac{55^2}{30(0.3 - G)}$$

$$x - 32 = \frac{55^2}{30(0.3 + G)}$$

$$32 = \frac{55^2}{30(0.3 - G)} - \frac{55^2}{30(0.3 + G)}$$

$$960(0.3 - G)(0.3 + G) = 55^2(0.3 + G) - 55^2(0.3 - G)$$

$$960(0.09 - G^2) = 55^2(2G)$$

$$86.4 - 960G^2 = 6050G$$

$$G^2 + 6.302G - 0.09 = 0$$

$$G = 0.014 \text{ (that is, 1.4\%)}$$

$$x = \frac{55^2}{30(0.3 - 0.014)} = 352.6 \text{ ft}$$

Example 3–3   Exit Ramp Stopping Distance

A motorist traveling at 55 mph on an expressway intends to exit from the expressway using an exit ramp with a maximum allowable speed of 30 mph. At what point on the expressway should the motorist step on her brakes in order to reduce her speed to the maximum allowable on the ramp just before entering the ramp. Assume that the coefficient of friction between the tires and the pavement is 0.3, and the alignment of this section of the road is horizontal.

- Use Eq. 3.20.

$$D_b = \frac{55^2 - 30^2}{30(0.3 - 0)} = 236.1 \text{ ft}$$

The brakes should be applied at least 236 ft from the ramp.

Example 3–4   Distance Required to Stop for an Obstacle in the Roadway

A motorist traveling at 55 mph down a grade of 5 percent on a highway observes an accident ahead of him involving an overturned truck that is completely blocking the road. If the motorist was able to stop his vehicle 30 ft from the overturned truck, what distance away from the truck was he when he first observed the accident. Assume perception-reaction time $= 2.5$ sec and $f = 0.3$.

- Use Eq. 3.21 to obtain the stopping distance.

$$S = 1.47ut + \frac{u^2}{30(0.3 - 0.05)} = 1.47 \times 55 \times 2.5 + \frac{55^2}{30 \times 0.25}$$

$$= 202.13 + 403.33$$

$$= 605.5 \text{ ft}$$

- Find the distance of the motorist when he first observed the accident.

$$S + 30 = 635.5 \text{ ft}$$

Estimate of Velocities. It is sometimes necessary to estimate the speed of a vehicle just before it is involved in an accident. This may be done by using the braking distance equations if skid marks can be seen on the pavement. The steps taken in making the speed estimate are as follows.

Step 1. Measure the length of skid marks for each tire and determine the average. The result is assumed to be the braking distance D_b of the vehicle.

Step 2. Determine the coefficient of friction f by performing trial runs at the site under similar weather conditions, using vehicles with tires in a state similar to those of the vehicle involved in the accident. This is done by driving the vehicle at a known speed u_k and measuring the distance traveled D_k while braking the vehicle to rest. The coefficient of friction f_k can then be estimated using Eq. 3.19 to give

$$f_k = \frac{u_k^2}{30D_k} \mp G$$

Alternatively, a tabulated value of f is used for f_k.

Step 3. Use the value of f_k obtained in step 2 to estimate the unknown velocity u_u just prior to impact, that is, the velocity at which the vehicle was traveling just before it was involved in the accident. This is done by using Eq. 3.20. If it can be assumed that the application of the brakes reduced the velocity u_u to zero, then u_u may be obtained from

$$D_b = \frac{u_u^2}{30(f_k \pm G)}$$

or

$$D_b = \frac{u_u^2}{30\left(\dfrac{u_k^2}{30D_k} \mp G \pm G\right)} = \left(\frac{u_u^2}{u_k^2}\right)D_k \qquad \textbf{(3.22)}$$

giving

$$u_u = \left(\frac{D_b}{D_k}\right)^{1/2} u_k \tag{3.23}$$

If, however, the vehicle involved in the accident was traveling at speed u_1 when impact took place and speed u_1 is known, then using Eq. 3.20, the unknown speed u_u just prior to the impact may be obtained from

$$D_b = \frac{u_u^2 - u_1^2}{30\left(\dfrac{u_k^2}{30D_k} \mp G \pm G\right)} = \frac{u_u^2 - u_1^2}{u_k^2} D_k$$

giving

$$u_u = \left(\frac{D_b}{D_k} u_k^2 + u_1^2\right)^{1/2} \tag{3.24}$$

Note that the unknown velocity just prior to the impact obtained from either Eq. 3.23 or Eq. 3.24 is only an estimate but will always be a conservative estimate as it will always be less than the actual speed at which the vehicle was traveling before impact. The reason is that some reduction of speed usually takes place before skidding commences and in using Eq. 3.23, the assumption of the initial velocity u_u being reduced to zero is always incorrect. The lengths of the measured skid marks do not reflect these factors.

Turning Radii

When a vehicle is moving round a circular curve, there are two main forces in the radial direction acting on it: an outward radial force (centrifugal force) and an inward radial force. The inward radial force is due to the frictional effect between the tires and the roadway. At high velocities, this force is not usually adequate to counterbalance the outward radial force. Thus, the road must be inclined toward the center of the curve, to provide an additional force, from the component of the vehicle weight down the incline. This additional force combines with the inward radial force to counterbalance the outward radial force. The inclination of the roadway toward the center of the curve is known as *superelevation*. The action of these forces on a vehicle moving round a circular curve is shown in Figure 3.7.

The minimum radius of a circular curve R for a vehicle traveling at speed u mph can be determined by considering the equilibrium of the vehicle with respect to its moving up or down the incline. If α is the angle of inclination of the highway, the component of the weight down the incline is $W \sin \alpha$, and the frictional force also acting down the incline is $Wf \cos \alpha$. The centrifugal force F_c is given as

$$F_c = \frac{Wa_c}{g}$$

Figure 3.7   Forces Acting on a Vehicle Traveling on a Horizontal Curve Section of a Road

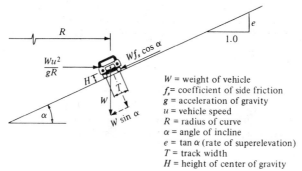

W = weight of vehicle
f_s = coefficient of side friction
g = acceleration of gravity
u = vehicle speed
R = radius of curve
α = angle of incline
e = tan α (rate of superelevation)
T = track width
H = height of center of gravity

Source: Redrawn from Donald R. Drew, *Traffic Flow Theory and Control*, copyright © 1968, McGraw-Hill Book Company.

where

a_c = acceleration for curvilinear motion = u^2/R
R = radius of the curve
W = weight of the vehicle
g = acceleration of gravity

When the vehicle is in equilibrium with respect to the incline (that is, vehicle moves forward but neither up nor down the incline), we may equate the three relevant forces and obtain

$$\frac{Wu^2}{gR} \cos \alpha = W \sin \alpha + Wf_s \cos \alpha \qquad (3.25)$$

where f_s = coefficient of side friction and $(u^2/g) = R(\tan \alpha + f_s)$, which gives

$$R = \frac{u^2}{g(\tan \alpha + f_s)} \qquad (3.26)$$

Tan α, the tangent of the angle of inclination of the roadway, is known as the rate of superelevation e. Eq. 3.26 can therefore be written as

$$R = \frac{u^2}{g(e + f_s)} \qquad (3.27)$$

Again, if g is taken as 32.2 ft/sec^2 and u is in mph, the minimum radius R is given in ft as

$$R = \frac{u^2}{15(e + f_s)} \qquad (3.28)$$

Table 3.5 Coefficient of Side Friction for Different Design Speeds

Design Speed (mph)	Coefficient of Side Friction, f_s
30	0.16
40	0.15
50	0.14
60	0.12
70	0.10

Source: Adapted from *A Policy on Geometric Design of Highways and Streets*, Washington, D.C.: The American Association of State Highway and Transportation Officials, copyright 1984. Used by permission.

Eq. 3.28 shows that to reduce R for a given velocity, either e or f_s or both should be increased.

There are, however, stipulated maximum values that should be used for either e or f_s. Several factors control the maximum value for the rate of superelevation. These include the location of the highway (that is, whether in an urban or rural area), weather conditions (such as the occurrence of snow), and the distribution of slow-moving traffic within the traffic stream. For highways located in rural areas with no snow or ice, a maximum superelevation rate of 0.10 is generally used. For highways located in areas with snow and ice, values ranging from 0.08 to 0.10 are used. For expressways in urban areas, a maximum superelevation rate of 0.08 is used. Because of the relatively low speeds on local urban roads, these roads usually are not superelevated.

The values used for side friction f_s generally vary with the design speed. Table 3.5 gives values generally used for design speeds up to 70 mph.

Example 3–5 Minimum Radius of a Highway Horizontal Curve

An existing horizontal curve on a highway has a radius of 268 ft, which restricts the maximum speed on this section of the road to only 60 percent of the design speed of the highway. If the curve is to be improved so that maximum speed will be that of the design speed of the highway, determine the minimum radius of the new curve. Assume the coefficient of side friction is 0.15 for the existing curve and the rate of superelevation is 0.08 for both the existing curve and the new curve to be designed.

- Use Eq. 3.28 to find the maximum permissible speed on the existing curve.

$$268 = \frac{u^2}{15(0.08 + 0.15)}$$

$$u = 30.4 \text{ mph}$$

- Determine the design speed of the highway.

$$30.4/0.6 = 50.67 \text{ mph}$$

- Find the radius of new curve by using Eq. 3.28

$$R = \frac{50.67^2}{15(0.08 + f_s)}$$

with the value of f_s for 50 mph from Table 3.5 ($f_s = 0.14$)

$$R = \frac{(50.67)^2}{15(0.08 + 0.14)} = 778 \text{ ft}$$

ROAD CHARACTERISTICS

The characteristics of the highway discussed in this section are related to stopping and passing since these have a more direct relationship to the characteristics of the driver and vehicle discussed earlier. Other characteristics of the highway are presented in detail in Chapter 14, where geometric design of the highway is discussed.

Sight Distance

Sight distance is the length of the roadway a driver can see ahead of him at any particular time. The sight distance available at each point of the highway must be such that when a driver is traveling at the highway's design speed, adequate time is given, after observing an object in the vehicle's path, to make the necessary evasive maneuvers without colliding with the object. The two types of sight distance are (1) stopping sight distance and (2) passing sight distance.

Stopping Sight Distance
The stopping sight distance (SSD) for design purposes is usually taken as the minimum sight distance required for a driver to stop a vehicle after seeing an object in the vehicle's path without hitting that object. This distance is the summation of the distance traveled during perception-reaction time and the distance traveled during braking. The SSD for a vehicle traveling at u mph is therefore the same as the stopping distance given in Eq. 3.21. The SSD is therefore given as

$$\text{SSD} = 1.47ut + \frac{u^2}{30(f \mp G)} \tag{3.29}$$

It is essential that highways are designed so that sight distance along the highway is at least equal to the SSD. Table 3.6 shows values for SSDs for different design

Table 3.6   SSDs for Different Design Speeds

Design Speed (mph)	Assumed Speed for Condition (mph)	Brake Reaction Time (sec)	Brake Reaction Distance (ft)	Coefficient of Friction, f	Braking Distance on Level[a] (ft)	SSD Computed[a] (ft)	SSD Rounded for Design (ft)
20	20–20	2.5	73.3– 73.3	0.40	33.3– 33.3	106.7–106.7	125–125
25	24–25	2.5	88.0– 91.7	0.38	50.5– 54.8	138.5–146.5	150–150
30	28–30	2.5	102.7–110.0	0.35	74.7– 85.7	177.3–195.7	200–200
35	32–35	2.5	117.3–128.3	0.34	100.4–120.1	217.7–248.4	225–250
40	36–40	2.5	132.0–146.7	0.32	135.0–166.7	267.0–313.3	275–325
45	40–45	2.5	146.7–165.0	0.31	172.0–217.7	318.7–382.7	325–400
50	44–50	2.5	161.3–183.3	0.30	215.1–277.8	376.4–461.1	400–475
55	48–55	2.5	176.0–201.7	0.30	256.0–336.1	432.0–537.8	450–550
60	52–60	2.5	190.7–220.0	0.29	310.8–413.8	501.5–633.8	525–650
65	55–65	2.5	201.7–238.3	0.29	347.7–485.6	549.4–724.0	550–725
70	58–70	2.5	212.7–256.7	0.28	400.5–583.3	613.1–840.0	625–850

[a]Different values for the same speed result from using unequal coefficients of friction.

Source: Adapted from *A Policy on the Geometric Design of Highways and Streets*, Washington, D.C.: The American Association of State Highway and Transportation Officials, copyright 1984. Used by permission.

speeds. The SSD requirements dictate the minimum lengths of vertical curves and minimum radii for horizontal curves that should be designed for any given highway.

Minimum Lengths of Vertical Curves. Highway vertical curves are usually designed as parabolic curves. The expressions developed for minimum lengths of vertical curves are therefore based on the properties of a parabola. The two types of vertical curves on a highway are (1) crest vertical curves and (2) sag vertical curves.

Crest Vertical Curves. Two conditions exist for the minimum length of crest vertical curves. These are when (1) the sight distance is greater than the length of the vertical curve, and (2) the sight distance is less than the length of the vertical curve. Let us first consider the case of the sight distance being greater than the length of the vertical curve. Figure 3.8 shows this condition. This figure schematically presents a vehicle on the grade at C with the driver's eye at height H_1 and an object of height H_2 located at D. If this object is seen by the driver, the line of sight is PN and the sight distance is S. Note that the line of sight is not necessarily horizontal, but in calculating the sight distance, the horizontal projection is considered.

From the properties of the parabola,

$$X_3 = \frac{L}{2}$$

The sight distance S is then given as

$$S = X_1 + \frac{L}{2} + X_2 \tag{3.30}$$

Figure 3.8   Sight Distance on Crest Vertical Curve ($S > L$)

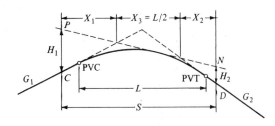

L = length of vertical curve (ft)
S = sight distance (ft)
H_1 = height of eye above roadway surface (ft)
H_2 = height of object above roadway surface (ft)
G_1 = slope of first tangent
G_2 = slope of second tangent
PVC = point of vertical curve
PVT = point of vertical tangent

X_1 and X_2 can be found in terms of the grades G_1 and G_2 and their algebraic difference A. The minimum length of the vertical curve for the required sight distance is obtained as

$$L_{min} = 2S - \frac{200(\sqrt{H_1} + \sqrt{H_2})^2}{A} \quad \text{(for } S > L) \quad \quad \textbf{(3.31)}$$

It has been the practice to assume the height of the driver H_1 to be 3.75 ft and the height of the object to be 0.5 ft. Because of the increasing number of compact automobiles on the nation's highways, the height of the driver's eye is now taken as 3.5 ft. With this assumption, Eq. 3.31 becomes

$$L_{min} = 2S - \frac{1329}{A} \quad \text{(for } S > L) \quad \quad \textbf{(3.32)}$$

When the sight distance is less than the length of the crest vertical curve, the configuration shown in Figure 3.9 applies. The properties of a parabola can also be used to show that the minimum length of the vertical curve is given as

$$L_{min} = \frac{AS^2}{200(\sqrt{H_1} + \sqrt{H_2})^2} \quad \text{(for } S < L) \quad \quad \textbf{(3.33)}$$

Substituting 3.5 ft for H_1 and 0.5 ft for H_2 gives

$$L_{min} = \frac{AS^2}{1329} \text{ (for } S < L) \quad \quad \textbf{(3.34)}$$

Figure 3.9   Sight Distance on Crest Vertical Curve ($S < L$)

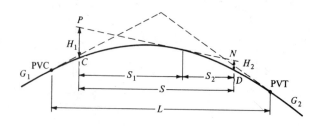

L = length of vertical curve (ft)
S = sight distance (ft)
H_1 = height of eye above roadway surface (ft)
H_2 = height of object above roadway surface (ft)
G_1 = slope of first tangent
G_2 = slope of second tangent
PVC = point of vertical curve
PVT = point of vertical tangent

Example 3–6 Minimum Length of a Crest Vertical Curve

A crest vertical curve is to be designed to join a $+2$ percent grade with a -2 percent grade at a section of a 2-lane highway. Determine the minimum length of the curve if the design speed of the highway is 60 mph and $S < L$. Assume $f = 0.29$ and the perception-reaction time is 2.5 sec.

■ Use Eq. 3.29 to determine the SSD required for the design conditions. (Since the grade changes constantly on a vertical curve, the worst case value for G is used to determine the braking distance.)

$$\text{SSD} = 1.47ut + \frac{u^2}{30(f-G)} = 1.47 \times 60 \times 2.5 + \frac{60^2}{30(0.29 - 0.02)}$$

$$= 220.50 + 444.44$$

$$= 664.94 \text{ ft}$$

■ Use Eq. 3.34 to obtain the minimum length of vertical curve.

$$L_{\min} = \frac{AS^2}{1329} = \frac{4 \times 664.94^2}{1329}$$

$$= 1330.8 \text{ ft}$$

Example 3–7 Maximum Safe Speed on a Crest Vertical Curve

An existing vertical curve on a highway joins a $+4$ percent grade with a -4 percent grade. If the length of the curve is 250 ft, what is the maximum safe speed on this curve? Assume $f = 0.4$ and perception-reaction time $= 2.5$ sec. Also assume $S < L$.

■ Determine the SSD using the length of the curve and Eq. 3.34.

$$250 = \frac{8 \times S^2}{1329}$$

$$S = 203.8 \text{ ft}$$

■ Now determine the maximum safe speed for this sight distance from Eq. 3.29.

$$203.8 = 1.47 \times 2.5u + \frac{u^2}{30(0.4 - 0.04)}$$

from which we obtain the quadratic equation

$$u^2 + 39.69u - 2201.04 = 0$$

- Solving the quadratic equation to find the maximum safe speed,

$$u = 31.1 \text{ mph}$$

The maximum safe speed for the SSD available is therefore 31.1 mph. However, if a speed limit is to be posted to satisfy this condition, 30 mph will be used since speed limits are usually set at 5 mph increments.

Sag Vertical Curves. The selection of the minimum length of a sag vertical curve is usually based on one of four different criteria: (1) sight distance provided by the headlight, (2) rider comfort, (3) control of drainage, and (4) general appearance.

The headlight sight distance requirement is based on the fact that as a vehicle is being driven on a sag vertical curve at night, the position of the headlight and the direction of the headlight beam dictate the stretch of highway ahead that is lighted and therefore the distance that can be seen by the driver. Figure 3.10 is a schematic of the situation when $S > L$. The headlight is located at a height H above the ground, and the headlight beam is inclined upward at an angle β to the horizontal. The headlight beam intersects the road at D, thereby restricting the available sight distance to S. The values used for H and β are usually 2 ft and 1°, respectively. Using the properties of the parabola, it can be shown that

$$L = 2S - \frac{200(H + S \tan \beta)}{A} \qquad \text{(for } S > L) \qquad \textbf{(3.35)}$$

Substituting 2 ft for H and 1° for β makes Eq. 3.35 become

$$L = 2S - \frac{(400 + 3.5S)}{A} \qquad \text{(for } S > L) \qquad \textbf{(3.36)}$$

Similarly, for the condition when $S < L$, it can be shown that

$$L = \frac{AS^2}{200(H + S \tan \beta)} \qquad \text{(for } S < L) \qquad \textbf{(3.37)}$$

Figure 3.10 Headlight Sight Distance on Sag Vertical Curves ($S > L$)

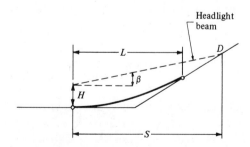

and substituting 2 ft for H and $1°$ for β gives

$$L = \frac{AS^2}{400 + 3.5S} \qquad \text{(for } S < L\text{)} \qquad (3.38)$$

To provide a safe condition on a sag vertical curve, the curve must be of such a length that it will make the light beam sight distance S be at least equal to the SSD. The SSDs for the appropriate design speeds are therefore used for S when Eqs. 3.37 and 3.38 are used to compute minimum lengths of sag vertical curves.

The comfort criterion for the design of sag vertical curves takes into consideration that, when a vehicle traverses a sag vertical curve, both the gravitational and centrifugal forces act in combination, resulting in a greater effect than on a crest vertical curve where these forces act in opposition to each other. Several factors such as weight carried, body suspension of the vehicle, and tire flexibility affect comfort due to change in vertical directions. This makes it difficult for comfort to be measured directly. It is, however, generally accepted that a comfortable ride will be provided if the radial acceleration is not greater than 1 ft/sec². An expression that has been used for the comfort criterion is

$$L = \frac{Au^2}{46.5} \qquad (3.39)$$

where u is the design speed in mph and L and A are the same as previously used. This length is usually about 75 percent of that obtained from the headlight sight distance requirement.

The drainage criterion for sag vertical curves is essential when the road is curbed. This criterion is different from the other criteria in that a maximum length requirement is given rather than a minimum length. The requirement usually specified to satisfy this criterion is that a minimum grade of 0.35 percent be provided within 50 ft of the level point of the curve. It has been observed that the maximum length for this criterion is usually greater than the minimum lengths for the other criteria for speeds up to 60 mph and about the same for a speed of 70 mph.

The criterion of general appearance is usually satisfied by the use of a rule of thumb expressed as

$$L = 100A \qquad (3.40)$$

where L is the minimum length of the sag vertical curve. Experience has shown, however, that longer curves are frequently necessary for high-type highways if the general appearance of these highways is to be improved.

Example 3–8   Minimum Length of a Sag Vertical Curve

A sag vertical curve is to be designed to join a -3 percent grade to a $+3$ percent grade. If the design speed is 40 mph, determine the minimum length of the curve that will satisfy all minimum criteria. Assume $f = 0.32$ and perception-reaction time $= 2.5$ sec.

- Find the stopping sight distance.

$$SSD = 1.47 \times 40 \times 2.5 + \frac{40^2}{30(0.32 - 0.03)} = 147.0 + 183.9$$

$$= 330.9 \text{ ft}$$

- Determine whether $S < L$ or $S > L$ for the headlight sight distance criterion.

$$L = 2 \times 330.9 - \frac{400 + 3.5 \times 330.9}{6} \qquad S > L$$

$$= 402.1 \text{ ft}$$

(This condition is not appropriate since $330.9 < 402.1$ and therefore $S \not> L$.)

$$L = \frac{6 \times 330.9^2}{400 + 3.5 \times 330.9} \qquad S < L$$

$$= 421.6 \text{ ft}$$

(This condition applies.)
- Determine minimum length for the comfort criterion.

$$L = \frac{6 \times 40^2}{46.5} = 206.5 \text{ ft}$$

- Determine minimum length for the general appearance criterion.

$$L = 100 \times 6 = 600 \text{ ft}$$

The minimum length to satisfy all criteria is 600 ft.

The above solution demonstrates a theoretical analysis to determine the minimum length of a sag vertical curve that satisfies the criteria discussed. AASHTO, however, suggests that the headlight criterion seems to be the most logical for general use, and therefore uses this criterion to establish minimum lengths of sag vertical curves.

AASHTO has developed charts based on SSD requirements from which minimum lengths of vertical curves can be determined. The use of these charts is discussed in detail in Chapter 14.

Minimum Radii of Horizontal Curves.   When a vehicle is being driven around a horizontal curve, an object located near the inside edge of the road may interfere with the view of the driver, which will result in a reduction of the driver's sight

Figure 3.11   Range of Lower Values for Stopping Sight Distances on Horizontal Curves

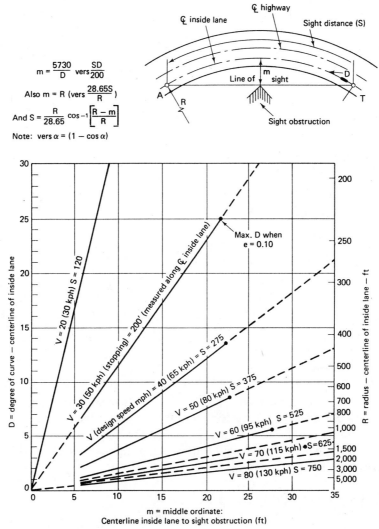

Source: Reproduced from *Transportation and Traffic Engineering Handbook*, 2nd ed., Institute of Transportation Engineers, Washington, D.C., 1982.

distance ahead of her. When such a situation exists, it is necessary to design the horizontal curve such that the available sight distance is at least equal to the safe stopping distance. Figure 3.11 is a schematic of this situation.

Consider a vehicle at point A and an object at point T. The line of sight that will just permit the driver to see the object is the chord AT. However, the horizontal distance traveled by the vehicle from point A to the object is arc AT. If 2θ (in degrees) is the angle subtended at the center of the circle by arc

AT, then

$$\frac{2R\theta\pi}{180} = S$$

where
   $R =$ radius of horizontal curve
   $S =$ sight distance $=$ length of arc AT

$$\theta = \frac{57.30}{2R} S$$

However,

$$\frac{R - m}{R} = \cos\theta$$

$$\frac{R - m}{R} = \cos\left(\frac{28.65}{R} S\right) \tag{3.41}$$

$$m = R\left(1 - \cos\frac{28.65}{R} S\right)$$

Eq. 3.41 can be used to determine m, R, or S, depending on the information known.

Example 3–9 Location of Object Near a Horizontal Curve

A horizontal curve having a radius of 800 ft forms part of a 2-lane highway that has a posted speed limit of 35 mph. If the highway is flat at this section, determine the minimum distance a large billboard can be placed from the center line of the inside lane of the curve, without reducing the required SSD. Assume perception-reaction time of 2.5 sec.

- Determine the required SSD.

$$(1.47 \times 35 \times 2.5) + \frac{35^2}{30(0.34)} = 248.72 \text{ ft}$$

Note that $f = 0.34$ is obtained from Table 3.4.
- Use Eq. 3.41.

$$m = 800\left[1 - \cos\left(\frac{28.65}{800}(248.72)\right)\right] = 800(1 - 0.988) \text{ ft}$$
$$= 9.6 \text{ ft}$$

Decision Sight Distance

The SSDs given in Table 3.6 are usually adequate for ordinary conditions, where the stimulus is expected by the driver. When, however, the stimulus is unexpected or it is necessary for the driver to make unusual maneuvers, longer SSDs are usually required since the perception-reaction time is much longer. This longer sight distance is the decision sight distance and it is defined as the "distance required for a driver to detect an unexpected or otherwise difficult-to-perceive information source or hazard in a roadway environment that may be visually cluttered, recognize the hazard of its threat potential, select an appropriate speed and path, and initiate and complete the required safety maneuvers safely and efficiently."[13] Table 3.7 shows recommended values for decision sight distances, which may be used at interchanges and intersections; toll plazas and lane drops and other locations where the cross section of the highway changes; sites at which unusual or unexpected maneuvers may be necessary; and areas where sources of information, such as traffic control devices, traffic, and advertisements, compete for the driver's attention.

Passing Sight Distance

The passing sight distance is the minimum sight distance required on a 2-lane, two-way highway that will permit a driver to complete a passing maneuver, without colliding with an opposing vehicle and without cutting off the passed vehicle. The passing sight distance will also provide for the driver to successfully abort the passing maneuver (that is, return to the right lane behind the vehicle being passed) if he or she so desires. In determining minimum passing sight distances for design purposes, only single passes are considered (that is, a single vehicle passing a single vehicle). Although it is possible for multiple passing maneuvers (that is, more than one vehicle pass or are passed in one maneuver) to occur, it is not pragmatic for minimum design criteria to be based on them.

In order to determine the minimum passing sight distance, certain assumptions have to be made regarding the movement of the passing vehicle during a passing maneuver.

1. The vehicle being passed (impeder) is traveling at a uniform speed.
2. The passing vehicle travels at the same speed as the impeder prior to the decision to begin the passing maneuver. If the passing maneuver is not undertaken, the vehicle continues at the same speed.
3. On arrival at a passing section, some time elapses during which the driver decides whether to undertake the passing maneuver.
4. If the decision is made to pass, the passing vehicle is accelerated during the passing maneuver, and the average passing speed is about 10 mph more than the speed of the impeder vehicle.
5. A suitable clearance exists between the passing vehicle and any opposing vehicle when the passing vehicle re-enters the right lane.

Table 3.7   Decision Sight Distances for Different Design Speeds

Design Speed (mph)	Time(s) Premaneuver		Maneuver (Lane Change)	Summation	Decision Sight Distance (ft)	
	Detection and Recognition	Decision and Response Initiation			Computed	Rounded for Design
30	1.5–3.0	4.2–6.5	4.5	10.2–14.0	449– 616	450– 625
40	1.5–3.0	4.2–6.5	4.5	10.2–14.0	598– 821	600– 825
50	1.5–3.0	4.2–6.5	4.5	10.2–14.0	748–1027	750–1025
60	2.0–3.0	4.7–7.0	4.5	11.2–14.5	986–1276	1000–1275
70	2.0–3.0	4.7–7.0	4.0	10.7–14.0	1098–1437	1100–1450

Source: Adapted from *A Policy on the Geometric Design of Highways and Streets*, Washington, D.C.: The American Association of State Highway and Transportation Officials, copyright 1984. Used by permission.

Figure 3.12 Elements of and Total Passing Sight Distance on 2-Lane Highways

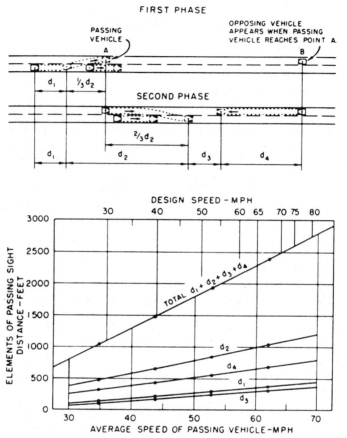

Source: Reproduced from *A Policy on Geometric Design of Highways and Streets*, Washington, D.C.: The American Association of State Highway and Transportation Officials, copyright 1984. Used by permission.

These assumptions have been used by AASHTO to develop a minimum passing sight distance requirement for 2-lane, two-way highways.

The minimum passing sight distance is the total of four components, as shown in Figure 3.12.

d_1 = distance traversed during perception-reaction time and during initial acceleration to the point where the passing vehicle just enters the left lane

d_2 = distance traveled during the time the passing vehicle is traveling on the left lane

d_3 = distance between the passing vehicle at the end of the passing maneuver and the opposing vehicle

d_4 = distance moved by the opposing vehicle during two-thirds of the time the passing vehicle is on the left lane (usually taken as $2/3d_2$)

The distance d_1 is obtained from the expression

$$d_1 = 1.47t_1\left(u - m + \frac{at_1}{2}\right)$$ **(3.42)**

where

t_1 = time for initial maneuver (sec)
a = average acceleration rate (mph/sec)
u = average speed of passing vehicle (mph)
m = difference in speeds of passing and impeder vehicles

The distance d_2 is obtained from

$$d_2 = 1.47ut_2$$

where

t_2 = time passing vehicle is traveling on left lane (sec)
u = average speed of passing vehicle (mph)

The clearance distance d_3 between the passing vehicle and the opposing vehicle at the completion of the passing maneuver has been found to vary between 110 ft and 300 ft.

Table 3.8 shows these components calculated for different speeds. It should

Table 3.8   Components of Safe Passing Sight Distances on 2-lane Highways

Component	Speed Range in mph (Average Passing Speed in mph)			
	30–40 (34.9)	40–50 (43.8)	50–60 (52.6)	60–70 (62.0)
Initial maneuver:				
a = average acceleration (mph/sec)[a]	1.40	1.43	1.47	1.50
t_1 = time (sec)[a]	3.6	4.0	4.3	4.5
d_1 = distance traveled (ft)	145	215	290	370
Occupation of left lane:				
t_2 = time (sec)[a]	9.3	10.0	10.7	11.3
d_2 = distance traveled (ft)	475	640	825	1030
Clearance length:				
d_3 = distance traveled (ft)[a]	100	180	250	300
Opposing vehicle:				
d_4 = distance traveled (ft)	315	425	550	680
Total distance, $d_1 + d_2 + d_3 + d_4$ (ft)	1035	1460	1915	2380

[a]For consistent speed relation, observed values are adjusted slightly.

Source: Adapted from *A Policy on Geometric Design of Highways and Streets*, Washington, D.C.: The American Association of State Highway and Transportation Officials, copyright 1984. Used by permission.

Table 3.9 Suggested Minimum Passing Zone and Passing Sight Distance Requirements for 2-Lane, Two-Way Highways in Mountainous Areas

85th Percentile Speed (mph)	Available Sight Distance (ft)	Minimum Passing Zone		Minimum Passing Sight Distance	
		Suggested (ft)	MUTCD* (ft)	Suggested (ft)	MUTCD* (ft)
30	600– 800	490		630	
	800–1000	530		690	
	1000–1200	580	400	750	500
	1200–1400	620		810	
35	600– 800	520		700	
	800–1000	560		760	
	1000–1200	610	400	820	550
	1200–1400	650		880	
40	600– 800	540		770	
	800–1000	590		830	
	1000–1200	630	400	890	600
	1200–1400	680		950	
45	600– 800	570		840	
	800–1000	610		900	
	1000–1200	660	400	960	700
	1200–1400	700		1020	
50	600– 800	590		910	
	800–1000	630		970	
	1000–1200	680	400	1030	800
	1200–1400	730		1090	

Manual on Uniform Traffic Control Devices, published by FHWA.

Source: Adapted from N. J. Garber and M. Saito, *Centerline Pavement Markings on Two-Lane Mountainous Highways*, Research Report No. VHTRC 84-R8, Virginia Highway and Transportation Research Council, Charlottesville, Va., March 1983.

be made clear that values given in Table 3.8 are for design purposes and cannot be used for marking passing and no-passing zones on completed highways. The values used for this purpose are obtained from different assumptions and are much shorter. Table 3.9 shows values recommended for this purpose in the *Manual of Uniform Control Devices*.[14] Recent studies have, however, shown that these values are inadequate.[15,16] Table 3.9 also shows minimum sight distance requirements obtained from a study on 2-lane, two-way roads in mountainous areas of Virginia.[17]

SUMMARY

The highway or traffic engineer needs to study and understand the fundamental elements that are important in the design of traffic control systems. This chapter has presented in a concise manner the basic characteristics of the fundamental elements that should be known and understood by transportation and/or traffic engineers. Although the characteristics are presented in terms of the highway mode, several of these are also used for other modes. For example, the driver and pedestrian characteristics also apply to other modes in which vehicles are manually driven and some possibility exists for interaction between the driven vehicle and pedestrians. It should be emphasized again that because of the wide range of capabilities among drivers and pedestrians, the use of average limitations of drivers and pedestrians in developing guidelines for design may result in the design of facilities that will not satisfy a significant percentage of the people using the facility. High percentile values (such as 85th or 95th percentile values) are therefore normally used for design purposes.

PROBLEMS

3–1 If the design speed of a multilane highway is 60 mph, what is the minimum sight distance that should be provided on the road if (a) the road is horizontal and (b) the road has a maximum grade of 5 percent? Assume the perception-reaction time = 2.5 sec.

3–2 The acceleration of a vehicle takes the form

$$\frac{dv}{dt} = 3.6 - 0.06v$$

where v is the vehicle speed in ft/sec. The vehicle is traveling at 45 ft/sec at time T_o.

(a) Determine the distance traveled by the vehicle, when accelerated to a speed of 55 ft/sec.

(b) Determine the time at which the vehicle attains the velocity of 55 ft/sec.

(c) Determine the acceleration of the vehicle after 3 sec.

3–3 A horizontal curve is to be designed for a section of a highway having a design speed of 60 mph. If the physical conditions restrict the radius of the curve to 500 ft, what value is required for the superelevation at this curve? Will this be a good design?

3–4 Determine the minimum radius required at a curved section of a highway if the design speed is 70 mph and the superelevation is 0.08.

3–5 Use the expression for the minimum length required for a crest vertical curve ($S < L$) in terms of minimum sight distance, algebraic difference of tangent gradients, height of driver's eyes above road pavement, and height of object above road pavement to determine the minimum length of a crest vertical curve formed by grades of +5 percent and +2 percent on a 2-lane highway with a design speed of 70 mph.

3–6 A curve of radius 200 ft and $e = 0.08$ is located at a section of an existing rural highway, which restricts the safe speed at this section of the highway to 55 percent of the design speed. This drastic reduction of safe speed resulted in a high accident rate at this section. To reduce the accident rate, a new alignment is to be designed with a crest vertical curve joining two tangents of $+3$ percent and -3 percent. Determine the minimum length of this vertical curve if the safe speed should be increased to the design speed of the highway. Assume $e = 0.08$, $f = 0.30$, and perception-reaction time $= 2.5$ sec.

3–7 A temporary diversion has been constructed on a highway of $+4$ percent gradient due to major repairs that are being undertaken on a bridge. The maximum speed allowed on the diversion is 10 mph. Determine the minimum distance from the diversion that a road sign should be located informing drivers of the temporary change on the highway.

>Maximum allowable speed on highway = 70 mph
>Letter height of road sign = 4 in.
>Coefficient of friction between tires and pavement = 0.2
>Perception-reaction time = 2.5 sec

Assume that a driver can read a road sign within his or her area of vision at a distance of 40 ft for each inch of letter height.

3–8 An elevated expressway goes through an urban area, and crosses a local street as shown in Figure 3.13. The ramp provided is on a 2 percent grade and all vehicles leaving the motorway should stop at the intersection with the local street before merging. Determine the minimum radius and length required for the ramp for the following conditions:

>Maximum speed on motorway = 60 mph
>Distance between exit sign and exit ramp = 260 ft
>Letter height of road sign = 3 in.
>Coefficient of friction between tires and pavement = 0.2
>Perception-reaction time = 2.5 sec
>Maximum superelevation = 0.08
>Highway grade = 0%

Assume that a driver can read a road sign within his or her area of vision at a distance of 50 ft for each inch of letter height, and the driver sees the stop sign immediately on entering the ramp.

FIGURE 3.13

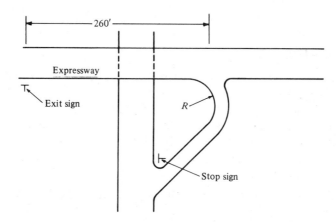

3–9 An exit ramp joins the right lane of a divided expressway to a frontage road, which is parallel to the expressway. (See Figure 3.14.) If the exit ramp is in the form of a reverse curve, develop the relationship between the radii R_1 and R_2 of the reverse curve, the distance d between the expressway and frontage road, and the distance D along the expressway center line between the points of tangency on the expressway and the frontage road. If $D = 350$ ft, $d = 75$ ft, and R_1 is restricted to 500 ft, determine the speed limit that should be specified on the ramp if the coefficient of side friction is 0.15 and the curvilinear acceleration is 0.08 g.

FIGURE 3.14

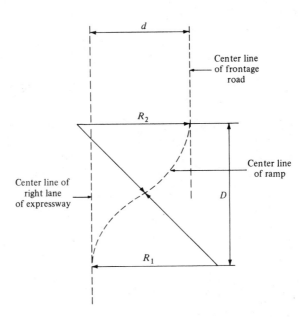

3–10 A corner of an existing building is 30 ft from the center line of a curved section of a rural highway in a mountainous area. If this section of the highway has a grade of 5 percent and the radius of the curve is only 300 ft, what speed limit will you recommend at this section of the highway?

REFERENCES

1. E. J. McCormick, *Human Factors Engineering*, McGraw-Hill Book Company, New York, 1957.

2. S. Salvatore, "Estimation of Vehicular Velocity Under Time Limitations and Restricted Conditions of Observations," *Highway Research Record* 195(1967):66–73.

3. E. Wolf, "Effects of Age on Peripheral Vision," *Highway Research Board Bulletin* 336(1962):26–31.

4. E. Wolf, "Glare Sensitivity in Relation to Age," *Highway Research Board Bulletin* 298(1961):18–22.

5. R. G. Mortimer, "The Effect of Glare in Simulated Night Driving," *Highway Research Record* 70(1965):57–62.

6. C. Johannsen and K. Rumar, "Driver's Brake Reaction Time," *Human Factors* 13(January 1971).

7. *A Policy on Geometric Design of Highways and Streets*, American Association of State Highway and Transportation Officials, Washington, D.C., 1984.

8. *Twin Trailer Trucks, Effects on Highways and Highway Safety*, National Research Council, Transportation Research Board, Washington, D.C., 1986.

9. *A Policy on Geometric Design of Highways and Streets.*

10. Ibid.

11. Paul J. Claffey, *Vehicle Operating Characteristics, Transportation and Traffic Engineering Handbook*, Institute of Transportation Engineers, Washington, D.C., 1975.

12. Ibid.

13. G. J. Alexander and H. Lunenfeld, *Positive Guidance in Traffic Control*, U.S. Department of Transportation, Federal Highway Administration, Washington, D.C., 1975.

14. *Manual on Uniform Traffic Control Devices*, U.S. Department of Transportation, Federal Highway Administration, Washington, D.C., 1977.

15. *Passing and No-Passing Zones: Signs, Markings and Warrants*, Report No. FHWA-RD-79-5, U.S. Department of Transportation, Federal Highway Administration, Washington, D.C., September 1977.

16. N. J. Garber and M. Saito, *Centerline Pavement Markings on Two-lane Mountainous Highways*, Research Report No. VHTRC 84-R8, Virginia Highway and Transportation Research Council, Charlottesville, Va., March 1983.

17. Ibid.

ADDITIONAL READINGS

Cohen, S. L., and W. R. Reilly, "Delay Measurement at Signalized Intersections," *Public Roads* 42(3):81–84 (December 1977).

Hulbert, Slade F., and Albert Burg, "Human Factors in Transportation Systems," in *Systems Psychology*, Kenyon B. DeGreene, ed., McGraw-Hill Book Company, New York, 1970.

Kent, P. M., et al., *1978 National Truck Characteristics Report*, U.S. Department of Transportation, Federal Highway Administration, Washington, D.C., November 1979.

Policy on Design of Urban Highways and Arterial Streets, American Association of State and Highway Transportation Officials, Washington, D.C., 1973.

Shina, D., *Psychology on the Road: The Human Factor in Traffic Safety*, John Wiley & Sons, New York, 1977.

Symposium on Geometric Design of Large Trucks, Transportation Research Record 1052, National Research Council, Transportation Research Board, Washington, D.C., 1986.

CHAPTER 4

Traffic Engineering Studies

The availability of highway transportation facilities in our communities has provided several positive qualities that have contributed to a high standard of living. At the same time, however, several problems related to the highway mode of transportation have developed. These problems include highway-related accidents, parking difficulties, congestion, and delay. To reduce the negative impact of highways, it is necessary to collect adequate information that describes the extent of the problems and identifies their locations. Such information is usually collected by organizing and conducting relevant traffic surveys and studies. This chapter introduces the reader to the different traffic engineering studies that are conducted to collect relevant traffic data. Brief descriptions of the methods of collecting and analyzing the data are also included.

Traffic studies may be grouped into three main categories: (1) static inventories, (2) administrative studies, and (3) dynamic studies. Inventories merely list or display graphically existing information, such as street widths, available parking spaces, transit routes, traffic regulations, and so forth. Some inventories, for example, available parking spaces and traffic regulations, change frequently and therefore require periodic updating, whereas others, such as street widths, do not. In carrying out administrative studies, use is made of existing engineering records available in government agencies and departments to obtain the necessary information. The information is then used to prepare an inventory of the relevant data. These inventories are sometimes recorded in files but frequently are recorded in automated data processing (ADP) systems. Administrative studies include field surveys, which may involve field measurements and/or aerial photography. Dynamic traffic studies involve the collection of data under operational conditions and include studies of speed, traffic volume, travel time and delay, parking, and accidents. These studies are all described in this chapter.

SPOT SPEED STUDIES

Spot speed studies are carried out to estimate the distribution of speeds of vehicles in a stream of traffic at a particular location on a highway. The speed of a vehicle is defined as the rate of movement of the vehicle and it is usually expressed in miles per hour (mph) or kilometers per hour (kph). A spot speed study is carried out by observing the instantaneous speeds of a sample of vehicles at the specified location. It is important to note that the speed characteristics identified by such a study will be true only for the traffic and environmental conditions that exist during the study. The speed characteristics identified may be used in

- Establishing speed zones
- Determining whether complaints on speeds are valid
- Establishing passing and no-passing zones
- Designing geometric alignment
- Analyzing accident data
- Evaluating the effect of physical improvements by conducting before and after studies
- Determining the effect of speed enforcement programs and speed control measures
- Determining speed trends by conducting periodic studies at the same location

Location of Spot Speed Studies

The location of spot speed studies depends on the anticipated use of the results obtained. The following locations are generally used for the different applications listed.

1. Locations that represent different traffic conditions on a highway or highways are used for *basic data collection*.
2. Midblocks of urban highways and straight and level sections of rural highways are sites for *speed trend analyses*.
3. Any location may be used for the solution of a *specific traffic engineering problem*.

When spot speed studies are being conducted, it is important that unbiased data are obtained. This requires that drivers should not be aware that such a study is being conducted. Equipment used should therefore be concealed from the driver. The observers conducting the study should be inconspicuous and a gathering of onlookers should be avoided. Since the speeds recorded will eventually be subjected to statistical analysis, it is important that an adequate number of vehicle speeds be recorded.

Time and Length of Spot Speed Studies

The time for conducting a speed study also depends on the purpose of the study. In general, when the study is to set speed limits, to observe speed trends, or to collect basic data, it is recommended that the study be conducted during off-peak hours at the following times:

- 9:00 a.m. to 11:30 a.m.
- 1:30 p.m. to 4:30 p.m.
- 7:00 p.m. to 10:00 p.m.

The length of time of the study should be such that at least the minimum number of vehicle speeds required for statistical analysis is recorded, but it is also useful for this time not to be less than one hr nor should the sample size be less than 30.

Sample Size for Spot Speed Studies

The representative speed value at any location is usually taken as the mean of the speeds recorded during a speed study and is assumed to be the true mean of all vehicle speeds at that location. The accuracy of the assumption depends on the sample size. The higher the sample size, the higher is the probability that the estimated mean $\bar{u}$ is not significantly different from the true mean. It is therefore necessary to select a sample size that will give an estimated mean within the acceptable error limits for the study. Statistical procedures are used to determine this minimum sample size. Before discussing these procedures, it is first necessary to define certain significant values that we need to describe the speed characteristics.

1. Average speed is the arithmetic mean of all vehicle speeds (that is, the sum of all spot speeds) divided by the number of speeds. It is given as

$$\bar{u} = \frac{\sum f_i u_i}{\sum f_i} \tag{4.1}$$

where

$\bar{u}$ = arithmetic mean
f_i = frequency for each speed group
u_i = midvalue for the ith speed group

2. Median speed is the speed representation of a middle value in a series of spot speeds that are arranged in ascending order. Fifty percent of the speed values will be greater than the median, and fifty percent will be less than the median.

3. Modal speed is the value of speed that occurs most frequently in a sample of spot speeds.

4. The ith percentile spot speed is the value of spot speed below which i percent of the vehicles travel; for example, 85th percentile spot speed is the speed below which 85 percent of the vehicles travel and above which 15 percent of the vehicles travel.

5. Pace is that range of speed, usually taken at 10 mph intervals, that has the greatest number of observations. For example, if a set of speed data includes speeds between 30 and 60 mph the speed intervals will be 30 to 40 mph, 40 to 50 mph, and 50 to 60 mph, assuming a range of 10 mph. The pace is 40 to 50 mph if this range of speed has the highest number of observations.

6. Standard deviation of speeds is a measure of the spread of the individual speeds. It is estimated from Eq. 4.2 as

$$S = \sqrt{\frac{\sum (u_j - \bar{u})^2}{N - 1}} \qquad (4.2)$$

where

S = standard deviation

$\bar{u}$ = estimated mean

$u_j = j$th observation

N = number of observations

Speed data are, however, frequently presented in classes where each class consists of a range of speeds. The standard deviation is computed for such cases from Eq. 4.3 as

$$S = \sqrt{\frac{\sum (f_i u_i^2) - (\sum f_i u_i)^2 / \sum f_i}{\sum f_i = 1}} \qquad (4.3)$$

where

u_i = midvalue of speed class i

f_i = frequency of speed class i

Standard probability sampling theories are used to determine the sample size for traffic engineering studies. Although a detailed discussion of these procedures is beyond the scope of this book, the simplest and most commonly used procedures are presented. Interested readers can find an in-depth treatment of the topic in publications listed in Additional Readings at the end of this chapter.

The minimum sample size depends on the precision level desired. The precision level is defined as "the degree of confidence that the sampling error of a produced estimate will fall within a desired fixed range."[1] Thus, for a precision level of 90–10, there is a 90 percent probability (confidence level) that the error of an estimate will not be greater than 10 percent of its true value. The confidence level is commonly given in terms of the level of significance (α) where

$\alpha = (100 - \text{confidence level})$. The commonly used confidence level for speed counts is 95 percent.

The basic assumption made in determining the minimum sample size for speed studies is that the normal distribution describes the speed distribution at a given section of highway. The normal distribution is given as

$$f(x) = \frac{1}{\sigma\sqrt{2\pi}} e^{-(x-\mu)^2/2\sigma^2} \qquad \text{for } -\infty < x < \infty \qquad (4.4)$$

where

    μ = true mean of the population

    σ = true standard deviation

    σ^2 = true variance

The properties of the normal distribution are then used to determine the minimum sample size for an acceptable error d of the estimated speed. The following basic properties are used and are shown in Figure 4.1.

1. The normal distribution is symmetrical about the mean.
2. The total area under the normal distribution curve is equal to 1 or 100 percent.
3. The area under the curve between $\mu + \sigma$ and $\mu - \sigma$ is 0.6827.
4. The area under the curve between $\mu + 1.96\sigma$ and $\mu - 1.96\sigma$ is 0.95.
5. The area under the curve between $\mu + 2\sigma$ and $\mu - 2\sigma$ is 0.9545.
6. The area under the curve between $\mu + 3\sigma$ and $\mu - 3\sigma$ is 0.9971.
7. The area under the curve between $\mu + \infty$ and $\mu - \infty$ is 1.000.

The last five properties are used to draw specific conclusions about speed data. For example, if it can be assumed that the true mean of the speeds in a section of highway is 50 mph and the true standard deviation is 4.5 mph, then

Figure 4.1 Shape of the Normal Distribution

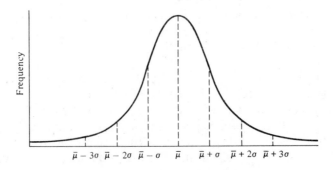

Table 4.1 Constant Corresponding to Level of Confidence

Confidence Level (%)	Constant Z
68.3	1.00
86.6	1.50
90.0	1.64
95.0	1.96
95.5	2.00
98.8	2.50
99.0	2.58
99.7	3.00

it can be concluded that 95 percent of all vehicle speeds will be between $(50 - 1.96 \times 4.5) = 41.2$ mph and $(50 + 1.96 \times 4.5) = 58.82$ mph. Similarly, if a vehicle is selected at random, there is a 95 percent chance that its speed is between 41.2 and 58.8 mph. The properties of the normal distribution have been used to develop an equation relating the sample size with the number of standard variations corresponding to a particular confidence level, the limits of tolerable error, and the standard deviation.

The formula is given as

$$N = \left(\frac{Z\sigma}{d}\right)^2 \tag{4.5}$$

where

N = minimum sample size

Z = number of standard deviations corresponding to the required confidence level

= 1.96 for 95 percent confidence level (Table 4.1)

σ = standard deviation (mph)

d = limit of acceptable error in the speed estimate (mph)

Table 4.2 Standard Deviation of Spot Speeds for Sample Size Determination

Traffic Area	Highway Type	Average Standard Deviation (mph)
Rural	2 lane	5.3
Rural	4 lane	4.2
Intermediate	2 lane	5.3
Intermediate	4 lane	5.3
Urban	2 lane	4.8
Urban	4 lane	4.9

Source: Adapted from *Manual of Traffic Engineering Studies*, 4th ed., Institute of Transportation Engineers, Washington, D.C., 1976.

The standard deviation can be estimated from previous data, or a small sample size can first be used. Table 4.2, which gives average standard deviations of spot speeds on different types of highways, may also be used to estimate the standard deviation.

Example 4-1   Minimum Sample Size for Spot Speed Study

Determine the minimum sample size required for a spot speed study on a rural 2-lane highway if the confidence level for the study is 95.0 percent and the tolerance is ± 1 mph.

- Use Eq. 4.5.

$$N = \left(\frac{Z\sigma}{d} \right)^2$$

where

$Z = 1.96$ (from Table 4.1)
$\sigma = \pm 5.30$ (from Table 4.2)
$d = \pm 1$

- Solve for N.

$$N = [(1.96 \times 5.3)/1]^2 = 108$$

The sample size should therefore be at least 108.

Methods of Conducting Spot Speed Studies

The methods used for conducting spot speed studies can generally be divided into two main categories: manual and automatic. Since the manual method is seldom used, automatic methods are described.

Automatic Methods

Several automatic devices that can be used to obtain the instantaneous speeds of vehicles at a location on a highway are now available on the market. These automatic devices can be grouped into two main categories: (1) those that use road detectors and (2) those that use Doppler principle meters (radar type).

Road Detectors.   Road detectors can be classified into two general categories: pneumatic road tubes and induction loops. These devices can be used to collect data on speeds at the same time that volume data are being collected. When road detectors are used to measure speed, they should be laid such that the probability of closing the connection of the meter by a passing vehicle during a speed measurement is reduced to a minimum. This is achieved by separating the road detectors by a distance of 3 to 15 ft.

The advantage of the detector meters is that human errors are considerably reduced. The disadvantages are that these devices tend to be rather expensive, and when pneumatic tubes are used, they are rather conspicuous and may, therefore, affect driver behavior that will result in a distortion of the speed distribution.

Pneumatic road tubes are laid across the lane at which data are to be collected. When a moving vehicle passes over the tube, an air impulse is transmitted through the tube to the counter. When used for speed measurements, two tubes are placed across the lane, usually about 6 ft apart. An impulse is recorded when the front wheels of a moving vehicle pass over the first tube, and shortly afterwards, a second impulse is recorded when the front wheels pass over the second tube. The time elapsed between the two impulses and the distance between the tubes are used to compute the speed of the vehicle.

An **inductive loop** is a rectangular wire loop buried under the roadway surface. It usually serves as the detector of a resonant circuit. It operates on the principle that a disturbance in the electrical field is created when a motor vehicle passes across it. This causes a change in potential that is amplified, resulting in an impulse being sent to the counter.

Doppler Principle Meters. Doppler meters work on the principle that when a signal is transmitted onto a moving vehicle, the change in frequency between the transmitted signal and the reflected signal is proportional to the speed of the moving vehicle. The difference between the frequency of the transmitted signal and that of the reflected signal is measured by the equipment, and this is converted to speed in mph. In setting up the equipment, care must be taken to reduce the angle between the direction of the moving vehicle and the line joining the center of the transmitter and the vehicle. The value of the speed recorded depends on that angle. If the angle is not zero, an error related to the cosine of that angle is introduced, resulting in a lower speed than that which would have been recorded if the angle was zero. However, this error is not very large as the cosine of small angles is not much less than 1.

The advantage of this method is that pneumatic tubes are not used, and if the equipment can be located at an inconspicuous position, the influence on driver behavior is considerably reduced.

Presentation and Analysis of Spot Speed Data

The data collected in spot speed studies are usually from a sample of vehicles using that section of the highway on which the study is conducted, but these data are used to determine the speed characteristics of the whole population of vehicles traveling on the study site. It is therefore necessary to use statistical methods in analyzing these data. Several characteristics are usually determined from the analysis of the data, some of which can be calculated directly from the data, whereas others can be determined from a graphical representation of the data. Thus, the data must be presented in a form suitable for analysis.

The presentation format most commonly used is the frequency distribution table. The first step in the preparation of a frequency distribution table is the selection of the number of classes, that is, the number of velocity ranges, into which the data are to be fitted. The number of classes chosen is usually between 8 and 20, depending on the data collected. One technique that can be used to determine the number of classes is to determine first the range for a class size of 8 and then of 20. Finding the difference between the maximum and minimum speeds in the data and dividing this number by 8 and then by 20 gives the maximum and minimum ranges in each class. A convenient range for each class is then selected and the number of classes determined. The midvalue of each class range is usually taken as the speed value for that class. Table 4.3 shows speed values that were obtained on a rural highway in Virginia during a speed study. The speeds range from 34.8 mph to 65.0 mph, giving a speed range of 30.2. For eight classes, the range per class is 3.75 mph, and for 20 classes, the range per class is 1.51 mph. It is convenient to choose a range of 2 mph per class, which will give 16 classes. A frequency distribution table can then be prepared, as shown in Table 4.4, in which the speed classes are listed in column 1 and the midvalues in column 2. The number of observations for each class is listed in column 3, and the cumulative percentages of all observations are listed in column 6.

Table 4.3 Speed Data Obtained on a Rural Highway

Car No.	Speed (mph)	Car No.	Speed (mph)	Car No.	Speed (mph)	Car No.	Speed (mph)
1	35.1	23	46.1	45	47.8	67	56.0
2	44.0	24	54.2	46	47.1	68	49.1
3	45.8	25	52.3	47	34.8	69	49.2
4	44.3	26	57.3	48	52.4	70	56.4
5	36.3	27	46.8	49	49.1	71	48.5
6	54.0	28	57.8	50	37.1	72	45.4
7	42.1	29	36.8	51	65.0	73	48.6
8	50.1	30	55.8	52	49.5	74	52.0
9	51.8	31	43.3	53	52.2	75	49.8
10	50.8	32	55.3	54	48.4	76	63.4
11	38.3	33	39.0	55	42.8	77	60.1
12	44.6	34	53.7	56	49.5	78	48.8
13	45.2	35	40.8	57	48.6	79	52.1
14	41.1	36	54.5	58	41.2	80	48.7
15	55.1	37	51.6	59	48.0	81	61.8
16	50.2	38	51.7	60	58.0	82	56.6
17	54.3	39	50.3	61	49.0	83	48.2
18	45.4	40	59.8	62	41.8	84	62.1
19	55.2	41	40.3	63	48.3	85	53.3
20	45.7	42	55.1	64	45.9	86	53.4
21	54.1	43	45.0	65	44.7		
22	54.0	44	48.3	66	49.5		

Another form of presentation used is the frequency histogram. The frequency histogram for the data shown in Table 4.4 is shown in Figure 4.2. The values in columns 2 and 3 of Table 4.4 are used to draw the frequency histogram, where the abscissa represents the speeds and the ordinate the observed frequency in each class.

The data can also be presented by preparing a frequency distribution curve as shown in Figure 4.3. In this case, a curve of percentage of observations against speed is drawn by plotting values from column 5 against the corresponding values in column 2 of Table 4.4 The total area under this curve is unity or 100 percent.

The cumulative frequency distribution curve is another form of presenting the data and is shown in Figure 4.4. In this case, the cumulative percentages in column 6 of Table 4.4 are plotted against the upper limit of each corresponding speed class. This curve, therefore, gives the percentages of vehicles that are traveling at or below a given speed.

The characteristics of the data can now be given in terms of the significant values defined earlier.

Table 4.4 Frequency Distribution Table for Set of Speed Data

1	2	3	4	5	6	7
Speed Class (mph)	Class Midvalue, u_i	Class Frequency (Number of Observations in Class), f_i	$f_i u_i$	Percentage of Observations in Class	Cumulative Percentage of All Observations	$f(u_i - \bar{u})^2$
34–35.9	35.0	2	70	2.3	2.30	420.5
36–37.9	37.0	3	111	3.5	5.80	468.75
38–39.9	39.0	2	78	2.3	8.10	220.50
40–41.9	41.0	5	205	5.8	13.90	361.25
42–43.9	43.0	3	129	3.5	17.40	126.75
44–45.9	45.0	11	495	12.8	30.20	222.75
46–47.9	47.0	4	188	4.7	34.90	25.00
48–49.9	49.0	18	882	21.0	55.90	9.0
50–51.9	51.0	7	357	8.1	64.0	15.75
52–53.9	53.0	8	424	9.3	73.3	98.00
54–55.9	55.0	11	605	12.8	86.1	332.75
56–57.9	57.0	5	285	5.8	91.9	281.25
58–59.9	59.0	2	118	2.3	94.2	180.50
60–61.9	61.0	2	122	2.3	96.5	264.50
62–63.9	63.0	2	126	2.3	98.8	364.50
64–65.9	65.0	1	65	1.2	100.0	240.25
		86	4260			3632.00

Figure 4.2 Histogram of Observed Vehicles' Speeds

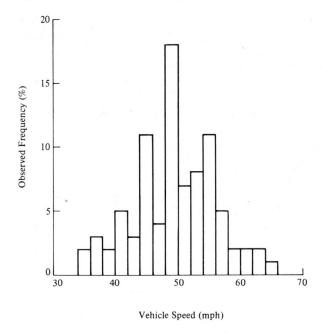

Figure 4.3 Frequency Distribution

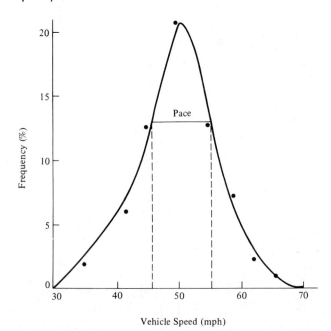

Figure 4.4 Cumulative Distribution

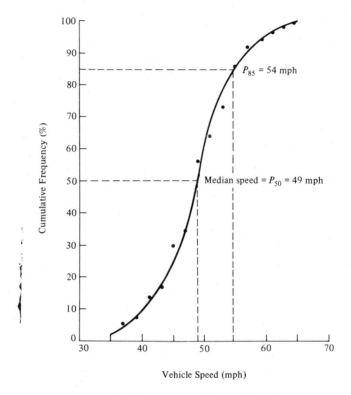

- The arithmetic mean speed is computed from Eq. 4.1.

$$\sum f_i = 86$$

$$\sum f_i u_i = 4260$$

$$\bar{u} = \frac{4260}{86} = 49.5 \text{ mph}$$

- The standard deviation is computed from Eq. 4.3.

$$\sum f_i(u_i - \bar{u})^2 = 3632$$

$$\sum f_i - 1 = 85$$

$$S^2 = \frac{3632}{85} = 42.73$$

$$S = \pm 6.5 \text{ mph}$$

- The median speed is obtained from the cumulative frequency distribution curve (Figure 4.4) as 49 mph, the 50th percentile speed.

- The pace is obtained from the frequency distribution curve (Figure 4.3) as 45 to 55 mph.
- The mode or modal speed is obtained from the frequency histogram as 49 mph (Figure 4.2). It may also be obtained from the frequency distribution curve shown in Figure 4.3, where the speed corresponding to the highest point on the curve is taken as an estimate of the modal speed.
- Eighty-fifth percentile speed is obtained from the cumulative frequency distribution curve as 54 mph (Figure 4.4).

Other Forms of Presentation and Analysis of Speed Data

Certain applications of speed study data may require a more complicated presentation and analysis of the speed data. For example, if the speed data is to be used in research on traffic flow theories, it may be necessary for the speed data to be fitted into a suitable theoretical frequency distribution, such as the normal distribution or the Gamma distribution. This is done by first assuming that the data fits a given distribution and testing this assumption using one of the methods of hypothesis testing such as chi-squared analysis. If the test suggests that the assumption can be accepted, then specific parameters of the distribution can be found using the speed data. The properties of this distribution are then used to describe the speed characteristics, and any form of mathematical computation can be carried out using the distribution. Detailed discussion of hypothesis testing is beyond the scope of this book, but interested readers will find additional information in any book on statistical methods for engineers.

It is also sometimes necessary to determine whether there is a significant difference between the mean speeds of two spot speed studies. This is done by comparing the absolute difference between the sample mean speeds and the product of the standard deviation of the difference in means and the factor Z for a given confidence level. If the absolute difference between the sample means is greater, it can then be concluded that there is a significant difference in sample means at that specific confidence level.

The standard deviation of the difference in means is given as

$$S_d = \sqrt{\frac{S_1^2}{n_1} + \frac{S_2^2}{n_2}} \tag{4.6}$$

where

n_1 = sample size for study 1
n_2 = sample size for study 2
S_d = square root of the variance of the difference in means
S_1^2 = variance about the mean for study 1
S_2^2 = variance about the mean for study 2

If $\bar{u}_1$ = mean speed of study 1, $\bar{u}_2$ = mean speed of study 2 and $|\bar{u}_1 - \bar{u}_2| > ZS_d$, where $|\bar{u}_1 - \bar{u}_2|$ is the absolute value of the difference in means, then it can be concluded that the mean speeds are significantly different at the confidence level corresponding to Z. This analysis assumes that $\bar{u}_1$ and $\bar{u}_2$ are estimated means from the same distribution. Since it is usual to use the 95 percent confidence level in traffic engineering studies, the conclusion will, therefore, be based on whether $|\bar{u}_1 - \bar{u}_2|$ is greater than $1.96S_d$.

Example 4–2 Significant Differences in Average Spot Speeds

Speed data were collected at a section of highway during and after utility maintenance work. The speed characteristics are given as $\bar{u}_1$, S_1 and $\bar{u}_2$, S_2 as shown in the accompanying data listing. Determine whether there was any significant difference between the average speed at the 95 percent confidence level.

$\bar{u}_1 = 35.5$ mph $\bar{u}_2 = 38.7$ mph
$S_1 = 7.5$ mph $S_2 = 7.4$ mph
$n_1 = 250$ $n_2 = 280$

- Use Eq. 4.6.

$$S_d = \sqrt{\frac{(7.5)^2}{250} + \frac{(7.4)^2}{280}} = 0.65$$

- Find the difference in means.

$$38.7 - 35.5 = 3.2 \text{ mph}$$
$$3.2 > (1.96)(0.65) = 1.3 \text{ mph}$$

It can, therefore, be concluded that the difference in mean speeds is significant at the 95 percent confidence level.

Several other statistical analyses may be carried out on speed studies data, but a discussion of them is beyond the scope of this book. Interested readers can find a detailed treatment of these methods in most statistical books.

VOLUME STUDIES

Traffic volume studies are carried out to collect data on the number of vehicles and/or pedestrians that pass a point on a highway facility during a specified time period. This time period varies from as little as 15 min to as much as a year, depending on the anticipated use of the data. The data collected may also be put into subclasses, which may include directional movement, occupancy rates, vehicle classification, and pedestrian age. Traffic volume studies are usually conducted when certain volume characteristics are needed, some of which follow.

1. Average Annual Daily Traffic (AADT) is the average of 24 hr counts collected every day in the year. AADTs are used in several traffic and transportation analyses for

  a. Estimation of highway user revenues

  b. Computation of accident rates in terms of 100 million vehicle miles

  c. Establishment of traffic volume trends

  d. Evaluation of the economic feasibility of highway projects

  e. Development of freeway and major arterial street systems

  f. Development of improvement and maintenance programs

2. Average Daily Traffic (ADT) is the average of 24 hr counts collected over a number of days greater than one but less than a year. ADTs may be used for

  a. Planning of highway activities

  b. Measurement of current demand

  c. Evaluation of existing traffic flow

3. Peak Hour Volume (PHV) is the maximum number of vehicles that pass a point on a highway during a period of sixty consecutive minutes. PHVs are used for

  a. Functional classification of highways

  b. Designing of the geometric characteristics of a highway; for example, number of lanes, intersection signalization, channelization, and so forth

  c. Capacity analysis

  d. Development of programs related to traffic operations; for example, one-way street systems or traffic routing

  e. Development of parking regulations

4. Vehicle Classification (VC) records volume with respect to the type of vehicles; for example, passenger cars, two-axle trucks, three-axle trucks, and so forth. VC is used in

  a. Design of geometric characteristics, with particular reference to turning radii requirements, maximum grades, lane widths, and so forth

  b. Capacity analyses, with respect to passenger car equivalents of trucks

  c. Adjustment of traffic counts obtained by machines

  d. Structural design of highway pavements, bridges, and so forth

5. Vehicle Miles of Travel (VMT) is a measure of travel along a section of road. It is the product of the traffic volume (that is, average weekday volume or

ADT) and the length of roadway in miles to which the volume is applicable. VMTs are used mainly as a base for allocating resources for maintenance and improvement of highways.

Methods of Conducting Volume Counts

Traffic volume counts are conducted using two basic methods: manual and mechanical.

Manual Method

The basic form of manual count involves a person recording each vehicle by making tally marks on a field sheet. Figure 4.5 shows a typical field sheet, which may be used by an individual to conduct traffic volume counts at an intersection. With this type of field sheet, both the turning movements at the intersection and the type of vehicles can be recorded. In this case, the vehicles have been classified into only two categories: passenger cars and similar vehicles (P) and trucks (T). Note that, in general, the inclusion of pickups and light trucks having four tires in the category of passenger cars does not create any significant deficiencies in the data collected, since the performance characteristics of these vehicles are similar to those of passenger cars. In some instances, however, a more detailed breakdown of commercial vehicles may be required, which would necessitate the collection of data by number of axles and/or weight. However, the degree of truck classification usually depends on the anticipated use of the data collected.

Mechanical hand counters may also be used for manual counts. Arranging the counters may be simple or complex, as shown in Figure 4.6 (*top*). The arrangement used depends on the type of data to be collected and the volume of traffic. At complex intersections with high volumes and several simultaneous flows, it would be difficult for a single person to count all the vehicles. In such cases, two or more people are used.

Advanced manual counting devices are now on the market that can be used to obtain several movements of traffic, pedestrians, bicycles, and so forth through an intersection. One such device is the TMC/48 electronic manual counter produced by Time Lapse, Inc., shown in Figure 4.6 (*bottom*). It is powered by two separate built-in rechargeable batteries, which may be recharged using a 120 v AC 60 HZ wall outlet. Several buttons are provided, each of which could be used to record volume data for different movements and different types of vehicles. The data for each movement is automatically separated into 48 distinct time intervals and stored in a semiconductor memory. The data may then be manually read out directly. The numbers are shown sequentially on a liquid crystal display at the front of the equipment. Alternatively, a time-lapse telephone transmission coupler could be used to transmit the data over a telephone line. A software package is then used to transfer the data to a computer where it is processed and printed. Figure 4.6 (*bottom*) shows the hookup of the equipment, the coupler, and the modem at the transmittal end.

Figure 4.5 Typical Field Sheet for Traffic Volume Counts at an Intersection by an Individual

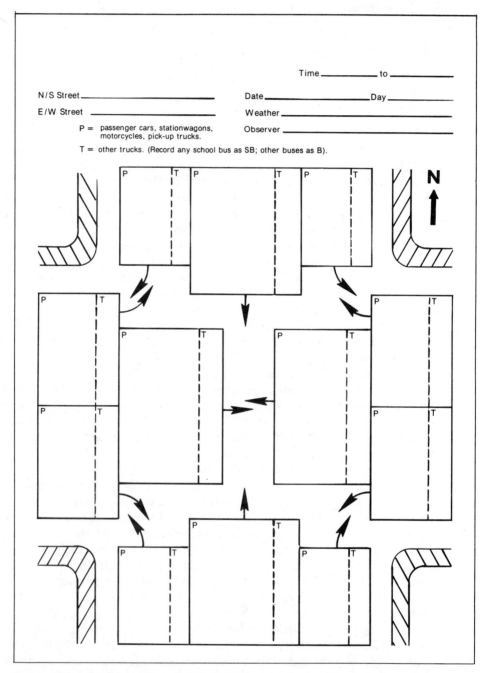

Source: Reproduced from *Manual of Traffic Engineering Studies*, 4th ed., Institute of Transportation Engineers, Washington, D.C., 1976.

Figure 4.6   Examples of Traffic Volume Hand Counters (Top: Layout of mechanical hand counters for counting turning movements at an intersection; bottom: TMC/48 manual counter)

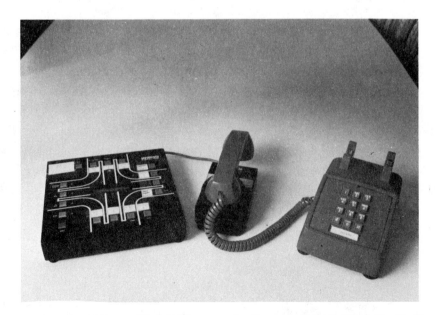

Source: Photos by Ed Deasy, Virginia Transportation Research Council, Charlottesville, Va.

The main disadvantages of the manual count method are that (1) it is labor intensive and can therefore be expensive, (2) it is subject to the limitations of human factors, and (3) it cannot be used for long periods of counting.

Mechanical Method

The mechanical counting method involves the laying of surface detectors such as pneumatic road tubes or subsurface detectors such as magnetic or electric contact devices on the road. These detect the passing vehicle and transmit the information to a recorder, which is connected to the detector located at the side of the road. Several types of recorders have been developed. These include junior counters, punch tape recorders, cassette tape recorders, and the counter/retriever combination.

Junior Counters. Figure 4.7 shows a typical junior counter, with a visible dial and a dry cell battery. This type of counter is usually connected to pneumatic tubes placed across the road. It determines the total number of vehicles passing over the pneumatic tube by recording one vehicle for every second axle that passes. In this case, the dial must be read at the beginning and end of every counting period or be reset to zero at the beginning of every counting period to obtain the volume for that counting period. It is useful for recording 24 hr counts. The main disadvantage of this type of counter is that it does not classify the vehicles, and the volumes obtained may have to be adjusted, by considering the percentage of vehicles having more than two axles, if this percentage is significant.

Figure 4.7 The Golden River Corporation Junior Counter

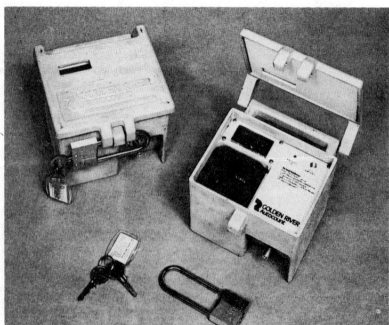

Source: Photo courtesy of Golden River Corporation, Rockville, Md.

Table 4.5 Manual Versus Mechanical Count Using a Junior Counter

Type of Vehicles	No. of Axles	Manual Count	Total Axles	No. of Vehicles Recorded by Machine
Passenger cars	2	1750	3500	1750
Trucks	3	670	2010	1005
Trucks	4	450	1800	900
Total		2870	7310	3655

For example, consider the data shown in Table 4.5 collected both manually and mechanically, using a junior counter. The total number of vehicles recorded by the junior counter will be 7310/2, that is, 3655 vehicles, while the actual count is only 2870. A truck adjustment factor can be calculated as 2870/3655, that is, 0.79 for this road. When junior counters are used to collect traffic volumes, the correct volumes should therefore be obtained by multiplying the recorded volume by the truck adjustment factor.

A special type of junior counter, known as the period counter, has also been developed. It can count vehicles for a specific length of time with the aid of a time clock that can be set to turn on and shut off at preselected times.

Cassette Tape Recorders.   These recorders use cassette tapes to record and store the impulses transmitted by the detectors. The cassettes are then removed and taken to the office, where a tape reader is used to retrieve and put the data into a form suitable for tabulation and analysis. Figure 4.8 shows the Leupold Stevens

Figure 4.8 Traffic Data Recorder (TDR)

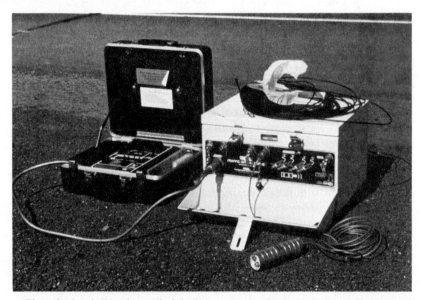

Source: Photo by Lewis Woodson, Virginia Transportation Research Council, Charlottesville, Va.

Figure 4.9   Sensor Set-up for the TDR

Source: Photo by Lewis Woodson, Virginia Transportation Research Council, Charlottesville, Va.

Traffic Data Recorder (TDR), which is a cassette tape recorder. Figure 4.9 shows the layout of the detectors for collecting data on traffic volume (including classification), velocity headway, and lateral placement. This equipment is capable of recording data for up to four different lanes at the same time, with the data for each lane being recorded on a different channel.

Counter/Retriever Combination.   This type of recorder basically consists of two components. The counter, which records and stores the vehicle counts at preset intervals (for example, every 15 min) and the retriever, which is used to first check the system and subsequently to collect the stored vehicle counts from the counter. Vehicles are detected by inductive loops placed in each traffic lane. Figure 4.10 shows an example of the counter/retriever combination devices.

Types of Volume Counts

Different types of traffic counts are carried out, depending on the anticipated use of the data to be collected. These different types are now discussed briefly.

Cordon Counts

When information is required on vehicle accumulation within an area, such as the central business district (CBD) of a city, particularly during a specific time, a cordon count is undertaken. The area for which the data is required is cordoned off by an imaginary closed loop, and the area enclosed within this loop is defined as the cordon area. Figure 4.11 shows such an area, where the CBD of a city is enclosed by the imaginary loop ABCA. The intersection of each street crossing the cordon line is taken as a count station, and volume counts of vehicles and/or persons entering and leaving the cordon area are carried out. The information obtained from such a count is useful for planning parking facilities, updating and

Figure 4.10 Golden River Corporation Marksman/Retriever Family

Source: Photos courtesy of Golden River Corporation, Rockville, Md.

Figure 4.11   Example of Station Locations for a Cordon Count

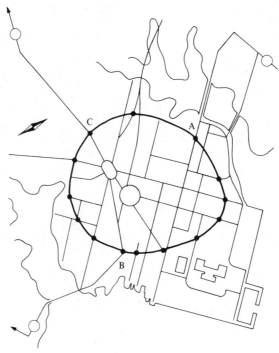

Source: Redrawn from *Manual of Traffic Engineering Studies*, 4th ed., Institute of Transportation Engineers, Washington, D.C., 1976.

evaluating traffic operational techniques, and long-range planning of freeway and arterial street systems.

Screen Line Counts

In screen line counts, the study area is divided into large sections by running imaginary lines, known as screen lines, across the study area. In some cases, natural and man-made barriers, such as rivers or railway tracks, are used as screen lines. Traffic counts are then taken at each crossing of a road and the screen line. It is usual for the screen lines to be such that they are not crossed more than once by the same street. Collection of data at these screen line stations at regular intervals facilitates the detection of variations in the traffic volume and traffic flow direction, due to changes in land-use pattern of the area.

Intersection Counts

Intersection counts are taken to determine vehicle classification, through movements, and turning movements at intersections. These data are used mainly in determining phase lengths and cycle times for signalized intersections, in the design of channelization at intersections, and in the design of general improvements to intersections.

Pedestrian Volume Counts

Volume counts of pedestrians are made at locations such as subway stations, midblocks, and cross walks. The counts usually are taken at these locations when the evaluation of existing or proposed pedestrian facilities are to be undertaken. Such facilities may include pedestrian overpasses or underpasses.

Periodic Volume Counts

In order to obtain certain traffic volume data, such as AADT, it is necessary to obtain data continuously. It is, however, not feasible to collect continuous data on all roads because of the cost involved. To make reasonable estimates of annual traffic volume characteristics on an areawide basis, different types of periodic counts with count durations ranging from 15 min to continuous are conducted, and the data from these different periodic counts are used to determine factors that are then used to estimate annual traffic characteristics. The periodic counts usually carried out are continuous, control, and coverage counts.

Continuous Counts. These counts are carried out continuously using mechanical counters. Stations at which continuous counts are carried out are known as permanent count stations. In selecting permanent count stations, the highways within the study area must first be properly classified. Each class should consist of highway links with similar traffic patterns and characteristics. A highway link for traffic count purposes is defined as a homogeneous section that has the same traffic characteristics such as AADT and daily, weekly, and seasonal variations in traffic volumes on each point. Broad classification systems for major roads may include freeways, expressways, and major arterials. For minor roads, classifications may include residential, commercial, and industrial streets.

Recent studies have, however, shown that a more detailed classification system is sometimes required since the traffic characteristics on an expressway in the northern part of a state may not be the same as those for the southern part of the state.[2] In developing highway classification systems, consideration must also be given to the location of the highways, the AADTs, and the primary uses of the highways. After developing a proper classification system for the highways, at least one permanent count station is selected for each class of highway. There are usually 30 to 60 permanent count stations in a state, which are located throughout the state so that continuous data on the traffic patterns and characteristics of a highway in each class are collected. These data are needed to expand shorter counts at other locations.

Control Counts. These counts are taken at stations known as control count stations, which are strategically located so that representative samples of traffic volume can be taken on each type of highway or street. The data obtained from control counts are used to determine seasonal and monthly variations of traffic characteristics, so that expansion factors can be determined. These expansion factors are used to determine year-round average values from short counts.

Control counts can be divided into two types: major control counts and minor control counts. Major control counts are taken monthly, with 24 hr directional counts being taken on at least three days during the week (Tuesday,

Wednesday, and Thursday) and on Saturday and Sunday to obtain information on weekend volumes. It is usual to locate at least one major control count station on every major street. The data collected give information regarding hourly, monthly, and seasonal variations of traffic characteristics. Minor control counts are nondirectional 48 hr counts carried out on weekdays on minor roads at least once every two years.

Coverage Counts.   These counts are used to estimate ADT, utilizing expansion factors developed from control counts. The study area is usually divided into zones having similar traffic characteristics. At least one coverage count station is located in each zone. A 24 hr nondirectional weekday count is taken at least once every 4 years at each coverage station. The data indicate changes in areawide traffic characteristics.

Traffic Volume Data Presentation

The data collected from traffic volume counts may be presented in one of several ways, depending on the type of count conducted and the primary use of the

Figure 4.12   Example of a Traffic Flow Map

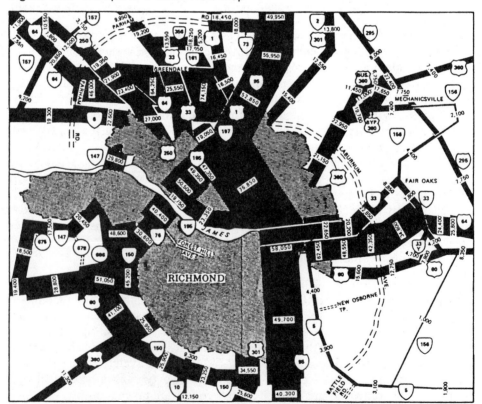

Source: Reproduced from *Manual of Traffic Engineering Studies*, 4th ed., Institute of Transportation Engineers, Washington, D.C., 1976.

data. A description of some of the conventional data presentation techniques follows.

Traffic Flow Maps. These maps show traffic volumes on individual routes. The volume of traffic on each route is represented by the width of a band, which is drawn in proportion to the traffic volume it represents, providing a graphic representation of the different volumes that facilitates easy visualization of the relative volumes of traffic on the different routes. When flows are significantly different in opposite directions of a particular street or highway, it is advisable to provide separate bands for each direction. In order to enhance the usefulness of such maps, the numerical value represented by each band is listed by the band. Figure 4.12 shows a typical traffic flow map.

Intersection Summary Sheets. These sheets are graphic representations of the volume and direction of all traffic movement through the intersection. These volumes can be either ADTs or PHVs, depending on the use of the data. Figure

Figure 4.13 Intersection Summary Sheet

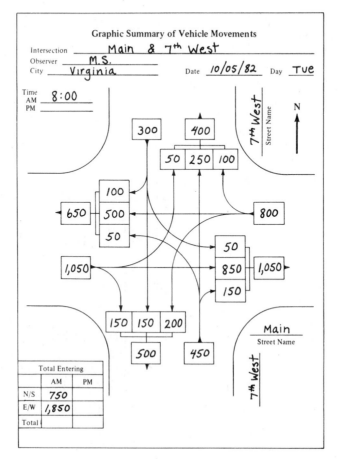

Figure 4.14   Traffic Volumes on an Urban Highway

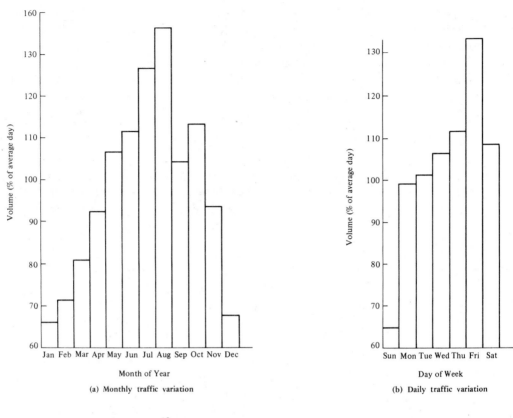

(a) Monthly traffic variation

(b) Daily traffic variation

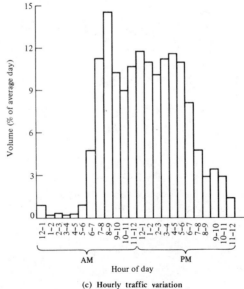

(c) Hourly traffic variation

Table 4.6   Summary of Traffic Volume Data

PHV	430
ADT	5375
VC	
Passenger cars	70%
Two-axle trucks	20%
Three-axle trucks	8%
Other trucks	2%

4.13 shows a typical intersection summary sheet showing peak hour traffic through the intersection.

Time-Based Distribution Charts.   These charts show the hourly, daily, monthly, or annual variations in traffic volume in an area or on a particular highway. The volumes are usually given as a percentage of the average volume. Figure 4.14 shows typical charts for monthly, daily, and hourly variations.

Summary Tables.   These tables give a summary of traffic volume data such as PHV, VC, and ADT in tabular form. Table 4.6 shows a typical summary table.

Traffic Volume Characteristics

A continuous count of traffic at a section of a road will show that traffic volume varies from hour to hour, day to day, and month to month. The regular observation of traffic volumes over the years has, however, identified certain characteristics that show that, although traffic volume at a section of a road varies from time to time, this variation is repetitive and rhythmic. These characteristics of traffic volumes are usually taken into consideration when traffic counts are being planned, so that volumes collected at a particular time or place can be related to volumes collected at other times and places. A knowledge of these characteristics can also be used to estimate the accuracy of traffic counts.

Monthly variations are shown in Figure 4.14a, where very low volumes are observed during January and February, mainly because of the winter weather, and the peak volume is observed during August, mainly due to holiday traffic. This suggests that traffic volumes taken either during the winter months of January or February or during summer months of July and August cannot be representative of the average annual traffic. If this information is presented for a number of consecutive years, the repetitive nature of the variation will be observed since the pattern of the variation will be similar for all years, although the actual volumes may not necessarily be the same.

Daily variations are shown in Figure 4.14b, where it is seen that traffic volumes on Tuesday, Wednesday, and Thursday are similar, but a peak is observed on Friday. This indicates that when short counts are being planned, it is useful to plan for the collection of weekday counts on Tuesday, Wednesday, and Thursday and, when necessary, to plan for the collection of weekend counts separately on Friday and Saturday.

Hourly variations in traffic volume are shown in Figure 4.14c, where the volume for each hour of the day is represented as a percentage of the ADT. It can be seen that there is hardly any traffic between the hours 1 a.m. to 5 a.m., and that peak volumes occur between 8 a.m. and 9 a.m., midday and 1 p.m., and 4 p.m. and 6 p.m. It can be inferred that work trips are primarily responsible for the peaks. If such data are collected every weekday for one week, the hourly variations will be similar to each other, although the actual volumes may not be the same from day to day.

Sample Size and Adjustment of Periodic Counts

The impracticality of collecting data continuously every day of the year at all counting stations makes it necessary to collect sample data from each class of highway and to estimate annual traffic volumes from periodic counts. This involves the determination of the minimum sample size (number of count stations) for a required level of accuracy and the determination of daily, monthly, and/or seasonal expansion factors for each class of highway.

Determination of Number of Count Stations

The minimum sample size depends on the precision level desired. The commonly used precision level for volume counts is 95–10. When the sample size is less than 30 and the selection of counting stations is random, a distribution known as the student's t distribution may be used to determine the sample size for each class of highway links. The student's t distribution is unbounded with a mean of zero and has a variance that depends on the scale parameter, commonly referred to as the *degrees of freedom* (v). The degrees of freedom (v) is a function of the sample size and $v = N - 1$, for the student's t distribution. The variance of the student's t distribution equals $v/(v - 2)$, which indicates that as v approaches infinity the variance approaches 1. The probabilities (confidence levels) for the student's t distribution for different degrees of freedom is given in Appendix A.

Assuming that the sampling locations are randomly selected, the minimum sample number is given as

$$n = \frac{t_{\alpha/2, N-1}^2 (S^2/d^2)}{1 + (1/N)(t_{\alpha/2, N-1})(S^2/d^2)} \tag{4.7}$$

where

n = minimum number of count locations required

t = value of the student's t distribution with $(1 - \alpha/2)$ confidence level $(N - 1$ degrees of freedom)

N = total number of links (population) from which a sample is to be selected

α = significance level

S = estimate of the spatial standard deviation of the link volumes

d = the allowable range of error

To use Eq. 4.7, estimates of the mean and standard deviation of the link volumes are required. These estimates can be obtained by taking volume counts at a few links or by using known values for other similar highways.

Example 4–3 Minimum Number of Count Stations

To determine a representative value for the ADT on 100 highway links having similar volume characteristics, it was decided to collect 24 hr volume counts on a sample of these links. Estimates of mean and standard deviation of the link volumes for the type of highways in which these links are located are 32,500 and 5500, respectively. Determine the minimum number of stations at which volume counts should be taken if a 95–10 precision level is required.

- Establish the data

  $\alpha = (100 - 95) = 5$ percent
  $S = 5500$
  $m = 32,500$
  $d = 0.1 \times 32,500 = 3250$ (allowable range of error)
  $v = 100 - 1 = 99$
  $t_{\alpha/2,99} \approx 1.984$ (from Appendix A)

- Use Eq. 4.7 to solve for n.

$$n = \frac{(1.984^2 \times 5500^2)/3250^2}{1 + (1/100)(1.984^2 \times 5500^2)/3250^2} = \frac{11.27}{1.11} = 10.1$$

Counts should be taken at a minimum of 11 stations. When sample sizes are greater than 30, the normal distribution is used instead of the student's t distribution.

The Federal Highway Administration (FHWA) has also presented a method for computing sample size, which uses the normal distribution and takes into consideration both the spatial variance S_1 and the temporal variance S_2.[3] The spatial variance is related to the variation of the link volumes (ADTs) within a given highway class, and the temporal variance is related to the variation of the ADT with time at a given link within a given cluster. This method is as follows:

$$N_h = \frac{N_o}{1 + (N_o/N)} \tag{4.8}$$

where

N_h = required number of count stations for a given cluster and a given precision level and corrected for finiteness

N_o = required sample size without finite adjustment

N = the total number of links in a given cluster

The formula for determining N_o is

$$N_o = \frac{Z^2(S_1^2 + S_2^2)}{d^2} \tag{4.9}$$

where

Z = value of the normal variate as applied to a specific confidence level

S_1^2 = spatial variance

S_2^2 = temporal variance

d = allowable range of error

The sample size obtained from either Eq. 4.7 or Eq. 4.8 is the number of links to be counted, as only one counting station is located on each link. An estimate of the spatial variance may be made by taking 24 hr volume counts on a small sample of highway links or by using available data. It is, however, more difficult to estimate temporal variance since this requires data from a given link over a period of time. It is therefore appropriate to use data from continuous count stations to determine the temporal variance. When no information on link volumes exists, representative values have been suggested by FHWA as shown in Tables 4.7 and 4.8.

Example 4–4 Minimum Sample Size for Highway Link Clusters

Determine the minimum sample size required for a cluster consisting of 100 arterial links, with an ADT range of 5,000–10,000, using both spatial and temporal variances if a count duration of 48 hr (2-day cluster) will be used. A 95–10 precision level is required.

- Establish the data.

$S_1 = 1500$ (from Table 4.7)
$S_2 = 980$ (from Table 4.8)
$Z = 1.96$

- Use Eq. 4.9, assuming that the mean is the midpoint of the range (mean = 7500).

$$N_o = \frac{1.96^2(1500^2 + 980^2)}{(0.1 \times 7500)^2} = 21.92 \approx 22$$

- Use Eq. 4.8.

$$N_h = \frac{22}{1 + (22/100)} = 18.03 \approx 18$$

Table 4.7 Suggested Spatial Measures of Variation

Range	Mean Volume	S_1	C(%)
Local streets			
0–2,000	1,000	300	30
Arterial streets			
0–5,000	2,500	1,500	60
5,000–10,000	7,500	1,500	20
10,000–15,000	12,500	1,500	12
15,000–20,000	17,500	1,500	9
20,000–25,000	22,500	1,500	7
25,000–30,000	27,500	1,500	5
30,000–35,000	32,500	1,500	5
35,000–40,000	37,500	1,500	4
Freeways			
20,000–60,000 (4 lanes)	40,000	12,000	30
40,000–120,000 (6 lanes)	80,000	24,000	30
80,000–200,000 (8 lanes)	130,000	42,000	32

Note: Data are rounded.

S_1 = spatial standard deviation.

C = coefficient of variation in percent.

Source: Adapted from *Urban Traffic Volume Counting Manual*, U.S. Department of Transportation, Washington, D.C., 1975.

The above examples demonstrate the use of a relatively simple expression to determine sample size. It must be pointed out, however, that more complex expressions have been developed for use in regional surveys. Some of the references listed at the end of this chapter contain some of these formulas.

Adjustment of Periodic Counts

Expansion factors are used to adjust periodic counts and are determined either from continuous count stations or from control count stations.

Expansions Factors from Continuous Count Stations. Hourly, daily, and monthly expansion factors can be determined using data obtained at continuous count stations.

Hourly expansion factors (HEFs) are determined as

$$\text{HEF} = \frac{\text{total volume for 24 hr period}}{\text{volume for particular hour}}$$

These factors are used to expand counts of durations shorter than 24 hr to 24 hr volumes by multiplying the hourly volume for each hour during the count period by the HEF for that hour and finding the mean of these products.

Table 4.8 Suggested Temporal Measures of Variation, Weekday Traffic

		Cluster (Day No.)				
		1	2	3	5	
	Mean Volume	C_1	S_2	S_2	S_2	S_2
Local streets						
0–2,000	1,000	30	300	290	280	250
Arterial streets						
0–5,000	2,500	20	500	480	440	400
5,000–10,000	7,500	14	1,050	980	900	810
10,000–15,000	12,500	12	1,500	1,350	1,230	1,080
15,000–20,000	17,500	10	1,750	1,680	1,510	1,300
20,000–25,000	22,500	10	2,250	1,940	1,760	1,490
25,000–30,000	27,500	9	2,470	2,200	1,980	1,700
30,000–35,000	32,500	9	2,520	2,470	2,210	1,890
35,000–40,000	37,500	8	3,000	2,700	2,400	2,000
Freeways						
Lanes, 20,000–60,000	40,000	8	3,200	2,800	2,500	2,100
Lanes, 40,000–120,000	80,000	6	4,800	4,800	3,850	3,200
Lanes, 60,000–200,000	130,000	5	6,500	6,250	5,350	4,200

Note: Data are rounded.

S_2 = temporal standard deviation of 1-, 2-, 3-, and 5-day clusters.

C_1 = coefficient of variation, 1-day cluster, in percent.

Mean volume is assumed as midpoint of range.

Source: Adapted from *Urban Traffic Volume Counting Manual*, U.S. Department of Transportation, Washington, D.C., 1975.

Daily expansion factors (DEFs) are computed as

$$DEF = \frac{\text{average total volume for week}}{\text{average volume for particular day}}$$

These factors are used to determine weekly volumes from counts of 24 hr duration by multiplying the 24 hr volume by the DEF.

Monthly expansion factors (MEFs) are computed as

$$MEF = \frac{\text{AADT}}{\text{ADT for particular month}}$$

The AADT for a given year may be obtained from the ADT for a given month by multiplying this volume by the MEF. Tables 4.9, 4.10, and 4.11 give expansion factors for a particular primary road in Virginia. Such expansion factors should be determined for each class of road in the classification system established for an area.

Table 4.9 Hourly Variation of Traffic Volume on a Rural Primary Road

Hour	Volume	HEF	Hour	Volume	HEF
6:00–7:00 a.m.	294	42.00	6:00–7.00 p.m.	743	16.62
7:00–8:00 a.m.	426	29.00	7:00–8:00 p.m.	706	17.49
8:00–9:00 a.m.	560	22.05	8:00–9:00 p.m.	606	20.38
9:00–10:00 a.m.	657	18.80	9:00–10:00 p.m.	489	25.26
10:00–11:00 a.m.	722	17.10	10:00–11:00 p.m.	396	31.19
11:00–12:00 noon	667	18.52	11:00–12:00 p.m.	360	34.31
12:00–1:00 p.m.	660	18.71	12:00–1:00 a.m.	241	51.24
1:00–2:00 p.m.	739	16.71	1:00–2:00 a.m.	150	82.33
2:00–3:00 p.m.	832	14.84	2:00–3:00 a.m.	100	123.50
3:00–4:00 p.m.	836	14.77	3:00–4:00 a.m.	90	137.22
4:00–5:00 p.m.	961	12.85	4:00–5:00 a.m.	86	143.60
5:00–6:00 p.m.	892	13.85	5:00–6:00 a.m.	137	90.14

Total daily volume = 12,350.

Example 4–5 Calculating AADT Using Expansion Factors

A traffic engineer urgently needs to determine the AADT on a rural primary road that has the volume distribution characteristics shown in Tables 4.9, 4.10, and 4.11. She collected the following data on a Tuesday during the month of May. Determine the AADT of the road.

7:00–8:00 a.m.	400
8:00–9:00 a.m.	535
9:00–10:00 a.m.	650
10:00–11:00 a.m.	710
11:00–12 noon	650

Table 4.10 Daily Expansion Factors for a Rural Primary Road

Day of Week	24 hr Daily Volume	DEF
Sunday	7,895	9.515
Monday	10,714	7.012
Tuesday	9,722	7.727
Wednesday	11,413	6.582
Thursday	10,714	7.012
Friday	13,125	5.724
Saturday	11,539	6.510
Totaly Weekly Volume	75,122	

Table 4.11   Monthly Expansion Factors for a Rural Primary Road

Month	ADT	MEF
January	1,350	1.756
February	1,200	1.975
March	1,450	1.635
April	1,600	1.481
May	1,700	1.394
June	2,500	0.948
July	4,100	0.578
August	4,550	0.521
September	3,750	0.632
October	2,500	0.948
November	2,000	1.185
December	1,750	1.354
Total	28,450	

Mean average daily volume = 2,370.

- Estimate the 24 hr volume for Tuesday using the factors given in Table 4.9.

$$\frac{(400 \times 29.0 + 535 \times 22.05 + 650 \times 18.80 + 710 \times 17.10 + 650 \times 18.52)}{5} \approx 11,959$$

- Adjust the 24 hr volume for Tuesday to an average volume for the week using the factors given in Table 4.10.

$$\text{total 7-day volume} = 11,959 \times 7.727$$

$$\text{average 24 hr volume} = \frac{11,959 \times 7.727}{7} = 13,201$$

- Since the data was collected in May, use the factor obtained in Table 4.11 for May to obtain the AADT.

$$\text{AADT} = 13,201 \times 1.394 = 18,402$$

TRAVEL TIME AND DELAY STUDIES

A travel time study determines the amount of time required to travel from one point to another on a given route. In carrying out such a study, information may also be collected on the location, duration, and causes of delays. When this is done, the study is known as a travel time and delay study. Data obtained from travel time and delay studies give a good indication of the level of service that exists on the study section. These data also aid the traffic engineer in identifying problem locations, which may require special attention in order to improve the overall flow of traffic on the route.

Applications of Travel Time and Delay Data

The data obtained from travel time and delay studies may be used in any one of the following traffic engineering tasks:

- Determination of the efficiency of a route with respect to its ability to carry traffic
- Identification of locations with relatively high delays and the causes for those delays
- Performance of before and after studies to evaluate the effectiveness of traffic operation improvements
- Determination of relative efficiency of a route by developing sufficiency ratings or congestion indices
- Determination of travel times on specific links for use in trip assignment models (See Chapter 10.)
- Compilation of travel time data that may be used in trend studies to evaluate the changes in efficiency and level of service with time
- Performance of economic studies in the evaluation of traffic operation alternatives that reduce travel time

Definition of Terms Related to Time and Delay Studies

Let us now define certain terms commonly used in travel time and delay studies.

1. Travel time is the time taken by a vehicle to traverse a given section of a highway.

2. Running time is the time a vehicle is actually in motion while traversing a given section of a highway.

3. Delay is the time lost by a vehicle due to causes beyond the control of the driver.

4. Operational delay is that part of the delay caused by the impedance of other traffic. This impedance can occur either as a side friction, where the stream flow is interfered with by other traffic (for example, parking or unparking vehicles) or as internal friction, where the interference is within the traffic stream (for example, reduction in capacity of the highway).

5. Stopped-time delay is that part of the delay during which the vehicle is not moving.

6. Fixed delay is that part of the delay caused by control devices such as traffic signals. This delay occurs regardless of the traffic volume or the impedance that may exist.

7. Travel-time delay is the difference between the actual travel time and the travel time that will be obtained by assuming that a vehicle traverses the study section at an average speed that is equal to that for an uncongested traffic flow on the section being studied.

Methods for Conducting Travel Time and Delay Studies

Several methods have been used to conduct travel time and delay studies. These methods can be grouped into two general categories: (1) those using a test vehicle and (2) those not requiring a test vehicle. The particular technique used for any specific study depends on the reason for conducting the study and the available personnel and equipment.

Methods Requiring a Test Vehicle

This category comprises the floating car, average-car, and moving-vehicle techniques.

Floating-Car Technique. In this method, the test car is driven by an observer along the test section such that the test car "floats" with the traffic. The driver of the test vehicle attempts to pass as many vehicles as pass his test vehicle. The time taken to traverse the study section is recorded. This is repeated for a minimum number of times, and the average time recorded as the travel time. The minimum number of test runs recommended for 95 percent confidence level and specified permitted errors are shown in Table 4.12.

Table 4.12 Approximate Minimum Sample Size Requirements for Travel Time and Delay Studies with Confidence Level of 95.0 Percent

Average Range in Travel Speed (mph)	Minimum Number of Runs for Specified Permitted Error				
	±1.0 mph	±2.0 mph	±3.0 mph	±4.0 mph	±5.0 mph
2.5	4	2	2	2	2
5.0	8	4	3	2	2
10.0	21	8	5	4	3
15.0	38	14	8	6	5
20.0	59	21	12	8	6

Average Range in Travel Speed (kph)	Minimum Number of Runs for Specified Permitted Error				
	±2.0 kph	±3.5 kph	±5.0 kph	±6.5 kph	±8.0 kph
5.0	4	3	2	2	2
10.0	8	4	3	3	2
15.0	14	7	5	3	3
20.0	21	9	6	5	4
25.0	28	13	8	6	5
30.0	38	16	10	7	6

Source: Adapted from *Manual of Traffic Engineering Studies*, 4th ed., Institute of Transportation Engineers, Washington, D.C., 1976.

Average-Speed Technique. This technique involves driving the test car along the length of the test section at a speed, that, in the opinion of the driver, is the average speed of the traffic stream. The time to traverse the test section is noted. The test run is repeated a minimum number of times, and the average time is recorded as the travel time.

In each of the above methods, it is first necessary to clearly identify the test section. The travel time is usually obtained by the observer starting a stop watch at the beginning point of the test section and stopping it at the end. Additional data may also be obtained by recording the time at which the test vehicle arrives at specific locations, which should have been identified before the start of the test runs. A second stop watch may also be used to determine the time that expires each time the vehicle is stopped. The summation of these times for any test run will give the stopped-time delay for that run. A typical field sheet for such a study is shown in Figure 4.15.

The Moving-Vehicle Technique. This technique is conducted by the observer making a round trip on the test section as shown in Figure 4.16, where it is

Figure 4.15 Typical Field Sheet for Time and Delay Study

DATE _____ WEATHER _____ TRIP NO. _____

ROUTE _____ DIRECTION _____

TRIP STARTED AT _____ AT _____ _____
(LOCATION) (MILEAGE)

TRIP ENDED AT _____ AT _____ _____
(LOCATION) (MILEAGE)

CONTROL POINTS		STOPS OR SLOWS		
LOCATION	TIME	LOCATION	SEC DELAY	CAUSE

TRIP LENGTH _____ TRIP TIME _____ TRAVEL SPEED _____

RUNNING TIME _____ STOPPED TIME _____ RUNNING SPEED _____

SYMBOLS OF DELAY CAUSE: S—TRAFFIC SIGNALS SS—STOP SIGN LT—LEFT TURNS
PK—PARKED CARS DP—DOUBLE PARKING T—GENERAL
PED—PEDESTRIANS BP—BUS PASSENGERS LOADING OR UNLOADING

COMMENTS _____

RECORDER _____

Source: Reproduced from *Manual of Traffic Engineering Studies*, 4th ed., Institute of Transportation Engineers, Washington, D.C., 1976.

Figure 4.16   Test Site for Moving-Vehicle Method

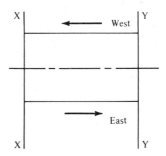

assumed that the road is in the east-west direction. The observer will start collecting the relevant data at section X-X, drive the car eastward to section Y-Y, then turn the vehicle around and drive westbound to section X-X again. The following data are collected as the test vehicle makes the round trip:

- The time it takes to travel from X-X to Y-Y (T_e) in minutes
- The time it takes to travel from Y-Y to X-X (T_w) in minutes
- The number of vehicles traveling west in the opposite lane while the test car is traveling east (N_e)
- The number of vehicles that overtake the test car while it is traveling from Y-Y to X-X, that is, in the westbound direction (O_w)
- The number of vehicles that the test car passes while it is traveling from Y-Y to X-X, that is, traveling westbound (P_w).

The volume (V_w) on the westbound direction can then be obtained from the expression

$$V_w = \frac{(N_e + O_w - P_w)60}{T_e + T_w} \qquad (4.10)$$

where ($N_e + O_w - P_w$) is the number of vehicles traveling westward and cross the line X-X during the time ($T_e + T_w$). Note that when the test vehicle starts at X-X traveling eastward, all vehicles traveling westward should get to X-X before the test vehicle, except those that are passed by the test vehicle when it is traveling westward. Similarly, all vehicles that pass the test vehicle when the test vehicle is traveling westward will get to X-X before the test vehicle. The test vehicle will also get to X-X before all vehicles it passes while traveling westward. These vehicles have, however, been counted as part of N_e or O_w and should therefore be subtracted from the sum of N_e and O_w to determine the number of vehicles that cross X-X traveling in the westbound direction during the time the test vehicle takes to travel from X-X to Y-Y and back to X-X. These considerations lead to Eq. 4.10.

Similarly, the average travel time $\bar{T}_w$ in the westbound direction is obtained from

$$\frac{\bar{T}_w}{60} = \frac{T_w}{60} - \frac{O_w - P_w}{V_w}$$

$$\bar{T}_w = T_w - \frac{60(O_w - P_w)}{V_w}$$

(4.11)

If the test car is traveling at the average speed of all vehicles, it will most likely pass the same number of vehicles that overtake it. Since it is probable that the test car will not be traveling at the average speed, the second term of Eq. 4.11 corrects for the difference between the number of vehicles that overtake the test car and that are overtaken by the test car.

Example 4–6 Volume and Travel Time Using Moving-Vehicle Technique

The data in Table 4.13 were obtained in a travel time study on a section of highway using the moving-vehicle technique. Determine the travel time and

Table 4.13 Data from Travel Time Study Using the Moving-Vehicle Technique

Run Direction/ Number	Travel Time (min)	No. of Vehicles Traveling in Opposite Direction	No. of Vehicles That Overtake Test Vehicle	No. of Vehicles Overtaken by Test Vehicle
Eastbound				
1	2.75	80	1	1
2	2.55	75	2	1
3	2.85	83	0	3
4	3.00	78	0	1
5	3.05	81	1	1
6	2.70	79	3	2
7	2.82	82	1	1
8	3.08	78	0	2
Average	2.85	79.50	1.00	1.50
Westbound				
1	2.95	78	2	0
2	3.15	83	1	1
3	3.20	89	1	1
4	2.83	86	1	0
5	3.30	80	2	1
6	3.00	79	1	2
7	3.22	82	2	1
8	2.91	81	0	1
Average	3.07	82.25	1.25	0.875

volume in each direction at this section of the highway.

> Mean time it takes to travel eastbound $T_e = 2.85$ min; westbound $T_w = 3.07$ min
>
> Average number of vehicles traveling westbound when test vehicle is traveling eastbound $(N_e) = 79.50$
>
> Average number of vehicles traveling eastbound when test vehicle is traveling westbound $(N_w) = 82.25$
>
> Average number of vehicles that overtake test vehicle while it is traveling westbound $(O_w) = 1.25$
>
> Average number of vehicles that overtake test vehicle while it is traveling eastbound $(O_e) = 1.00$
>
> Average number of vehicles that the test vehicle passes while traveling westbound $(P_w) = 0.875$
>
> Average number of vehicles that the test vehicle passes while traveling eastbound $(P_e) = 1.5$.

■ From Eq. 4.10, find the volume in the westbound direction.

$$V_w^* = \frac{79.50 + 1.25 - 0.875}{(2.85/60) + (3.07/60)} = 809.5 \qquad \text{(or 810 vph)}$$

■ Similarly, calculate the volume in the eastbound direction.

$$V_e = \frac{82.25 + 1.00 - 1.50}{(2.85/60) + (3.07/60)} = 828.5 \qquad \text{(or 829 vph)}$$

■ Find the average travel time in the west direction.

$$\bar{T}_w = 3.07 - \frac{(1.25 - 0.875)}{810}60 = 3.0 \text{ min}$$

■ Find the average travel time in the east direction.

$$\bar{T}_e = 2.85 - \frac{(1.00 - 1.50)}{829}60 = 2.9 \text{ min}$$

Methods Not Requiring a Test Vehicle

This category includes the license plate method and interview method.

License Plate Observation.   The license plate method requires that observers be positioned at the beginning and end of the test section. Observers can also be positioned at other designated locations if elapsed times to these locations are required. Each observer records the last three or four digits of the license plate of each car that passes together with the time at which the car passes. The reduction of the data is accomplished in the office by matching the times of arrival at the beginning and end of the test section of each license plate recorded. The difference between these times is the traveling time of each vehicle. The average of these is the average traveling time on the test section. It has been suggested that a sample size of 50 matched license plates will give reasonably accurate results.

Interviews.   The interviewing method is carried out by obtaining information from individuals who drive on the study site regarding their travel times, their experience of delays, and so forth. This method facilitates the collection of a large amount of data in a relatively short time. However, it requires the cooperation of the individuals contacted since the result depends entirely on the information given by the individuals interviewed.

PARKING STUDIES

Any vehicle traveling on a highway will at one time or another be parked for either a relatively short time or a much longer time, depending on the reason for parking the vehicle. The provision of parking facilities is therefore an essential element of the highway mode of transportation. The need for parking spaces is usually very great in areas where the land use includes business, residential, or commercial activities. The growing use of the automobile as a personal feeder service to transit systems (park-and-ride) has also increased the demand for parking spaces at transit stations. In areas of high density, where space is very expensive, the space provided for the automobile usually has to be divided between that allocated for the movement of the automobile and that allocated for the parking of the automobile.

Providing adequate parking spaces to meet the demand for parking in the CBD may necessitate the provision of parking bays along curbs, which reduces the capacity of the highways and may affect the levels of service on them. This problem usually confronts a city traffic engineer. The solution is not simple since the allocation of available space will depend on the goals of the community, which the traffic engineer must take into consideration when trying to solve the problem. Parking studies are therefore used to determine the demand and supply of parking facilities in an area, the projection of the demand, and the collection of the views of various interest groups on how best to solve the problem. Before discussing the details of parking studies, it is necessary to discuss the different types of parking facilities. The design of these facilities will be discussed in Chapter 14.

Types of Parking Facilities

Parking facilities can be divided into two main groups: on-street and off-street.

On-Street Parking Facilities.   These are also known as curb facilities. Parking bays are provided alongside the curb on either one or both sides of the road. These bays can be unrestricted parking facilities if the duration of parking is unlimited and parking is free, or they can be restricted parking facilities, if parking is limited to specific times of the day for a maximum duration. Parking on restricted facilities may or may not be free. Restricted facilities may also be provided for specific purposes such as for handicapped parking or as bus stops or loading bays.

Off-Street Parking Facilities.   These facilities may be privately or publicly owned and include surface lots and garages. Self-parking garages require that drivers park their own automobiles, whereas attendant-parking garages provide attendants who park the automobiles.

Definition of Parking Terms

Before discussing the different methods for conducting a parking study, it is necessary to define some commonly used terms in parking studies, such as space hour, parking volume, parking accumulation, parking load, parking duration, and parking turnover.

   1. Space hour is a unit of parking that defines the use of a single space of parking for a period of 1 hr.

   2. Parking volume is the number of vehicles involved in parking in a study area during a specific length of time, usually a day.

   3. Parking accumulation is the number of parked vehicles in a study area at any specified time. This data can be plotted as a curve of parking accumulation against time, which shows the variation of the parking accumulation during the day. Such a curve is known as a parking accumulation curve. (See Figure 4.17.)

   4. Parking load is the area under the accumulation curve between two specific times. It is usually given as the number of space hours used during the specified period of time.

   5. Parking duration is the length of time a vehicle is parked at a parking bay. When the parking duration is given as an average, it gives an indication of how frequently a parking space becomes available. For example, if n parking spaces are available and the average parking duration is t hrs, then the number of cars that can park during a period of T hrs is given by nTf/t, where f is a factor that corrects for the time it takes to park and remove a vehicle from the parking spot, which is usually between 0.85 and 0.95.

   6. Parking turnover is the rate of use of a parking space. It is obtained by dividing the parking volume for a specified period by the number of parking spaces.

Methodology of Parking Studies

A comprehensive parking study usually involves (1) inventorying existing parking facilities, (2) collecting data on parking accumulation, parking turnover, and parking duration, (3) identifying parking generators, and (4) obtaining information on parking demand. Information on related factors such as financial, legal, and administrative may also be collected.

Inventory of Existing Parking Facilities

An inventory of existing parking facilities is a detailed listing of the location and all other relevant characteristics of each legal parking facility, private and public, in the study area. The inventory includes both on- and off-street facilities. The relevant characteristics usually listed include

- Type and number of parking spaces at each parking facility
- Times of operation and limit on duration of parking, if any
- Type of ownership (that is, whether private or public)
- Parking fees, if any, and method of collection
- Restrictions on usage (that is, whether or not open to the public)
- Other restrictions, if any (such as loading and unloading zones, bus stops, taxi ranks, and so forth)
- Probable degree of permanency, that is, can the facility be regarded as permanent or is it just a temporary facility

The information obtained from an inventory of parking facilities is useful to both the traffic engineer and other public agencies, such as zoning commissions and planning departments. The inventory should be updated at regular intervals of about 4 to 5 years.

Collection of Parking Data

Accumulation. The accumulation data are obtained by checking the amount of parking during regular intervals on different days of the week. The checks are usually carried out on an hourly or 2 hr basis between 6:00 a.m. and 8:00 p.m. The selection of the times depends on the operating times of land-use activities that act as parking generators. For example, if a commercial zone is included, checks should be made during the times retail shops are open, which may include periods up to 9:30 p.m. on some days. The information obtained is used to determine hourly variations of parking and peak periods of parking demand. (See Figure 4.17.)

Turnover and Duration. The information on turnover and duration are usually obtained by collecting data on a sample of parking spaces in a given block. This is done by recording the license plate of the vehicle parked on each parking space in the sample at the end of fixed intervals during the study period. The length of the fixed interval depends on the maximum permissible duration. For example, if the maximum permissible duration of parking at a curb face is 1 hr, a suitable

Figure 4.17   Parking Accumulation at a Parking Lot

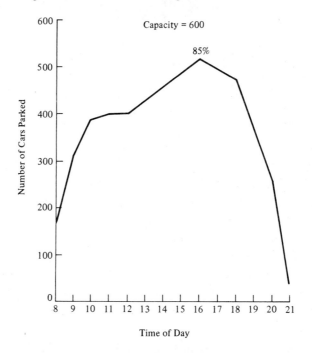

interval is every 20 min. If the length duration is 2 hr, checking every 30 min would be appropriate. Turnover is then obtained from the equation

$$T = \frac{\text{number of different vehicles parked}}{\text{number of parking spaces}} \qquad \textbf{(4.12)}$$

Identification of Parking Generators

This phase involves identifying parking generators (for example, shopping centers or transit terminals) and locating these on a well-prepared map of the study area.

Parking Demand

The information on parking demand is obtained by interviewing drivers at the various parking facilities listed during the inventory. An effort should be made to interview all drivers using the parking facilities on a typical weekday between 8:00 a.m. and 10:00 p.m. Information sought should include (1) trip origin, (2) purpose of trip, and (3) driver's destination reached on foot after parking. The interviewer must also note the location of the parking facility, the times of arrival and departure, and the vehicle type.

Parking interviews can also be carried out, using the post card technique, where stamped post cards with a return address and the appropriate questions are handed to drivers or placed under windshield wipers. When this technique is used, usually only about 30 to 50 percent of the cards distributed are returned. It is therefore necessary to record the time and the number of cards distributed at each location, as this information is required to develop expansion factors, which are later used to expand the sample.

Analysis of Parking Data

Analysis of parking data includes summarizing, coding, and interpreting the data so that the relevant information required for decision making can be obtained. The relevant information includes

- Number and duration of vehicles legally parked
- Number and duration of vehicles illegally parked
- Space hours of demand for parking
- Supply of parking facilities

The analysis required to obtain information on the first two items is simple and straightforward and usually involves simple arithmetical and statistical calculations. Data obtained from these items are then used to determine parking space hours.

The space hours of demand for parking are obtained from the expression

$$D = \sum_{i=1}^{N} (n_i t_i) \tag{4.13}$$

where

D = space vehicle hours demand for a specific period of time
N = number of classes of parking duration ranges
t_i = midparking duration of the ith class
n_i = number of vehicles parked for the ith duration range

The space hours of supply are obtained from the expression:

$$S = f \sum_{i=1}^{N} (t_i) \tag{4.14}$$

where

S = the practical number of space hours of supply for a specific period of time
N = number of parking spaces available
t_i = the total length of time in hours the ith space can be legally parked on during the specific period
f = efficiency factor

The efficiency factor f is used to correct for time lost in each turnover. It is determined on the basis of the best performance a parking facility is expected to operate on. Efficiency factors should therefore be determined for different types of parking facilities, for example, surface lots, curb parking, and garages. Efficiency factors for curb parking, during highest demand, varies from 78 percent to 96 percent and for surface lots and garages from 75 percent to 92 percent. Average values of f are 90 percent for curb parking, 80 percent for garages, and 85 percent for surface lots.

Example 4–7   Space Requirements for a Parking Garage

The owner of a parking garage located in a CBD has observed that 20 percent of those wishing to park are turned back every day during the open hours of 8 a.m. to 6 p.m. because of lack of parking spaces. An analysis of data collected at the garage indicated that 60 percent of those who park are commuters, with an average parking duration of 9 hr and the remaining are shoppers with average parking duration of 2 hr. If 20 percent of those who cannot park are commuters and the rest are shoppers, and a total of 200 vehicles currently park daily in the garage, determine the number of additional spaces required to meet the excess demand. Assume parking efficiency is 0.80.

- Calculate the space hours of demand using Eq. 4.13.

   Commuters now being served $= 0.6 \times 200 \times 9 = 1080$ space hr

   Shoppers now being served $= 0.4 \times 200 \times 2 = 160$ space hr

   Total number of vehicles turned away $= \dfrac{200}{0.8} - 200 = 50$

   Commuters not being served $= 0.2 \times 50 \times 9 = 90$ space hr

   Shoppers not being served $= 0.8 \times 50 \times 2 = 80$ space hr

   Total space hours of demand $= (1080 + 160 + 90 + 80) = 1410$

   Total space hours served $= 1080 + 160 = 1240$

   Number of space hours required $= 1410 - 1240 = 170$

- Determine the number of parking spaces required from Eq. 4.14.

$$S = f \sum_{i=1}^{N} t_i = 170 \text{ space hr}$$

Use the length of time each space can be legally parked (8 a.m. through 6 p.m. = 10 hours) to determine the number of additional spaces,

$$0.8 \times 10 \times N = 170$$

$$N = 21.25$$

At least 22 additional spaces will be required, as a fraction of a space is not feasible.

ACCIDENT STUDIES

The high accident rate on our nation's highways results in considerable loss of human and economic resources. It is estimated that the United States loses over $10 billion every year as a result of motor vehicle accidents. Traffic and highway engineers are therefore continually engaged in the design and/or operation of traffic control devices on the nation's highways, with the aim of reducing the high accident rate. This effort also involves the redesigning and reconstruction of specific highways, which have high potential for accidents. To evaluate the success or failure of these efforts in reducing accidents, data on the frequency and severity of accidents are needed. The need for adequate accident data and the necessity to reduce the high accident rate have led to emphasis on highway safety programs. During the late 1960s and early 1970s, the FHWA took several legislative actions that led to a significant growth in research on highway safety. For example, in the mid-sixties, the FHWA introduced the Spot Improvement Program, which attempted to identify highway locations with high accident potentials. The program also provided funds for the improvement of these locations. Also in the mid-sixties, the Highway Safety Act (23 U.S.C. 402) was passed by Congress. This act set the requirements for states to develop and maintain a safety program through the Highway Safety Program Standards. The maintenance of such a program was assisted by the publication of the *Yellow Book* by AASHTO, which was subsequently revised. These publications described safety design practices and policies.

Funds were made available by the federal government to support the efforts of state and local governments in the application of new procedures in highway safety programs. The objectives of these procedures included the development of better methodologies for the collection, analysis, and evaluation of accident data. This led to the publication of *Highway Safety Improvement Program* (Federal-Aid Highway Program Manual 6-8-2-1), which was superseded by *Highway Safety Improvement Program* (Federal-Aid Highway Program Manual 8-2-3) in the late seventies. This manual advocates that each state should "develop and implement on a continuing basis, a highway safety improvement program (HSIP), which has the overall objective of reducing the number and severity of accidents and decreasing the potential for accidents on all highways." The HSIP consists of the planning component, the implementation component, and the evaluation component. Each of these components consists of one or more processes as shown in Figure 4.18. In this section, the tasks involved in each of these processes will be described briefly. Readers interested in an in-depth study of the program may refer to *Highway Safety Engineering Studies Procedural Guide*, published by the FHWA.[4]

Planning Component of the Highway Safety Improvement Program

The planning component of the program consists of four processes as shown in Figure 4.18. These are (1) collecting and maintaining data, (2) identifying

Figure 4.18   Highway Safety Improvement Program at the Process Level

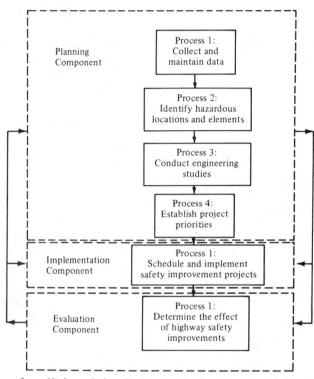

Source: Redrawn from *Highway Safety Engineering Studies Procedural Guide*, U.S. Department of Transportation, Washington, D.C., June 1981.

hazardous locations and elements, (3) conducting engineering studies, and (4) establishing project priorities. Figure 4.18 shows that the information obtained under the planning component serves as input to the two other components and results obtained from the evaluation component may also serve as input to the planning component.

Collecting and Maintaining Data

Accident data are usually obtained from state and local transportation and police agencies. All relevant information on a reported accident is usually recorded by the police on an accident report form. The type of accident form used differs from state to state, but a typical completed accident form will include information on the location of the accident, the time of occurrence, roadway and environmental conditions at time of accident, types and number of vehicles involved, a sketch showing the original paths of the maneuver or maneuvers of the vehicles involved, and the severity of the accident (fatal, injury, or property damage only). Figure 4.19 shows the Virginia accident report form, which is completed by the police officer investigating the accident. Information on minor accidents that do not

Figure 4.19 Virginia Police Accident Report Form

COMMONWEALTH OF VIRGINIA · DIVISION OF MOTOR VEHICLES

POLICE AGENCY COPY

PAGE____ OF____ PAGES

POLICE ACCIDENT REPORT

FR 300P 1/78

ACCIDENT DATE Month Day Year	DAY OF WEEK	TIME AM PM	COUNTY OF ACCIDENT	MILE POST NUMBER	RAILROAD CROSSING ID. NO. IF WITHIN 150 FEET

1 CITY OR TOWN ☐ ☐ OF | LANDMARKS AT SCENE | NUMBER OF VEHICLES | OFFICIAL USE ONLY 17

ROUTE NO. OR STREET NAME AT SCENE 18

☐ AT INTERSECTION WITH OR _____/☐ MILES ☐ FEET ☐ N ☐ S ☐ E ☐ W OF ROUTE NUMBER OR STREET NAME

2

VEHICLE NO. 1			VEHICLE NO. 2 (OR PEDESTRIAN)		
DRIVER'S NAME (LAST, FIRST, MIDDLE)		OCCUPATION	DRIVER'S NAME (LAST, FIRST, MIDDLE)		OCCUPATION
3 ADDRESS (STREET & NO.)		YEARS OF DRIVING EXPERIENCE	ADDRESS (STREET & NO.)		YEARS OF DRIVING EXPERIENCE 19
CITY	STATE	ZIP CODE	CITY	STATE	ZIP CODE 20
4 DATE OF BIRTH Month Day Year / SEX	DRIVER'S LICENSE NUMBER	STATE	DATE OF BIRTH Month Day Year / SEX	DRIVER'S LICENSE NUMBER	STATE 21
VEHICLE OWNER'S NAME (LAST, FIRST, MIDDLE)			VEHICLE OWNER'S NAME (LAST, FIRST, MIDDLE)		22
5 ADDRESS (STREET & NO.)			ADDRESS (STREET & NO.)		23
CITY	STATE	ZIP CODE	CITY	STATE	ZIP CODE
MAKE & TYPE OF VEHICLE (SHOW MOPED, MOTORCYCLE, AMBULANCE, ETC....)	YEAR	REPAIR COST	MAKE & TYPE OF VEHICLE (SHOW MOPED, MOTORCYCLE, AMBULANCE, ETC....)	YEAR	REPAIR COST 24
6 LICENSE PLATE NUMBER	STATE	NAME OF INSURANCE CO. (NOT AGENT)	LICENSE PLATE NUMBER	STATE	NAME OF INSURANCE CO. (NOT AGENT) 25

DAMAGE TO PROPERTY OTHER THAN VEHICLES | OBJECT STRUCK (TREE, FENCE, ETC.) | OWNER'S NAME (LAST, FIRST, MIDDLE) | ADDRESS | REPAIR COST 26

7 **VEHICLE NO. 1 DAMAGE** CHECK POINTS OF IMPACT | ACCIDENT DIAGRAM | **VEHICLE NO. 2 DAMAGE** CHECK POINTS OF IMPACT 27

8

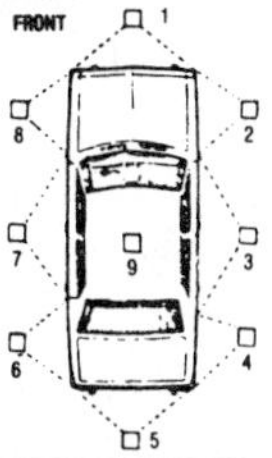

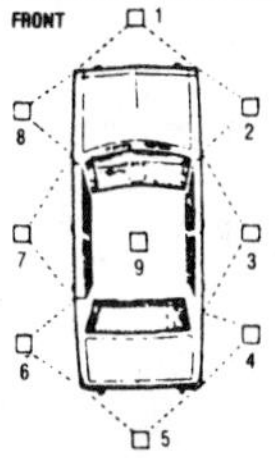

28 29 30 31 32 33

SPEED		
BEFORE ACCIDENT	LIMIT	MAXIMUM SAFE

INDICATE NORTH BY ARROW

SPEED		
BEFORE ACCIDENT	LIMIT	MAXIMUM SAFE

VEHICLE NO. 1 DAMAGES: ☐ UNKNOWN ☐ NO DAMAGE ☐ OVERTURNED ☐ MOTOR ☐ UNDERCARRIAGE ☐ TOTALED ☐ BY FIRE ☐ OTHER

VEHICLE NO. 2 DAMAGES: ☐ UNKNOWN ☐ NO DAMAGE ☐ OVERTURNED ☐ MOTOR ☐ UNDERCARRIAGE ☐ TOTALED ☐ BY FIRE ☐ OTHER 34

ACCIDENT DESCRIPTION: 35

36

37

OFFENSES CHARGED DRIVER:

ALL INJURED	9	10	11	12	13	14	15	16	NAMES OF INJURED · IF DECEASED, INCLUDE DATE OF DEATH
A									
B									
C									
D									
E									

TROOPER/OFFICER'S NAME	BADGE/CODE NUMBER	DEPARTMENT NAME AND CODE NUMBER	REVIEWING OFFICER	DATE REPORT FILED

Source: Reproduced from Commonwealth of Virginia, Division of Motor Vehicles, Richmond, Va.

Figure 4.20 Virginia Citizen Accident Report Form

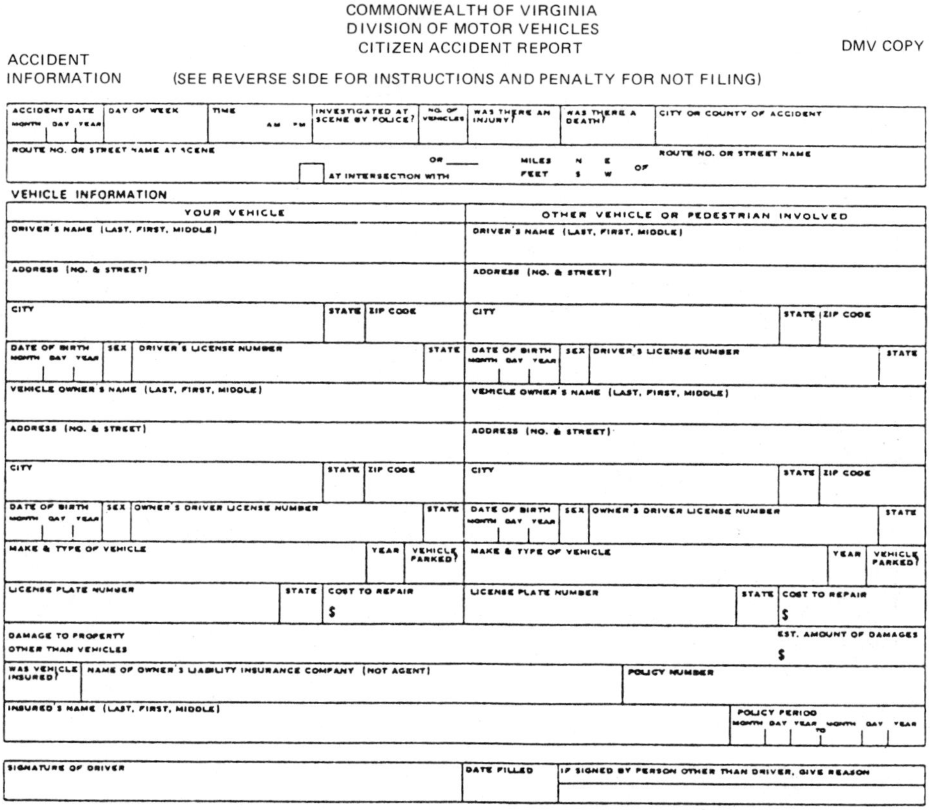
COMMONWEALTH OF VIRGINIA
DIVISION OF MOTOR VEHICLES
CITIZEN ACCIDENT REPORT
DMV COPY
ACCIDENT INFORMATION
(SEE REVERSE SIDE FOR INSTRUCTIONS AND PENALTY FOR NOT FILING)

ACCIDENT DATE (MONTH, DAY, YEAR) | DAY OF WEEK | TIME (AM, PM) | INVESTIGATED AT SCENE BY POLICE? | NO. OF VEHICLES | WAS THERE AN INJURY? | WAS THERE A DEATH? | CITY OR COUNTY OF ACCIDENT
ROUTE NO. OR STREET NAME AT SCENE | AT INTERSECTION WITH | OR ___ MILES / FEET N S E W OF | ROUTE NO. OR STREET NAME

VEHICLE INFORMATION

YOUR VEHICLE
DRIVER'S NAME (LAST, FIRST, MIDDLE)
ADDRESS (NO. & STREET)
CITY | STATE | ZIP CODE
DATE OF BIRTH (MONTH, DAY, YEAR) | SEX | DRIVER'S LICENSE NUMBER | STATE
VEHICLE OWNER'S NAME (LAST, FIRST, MIDDLE)
ADDRESS (NO. & STREET)
CITY | STATE | ZIP CODE
DATE OF BIRTH (MONTH, DAY, YEAR) | SEX | OWNER'S DRIVER LICENSE NUMBER | STATE
MAKE & TYPE OF VEHICLE | YEAR | VEHICLE PARKED?
LICENSE PLATE NUMBER | STATE | COST TO REPAIR $

OTHER VEHICLE OR PEDESTRIAN INVOLVED
DRIVER'S NAME (LAST, FIRST, MIDDLE)
ADDRESS (NO. & STREET)
CITY | STATE | ZIP CODE
DATE OF BIRTH (MONTH, DAY, YEAR) | SEX | DRIVER'S LICENSE NUMBER | STATE
VEHICLE OWNER'S NAME (LAST, FIRST, MIDDLE)
ADDRESS (NO. & STREET)
CITY | STATE | ZIP CODE
DATE OF BIRTH (MONTH, DAY, YEAR) | SEX | OWNER'S DRIVER LICENSE NUMBER | STATE
MAKE & TYPE OF VEHICLE | YEAR | VEHICLE PARKED?
LICENSE PLATE NUMBER | STATE | COST TO REPAIR $

DAMAGE TO PROPERTY OTHER THAN VEHICLES | EST. AMOUNT OF DAMAGES $
WAS VEHICLE INSURED? | NAME OF OWNER'S LIABILITY INSURANCE COMPANY (NOT AGENT) | POLICY NUMBER
INSURED'S NAME (LAST, FIRST, MIDDLE) | POLICY PERIOD (MONTH DAY YEAR TO MONTH DAY YEAR)
SIGNATURE OF DRIVER | DATE FILLED | IF SIGNED BY PERSON OTHER THAN DRIVER, GIVE REASON

Source: Reproduced from Commonwealth of Virginia, Division of Motor Vehicles, Richmond, Va.

involve police investigation may be obtained from routine reports given at the police stations by the drivers involved as is required in some states. Drivers involved in accidents are also sometimes required to complete accident report forms, even though the accident is investigated by the police. Figure 4.20 shows the type of form used for this purpose in Virginia.

Storage and Retrieval of Accident Data. Basically, two techniques are used in the storage of accident data. The first technique involves the manual filing of each completed accident report form in the offices of the appropriate police agency. These forms are usually filed by either date of the accident, the name or number of the routes on which the accident occurred, or location, which may be identified by intersection and roadway links. Summary tables, which give the number and percentage of each type of accident occurring during a given year at a given location, are also prepared. The location can be a specific spot on the highway or an identifiable length of the highway. This technique is suitable for areas where the total number of accidents is less than 500 a year, although it may be used when the total number of accidents is between 500 and 1000 annually. This technique, however, becomes time consuming and inefficient when accidents total more than 1000 a year.

The second technique involves the use of a computer, where each item of information on the accident report form is coded and stored in a computer file. This technique is suitable for areas where total number of accidents per year is higher than 500. With this technique, facilities are provided for storing a large amount of data in a small space. The technique also facilitates flexibility in the choice of methods used for data analysis and permits the study of a large number of accident locations in a short time. There are, however, some disadvantages associated with this technique. These include the high cost of equipment and the requirement of trained computer personnel for the operation of the system. Several agencies are avoiding the necessity of investing large amounts to purchase computer systems by sharing computer facilities. The advent of microcomputers has also made it feasible for relatively small agencies to purchase individual systems.

Several national accident data banks use computerized systems to store data on national accident statistics. These include

- Highway Performance Monitoring System (HPMS), compiled by the FHWA
- Fatal Accident Reporting System (FARS), compiled by the National Center for Statistics and Analysis (NCSA) of the National Highway Safety Administration
- National Injury Surveillance System, which compiles accident data collected at emergency rooms of hospitals
- Bureau of motor carrier series, which contains data on accidents involving passengers and properties of members of the Bureau of Motor Carriers.

Information provided by these data banks may be retrieved by computer techniques for research purposes.

The techniques used for retrieving specific accident data depends on the method of storage of the data. When the data is manually stored, the retrieval is done manually. In this case, the file is examined by a trained technician, who then retrieves the appropriate accident report forms. When the data is stored by the computer technique, retrieval requires only the input of appropriate commands into the computer for any specific data required and this is immediately given as output.

Collision Diagrams.   These diagrams present pictorial information on individual accidents at a location. Different symbols are used to represent different types of maneuvers, types of accidents, and severity of accidents. The date and time (day or night) at which the accident occurs are also indicated. Figure 4.21 shows a typical collision diagram. One advantage of collision diagrams is that they give information on the location of the accidents, which statistical summaries do not give. Collision diagrams may be prepared manually by retrieving data filed manually or by a computer when data are stored in a computer file.

Figure 4.21 Collision Diagram

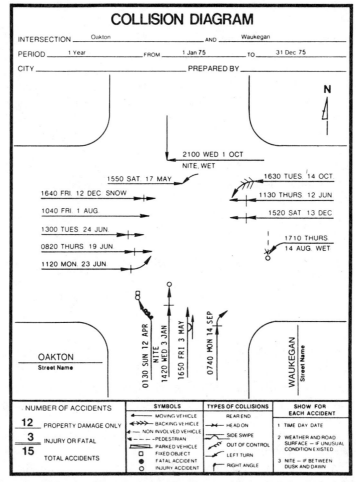

Source: Reproduced from *Transportation and Traffic Engineering Handbook*, 2nd ed., Institute of Transportation Engineers, Washington, D.C., 1982.

Analysis of Accident Data. The reasons for analyzing traffic data are (1) to identify any accident pattern that may exist, (2) to determine the probable causes of accidents with respect to drivers, highways, and vehicles, and (3) to develop countermeasures that will reduce the rate and severity of accidents. To facilitate the comparison of results obtained from the analysis of accidents at a particular location with those of other locations, use is made of one or more accident rates. These accident rates are determined on the basis of exposure data, such as traffic volume, and length of road section being considered. Commonly used rates are rate per million of entering vehicles and rate per 100 million vehicle miles.

The **rate per million of entering vehicles** (RMEVs) is the number of accidents per million vehicles entering the study location during the study period. It is expressed as

$$RMEV = \frac{A \times 1,000,000}{V} \qquad (4.15)$$

where

$RMEV$ = accident rate per million entering vehicles

A = total numbers of accidents or number of accidents by type occurring in 1 yr at the location

$V = ADT \times 365$

This rate is very often used as a measure of accident rates at intersections.

Example 4–8   Computing Accident Rates at Intersections

The number of all accidents recorded at an intersection in a year was 23, and the average 24 hr volume entering from all approaches was 6500. Determine the accident RMEV.

$$RMEV = \frac{23 \times 1,000,000}{6500 \times 365} = 9.69 \text{ accidents/million entering vehicles}$$

The **rate per 100 million vehicle miles** (RMVM) is the number of accidents per 100 million vehicle miles of travel. It is obtained from the expression

$$RMVM = \frac{A \times 100,000,000}{VMT} \qquad (4.16)$$

where

A = number of total accidents or number of accidents by type at the study location, during a given period

VMT = total vehicle miles of travel during the given period

= $ADT \times$ (number of days in study period) $\times$ (length of road)

This rate is very often used as a measure of accident rates on a stretch of highway with similar traffic and geometric characteristics.

Example 4–9   Computing Accident Rates on Roadway Sections

It is observed that 40 traffic accidents occurred on a 17.5 mi long section of a highway in one year. If the ADT on the section was 5000 vehicles, determine

(a) the rate of total accidents per 100 million vehicle miles, and (b) the rate of fatal accidents per 100 million vehicle miles, if 5 percent of the accidents were fatal.

(a)
$$RMVM_T = \frac{40 \times 100,000,000}{17.5 \times 5000 \times 365} = 125.2$$

(b)
$$RMVM_F = \frac{0.05 \times 40 \times 100,000,000}{17.5 \times 5000 \times 365} = 6.26$$

Note that any accident rate may be given in terms of the total accidents occurring or in terms of a specific type of accident. Therefore, it is important that the basis on which accident rates are determined be clearly stated; comparisons between two locations can only be made using results obtained from an analysis based on similar exposure data.

Accident Patterns.   Two commonly used techniques to determine accident patterns are (1) expected value analysis and (2) cluster analysis. A suitable summary of accident data can also be used to determine accident patterns.

   Expected value analysis is a mathematical method used to identify locations with abnormal accident characteristics. It should only be used to compare sites with similar characteristics (for example, geometrics, volume, traffic control), as the analysis does not consider exposure levels. The analysis is carried out by determining the average number of a specific type of accident occurring at several locations with similar geometric and traffic characteristics. This average, adjusted for a given level of confidence, indicates the "expected" value for the specific type of accident. Locations with accident values higher than the expected value are considered as overrepresenting that specific type of accident. The expected value can be obtained from the expression

$$EV = \bar{x} \pm ZS \tag{4.17}$$

where

   EV = expected range of accident frequency
   $\bar{x}$ = average number of accidents per location
   S = estimated standard deviation of accident frequencies
   Z = the number of standard deviations corresponding to the required
       confidence level

Example 4–10   Identifying High Accident Rate Locations

Data collected for 3 consecutive years at an intersection study site show that 14 rear-end collisions and 10 left-turn collisions occurred during the 3 yr period. Data collected at 10 other intersections with similar geometric and traffic characteristics give the information shown in Table 4.14. Determine whether any type of accident is overrepresented at the study site for a 95.0 percent confidence level.

Table 4.14 Number of Rear-End and Left-Turn Collisions at Ten Control Stations for Three Consecutive Years

Control Site	Rear-End Collisions	Left-Turn Collisions
1	8	11
2	5	12
3	7	4
4	8	5
5	6	8
6	8	3
7	9	4
8	10	9
9	6	7
10	7	6
Mean	7.40	6.90
Standard deviation	1.50	3.07

Mean rear-end collisions at 10 control sites = 7.4

Standard deviation of rear-end accidents at 10 control sites = 1.50

Expected range (95% confidence level) = $7.4 \pm 1.5 \times 1.96 = 10.3$ max.

Number of rear-end collisions at study site = 14 (greater than the maximum expected)

Rear-end accidents are, therefore, overrepresented at the study site at 95 percent confidence level.

Mean of left-turn collisions at 10 control sites = 6.9

Standard deviation of left-turn collisions at 10 control sites = 3.07

Expected range (95% confidence level) = $6.9 \pm 3.07 \times 1.96 = 6 \pm 6.02 = 12.9$ max.

Number of left-turn collisions at study site = 10 (within the exposed range)

Left-turn collisions are, therefore, not overrepresented at the study site.

Cluster analysis involves the identification of a particular accident characteristic from the accident data obtained at a site. It identifies any abnormal occurrence of a specific accident type in comparison with other types of accidents at the site. For example, if there are two rear-end collisions, one right-angle accident, and six left-turn accidents at an intersection during a given year, the left-turn accidents could be defined as a *cluster* or grouping, with abnormal occurrence at the site.

It is, however, very difficult to assign discrete values, which can be used to identify accident patterns. This is because accident frequencies, which are the basis for determining accident patterns, differ considerably from site to site. The use of accident rates using exposure data, such as traffic volumes, to define patterns is sometimes useful. Care must be taken, however, to use the correct exposure data. For example, if total intersection volume is used to determine left-turn accident rates at different sites, these rates are not directly comparable as the percentages of left-turn vehicles at these sites may be significantly different. Because of these difficulties, it is desirable to use good engineering judgment when this approach is being used.

Methods of Summarizing Accident Data.　A summary of accidents can be used to identify safety problems that may exist at a particular site. It can also be used to identify the accident pattern at a site, from which possible causes of accidents may be identified, which may lead to the identification of possible remedial actions (countermeasures).

There are five different ways in which the accidents at a site can be summarized:

- Type of accident
- Severity of accidents
- Contributing circumstances
- Environmental conditions
- Time periods of accidents

Accident Summary by Type of Accidents.　This method of summarizing accidents involves the identification of the pattern of accidents at a site, based on the specific types of accidents occurring at the site. The types of accidents commonly used are

- Rear-end
- Right-angle
- Left-turn
- Fixed object
- Sideswipes
- Pedestrian-related
- Run off road
- Head-on
- Parked vehicle
- Bicycle-related

Accident Summary by Severity.　This method involves listing each accident occurring at a site under one of three severity classes: fatal (F), personal injury

(PI), and property damage (PD). Fatal accidents result in at least one death. Accidents that result in injuries, but no deaths, are classified as personal injury accidents. Accidents that result in neither death nor injuries but do involve damage to property are classified as property damage accidents.

This method of summarizing accidents is commonly used to compare accidents at different locations by assigning a weighted scale to each accident based on the severity of the accident. Several weighting scales have been used, but a typical one is given as

Fatality = 12
Personal injury = 3
Property damage only = 1

For example, if 1 fatal accident, 3 personal injury accidents, and 5 property damage accidents occurred during a year at a particular site, the severity number is obtained by

$$12 \times 1 + 3 \times 3 + 1 \times 5 = 26$$

The greatest disadvantage in using this scale is the large difference between the severity scales for fatal and property damage accidents. This may overemphasize the seriousness of the fatal accidents over several property damage accidents. For example, a site with only one fatal accident will be considered more dangerous than a site with nine property damage accidents. This effect can be reduced by using a lower weighting, for example, 8 for fatal accidents, especially at locations where fatal accidents are very rare in comparison with other accidents.

Accident Summary by Contributing Circumstances. In this method, each accident occurring at a site is listed under one of three contributing factors: (1) human (driver) factors, (2) environmental factors, and (3) vehicle-related factors. The necessary information is usually obtained from accident reports.

Accident Summary by Environmental Conditions. This method categorizes accidents based on the environmental conditions that existed at the time of the accidents. Two main categories of environmental conditions are (1) lighting condition, that is, daylight, dusk, dawn, or dark, and (2) roadway surface condition, that is, dry, wet, snow/icy. This method of summarizing accidents facilitates the identification of possible causes of accidents and safety deficiencies that may exist at a particular location. The expected value method may be used to ascertain whether accident rates under a particular environmental condition are significantly greater at one site than at other similar sites.

Accident Summary by Time Periods. This method categorizes all accidents under different time periods to identify whether accident rates are significantly higher during any specific time periods. Three different time periods can be used: (1) time, (2) day, and (3) month. This method of summarizing accident data also facilitates the use of the expected value method to identify time periods during which accident occurrences are overrepresented.

Determination of Accident Causes.   Having identified the hazardous locations and the accident pattern, the next stage in the data analysis is to determine possible causes of accidents. The types of accidents identified are matched with a list of possible causes from which several probable causes are identified. Table 4.15 shows a list of possible causes for different types of accidents. The environmental conditions existing at the times of accidents may also help in the identification of possible causes of accidents.

Identifying Hazardous Locations and Elements

Hazardous locations are sites where accident frequencies, calculated on the basis of the same exposure data, are higher than the expected value for other similar locations or conditions. Any of the accident rates or accident summaries described earlier may be used to identify hazardous locations. A common method of analysis involves the determination of accident rates based on the same exposure data for the study site with apparent high accident rates and several other sites with similar traffic and geometric characteristics. An appropriate statistical test such as the expected value analysis is then performed to determine whether the apparent high accident rate at the study site is actually significantly higher. If the statistical test shows that the apparent high accident rate is significantly higher, this indicates an abnormal rate of accidents at the test location and that location is considered a hazardous location.

Conducting Engineering Studies

After a particular location has been identified as hazardous, a detailed engineering study is performed to identify the safety problem. Once the safety problem is identified, suitable safety-related countermeasures can be developed.

The first task in this subprocess is an in-depth study of the accident data obtained at the hazardous site. The results of the analysis will indicate the type or types of accidents that predominate or that have abnormal frequency rates. Possible causes for the accidents can then be identified from Table 4.15. Note, however, that the list of possible causes obtained at this stage is preliminary, and personal knowledge of the site, field conditions, and police accident reports should all be used to improve this list.

The next task is to conduct a field review of the study site. This review involves an inspection of the physical condition of the site and an observation of traffic operation at the site. Information obtained from this field review is then used to confirm the existence of physical deficiencies suspected, based on the pattern of accidents, and to refine the list of possible accident causes first obtained by adding or deleting possible accident causes on the observation.

The refined list of possible accidents is used to determine what data will be required to identify the safety deficiencies at the study site. Table 4.16 gives a partial list of data needs for different possible causes of accidents. A complete list is given in *Highway Safety Engineering Studies Procedural Guide*.[5] After identifying the data needs, existing records will then be reviewed to determine whether the data required are available. Care must be taken to ensure that any existing data

(text continues on p. 147)

Table 4.15 Probable Accident Causes for Different Types of Accidents

Accident Pattern	Probable Cause
Left-turn head-on collisions	▪ Large volume of left-turns ▪ Restricted sight distance ▪ Too short amber phase ▪ Absence of special left-turning phase ▪ Excessive speed on approaches
Right-angle collisions at signalized intersections	▪ Restricted sight distance ▪ Excessive speed on approaches ▪ Poor visibility of signal ▪ Inadequate signal timing ▪ Inadequate roadway lighting ▪ Inadequate advance intersection warning signs ▪ Large total intersection volume
Right-angle collisions at unsignalized intersections	▪ Restricted sight distance ▪ Large total intersection volume ▪ Excessive speed on approaches ▪ Inadequate roadway lighting ▪ Inadequate advance intersection warning signals ▪ Inadequate traffic control devices
Rear-end collisions at unsignalized intersections	▪ Driver not aware of intersection ▪ Excessive speed on approach ▪ Slippery surface ▪ Lack of adequate gaps ▪ Large number of turning vehicles ▪ Crossing pedestrians ▪ Inadequate roadway lighting
Rear-end collisions at signalized intersections	▪ Slippery surface ▪ Large number of turning vehicles ▪ Poor visibility of signals ▪ Inadequate signal timing ▪ Unwarranted signals ▪ Inadequate roadway lighting
Pedestrian-vehicle collisions	▪ Restricted sight distance ▪ Inadequate protection for pedestrians ▪ School crossing area ▪ Inadequate signals ▪ Inadequate phasing signal

Source: Adapted from *Highway Safety Engineering Studies Procedural Guide*, U.S. Department of Transportation, Washington, D.C., June 1981.

Table 4.16 Data Needs for Different Possible Causes of Accidents

Possible Causes	Data Needs	Procedures to Be Performed
Rear-end collisions at unsignalized intersections		
Drivers not aware of intersection	• Roadway inventory • Sight distance characteristics • Speed characteristics	• Roadway Inventory Study • Sight Distance Study • Spot Speed Study
Slippery surface	• Pavement skid resistance characteristics • Conflicts resulting from slippery surface	• Skid Resistance Study • Weather-Related Study • Traffic Conflict Study
Large number of turning vehicles	• Volume data • Roadway inventory • Conflict data	• Volume Study • Roadway Inventory Study • Traffic Conflict Study
Inadequate roadway lighting	• Roadway inventory • Volume data • Data on existing lighting	• Roadway Inventory Study • Volume Study • Highway Lighting Study
Excessive speed on approaches	• Speed characteristics	• Spot Speed Study
Lack of adequate gaps	• Roadway inventory • Volume data • Gap data	• Roadway Inventory Study • Volume Study • Gap Study
Crossing pedestrians	• Pedestrian volumes • Pedestrian/vehicle conflicts • Signal inventory	• Volume Study • Pedestrian Study • Roadway Inventory Study
Rear-end collisions at signalized intersections		
Slippery surface	• Pavement skid resistance characteristics • Conflicts resulting from slippery surface	• Skid Resistance Study • Weather-Related Study • Traffic Conflict Study

Continued

Table 4.16—_Continued_

Possible Causes	Data Needs	Procedures to Be Performed
Rear-end collisions at signalized intersections		
Large number of turning vehicles	• Volume data • Roadway inventory • Conflict data • Travel time and delay data	• Volume Study • Roadway Inventory Study • Traffic Conflict Study • Delay Study
Poor visibility of signals	• Roadway inventory • Signal review • Traffic conflicts	• Roadway Inventory Study • Traffic Control Device Study • Traffic Conflict Study
Left-turn head-on collisions		
Large volume of left-turns	• Volume data • Vehicle conflicts • Roadway inventory • Signal timing and phasing • Travel time and delay data	• Volume Study • Traffic Conflict Study • Roadway Inventory Study • Capacity Study • Travel Time and Delay Study
Restricted sight distance	• Roadway inventory • Sight distance characteristics • Speed characteristics	• Roadway Inventory Study • Sight Distance Study • Spot Speed Study
Too short amber phase	• Speed characteristics • Volume data • Roadway inventory • Signal timing and phasing	• Spot Speed Study • Volume Study • Roadway Inventory Study • Capacity Study
Absence of special left-turning phase	• Volume data • Roadway inventory • Signal timing and phasing • Delay data	• Volume Study • Roadway Inventory Study • Capacity Study • Travel Time and Delay Study
Excessive speed on approaches	• Speed characteristics	• Spot Speed Study

Source: Adapted from _Highway Safety Engineering Studies Procedural Guide_, U.S. Department of Transportation, Washington, D.C., June 1981.

Table 4.17   General Countermeasures for Different Safety Deficiencies

Probable Cause	*General Countermeasure*
Left-turn head-on collisions	
Large volume of left-turns	▪ Create one-way street ▪ Widen road ▪ Provide left-turn signal phases ▪ Prohibit left-turns ▪ Reroute left-turn traffic ▪ Channelize intersection ▪ Install stop signs (see MUTCD)* ▪ Revise signal sequence ▪ Provide turning guidelines (if there is a dual left-turn lane) ▪ Provide traffic signal if warranted by MUTCD* ▪ Retime signals
Restricted sight distance	▪ Remove obstacles ▪ Provide adequate channelization ▪ Provide special phase for left-turning traffic ▪ Provide left-turn slots ▪ Install warning signs ▪ Reduce speed limit on approaches
Too short amber phase	▪ Increase amber phase ▪ Provide all red phase
Absence of special left-turning phase	▪ Provide special phase for left-turning traffic
Excessive speed on approaches	▪ Reduce speed limit on approaches
Rear-end collisions at unsignalized intersections	
Driver not aware of intersection	▪ Install/improve warning signs
Slippery surface	▪ Overlay pavement ▪ Provide adequate drainage ▪ Groove pavement ▪ Reduce speed limit on approaches ▪ Provide "slippery when wet" signs
Large numbers of turning vehicles	▪ Create left- or right-turn lanes ▪ Prohibit turns ▪ Increase curb radii
Inadequate roadway lighting	▪ Improve roadway lighting
Excessive speed on approach	▪ Reduce speed limit on approaches
Lack of adequate gaps	▪ Provide traffic signal if warranted (see MUTCD)* ▪ Provide stop signs
Crossing pedestrians	▪ Install/improve signing or marking of pedestrian crosswalks

Continued

Table 4.17—Continued

Probable Cause	General Countermeasure
Rear-end collisions at signalized intersections	
Slippery surface	• Overlay pavement • Provide adequate drainage • Groove pavement • Reduce speed limit on approaches • Provide "slippery when wet" signs
Large number of turning vehicles	• Create left- or right-turn lanes • Prohibit turns • Increase curb radii • Provide special phase for left-turning traffic
Poor visibility of signals	• Install/improve advance warning devices • Install overhead signals • Install 12-in. signal lenses (see MUTCD)* • Install visors • Install back plates • Relocate signals • Add additional signal heads • Remove obstacles • Reduce speed limit on approaches
Inadequate signal timing	• Adjust amber phase • Provide progression through a set of signalized intersections • Add all-red clearance
Unwarranted signals	• Remove signals (see MUTCD)*
Inadequate roadway lighting	• Improve roadway lighting

*Manual on Uniform Traffic Control Devices, published by FHWA.

Source: Adapted from *Highway Safety Engineering Studies Procedural Guide*, U.S. Department of Transportation, Washington, D.C., June 1981.

are current and are related to the time for which the study is being conducted. In cases where the necessary data are available, it will not be necessary to carry out specific engineering studies. When the appropriate data are not available, the engineering studies identified from Table 4.16 will then be conducted. Some of these studies are described earlier in this chapter.

The results of these studies are used to determine traffic characteristics of the study site, through which specific safety deficiencies at the study site are determined. For example, a sight distance study at an intersection may reveal inadequate sight distance at that intersection that results in an abnormal rate of left-turn head-on collisions. Similarly, a volume study, which includes turning movements at an intersection with no separate left-turn phase, may indicate a high volume of left-turn vehicles, which suggests that a deficiency is the absence of a special left-turn phase.

Having identified the safety deficiencies at the study site, the next task is to develop alternative countermeasures to alleviate the identified safety deficiencies. A partial list of general countermeasures for different types of possible causes is shown in Table 4.17. The selection of countermeasures should be carefully made by the traffic engineer based on his or her personal knowledge of the effectiveness of each countermeasure considered in reducing the rate at similar sites for the specific type of accidents being considered. Note that countermeasures that are very successful in achieving significant major benefits in one part of the country may not be that successful in another part of the country due to the complexity of the interrelationship that exists among the traffic variables.

Accident Reduction Capabilities of Countermeasures. Accident reduction capabilities are used to estimate the expected reduction in accidents that will occur during a given period as a result of implementing a proposed countermeasure. This estimate can be used to carry out an economic evaluation of alternate countermeasures. Accident reduction capabilities are given in terms of factors that have been developed by several states and agencies. These factors are known as *accident reduction* (AR) factors, and they are usually based on the evaluation of data obtained from safety projects. AR factors can be obtained from state agencies involved in accident analysis.

In using the AR factor to determine the reduction in accidents due to the implementation of a specific countermeasure, the following equation is used.

$$\text{Accidents prevented} = N \times AR \, \frac{(ADT \text{ after period})}{(ADT \text{ before period})} \qquad \textbf{(4.18)}$$

where

N = expected number of accidents if countermeasure is not implemented and if the traffic volume remains the same

AR = accident reduction factor for a specific countermeasure

ADT = average daily traffic

Example 4–11   Expected Reduction in Accident Due to a Safety Improvement

The AR factor for a specific type of countermeasure = 30 percent; the ADT before period = 7850 (average over 3 yr period); and the ADT after period = 9000. Over the 3 yr period, the number of specific types of accidents occurring per year = 12, 14, 13. Determine the expected reduction in number of accidents occurring after the implementation of the countermeasure.

Average number of accidents/year $= 13$.

$$\text{Accidents prevented} = \frac{13 \times 0.30 \times 9000}{7850}$$

$$= 4.47 \, (4 \text{ accidents})$$

This indicates that, if the ADT increases to 9000, the expected number of accidents with the implementation of the countermeasure will be less by 4 every year than the expected number of accidents without the countermeasure. Note that in using Eq. 4.18 the value of N is the average result of the study period. The value used for the ADT before period is also the average of the ADTs throughout the before period.

It is also sometimes necessary to consider multiple countermeasures at a particular site. In such cases, the overall accident reduction factor is obtained from the individual accident reduction factors by using Eq. 4.19, which was proposed by Roy Jorgensen and Associates.

$$AR = AR_1 + (1 - AR_1)AR_2 + (1 - AR_1)(1 - AR_2)AR_3 + \cdots$$
$$+ (1 - AR_1), \ldots, (1 - AR_{m-1})AR_m \qquad \textbf{(4.19)}$$

where

AR = overall accident reduction factor for multiple mutually exclusive improvements at a single site

AR_i = accident reduction factor for a specific countermeasure

m = number of countermeasures at the site

In using Eq. 4.19, it is first necessary to list all the individual countermeasures in order of importance. The countermeasure with the highest reduction factor will be listed first, and its reduction factor will be designated AR_1, the countermeasure with the second highest reduction factor will be listed second and its reduction factor designated AR_2 and so on.

Example 4–12 Accident Reduction Due to Multiple Safety Improvements

At a single location three countermeasures with ARs of 40 percent, 28 percent, and 20 percent are proposed. Determine the overall AR factor.

$AR_1 = 0.40$

$AR_2 = 0.28$

$AR_3 = 0.20$

$$AR = 0.4 + (1 - 0.4)0.28 + (1 - 0.4)(1 - 0.28)0.2$$

$$= 0.4 + 0.17 + 0.09$$

$$= \underline{0.66}$$

Establishing Project Priorities

Economic Analysis. The purpose of this task is to determine the economic feasibility of each set of countermeasures and to determine the best alternative among feasible mutually exclusive countermeasures. This involves the use of many of the techniques discussed in Chapter 11. Benefits are determined on the basis of expected number of accidents that will be prevented if a specific proposal is implemented, and costs are the capital and continuing costs for constructing and operating the proposed countermeasure.

The benefit may be obtained in monetary terms, by multiplying the expected number of accidents prevented by an assigned cost for each type of accident severity. Table 4.18 shows costs proposed by the National Safety Council (NSC) and the National Highway Traffic Safety Administration (NHTSA). Note, however, that individual states have determined accident costs that are applicable to them. When such information is available, it is advisable to use it.

Implementation and Evaluation. The scheduling and implementation of the proposal selected is the next step. The evaluation component involves the determination of the effect of the highway safety improvement. This includes the collection of data for a period after the implementation of the improvement to determine whether the anticipated benefits are actually accrued. This task is important, since the information obtained will provide valuable data for other similar projects. This is discussed further in Chapter 12 under Transportation Systems Management (TSM), since a common reason for implementing a TSM action is to reduce accident rates at a particular location or stretch of highway.

Table 4.18 National Safety Council and National Highway Traffic Safety Administration Accident Costs

Source	Accident Severity	Cost per Involvement
NSC (1979)	Fatal	$160,000
	Nonfatal disabling injury	6,200
	Property damage (including minor injuries)	870
NHTSA (1975)	Fatality	$287,175
	Critical injury	192,240
	Severe injury—life threatening	89,955
	Severe injury—not life threatening	8,085
	Moderate injury	4,350
	Minor injury	2,190
	Average injury	3,185
	Property damage only	520

Source: Adapted from *Highway Safety Engineering Studies Procedural Guide*, U.S. Department of Transportation, Washington, D.C., June 1981.

SUMMARY

Highway transportation has provided considerable opportunities for humans, particularly provision of the freedom to move from place to place at one's will and convenience. The positive aspects of the highway mode, however, go hand in hand with numerous negative aspects, which include traffic congestion, accidents, pollution, and parking difficulties. Traffic and transportation engineers are continually involved in determining ways to reduce the effect of these negative aspects. The effective reduction of the negative impact of the highway mode of transportation at any location can be achieved only after adequate information is obtained to define the problem and the extent to which the problem has a negative impact on the highway system. This information is obtained by conducting appropriate studies to collect and analyze the relevant data. These studies are generally referred to as traffic engineering studies.

This chapter has presented the basic concepts of the different traffic engineering studies under the subheadings spot speed studies, volume studies, travel time and delay studies, parking studies, and accident studies. It should be emphasized here that no attempt has been made to present an in-depth discussion of any of these studies, as such a discussion is beyond the scope of this book. However, enough material has been provided to introduce the reader to the subject so he or she will be able to understand the more advanced literature on the subject. Additional Readings are provided at the end of this chapter; the interested reader can use it for further reading on traffic engineering studies.

PROBLEMS

4–1 Describe in detail the different traffic count programs carried out in your state, indicating the use of the data collected under each program.

4–2 Define the following terms and give examples of uses for each.

Average annual daily traffic (AADT)
Average daily traffic (ADT)
Vehicle miles of travel (VMT)
Peak hour volume (PHV)

4–3 An engineer, wishing to determine a truck adjustment factor to be applied to traffic volumes collected with a junior counter, collected the accompanying data. Both manual and machine counts were collected simultaneously. Determine the truck adjustment factor that the engineer should apply to a machine count in order to obtain the number of vehicles.

Type	No. of Axles	Manual Count	No. of Vehicles Recorded by Machine
Passenger cars	2	3850	3850
Trucks	3	1370	2055
Trucks	4	520	1040

4–4 Using the data furnished in problem 4–5, draw the histogram frequency distribution and cumulative percentage distribution of your data and determine: (a) average speed, (b) eighty-fifth percentile speed, (c) fifteenth percentile speed, (d) mode, (e) median, and (f) pace.

4–5 The accompanying data show spot speeds collected at a section of highway located in a residential area before and after a posted speed limit was reduced from 35 mph to 25 mph. Using the student's t test, determine whether there was a significant difference in the average speeds at the 95 percent confidence level.

Before	After	Before	After
40	23	38	25
35	33	35	21
38	25	30	35
37	36	30	30
33	37	38	33
30	34	39	21
28	23	35	28
35	28	36	23
35	24	34	24
40	31	33	27
33	24	31	20
35	20	36	20
36	21	35	30
36	28	33	32
40	35	39	33

4–6 Describe each of the following types of traffic volume counts and state when you would use each of them: (a) screen line counts, (b) cordon counts, (c) intersection counts, and (d) control counts.

4–7 What are travel time and delay studies used for? Describe one method for collecting data on travel time and delay at a section of a highway and explain how you will obtain the following from the data collected: (a) travel time, (b) operational delay, (c) stopped time delay, (d) fixed delay, and (e) travel time delay.

4–8 Table 4.19 shows data obtained in a travel time study on a section of highway using the moving-vehicle technique. Estimate the travel time and volume in each direction at this section of the highway.

4–9 Briefly describe the tasks you would include in a comprehensive parking study for your college campus, indicating how you would perform each task and the way you would present the data collected.

4–10 Data collected at a parking lot indicated that a total of 300 cars park between 8 a.m. and 6 p.m. Ten percent of these cars are parked for an average of 2 hr, 30 percent for an average of 4 hr, and the remaining cars are parked for an average of 10 hr. Determine the space hours of demand at the lot.

Table 4.19

Run Direction/ Number	Travel Time (min)	No. of Vehicles Traveling in Opposite Direction	No. of Vehicles That Overtook Test Vehicles	No. of Vehicles Overtaken by Vehicle
Northbound				
1	5.25	100	2	2
2	5.08	105	2	1
3	5.30	103	3	1
4	5.15	110	1	0
5	5.00	101	0	0
6	5.51	98	2	2
7	5.38	97	1	1
8	5.41	112	2	3
9	5.12	109	3	1
10	5.31	107	0	0
Southbound				
1	4.95	85	1	0
2	4.85	88	0	1
3	5.00	95	0	1
4	4.91	100	2	1
5	4.63	102	1	2
6	5.11	90	1	1
7	4.83	95	2	0
8	4.91	96	3	1
9	4.95	98	1	2
10	4.83	90	0	1

4–11 If 10 percent of the parking bays are vacant on an average at the parking lot of problem 4–10, determine the number of parking bays in the parking lot. Assume an efficiency factor of 0.85.

4–12 The owner of the parking lot of problems 4–10 and 4–11 is planning an expansion of her lot to provide adequate demand for the following 5 yr. If she has estimated that parking demand for all categories will increase by 5 percent a year, determine the number of additional parking bays that will be required.

4–13 Briefly describe the different ways accidents at a site may be summarized, stating when you would use each of these.

4–14 An engineer has proposed four countermeasures to be implemented to reduce the high accident rate at an intersection. AR factors for these countermeasures are 0.25, 0.30, 0.17, and 0.28. The number of accidents occurring at the intersection during the past 3 yr were 28, 30, and 31 and the AADTs during those years were 8450, 9150, and 9850, respectively. Determine the expected reduction in number of accidents during the first 3 yr after the implementation of the countermeasures if the AADT during the first year of implementation is 10,850 and the estimated traffic growth rate is 4 percent per annum during the first 3 yr after implementation of the countermeasures.

REFERENCES

1. *Transportation and Traffic Engineering Handbook*, 2nd ed., Institute of Transportation Engineers, Washington, D.C., 1982.

2. *Traffic Monitoring Guide*, U.S. Department of Transportation, Federal Highway Administration, Washington, D.C., June 1985.

3. *Transportation and Traffic Engineering Handbook*.

4. *Highway Safety Engineering Studies Procedural Guide*, U.S. Department of Transportation, Federal Highway Administration, Washington, D.C., November 1981.

5. Ibid.

ADDITIONAL READINGS

Accident Research Manual, University of North Carolina, Highway Safety Research Center, Chapel Hill, N.C., January 1980.

Brown, A. M., and A. Churly, "Parking Surveys and the Use of VISTA in West Sussex," *Traffic Engineering and Control* 23 (December 1982).

DART (Data Analysis and Reporting Techniques) Software System, U.S. Department of Transportation, National Highway Traffic Safety Administration, Washington, D.C., April 1978.

Guide to Urban Traffic Volume Counting, U.S. Department of Transportation, Federal Highway Administration, Washington, D.C., September 1981.

Hague, P. W., "Photographic Parking Duration Surveys," *Traffic Engineering and Control* 21(3):123–126 (March 1980).

Harrison, Ian, and Mick Roberts, "A Model for Predicting Seasonal Traffic," *Traffic Engineering and Control* 22(3):122–126 (March 1981).

Highway Performance Monitoring System, U.S. Department of Transportation, Federal Highway Administration, Washington, D.C., January 1979.

Hoang, Lap T., and Victor P. Poteat, "Estimating VMT by Using Random Sampling Technique," *Transportation Research Record* 779(1980):6–10.

Laughland, J. C., *Accident Reduction Factors Developed at the National Level*, U.S. Department of Transportation, Federal Highway Administration, Washington, D.C., June 1980.

Laughland, J. C., L. E. Haefner, J. W. Hall, and D. R. Clough, *Methods for Evaluating Highway Safety Improvements*, NCHRP Report 162, National Research Council, Transportation Research Board, 1975.

Manual on Identification, Analysis and Correction of High Accident Location, U.S. Department of Transportation, Federal Highway Administration, Washington, D.C., 1976.

Oppenlander, Joseph C., "Simple Size Determination for Travel Time and Delay Studies," *Traffic Engineering* 46(9):25–29 (September 1976).

Phillips, Garwyn, *Accuracy of Annual Traffic Flow Estimates from Automatic Counts*, TRRL Special Report 515, Transport and Road Research Laboratory, Crowthorne, Berkshire, England, 1979.

Phillips, Garwyn, and Philip Blake, "Estimating Total Annual Traffic Flow from Short Period Count," *Transportation Planning and Technology* 6(1980):169–174.

Priestas, E. L., and T. E. Mulinzazzi, "Traffic Volume Counting Recorders," *Transportation Engineering Journal* 101(May 1975):211–224.

Procedure for the Analysis of High Accident Locations, Wayne State University, Detroit, Mich., December 1976.

Reilly, William R., and Craig C. Gardner, "Technique for Measuring Delay at Intersections," *Transportation Research Record* 644(1977):1–7.

Russell, Eugene R., Thomas Butcher, and Harold L. Michael, "A Quick, Inexpensive Data Collection Technique for Approach Speed Analysis," *Traffic Engineering* 46(5):13–17 (May 1976).

Sharma, Satish C., and Al Werner, "Improved Method of Grouping Provincewide Permanent Traffic Counters," *Transportation Research Record* 815(1981):12–18.

Sosslau, Arthur R., and Gary E. Honts, "Traffic Characteristics Measurements Using License Matching Techniques," *Traffic Engineering* 45(10):16–20 (October 1975).

Technical Council Committee 4H-M, "Areawide Traffic Performance Evaluation Measures," *Traffic Engineering* 45(9):34 (September 1975).

Traffic Control Systems Handbook, U.S. Department of Transportation, Federal Highway Administration, Washington, D.C., June 1976.

CHAPTER 5

Fundamental Principles of Traffic Flow

In Chapter 4, we described field studies for collecting data on several elements of a traffic stream, including speed and volume. The data collected can be used to determine the relationships among these elements to more fully understand the resultant effect on the characteristics of the traffic stream due to changes in the value of any of these elements. Determination of these relationships is of primary interest in traffic flow theory, which involves the development of mathematical relationships among the primary elements of a traffic stream: flow, density, and speed. The relationships help the traffic engineer in planning, designing, and evaluating the effectiveness of implementing traffic engineering measures on a highway system.

One example of the use of traffic flow theory in design is the determination of adequate lane lengths for storing left-turn vehicles on separate left-turn lanes. The determination of the average delay at intersections and freeway ramp merging areas is another example of the application of traffic flow theory. An example relating to the evaluation of traffic engineering measures is the determination of changes in the level of freeway performance due to the installation of improved vehicular control devices on ramps. Another important application of traffic flow theory is simulation, where mathematical algorithms are used to study the complex interrelationships that exist among the elements of a traffic stream or network to estimate the effect of changes in traffic flow on factors such as accidents, travel time, air pollution, and gasoline consumption.

Methods ranging from physical models to empirical models have been used in studies related to the description and quantification of traffic flow. This chapter, however, will introduce only those aspects of traffic flow theory that can be used in the planning, designing, and operation of highway systems. Readers interested in a more detailed treatment of the topic are referred to the Additional Readings at the end of this chapter.

TRAFFIC FLOW ELEMENTS

Let us first define the elements of traffic flow before discussing the relationships among them. Before we do that, though, we will describe the time–space diagram, which serves as a useful device for defining the elements of traffic flow.

The time–space diagram is a graph that describes the relationship between the location of vehicles in a traffic stream and time as the vehicles progress along the highway. Figure 5.1 shows a time–space diagram for six vehicles, with the distance plotted on the vertical axis and time on the horizontal axis. At time zero, vehicles 1, 2, 3, and 4 are at distances d_1, d_2, d_3, and d_4, respectively, from a reference point, whereas vehicles 5 and 6 cross the reference point later at times t_5 and t_6, respectively.

We earlier referred to the primary elements of traffic flow as being the flow, density, and speed. Another element, associated with density, is the gap or headway between two vehicles in a traffic stream. The definitions of these elements follow.

Volume (V) is the total number of vehicles that pass a point on a highway during a given time interval. When the time interval is an hour, the unit of volume is vehicles per hour (vph). Volume can also be measured by day (vpd) or by year (vpy). In some circumstances, volume is also given as the number of passenger cars per hour, with all trucks and buses counted as an equivalent number of passenger cars. The *Highway Capacity Manual* gives equivalent factors for trucks and buses for varying types of roads, terrain conditions, and velocities.[1]

Flow (q) is the equivalent hourly rate at which vehicles pass a point on a highway lane during a time period less than 1 hr.

Density (k), sometimes referred to as concentration, is the number of vehicles traveling over a unit length of highway at an instant in time. The unit length is usually 1 mi, thereby making the vehicles per mile (vpm) unit of density.

Figure 5.1 Time–Space Diagram

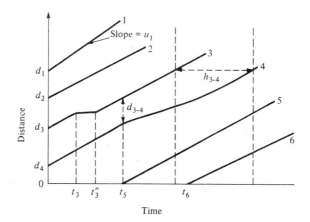

Speed (u) is the distance traveled by a vehicle during a unit of time. It can be expressed in miles per hour (mph), kilometers per hour (kph), or in feet per second (fps). The speed of a vehicle at any time t is the slope of the time–space diagram for that vehicle at time t. Vehicles 1 and 2 in Figure 5.1, for example, are moving at constant speeds because the slopes of the associated graphs are constant. Vehicle 3, however, moves at a constant speed between time zero and time t_3, then stops for the period t_3 to t_3'' (slope of graph equals zero), and then accelerates and eventually moves at a constant speed. There are two mean speeds, namely, time mean speed and space mean speed.

Time mean speed ($\bar{u}_t$) is the arithmetic mean of the speeds of vehicles passing a point on a highway during an interval of time.

Space mean speed ($\bar{u}_s$) is the harmonic mean of the speeds of vehicles passing a point on a highway during an interval of time. It is obtained by dividing the total distance traveled by two or more vehicles on a section of highway by the total time required by these vehicles to travel that distance. This is the speed that is involved in flow–density relationships.

Time headway (h) is the difference between the time the front of a vehicle arrives at a point on the highway and the time the front of the next vehicle arrives at that same point. Time headway usually is expressed in seconds. For example, in the time–space diagram (Figure 5.1), the time headway between vehicles 3 and 4 at d_1 is h_{3-4}.

Space headway (d) is the distance between the front of a vehicle and the front of the following vehicle. It is usually expressed in feet. The space headway between vehicles 3 and 4 at time t_5 is d_{3-4} (see Figure 5.1).

To demonstrate the physical interpretation of traffic flow elements, consider the situation shown in Figure 5.2, which shows vehicles traveling at constant speeds on a 2-lane highway between sections X and Y with their positions and speeds obtained at an instant of time by photography. An observer located at point X will observe the four vehicles passing point X during a period of time, T sec, with velocities 45, 40, 35, and 20 mph, respectively.

The flow is calculated by

$$q = \frac{n \times 3600}{T} \tag{5.1}$$

$$= \frac{4 \times 3600}{T} = \frac{14,400}{T} \text{ vph}$$

With L equal to the distance between X and Y (ft), density is obtained by

$$k = \frac{n}{L}$$

$$= \frac{4}{300} \times 5280 = 70.4 \text{ vpm}$$

The time mean speed is found by

$$\bar{u}_t = \frac{1}{n} \sum_{i=1}^{n} u_i \tag{5.2}$$

With n as the number of vehicles passing point X and u_i as the speed of the ith vehicle, we have

$$\bar{u}_t = \frac{20 + 35 + 40 + 45}{4} \text{ mph} = 35 \text{ mph}$$

The space mean speed is found by

$$\bar{u}_s = \frac{n}{\sum_{i=1}^{n} (1/u_i)} \tag{5.3}$$

$$= \frac{Ln}{\sum_{i=1}^{n} t_i}$$

$$= \frac{300n}{\sum_{i=1}^{n} t_i}$$

where t_i is the time it takes the ith vehicle to travel from X to Y at speed u_i and L (ft) is the distance between X and Y.

$$t_i = \frac{L}{1.47u_i} \text{ sec}$$

$$t_A = \frac{300}{1.47 \times 45} = 4.54 \text{ sec}$$

$$t_B = \frac{300}{1.47 \times 40} = 5.10 \text{ sec}$$

$$t_C = \frac{300}{1.47 \times 35} = 5.83 \text{ sec}$$

$$t_D = \frac{300}{1.47 \times 20} = 10.20 \text{ sec}$$

$$\bar{u}_s = \frac{4 \times 300}{4.54 + 5.10 + 5.83 + 10.20} = 46.75 \text{ ft/sec}$$

$$= 31.90 \text{ mph}$$

Figure 5.2 Locations and Speeds of Four Vehicles on a 2-Lane Highway at an Instant of Time

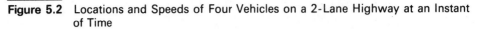
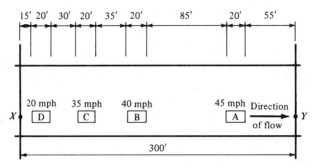

FLOW DENSITY RELATIONSHIPS

The general equation relating flow, density, and space mean speed is given as

$$\text{flow} = \text{density} \times \text{space mean speed}$$
$$q = k\bar{u}_s \tag{5.4}$$

Each of the variables in Eq. 5.4 also depends on several other factors, such as the characteristics of the roadway, the characteristics of the vehicle, the characteristics of the driver, and environmental factors such as the weather.

Other relationships that exist among the traffic flow variables are given as follows.

$$\text{space mean speed} = (\text{flow}) \times (\text{space headway})$$
$$\bar{u}_s = q\bar{d} \tag{5.5}$$

where
$$\bar{d} = (1/k) = \text{average space headway} \tag{5.6}$$

$$\text{density} = (\text{flow}) \times (\text{travel time for unit distance})$$
$$k = q\bar{t} \tag{5.7}$$

where $\bar{t}$ is the average time for unit distance.

$$\text{average space headway} = (\text{space mean speed}) \times (\text{average time headway})$$
$$\bar{d} = \bar{u}_s\bar{h} \tag{5.8}$$

$$\text{average time headway} = (\text{average travel time for unit distance})$$
$$\times (\text{average space headway})$$
$$\bar{h} = \bar{t}\bar{d} \tag{5.9}$$

Fundamental Diagram of Traffic Flow

The relationship between the density in number of vpm on a highway and the corresponding flow of traffic generally is referred to as the fundamental diagram of traffic flow. The following theory has been postulated with respect to the shape of the curve depicting this relationship.

1. When the density on the highway is zero, the flow is also zero as there are no vehicles on the highway.

2. As the density increases, the flow also increases.

3. However, when the density reaches its maximum, generally referred to as the jam density (k_j), the flow must be zero as vehicles will tend to line up end to end.

4. It follows, therefore, that as density increases from zero, the flow will also initially increase from zero to a maximum value. Further continuous increase in density, however, will result in continuous reduction of the flow, which will eventually be zero when the density is the jam density. The shape of the curve will, therefore, take the form in Figure 5.3a.

Data have been collected that tend to confirm the argument postulated above, but there is some controversy regarding the exact shape of the curve. A similar argument can be postulated for the general relationship between the space mean speed and the flow. When the flow is very low, there is very little interaction between individual vehicles. Drivers are therefore free to travel at the maximum possible speed. The absolute maximum speed is obtained as the flow tends to zero, and it is known as the mean free speed (u_f). The magnitude of the mean free speed depends on the physical characteristics of the highway. Continuous increase in flow will result in a continuous decrease in speed. A point will be reached, however, when further addition of vehicles will result in the reduction of actual number of vehicles that pass a point on the highway (that is, reduction of flow). This results in congestion and eventually both the speed and the flow become zero. Figure 5.3c shows this general relationship. Figure 5.3b shows the direct relationship between speed and density.

From Eq. 5.4, we know that space mean speed is flow divided by density, which makes the slopes of lines 0B, 0C, and 0E in Figure 5.3a represent the space mean speeds at densities k_a, k_c, and k_e, respectively. The slope of line 0A is the speed as the density tends to zero and there is little interaction between vehicles. The slope of this line is therefore the mean free speed (u_f); it is the maximum speed that can be attained on the highway. The slope of line 0E is the space mean speed for maximum flow. This maximum flow is the capacity of the highway. Thus it can be seen that it is desirable for highways to operate at densities not greater than that required for maximum flow.

Figure 5.3 Fundamental Diagrams of Traffic Flow: (a) Flow vs. Density, (b) Space Mean Speed vs. Density, (c) Space Mean Speed vs. Volume

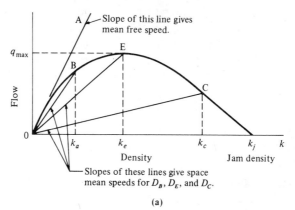

(a)

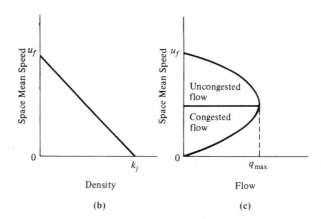

(b)   (c)

Mathematical Relationships Describing Traffic Flow

Mathematical relationships describing traffic flow can be classified into two general categories—macroscopic approach and microscopic approach—depending on the approach used in the development of these relationships. The macroscopic approach considers flow-density relationships, whereas the microscopic approach considers spacings and speeds of individual vehicles.

Macroscopic Approach

The macroscopic approach considers traffic streams and develops algorithms that relate the flow to the density and space mean speeds. The two most commonly used macroscopic models are the Greenshields and Greenberg models.

Greenshields Model. Greenshields carried out one of the earliest recorded works, in which he studied the relationship between speed and density.[2] Using a set of data, he hypothesized that a linear relationship existed between speed and density, which he expressed as

$$\bar{u}_s = u_f - \frac{u_f}{k_j} k \tag{5.10}$$

Corresponding relationships for flow and density and for flow and speed can be developed. Since $q = \bar{u}_s k$, substituting $q/\bar{u}_s$ for k in Eq. 5.10 gives

$$\bar{u}_s^2 = u_f \bar{u}_s - \frac{u_f}{k_j} q \tag{5.11}$$

Also substituting q/k for $\bar{u}_s$ in Eq. 5.10 gives

$$q = u_f k - \frac{u_f}{k_j} k^2 \tag{5.12}$$

Eqs. 5.11 and 5.12 indicate that if a linear relationship in the form of Eq. 5.10 is assumed for speed and density, then a parabolic relationship is obtained between flow and density and between flow and speed. The shape of the curve shown in Figure 5.3a will therefore be a parabola. Also, Eqs. 5.11 and 5.12 can be used to determine the corresponding speed and the corresponding density for maximum flow.

Consider Eq. 5.11.

$$\bar{u}_s^2 = u_f \bar{u}_s - \frac{u_f}{k_j} q$$

Differentiating q with respect to $\bar{u}_s$, we obtain

$$2\bar{u}_s = u_f - \frac{u_f}{k_j} \frac{dq}{du_s}$$

that is,

$$\frac{dq}{du_s} = u_f \frac{k_j}{u_f} - 2\bar{u}_s \frac{k_j}{u_f} = k_j - 2\bar{u}_s \frac{k_j}{u_f}$$

For maximum flow,

$$\frac{dq}{du_s} = 0 \qquad k_j = 2\bar{u}_s \frac{k_j}{u_f} \qquad u_o = \frac{u_f}{2} \tag{5.13}$$

Thus, the space mean speed u_o at which the volume is maximum is equal to half the free mean speed.

Consider Eq. 5.12.

$$q = u_f k - \frac{u_f}{k_j} k^2$$

Differentiating q with respect to k, we obtain

$$\frac{dq}{dk} = u_f - 2k \frac{u_f}{k_j}$$

For maximum flow,

$$\frac{dq}{dk} = 0 \qquad u_f = 2k \frac{u_f}{k_j} \qquad \frac{k_j}{2} = k_o \tag{5.14}$$

Thus, at the maximum flow, the density k_o is half the jam density. The maximum flow for the Greenshields relationship can therefore be obtained from Eqs. 5.4, 5.13, and 5.14, as shown in Eq. 5.15.

$$q_{max} = \frac{k_j u_f}{4} \tag{5.15}$$

Greenberg Model. Several researchers have used the analogy of the flow of fluid to describe the flow of traffic to develop macroscopic relationships for traffic flow. One of the major contributions using the fluid-flow analogy was developed by Greenberg[3] in the form

$$\bar{u}_s = c \ln \frac{k_j}{k} \tag{5.16}$$

$$q = ck \ln \frac{k_j}{k} \tag{5.17}$$

Differentiating q with respect to k, we obtain

$$\frac{dq}{dk} = c \ln \frac{k_j}{k} - c$$

For maximum flow,

$$\frac{dq}{dk} = 0 \qquad \ln \frac{k_j}{k} = 1$$

giving

$$\ln k_j = 1 + \ln k_o \tag{5.18}$$

That is,

$$\ln \frac{k_j}{k_o} = 1$$

Substituting 1 for ln (k_j/k_o) in Eq. 5.16 gives

$$u_o = c$$

Thus, the value of c is the speed at maximum flow.

Model Application

Use of these macroscopic models depends on whether they satisfy the boundary criteria of the fundamental diagram of traffic flow at the region that describes the traffic conditions. For example, the Greenshields model satisfies the boundary conditions when the density (k) is approaching zero as well as when the density k is approaching the jam density (k_j). The Greenshields model can therefore be used for light or dense traffic. The Greenberg model, on the other hand, satisfies the boundary conditions when the density (k) is approaching the jam density k_j but does not satisfy the boundary conditions when k is approaching zero. The Greenberg model is therefore useful only for dense traffic conditions.

Calibration of Macroscopic Traffic Flow Models. The traffic models discussed thus far can be used to determine specific characteristics such as the speed and density at which maximum flow occurs and the jam density of a facility. This usually involves the collection of appropriate data on the particular facility of interest and fitting the data points obtained to a suitable model. The most common method of approach is regression analysis. This is done by minimizing the squares of the differences between the observed and expected values of a dependent variable. When the dependent variable is linearly related to the independent variable, the process is known as *linear regression analysis*, and when the relationship is with two or more independent variables, the process is known as *multiple linear regression analysis*.

If a dependent variable y and an independent variable x are related by an estimated regression function, then

$$y = a + bx \tag{5.19}$$

The constants a and b could be determined from Eqs. 5.20 and 5.21. For development of these equations, see Appendix B.

$$a = \frac{1}{n}\sum_{i=1}^{n} y_i - \frac{b}{n}\sum_{i=1}^{n} x_i = \bar{y} - b\bar{x} \tag{5.20}$$

and

$$b = \frac{\sum_{i=1}^{n} x_i y_i - \frac{1}{n}\left(\sum_{i=1}^{n} x_i\right)\left(\sum_{i=1}^{n} y_i\right)}{\sum_{i=1}^{n} x_i^2 - \frac{1}{n}\left(\sum_{i=1}^{n} x_i\right)^2} \tag{5.21}$$

where

n = number of sets of observations

x_i = ith observation for x

y_i = ith observation for y

A measure commonly used to determine the suitability of an estimated regression function is the coefficient of determination (or square of the estimated correlation coefficient) R^2, which is given by

$$R^2 = \frac{\sum_{i=1}^{n} (Y_i - \bar{y})^2}{\sum_{i=1}^{n} (y_i - \bar{y})^2} \tag{5.22}$$

where Y_i = the value of the dependent variable as computed from the regression equations. The closer R^2 is to 1, the better is the regression fit.

Example 5–1   Fitting Speed and Density Data to the Greenshields Model

Let us now use the data shown in Table 5.1 (Columns 1 and 2) to demonstrate the use of the method of regression analysis in fitting speed and density data to the macroscopic models discussed earlier. Let us first consider the Greenshields expression

$$\bar{u}_s = u_f - \frac{u_f}{k_j} k$$

- Comparing this expression with our estimated regression function, Eq. 5.19, we see that the speed $\bar{u}_s$ in the Greenshields expression is represented by Y in the estimated regression function, the mean free speed u_f is represented by a, and the value of the mean free speed u_f divided by jam density k_j is represented by $-b$. We therefore obtain

$$\sum y_i = 404.8 \qquad \sum x_i = 892 \qquad \bar{y} = 28.91$$

$$\sum x_i y_i = 20{,}619.8 \qquad \sum x_i^2 = 66{,}628 \qquad \bar{x} = 63.71$$

- Using Eqs. 5.20 and 5.21, we obtain

$$a = 28.91 - 63.71b$$

$$b = \frac{20{,}619.8 - \dfrac{(892)(404.8)}{14}}{66{,}628 - \dfrac{(892)^2}{14}} = -0.53$$

or $a = 28.91 - 63.71(-0.53) = 62.68$

Table 5.1   Speed and Density Observations at a Rural Road

(a) Computations for Example 5-1

Speed, u_s (mph) y_i	Density, k (vpm) x_i	x_iy_i	x_i^2
53.2	20	1064.0	400
48.1	27	1298.7	729
44.8	35	1568.0	1,225
40.1	44	1764.4	1,936
37.3	52	1939.6	2,704
35.2	58	2041.6	3,364
34.1	60	2046.0	3,600
27.2	64	1740.8	4,096
20.4	70	1428.0	4,900
17.5	75	1312.5	5,625
14.6	82	1197.2	6,724
13.1	90	1179.0	8,100
11.2	100	1120.0	10,000
8.0	115	920.0	13,225
$\sum = 404.8$	$\sum = 892$	$\sum = 20,619.8$	$\sum = 66,628.0$
$\bar{y} = 28.91$	$\bar{x} = 63.71$		

(b) Computations for Example 5-2

Speed, u_s (mph) y_i	Density, k (vpm)	ln k x_i	x_iy_i	x_i^2
53.2	20	2.995732	159.3730	8.974412
48.1	27	3.295837	158.5298	10.86254
44.8	35	3.555348	159.2796	12.64050
40.1	44	3.784190	151.746	14.32009
37.3	52	3.951244	147.3814	15.61233
35.2	58	4.060443	142.9276	16.48720
34.1	60	4.094344	139.6171	16.76365
27.2	64	4.158883	113.1216	17.29631
20.4	70	4.248495	86.66929	18.04971
17.5	75	4.317488	75.55605	18.64071
14.6	82	4.406719	64.33811	19.41917
13.1	90	4.499810	58.94750	20.24828
11.2	100	4.605170	51.57791	21.20759
8	115	4.744932	37.95946	22.51438
$\sum = 404.8001$		$\sum = 56.71864$	$\sum = 1547.024$	$\sum = 233.0369$
$\bar{y} = 28.91$		$\bar{x} = 4.05$		

Since, $a = 62.68$ and $b = -0.53$, then $u_f = 62.56$ mph, and $u_f/k_j = 0.53$, $k_j = 118$ vpm and $\bar{u}_s = 62.68 - 0.53k$.

- Using Eq. 5.22 to determine the value of R^2, we obtain $R^2 = 0.95$.
- Using the above estimated values for u_f and k_j, we can determine the maximum flow from Eq. 5.15 as

$$q_{max} = \frac{k_j u_f}{4} = \frac{118 \times 62.68}{4}$$
$$= 1849 \text{ vph}$$

- Using Eq. 5.13 we also obtain the velocity at which flow is maximum—$(62.68/2) = 31.3$ mph—and using Eq. 5.14, the density at which flow is maximum—$(118/2) = 59$ vph.

Example 5–2 Fitting Speed and Density Data to the Greenberg Model

The data in Table 5.1 can also be fitted into the Greenberg model shown in Eq. 5.16

$$\bar{u}_s = c \ln \frac{k_j}{k}$$

which can be written as

$$\bar{u}_s = c \ln k_j - c \ln k \tag{5.23}$$

- Comparing Eq. 5.23 and the estimated regression function Eq. 5.19, we see that $\bar{u}_s$ in the Greenberg expression is represented by Y in the estimated regression function, $c \ln k_j$ is represented by a, c is represented by $-b$, and $\ln k$ is represented by x. Table 5.1b shows values for x_i, $x_i y_i$, and x_i^2. (Note that these values are computed to a higher degree of accuracy as they involve logarithmic values.) We therefore obtain

$$\sum y_i = 404.8 \qquad \sum x_i = 56.72 \qquad \bar{y} = 28.91$$

$$\sum x_i y_i = 1547.02 \qquad \sum x_i^2 = 233.04 \qquad \bar{x} = 4.05$$

- Using Eqs. 5.20 and 5.21, we obtain

$$a = 28.91 - 4.05b$$

$$b = \frac{1547.02 - \dfrac{(56.72)(404.8)}{14}}{233.04 - \dfrac{(56.72)^2}{14}} = -28.68$$

or $\qquad a = 28.91 - 4.05(-28.68) = 145.06$

Since $a = 145.06$ and $b = -28.68$, then the speed for maximum flow is $c = 28.68$ mph.

$$c \ln k_j = 145.06$$
$$\ln k_j = 5.06$$
$$k_j = 157 \text{ vpm}$$

give

$$\bar{u}_s = 28.68 \ln \frac{157}{k}$$

- Obtaining k_o, the density for maximum flow from Eq. 5.18, we then use Eq. 5.17 to determine the value of the maximum flow.

$$\ln k_j = 1 + \ln k_o$$
$$\ln 158 = 1 + \ln k_o$$
$$5.06 = 1 + \ln k_o$$
$$58.0 = k_o$$
$$q_{max} = 58.0 \times 28.68 \text{ vph}$$
$$q_{max} = 1663 \text{ vph}$$

The R^2 based on the Greenberg expression is 0.9, which indicates that the Green-shields expression is a better fit for the data in Table 5.1. Figure 5.4 shows plots

Figure 5.4   Speed Versus Density

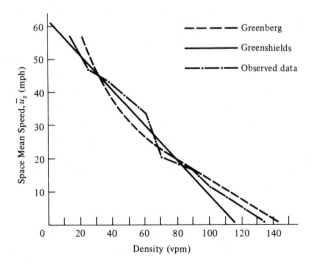

Figure 5.5 Volume Versus Density

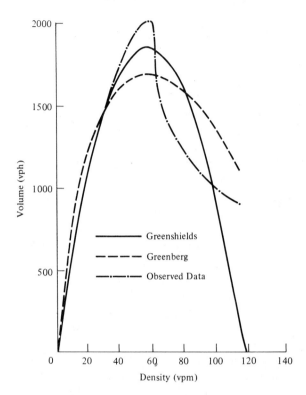

of speed versus density for the two estimated regression functions obtained and also for the actual data points. Figure 5.5 also shows similar plots for the flow against speed.

Microscopic Approach

The microscopic approach, which is sometimes referred to as car-following theory or follow-the-leader theory, considers spacings and speeds of individual vehicles. Consider two consecutive vehicles, A and B, on a single lane of a highway as shown in Figure 5.6. If the leading vehicle is considered as the nth vehicle and the following vehicle the $(n + 1)$th vehicle, then the distances of these vehicles from a fixed section at any time t can be taken as x_n and x_{n+1}, respectively.

If it is assumed that the driver of vehicle B maintains an additional separation distance P above the separation distance at rest S and that P is proportional to the speed of vehicle B, then

$$P = \rho \dot{x}_{n+1} \tag{5.24}$$

where

    ρ = factor of proportionality with units of time
    $\dot{x}_{n+1}$ = speed of the $(n + 1)$th vehicle

Figure 5.6 Basic Assumptions in Follow-the-Leader Theory

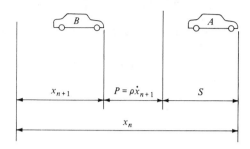

We can write

$$x_n - x_{n+1} = \rho \dot{x}_{n+1} + S \tag{5.25}$$

where

$S = $ distance between front bumpers of vehicles at rest

Differentiating Eq. 5.25 gives

$$\ddot{x}_{n+1} = \frac{1}{\rho}[\dot{x}_n - \dot{x}_{n+1}] \tag{5.26}$$

Equation 5.26 is the basic equation of the microscopic models and it describes the stimulus–response of the models. Researchers have shown that a time lag exists for a driver to respond to any stimulus that is induced by the vehicle just ahead,[4] and Eq. 5.26 can therefore be written as

$$\ddot{x}_{n+1}(t + T) = \lambda[\dot{x}_n(t) - \dot{x}_{n+1}(t)] \tag{5.27}$$

where

$T = $ time lag of response to the stimulus
$\lambda = (1/\rho)$ (sometimes called the sensitivity)

A general expression for λ is given in the form

$$\lambda = a \frac{\dot{x}^m_{n+1}(t + T)}{[x_n(t) - x_{n+1}(t)]^\ell} \tag{5.28}$$

The general expression for the microscopic models can then be written as

$$\ddot{x}_{n+1}(t + T) = a \frac{\dot{x}^m_{n+1}(t + T)}{[x_n(t) - x_{n+1}(t)]^\ell}[\dot{x}_n(t) - \dot{x}_{n+1}(t)] \tag{5.29}$$

where a, ℓ, and m are constants.

The microscopic model (Eq. 5.29) can be used to determine the velocity, flow, and density of a traffic stream when the traffic stream is moving in a steady state. The direct analytical solution of either Eq. 5.27 or Eq. 5.29 is not easy. It can be shown, however, that the macroscopic models discussed earlier can all be obtained from Eq. 5.29.

For example, if $m = 0$ and $\ell = 1$, the acceleration of the $(n + 1)$th vehicle is given as

$$\ddot{x}_{n+1}(t + T) = a\frac{\dot{x}_n(t) - \dot{x}_{n+1}(t)}{x_n(t) - x_{n+1}(t)}$$

Integrating the above expression, we obtain the velocity of the $(n + 1)$th vehicle, as

$$\dot{x}_{n+1}(t + T) = a \ln\left[x_n(t) - x_{n+1}(t + 1)\right] + C$$

Since we are considering the steady state condition,

$$\dot{x}_n(t + T) = \dot{x}_n(t) = u$$

$$u = a \ln\left[x_n - x_{n+1}\right] + C$$

Also,

$$x_n - x_{n+1} = \text{average space headway} = \frac{1}{k}$$

$$u = a \ln\left(\frac{1}{k}\right) + C$$

Using the boundary condition,

$$u = 0 \qquad \text{when } k = k_j$$

$$0 = a \ln\left(\frac{1}{k_j}\right) + C$$

$$C = -a \ln\left(\frac{1}{k_j}\right)$$

Substituting for C in the equation for u, we obtain

$$u = a \ln\left(\frac{1}{k}\right) - a \ln\left(\frac{1}{k_j}\right)$$

$$= a \ln\left(\frac{k_j}{k}\right)$$

which is the Greenberg model given in Eq. 5.16. Similarly, if m is allowed to be 0 and $\ell = 2$, we obtain the Greenshields model.

SHOCK WAVES IN TRAFFIC STREAMS

The fundamental diagram of traffic flow for two adjacent sections of a highway with different capacities (maximum flows) are shown in Figure 5.7. This figure describes the phenomenon of backups and queuing on a highway due to a sudden reduction of the capacity of the highway (bottleneck condition). The sudden reduction in capacity could be due to reduction in the number of lanes, accidents, restricted bridge sizes, work zones, a signal turning red, and so forth, creating a situation where the capacity on the highway suddenly changes from C_1 to a lower value of C_2, with a corresponding change in optimum density from k_o^a to a value of k_o^b.

When such a condition exists and the flow and density on the highway are relatively large, the speeds of the vehicles will have to be reduced while passing the bottleneck. The point at which the speed reduction takes place can be noted by the brake lights of the vehicles coming on. An observer will see that this point moves upstream as traffic continues to approach the vicinity of the bottleneck indicating an upstream movement of the point at which changes in flow and density occur. This phenomenon is usually referred to as a *shock wave* in the traffic stream.

Let us consider two different densities of traffic, k_1 and k_2, along a straight highway as shown in Figure 5.8, where $k_1 > k_2$. Let us also assume that these densities are separated by the line w representing the shock wave moving at a speed u_w. If the line w moves in the direction of the arrow (that is, in the direction of the traffic flow) u_w is positive.

Figure 5.7 Kinematic and Shock Wave Measurements Related to Flow–Density Curve

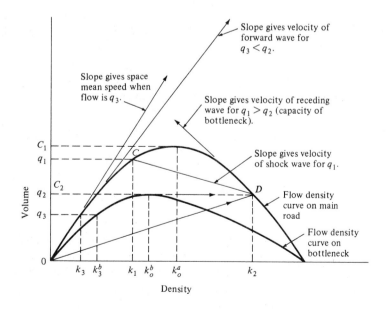

Figure 5.8   Movement of Shock Wave Due to Change in Densities

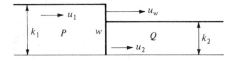

With u_1 equal to the space mean speed of vehicles in the area with density k_1 (section P), the speed of the vehicle in this area relative to the line w is

$$u_{r_1} = (u_1 - u_w)$$

The number of vehicles crossing line w from area P during a time period t is

$$N_1 = u_{r_1} k_1 t$$

Similarly, the speed of vehicles in the area with density k_2 (section Q) relative to line w is

$$u_{r_2} = (u_2 - u_w)$$

and the number of vehicles crossing line w during a time period t is

$$N_2 = u_{r_2} k_2 t$$

Since the net change is zero—$N_1 = N_2$ and $(u_1 - u_w)k_1 = (u_2 - u_w)k_2$—we have

$$u_2 k_2 - u_1 k_1 = u_w(k_2 - k_1) \tag{5.30}$$

If the flow rates are q_1 and q_2 in sections P and Q, respectively, then

$$q_1 = k_1 u_1 \qquad q_2 = k_2 u_2$$

Substituting q_1 and q_2 for $k_1 u_1$ and $k_2 u_2$ in Eq. 5.30 gives

$$q_2 - q_1 = u_w(k_2 - k_1)$$

That is,

$$u_w = \frac{q_2 - q_1}{k_2 - k_1} \tag{5.31}$$

which is also the slope of the line CD shown in Figure 5.7. This indicates that the velocity of the shock wave created by a sudden change of density from k_1 to k_2 on a traffic stream is the slope of the chord joining the points associated with k_1 and k_2 on the volume density curve for that traffic stream.

Example 5–3 Length of Queue Due to a Moving Shock Wave

The volume at a section of a 2-lane highway is 1000 vph in each direction. The space mean speed in the upgrade direction is estimated to be 50 mph when the density is about 20 vpm. A large dump truck loaded with soil from an adjacent construction site joins the traffic stream and travels at a speed of 15 mph for a length of 1.5 mi along the upgrade before turning off onto a dump site. Due to the relatively high flow in the opposite direction, it is impossible for any car to pass the truck. Vehicles just behind the truck therefore have to travel at the speed of the truck, which results in the formation of a platoon having a density of 85 vpm and a flow of 1275 vph. Determine how many vehicles will be in the platoon by the time the truck leaves the highway.

- Use Eq. 5.31 to obtain the wave velocity.

$$u_w = \frac{1275 - 1000}{85 - 20} \text{ mph} \approx 4.2 \text{ mph}$$

- Knowing that the truck is traveling at 15 mph and that the shock wave is moving forward relative to the road at 4.2 mph, determine the growth rate of the platoon.

$$(15 - 4.2)\text{mph} = 10.8 \text{ mph}$$

- Calculate the time spent by the truck on the highway—$(1.5/15) = 0.1$ hr—to determine the length of the platoon by the time the truck leaves the highway.

$$0.1 \times 10.8 = 1.08 \text{ mi}$$

- Use the density of 85 vpm to calculate the number of vehicles in the platoon.

$$85 \times 1.08 \approx 92 \text{ vehicles}$$

Special Cases of Shock Wave Propagation

The shock wave phenomenon can also be explained by considering a continuous change of flow and density in the traffic stream. If the change in flow and the change in density are very small, we can write

$$(q_2 - q_1) = \Delta q \qquad (k_2 - k_1) = \Delta k$$

The wave velocity can then be written as

$$u_w = \frac{\Delta q}{\Delta k} = \frac{dq}{dk} \tag{5.32}$$

Since $q = k\bar{u}_s$, substituting $k\bar{u}_s$ for q in Eq. 5.32 gives

$$u_w = \frac{d(k\bar{u}_s)}{dk} \tag{5.33}$$

$$= \bar{u}_s + k\frac{d\bar{u}_s}{dk} \tag{5.34}$$

When such a continuous change of volume occurs in a vehicular flow, a phenomenon similar to that of fluid flow exists, in which the waves created in the traffic stream transport the continuous changes of flow and density. The speed of these waves is dq/dk and it is given by the Eq. 5.34.

We have already seen that as density increases, the space mean speed decreases (see Eq. 5.4), giving a negative value for $d\bar{u}_s/dk$. This shows that at any point on the fundamental diagram, the speed of the wave is theoretically less than the space mean speed of the traffic stream. Thus, the wave moves in the opposite direction relative to that of the traffic stream. The actual direction and speed of the wave will depend on the point at which we are on the curve (that is, the flow and density on the highway) and the resultant effect on the traffic downstream will depend on the capacity of the restricted area (bottleneck). When both the flow and the density of the traffic stream are very low, that is, approaching zero, the flow is much lower than the capacity of the restricted area and there is very little interaction between the vehicles. The differential of $\bar{u}_s$ with respect to k ($d\bar{u}_s/dk$) then tends to zero, and the wave velocity approximately equals the space mean speed. The wave therefore moves forward with respect to the road and no backups result. As the flow of the traffic stream increases to a value much higher than zero but still lower than the capacity of the restricted area (say, q_3 in Figure 5.7), the wave velocity is still less than the space mean speed of the traffic stream, and the wave moves forward relative to the road. This results in a reduction in speed and an increase in the density from k_3 to k_3^b as vehicles enter the bottleneck but no backups occur. When the volume on the highway is equal to the capacity of the restricted area (C_2 in Figure 5.7), the speed of the wave is zero and the wave does not move. This results in a much slower speed and a greater increase in the density to k_o^b as the vehicles enter the restricted area. Again, delay occurs but there are no backups. When, however, the flow on the highway is greater than the capacity of the restricted area, for example, q_1, the flow through the bottleneck is only q_2, the speed of the wave is not only less than the space mean speed of the vehicle stream, but it moves backward relative to the road. As vehicles enter the restricted area, a complex queuing condition arises, resulting in an immediate increase in the density from k_1 to k_2 in the upstream section of the road and a considerable decrease in speed. The movement of the wave toward the upstream section of the traffic stream creates a shock wave in the traffic stream, eventually resulting in backups, which gradually moves upstream of the traffic stream.

The expressions developed for the speed of the shock wave, Eqs. 5.31 and 5.34, can be applied to any of the specific models described earlier. For example, the Greenshields model can be written as

$$\bar{u}_{si} = u_f\left(1 - \frac{k_i}{k_j}\right) \qquad \bar{u}_{si} = u_f(1 - \eta_1) \tag{5.35}$$

where $\eta_i = (k_i/k_j)$ (normalized density).

If the Greenshields model fits the flow density relationship for a particular traffic stream, Eq. 5.31 can be used to determine the speed of a shock wave as

$$
\begin{aligned}
u_w &= \frac{\left[k_2 u_f\left(1 - \frac{k_2}{k_j}\right)\right] - \left(k_1 u_f\left(1 - \frac{k_1}{k_j}\right)\right)}{k_2 - k_1} \\
&= \frac{k_2 u_f(1 - \eta_2) - k_1 u_f(1 - \eta_1)}{k_2 - k_1} \\
&= \frac{u_f(k_2 - k_1) - k_2 u_f \eta_2 + k_1 u_f \eta_1}{k_2 - k_1} \\
&= \frac{u_f(k_2 - k_1) - \frac{u_f}{k_j}(k_2^2 - k_1^2)}{k_2 - k_1} \\
&= \frac{u_f(k_2 - k_1) - \frac{u_f}{k_j}(k_2 - k_1)(k_2 + k_1)}{k_2 - k_1} \\
&= u_f[1 - (\eta_1 + \eta_2)]
\end{aligned}
$$

The speed of a shock wave for the Greenshields model is therefore given as

$$u_w = u_f[1 - (\eta_1 + \eta_2)] \tag{5.36}$$

Density Nearly Equal

When there is only a small difference between k_1 and k_2 (that is, $\eta_1 \approx \eta_2$),

$$u_w = u_f[1 - (\eta_1 + \eta_2)] \qquad \text{(neglecting the small change in } \eta_1)$$
$$= u_f[1 - 2\eta_1]$$

Stopping Waves

Eq. 5.36 can also be used to determine the velocity of the shock wave due to the change from green to red of a signal at an intersection approach if the Greenshields model is applicable. During the green phase, the normalized density is η_1. When the traffic signal changes to red, the traffic at the stop line x_o of the approach comes to a halt, which results in a density equal to the jam density. The value of η_2 is then equal to 1.

The speed of the shock wave, which in this case is a stopping wave, can be obtained by

$$u_w = u_f[1 - (\eta_1 + 1)] = -u_f\eta_1 \qquad (5.37)$$

Eq. 5.37 indicates that in this case the shock wave travels upstream of the traffic with a velocity of $u_f\eta_1$. If the length of the red phase is t sec, then the length of the line of cars upstream of the stopline is $u_f\eta_1 t$.

Starting Waves

At the instant when the signal again changes from red to green, η_1 equals 1. Vehicles will then move forward at a speed of $\bar{u}_{s2}$ resulting in a density of η_2. The speed of the shock wave, which in this case is a starting wave, is obtained by

$$u_w = u_f[1 - (1 + \eta_2)] = -u_f\eta_2 \qquad (5.38)$$

Eq. 5.35, $\bar{u}_{s2} = u_f(1 - \eta_2)$, gives

$$\eta_2 = 1 - \frac{\bar{u}_{s2}}{u_f}$$

The velocity of the shock wave is then obtained as

$$u_w = -u_f + \bar{u}_{s2}$$

Since the starting velocity $\bar{u}_{s2}$ just after the signal changes to green is usually small, the velocity of the starting shock wave approximately equals $-u_f$.

Example 5–4   Length of Queue Due to a Stopping Shock Wave

Studies have shown that the traffic flow on a single-lane approach to a signalized intersection can be described by the Greenshields model. If the jam density on the approach is 120 vpm, determine the velocity of the stopping wave, when the approach signal changes to red, if the density on the approach is 50 vpm and the space mean speed is 36 mph. Up to what length of the approach upstream from the stop line will vehicles be affected by the red signal if the red phase is 30 sec?

- Use the Greenshields model.

$$\bar{u}_s = u_f - \frac{u_f}{k_j}k$$

$$36 = u_f - \frac{u_f}{120}50$$

$$4320 = 120u_f - 50u_f$$

$$61.7 \text{ mph} = u_f$$

- Use Eq. 5.37 for a stopping wave.

$$u_w = -u_f \eta_1$$

$$= -61.7 \times \frac{50}{120}$$

$$= -25.7 \text{ mph}$$

Since u_w is negative, the wave moves upstream.

- Determine the approach length that will be affected in 30 sec.

$$25.7 \times 1.47 \times 30 \text{ ft} = 1133.4 \text{ ft}$$

GAP AND GAP ACCEPTANCE

Thus far we have been considering the theory of traffic flow as it relates to the flow of vehicles in a single stream. Another important aspect of traffic flow is the interaction of vehicles as they join, leave, or cross a traffic stream. Examples of these include ramp vehicles merging onto an expressway stream, freeway vehicles leaving the freeway onto frontage roads, and the changing of lanes by vehicles on a multilane highway. The most important factor a driver considers in making any one of these maneuvers is the availability of a gap between two vehicles that, in the driver's judgment, is adequate for him or her to complete the maneuver. The evaluation of available gaps and the decision to carry out a specific maneuver within a particular gap are inherent in the concept of gap acceptance.

Following are the important measures that involve the concept of gap acceptance.

1. **Merging** is the process by which a vehicle in one traffic stream joins another traffic stream moving in the same direction, such as a ramp vehicle joining a freeway stream.
2. **Diverging** is the process by which a vehicle in the traffic stream leaves the traffic stream, such as a vehicle leaving the outside lane of an expressway.
3. **Weaving** is the process by which a vehicle first merges into a stream of traffic, obliquely crosses that stream, and then merges into a second stream moving in the same direction; for example, the maneuver required for a ramp vehicle to join the far side stream of flow on an expressway.

The important variables in merging, diverging, and weaving maneuvers are time headway, space headway (both of which have already been defined), gap, time lag, and space lag. The *gap* is expressed either in units of time (time gap) or in units of distance (space gap), and is defined as the headway in a major stream, which is evaluated by a driver in a minor stream vehicle wishing to merge into the major stream. *Time lag* is defined as the difference between the time a vehicle that merges into a main traffic stream reaches a point on the highway in the area of merge and the time a vehicle in the main stream reaches the same point. *Space lag* is defined as the difference, at an instant of time, between the distance a merging vehicle is away from a reference point in the area of merge and the distance a main stream vehicle is away from the same point. Figure 5.9 depicts the time–distance relationships for a vehicle at a stop sign waiting to merge and for vehicles on the near lane of the main stream traffic.

A driver who intends to merge first has to evaluate the gaps that become available to determine one that, in his or her opinion, is large enough to accept. In accepting that gap, the driver feels that he or she will be able to complete the merging maneuver and safely join the main stream within the length of the gap. This phenomenon is generally referred to as *gap acceptance*. It is of importance when engineers are considering the delay of vehicles on minor roads wishing to join a major road traffic stream, at unsignalized intersections, and the delay of ramp vehicles wishing to join expressways. It can also be used in timing the release of vehicles at an on-ramp of an expressway, such that the probability of the released vehicle finding an acceptable gap in arriving at the freeway shoulder lane is maximum. To use the phenomenon of gap acceptance in evaluating delays, waiting times, queue lengths, and so forth at on-ramps and at unsignalized intersections, the average minimum gap length that will be accepted by drivers should first be determined. Several definitions have been given to this "critical" value. Greenshields referred to it as the "acceptable average minimum time gap" and defined it as the gap accepted by 50 percent of the drivers.[5] The concept of "critical gap" was used by Raff, who defined it as the gap that has the number of accepted gaps shorter than it being the same as the number of rejected gaps longer than it.[6] The data in Table 5.2 are used to demonstrate the determination of the critical gap using Raff's definition. Either a graphical or an algebraic method can

Figure 5.9   Time–Space Diagrams for Vehicles in the Vicinity of a Stop Sign

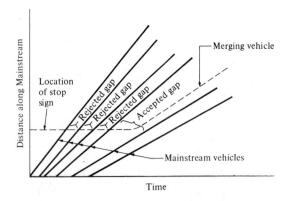

Table 5.2 Computation of Critical Gap (t_c)

(a) Gaps Accepted and Rejected

1 Length of Gap (t sec)	2 Number of Accepted Gaps (less than t sec)	3 Number of Rejected Gaps (greater than t sec)
0.0	0	116
1.0	2	103
2.0	12	66
3.0	$m = 32$	$r = 38$
4.0	$n = 57$	$p = 19$
5.0	84	6
6.0	116	0

(b) Difference in Gaps Accepted and Rejected

1 Consecutive Gap Lengths (t sec)	2 Change in Number of Accepted Gaps (less than t sec)	3 Change in Number of Gaps (greater than t sec)	4 Difference Between Columns 1 and 2
0.0–1.0	2	13	11
1.0–2.0	10	37	27
2.0–3.0	20	28	8
3.0–4.0	25	19	6
4.0–5.0	27	13	14
5.0–6.0	32	6	26

be used. In using the graphical method, two cumulative distribution curves are drawn as shown in Figure 5.10. One of them relates gap lengths t with the number of accepted gaps less than t and the other relates t with the number of rejected gaps greater than t. The intersection of these two curves gives the value of t for the critical gap.

In using the algebraic method, it is first necessary to identify the gap lengths between which the critical gap lies. This is done by comparing the change in number of accepted gaps less than t sec (column 2 of Table 5.2b) for two consecutive gap lengths, with the change in number of rejected gaps greater than t sec (column 3 of Table 5.2b) for the same two consecutive gap lengths. The critical gap length lies between the two consecutive gap lengths, where the difference between the two changes is minimum. Table 5.2b shows the computation and indicates that the critical gap for this case lies between 3 and 4 sec.

In the example of Figure 5.10, with Δt equal to time increment used for gap analysis, the critical gap lies between

$$t_1 \text{ and } t_2 = t_1 + \Delta t$$

Figure 5.10   Cumulative Distribution Curves for Accepted and Rejected Gaps

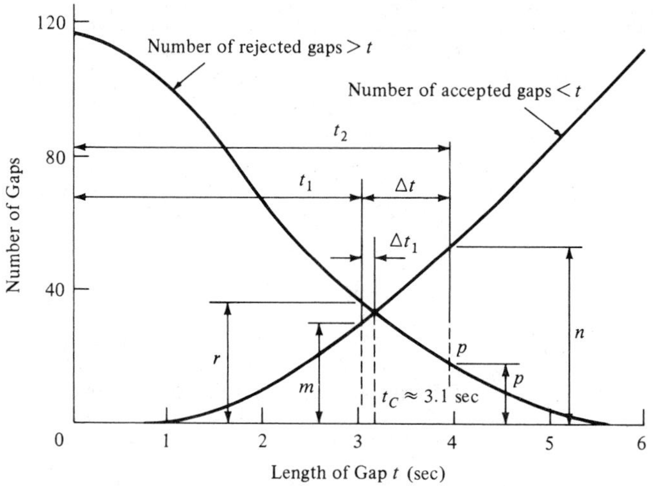

and

m = number of accepted gaps less than t_1
r = number of rejected gaps greater than t_1
n = number of accepted gaps less than t_2
p = number of rejected gaps greater than t_2

Assuming that the curves are linear between t_1 and t_2, the point of intersection of these two lines gives the critical gap. From Figure 5.10, the critical gap expression can be written as

$$t_c = t_1 + \Delta t_1$$

Using the properties of similar triangles,

$$\frac{\Delta t_1}{r - m} = \frac{\Delta t - \Delta t_1}{n - p}$$

$$\Delta t_1 = \frac{\Delta t(r - m)}{(n - p) + (r - m)}$$

we obtain

$$t_c = t_1 + \frac{\Delta t(r - m)}{(n - p) + (r - m)} \tag{5.39}$$

For the data given in Table 5.2, we thus have

$$t_c = 3 + \frac{1(38 - 32)}{(57 - 19) + (38 - 32)} = 3 + \frac{6}{38 + 6}$$

$$\approx 3.14 \text{ sec}$$

STOCHASTIC APPROACH TO GAP AND GAP ACCEPTANCE PROBLEMS

The use of the phenomenon of gap acceptance to determine the delay of vehicles in minor streams wishing to merge onto major streams requires a knowledge of the frequency of arrivals of gaps that are at least equal to the critical gap. This in turn depends on the distribution of arrivals of main stream vehicles at the area of merge. It is generally accepted that, for light to medium traffic flow on a highway, the arrival of vehicles is randomly distributed. It is therefore important that the probabilistic approach to the subject be discussed. It is usually assumed that for light to medium traffic the distribution is Poisson, although assumptions of gamma and exponential distributions have also been made.

Assuming that the distribution of main stream arrival is Poisson, then the probability of x arrivals in any interval of time t sec can be obtained from the expression

$$P(x) = \frac{\mu^x e^{-\mu}}{x!} \qquad \text{for } x = 0, 1, 2 \ldots \infty \qquad \textbf{(5.40)}$$

where

$P(x)$ = the probability of x vehicles arriving in time t sec

μ = average number of vehicles arriving in time t

If V = total number of vehicles arriving in time T sec, then the average number of vehicles arriving per second is

$$\lambda = \frac{V}{T} \qquad \mu = \lambda t$$

We can therefore write Eq. 5.40 as

$$P(x) = \frac{(\lambda t)^x e^{-\lambda t}}{x!} \qquad \textbf{(5.41)}$$

Now consider a vehicle waiting at an unsignalized intersection or at a ramp to merge into the main stream flow, arrivals of which can be described by Eq. 5.41. The minor stream vehicle will merge only if there is a gap of t sec equal to or greater than its critical gap. This will occur when no vehicles arrive during a period t sec long. The probability of this is the probability of zero car arriving (that is, when x is zero in Eq. 5.41). Substituting zero for x in Eq. 5.41 will therefore give a probability of a gap ($h \geq t$) occurring. Thus,

$$P(o) = P(h \geq t) = e^{-\lambda t} \qquad \text{for } t \geq 0 \qquad \textbf{(5.42)}$$

$$P(h < t) = 1 - e^{-\lambda t} \qquad \text{for } t \geq 0 \qquad \textbf{(5.43)}$$

Since

$$P(h < t) + P(\geq t) = 1$$

It can be seen that t can take all values from 0 to ∞, which therefore makes Eqs. 5.42 and 5.43 continuous functions. The probability function described by Eq. 5.42 is known as the *exponential distribution*.

Eq. 5.42 can be used to determine the expected number of acceptable gaps that will occur at an unsignalized intersection or at the merging area of an expressway on-ramp during a period T, if the Poisson distribution is assumed for the main stream flow and the volume V is also known. Let us assume that T is equal to 1 hr, then V is the volume in number of vph on the main stream flow. Since $(V - 1)$ gaps occur between V successive vehicles in a stream of vehicles, then the expected number of gaps greater or equal to t is given as

$$\text{Freq. } (h \geq t) = (V - 1)e^{-\lambda t} \tag{5.44}$$

and the expected number of gaps less than t is given as

$$\text{Freq. } (h < t) = (V - 1)(1 - e^{-\lambda t}) \tag{5.45}$$

Example 5–5   Number of Acceptable Gaps for Vehicles on an Expressway
Ramp

The peak hour volume on an expressway at the vicinity of the merging area of an on-ramp was determined to be 1800 vph. If it can be assumed that the arrival of expressway vehicles can be described by a Poisson distribution and the critical gap for merging vehicles is 3.5 sec, determine the expected number of acceptable gaps for ramp vehicles that will occur on the expressway during the peak hour.

- List the data

  $V = 1800$
  $T = 3600$ sec
  $\lambda = (1800/3600) = 0.5$ vps

- Calculate the expected number of acceptable gaps in 1 hr using Eq. 5.44.

$$(h \geq t) = (1800 - 1)e^{(-0.5 \times 3.5)} = 1799e^{-1.75}$$
$$= 312$$

The expected number of occurrences of different gaps t for the above example have been calculated and are shown in Table 5.3.

The basic assumption made in the above analysis is that the arrival of main stream vehicles can be described by a Poisson distribution. This assumption is reasonable for light to medium traffic but may not be acceptable for conditions of heavy traffic. Analyses of the occurrence of different gap sizes when traffic volume

Table 5.3 Number of Different Lengths of Gaps Occurring During a Period of 1 hr for $V = 1800$ vph and an Assumed Distribution of Poissons for Arrivals

Gap (t sec)	Probability		No. of Gaps	
	$P(h \geq t)$	$P(h < t)$	$h \geq t$	$h \leq t$
0	1.0000	0.0000	1799	0
0.5	0.7788	0.2212	1401	398
1.0	0.6065	0.3935	1091	708
1.5	0.4724	0.5276	849	950
2.0	0.3679	0.6321	661	1138
2.5	0.2865	0.7135	515	1284
3.0	0.2231	0.7769	401	1398
3.5	0.1738	0.8262	312	1487
4.0	0.1353	0.8647	243	1556
4.5	0.1054	0.8946	189	1610
5.0	0.0821	0.9179	147	1652

is heavy have shown that the main discrepancies occur at gaps of short lengths (that is, less than 1 sec). The reason for this is that, although theoretically there are definite probabilities for the occurrence of gaps between zero and 1 sec, in reality, these gaps very rarely occur, as a driver will tend to keep a safe distance between his or her vehicle and the vehicle immediately in front. One alternative used to deal with this situation is to restrict the range of headways by introducing a minimum gap. Eqs. 5.42 and 5.43 can then be written as

$$P(h \geq t) = e^{-\lambda(t - \tau)} \qquad \text{for } t \geq 0 \qquad \textbf{(5.46)}$$

$$P(h < t) = 1 - e^{-\lambda(t - \tau)} \qquad \text{for } t \geq 0 \qquad \textbf{(5.47)}$$

where τ is the minimum headway introduced.

Example 5–6 Number of Acceptable Gaps with a Restrictive Range, for Vehicles on an Expressway Ramp

Repeat Example 5–4 using a minimum gap in the expressway traffic stream of 1.0 sec and the following data.

$V = 1800$
$T = 3600$
$\lambda = (1800/3600) = 0.5$ vps

- Calculate the expected number of acceptable gaps in 1 hr.

$$(h \geq t) = (1800 - 1)e^{-0.5(3.5 - 1.0)} = 1799e^{-0.5 \times 2.5}$$

$$= 515$$

INTRODUCTION TO QUEUING THEORY

One of the greatest concerns of traffic engineers is the serious congestion that exists on urban highways, especially during peak hours. This congestion results in the formation of queues on ramps leading onto expressways, at signalized and unsignalized intersections, and on arterials, where moving queues may occur. An understanding of the processes that lead to the occurrence of queues and the subsequent delays on highways is essential for the proper analysis of the effects of queuing. The theory of queuing therefore concerns the use of mathematical algorithms to describe the processes that result in the formation of queues, so that a detailed analysis of the effects of queues can be undertaken. These mathematical algorithms can be used to determine the probability that an arrival will be delayed, the expected waiting time for all arrivals, the expected waiting time of an arrival that waits, and so forth.

Several models have been developed that can be applied to traffic situations such as merging of ramp traffic to freeway traffic, pedestrian crossings, and sudden reduction of capacity on freeways. This section will give only the elementary queuing theory relationships for a specific type of queue, that is, the single-channel queue. The theoretical development of these relationships is not included here. Interested readers are referred to the Additional Readings at the end of this chapter for a more detailed treatment of the topic.

A queue is formed when arrivals wait at a service area for service. This service can be the arrival of an accepted gap in a main traffic stream, the collection of tolls at a toll booth, the payment of parking fees at a parking garage, and so forth. The service can be provided in a single channel or several channels. Proper analysis of the effects of such a queue can be carried out only if the queue is fully specified. This requires that the following characteristics of the queue be given: (1) the distribution of arrivals, that is, uniform, Poisson, and so forth; (2) the method of service, that is, first come-first served, random, priority, and so forth; (3) the characteristic of the queue length, that is, whether finite or infinite; (4) the distribution of service times; and (5) the channel layout, that is, whether single or multiple channels and, in the case of multiple channels, whether they are in series or parallel. Several methods for the classification of queues based on the above characteristics have been used, some of which are discussed below.

Arrival Distribution.   The arrivals can be described as either a deterministic or a random distribution. Light to medium traffic is usually described by a Poisson distribution, and this is generally used in queuing theories related to traffic flow.

Service Method.   Queues can also be classified by the method used in servicing the arrivals. These include first come-first served, where units are served in order of their arrivals, last in-first served where the service is reversed to the order of arrival. The service method can also be based on priority, where arrivals are directed to specific queues of appropriate priority levels. Queues are then serviced in order of their priority level.

Characteristics of the Queue Length. The maximum length of the queue, that is, the maximum number of units in the queue, is specified, in which case the queue is a finite or truncated queue or there may be no restriction on the length of the queue. Finite queues are sometimes necessary when the waiting area is limited.

Service Distribution. This distribution is also usually considered as random, and the Poisson and negative exponential distributions have been used.

Number of Channels. The number of channels usually corresponds to the number of waiting lines and is therefore used to classify queues, for example, single-channel queue or multichannel queue.

Oversaturated and Undersaturated Queues. Oversaturated queues are those in which the arrival rate is higher than the service rate, and undersaturated queues are those in which the arrival rate is less than the service rate. The length of an undersaturated queue may vary but will reach a steady state with the arrival of units. The length of an oversaturated queue will, however, never reach a steady state but will continue to increase with the arrival of units.

Single-Channel, Undersaturated, Infinite Queues

Figure 5.11 is a schematic of a single-channel queue in which the rate of arrival is q vph and the service rate is Q vph. For an undersaturated queue, $Q > q$, assuming that both the rate of arrivals and the rate of service are random, the following relationships can be developed.

1. Probability of n units in the system $P(n)$

$$P(n) = \left(\frac{q}{Q}\right)^n \left(1 - \frac{q}{Q}\right) \qquad (5.48)$$

   where n is the number of units in the system, including that being serviced.

2. The expected number of units in the system $E(n)$

$$E(n) = \frac{q}{Q - q} \qquad (5.49)$$

Figure 5.11 A Single-Channel Queue

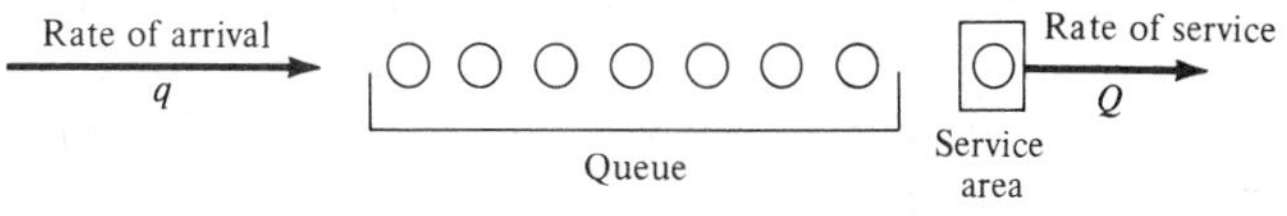

3. The expected number of units waiting to be served (that is, the mean queue length) in the system $E(m)$

$$E(m) = \frac{q^2}{Q(Q-q)} \tag{5.50}$$

Note that $E(m)$ is not exactly equal to $E(n) - 1$, the reason being that there is a definite probability of zero units being in the system $P(o)$.

4. Average waiting time in the queue $E(w)$

$$E(w) = \frac{q}{Q(Q-q)} \tag{5.51}$$

5. Average waiting time of an arrival, including queue and service $E(v)$

$$E(v) = \frac{1}{Q-q} \tag{5.52}$$

6. Probability of spending time t or less in the system

$$P(v \le t) = 1 - e^{-\left(1-\frac{q}{Q}\right)qt} \tag{5.53}$$

7. Probability of waiting for time t or less in the queue

$$P(w \le t) = 1 - \frac{q}{Q} e^{-\left(1-\frac{q}{Q}\right)qt} \tag{5.54}$$

8. Probability of more than N vehicles being in the queue; that is, $P(n > N)$

$$P(n > N) = \left(\frac{q}{Q}\right)^{N+1} \tag{5.55}$$

Eq. 5.49 can be used to produce a graph of the relationship between the expected number of units in the system $E(n)$ and the ratio of the rate of arrival to the rate of service $\rho = q/Q$. Figure 5.12 is such a representation for different values of $\rho = q/Q$. It should be noted that as this ratio tends to 1 (that is, approaching saturation), the expected number of vehicles in the system tends to infinity. This shows that q/Q, which is usually referred to as the traffic intensity (ρ), is an important factor in the queuing process. The figure also indicates that queuing is of no significance when ρ is less than 0.5, but at values of 0.75 and above, the average queue lengths tend to increase rapidly. Figure 5.13 also is a graph of the probability of n units being in the system versus q/Q. Eq. 5.49 can also be used to produce this graph.

Figure 5.12 Expected Number of Vehicles in the System $E(n)$ Versus Traffic Intensity (ρ)

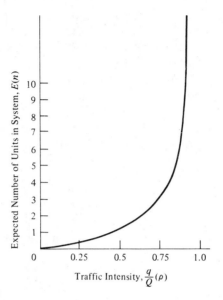

Figure 5.13 Probability of n Vehicles Being in the System for Different Traffic Intensities (ρ)

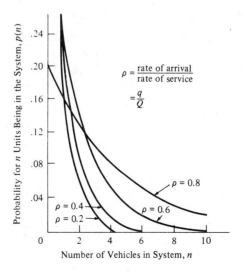

Example 5–7 Application of the Single-Channel, Undersaturated, Infinite Queue Theory to a Toll Booth Operation

On a given day, 375 vehicles per hour arrive at a toll booth located at the end of an off-ramp of a rural expressway. If the vehicles can be serviced by only a single channel at the service rate of 500 vph, determine **(a)** the percentage of time the operator of the toll booth will be free, **(b)** the average number of vehicles in the system, and **(c)** the average waiting time for the vehicles that wait. (Assume Poisson arrival and negative exponential service time.)

(a) $q = 375$ and $Q = 500$. For the operator to be free, the number of vehicles in the system must be zero. From Eq. 5.48,

$$P(o) = 1 - \frac{q}{Q} = 1 - \frac{375}{500}$$

$$= 0.25$$

The operator will be free 25 percent of the time.

(b) From Eq. 5.49,

$$E(n) = \frac{375}{500 - 375}$$

$$= 3$$

(c) From Eq. 5.52,

$$E(v) = \frac{1}{500 - 375} \text{ hr} = 0.008 \text{ hr}$$

$$= 28.8 \text{ sec}$$

Single-Channel, Undersaturated, Finite Queues

In the case of a finite queue, the maximum number of units in the system is specified. Let this number be N. Let the rate of arrival be q and the service rate be Q. If it is also assumed that both the rate of arrival and the rate of service are random, the following relationships can be developed for the finite queue.

1. Probability of n units in the system

$$P(n) = \frac{1 + \rho}{1 - \rho^{N+1}} \rho^n \tag{5.56}$$

where $\rho = q/Q$.

2. The expected number of units in the system

$$E(n) = \frac{\rho}{1 - \rho} \frac{1 - (N+1)\rho^N + N\rho^{N+1}}{1 - \rho^{N+1}} \tag{5.57}$$

Example 5–8 Application of the Single-Channel, Undersaturated, Finite Queue Theory to an Expressway Ramp

The number of vehicles that can enter the on-ramp of an expressway is controlled by a metering system, which allows a maximum of 10 vehicles to be on the ramp at any one time. If the vehicles can enter the expressway at a rate of 500 vph and the rate of arrival of vehicles at the on-ramp is 400 vph during the peak hour, determine **(a)** the probability of 5 cars being on the on-ramp, **(b)** the percent of time the ramp is full, and **(c)** the expected number of vehicles on the ramp during the peak hour.

(a) Probability of 5 cars being on the on-ramp: $q = 400$, $Q = 500$, and $\rho = (400/500) = 0.8$. From Eq. 5.56,

$$P(5) = \frac{(1 - 0.8)}{1 - (0.8)^{11}} (0.8)^5$$

$$= 0.072$$

(b) From Eq. 5.56,

$$P(10) = \frac{1 - 0.8}{1 - (0.8)^{11}} (0.8)^{10}$$

$$= 0.023$$

that is, the ramp is full only 2.3 percent of the time.

(c) The expected number of vehicles on the ramp is obtained from Eq. 5.57.

$$E(n) = \frac{0.8}{1 - 0.8} \frac{1 - (11)(0.8)^{10} + 10(0.8)^{11}}{1 - (0.8)^{11}} = 2.97$$

The expected number of vehicles on the ramp is 3.

SUMMARY

One of the most important current functions of a traffic engineer is to implement traffic control measures that will facilitate the efficient use of existing highway facilities, since extensive highway construction is no longer taking place at the rate it once was. Efficient use of any highway system entails the flow of the maximum volume of traffic without causing excessive delay to the traffic and inconvenience to the motorist. It is therefore essential that the traffic engineer understand the basic characteristics of the elements of a traffic stream since these characteristics play an important role in the success or failure of any traffic engineering action to achieve an efficient use of the existing highway system.

This chapter has furnished the fundamental theories that are used to determine the effect of these characteristics. The definitions of the different elements have been presented, together with mathematical relationships of these elements. These relationships are given in the form of macroscopic models, which consider the traffic stream as a whole, and microscopic models, which deal with individual vehicles in the traffic stream. Using the appropriate model for a traffic flow will facilitate the computation of any change in one or more elements due to a change in another element. An introduction to queuing theory is also presented to provide the reader with simple equations that can be used to determine delay and queue lengths in simple traffic queuing systems.

PROBLEMS

5–1 Observers stationed at two sections XX and YY, 500 ft apart on a highway, record the arrival times of four vehicles as shown in the accompanying table. If the total time of observation at XX was 15 sec, determine **(a)** the time mean speed, **(b)** the space mean speed, and **(c)** the flow at section XX.

Time of Arrival

Vehicle	Section XX	Section YY
A	T_0	$T_0 + (7.58 \text{ sec})$
B	$T_0 + (3 \text{ sec})$	$T_0 + (9.18 \text{ sec})$
C	$T_0 + (6 \text{ sec})$	$T_0 + (12.36 \text{ sec})$
D	$T_0 + (12 \text{ sec})$	$T_0 + (21.74 \text{ sec})$

5–2 Data obtained from aerial photography showed six vehicles on a 600 ft long section of road. Traffic data collected at the same time indicated an average time headway of 4 sec. Determine **(a)** the density on the highway, **(b)** the flow on the road, and **(c)** the space mean speed.

5–3 The data shown below was obtained by time-lapse photography on a highway. Use regression analysis to fit this data to the Greenshields model and determine **(a)** the mean free speed, **(b)** the jam density, **(c)** the capacity, and **(d)** the speed at maximum flow.

Speed (mph)	Density (vpm)
14.2	85
24.1	70
30.3	55
40.1	41
50.6	20
55.0	15

5–4 Two sets of students are collecting traffic data at two sections, *xx* and *yy*, of a highway 1500 ft apart. Observations at *xx* show that five vehicles passed that section at intervals of 3, 4, 3, and 5 sec, respectively. If the velocities of the vehicles were 50, 45, 40,

35, and 30 mph, respectively, draw a schematic showing the locations of the vehicles 20 sec after the first vehicle passed section xx. Also determine (a) the time mean speed, (b) the space mean speed, and (c) the density on the highway.

5–5 Researchers have used analogies between the flow of fluids and the movement of vehicular traffic to develop mathematical algorithms describing the relationship among traffic flow elements. Discuss briefly in one or two paragraphs the main deficiencies in this approach.

5–6 Assuming that the expression

$$\bar{u}_s = u_f e^{-k/k_j}$$

can be used to describe the speed density relationship of a highway, determine the capacity of the highway from the following data using regression analysis

k (vpm)	$\bar{u}_s$ (mph)
43	38.4
50	33.8
8	53.2
31	42.3

Under what flow conditions is the above model valid?

5–7 Results of traffic flow studies on a highway indicate that the flow–density relationship can be described by the expression

$$q = u_f k - \frac{u_f}{k_j} k^2$$

If speed and density observations give the data shown below, develop an appropriate expression for speed versus density for this highway and determine the density at which the maximum volume will occur as well as the value of the maximum volume. Also plot the speed versus density and the volume versus speed for both the expression developed and the data shown. Comment on the differences between the two sets of curves.

Speed (mph)	Density (vpm)
50	18
45	25
40	41
34	58
22	71
13	88
12	99

5–8 Studies have shown that the traffic flow on a 2-lane road adjacent to a school can be described by the Greenshields model. A length of 0.5 mi adjacent to a school is described as a school zone (see Figure 5.14). The school zone operates for only 20 min. Data

FIGURE 5.14

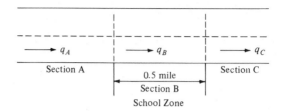

collected at the site when the school zone is in operation are given below. Determine the speed of the shock waves AB and BC. Also determine the length of the queue after the 20 min period and the time it will take the queue to dissipate after the 20 min period.

q_A (one direction) = 1750 vph $\bar{u}_a$ = 35 mph
q_B (one direction) = 1543 vph $\bar{u}_b$ = 18 mph
q_C (one direction) = 1714 vph $\bar{u}_c$ = 30 mph

5–9 Briefly discuss the phenomenon of gap acceptance with respect to merging and weaving maneuvers in traffic streams.

5–10 The table below gives data on accepted and rejected gaps of vehicles on the minor road of an unsignalized intersection. If the arrival of major road vehicles can be described by the Poisson distribution, and the peak hour volume is 1100 vph, determine the expected number of accepted gaps that will be available for minor road vehicles during the peak hour.

Gap (t)	Number of Rejected Gaps > t	Number of Accepted Gaps < t
1.5	92	3
2.5	52	18
3.5	30	35
4.5	10	62
5.5	2	100

5–11 Using appropriate diagrams, describe the resultant effect of a sudden reduction of the capacity (bottleneck) on a highway on the traffic flow both upstream and downstream of the bottleneck.

5–12 The capacity of a highway is suddenly reduced to 60 percent of its normal capacity due to closure of certain lanes in a work zone. If it can be assumed that the Greenshields model describes the relationship between speed and density on the highway, the jam density of the highway is 112 vpm, and the mean free speed is 64.5 mph, determine by what percentage the space mean speed at the vicinity of the work zone will be reduced if the flow upstream is 80 percent of the capacity of the highway.

5–13 The arrival times of vehicles at the ticket gate of a sports stadium may be assumed to be Poisson with a mean of 30 vph. It takes an average of 1.5 min for the necessary tickets to be bought for occupants of each car.

(a) What is the expected length of queue at the ticket gate, not including the vehicle being served?

(b) What is the probability that there are no more than 5 cars at the gate, including the vehicle being served?

(c) What will be the average waiting time of a vehicle?

5–14 An expressway off-ramp consisting of a single lane leads directly to a toll booth. The rate of arrival of vehicles at the expressway can be considered to be Poisson with a mean of 50 vph, and the rate of service to vehicles can be assumed to be exponentially distributed with a mean of 1 min.

(a) What is the average number of vehicles waiting to be served at the booth (that is, number of vehicles in queue, not including the vehicle being served)?

(b) What is the length of the ramp required to provide storage for all exiting vehicles 85 percent of the time? Assume average length of vehicle is 20 ft and an average space of 5 ft between consecutive vehicles waiting to be served.

(c) What is the average waiting time a driver waits before being served at the toll booth (that is, average waiting time in the queue)?

REFERENCES

1. *Highway Capacity Manual*, Special Report 209, National Research Council, Transportation Research Board, Washington, D.C., 1985.

2. B. D. Greenshields, "A Study of Highway Capacity," *Highway Research Board Proceedings*, 14(1934).

3. H. Greenberg, "An Analysis of Traffic Flow," *Operations Research* 7(1959).

4. R. E. Chandler, R. Herman, and E. Motroll, "Traffic Dynamics—Studies in Car Following," *Operations Research* 6(2)(1958):165–186.

5. B. D. Greenshields, D. Shapiro, and E. L. Erickson, *Traffic Performance at Urban Street Intersections*, Technical Report 1, Bureau of Highway Traffic, Yale University, New Haven, Conn., 1947.

6. M. S. Raff, and J. W. Hart, *A Volume Warrant for Urban Stop Signs*, The Eno Foundation for Highway Traffic Control, Saugatuck, Conn., 1950.

ADDITIONAL READINGS

Gazis, D. C., R. Herman, and R. W. Rothery, "Follow the Leader Models of Traffic Flow," *Operations Research* 9(1959).

Gerlough, Daniel L., and Matthew J. Huber, *Traffic Flow Theory*, Special Report 165, National Research Council, Transportation Research Board, Washington, D.C., 1975.

Haight, Frank A., B. F. Whister, and W. W. Mosher, Jr., "New Statistical Methods for Describing Highway Distribution of Cars," *Highway Research Board Proceedings* 40(1961):557–564.

Herman, R., E. W. Montroll, R. B. Potts, and R. W. Rothery, "Traffic Dynamics: Analysis of Stability in Car Following," *Operations Research* 7(1959).

Lieberman, E. A., "Dynamic Analysis of Freeway Corridor Traffic," presented at 1970 Joint Transportation Engineering Conference, Chicago, Ill., October 12–14, 1970.

Munjal, P. K., and L. A. Pipes, "Propagation of On-Ramp Density Perturbations on Uni-Directional and Two- and Three-Lane Freeways," *Transportation Research* 5(4)(1971):241–255.

Pipes, L. A., "Car Following Models and the Fundamental Diagram of Road Traffic," *Transportation Research* 1(2)(1967):21–29.

Yager, S., CORQ—A Model for Predicting Flows and Queues in a Road Corridor, Transportation Research Record 533, National Research Council, Transportation Research Board, Washington, D.C., 1975.

Yager, S., *Applications of Traffic Flow Theory in Modeling Network Operations*, Transportation Research Record 567, National Research Council, Transportation Research Board, Washington, D.C., 1976.

CHAPTER 6

Intersection Control

An intersection is an area shared by two or more roads, and its main function is to provide for the change of route directions. A simple intersection may consist of only two intersecting roads, whereas a complex intersection may have several intersecting roads within the same area. The intersection is therefore an area of decision for all drivers; each driver must select one of the available alternative choices to proceed. This requires an additional effort by the driver, which is not necessary at nonintersection areas of a highway. The flow of traffic on any highway is greatly affected by the flow of traffic through the intersection points on that highway because the intersection usually performs at a level below that of any other section of the road.

Intersections can be classified as *grade-separated* or *at grade*. Grade-separated intersections are commonly known as interchanges and usually consist of structures that provide for the cross flow of traffic at different levels without interruption. This reduces delay, particularly when volumes are high. Figure 6.1 shows different types of at-grade intersections and Figure 6.2 shows different types of interchanges. Several types of traffic control systems are used to reduce traffic delays and accidents at at-grade intersections and to increase the capacity of highways and streets. However, the appropriate regulations must be enforced if these systems are to be effective. This chapter discusses the different methods of controlling traffic at at-grade intersections and freeway ramps.

GENERAL CONCEPTS OF TRAFFIC CONTROL

The purpose of traffic control is to facilitate highway safety by ensuring the orderly and predictable movement of all traffic on highways. The control may be achieved by using traffic signals, signs, or markings that regulate, guide, warn, and/or channel traffic. The more complex the maneuvering area, the higher is the need for a properly designed traffic control system. Intersections, being complex maneuvering areas, therefore require properly designed traffic control systems. Guidelines for determining whether a particular control type is suitable for a

Figure 6.1   Types of Intersections at Grade

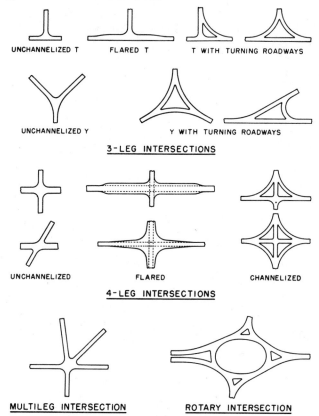

Source: Reproduced from *A Policy on Geometric Design of Rural Highways 1965*, Washington, D.C.: The American Association of State Highway and Transportation Officials, copyright 1966. Used by permission.

given intersection have been developed and are given in the *Manual on Uniform Traffic Control Devices* (MUTCD).[1]

To be effective, a traffic control device must

- Fulfill a need
- Command attention
- Convey a clear simple meaning
- Command respect of road users
- Give adequate time for proper response

To ensure that a traffic control device possesses these five properties, the MUTCD recommends that engineers consider the following five factors.

Figure 6.2 Types of Interchanges

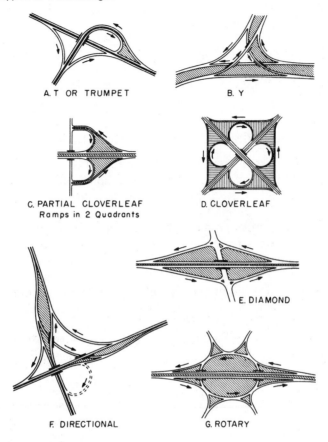

Source: Reproduced from *A Policy on Geometric Design of Rural Highways and Streets 1965,* Washington, D.C.: The American Association of State Highways and Transportation Officials, copyright 1966. Used by permission.

1. **Design.** The design should be such that a suitable combination of size, color, and shape is obtained that will convey a message and command the respect and attention of the driver.

2. **Placement.** The device should be located so it is within the cone of vision of the viewer and the driver has adequate response time when driving at normal speed.

3. **Operation.** The device should be used in a manner that ensures the fulfillment of traffic requirements in a consistent and uniform way.

4. **Maintenance.** The device must be regularly maintained to ensure that legibility is sustained.

5. **Uniformity.** Similar devices should be used at locations with similar
    traffic and geometric characteristics to facilitate the recognition and under-
    standing of these devices by drivers.

    In addition to these considerations, it is essential that engineers avoid using
control devices at locations that conflict with one another. It is imperative that
control devices aid each other in transmitting the required message to the driver.

CONFLICT POINTS AT INTERSECTIONS

A number of conflicts can occur at intersections. Traffic conflicts occur when
traffic streams moving in different directions interfere with each other. Three
types of conflicts are merging, diverging, and crossing. Figure 6.3 shows the
different conflict points that exist at a four-approach, unsignalized intersection.
There are 32 conflict points in this case. The number of possible conflict points at
any intersection depends on the number of approaches, the turning movements,
and the type of traffic control at the intersection.

The primary objective in the design of a traffic control system at an intersection is to reduce the number of significant conflict points. In designing such a
system, it is first necessary to undertake an analysis of the turning movements at
the intersection, which will indicate the significant types of conflicts. Factors that
influence the significance of a conflict include the type of conflict, the number of
vehicles in each of the conflicting streams, and the speeds of the vehicles in these

Figure 6.3   Conflict Points at a Four-Approach Unsignalized Intersection

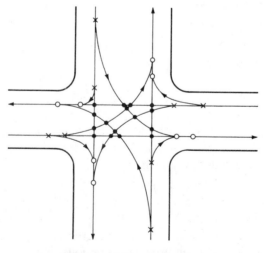

o Merging conflict points = 8
× Diverging conflict points = 8
• Crossing conflict points = 16

streams. Crossing conflicts, however, tend to have the most severe effect on traffic flow and should be reduced to a minimum whenever possible.

TYPES OF INTERSECTION CONTROL

Several methods of controlling conflicting streams of vehicles at intersections are in use. The use of any one of these methods depends on the type of intersection and the volume of traffic in each of the conflicting streams. Guidelines for determining whether a particular control type is suitable for a given intersection have been developed and are given in the MUTCD.[2] These guidelines are given in the form of warrants, which have to be compared with the traffic and geometric characteristics at the intersection being considered. The different types of intersection control are presented next.

Yield Signs

All drivers on approaches with yield signs are required to slow down and yield the right of way to all other vehicles at the intersection. Stopping at yield signs is not mandatory, but drivers are required to stop when necessary to avoid interfering with a traffic stream that has the right of way. Yield signs are therefore usually placed on minor road approaches, where it is necessary to yield the right of way to the major road traffic. Figure 6.4 shows the regulated shape and dimensions for a yield sign. The most significant factor included in the warrant for yield signs is the approach speed on the minor road. The yield sign is not warranted when the approach speed is less than 10 mph. This sign is warranted at intersections where there is a separate or channelized right-turn lane, without an adequate acceleration lane.

Stop Signs

A stop sign is used where an approaching vehicle is required to stop before entering the intersection. Figure 6.4 also shows the regulated shape and dimensions of a stop sign. Stop signs should be used only when they are warranted since the use of this sign results in considerable inconvenience to motorists. Stop signs

Figure 6.4 Stop Sign and Yield Sign

R1-1
30″ X 30″

R1-2
36″ X 36″ X 36″

should not be used at signalized intersections or on through roadways of express-ways. The warrants for stop signs suggest that a stop sign may be used on a minor road when it intersects a major road, at an unsignalized intersection, and where a combination of high speed, restricted view, and serious accidents indicate the necessity for such a control.

Multiway Stop Signs

Multiway stop signs require that all vehicles approaching the intersection stop before entering that intersection. They are used as a safety measure at some intersections and are normally used when the traffic volumes on all the approaches are approximately equal. When traffic volumes are high, however, the use of signalization is recommended.

The warrants for this control specify that total intersection approach volume should not be less than 500 vph for 8 hours of an average day, nor should the combined volume of vehicles and pedestrians from the minor approach be less than 200 vph for the same 8 hours. The average delay of the vehicles on the minor street should also not be less than 30 sec per vehicle during the maximum hour. The minimum requirement for vehicular volume can be reduced by 30 percent if the 85th percentile approach speed on the major approach is greater than 40 mph.

Intersection Channelization

Intersection channelization is used mainly to separate turn lanes from through lanes. A channelized intersection consists of solid white lines or raised barriers, which guide traffic within a lane so that vehicles can safely negotiate a complex intersection. When raised islands are used, they can also serve as a refuge for pedestrians.

Channelization design criteria have been developed by many states individu-ally to provide guidelines for the cost-effective design of channelized intersections. A detailed description of the techniques that have proven effective for both simple and complicated intersections is given in *Intersection Channelization Design Guide*.[3] Guidelines for the use of channels at intersections include

- Islands or channel lines laid out such that the driving channels provide for a natural convenient flow of traffic
- Avoiding confusion by using a few well-located islands
- Adequate radii of curves and width of lanes for the prevailing type of vehicle

Traffic Signals

One of the most effective ways of controlling traffic at an intersection is the use of traffic signals. Traffic signals can be used to eliminate many conflicts, as differ-ent traffic streams can be assigned the use of the intersection at different times. Since this results in a delay to vehicles in all streams, it is important that traffic signals be used only when necessary. The most important factor that indicates when use of traffic signals at a particular intersection is necessary is the intersection approach traffic volume, although other factors such as pedestrian volume and accident experience may also play a significant role. The *Manual on Traffic Signal Design* gives the fundamental concepts and standard practices used in the design of traffic signals.[4] In addition, the MUTCD describes in detail eleven warrants, at least one of which should be satisfied for an intersection to be signalized.[5] However, these warrants should be considered only as a guide; professional judgment based on experience should also be used to decide whether or not an intersection should be signalized. The factors considered in the warrants given are

- Minimum vehicular volume
- Interruption of continuous traffic
- Minimum pedestrian volume
- School crossing
- Progressive movement
- Accident experience
- Systems
- Combination of warrants
- Four-hour volume
- Peak-hour delay
- Peak-hour volume

A brief discussion of each of these is given below. Interested readers are referred to the MUTCD for details.[6]

Minimum Vehicle Volume. This warrant is applied when the principal factor for considering signalization is the intersection traffic volume. The warrant is satisfied when traffic volumes on major streets and on higher-volume minor streets for each of any 8 hours of an average day are at least equal to specified volumes.

Interruptions of Continuous Traffic. This warrant should be considered when traffic on a minor street suffers excessive delay due to the heavy volume of traffic on a major street. Heavy major street traffic may also make it hazardous for minor street traffic to enter or cross the major street. The warrant is satisfied when the traffic volume on the major street and on the higher-volume minor street for each of any 8 hours of an average day is at least equal to specified volumes.

Minimum Pedestrian Volume. This warrant is based on a combination of pedestrian and vehicle volumes. It is satisfied when the traffic volume on the major street and the pedestrian volume on the highest volume crosswalk crossing the major street for each of any 8 hours of an average day are at least equal to specified volumes. The 8 hours should be the same for both the vehicle and pedestrian volumes. When the decision to install a signal at an isolated intersection is based on this warrant, the signal should be of the traffic-actuated type and should also be equipped with push buttons for pedestrians crossing the major street.

School Crossing.   When an analysis of gap data at an established school zone shows that the frequency of occurrence of gaps and lengths of gaps are inadequate for safe crossing of the street by school children, this warrant is applied. It stipulates that, if during the period when school children are using the crossing the number of acceptable gaps is less than the number of minutes in that period, the use of traffic signals is warranted. The signal in this case should be pedestrian actuated, and all obstructions to view, such as parked vehicles, should be prohibited for at least 100 ft before and 20 ft after the crosswalk.

Progressive Movement.   This warrant may justify the installation of traffic lights at an intersection where lights would not otherwise have been installed. It justifies the installation of traffic lights when such an installation will help maintain a proper grouping of vehicles and effectively regulate group speed.

Accident Experience.   This warrant justifies signalization of an intersection when accident frequency has not been reduced by adequate trial of less restrictive measures.

Systems.   This warrant justifies the installation of signals at some intersections when such an installation will help to encourage concentration and organization of traffic networks. The warrant can be applied when the total intersection vehicle volume is at least 800 during the peak hour of a typical weekday or each of any 5 hours of a Saturday or Sunday.

Combination of Warrants.   This warrant, in exceptional cases, justifies the installation of signals when none of the above warrants is satisfied but when two or more of the first three warrants are satisfied to the extent of 80 percent of the stipulated volumes.

Four-Hour Volume.   This warrant is based on the comparison of standard graphs given in the MUTCD with a plot on the same graph of the total volume of vph on the approach with the higher volume in the minor street (vertical scale) against the total vph on both approaches of the major street (horizontal scale) at the intersection under consideration. When the plot for each of any 4 hours of an average day falls above the standard graph, this warrant is satisfied. Standard graphs are given for different types of lane configurations at the intersection. A different set of standard plots is also given for intersections where the 85th percentile speed of the major street traffic is higher than 40 mph or where the intersection is located in a built-up area of an isolated community having a population less than 10,000.

Peak-Hour Delay.   This warrant is used to justify the installation of traffic signals at intersections where traffic conditions during 1 hour of the day result in undue delay to traffic on the minor street. The warrant is satisfied, when the delay during any four consecutive 15 minute periods on one of the minor street approaches (one direction only) controlled by a stop sign is equal to or higher than specified levels, and the same minor street approach (one direction only) volume and the total intersection entering volume are equal to or higher than the specified levels.

Peak-Hour Volume.   This warrant is also used to justify the installation of traffic signals at intersections where traffic conditions during 1 hour of the day result in undue delay to traffic on the minor street. It is based on the comparison of the standard graphs given in the MUTCD with a plot of the peak-hour volume on the approach with the higher volume on the minor street against the corresponding peak-hour volume on the major street (both directions) of the intersection under consideration. When this plot for an average day' lies above the appropriate standard curve, this warrant is satisfied.

SIGNAL TIMING FOR DIFFERENT COLOR INDICATIONS

The warrants given earlier will only help the engineer to decide whether a traffic signal should be used at an intersection. The efficient operation of the signal, however, requires proper timing of the different color indications, which is obtained by carrying out the necessary design. Before presenting the different methods of signal timing design, however, it is first necessary to define a number of terms commonly used in the design of signal times.

Controller is a device in a traffic signal installation that changes the colors indicated by the signal lamps with respect to a fixed or variable plan. It assigns the right of way to different approaches at appropriate times.

Cycle (*cycle length*) is the time in seconds required for one complete color sequence. Figure 6.5 is a schematic of a cycle.

Phase (*signal phase*) is that part of a cycle allocated to a stream of traffic or a combination of two or more streams of traffic having the right of way simultaneously during one or more intervals. See Figure 6.5.

Interval is any part of the cycle length during which signal indications do not change.

Offset is the time lapse in seconds or percent of the cycle length between the beginning of a green phase at the intersection and the beginning of a corresponding green phase at the next intersection. It is the time base of the system controller.

Clearance interval is the length of time in seconds of the yellow signal indication. This time is provided for vehicles to clear the intersection after the green interval.

All red interval is the display time of a red indication for all approaches. It is sometimes used as a phase exclusively for pedestrian crossing or to allow vehicles .

Fgiure 6.5 Two-Phase Signal System

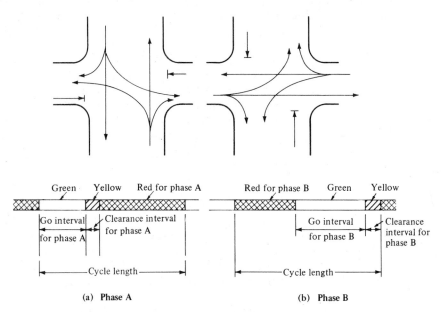

(a) **Phase A**   (b) **Phase B**

and pedestrians to clear very large intersections before opposing approaches are given the green indication.

Split phase is that part of a phase that is set apart from the primary movement, thus forming a special phase that relates to the parent phase.

Peak-hour factor (PHF) is a measure of the variability of demand during the peak hour. It is the ratio of the volume during the peak hour to the maximum rate of flow during a given time period within the peak hour. For intersections, the time period used is 15 min and the PHF is given as

$$\text{PHF} = \frac{\text{volume during peak hour}}{4 \times \text{volume during peak 15 min within peak hour}}$$

The PHF may be used in signal timing design to compensate for the possibility that peak arrival rates for short periods during the peak hour may be much higher than the average for the full hour. Design hourly volume (DHV) can then be obtained as

$$\text{DHV} = \frac{\text{peak-hour volume}}{\text{PHF}}$$

Not all factors that affect PHF have been identified, but it is generally known that the PHF is a function of the traffic generators being served by the highway, the distances between these generators and the highway, and the population of the metropolitan area in which the highway is located.

Passenger car equivalent (PCE) is a factor used to convert straight-through volumes of buses and trucks to straight-through volumes of passenger cars. This conversion is necessary since buses and trucks require more time to cross the intersection than do passenger cars. PCE factors commonly used for both buses and trucks range from 1.4 to 1.6, depending on the predominant type of truck.

Turning movement factors are required since turning vehicles generally require a longer green time than straight-through vehicles. These factors, which are used to convert turning vehicles to equivalent straight-through vehicles, range from 1.4 to 1.6 for left-turning vehicles and 1.0 to 1.4 for right-turning vehicles.

Critical lane volume is the maximum lane volume (vph) in a phase.

Objectives of Signal Timing

The main objectives of signal timing at an intersection are to reduce the average delay of all vehicles and the probability of accidents by minimizing the possible conflict points. This is achieved by assigning the right of way to different traffic streams at different times. The objective of reducing delay, however, sometimes conflicts with that of accident reduction. This is because the number of distinct phases should be kept to a minimum to reduce average delay, whereas many more distinct phases may be required to separate all traffic streams from each other. When this situation exists, it is essential that engineering judgment be used to determine a compromise solution. In general, however, it is usual to adapt a two-phase system whenever possible, using the shortest practical cycle length that is consistent with the demand. At a complex intersection, though, it may be necessary to use a multiphase (three or more phases) system to achieve the main design objectives.

Signal Timing at Isolated Intersections

An isolated intersection is one at which the signal time is not coordinated with that of any other intersection and therefore operates independently of any other intersection. The cycle length for an intersection of this type should be short, preferably between 35 and 60 sec, although it may be necessary to use longer cycles when approach volumes are very high. However, cycle lengths should be kept below 120 sec since very long cycle lengths will result in excessive delay. Several methods have been developed for determining the optimal cycle length at an intersection and, in most cases, the clearance interval is considered as a component of the cycle length. Before discussing a few of these methods, we will discuss the basis for selecting the clearance interval at an intersection.

Clearance Interval

The main purpose of the yellow indication after the green is to alert motorists to the fact that the green light is about to change to red and to allow vehicles already at the intersection to cross it. A bad choice of clearance interval may lead to the creation of a *dilemma zone*, an area close to an intersection in which a vehicle can neither stop safely before the intersection, nor clear the intersection without speeding before the red signal comes on. The required clearance interval is that time period that guarantees that an approaching vehicle can either stop safely or proceed through the intersection without speeding.

Figure 6.6 is a schematic of a dilemma zone. For the dilemma zone to be eliminated, the distance X_o should be equal to the distance X_c. Let τ_{min} = clearance interval (sec) and the distance traveled during the clearance interval without accelerating = $u_o(\tau_{min})$, with u_o = speed limit on approach (ft/sec). If the vehicle just clears the intersection, then

$$X_o = u_o(\tau_{min}) - (W + L)$$

where

W = width of intersection (ft)
L = length of vehicle (ft)

For vehicles to be able to stop, however,

$$X_c = u_o\delta + \frac{u_o^2}{2a}$$

where

δ = perception-reaction time
a = constant rate of braking deceleration (ft/sec²)

Figure 6.6   Schematic of a Dilemma Zone at an Intersection

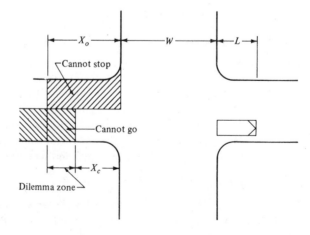

For the dilemma zones to be eliminated, X_o must be equal to X_c. Accordingly,

$$u_o\tau_{\min} - (W + L) = u_o\delta + \frac{u_0^2}{2a}$$

$$\tau_{\min} = \delta + \frac{W + L}{u_o} + \frac{u_o}{2a}$$

(6.1)

Safety considerations, however, normally preclude clearance intervals less than 3 sec, and to encourage motorists' respect for the clearance interval, it is usually not made longer than 5 sec. When longer clearance intervals are required as computed from Eq. 6.1, an all red phase can be inserted to follow the clearance interval.

Example 6–1   Determining the Minimum Clearance Interval at an Intersection

Determine the minimum clearance interval at an intersection having a width of 40 ft if the maximum allowable speed on the approach roads is 30 mph. Assume average length of vehicle is 20 ft.

To solve this problem, we must first decide on a deceleration rate. Research on this topic has shown that a comfortable deceleration rate can be taken as $0.27g$—that is, 27 percent of the gravitational acceleration (32.2). Assuming this value for a and taking δ as 1.0 sec, we obtain

$$\tau_{\min} = 1.0 + \frac{40 + 20}{30 \times 1.47} + \frac{30 \times 1.47}{2 \times 0.27 \times 32.2}$$

$$= 4.9 \text{ sec}$$

In this case, an amber period of 5 sec will be needed.

Cycle Lengths of Fixed (Pretimed) Signals

The signals at isolated intersections can either be pretimed (fixed), semiactuated, or fully actuated. Pretimed signals assign the right of way to different traffic streams in accordance with a preset timing program. Each signal has a preset cycle length that remains fixed for a specific period of the day or for the entire day. Each phase also consists of a fixed proportion of the cycle for the green indication. This proportion may be fixed for specific periods of the day or for the entire day. Several design methods have been developed to determine the optimum cycle length, three of which—the failure rate, Webster, and Pignataro methods—are presented here. A fourth method, described in the *Highway Capacity Manual* of 1985,[7] is presented in Chapter 8, under level of service of signalized intersections, as it is directly linked with the level of service at the intersection.

Failure Rate Method. This method was developed by Drew and Pinnell and is based on the critical lane volume for each phase, which is the maximum hourly volume per lane that can go through the intersection during a given phase.[8] Figure 6.7 shows a time–space diagram of the conditions that exist on a simple two-phase system. Although the diagram shows only a two-phase system, the method can be used for a complex intersection with many phases.

Noting that the ordinate represents distance and the abscissa time, it can be seen that the progress of vehicles as they approach and leave the intersection is represented by the lighter lines. When the green indicator appears on an approach, some time elapses before all vehicles in the queue start to move. This elapsed time is usually referred to as the starting delay.

Let

K_1 = starting delay (sec)
K_2 = time taken by the last vehicle to go through the intersection (sec)
$K = K_1 + K_2$
G_i = green phase + clearance interval for phase i (sec)
X_i = maximum number of vehicles per lane that go through the intersection during the time G_i

During a time period of $G_i - (K_1 + K_2)$ the number of vehicles that cross the stop line is $(X_i - 1)$. The average minimum headway (h_i) can be obtained from

$$\text{average minimum headway} = \frac{\text{time}}{\text{volume}}$$

That is,

$$h_i = \frac{G_i - (K_1 + K_2)}{X_i - 1} = \frac{G_i - K}{X_i - 1}$$

from which we obtain

$$G_i = (X_i - 1)h_i + K$$

and

$$X_i = \frac{G_i - (K - h_i)}{h_i} \tag{6.2}$$

If the critical lane volume for phase $i = V_i$ vph, then

$$V_i = \frac{3600}{C}X_i$$

Figure 6.7 Time–Space Relationship for Two-Phase System

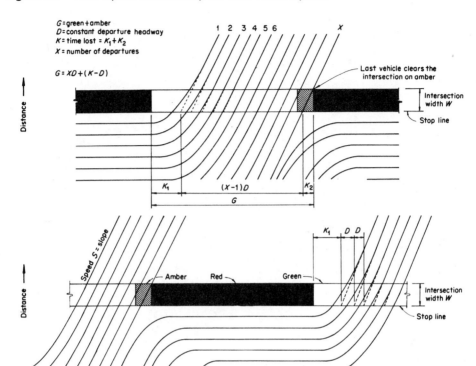

Source: Reproduced from Donald R. Drew, *Traffic Flow Theory and Control,* copyright © 1968, McGraw-Hill Book Company.

where C = minimum cycle length = $\Sigma\, G_i$, assuming no all-red phase. (Note that in this method the yellow is considered as part of the green phase.)

The sum of the critical lane volumes for all phases is therefore given as

$$\sum_{i=1}^{\phi} V_i = \frac{3600}{C} \sum_{i=1}^{\phi} X_i$$

Substituting the expression in Eq. 6.2 for X_i, we obtain

$$\sum_{i=1}^{\phi} V_i = \frac{3600}{C} \sum_{i=1}^{\phi} \frac{G_i - (K_i - h_i)}{h_i} \qquad (6.3)$$

The minimum headway can be assumed to be constant, and Eq. 6.3 can then be written as

$$\sum_{i=1}^{\phi} V_i = \frac{3600}{C} \frac{\sum_{i=1}^{\phi} G_i - \left(\sum_{i=1}^{\phi} K_i + \phi h \right)}{h} \qquad (6.4)$$

where ϕ is the number of phases per cycle.

If K_i is also taken as a constant, K, we obtain

$$\sum_{i=1}^{\phi} V_i = \left(\frac{3600}{C}\right) \frac{\sum G_i - \phi(K-h)}{h} \tag{6.5}$$

and

$$C = \left(\frac{3600}{\displaystyle\sum_{i=1}^{\phi} V_i}\right) \frac{\displaystyle\sum_{i=1}^{\phi} G_i - \phi(K-h)}{h} \tag{6.6}$$

but

$$\sum_{i=1}^{\phi} G_i = C$$

$$C = \frac{3600 \times \phi(K-h)}{3600 - h \displaystyle\sum_{i=1}^{\phi} V_i} \tag{6.7}$$

Studies have shown that reasonable and representative values are 6 sec for K, 2 sec for h, and 3.5 sec for the starting delay K_1.[9]

The assumption that is implied in the development of the above is that the arrival of vehicles is uniform during the period of 60 min. This, however, is true only when traffic volume is very high. For light to medium traffic, the assumption of a Poisson distribution for the arrival of vehicles gives the best estimate of demand. Using this assumption, the probability of $(X_i + 1)$ or more vehicles arriving during the ith phase can be determined. This probability gives the percentage of failing cycles, where a failing cycle is one during which the number of vehicles arriving on the approach is greater than the number that can get through the intersection. Based on the assumption of a Poisson distribution of arrivals, the probability of failure of phase i is given as

$$P(X \geq (X_i + 1)) = \sum_{X_i+1}^{\infty} \frac{m^{X_i+1} e^{-m}}{(X_i+1)!} \tag{6.8}$$

where

$$m_i = V_i/(3600/C) = \text{average arrivals per cycle} \tag{6.9}$$

$$X_i = \frac{G_i - (K-h)}{h}$$

$$C = \sum_{i=1}^{\phi} G_i$$

The failure rate can then be determined by solving these equations successively. The failure rate thus obtained gives a means of evaluating the effectiveness of the designed cycle length. Figure 6.8 shows a design chart that can be used to simplify the computation procedure. The following steps are involved in using the design charts.

Step 1. Determine the hourly rate of flow equivalent to the arrivals at each approach during the peak period.

Step 2. Select a suitable phasing system.

Step 3. Determine the critical lane volume for each phase, V_i.

Step 4. Determine the minimum cycle length, C, using Eq. 6.7, and use the result as the assumed cycle length.

Step 5. Determine the average arrivals per cycle, m, using Eq. 6.9.

Step 6. Determine the phase length, G_i, for each phase and for various probabilities of failure using the design chart. (See Problem 6–2.)

Step 7. The designed failure rate is that for which the summation of all green phases is equal to the assumed cycle length—that is, $\Sigma\, G_i = C$.

Drew and Pinnell suggested that a practical design failure rate should be about 30 percent to 35 percent during the peak period, which will be only about 10 percent to 15 percent during the off-peak period.[10]

Figure 6.8 Failure Rate Method Design Chart

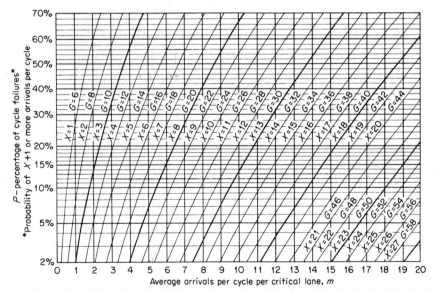

Source: Reproduced from Donald R. Drew, *Traffic Flow Theory and Control*, copyright © 1968, McGraw-Hill Book Company.

Example 6–2 Signal Timing Using the Failure Rate Method

Figure 6.9a shows peak hour volumes for a major intersection on an expressway. Determine a suitable signal timing for the intersection using a four-phase system and the additional data given in the figure.

- First convert the mixed volumes to equivalent straight-through passenger cars. The equivalent volumes are shown in Figure 6.9b. The volumes were obtained by dividing by the PHF, and then applying the relevant factors for trucks and left-turning vehicles as necessary. No factors for right-turning vehicles were used, as those volumes were very low.

- Assume the following phasing system, where the arrows indicate traffic streams that have the right of way.

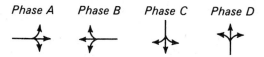

| Phase A | Phase B | Phase C | Phase D |

- Use the following critical lane volumes.

Phase, ϕ	Critical Lane Volume
A	499
B	338
C	115
D	519
	$\sum$ 1471

- Use Eq. 6.7—$C = [3600 \times \phi(K-h)]/(3600 - h \sum V)$—with $\phi = 4$, $K = 6.0$, and $h = 2.0$ to find the cycle length.

$$C = \frac{3600 \times 4(6-2)}{3600 - 2 \times 1471} = \frac{57,600}{658} = 87.5 \text{ sec}$$

- Assuming a cycle length of 90 sec, determine green time required for each phase and for different percentages of failure, using Figure 6.8. For example, knowing the average arrivals per cycle m for phase A is 12.5, we enter the chart at $m = 12.5$ on the horizontal scale and project a line upward vertically to intersect the 30 percent failure rate line, which will give a required green time of 31 sec for a 30 percent failure rate. In a similar manner, we can determine the following values.

Phase, ϕ	Average Arrivals per Cycle, m	Length of Green Phase for Various Percentages of Failure				
		15	25	30	35	40
A	12.5	35	33	31	31	30
B	8.5	26	24	23	22	21
C	2.9	12	11	10.	10	9
D	13.0	36	34	32	32	30
Total		112	102	96	95	90

Figure 6.9 Signal Timing Using the Failure Rate Method

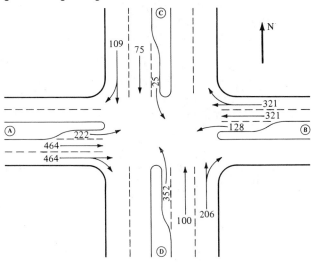

PHF = 0.95
Left-turn factor = 1.4
PCE for buses and trucks = 1.6
Truck percentages

North Approach		South Approach		West Approach		East Approach	
Through	Left	Through	Left	Through	Left	Through	Left
0	0	0	0	4	4	0	0

Pedestrian volume is negligible.

(a) Data for Example 6.2

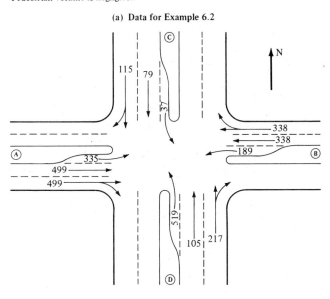

DHV for EB (West approach) through traffic = $\dfrac{464}{0.95}$ = 488 vehicles

PCE = [488 − 0.04 × 488] + 0.04 × 488 × 1.6 = 468 + 31 = 499

(b) Equivalent straight-through passenger cars

- Determine the designed failure rate—that is, the rate at which the sum of the green times obtained equals the cycle length. We can see that $\Sigma \, G_i = 90$ gives a 40 percent failure rate, which indicates that with a cycle length of 90 sec, the failure rate will be 40 percent during the peak hour.

At this point the designer has to decide whether the designed failure rate obtained is acceptable. If it is unacceptable, then the computation has to be repeated using a different cycle length. For example, if the failure rate is to be reduced in this problem, the cycle length will have to be increased and the computation repeated.

Webster Method.   Webster[11] has shown that, for a wide range of practical conditions, minimum intersection delay is obtained when the cycle length is obtained by the equation

$$C_o = \frac{1.5L + 5}{1 - \sum\limits_{i=1}^{\phi} Y_i} \qquad\qquad (6.10)$$

where

C_o = optimum cycle length in seconds

L = total lost time per cycle in seconds

Y_i = maximum value of the ratios of approach volumes to saturation flows for phase i

ϕ = number of phases

Total Lost Time.   Figure 6.10 shows a graph of rate of discharge of vehicles at various times during a green phase of a signal cycle at an intersection. Initially, some time is lost before the vehicles start moving, then the rate of discharge increases to a maximum. This maximum rate of discharge is the saturation flow. If there are sufficient vehicles in the queue to use the available green time, the maximum rate of discharge will be sustained until the clearance interval (amber phase). The rate of discharge will then fall to zero as the amber

Figure 6.10   Discharge of Vehicles at Various Times During a Green Phase

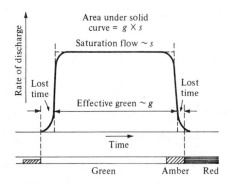

signal changes to red in the manner shown. The number of vehicles that go through the intersection is represented by the area under the curve. Dividing the number of vehicles that go through the intersection by the saturation flow will give the effective green time, which is less than the sum of the green and amber times. This difference is considered lost time as it is not used by any other phase for the discharge of vehicles; it can be expressed as

$$\ell_i = G_{iw} + \tau_i - g_i \qquad (6.11)$$

where

ℓ_i = lost time for phase i

G_{iw} = total green time for phase i (not including amber time)

τ_i = amber time for phase i

g_i = effective green time

V_i = number of vehicles discharged on the average during a saturated phase i

s_i = saturation flow for phase i

Total lost time is given as

$$L = \sum_{i=1}^{\phi} \ell_i + R \qquad (6.12)$$

where R is the total all red time during the cycle.

Allocation of Green Times.   In general, the total effective green time available per cycle is given by

$$g_T = C - L = C - \left(\sum_{i=1}^{\phi} \ell_i + R \right)$$

where

C = actual cycle length used (usually obtained by rounding off C_o to the nearest 5 sec)

g_T = total effective green time per cycle

To obtain minimum overall delay, the total effective green time should be distributed among the different phases in proportion to their Y values.

$$g_i = \frac{Y_i}{Y_1 + Y_2 + \cdots Y_\phi} g_T \qquad (6.13)$$

and the green time for each phase is obtained as

$$G_{1w} = g_1 + \ell_1 - \tau_1$$
$$G_{2w} = g_2 + \ell_2 - \tau_2$$
$$G_{iw} = g_i + \ell_i - \tau_i \tag{6.14}$$
$$G_{nw} = g_n + \ell_n - \tau_n$$

Example 6–3   Signal Timing Using Webster Method

Solve Example 6–2 using the Webster method and assuming that the saturation flow is 2000 passenger cars per hour for each lane and the lost time per phase is 3.5 sec.

- Compute the total lost time using Eq. 6.12. Since there is not an all red phase—that is, $R = 0$—and there are 4 phases,

$$L = \sum \ell_i = 4 \times 3.5 = 14 \text{ sec}$$

- Determine Y_i and $\Sigma \ Y_i$.

Lanes:	Phase A (EB)			Phase B (WB)			Phase C (SB)			Phase D (NB)		
	1	2	3	1	2	3	1	2	3	1	2	3
V_i	335	499	499	189	338	338	115	79	37	519	105	217
s_i	2000	2000	2000	2000	2000	2000	2000	2000	2000	2000	2000	2000
V_i/s_i	0.17	0.25	0.25	0.09	0.17	0.17	0.06	0.04	0.019	0.26	0.05	0.11
Y_i		0.25			0.17			0.06			0.26	

$$\sum Y_i = 0.74$$

- Determine the optimum cycle length using Eq. 6.10.

$$C_o = \frac{(1.5 \times 14) + 5}{1 - 0.74} = 100 \text{ sec.}$$

- Find the total effective green time.

$$g_T = (100 - 14) \text{ sec} = 86 \text{ sec.}$$

Assuming amber time $\tau = 3.0$ sec, the green time G_{iw} for each phase is obtained from Eq. 6.14 as

$$G_{iw} = g_i + l_i - 3.0$$

$$\text{Total green time for Phase A } (G_A) = \frac{0.25}{0.74} \times 86 + 3.5 - 3.0 \approx 30 \text{ sec}$$

$$\text{Total green time for Phase B } (G_B) = \frac{0.17}{0.74} \times 86 + 3.5 - 3.0$$

$$\approx 20 \text{ sec}$$

$$\text{Total green time for Phase C } (G_C) = \frac{0.06}{0.74} \times 86 + 3.5 - 3.0$$

$$\approx 7 \text{ sec}$$

$$\text{Total green time for Phase D } (G_D) = \frac{0.26}{0.74} \times 86 + 3.5 - 3.0$$

$$\approx 31 \text{ sec}$$

Pignataro Method. Pignataro has also presented a method to determine the minimum cycle length of a pretimed signal system, based on the total time required for all vehicles to pass through the intersection during the peak 15 min period.[12] The volume going through the intersection is represented by the summation of the number of vehicles on the critical lane for each phase.

Let

V_i = critical lane volume for phase i

τ_i = clearance interval for phase i

n = number of signal cycles for a 15 min period

ϕ' = number of phases for a 15 min period

T = total time in seconds required for all vehicles to go through the intersection during the peak 15 min period

h_i = average headway for phase i on critical lane (lane that carries the critical volume V_i)

C = cycle length (sec)

To obtain T, we first have to compute the peak volume for a 15 min period for each phase. This is obtained by dividing the hourly volume by 4. This volume is then corrected for the peak 15 min period by dividing it by the PHF.
Then,

$$T = \frac{\sum\limits_{i=1}^{\phi'} V_i h_i}{4 \text{ (PHF)}} \tag{6.15}$$

If total clearance time per cycle is $\sum_{i=1}^{\phi} \tau_i$, then total clearance time during a 15 min period can be found.

$$n \sum_{i=1}^{\phi'} \tau_i = n(\tau_1 + \tau_2 + \tau_3 + \cdots \tau_\phi)$$

Since the sum of time T and the total clearance time cannot be greater than 15 min,

$$T + n \sum_{i=1}^{\phi'} \tau_i \leq 900 \qquad (6.16)$$

The limiting condition for Eq. 6.16 is

$$n = \frac{900 - T}{\displaystyle\sum_{i=1}^{\phi'} \tau_i} \qquad (6.17)$$

The number of cycles in a 15 min period, where C is the cycle length, is

$$n = \frac{900}{C} \qquad (6.18)$$

Combining Eqs. 6.17 and 6.18 gives

$$\frac{900}{C} = \frac{900 - T}{\displaystyle\sum_{i=1}^{\phi'} \tau_i} \qquad (6.19)$$

from which we obtain

$$C = \frac{900 \displaystyle\sum_{i=1}^{\phi'} \tau_i}{900 - T}$$

$$= \frac{900 \displaystyle\sum_{i=1}^{\phi'} \tau_i}{900 - \left(\displaystyle\sum_{i=1}^{\phi'} V_i h_i \right) / 4\,(\text{PHF})}$$

Thus,

$$C = \frac{\displaystyle\sum_{i=1}^{\phi'} \tau_i}{1 - \left(\displaystyle\sum_{i=1}^{\phi'} V_i h_i \right) / 3600\,(\text{PHF})} \qquad (6.20)$$

This expression can also be used to determine the minimum cycle length when an all red phase is provided in each cycle. Hence,

$$C = \frac{\displaystyle\sum_{i=1}^{\phi'} \tau_i + R}{1 - \left(\displaystyle\sum_{i=1}^{\phi'} V_i h_i \right) / 3600\,(\text{PHF})} \qquad (6.21)$$

where R is the time of the all red phase during the cycle (sec).

The total available green time is then distributed among the different green phases in proportion to their critical lane volumes as follows:

$$G_i = \frac{V_i}{\sum\limits_{i=1}^{\phi'} V_i} \left[C - \left(\sum\limits_{i=1}^{\phi'} \tau_i + R \right) \right] \tag{6.22}$$

This calculation assumes that the average headway between vehicles is approximately the same for each approach.

It must be noted, however, that the cycle lengths obtained from Eqs. 6.7, 6.10, 6.20, or 6.21 are approximate values. Other factors, such as pedestrian crossing requirements, should be used to check the cycle length and, if necessary, to adjust it to obtain a satisfactory cycle length. For example, if an all red phase is not provided, the green time for a given phase should be at least equal to the time it takes pedestrians to cross the intersection in the direction that the traffic flows for that phase. Similarly, when an all red phase is provided, the length of this phase should be at least equal to the time it takes pedestrians to cross the widest approach. However, pedestrian refuge islands can be provided on very wide approaches to facilitate pedestrian crossing in two stages, thereby reducing the time allocated for pedestrian crossing. The time required for pedestrians to cross any approach of an intersection depends on (1) the volume of pedestrians crossing that approach, (2) the average walking speed of pedestrians, (3) the width of the approach, and (4) pedestrian start up and delay time, which includes the perception-reaction time and the time that passes before all pedestrians waiting to cross the approach start moving.

Example 6–4   Signal Timing Using Pignataro's Method

Figure 6.11(a) shows peak hour volumes and other traffic characteristics for an intersection. Design a suitable signal timing for a two-phase system using the information given.

- Since there is moderately high pedestrian volume, consider the minimum times required to cross the intersection, although it may not be necessary to provide an all red phase.

Constant rate of braking deceleration $a = 0.27$ g
Pedestrian start up and delay time $= 5$ sec
Average length of vehicle $= 20$ ft
Average walking speed $= 4.0$ ft/sec
Average headway $= 2.5$ sec
Perception-reaction time $\delta = 2.5$ sec

Figure 6.11   Signal Timing Using Pignataro Method

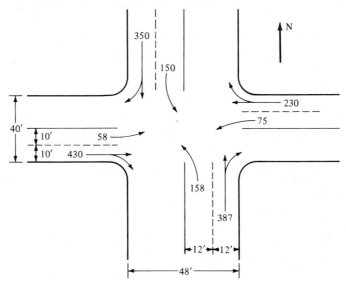

PHF = 0.90
Left-turn factor = 1.6
Truck factor = 1.5
Approach speed on all approaches = 30 mph
Truck percentages

South Approach		North Approach		East Approach		West Approach	
Through	Left	Through	Left	Through	Left	Through	Left
0	0	2	2	0	0	3	3

Pedestrian volume is moderately high.

(a) **Data for Example 6.4**

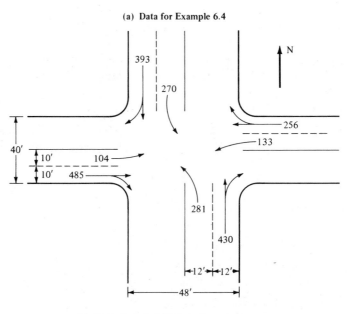

(b) **Equivalent straight- through passenger cars**

- Find the time required to cross a 48 ft wide intersection.

$$5 + \frac{48}{4} = 17 \text{ sec}$$

- Find the time required to cross a 40 ft wide intersection.

$$5 + \frac{40}{4} = 15 \text{ sec}$$

- Find the minimum yellow interval by using Eq. 6.1. For the N–S movement,

$$\tau_{min} = \delta + \frac{W + L}{V_o} + \frac{V_o}{2a}$$

$$= 2.5 + \frac{40 + 20}{30 \times 1.47} + \frac{30 \times 1.47}{2 \times 0.27 \times 32.2}$$

$$= 2.5 + 1.36 + 2.54$$

$$= 6.40 \text{ sec}$$

For the E–W movement,

$$\tau_{min} = 2.5 + \frac{48 + 20}{30 \times 1.47} + \frac{30 \times 1.47}{2 \times 0.27 \times 32.2}$$

$$= 2.5 + 1.54 + 2.54$$

$$= 6.58 \text{ sec}$$

- Allow a clearance interval of 5 sec for each phase (and a 2 sec all red phase). Note that the green time for traffic in the N–S direction must be at least equal to the time required for pedestrians to cross the 40 ft wide road—that is, minimum green time for the N–W direction is $15 - 5 = 10$ sec. Similarly, minimum time for the E–W direction is $17 - 5 = 12$ sec
- Find the minimum cycle length using Eq. 6.21 and Figure 6.11(b).

$$C_{min} = \frac{5 + 5 + 2}{1 - \dfrac{1}{3600} (430 \times 2.5 + 485 \times 2.5)}$$

$$= 33.0 \text{ sec} \quad (\text{say, } 45 \text{ sec})$$

$$\sum V_i = 485 + 430$$

$$= 915$$

- Find the minimum total green time. For the N–S direction,

$$G_{NS} = \frac{430}{915}(45 - 5 - 5 - 2) = 16 \text{ sec}$$

For the E–W direction,

$$G_{EW} = \frac{485}{915}(45 - 5 - 5 - 2) = 17 \text{ sec}$$

These green times satisfy the pedestrian requirements.

Delay at Pretimed Signalized Intersections

One of the main objectives in installing a signal system at an intersection is to reduce the average delay of vehicles at the intersection. Delay is therefore an important measure of effectiveness used in the evaluation of a signalized intersection. Delay at a signalized intersection can be determined by using an expression developed by Webster and given in Eq. 6.23.[13] It gives the average delay experienced per vehicle on the jth approach during the ith phase.

$$d_j = \left(CA + \frac{B}{V_j}\right)\frac{100 - P}{100} \qquad (6.23)$$

where

d_j = average delay per vehicle on jth approach during ith phase

$A = \dfrac{(1 - \lambda_i)^2}{2(1 - \lambda_i x_j)}$

$B = \dfrac{x_j^2}{2(1 - x_j)}$

C = cycle length (sec)

V_j = actual volume on jth approach during ith phase (vehicles/lane/sec)

λ_i = proportion of cycle length that is effectively green (that is, g_i/C, where g_i is effective green time)

x_j = degree of saturation for the jth approach = $V_j/\lambda_i s$

s = saturation flow (vehicles/lane/sec)

P = percentage correction, ranging from 5 percent to 15 percent for normal conditions (see Table 6.3)

The total hourly delay for any approach can be estimated by determining the d_j for each lane on that approach, multiplying each d_j by the corresponding lane volumes, and then summing these. The total intersection hourly delay can then be determined by summing the total delay for each approach. Values for A, B, and P have been calculated and are shown in Tables 6.1, 6.2, and 6.3.

Table 6.1 A Values for Use in Webster Delay Model

x_j	λ_i												
	0.1	0.2	0.3	0.35	0.40	0.45	0.50	0.55	0.60	0.65	0.70	0.80	0.90
0.1	0.409	0.327	0.253	0.219	0.188	0.158	0.132	0.107	0.085	0.066	0.048	0.022	0.005
0.2	0.413	0.333	0.261	0.227	0.196	0.166	0.139	0.114	0.091	0.070	0.052	0.024	0.006
0.3	0.418	0.340	0.269	0.236	0.205	0.175	0.147	0.121	0.098	0.076	0.057	0.026	0.007
0.4	0.422	0.348	0.278	0.246	0.214	0.184	0.156	0.130	0.105	0.083	0.063	0.029	0.008
0.5	0.426	0.356	0.288	0.256	0.225	0.195	0.167	0.140	0.114	0.091	0.069	0.033	0.009
0.55	0.429	0.360	0.293	0.262	0.231	0.201	0.172	0.145	0.119	0.095	0.073	0.036	0.010
0.60	0.431	0.364	0.299	0.267	0.237	0.207	0.179	0.151	0.125	0.100	0.078	0.038	0.011
0.65	0.433	0.368	0.304	0.273	0.243	0.214	0.185	0.158	0.131	0.106	0.083	0.042	0.012
0.70	0.435	0.372	0.310	0.280	0.250	0.221	0.192	0.165	0.138	0.112	0.088	0.045	0.014
0.75	0.438	0.376	0.316	0.286	0.257	0.228	0.200	0.172	0.145	0.120	0.095	0.050	0.015
0.80	0.440	0.381	0.322	0.293	0.265	0.236	0.208	0.181	0.154	0.128	0.102	0.056	0.018
0.85	0.443	0.386	0.329	0.301	0.273	0.245	0.217	0.190	0.163	0.137	0.111	0.063	0.021
0.90	0.445	0.390	0.336	0.308	0.281	0.254	0.227	0.200	0.174	0.148	0.122	0.071	0.026
0.92	0.446	0.392	0.338	0.312	0.285	0.258	0.231	0.205	0.179	0.152	0.126	0.076	0.029
0.94	0.447	0.394	0.341	0.315	0.288	0.262	0.236	0.210	0.183	0.157	0.132	0.081	0.032
0.96	0.448	0.396	0.344	0.318	0.292	0.266	0.240	0.215	0.189	0.163	0.137	0.086	0.037
0.98	0.449	0.398	0.347	0.322	0.296	0.271	0.245	0.220	0.194	0.169	0.143	0.093	0.042

Note: Values of A calculated from

$$A = \frac{(1 - \lambda_i)^2}{2(1 - \lambda_i x_j)}$$

Source: Adapted from F. V. Webster and B. M. Cobbe, *Traffic Signals*, Road Research Technical Paper No. 39, Her Majesty's Stationery Office, London, 1958.

Table 6.2 B Values for Use in Webster Model

x_i	0.00	0.01	0.02	0.03	0.04	0.05	0.06	0.07	0.08	0.09
0.1	0.006	0.007	0.008	0.010	0.011	0.013	0.015	0.017	0.020	0.022
0.2	0.025	0.028	0.031	0.034	0.038	0.042	0.046	0.050	0.054	0.059
0.3	0.064	0.070	0.075	0.081	0.088	0.094	0.101	0.109	0.116	0.125
0.4	0.133	0.142	0.152	0.162	0.173	0.184	0.196	0.208	0.222	0.235
0.5	0.250	0.265	0.282	0.299	0.317	0.336	0.356	0.378	0.400	0.425
0.6	0.450	0.477	0.506	0.536	0.569	0.604	0.641	0.680	0.723	0.768
0.7	0.817	0.869	0.926	0.987	1.05	1.13	1.20	1.29	1.38	1.49
0.8	1.60	1.73	1.87	2.03	2.21	2.41	2.64	2.91	3.23	3.60
0.9	4.05	4.60	5.28	6.18	7.36	9.03	11.5	15.7	24.0	49.0

Note: Values of B calculated from

$$B = \frac{x_j^2}{2(1 - x_j)}$$

Source: Adapted from F. V. Webster and B. M. Cobbe, *Traffic Signals*, Road Research Technical Paper No. 39, Her Majesty's Stationery Office, London, 1958.

Example 6–5 Computing Delay Using the Webster Delay Model

Using the solution obtained for Example 6–3 and Eq. 6.23, determine the total hourly delay of vehicles in lane 1 of the eastbound approach (Phase A).

$$d_j = \left(CA + \frac{B}{V_j} \right) \frac{100 - P}{100}$$

- Use the following values for lane 1.

$$g_1 = \frac{0.25}{0.74} \times 86 = 29.05 \text{ sec}$$

$V_1 = 335$ passenger cars per hour (pc/hr) $= 0.093$ pc/sec
$s = 2000$ pc/hr $= 0.556$ pc/sec

$$\lambda_1 = \frac{29.05}{100} = 0.291$$

$$x_1 = \frac{0.093}{0.291 \times 0.556} = 0.575$$

- Determine A and B either by using Tables 6.1 and 6.2, respectively, or by computing them from the appropriate equations. Using the appropriate equations,

$$A = \frac{(1 - \lambda_i)^2}{2(1 - \lambda_i x_i)} = \frac{(1 - 0.291)^2}{2[1 - (0.291)(0.575)]} = 0.302$$

$$B = \frac{x_1^2}{2(1 - x_j)} = \frac{0.575^2}{2(1 - 0.575)} = 0.389$$

Table 6.3 *P* Values for Use in Webster Model

		M_j							M_j			
x_i	λ_i	2.5	5	10	20	40	x_i	λ_i	2.5	5	10	20	40
0.3	0.2	2	2	1	1	0	0.8	0.2	18	17	13	10	7
	0.4	2	1	1	0	0		0.4	16	15	13	10	9
	0.6	0	0	0	0	0		0.6	15	15	14	12	9
	0.8	0	0	0	0	0		0.8	14	15	17	17	15
0.4	0.2	6	4	3	2	1	0.9	0.2	13	14	13	11	8
	0.4	3	2	2	1	1		0.4	12	13	13	11	9
	0.6	2	2	1	1	0		0.6	12	13	14	14	12
	0.8	2	1	1	1	1		0.8	13	13	16	17	17
0.5	0.2	10	7	5	3	2	0.95	0.2	8	9	9	9	8
	0.4	6	5	4	2	1		0.4	7	9	9	10	9
	0.6	6	4	3	2	2		0.6	7	9	10	11	10
	0.8	3	4	3	3	2		0.8	7	9	10	12	13
0.6	0.2	14	11	8	5	3	0.975	0.2	8	9	10	9	8
	0.4	11	9	7	4	3		0.4	8	9	10	10	9
	0.6	9	8	6	5	3		0.6	8	9	11	12	11
	0.8	7	8	8	7	5		0.8	8	10	12	13	14
0.7	0.2	18	14	11	7	5							
	0.4	15	13	10	7	5							
	0.6	13	12	10	8	6							
	0.8	11	12	13	12	10							

Note: M_j is the average actual flow per lane per cycle for the jth approach.

$$M_j = V_j C$$

Source: Adapted from F. V. Webster and B. M. Cobbe, *Traffic Signals*, Road Research Technical Paper No. 39, Her Majesty's Stationery Office, London, 1958.

- Find *M* in order to compute *P*.

$$M_j = V_j C = 0.093 \times 100 = 9.3$$

From Table 6.3, we obtain $P = 7$ by interpolating and rounding.
- Use Eq. 6.23.

$$d_1 = \left(100 \times 0.302 + \frac{0.389}{0.093} \right) \frac{100 - 7}{100}$$

$$= 31.98 \text{ sec/vehicle}$$

- Calculate the total hourly delay in lane 1.

$$d_1 = d_{T1} = d_1 \times \text{hourly volume in lane 1}$$

$$= 31.98 \times 222 \text{ sec [note hourly volume is given in Figure 6.9(a)]}$$

$$= 1.97 \text{ hr}$$

Cycle Lengths of Actuated Traffic Signals

A major disadvantage of fixed or pretimed signals is that they cannot adjust themselves to handle fluctuating volumes. When the fluctuation of traffic volumes warrants it, a vehicle-actuated signal is used. These signals are capable of adjusting themselves and therefore do not have the same disadvantage as the pretimed signals. When a vehicle-actuated signal is used, vehicles arriving at the intersection are detected by detectors, which then transmit this information to a controller. The controller then adjusts the phase lengths to meet the requirements of the prevailing traffic condition.

Let us first define certain terms associated with actuated signals before discussing the different design procedures.

Initial portion is that first portion of the green phase that an actuated controller has timed out for vehicles waiting between the detector and the stop line during the red phase.

Extendible portion is that portion of the green phase that follows the initial portion, to allow for more vehicles arriving between the detector and the stop line during the green phase to go through the intersection.

Unit extension is the minimum time by which a green phase could be increased during the extendible portion after an actuation on that phase. The total extendible portion should, however, not exceed the extension limit.

Extension limit is the maximum additional time that can be given to the extendible portion of a phase after actuation on another phase.

Minimum period is the shortest time that should be provided for a green interval during any traffic phase.

Demand is a request for the right of way by a traffic stream through the controller.

Semiactuated Signals. Actuated signals can be either semiactuated or fully actuated. A semiactuated signal uses detectors only in the minor stream flow. This system can be installed even when the minor stream volume does not satisfy the volume requirements for signalization. The operation of the semiactuated signal is based on the ability of the controllers to vary the lengths of the different phases to meet the demand on the minor approach. The signals are set as follows.

1. The green signal on the major approach is preset for a minimum period, but it will continue to be on until the signal is actuated by a minor stream vehicle.

2. If the green signal on the major approach has been on for a period equal to or greater than the preset minimum, then the signal will change to red in response to the actuation caused by the minor street vehicle.

3. The green signal on the minor stream will then come on for at least a period equal to the preset minimum for this stream. This minimum is given an extendible green for each vehicle arriving up to a preset extension limit.

4. The signal on the minor stream then changes to red and that on the major stream changes to green.

Note that when the volume is high on the minor stream, the signal becomes a pretimed one.

The operation of a semiactuated signal requires certain times to be set for both the minor and major streams. For the minor streams, times should be set for the initial portion, unit extension, maximum green (sum of initial and extendible portions), and clearance interval. For the major streams, times should be set for the minimum green and clearance interval. When pedestrian actuators are installed, it is also necessary to set a time for pedestrian clearance. Several factors should be taken into consideration when setting these times. The major factor, however, is that a semiactuated signal works as a pretimed signal during peak periods. It is therefore important that the time set for the maximum green in the minor stream be adequate to meet the demand during the peak period. Similarly, the time set for the minimum green on the major approach should be adequate to provide for the movement through the intersection of the expected number of vehicles waiting between the stop line and the detector whenever the signals change to green during the peak period. These settings should, however, not be so large that the resulting cycle length becomes undesirable. In general, the following procedures can be used to obtain some indication of the required lengths of the different set times.

Unit Extension. The unit extension depends on the average speed of the approaching vehicles and the distance between the detectors and the stop line. The unit extension should be at least the time it takes a vehicle traveling at the average speed to travel from the location of the detectors to the stop line. Let u = average speed (mph) and X = distance between detectors and stop line (ft). Therefore,

$$\text{unit extension} = \frac{X}{1.47u} \, (\text{sec})$$

This time will allow a vehicle detected at the end of the initial portion to arrive at the stop line just as the signal is changing to amber and to clear the intersection during the clearance interval.

If, however, the desire is to provide a unit extension time that will also allow the vehicle to clear the intersection, then

$$\text{unit extension time} = \frac{X + W + L}{1.47u} \, (\text{sec})$$

where
W = width of the cross street (ft)
L = length of the vehicle

Initial Portion. This time should be adequate to allow vehicles waiting between the stop line and detector during the red phase to clear the intersection. This time depends on the number of vehicles waiting, the average headway, and the starting delay. The time for the initial portion can be obtained as

$$\text{initial portion} = (hn + K_1)$$

where

h = average headway

n = number of vehicles waiting between the detectors and the stopline

K_1 = starting delay

Suitable values for h and K_1 are 2 sec and 3.5 sec, respectively.

Minimum Green. This is given as the sum of the initial portion and the unit extension.

Fully Actuated Signals. Fully actuated signals are suitable for intersections at which large fluctuations of traffic volumes exist on all approaches during the day. Maximum and minimum green times are set for each approach. The basic operation of the fully actuated signal can be described using an intersection with four approaches and a two-phase signal system. Let phase A be assigned to the north–south direction and phase B to the east–west direction. If phase A is given the right of way, the green signal will continue to be on until the minimum green time is expired and an approaching vehicle in the east–west direction actuates one of the detectors for phase B. A demand for the right of way in the east–west direction is then registered. If there is no traffic on the north–south direction, the red signal will come on for phase A and the right of way will be given to vehicles on the east–west direction—that is, the green indicator will come on for phase B. This right of way will be held by phase B until at least the minimum green time is expired. At the expiration of the minimum green time, the right of way will be given to phase A, that is, the north–south direction, only if during the period of the minimum green, a demand is registered by an approaching vehicle in this direction. If no demand is registered in the north–south direction and vehicles continue to arrive at the east–west direction, the right of way will continue to be given to phase B until the maximum green is reached. At this time the right of way is given to phase A and so on. The procedures used to determine the lengths of the different set times are similar to that described for semiactuated signals.

Signal Timing of Arterial Routes

In urban areas where two or more intersections are adjacent to each other, the signals should be timed so that, when a queue of vehicles is released by receiving the right of way at an intersection, these vehicles will also have the right of way at the adjacent intersections. This coordination will reduce the delay vehicles experience on the arterial. To obtain this coordination, all intersections in the system must have the same cycle length. In rare instances, however, some intersections in the system may have cycle lengths equal to half or twice the common cycle length. It is usual for a common cycle length to be set, with an offset that is suitable for the main street. Traffic conditions at a given intersection are used to determine the appropriate phases of green, red, and amber times for that intersection. The methods used to achieve the required coordination are simultaneous system, alternate system, and progressive system. Before discussing these different systems, it is first necessary to define certain terms that are used frequently in arterial signal timing.

Speed of progression (*band speed*) is the speed at which a platoon of vehicles released at an intersection will proceed along the arterial. It is usually taken as the mean operating speed of vehicles on the arterial for the specific time of day being considered. This speed is represented by the ratio of the distance between the traffic signals and the corresponding travel time.

Through band is the space delineated by two progressive speed lines on a time–space diagram as shown in Figure 6.12.

Figure 6.12 Typical Time–Space Diagram

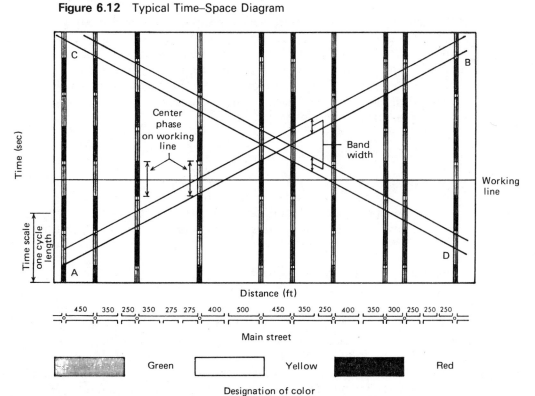

Source: Reproduced from *Transportation and Traffic Engineering Handbook*, 2nd ed., Institute of Transportation Engineers, Washington, D.C., 1982.

Let us now describe the three methods normally used for traffic signal coordination.

Simultaneous System

In a simultaneous system, all signals along a given arterial have the same cycle length and have the green phase showing at the same time. When given the right of way, all vehicles move at the same time along the arterial and stop at the nearest signalized intersection when the right of way is given to the side streets. A simple approximate mathematical relationship can be developed for this system.

$$u = \frac{X}{1.47C} \tag{6.24}$$

where

X = average spacing for signals (ft)
u = progression speed (mph)
C = cycle length (sec)

Alternate System

With the alternate system, intersections on the arterial are formed into groups of one or more adjacent intersections. The signals are then set such that successive groups of signals are given the right of way alternatively. This system is known as the single alternate when the groups are made up of individual signals—that is, each signal alternates with those immediately adjacent to it. It is known as double alternate when the groups are made up of two adjacent signals, and so on. The band speed in a single alternate system is given as

$$u = \frac{2X}{1.47C} \tag{6.25}$$

The same expression is used for the double alternate, but in this case the distance X should represent the midpoints of adjacent pairs. The alternate system is most effective when the intersections are at equal distance from each other.

Example 6–6   Choosing an Appropriate Alternate System for an
                  Urban Arterial

The traffic signals on an urban arterial are to be coordinated to facilitate the flow of traffic. The intersections are spaced at approximately 500 ft intervals, with at least one intersection being part of another coordinated system, having a cycle length of 60 sec. Determine whether a single-, double-, or triple-alternate system is preferable for this arterial if the mean velocity on the arterial is 35 mph.

Since there is one intersection in another coordinated system, it is necessary to use the cycle length at that intersection—that is, $C = 60$ sec. We also want to use Eq. 6.24.

- Find the mean speed for the single-alternate system.

$$u = \frac{2 \times 500}{1.47 \times 60} = 11.34 \text{ mph}$$

- Find the mean speed for the double alternate system.

$$u = \frac{4 \times 500}{1.47 \times 60} = 22.68 \text{ mph}$$

- Find the mean speed for the triple alternate system.

$$u = \frac{6 \times 500}{1.47 \times 60} = 34.01 \text{ mph}$$

Since the mean speed is currently 35 mph, the triple alternate system is preferable as this requires a progressive speed that is approximately equal to the existing mean speed.

Progressive System

The progressive system provides for a continuous flow of traffic through all intersections under the system when traffic moves at the speed of progression. The same cycle length is used for all intersections, but the green indication for each succeeding intersection is offset a given time from that of the preceding intersection, depending on the distance from the preceding intersection and the speed of progression for that section of the street. When the offset and cycle length are fixed, the system is known as the limited or simple progressive system, and when the offset and cycle length can be changed to meet the demands of fluctuating traffic at different times of the day, it is known as a flexible progressive system.

Design of Progressive Signal System. The design of a progressive signal system involves the selection of the best cycle length, using the criterion that the band speeds are approximately equal to the mean operating speed of the vehicles on the arterial street. This selection is accomplished by a trial-and-error procedure. Eq. 6.24 can be used to obtain an indication of a suitable cycle length by using the mean operating speed of the arterial for u and the measured distance between intersections as X. In addition, the required cycle length for each intersection should be computed and compared with that obtained from Eq. 6.24. If this cycle length is approximately equal to those obtained for the majority of the intersections, it can be selected on a trial basis. It is the usual practice, however, to use cycle lengths that have been established for intersecting or adjacent systems as a guide for selecting a suitable cycle length. The actual design of the progressive system is normally done by using one of several available computer software packages. The first step in any method used is the collection of adequate data on traffic volumes, intersection spacings, speed limits, on-street parking, operating speed, and street geometrics.

Computer programs have been developed to cope with several problems associated with progressive signal timing, such as large variation in distance between intersections, difference in speeds between traffic streams in opposite directions, variable speed patterns at different sections of the system, and the requirement of a high level of computational effort. The use of computers to reduce the computational effort and increase analysis flexibility has made the design of signalized arterial systems less taxing.

The earliest computer programs were based on the time–space relation, with the through band being maximized. One of the first programs in this group was SIGART.[14] This program determines all possible through bands that are greater than a specified minimum. The inputs include the cycle length (sec), the number and spacings of the signalized intersections being considered, and three speed values—the desired speed, the maximum speed to be considered, and the minimum speed that will provide for an acceptable minimum band width.

The same primary objective—that is, maximization of through band width—is used in later programs, such as Maximum Band Width (MAXBAND) and Progression Analysis and Signal System Evaluation Routine (PASSER).[15] These

---

Figure 6.13 AAP/M Output and Space Diagram for a Seven-Intersection Arterial

ROUTE: ROUTE 29N CHARLOTTESVILLE

INTERSECTIONS: 7   CYCLE LENGTH: 60   SYSTEM OFFSET: 0

0.	.... TIME-LOCATION DIAGRAM	DISTANCE		SPEED	
	RIGHTBOUND ... READ DOWN	LEFT	RIGHT	LEFT	RIGHT
1	XXXXXXXXXXXXXXX	300	0	35	35
2	XXXXXXXXXXXXXXX	500	300	35	35
3	XXXXXXXXXXXXXXX	350	500	35	35
4	XXXXXXXXXXXXXXX	550	350	35	35
5	XXXXXXXXXXXXXXX	400	550	35	35
6	XXXXXXXXXXXXXXX	350	400	35	35
7	XXXXXXXXXXXXXXX	0	350	35	35

0. OFFSET	.... TIME-LOCATION DIAGRAM	PHASE SPLITS (%)							
(PERCENT)	LEFTBOUND ... READ UP	1	2	3	4	5	6	7	8
1 44	XXXXXXXXXXXXXXX	60	40						
2 54	XXXXXXXXXXXXXXX	60	40						
3 70	XXXXXX XXXXXXXXX	60	40						
4 80	XXXXXXXXXXXXXXX	60	40						
5 0	XXXXXXXXXXXXXX	60	40						
6 13	X XXXXXXXXXXXXXX	60	40						
7 24	XXXXXXXXXX XXXXX	60	40						

Continued

Figure 6.13—*Continued*

TIME-SPACE DIAGRAM

ROUTE: ROUTE 29N CHARLOTTESVILLE

COMMENT: JACK'S PROJECT
CYCLE LENGTH 60 SECONDS: SCALE 1 INCH = 40% OF CYCLE: 1 LINE = 61 FT

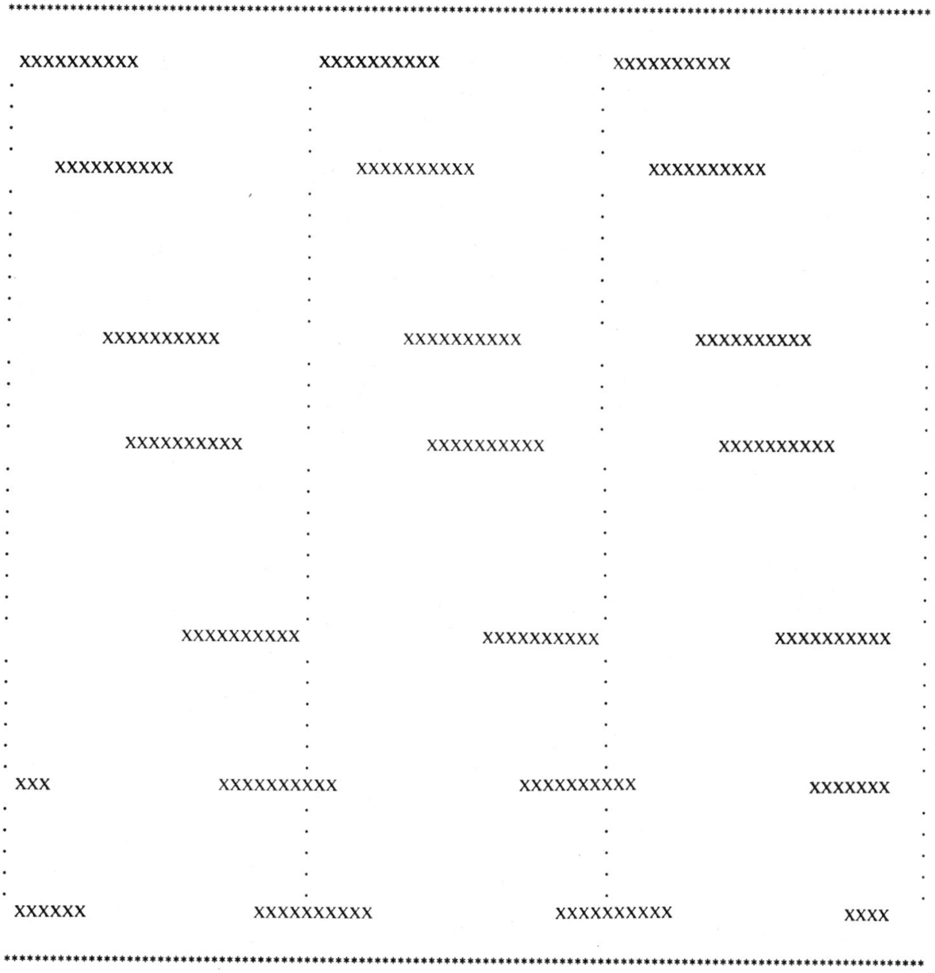

later models take into consideration traffic demands in the determination of cycle lengths and green phases. They also optimize phase lengths for a range of cycle lengths and alternative phase sequences.

With the advent of microcomputers, several programs have been developed that can be used to design an arterial signalized system. One such program, Arterial Analysis Package/Microcomputer (AAP/M), was developed by the University of Florida Transportation Research Center for Tampa's Division of

Traffic Engineering.[16] This program can be used as a computerized tool to design and evaluate signal timing for arterial traffic control systems. Three existing traffic signal timing programs are combined in this single package: SOAP, which deals with individual intersections; PASSER II, which is used to optimize arterial progression; and TRANSYT 7F, which optimizes the stops and delay performance of a coordinated traffic control system.

The main feature of the AAP/M is that it provides for the microcomputer to be interfaced with a mainframe computer system, although the interfacing with the mainframe is really only needed for the TRANSYT program. The design and analysis of simple problems can also be carried out without the mainframe computer. When the microcomputer is interfaced with the mainframe, the AAP/M acts as an intelligent terminal to the mainframe, which allows for the conversational entry and retrieval of data, as well as graphical display of the system operation.

Inputs for the AAP/M include the common cycle length, the number of intersections, and the distances between successive intersections. Output includes the offset for each intersection as a percentage of the cycle length. Figure 6.13 shows the output for an arterial consisting of seven intersections. The figure also shows the time–space diagram obtained. There is no doubt that other signal optimization programs will be developed for microcomputers as the use of these computers becomes more widespread.

FREEWAY RAMPS

Ramps are usually part of grade-separated intersections, where they serve as interconnecting roadways for traffic streams at different levels. They are also sometimes constructed between two parallel highways to allow vehicles to change from one highway to the other. Freeway ramps can be divided into two groups: entrance ramps, which provide for the merging of vehicles into the freeway stream; and exit ramps, which provide for vehicles to leave the freeway stream. When it becomes necessary to control the number of vehicles entering or leaving a freeway at a particular location, the access to the entrance or exit ramp is controlled in one of several ways.

Freeway Entrance Ramp Control

The control of entrance ramps is essential to the efficient operation of freeways, particularly when volumes are high. The main objective in controlling entrance ramps is to regulate the number of vehicles entering the freeway so that the volume is kept lower than the capacity of the freeway. This will ensure that freeway traffic moves at a speed approximately equal to the optimum speed that will result in maximum flow rates. The control of entrance ramps also provides for a better level of service on the freeway and a safer overall operation of both the freeway and the ramp. On the other hand, entrance ramp control may lead to long queues on the ramps, formed by vehicles waiting to join the freeway traffic stream, or to the diversion of traffic to local roads, which may result in serious congestion on the local roads. It is therefore essential that the control of freeway entrance ramps be undertaken only when certain conditions are satisfied. The MUTCD gives general guidelines for the successful application of ramp control.[17] The following guidelines are mainly qualitative due to the lack of adequate data on which numerical warrants can be based.

1. Installation of freeway entrance ramp control signals may be warranted when:

    a. The expected reduction in delay to freeway traffic exceeds the expected delay to ramp users and added travel time for diverted traffic and traffic on alternative surface routes.

    b. There is adequate storage space for the vehicles that will be delayed.

    c. There are suitable alternate surface routes available having capacity for traffic diverted from the freeway ramps.

    d. There is recurring congestion on the freeway due to traffic demand in excess of the capacity; or there is recurring congestion or a severe accident hazard at the freeway entrance because of an inadequate ramp-merging area.

2. Installation of freeway entrance ramp control signals may be warranted to reduce sporadic congestion on isolated sections of the freeway caused by short period peak traffic loads from special events or from severe peak loads of recreational traffic.

    Guidelines for the design, implementation, and operation of entrance ramp control systems were also produced by the Stanford Research Institute, based on the results of a National Cooperative Highway Research Program study.[18]

Methods for Controlling Freeway Entrance Ramps

The methods used in controlling freeway entrance ramps are

- Closure
- Simple metering
- Traffic response metering
- Gap acceptance
- Integrated systems control

Closure.   Closure entails the physical closure of the ramp by using Do Not Enter signs or by placing barriers at the entrance to the ramp. This is the simplest form of ramp control, but it is unfortunately the most restrictive. It should therefore be used only when absolutely necessary. Factors that suggest the use of closure include inadequacy of storage area on the ramp for entering vehicles, improper design of merging area connecting ramp traffic and freeway traffic, and volume of traffic on freeway being at or approaching capacity. Experience has shown that the use of signs is relatively ineffective when compared to barriers. Barriers can be either manually placed or automated. Manually placed barriers are labor intensive and therefore are not very efficient for closure at specific times of the day, for example, peak hours over a long period of time. Automatic barriers, however, provide the flexibility of opening and closing of the ramp in response to traffic conditions.

Simple Metering.   This form of control consists of setting up a pretimed signal with extremely short cycles at the ramp entrance. The time settings are usually made for different times of the day and/or days of the week. Simple metering can be used to reduce the flow of traffic on the ramp from about 1200 vph, which is the normal capacity of a properly designed ramp, to about 250 vph. Figure 6.14 shows the layout of a typical simple metering system for an entrance ramp, including some optional features that can be added to the basic system. The basic system consists of a traffic signal, with a two-section (green–red) or three-section (green–yellow–red) indicator located on the ramp, a warning sign, which informs motorists that the ramp is being metered, and a controller, which is actuated by a time clock. The detectors shown are optional but when used will enhance the efficient operation of the system. For example, the check-in detector provides for the signal to change to green only when a vehicle is waiting, which means that the signal will continue to be red until a vehicle is detected by the check-in detector and the minimum red time has elapsed. The check-out detector is useful when a single entry system is desired. The green interval is terminated immediately when a vehicle is detected by the check-out detector.

The calculation of the metering rate depends on the primary objective of the control. When this objective is to reduce congestion on the highway, the difference between the downstream volume and the upstream capacity is used as the metering rate. This is demonstrated in Figure 6.15, where the metering rate is 400 vph. It must be remembered, however, that the guidelines given by the MUTCD must be taken into consideration.[19] For example, if the storage space on

Figure 6.14   Layout of Pretimed Entrance Ramp Metering System

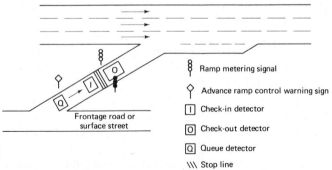

Source: Reproduced from *Transportation and Traffic Engineering Handbook*, 2nd ed., Institute of Transportation Engineers, Washington, D.C., 1982.

Figure 6.15 Relations Between Metering Rate, Upstream Demand, and Downstream Capacity

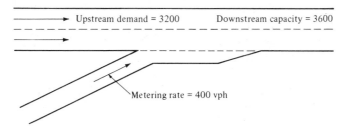

Upstream demand = 3200   Downstream capacity = 3600

Metering rate = 400 vph

the ramp is not adequate to accommodate this volume, the signal should be pretimed for a metering rate that can be accommodated on the ramp. When the objective is to enhance safety at the merging area of the ramp and the freeway, the metering rate will be such that only one vehicle at a time is within the merging area. This will allow for an individual vehicle to merge into the freeway traffic stream before the following vehicle reaches the merging area. The metering rate in this case will depend on the average time it takes a stopped vehicle to merge, which in turn depends on the ramp geometry and the probability of an acceptable gap occurring in the freeway stream. If it is estimated that it takes 9 sec to merge on the average, then the metering rate is 3600/9—that is, 400 vph.

One of two methods of metering can be used in this system, depending on the magnitude of the metering rate. When the rate is below 900 vph, single entry is used; platoon entry is used for rates higher than 900 vph. Single entry allows for only one vehicle to merge into the freeway stream during the green interval, whereas the platoon entry allows for the release of two or more vehicles per cycle. Platoon entry can either be parallel, where two vehicles abreast of each other on two parallel lanes are released, or tandem, where vehicles are released one behind the other. Care should be taken in designing the green interval for tandem platooning; it must be long enough to allow all the vehicles in the platoon to pass the signal.

Traffic Response Metering. This control system is based on the same principles as the simple metering system, but the traffic response system uses actual current information on traffic conditions to determine the metering rates, whereas historical data on traffic volumes are used to determine metering rates in the simple metering system. The traffic response system therefore has the advantage of being capable of responding to short-term changes in traffic conditions. Figure 6.16 shows the basic requirements for a traffic responsive metering system, with some optional features that can be added to the system. The basic requirements include a traffic signal, detectors, a ramp control sign, and a controller that can monitor the variation of traffic conditions. The optional features include the queue detector, which when continuously actuated indicates that the vehicles queued on the ramp may interfere with traffic on the local road; a merging detector, which indicates whether a merging vehicle is still in the merging area; and a check-out detector, which indicates whether a vehicle uses the green interval to proceed to

Figure 6.16 Layout of Traffic-Responsive Ramp Metering System

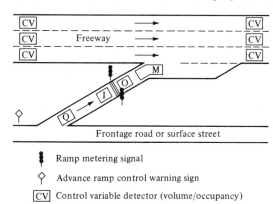

![signal]	Ramp metering signal
![sign]	Advance ramp control warning sign
CV	Control variable detector (volume/occupancy)
Q	Queue detector
I	Check-in detector
O	Check-out detector
M	Merge detector (used primarily where merge geometrics are substandard)
\\\	Stop line

Source: Redrawn from D. R. Drew, K. A. Brewer, J. H. Buhr, and R. H. Whitson, "Multilevel Approach to the Design of a Freeway Control System," *Highway Research Record* 9, 1969.

the merging area. Two methodologies are used to determine the metering rate: (1) the demand capacity control and (2) the occupancy control.

Demand Capacity Control.   In this method, the actual upstream volume is measured at regular short intervals and then is compared with the downstream capacity, which may be preset or calculated using downstream traffic conditions. To ascertain whether the freeway is operating under congested or free-flow conditions, occupancy measurements are also made from at least one upstream detector.

Occupancy Control.   This method uses a predetermined relationship between occupancy rate and lane volume developed from data previously collected at the freeway adjacent to the ramp being considered. An example of such a relationship is shown in Figure 6.17. This relationship also gives the occupancy rate at capacity. A metering rate is selected for the subsequent control period (usually 1 min), based on the occupancy rate measured during the current period either upstream or downstream of the ramp. The metering rate selected is the difference between the previously determined capacity of the freeway and the current volume. The single entry metering system is normally used, except when metering rates are higher than 900 vph.

Gap Acceptance.   This control system is based on the maximum use of acceptable gaps by ramp vehicles. The objective in the method is to minimize the conflict of vehicles at the merging area by ensuring that a ramp vehicle arrives at the

Figure 6.17   Plot of Lane Volume Versus Lane Occupancy

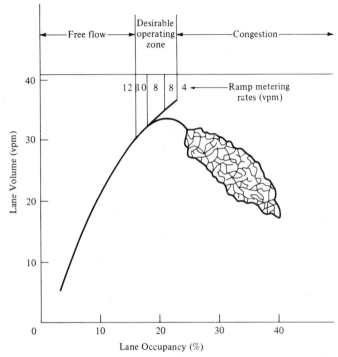

Source: Redrawn from J. M. McDermott et al., *Chicago Area Expressway Surveillance and Control*, Final Report, State of Illinois, Department of Transportation, March 1979.

merge area at the same time an acceptable gap occurs. The process starts with the detection of an acceptable gap upstream of the ramp and the determination of the time this gap will arrive at the merging area (see Figure 6.18). A ramp vehicle is then released so that it arrives at the merging area at the same time as the detected acceptable gap, as shown in Figure 6.18. this system of ramp control requires expensive ·equipment, therefore, it is not widely used since it is questionable whether the additional benefits warrant the added cost over that of other methods described earlier.

Integrated System Control.   The philosophy used in this system is that it is more effective to integrate the control on several ramps rather than dealing with individual ramps independently. The metering rates are therefore based on the consideration of the demand and the capacity of the entire stretch of the freeway being considered, rather than the traffic considerations at individual ramps. A detailed description of the design of an integrated control system is beyond the scope of this book, but interested readers may refer to the Stanford Research Institute publication for a detailed treatment of the subject.[20] This system is usually applied to a group of ramps that are installed with either the traffic-response or pretimed metering devices. When the traffic-response system is

Figure 6.18   Gap Acceptance Mode of Ramp Control

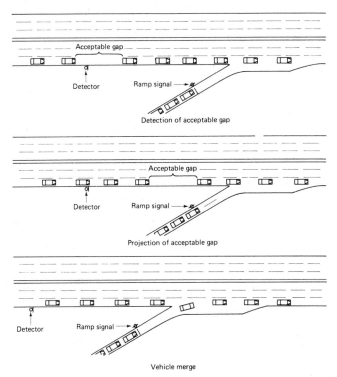

Source: Reproduced from *Transportation and Traffic Engineering Handbook*, 2nd ed., Institute of Transportation Engineers, Washington, D.C., 1982.

installed, linear programming is used to determine a set of integrated metering rates for each ramp, based on the expected range of capacity and demand conditions. The proper metering rate at each intersection is then selected from the appropriate set of metering rates, based on current traffic conditions on the freeway.

Freeway Exit Ramp Control

The control of exit ramps can be used to reduce the flow of traffic from the freeway to congested local streets, to reduce weaving when the distance between an entrance ramp and an exit ramp is short, and to reduce the volume of traffic on the freeway beyond the point of lane-drop by encouraging more traffic to leave the freeway before the lane-drop. Care should be taken in using the control of exit ramps as this can result in queuing of vehicles onto the freeway and increasing the risk of rear-end collisions. The control of exit ramps can be achieved by using either a metering system or by closing the ramp. Closure of exit ramps usually results in an increase of travel time for some motorists, a situation not likely to be appreciated by them.

SUMMARY

The control of traffic at highway intersections is of fundamental importance to traffic engineers engaged in finding ways to achieve efficient operation of any highway system. Several methods of controlling traffic at intersections are presented in this chapter. When traffic conditions warrant it, signalization may be considered an effective means of traffic control at intersections since the number of conflict points at an intersection could be significantly reduced with signalization. Note, however, that an attempt to significantly reduce the number of conflict points will increase the number of phases required, which may result in increased delay. The methods commonly used to determine the cycle and phase lengths of signal systems are discussed, together with the methodologies for determining the important parameters for coordinated signals. Any of these methods could be used to obtain reasonable values of cycle lengths for different traffic conditions at intersections. The method described in the *Highway Capacity Manual of 1985* is presented in Chapter 8 (Level of Service at Signalized Intersections) because that method is used primarily for determining the level of service at an intersection, although it could be used for determining the cycle length and phase lengths for a signalized system.

Ramp control can be used to limit the number of vehicles entering or leaving an expressway at an off- or on-ramp. Care should be taken, however, in using ramp control, as this may lead to longer travel times or congestion on the local streets.

Although mathematical algorithms are presented for computing the important parameters required for signalization and ramp control, traffic engineers must always be aware that a good design is based on both correct mathematical computation and use of engineering judgment.

PROBLEMS

6–1 Using an appropriate diagram, identify all the possible conflict points at an unsignalized *T*-intersection.

6–2 Assuming that a two-phase signal system is installed at the intersection of problem 6–1, with channelized left-turn lanes and shared through and right-turn lanes, using a suitable diagram, determine the possible conflict points. Indicate the phasing system used.

6–3 Under what conditions will you recommend the use of each of the following intersection control devices at urban intersections: (a) yield sign, (b) stop sign, and (c) multiway stop sign?

6–4 Both accident rates and traffic volumes at an unsignalized urban intersection have steadily increased during the past few years. Briefly describe the types of data you will collect, and how you will use that data to justify the installation of signal lights at the intersection.

6–5 For the geometric and traffic characteristics shown below, determine a suitable signal phasing system and phase lengths for the intersection using the failure-rate method, assuming regular arrivals. Show a detailed layout of the phasing system and the intersection geometry used.

| | North Approach 56 ft | South Approach 56 ft | East Approach 68 ft | West Approach 68 ft |
Approach width:	56 ft	56 ft	68 ft	68 ft
Peak-hour approach volume				
Left turn	133	73	168	134
Through movement	420	373	563	516
Right turn	140	135	169	178
PHF	0.95	0.95	0.95	0.95
Vehicle mix				
Passenger cars	97%	97%	98%	98%
Trucks	3%	3%	2%	2%
Conflicting pedestrian volume	300 peds/hr	350 peds/hr	475 peds/hr	450 peds/hr

Left turn factor $= 1.4$ $\tau = 4$ sec $l = 3.5$ sec

PCE for trucks $= 1.6$ $K = 6$ sec $h = 2$ sec $s = 2000$ vph

6–6 Repeat problem 6–5, assuming a Poisson distribution of vehicle arrivals and that failure rate should not be greater than 35 percent during the peak hour.

6–7 Repeat problem 6–5, using the Webster method.

6–8 Repeat problem 6–5, using the Pignataro method.

6–9 Using the Webster delay model, determine the total hourly delay for each of the solutions of problems 6–5, 6–6, and 6–7.

6–10 Using the results for problems 6–5, 6–7, and 6–8, compare the three different approaches used for computing the cycle length. Assume lost time per phase $= 2$ sec.

6–11 A four-leg signalized intersection has a two-phase signal system, with a cycle length of 60 sec and the following critical lane volume.

North approach: 450 vph
South approach: 500 vph
East approach: 350 vph
West approach: 475 vph

If the demand and the capacity are equal, determine the value of K in Eq. 6.7.

6–12 Briefly describe the different ways the signal lights at the intersection of an arterial route could be coordinated, stating under which condition you will use each of them.

6–13 Briefly discuss the different methods by which freeway entrance ramps can be controlled. Clearly indicate the advantages and/or disadvantages of each method and give the conditions under which each of them can be used.

REFERENCES

1. *Manual on Uniform Traffic Control Devices*, U.S. Department of Transportation, Federal Highway Administration, Washington, D.C., 1978.

2. Ibid.

3. *Intersection Channelization Design Guide*, National Cooperative Highway Research Program Report 279, National Research Council, Transportation Research Board, Washington, D.C., November 1985.

4. *Manual of Traffic Signal Design*, Institute of Transportation Engineers, Washington, D.C., 1982.

5. *Manual on Uniform Traffic Control Devices.*

6. Ibid.

7. *Highway Capacity Manual*, Special Report 209, National Research Council, Transportation Research Board, Washington, D.C., 1985.

8. D. Drew and C. Pinnell, "A Study of Peaking Characteristics of Signalized Urban Intersections as Related to Capacity Design," *Highway Research Board Bulletin* 352(1962).

9. Ibid.

10. Ibid.

11. F. V. Webster and B. M. Cobbe, *Traffic Signals*, Road Research Technical Paper No. 39, Her Majesty's Stationery Office, London, 1958.

12. Louis J. Pignataro, *Traffic Engineering Theory and Practice*, Prentice-Hall, Englewood Cliffs, N.J., 1973.

13. F. V. Webster and B. M. Cobbe, *Traffic Signals.*

14. *SIGART Program Documentation Manual*, Metropolitan Toronto Roads and Traffic Department, Toronto, April 1965.

15. C. J. Messer, H. E. Haenel, and E. A. Leoppe, *A Report on the User's Manual for Progression Analysis and Signal System Evaluation Routine PASSER II*, College Station Report 165–14, Texas Transportation Institute, College Station, Tex., August 1974.

16. University of Florida, Transportation Research Center, *Arterial Analysis Package/Microcomputer Version AAP/M*, Mctrans Microcomputer Applications in Transportation Engineering, vol. 1, Center for Microcomputers in Transportation, University of Florida, Gainesville, Fla., January 1983.

17. *Manual on Uniform Traffic Control Devices.*

18. Stanford Research Institute, *Guidelines for the Design and Operation of Ramp Control Systems*, NCHRP Project 3–22, National Research Council, Transportation Research Board, Washington, D.C., 1975.

19. *Manual on Uniform Traffic Control Devices.*

20. Stanford Research Institute, *Guidelines.*

CHAPTER 7

Highway Capacity and Level of Service

In Chapter 5 we used the fundamental diagram of traffic flow to show the relationship between flow and density. It was shown that traffic flows reasonably well when the flow rate is less than optimum (capacity), but excessive delay and congestion occur when the flow rate is at or near capacity. This phenomenon is a primary consideration in the planning and design of highway facilities, in that a main objective is to design or plan facilities that will operate at flow rates below their optimum rates. This objective can be achieved only if a good estimate of the optimum flow of a facility can be made. Capacity analysis therefore involves the quantitative evaluation of the capability of a road section to carry traffic. It uses a set of procedures to determine the maximum flow of traffic that a given section of highway will carry under prevailing traffic and control conditions.

It was also shown in Chapter 5 that the maximum safe speed that can be achieved on a section of highway is the mean free speed, which depends solely on the physical characteristics of the highway. This speed can be achieved only when traffic demand volume tends to zero and there is no interaction between any two vehicles on the highway segment. A driver is then free to drive at his or her desired speed up to the mean free speed. Under these conditions, motorists have a very high perception of the quality of flow on the highway. As the demand volume increases, however, vehicle interaction increases, resulting in the gradual lowering of the speed that can be safely achieved by the drivers. As the interaction among vehicles increases, the drivers are increasingly influenced by the action of other drivers, and individual drivers find it more difficult to achieve their desired speeds. Motorists therefore perceive a lowering of the quality of flow as demand volume on the highway increases. In other words, for a given capacity, the level of operating performance—that is, the quality of flow—changes with the demand traffic volume on the highway. The level of operating performance is indicated by the level of service at which the highway segment operates under a set of traffic and control conditions. This chapter presents procedures for determining the level of service on segments of freeways and segments of 2-lane, two-way rural highways.

246

HIGHWAY CAPACITY

The capacity of a highway segment is the maximum flow in vehicles per hour (vph) that can be reasonably expected on that segment of highway during a given time period under prevailing roadway, traffic, and control conditions. A 15 min time period is normally used because this period is regarded as the longest length of time during which flow is stable. The roadway conditions are associated with the geometric characteristics, which include number of lanes by direction, lane width, shoulder width, horizontal and vertical alignments, lateral clearance, and design speed. The traffic conditions are associated with the characteristics of the traffic stream on the segment of the highway. These include the distribution of the different types of vehicles in the traffic stream, the distribution of the local traffic on the different lanes, and the directional distribution (directional split) of the traffic volume on the highway segments. The control conditions are associated with the types of traffic control devices in operation and traffic regulations that may apply to the highway segment.

LEVEL OF SERVICE

Level of service (LOS) qualitatively measures the operating conditions within a traffic system and how these conditions are perceived by drivers and passengers. It is related to the physical characteristics of the highway and the different operating characteristics that can occur when the highway carries different traffic volumes. Although traffic volume is the principal factor affecting the level of service of a highway segment, factors such as lane width, lateral obstruction, traffic composition, grade, and speed also have some effect, since the maximum flow on a given highway segment depends on each of these factors. The effects of each of these factors on flow is briefly discussed.

   1. Lane Width. Traffic flow tends to be restricted when lane widths are narrower than 12 ft, the reason being that vehicles have to travel closer together in the lateral direction, and motorists tend to compensate for this by driving more cautiously and increasing the spacing between vehicles, which reduces the maximum flow on the highway.

   2. Lateral Obstruction. In general, when objects are located near the edge of the roadway or in the median, motorists in lanes adjacent to the object tend to drive away from the object with the result that the lateral distance between vehicles is reduced. This lateral reduction in space also results in longer spacings between vehicles and reduction in the maximum flow on the highway. This effect is eliminated if the object is less than 6 in. high or if it is located at least 6 ft from the edge of the roadway. Note, however, that lateral clearances are based mainly on safety considerations and not on flow consideration. In fact, current design criteria for freeways and other expressways call for the provision of shoulder widths wider than 6 ft (see Chapter 14), and obstructions are normally not allowed on the shoulders. A lateral clearance of less than 6 ft may, however, be found in very old highways or at construction or maintenance work zones.

3. Traffic Composition. The presence of trucks, buses, and recreational vehicles in a traffic stream reduces the maximum flow on the highway because of their size, operating characteristics, and interaction with other vehicles.

4. Grade. The effect of a grade depends on both the length and the slope of the grade. Traffic operations are significantly affected when grades with slopes of 3 percent or higher are 1/4 mi in length or longer and when grades less than 3 percent are 1/2 mi or longer in length. The effect of trucks on such grades is magnified.

5. Speed. Space mean speed, defined in Chapter 5, is used in level-of-service analysis because flow has a significant effect on speed.

Because these factors affect traffic operations on the highway, it is essential that they be considered in any LOS analysis. Computational methodologies used to take into consideration the effect of these factors are presented later in this chapter.

Six levels of service, designated A through F, have been established, with level of service A designating the best service and level of service F designating the worst service in terms of motorist satisfaction. Procedures given in the 1965 *Highway Capacity Manual* (HCM) of the Transportation Research Board have served as the principal means of determining level of service since 1965. These procedures are based primarily on traffic flow relationships between operating speeds and volumes.

Since the development of the 1965 HCM, however, several studies, including controlled experiments by computer simulation, have indicated that the procedures developed in the 1965 HCM needed updating. The Transportation Research Board (TRB), through the National Cooperative Highway Research Program (NCHRP), funded the NCHRP projects 3–28[1] and 3–28A[2] with the main objective of updating the 1965 HCM. The results of these studies have been used to develop procedures and methodologies that are presented in the 1985 *Highway Capacity Manual*.[3] The procedures for 2-lane, two-way rural highways were developed from the results obtained from using a detailed microscopic simulation model, combined with theoretical and empirical evidence. The model was validated using data collected from four states and Alberta, Canada. The procedures for freeways were based on the results of several major contracts for research in highway capacity.[4-9] The procedures for freeways were developed separately for each freeway component subsection: basic freeway segments, weaving areas, and ramp junctions. The procedures presented here are for basic freeway segments and for 2-lane two-way rural highways.

BASIC FREEWAY SEGMENTS

Basic freeway segments are sections of a freeway that are not influenced by weaving, diverging, and merging maneuvers caused by motorists entering or leaving the freeway. The 1985 *Highway Capacity Manual* gives the zones of influence of weaving areas and ramp junctions as

- On-ramp terminals: 500 ft upstream and 2500 ft downstream
- Off-ramp terminals: 2500 ft upstream and 500 ft downstream
- Weaving areas: 500 ft upstream and downstream (in addition to the weaving area itself)

At least one study has shown that the zones of influence may extend much more than those given in the HCM, as the extent of influence depends on several factors, such as geometrics, local conditions, and relative traffic volumes.[10] The influence zones given above should therefore be regarded as general guidelines. Engineering judgment must always be used when identifying basic segments of a specific freeway.

The following terms are used in the analysis of basic freeway segments.

Rate of flow is the equivalent hourly rate at which vehicles pass a point on the freeway lane during a given period of time less than one hour. The time period usually taken for flow is 15 min. The difference between volume (defined in Chapter 5) and rate of flow should be noted. Volume gives the total number of vehicles that pass a point during a given period of time, which may be 1 hr, a day, a week, and so forth.

Freeway capacity is the maximum rate of flow that can be sustained over a 15 min period on a uniform freeway segment in one direction under prevailing roadway and traffic conditions. This rate, however, is converted to hourly flow as the capacity is given in terms of vph.

Road characteristics are the geometric characteristics such as grades, number of lanes, design speeds, width of lanes, curvature, and so forth of the freeway segment.

Traffic characteristics include the composition of traffic (for example, percentages of trucks, buses, and recreational vehicles) distribution of traffic in one direction among the lanes in the same direction, and characteristics of the drivers (for example, weekday commuters versus recreational drivers).

Design speed relates to freeway segments. While the design speed for freeways is usually 70 mph, isolated sections, such as a sharp horizontal curve, may have design speeds less than 70 mph. It is, therefore, necessary to consider these isolated sections separately. A long freeway segment having a predominant number of sections with a lower design speed may however be treated as a unit with the lower design speed.

Level of Service of Basic Freeway Segments

The LOS for a basic freeway segment is based on the density on that segment, since the density is directly related to the freedom of the drivers to maneuver within the traffic stream. In addition, rate of flow increases as density increases for flows less than the capacity of the freeway segment. Six levels of service have been defined for the basic freeway segment.

Level of Service A. At this level of service, the speed of an individual vehicle is controlled only by the desires of the driver and the prevailing conditions. There is no interference from other vehicles and free-flow conditions exist. The average running speed is about 60 mph on freeways with design speed of 70 mph. The density is not greater than 12 passenger cars per mile, per lane (pc/mi/ln), with an average spacing of 440 ft between consecutive vehicles. Under these conditions, an incident on the freeway may result in the lowering of the level of service at the vicinity of the incident, but traffic quickly returns to operating at level of service A after passing the incident.

Level of Service B. Traffic is moving under free-flow conditions, and average speeds higher than 57 mph are sustained on freeways with design speed of 70 mph. The maximum density, however, increases to about 20 pc/mi/ln, with an average spacing of about 260 ft between consecutive vehicles. There is slight restriction on the ability to make lane changes or to leave or enter the traffic stream. Drivers, however, do not find it difficult to make such maneuvers, and a high level of physical and psychological comfort is still provided to the drivers. Minor incidents can result in a more severe local deterioration of level of service than at level of service A, but level of service B is easily obtained after passing the incident.

Level of Service C. At level of service C, stable operation is provided, but flows approach the range at which an increase in volume immediately results in a deterioration in service. Average running speed is still over 54 mph for freeways with design speed of 70 mph, but drivers are definitely restricted in making maneuvers such as lane changes. Density increases to about 30 pc/mi/ln, with an average spacing of about 175 ft between consecutive vehicles. Significant incidents and point breakdowns are not easily absorbed and may result in the formation of queues. Additional vigilance is required of drivers in order to maintain safe operation, which results in drivers experiencing a significant increase in tension.

Level of Service D. At this level of service, operation is approaching unstable flow, and average running speed when no incidents occur is about 46 mph on highways with a design speed of 70 mph. Maximum density increases to about 42 pc/mi/ln, with an average spacing of 125 ft between consecutive vehicles. Motorists are severely restricted in carrying out maneuvers such as lane change, and there is a significant reduction of the driver's physical and psychological comfort. The occurrence of a minor incident will result in extensive queuing.

Level of Service E. Flow at this level of service is unstable. The average spacing between consecutive vehicles decreases to 80 ft, and for all practical purposes, no usable gaps exist between consecutive vehicles. Maneuvers such as lane change or merging of traffic from entrance ramps will result in a disturbance of the traffic stream. Minor incidents result in immediate extensive queue build up. When operating conditions are at level of service E, the freeway segment is operating at or near capacity. Average speeds at capacity are approximately 30 mph.

Level of Service F. Operation under this level of service is under forced or breakdown conditions and uniform moving flow cannot be maintained. These conditions prevail in queues behind the sections of freeways, with a temporary reduction in capacity. The flow conditions are such that the number of vehicles that can pass a point is less than the number of vehicles arriving at the point.

Maximum Service Flow Rate per Lane

The maximum service flow rate at level of service i (MSF_i) is the maximum flow that a section of the freeway can maintain at level of service i under ideal conditions. Ideal conditions are defined as follows.

1. No trucks, buses, or recreational vehicles are in the traffic stream—that is, only passenger cars in the traffic stream.
2. Lanes are 12 ft or wider.
3. Lateral obstructions are no closer than 6 ft to the edge of the pavement.
4. In urban areas, typical characteristics of weekday commuter drivers prevail, and in other areas typical characteristics of regular users prevail.

The MSF_i is determined as the product of the capacity under ideal conditions and the maximum volume to capacity ratio for the level of service i, as shown in Eq. 7.1[11]

$$MSF_i = C_j(v/c)_i \qquad (7.1)$$

where

MSF_i = maximum service flow rate per lane for level of service i under ideal conditions in passenger cars per hour, per lane (pc/hr/ln)

$(v/c)_i$ = maximum volume to capacity ratio for level of service i

C_j = capacity under ideal conditions for the freeway segment having design speed j

= 2000 pc/hr/ln for freeway segments with design speeds of 60 and 70 mph

= 1900 pc/hr/ln for freeway segments with design speed of 50 mph.

Table 7.1* gives maximum service flow rates for different design speeds and levels of service. Note that, since operating at level of service E is the same as operating at capacity, the maximum service flow rate at level of service E equals the capacity of the freeway segment.

---

*For convenience in looking up values, all tables referenced in this chapter are gathered in an appendix to the chapter beginning on page 280.

Service Flow Rate for Level of Service i

The service flow rate for level of service i is the maximum flow rate in one direction that can be maintained while operating at level of service i, under prevailing conditions. Since the prevailing conditions are usually not the same as the ideal conditions, the service flow rate is obtained by adjusting the MSF_i to reflect the number of lanes and the prevailing conditions. This is done by multiplying MSF_i by different adjustment factors for each condition. The service flow rate is therefore obtained as

$$SF_i = MSF_i(N)(f_w)(f_{HV})(f_p) \tag{7.2}$$

Substituting for MSF_i using Eq. 7.1.

$$SF_i = C_j(v/c)_i(N)(f_w)(f_{HV})(f_p) \tag{7.3}$$

where

SF_i = service flow rate under prevailing traffic and roadway condition for the level of service i (vph)

MSF_i = maximum service flow rate under ideal conditions for level of service i

f_w = adjustment factor for the effect of restricted lane widths and/or lateral clearance

f_{HV} = adjustment factor for the combined effect of trucks, buses, and recreational vehicles in the traffic stream

f_p = adjustment factor for the effect of driver population

N = number of lanes in one direction

Note that the adjusted service flow rate obtained from either Eq. 7.2 or Eq. 7.3 will be achieved only if good pavement and weather conditions exist and there are no incidents on the freeway segments. If these conditions do not exist, the actual service flow that will be achieved may be less. Let us now discuss how each factor is obtained for use in Eq. 7.2 or Eq. 7.3.

Lane Width and Lateral Clearance Factor, f_w

This factor is used to adjust for lane widths less than 12 ft and/or lateral clearance less than 6 ft. Table 7.2* gives the appropriate values for this factor for different lane widths and lateral clearances. In cases where there are obstructions on both sides of the road at different distances from the edge, the average of the two distances is used to obtain the adjustment factor from the table. The lane width condition is relatively simple to determine since the lane width of a freeway segment is easily obtained. The determination of the lateral clearance condition

---

*See the appendix to this chapter.

is, however, not that simple because not all objects located less than 6 ft from the pavement will have a significant impact on the traffic. For example, when drivers become familiar with a certain type of obstruction, the effect of such an obstruction is negligible even when it is located within 6 ft of the pavement. Studies have also shown that certain common types of barriers have very little effect on traffic. Good judgment must be used when adjustment for lateral clearance is made.

Combined Effects of Trucks, Buses, and Recreational Vehicles

To adjust the maximum service flow for the composition of the traffic stream, it is first necessary to determine the passenger car equivalent (PCE) of each truck, bus, or recreational vehicle (E_T, E_B, or E_R) for the prevailing traffic and roadway conditions. These factors give the number of passenger cars that will use up the same space on the freeway as one truck, bus, or recreational vehicle under the prevailing conditions being considered. The next step is to use these PCEs and fractional distribution, or proportion, of each type of vehicle (P_T, P_B, and P_R) to determine a single multiplicative correction factor, f_{HV}, for the combined effect.

Since the extent to which the presence of a truck affects the traffic stream also depends on the vertical alignment (grade) of the segment being considered, PCEs for trucks are given for two grade conditions: extended general freeway segments and specific grades of significant length.

PCEs for Extended General Freeway Segments. In cases where the freeway segment is not dominated by a specific grade but consists of upgrades, downgrades, and level sections, PCEs can be determined for extended general freeway segments. In general, this can be done for freeway segments where no one grade of 3 percent or greater is longer than 1/2 mi or no one grade less than 3 percent is longer than 1 mi. PCEs for these cases are given in Table 7.3. The PCEs are given in terms of the type of terrain, classified as level, rolling, or mountainous, defined as follows.

Level Terrain. Freeway segments are considered to be on level terrain if the combination of grades and horizontal alignment does not restrict the speed of trucks and permits such trucks to travel at approximately the speed of passenger cars. Grades are generally not higher than 1 or 2 percent for level terrain.

Rolling Terrain. Freeway segments are considered to be on rolling terrain if the combination of grades and horizontal alignment restricts the speed of trucks to values substantially lower than that of passenger cars but such trucks do not travel at crawl speed for a significant length of time.

Mountainous Terrain. Freeway segments are considered to be in mountainous terrain if the combination of grades and horizontal alignment causes trucks to travel at crawl speeds for a significant distance or at frequent intervals.

PCEs for Specific Grades of Significant Length. When a grade of 3 percent or greater is longer than 1/2 mi or a grade of less than 3 percent is longer than 1 mi, it should be considered separately as a specific grade of significant length. Specific grades are analyzed separately for downgrade and upgrade conditions. In addition, the specific grade is considered either as a single isolated grade with a fixed percentage or as a number of consecutive grades forming a composite segment.

The variety of trucks and recreational vehicles with varying characteristics results in a wide range of performance capabilities on specific grades. Although studies have shown that the truck population on freeways has an average weight-to-horse power ratio (lb/hp) of 125 to 150, the procedure suggests the use of a lb/hp ratio of 200 since the heavier trucks (that is, trucks with higher ratios of lb/hp) have a greater effect on traffic flow than do lighter trucks. PCE factors for trucks are, however, presented for average lb/hp ratios of 100, 200, and 300 to allow the use of the procedure when the trucks in the traffic stream are more or less heavy than the suggested value of 200 lb/hp.

Tables 7.4, 7.5, and 7.6 give PCEs for trucks traveling up grades for different grade percentages, lengths of grade, and truck percentages within the traffic stream, since all these factors affect the extent to which the presence of a heavy truck influences the traffic stream. For example, the effect of a truck traveling up a grade increases with increase in grade length because the truck experiences a longer decrease in speed for a longer upgrade. The PCEs selected should be associated with the point where the effect is greatest, which is usually at the end of the grade. It has therefore been suggested that the grade length be obtained from a profile of the road and include the tangent portion, plus 25 percent of the vertical curves of the beginning and end of the grade. In cases where two consecutive upgrades are connected by a vertical curve, 50 percent of the curve length is added to each portion of the grade.[12] This guideline may, however, vary for some specific conditions. For example, if an analysis is to be carried out to determine the effect of an on-ramp at an upgrade section of the expressway, the length to be used is that up to the ramp junction.

Note that PCE factors are not given for specific downgrades. The general approach recommended is to consider a downgrade of less than 4 percent or shorter than 3000 ft as an extended section of a level terrain and to select PCEs from Table 7.3. When, however, it is necessary for trucks to engage low gears because of severe downgrades, it is advisable to collect field data on speed and use these data to determine a comparable upgrade condition, from which the PCE is determined. Alternative charts provided are used, as demonstrated in the next section for composite upgrades.

PCEs for Trucks on Composite Upgrades. When a segment of highway consists of two or more consecutive upgrades with different slopes, the passenger car equivalent of trucks is determined by using either one of two techniques. The first technique is simple but only gives an approximate value of the passenger car equivalent. This technique consists of determining the average slope of the segment by finding the total rise in elevation and dividing it by the total horizontal distance. This average slope is then used to enter either Table 7.4, Table 7.5, or Table 7.6.

Example 7–1 Computation of PCE for a Composite Upgrade

A segment of 4-lane freeway consists of two consecutive upgrades of 4 percent, 3/4 mi long and 5 percent, 1/2 mi long. Determine the PCE of trucks on this composite upgrade if 2 percent of the vehicles are trucks. Assume typical trucks of 200 lb/hp ratios.

$$\text{total rise} = 0.04 \times 0.75 \times 5280 + 0.05 \times 0.5 \times 5280$$

$$= 290.4 \text{ ft}$$

$$\text{average grade} = 290.4/(3960 + 2640)$$

$$= 0.444 \quad (\text{that is, } 4.4\%)$$

$$\text{total length} = 1.25 \text{ mi}$$

Entering Table 7.4, with 5 percent grade (since 4.4 > 4, see note on bottom of Table 7.4) and length > 1 mi, we obtain a PCE of 14.0.

The second technique is a modified version of one developed by Leisch,[13] based on trucks' acceleration/deceleration curves as shown in Figures 7.1, 7.2, and 7.3. The curves are used to determine an equivalent grade, having the same length as the total length of the composite grade under consideration, that will cause the final speed of trucks to be the same as that on the composite grade. The use of this technique is demonstrated by resolving Example 7–1, using the following steps.

Step 1. Using the grade and length of the initial grade, determine the final speed of trucks from Figure 7.1. Trucks will enter the second grade at this speed. Initial grade is 4 percent and length of initial grade is 3/4 mi (3960 ft). Entering the graph at 3960 ft (A) and projecting vertically to the 4 percent curve, we find final speed = 33 mph (B).

Step 2. Determine the length along the next grade that will give the same speed as that determined in step 1—that is, 33 mph, which is taken as the starting point for the length of the second grade. This length is obtained by locating the point where the 33 mph line intersects the 5 percent curve (C) and projecting vertically to the horizontal side (D). This step gives a length of about 1700 ft.

Step 3. Determine the sum of the length of the second grade and the length determined in step 2: (1700 + 2640) ft = 4340 ft. This addition is necessary because the effect is that, at a speed of 33 mph, a truck will enter the 5 percent grade as if it has already traveled a distance of 1700 ft on that grade starting from a level terrain.

Step 4. Using the length determined in step 3, enter the curve at 4340 ft (G), project vertically to the curve of the slope of the second grade, and determine the resultant speed (E)—that is, length = 4340 ft and grade = 5 percent, giving a speed of 26 mph.

Figure 7.1 Performance Curves for Typical Trucks (200 lb/hp)

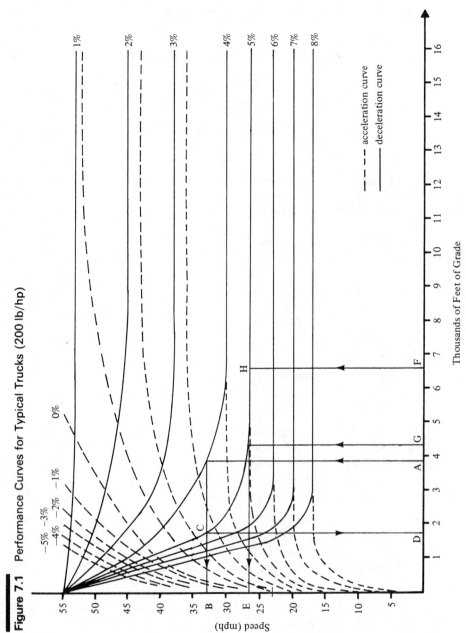

Source: Redrawn from *Highway Capacity Manual*, Special Report 209, Transportation Research Board, Washington, D.C., 1985.

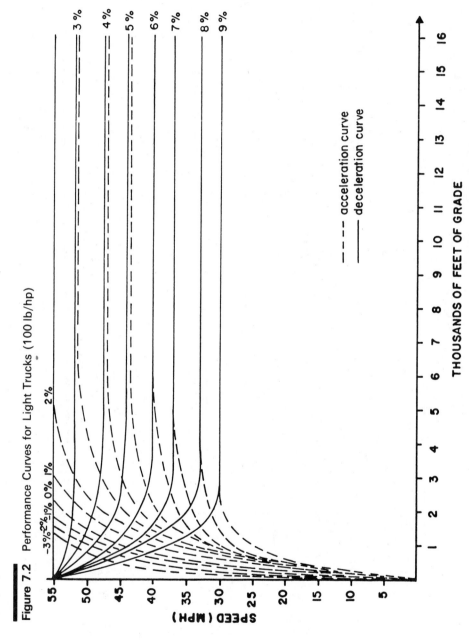

Figure 7.2 Performance Curves for Light Trucks (100 lb/hp)

Source: Reproduced from *Highway Capacity Manual*, Special Report 209, Transportation Research Board, Washington, D.C., 1985.

Figure 7.3   Performance Curves for Heavy Trucks (300 lb/hp)

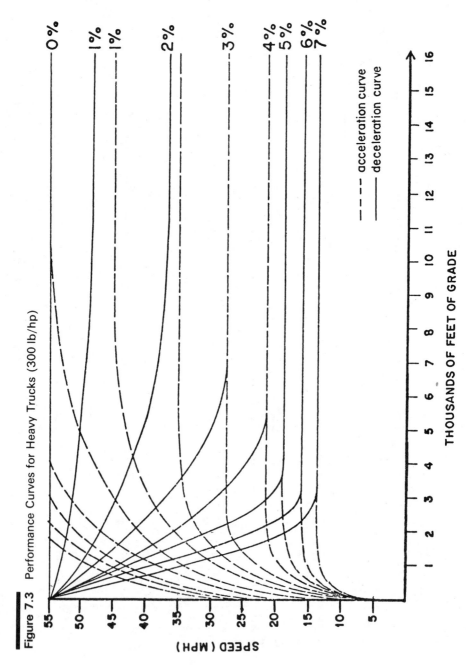

Source: Reproduced from *Highway Capacity Manual*, Special Report 209, Transportation Research Board, Washington, D.C., 1985.

Step 5. Using the final speed determined in step 4 and the total length of the composite grade (F), determine the equivalent grade (H)—that is, speed = 26 mph and length = (3960 + 2640) = 6600 ft, giving an equivalent grade of 5 percent.

Step 6. Determine the PCE using the equivalent grade from Table 7.4: PCE = 14.

Although the same PCE value is obtained for both methods in this case, it should not be assumed that this is always true. Significantly different values of PCE may be obtained, depending on the lengths and grades being considered. The difference is attributed mainly to the second technique being more accurate since it takes the performance of the truck on the individual grades into consideration.

When the composite grade consists of more than two grades, steps 2, 3, and 4 are repeated until the final speed is obtained. When downgrades are part of composite grades, the same procedure applies, but the curves for negative grades are used. Note, however, that this technique should not be used for buses because of the form the PCEs take, nor should it be used for recreational vehicles because of the approximate nature of the PCEs. It is also unnecessary to use this technique for grades having lengths less than 2500 ft as the average technique is adequate for such cases.

PCEs for Recreational Vehicles. The characteristics of recreational vehicles (RVs) also vary considerably, ranging from cars towing different types of trailers to self-contained mobile campers. The typical lb/hp ratio for these vehicles ranges only between 30 and 60. Therefore, only one set of PCEs for RVs, E_R is given in Table 7.7

PCEs for Buses. Table 7.8 gives the PCEs for buses, E_B. The PCE values are based only on the grade percent. This is because the characteristics of intercity buses vary only slightly and because relatively little research has been undertaken on performance characteristics of buses since the early 1960s. The factors given are thus based on information available in the early sixties.

Having determined the PCEs for trucks, buses, and recreational vehicles, the combined correction factor f_{HV} is determined using Eq. 7.4.

$$f_{HV} = \frac{1}{1 + P_T(E_T - 1) + P_B(E_B - 1) + P_R(E_R - 1)} \qquad (7.4)$$

where

f_{HV} = combined correction factor

P_T, P_B, P_R = proportion of trucks, buses, and RVs, respectively, in the traffic stream

E_T, E_B, E_R = PCEs for trucks, buses, and RVs, respectively

This procedure is used to determine the combined correction factor when there is a significant percentage of trucks, buses, or recreational vehicles. When the percentage of trucks is at least 5 times the total percentage of buses and recreational vehicles, it can be assumed that the traffic stream consists of passenger cars and trucks only. For example, a traffic stream consisting of 16 percent trucks, 1 percent buses, and 1 percent RVs can be assumed to contain 18 percent trucks, and Tables 7.4, 7.5, or 7.6 can be used directly.

Driver Population Adjustment Factor, f_p

The adjustment factors f_p for the characteristics of the driver population are given in Table 7.9. These adjustment factors reflect the fact that drivers unfamiliar with a particular freeway will use that freeway less efficiently than weekday commuter drivers who are familiar with the freeway. Since the "ideal" conditions discussed earlier include weekday commuter traffic, it is necessary to correct for the case when noncommuter drivers are prevalent in the traffic stream. It is suggested, however, that these values be used with caution, and that users of the procedure apply their general knowledge of the area in selecting a specific value of f_p. In cases where considerable accuracy is required, it is necessary to compare speed and flows on regular commuting days with those occurring on weekends.

Design Service Volume

When the LOS on a road segment is to be determined, the variation of traffic flow within the hour is taken into consideration by converting the directional design hourly volume (DDHV) on the highway segment to the peak 15 min flow rate, using the peak hour factor (PHF). The demand service flow (DSF) used for analysis or design purposes is therefore given as

$$DSF = \frac{DDHV}{PHF}$$

where

    DSF = demand service flow rate for the highway segment being considered (vph)

$$PHF = \frac{\text{Volume during peak hour}}{4 \times \text{volume during peak 15 min within peak hour}}$$

    DDHV = actual hourly demand volume on the highway segment being considered

---

Example 7–2   Computation of Level of Service at an Extended Section of a Freeway

Determine the LOS on a regular weekday on an extended section of a 6-lane freeway.

- Use the following data.

  DDHV = 3500 vph
  PHF = 0.85
  Traffic composition:
      Trucks = 10 percent
      Buses = 2 percent
      RVs = 2 percent
  Lane width = 11 ft
  Terrain = level
  Design speed = 70 mph
  Lateral obstruction = none
  DSF = 3500/0.85 = 4118 vph

- Determine the correction factors:

  Restricted lane width and lateral clearance $f_w = 0.97$ (from Table 7.2)
  PCEs:
      Trucks = 1.7 (from Table 7.3)
      Buses = 1.5 (from Table 7.3)
      RVs = 1.6 (from Table 7.3)
  Driver population adjustment factor $f_p = 1.0$ (from Table 7.9)

- Calculate the combined correction factor.

$$f_{HV} = \frac{1}{1 + 0.1(1.7 - 1) + 0.02(1.5 - 1) + 0.02(1.6 - 1)}$$

$$= \frac{1}{1 + 0.07 + 0.01 + 0.01}$$

$$= 0.92.$$

- Use Eq. 7.3 to determine the maximum volume to capacity ratio.

$$(v/c) = \frac{SF}{(C_j)N(f_w)(f_{HV})(f_p)}$$

with

C_j for design speed of 70 mph = 2000 pc/hr/ln

Thus,

$$(v/c) = \frac{4118}{2000 \times 3 \times 0.97 \times 0.92 \times 1}$$

$$= \frac{4118}{5354}$$

$$= 0.77.$$

- Use Table 7.1 to obtain level of service C.

Note that 0.77 is the maximum value of v/c for a level of service C for a design speed of 70 mph. Any value higher than 0.77 will give a worse level of service.

Example 7–3 Computation of the Required Number of Lanes for a Given Level of Service

A 1/2 mi, 4 percent grade section of a 4-lane freeway is to be widened to improve the LOS to level C. Determine the number of lanes that should be added in each direction.

- Use the following design features.

    DDHV = 2500 vph (weekday commuter traffic)
    PHF = 0.95
    Traffic composition:
        Trucks = 8 percent
        Buses = 3 percent
        RVs = 2 percent
    Design speed = 70 mph
    Lane width = 12 ft
    Lateral obstruction = none
    DSF = 2500/0.95 = 2632

- Determine the correction factors (assuming typical trucks and considering the upgrade direction).

    $f_w = 1$ (from Table 7.2)
    $E_T = 6$ (from Table 7.4)
    $E_R = 4$ (from Table 7.7)
    $E_B = 1.6$ (from Table 7.8)
    $f_p = 1$ (from Table 7.9)

- Calculate the combined correction factor.

$$f_{HV} = \frac{1}{1 + P_T(E_T - 1) + P_B(E_B - 1) + P_R(E_R - 1)}$$

$$= \frac{1}{1 + 0.08(6 - 1) + 0.03(1.6 - 1) + 0.02(4.0 - 1)}$$

$$= \frac{1}{1 + 0.40 + 0.02 + 0.06}$$

$$= 0.68$$

- Use Eq. 7.3 to determine the number of lanes.

$$SV_C = 2000(N)(v/c)(f_w)(f_{HV})(f_p)$$

with

$$(v/c) = 0.77 \text{ (for level of service C from Table 7.1)}$$

Thus,

$$2632 = 2000(N)(0.77)(1)(0.68)(1)$$

$$2.51 = N$$

One more lane must be added in the upgrade.

- Repeat the computations considering the downgrade direction and assuming typical trucks. Since the grade length is 2640 ft (which is less than 3000 ft), the downgrade is treated as an extended section of a level terrain. From table 7.3,

$$E_T = 1.7$$
$$E_B = 1.5$$
$$E_R = 1.6$$

$$f_{HV} = \frac{1}{1 + P_T(E_T - 1) + P_B(E_B - 1) + P_R(E_R - 1)}$$

$$= \frac{1}{1 + 0.08(1.7 - 1) + 0.03(1.5 - 1) + 0.02(1.6 - 1)}$$

$$= \frac{1}{1 + 0.06 + 0.02 + 0.01}$$

$$= 0.92$$

$$2632 = 2000(N)(v/c)(f_w)(f_{HV})(f_p)$$

$$= 2000(N)(0.77)(1)(0.92)(1)$$

$$1.85 = N$$

This result indicates that in the downgrade direction no additional lane is necessary as the operating LOS is currently C, which is confirmed by determining the (v/c) ratio in the downgrade direction.

$$(v/c) = \frac{2632}{2000(2)(1)(0.92)(1)} = 0.72$$

$0.54 < 0.72 < 0.77$ (that is, level of service C)

TWO-LANE, TWO-WAY RURAL HIGHWAYS

The procedure developed for 2-lane, two-way rural highways provide for evaluating the level of service and capacity of these roads at two levels of analysis: operational and system planning.

At the *operational level of analysis*, the level of service is determined for an existing 2-lane highway with a known set of traffic and roadway characteristics. The traffic characteristics can be either current or projected future characteristics. This analysis may be carried out for general terrain segments or for specific grades.

At the *system planning level of analysis*, the average daily traffic (ADT) that can be accommodated on 2-lane highways for different levels of service can be quickly estimated for planning purposes.

Operational Level of Analysis

Factors of primary importance at this level of analysis are average travel speed, percent time delay, capacity utilization, and percentage of marked no-passing zones.

Average travel speed is represented by the average speed of the traffic streams traveling in both directions over the section of the highway.

Percent time delay is the average delay of all motorists in a platoon, expressed as a percentage of the total travel time on a given section of highway. Delay is considered to have occurred if motorists are traveling behind a platoon leader at headways less than 5 sec and at speeds less than their desired speeds. The level of service is strongly reflected in the average percent time delay. When traffic volume is low, average headways are high, demand for passing is very low, and percent of time delay tends to 0 percent. When traffic volume is very high, however, the roadway capacity is approached, and the demand for passing becomes higher than the passing capacity. This results in long platoons of traffic and percent of time delay tends to 100 percent.

Figure 7.4 shows established basic relationships between average speed and percent time delay and traffic volumes. These relationships are based on ideal traffic and roadway conditions, which are defined as highways with no restrictive geometric, traffic, or environmental conditions, with the following characteristics:

- Level grade
- Horizontal alignment having a design speed of at least 60 mph
- No passing restrictions
- Lane widths of 12 ft or greater
- Usable shoulders of 6 ft or greater
- No restriction to through traffic either by turning movements or traffic control devices
- Same volume of traffic in both directions (50/50 directional split)
- Only passenger cars in traffic streams

Capacity utilization indicates to what extent additional traffic can be accommodated by the highway section, and it is expressed as the ratio of the flow rate of the demand to the capacity of the highway section.

Average percentage of marked no-passing zones in both directions of the highway indicates the frequency of no-passing zones on the 2-lane highway, which in turn characterizes the design of the highway and the type of traffic condition that will be expected on it. As an alternative to marked no-passing

Figure 7.4 Speed Flow and Percent Time Delay Flow Relationships for 2-Lane Rural Highways (Ideal Conditions)

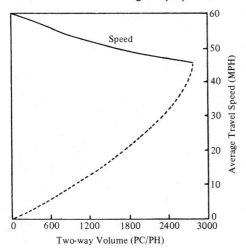

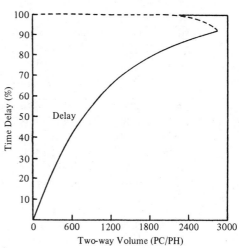

(a) Relation between average speed and flow on two-lane highways

(b) Relation between percent time delay and flow on two-lane highways

Source: Redrawn from *Highway Capacity Manual*, Special Report 209, Transportation Research Board, Washington, D.C., 1985.

zones, sections with passing sight distances less than 1500 ft are considered as no-passing zones. No-passing zones on 2-lane highways typically range from 20 to 50 percent, although the value can be as high as 100 percent on winding mountainous highways. The impact of no-passing zones is greater on highways located on mountainous terrain than on those located on level terrain.

Levels of Service at Operational Analysis Level

Six levels of service have been defined in the recommended procedures for the operational level of analysis.

Level of Service A. This level of service gives the highest quality of traffic flow. Drivers are free to drive at their desired speed. If there is no enforcement of a speed limit, average speed will approach 60 mph, and the demand for passing will be less than the passing capacity. The number of platoons consisting of three or more vehicles will be negligible, the average percent time delay will be less than 30 percent, and the maximum flow rate (service flow rate) total in both directions will not be greater than 420 pc/hr under ideal conditions.

Level of Service B. Average speeds on level terrain are approximately 55 mph. To maintain desired speeds, passing demand becomes significant and approaches the passing capacity at the lower boundary of this level of service. Average percent time delay is about 45 percent, and service flow rates approach 750 pc/hr total in both directions under ideal traffic conditions.

Level of Service C. Traffic flow is still stable under this level of service, but turning and slow-moving traffic tends to create some congestion. Average speeds on level terrain are higher than 52 mph, but unrestricted passing demand is higher than the passing capacity. Service flow rates can be as high as 1200 pc/hr in both directions under ideal traffic conditions with average percent time delay getting up to 60 percent. When 2-lane, two-way rural highways are designed for level of service C, intersections with significant turning movements must be provided with special treatment in order to avoid significant delay.

Level of Service D. Traffic flow approaches unstable conditions at this level of service. The passing capacity tends to zero, while passing demand is very high. Average speeds of about 50 mph can be achieved, with an average of 5 to 10 vehicles in a platoon. Major shock waves are caused in the traffic stream by turning vehicles, and the average percent time delay gets up to about 75 percent. The maximum service flow rate under ideal traffic conditions is 1800 pc/hr in both directions.

Level of Service E. Average speeds at this level of service are less than 50 mph under ideal conditions but may reduce to 25 mph on long grades. Operating conditions on the highway become unstable and difficult to predict. The maximum service volume that can be attained at this level of service is the capacity of the highway, which is 2800 pc/hr in both directions under ideal conditions. The maximum service volume, is, however, affected by the percentage of the total traffic traveling in each direction (directional split) as shown in Table 7.10.

Level of Service F. At this level of service traffic demand is greater than the capacity, resulting in heavily congested flows. Average speeds are less than capacity speeds, and actual flows are less than the capacity.

Table 7.11 gives the criteria for the different levels of service on a general terrain segment. The criteria are given for level terrain, rolling terrain, and mountainous terrain separately and for different percentages of no-passing zones. The definitions for the different types of terrain are the same as those given earlier for basic freeway sections. Note that maximum values for (v/c) ratios are slightly different than those given for basic freeway sections, because the values in this case are based on the ideal capacity of 2800 pc/hr on a level terrain segment having an ideal geometric alignment.

LOS criteria for specific grade segments are given in Table 7.12, which relates the level of service with only the average upgrade travel speed. The reason for using only travel speed is that the determination of the capacity of a specific grade is rather difficult, as the capacity speed for a given grade is strongly influenced by the volume of traffic on the grade and the percentage and length of the grade.

The procedures do not give a specific method of analysis for downgrades. It may be assumed, however, that operation conditions on downgrades less than 3 percent are similar to those on level sections, and on steeper grades, operation conditions are about midway between level conditions and upgrade conditions for similar traffic and roadway characteristics.

Procedures for Evaluating Levels of Service at Operational Analysis Level

The procedures for evaluating the LOS of a 2-lane, two-way rural highway at the operational level of analysis are given separately for general terrain segments and specific grades.

General Terrain Segments. This procedure is used to determine the LOS at which a given highway will operate for a known set of traffic data or to determine the permissible flow rate for a given level of service. The capacity of a segment of a 2-lane highway can also be determined by using this procedure. The study period in this case is the design hour. The flow rate during the design hour is adjusted for fluctuations by using the peak hour factor, which is based on the peak 15 min of the study hour.

The factors considered in this methodology include highway geometrics, average terrain, and traffic volume. Characteristics considered under highway geometrics are the cross-section data of the highway, such as lane width and usable shoulder width, and a general description of the longitudinal section, based on the proportion of the length of the highway with no-passing zones. When no-passing zone data are unavailable, sections of road with sight distances less than 1500 ft are considered no-passing zones. The terrain is classified as flat, rolling, or mountainous. The traffic data required are two-way hourly volume, PHF, directional distribution of traffic flow, and traffic mix. Table 7.13 gives suggested PHFs, which can be used if field measurements are not readily available. The traffic mix is given in terms of the fractions of trucks, recreational vehicles, and buses on the section of highway being considered. The following default values are suggested for rural primary roads if actual data are not available.

Trucks $P_T = 0.14$
RVs $P_R = 0.04$
Buses $P_B = 0.00$

Note that the proportion for RVs on recreational routes will usually be higher than 0.04.

The service flow rate SF for any level of service or speed is determined from Eq. 7.5.

$$SF_i = 2800(f_d)(f_w)(f_{HV})(v/c)_i \qquad \textbf{(7.5)}$$

where

$SF_i =$ total service flow rate for a given level of service under prevailing conditions in vph (mixed vehicles per hour, total of both directions)

$f_d =$ ratio of capacity of highway section to ideal conditions as related with directional distribution (see Table 7.10)

$f_w =$ lane width and shoulder width adjustment factor for a given level of service (see Table 7.14)

$(v/c)_i =$ service flow rate to ideal capacity ratio for a given level of service (see Table 7.11)

$f_{HV} =$ traffic mix adjustment factor for trucks, RVs, and buses along a highway section for given terrain and level of service

The fractions of trucks, RVs, and buses in the traffic stream are used in Eq. 7.6 to determine the traffic mix adjustment factor.

$$f_{HV} = \frac{1}{1 + P_T(E_T - 1) + P_R(E_R - 1) + P_B(E_B - 1)} \qquad \textbf{(7.6)}$$

where

$P_T =$ decimal fraction of trucks and tractor-trailer combinations in the traffic stream

$P_R =$ decimal fraction of recreational vehicles and auto-trailer combinations in the traffic stream

$P_B =$ decimal fraction of buses in the traffic stream

$E_T =$ PCE of trucks (see Table 7.15)

$E_R =$ PCE of RVs and auto-trailer combinations (see Table 7.15)

$E_B =$ PCE of buses (see Table 7.15)

Example 7–4 Determining Level of Service for a 2-Lane Highway Using the General Terrain Concept

A section of rural 2-lane highway is to be designed to carry a peak hour volume of 1500 vehicles on a level terrain. Determine the level of service at which it will operate.

- Use the following design features.

    Lane width = 11 ft
    Shoulder width = 6 ft
    No-passing zones = 20 percent
    Directional split = 60/40
    Decimal fraction of trucks = 0.1
    Decimal fraction of recreational vehicles = 0.04
    Decimal fraction of buses = 0.01
    PHF = 0.95

- Use Eq. 7.5 to determine the service volumes for different levels of service and compare these with the demand service volume to obtain the LOS.

$$SF_i = 2800(f_d)(f_w)(f_{HV})(v/c)_i$$

with

$f_d = 0.94$ (from Table 7.10)
$f_w = 0.93$ for level of service A–D
$\quad = 0.94$ for level of service E   (from Table 7.14)

- Calculate the truck mix factor from Eq. 7.6 as

$$f_{HV} = \frac{1}{1 + P_T(E_T - 1) + P_R(E_R - 1) + P_B(E_B - 1)}$$

- Obtain PCEs from Table 7.15.

Level of Service	B	C	D	E
$E_T =$	2.2	2.2	2.0	2.0
$E_R =$	2.5	2.5	1.6	1.6
$E_B =$	2.0	2.0	1.6	1.6

- Determine the correction factor.

    f_{HV} for level of service B = 0.84
    f_{HV} for level of service C = 0.84
    f_{HV} for level of service D = 0.88
    f_{HV} for level of service E = 0.88

- Use Table 7.11 to obtain the (v/c) ratios.

  (v/c) for level of service B $= 0.24$
  (v/c) for level of service C $= 0.39$
  (v/c) for level of service D $= 0.62$
  (v/c) for level of service E $= 1.00$

- Determine service flow rates.

  $SF_B = 2800(0.94)(0.93)(0.84)(0.24) = 493$
  $SF_C = 2800(0.94)(0.93)(0.84)(0.39) = 802$
  $SF_D = 2800(0.94)(0.93)(0.88)(0.62) = 1335$
  $SF_E = 2800(0.94)(0.94)(0.88)(1) = 2177$

- Convert the full-hour demand volume to an equivalent peak 15 min flow rate using the PHF.

$$DSF = \frac{1500}{0.95} = 1579$$

Thus,

$SF_D < DSF < SF_E$ (that is, level of service is E)

Specific Upgrades.   This level of analysis is used to evaluate the operating conditions on a specific section of a 2-lane highway with a steep grade. Results of recent studies have indicated that the effect of grades on the operation of 2-lane highways is more significant than on multilane highways with similar grades. One reason is that any driver wishing to pass a slow-moving vehicle can do so only by using the opposing lane, which may be very difficult if the volume is high. Thus platoons are difficult to dissipate. In addition, most passenger vehicles operate less efficiently on grade than on level terrain, even when no trucks are present. Therefore, geometric characteristics are important factors that affect the level of service at a specific grade. Geometric characteristics are defined by the percent grade, length of the grade, and percentage of no-passing zones. It is assumed that the approach to the grade is level. The level of service is based on the average speed of the traffic stream moving up the grade. The average speed is defined as the predicted average traffic speed for all vehicles operating over the length of upgrade being evaluated. Table 7.12 gives the relationship between minimum average speed and level of service. It is seen that the speed at capacity varies between 25 and 40 mph. The actual speed at capacity for a given specific grade depends on the geometric characteristics, the volume, and traffic composition. An expression that can be used to determine this is presented later. Note, however, that the procedure does not consider factors such as the effect of snow and ice, nor does it consider operational problems resulting from accidents or stalled vehicles, which may occur on steep grades when volumes are high. The procedure can be used to determine the maximum service flow rate (two-way) that will allow for the minimum average upgrade speed for a given level of service. This maximum service volume is obtained from Eq. 7.7.

$$SF_i = 2800(f_d)(f_w)(f_g)(f_{HV})(v/c)_i \tag{7.7}$$

where

SF_i = maximum service flow rate in vph (mixed vehicles per hour in both directions under prevailing conditions) for a given level of service or speed

f_d = directional distribution factor for upgrade direction as related to directional distribution (see Table 7.16)

f_w = lane width and shoulder width adjustment factor for a given level of service (see Table 7.14)

f_g = adjustment factor for passenger cars on specific grades for level of service computed as shown below

f_{HV} = traffic mix adjustment factor for given traffic mix, upgrades, and given level of service computed as shown below

$(v/c)_i$ = service volume to capacity ratio for a given percent no-passing zones, percent grade, and average upgrade speed (see Table 7.17)

This procedure is generally restricted to grades higher than 3 percent or longer than 1/2 mi. The adjustment factor for passenger cars on specific upgrades (f_g) is used to correct for the effect of the slight reduction of average speeds experienced by an all-passenger-car traffic stream on high upgrades. This factor is determined from Eq. 7.8.

$$f_g = \frac{1}{1 + P_P I_P} \tag{7.8}$$

where

P_P = decimal fraction of total volume composed of passenger vehicles (cars, vans, pickups)

I_P = specific upgrade impedance factor for passenger vehicles
    = $0.02(E - E_o)$

E = base passenger car equivalent for given percent grade, length of grade, and average upgrade speed (see Table 7.18)

E_o = base PCE for 0 percent grade and average speed (see Table 7.18)

The traffic mix adjustment factor f_{HV} is obtained from Eq. 7.9.

$$f_{HV} = \frac{1}{1 + P_{HV}(E_{HV} - 1)} \tag{7.9}$$

where

P_{HV} = decimal fraction of total volume composed of trucks, RVs, and buses

= $P_T + P_R + P_B$

E_{HV} = average passenger car equivalent of trucks, RVs, and buses

= $1 + (0.25 + f_t)(E - 1)$

f_t = decimal fraction of trucks in the truck, recreational vehicle, and bus mix (P_T/P_{HV})

The procedure described so far can be used to determine the service flow rates on specific grades from level of service A to level of service D. However, it is not that simple to determine the capacity, which is the service flow rate at level of service E, because the speed at which the capacity is obtained depends on the percent and length of grade being considered. For normal grades varying between 3 percent and 7 percent and not more than 4 mi long, the speed at which capacity is achieved usually varies between 25 and 40 mph and can be obtained from Eq. 7.10.[14]

$$S_C = 25 + 3.75 \left(\frac{q_c}{1000} \right)^2 \tag{7.10}$$

where

S_C = average upgrade traffic speed at capacity (mph)

q_c = total two-way flow rate at capacity (vph, in mixed vehicles per hour)

The capacity is determined by plotting the service flow rate against speed as obtained from Eq. 7.7 and capacity flow against capacity speed as obtained from Eq. 7.10 on the same graph sheet; the intersection of the two curves gives the flow rate and speed at capacity. Note, however, that Eq. 7.10 can only be used for upgrade speeds between 25 mph and 45 mph.

Example 7-5 Computing Service Flow of a Specific Grade on a 2-Lane Highway

A specific upgrade on a rural 2-lane highway is 3 mi long and has a grade of 5 percent. Determine the maximum hourly volume that can maintain an average speed of at least 50 mph on this upgrade.

- Use the characteristics.

  Width of each lane = 11 ft
  Width of stabilized shoulder = 7 ft
  No-passing zone rate = 50 percent
  Directional split (upgrade) = 60/40
  Decimal fraction of trucks = 0.10
  Decimal fraction of recreational vehicles = 0.06
  Decimal fraction of passenger cars = 0.84

- Use Eq. 7.7 to estimate the maximum service flow rate. Also note that Table 7.12 shows the level of service B will operate if a minimum average speed of 50 mph is maintained at the maximum volume.

$$SV_B = 2800(f_d)(f_w)(f_g)(f_{HV})(v/c)_B$$

where

  $f_d = 0.87$ (from Table 7.16, directional split = 60/40)
  $f_w = 0.93$ (from Table 7.14)

- Determine f_g by first determining the specific upgrade impedance factor.

$$I_P = 0.02(E - E_o)$$

where

  E for 5 percent grade, 3 mi long, and average speed of 50 mph = 37.0
  (from Table 7.18)
  E_o for average speed of 50 mph = 1.6 (from Table 7.18)

Thus,

$$I_P = 0.02(37.0 - 1.6) = 0.71$$

- Find the adjustment factor f_g using Eq. 7.8.

$$f_g = \frac{1}{1 + P_P I_P}$$

with

$$P_P = 0.84$$

Thus,

$$f_g = \frac{1}{1 + (0.84)(0.71)} = 0.63$$

■ Determine the traffic mix adjustment factor f_{HV} by first determining the average PCE of trucks, recreational vehicles, and buses (E_{HV}) using the equation

$$E_{HV} = 1 + (0.25 + f_t)(E - 1)$$

where f_t is the decimal fraction of trucks in the truck, recreational vehicle, and bus mix and is given as

$$f_t = \frac{P_t}{P_{HV}}$$

where P_t is the decimal fraction of trucks and tractor-trailer combinations.

$$P_{HV} = P_T + P_R + P_C = 0.10 + 0.06 + 0 = 0.16$$

Fraction of trucks in heavy vehicle flow is

$$f_t = \frac{0.10}{0.16} = 0.63$$

Thus,

$$E_{HV} = 1 + (0.25 + 0.63)(37 - 1) = 32.68$$

■ Now compute the traffic mix adjustment factor.

$$f_{HV} = \frac{1}{1 + P_{HV}(E_{HV} - 1)}$$

$$= \frac{1}{1 + 0.16(32.68 - 1)}$$

$$= 0.165$$

■ Obtain the $(v/c)_B$ ratio from Table 7.17.

$$(v/c)_B = 0.43 \quad \text{(by interpolation)}$$

■ Now determine the total two-way flow with a 60/40 directional split that will maintain an average upgrade speed of at least 50 mph as

$$SV_B = (2800)(0.87)(0.93)(0.63)(0.165)(0.43)$$

$$= 101 \text{ vph}$$

However, maximum hourly volume/PHF = SV. Assuming PHF = 0.92 for level of service B (Table 7.13), maximum hourly volume = $0.92 \times 101 = 93$ vph.

The problem demonstrates the difficulty in achieving a level of service as high as B on a specific grade of this type, which will be achieved only when the demand service flow is very low.

Highway Systems Planning

This level of analysis can be used for planning and policy formulation studies where detailed traffic and geometric data are unknown. Projected ADT volumes for some future planning year are used as the traffic demand. Estimated maximum ADT volumes have been established for different levels of service and design hour factors K for different terrains (see Table 7.19). The ADT volumes were obtained from Eq. 7.11.

$$\text{ADT}_i = \frac{SF_i}{K} \text{ (PHF)} \qquad \qquad (7.11)$$

where

ADT$_i$ = maximum average daily traffic permitted for level of service i, in number of vehicles per day (vpd)

PHF = peak hour factor

SF_i = service volume limit for level of service i in vph (mixed vehicles per hour total for both directions); that is, the maximum volume in number of vehicles for which the level of service i can be maintained

K = 30th highest hourly volume factor (see Chapter 14 for definition of 30th highest hourly volume, where K = 30th highest hourly volume/ADT.

The levels of service given in Table 7.19 are based on the peak 15 min period of the design hour and the following assumptions were made to develop the ADT volumes.

Traffic mix: 14 percent trucks, 4 percent recreational vehicles, and 0 percent buses

Directional split: 60/40

Percentage of no-passing zones: 20 percent for level terrain, 40 percent for rolling terrain, and 60 percent for mountainous terrain

Lane width: 12 ft

Shoulder width: 6 ft

Design speed: 60 mph minimum

Example 7–6 Computing Level of Service Using the Planning Level of Analysis

The traffic growth rate on an existing 2-lane rural highway is 5 percent and the present ADT is 2000 vpd. If the highway is located on a rolling terrain and the K

factor for the road is 0.13, what will be the level of service 10 years from now? If the policy of the state does not permit a level of service E on its rural highways, determine by which year the road should be widened.

- Determine the ADT in 10 years by calculating the increase annually over the period of 10 years using the compound factor, which is given as $(1 + i)^n$, where i is the annual growth rate and n is the number of years.

$$\text{Present ADT} = 2000$$

$$\text{ADT 10 years from now} = 2000(1 + 0.05)^{10} = 3258$$

- Use Table 7.19.

$$2100 \ (\text{LOSB}) < \text{ADT in 10 years} < 4000 \ (\text{LOSC})$$

In 10 years time level of service will be C.
- Determine the number of years in which the highway should be widened to operate at level of service D (ADT = 6100).

$$6100 = 2000(1 + 0.05)^n$$

$$3.05 = (1.05)^n$$

$$n = 22.85 \text{ years}$$

SUMMARY

Traffic engineers frequently are engaged in evaluating the performance of different facilities of the highway system. These facilities include highway segments, intersections, ramps, and so forth. The level of service is usually taken as a good indication of how well the particular component is operating. It is a qualitative measure of motorists' perception of the operational conditions existing on the facility. A primary objective in traffic engineering is to provide highway facilities that operate at levels of service acceptable by the users of those facilities. Regular evaluation of the level of service at the facilities will help the engineers to determine whether acceptable conditions exist and to identify those locations where improvements may be necessary. Levels of different operating conditions are assigned to different levels of service, ranging from level of service A to level of service F for each facility. The levels of different operating conditions are also related to the volume of traffic that can be accommodated by the specific component. This amount of traffic is also related to the capacity of the facility.

This chapter discusses the different procedures for determining the level of service on highway segments as presented in the *Highway Capacity Manual* of 1985. The procedures are based on the results of several major studies, from which several empirical expressions have been developed to determine the maximum flow rate at a specific facility under prevailing conditions. It is, however, suggested that in cases where specific information is available or can be easily obtained, that this information be used.

PROBLEMS

7–1 Briefly describe the traffic characteristics associated with each of the six levels of service for basic freeway segments.

7–2 A freeway is being designed to carry a DDHV of 5000 vph on a regular weekday in a rolling terrain. If the PHF is 0.9 and the traffic consists of 90 percent passenger cars and 10 percent trucks, determine the number of 12 ft lanes required in each direction if the highway is to operate at level of service C. The design speed is 70 mph and there is no lateral obstruction.

7–3 A section of a 4-lane (2 lanes in each direction) freeway, 2 mi long and having a sustained grade of 4 percent is to be improved to carry a DDHV of 3000 vph, consisting of 85 percent passenger cars, 10 percent trucks, 2 percent buses, and 3 percent recreational vehicles on a regular weekday. The PHF is 0.95. Determine the additional number of 12 ft lanes required in each direction if the road is to operate at level of service B. The design speed is 70 mph and there is a lateral obstruction 5 ft from the pavement on one side of the road.

7–4 Describe the three types of level of service analysis that can be carried out for a 2-lane, two-way highway, giving examples to show when each type is suitable.

7–5 A 3 mi section of a 2-lane, two-way highway located in a mountainous terrain is to be designed to carry an ADT of 750 vehicles. Using the operational level of analysis, determine the level of service at which the highway will operate for the following traffic and geometric characteristics.

    Lane width = 12 ft
    Shoulder width = 6 ft
    No-passing zones = 10 percent
    Directional split = 60/40
    Decimal fraction of recreational vehicles = 0.00
    Decimal fraction of buses = 0.00
    Decimal fraction of trucks = 0.08
    PHF = 0.95

7–6 A 4 mi section of a rural 2-lane, two-way highway is being designed in a rolling terrain. Design features include 11 ft lanes, 6 ft paved shoulder, and 10 percent no-passing zones. The traffic will consist of 85 percent passenger cars and 15 percent trucks, and directional distribution during peak traffic conditions will be 50/50. What will be the average general speed and percent time delay during the peak hour at this section of the highway if the estimated peak hour volume is 1300 vehicles. Also determine the capacity of this stretch of the highway. PHF = 0.95.

7–7 A specific 2 mi stretch of a 2-lane rural highway has a 4 percent grade. The highway has 11 ft lanes and 6 ft stabilized shoulders. The average no-passing zone rate along this grade is 40 percent and studies have shown that the direction split during peak traffic conditions is 70/30 in the upgrade direction. The traffic consists of 80 percent passenger cars, 15 percent trucks, and 5 percent recreational vehicles. Determine the maximum hourly volume that can be maintained on this stretch of highway if the average upgrade speed should not be less than 45 mph. PHF = 0.95.

7–8 If the specific grade in problem 7–7 is to operate at capacity, what will be the maximum hourly volume that can be maintained?

7–9 A 2-lane, two-way highway located in a mountainous terrain is currently carrying a maximum ADT of 900 vpd. The 30th highest hour factor is estimated to be 0.12. Using highway systems level of analysis, determine the LOS at which the road will be operating in 15 yr if the growth of traffic is 6 percent annually. PHF = 0.95.

7–10 An existing 2-lane, two-way rural highway located in a rolling terrain is operating at level of service C. Studies have shown that the growth rate of traffic on the highway is 5 percent per annum, and it is believed that this rate of traffic will continue for several years. The K factor for the road is 0.13. In how many years will the number of lanes have to be increased if the level of service on the road must not be worse than D?

7–11 Plans are being developed for a 2-lane, two-way highway, which will be located on a rural rolling terrain. The 30th highest hour factor is estimated to be 0.12. If the highway agency uses level of service D for planning purposes, what is the service volume the highway can handle?

REFERENCES

1. *Interim Material on Highway Capacity*, TRB Circular No. 212, Transportation Research Board, National Research Council, Washington, D.C., January 1980.

2. C. J. Messler, *Two-Lane, Two-Way Rural Highway Capacity*, NCHRP Project 3–28A Final Report, Transportation Research Board, National Research Council, Washington, D.C., March 1983.

3. *Highway Capacity Manual*, Special Report 209, Transportation Research Board, National Research Council, Washington, D.C., 1985.

4. L. J. Pignataro et al., *Weaving Areas: Design and Evaluation*, National Cooperative Highway Research Programs Report 159, Transportation Research Board, National Research Council, Washington, D.C., 1975.

5. A. D. St. John et al., *Freeway Design and Control Strategies as Affected by Trucks and Traffic Regulations*, Report No. FHWA–RD–75–42, Midwest Research Institute, April 1975.

6. A. D. St. John, *Grade Effects on Traffic Flow Stability and Capacity*, NCHRP 3–19, Midwest Research Institute, August 1974.

7. J. Leish, *Capacity Analysis Techniques for Design and Operation of Freeway Facilities*, U.S. Department of Transportation, Federal Highway Administration, Washington, D.C., 1974.

8. *Review of Vehicle Weight/Horse Power Ratio as Related to Passing Lane Design*, NCHRP Project 20–7, Pennsylvania State University, State College, Pa., 1978.

9. Leish, *Capacity Analysis Techniques*.

10. Ibid.

11. *Highway Capacity Manual.*

12. Ibid.

13. Leish, *Capacity Analysis Techniques.*

14. *Highway Capacity Manual.*

15. Ibid.

CHAPTER 7 APPENDIX: TABLES

All tables in this appendix are reprinted from *Highway Capacity Manual*, Special Report 209, Transportation Research Board, Washington, D.C., 1985.

Table 7.1 Level of Service (LOS) for Basic Freeway Sections

LOS	Density (pc/mi/ln)	70 mph Design Speed			60 mph Design Speed			50 mph Design Speed		
		Speed[b] (mph)	v/c	MSF[a] (pc/hr/ln)	Speed[b] (mph)	v/c	MSF[a] (pc/hr/ln)	Speed[b] (mph)	v/c	MSF[a] (pc/hr/ln)
A	≤12	≥60	0.35	700	—	—	—	—	—	—
B	≤20	≥57	0.54	1,100	≥50	0.49	1,000	—	—	—
C	≤30	≥54	0.77	1,550	≥47	0.69	1,400	≥43	0.67	1,300
D	≤42	≥46	0.93	1,850	≥42	0.84	1,700	≥40	0.83	1,600
E	≤67	≥30	1.00	2,000	≥30	1.00	2,000	≥28	1.00	1,900
F	>67	<30	c	c	<30	c	c	<28	c	c

[a]Maximum service flow rate per lane under ideal conditions.
[b]Average travel speed.
[c]Highly variable, unstable.
Note: All values of MSF rounded to the nearest 50 pc/hr.

Table 7.2   Adjusted Factor for Restricted Lane Width and Lateral Clearance

DISTANCE FROM TRAVELED PAVEMENT[a] (FT)	ADJUSTMENT FACTOR, f_w							
	OBSTRUCTIONS ON ONE SIDE OF THE ROADWAY				OBSTRUCTIONS ON BOTH SIDES OF THE ROADWAY			
	LANE WIDTH (FT)							
	12	11	10	9	11	12	10	9
4-LANE FREEWAY (2 LANES EACH DIRECTION)								
≥ 6	1.00	0.97	0.91	0.81	0.97	1.00	0.91	0.81
5	0.99	0.96	0.90	0.80	0.96	0.99	0.90	0.80
4	0.99	0.96	0.90	0.80	0.95	0.98	0.89	0.79
3	0.98	0.95	0.89	0.79	0.93	0.96	0.87	0.77
2	0.97	0.94	0.88	0.79	0.91	0.94	0.86	0.76
1	0.93	0.90	0.85	0.76	0.85	0.87	0.80	0.71
0	0.90	0.87	0.82	0.73	0.79	0.81	0.74	0.66
6- or 8- LANE FREEWAY (3 or 4 LANES EACH DIRECTION)								
≥ 6	1.00	0.96	0.89	0.78	0.96	1.00	0.89	0.78
5	0.99	0.95	0.88	0.77	0.95	0.99	0.88	0.77
4	0.99	0.95	0.88	0.77	0.94	0.98	0.87	0.77
3	0.98	0.94	0.87	0.76	0.93	0.97	0.86	0.76
2	0.97	0.93	0.87	0.76	0.92	0.96	0.85	0.75
1	0.95	0.92	0.86	0.75	0.89	0.93	0.83	0.72
0	0.94	0.91	0.85	0.74	0.87	0.91	0.81	0.70

[a] Certain types of obstructions, high-type median barriers in particular, do not cause any deleterious effect on traffic flow. Judgment should be exercised in applying these factors.

Table 7.3   PCEs in Extended Basic Freeway Segments

FACTOR	TYPE OF TERRAIN		
	LEVEL	ROLLING	MOUNTAINOUS
E_T for Trucks	1.7	4.0	8.0
E_B for Buses	1.5	3.0	5.0
E_R for RV's	1.6	3.0	4.0

Table 7.4 PCEs for Typical Trucks (200 lb/hp) on Specific Freeway Upgrades

GRADE (%)	LENGTH (MI)	PASSENGER-CAR EQUIVALENT, E_T															
		4-LANE FREEWAYS								6–8 LANE FREEWAYS							
PERCENT TRUCKS		2	4	5	6	8	10	15	20	2	4	5	6	8	10	15	20
<1	All	2	2	2	2	2	2	2	2	2	2	2	2	2	2	2	2
1	0–1/2	2	2	2	2	2	2	2	2	2	2	2	2	2	2	2	2
	1/2–1	3	3	3	3	3	3	3	3	3	3	3	3	3	3	3	3
	≥1	4	3	3	3	3	3	3	3	4	3	3	3	3	3	3	3
2	0–1/4	4	4	4	3	3	3	3	3	4	4	4	3	3	3	3	3
	1/4–1/2	5	4	4	4	3	3	3	3	5	5	4	4	3	3	3	3
	1/2–3/4	6	5	5	4	4	4	4	4	6	5	5	4	4	4	4	4
	3/4–1½	7	6	6	5	4	4	4	4	7	6	5	5	4	4	4	4
	≥1½	8	6	6	6	5	5	4	4	8	6	6	5	5	5	4	4
3	0–1/4	6	5	5	5	4	4	4	3	6	5	5	5	4	4	4	3
	1/4–1/2	8	6	6	6	5	5	5	4	7	6	6	6	5	5	5	4
	1/2–1	9	7	7	6	5	5	5	5	9	7	7	6	5	5	5	5
	1–1½	9	7	7	7	6	6	5	5	9	7	7	6	5	5	5	5
	≥1½	10	7	7	7	6	6	5	5	10	7	7	6	6	6	5	5
4	0–1/4	7	6	6	5	4	4	4	4	7	6	6	5	4	4	4	4
	1/4–1/2	10	7	7	6	5	5	5	5	9	7	7	6	5	5	5	5
	1/2–1	12	8	8	7	6	6	6	6	10	8	7	6	5	5	5	6
	≥1	13	9	9	9	8	8	7	7	11	9	9	8	7	6	6	7
5	0–1/4	8	6	6	6	5	5	5	5	8	6	6	6	5	5	5	5
	1/4–1/2	10	8	8	7	6	6	6	6	8	7	7	6	5	6	6	6
	1/2–1	12	11	11	10	8	8	8	8	12	10	9	8	7	7	7	8
	≥1	14	11	11	10	8	8	8	8	12	10	9	8	7	7	7	8
6	0–1/4	9	7	7	7	6	6	6	6	9	7	7	6	5	5	5	5
	1/4–1/2	13	9	9	8	7	7	7	7	11	8	8	7	6	6	6	6
	1/2–3/4	13	9	9	8	7	7	7	7	11	9	9	8	7	6	6	6
	≥3/4	17	12	12	11	9	9	9	9	13	10	10	9	8	8	8	8

NOTE: If a length of grade falls on a boundary condition, the equivalent for the longer grade category is used. For any grade steeper than the percentage shown, use the next higher grade category.

Table 7.5   PCEs for Light Trucks (100 lb./hp) on Specific Freeway Upgrades

GRADE (%)	LENGTH (MI)	PASSENGER-CAR EQUIVALENT, E_T															
		4-LANE FREEWAYS								6–8 LANE FREEWAYS							
	PERCENT TRUCKS	2	4	5	6	8	10	15	20	2	4	5	6	8	10	15	20
≤2	All	2	2	2	2	2	2	2	2	2	2	2	2	2	2	2	2
3	0-1/4	3	3	3	3	3	3	3	3	3	3	3	3	3	3	3	3
	1/4-1/2	4	4	4	3	3	3	3	3	4	4	4	3	3	3	3	3
	1/2-3/4	4	4	4	4	3	3	3	3	4	4	4	3	3	3	3	3
	3/4-1	5	4	4	4	3	3	3	3	5	4	4	4	3	3	3	3
	>1	6	5	5	5	4	4	4	3	6	5	5	4	4	4	3	3
4	0-1/4	4	4	4	3	3	3	3	3	5	4	4	4	3	3	3	3
	1/4-1/2	5	5	5	4	4	4	4	4	5	5	4	4	4	4	4	4
	1/2-1	6	5	5	5	4	4	4	4	6	5	5	4	4	4	4	4
	>1	7	6	6	5	4	4	4	4	7	5	5	5	4	4	4	4
5	0-1/4	6	5	5	5	4	4	4	3	6	5	5	5	4	4	4	3
	1/4-1	8	7	7	6	5	5	5	5	8	7	7	6	5	5	5	5
	>1	9	7	7	6	5	5	5	5	8	7	7	6	5	5	5	5
6	0-1/4	7	5	5	5	4	4	4	4	7	5	5	5	4	4	3	3
	1/4-1	9	7	7	6	5	5	5	5	8	7	7	6	5	5	5	5
	>1	9	7	7	7	6	6	5	5	9	7	7	6	5	5	5	5

NOTE: If a length of grade falls on a boundary condition, the equivalent from the longer grade category is used. For any grade steeper than the percentage shown, use the next higher grade category.

Table 7.6 PCEs for Heavy Trucks (300 lb/hp) on Specific Freeway Upgrades

GRADE (%)	LENGTH (MI)	PASSENGER-CAR EQUIVALENT, E_T															
		4-LANE FREEWAYS								6–8 LANE FREEWAYS							
PERCENT TRUCKS		2	4	5	6	8	10	15	20	2	4	5	6	8	10	15	20
<1	All	2	2	2	2	2	2	2	2	2	2	2	2	2	2	2	2
1	0-1/4	2	2	2	2	2	2	2	2	2	2	2	2	2	2	2	2
	1/4-1/2	3	3	3	3	3	3	3	3	3	3	3	3	3	3	3	3
	1/2-3/4	4	4	4	4	3	3	3	3	4	4	4	3	3	3	3	3
	3/4-1	5	4	4	4	4	3	3	3	5	4	4	4	3	3	3	3
	1-1½	6	5	5	5	4	4	4	3	6	5	5	4	4	4	3	3
	>1½	7	5	5	5	4	4	4	3	7	5	5	5	4	4	3	3
2	0-1/4	4	4	4	3	3	3	3	3	4	4	4	3	3	3	3	3
	1/4-1/2	7	6	6	5	4	4	4	4	7	5	5	5	4	4	4	4
	1/2-3/4	8	6	6	5	5	4	4	4	8	6	6	6	5	5	4	4
	3/4-1	8	6	6	6	5	5	5	5	8	6	6	6	5	5	5	5
	1-1½	9	7	7	7	6	6	5	5	9	7	7	6	5	5	5	5
	>1½	10	7	7	7	6	6	5	5	10	7	7	6	5	5	5	5
3	0-1/4	6	5	5	5	4	4	4	3	6	5	5	5	4	4	4	3
	1/4-1/2	9	7	7	6	5	5	5	5	8	7	7	6	5	5	5	5
	1/2-3/4	12	8	8	7	6	6	6	6	10	8	7	7	6	6	5	5
	3/4-1	13	9	9	8	7	7	7	7	11	8	8	7	6	6	6	6
	>1	14	10	10	9	8	8	7	7	12	9	9	8	7	7	7	7
4	0-1/4	7	5	5	5	4	4	4	4	7	5	5	5	4	4	3	3
	1/4-1/2	12	8	8	7	6	6	6	6	10	8	7	6	5	5	5	5
	1/2-3/4	13	9	9	8	7	7	7	7	11	9	9	8	7	6	6	6
	3/4-1	15	10	10	9	8	8	8	8	12	10	10	9	8	7	7	7
	>1	17	12	12	10	9	9	9	9	13	10	10	9	8	8	8	8
5	0-1/4	8	6	6	6	5	5	5	5	8	6	6	6	5	5	5	5
	1/4-1/2	13	9	9	8	7	7	7	7	11	8	8	7	6	6	6	6
	1/2-3/4	20	15	15	14	11	11	11	11	14	11	11	10	9	9	9	9
	>3/4	22	17	17	16	13	13	13	13	17	14	14	13	12	11	11	11
6	0-1/4	9	7	7	7	6	6	6	6	9	7	7	6	5	5	5	5
	1/4-1/2	17	12	12	11	9	9	9	9	13	10	10	9	8	8	8	8
	>1/2	28	22	22	21	18	18	18	18	20	17	17	16	15	14	14	14

NOTE: If a length of grade falls on a boundary condition, the equivalent from the longer grade condition is used. For any grade steeper than the percent shown, use the next higher grade category.

Table 7.7 PCEs for Recreational Vehicles on Specific Freeway Upgrades

GRADE (%)	LENGTH (MI)	PASSENGER-CAR EQUIVALENT, E_R															
		4-LANE FREEWAYS								6-8 LANE FREEWAYS							
	PERCENT RV's	2	4	5	6	8	10	15	20	2	4	5	6	8	10	15	20
<2	All	2	2	2	2	2	2	2	2	2	2	2	2	2	2	2	2
3	0–1/2	3	2	2	2	2	2	2	2	2	2	2	2	2	2	2	2
	≥1/2	4	3	3	3	3	3	3	3	4	3	3	3	3	3	3	3
4	0–1/4	3	2	2	2	2	2	2	2	3	2	2	2	2	2	2	2
	1/4–3/4	4	3	3	3	3	3	3	3	4	3	3	3	3	3	3	3
	≥3/4	5	4	4	4	3	3	3	3	4	4	4	4	3	3	3	3
5	0–1/4	4	3	3	3	3	3	3	3	4	3	3	3	3	3	3	3
	1/4–3/4	5	4	4	4	4	4	4	4	5	4	4	4	4	4	4	4
	≥3/4	6	5	4	4	4	4	4	4	5	4	4	4	4	4	4	4
6	0–1/4	5	4	4	4	3	3	3	3	5	4	4	4	3	3	3	3
	1/4–3/4	6	5	5	4	4	4	4	4	6	5	5	4	4	4	4	4
	≥3/4	7	6	6	6	5	5	5	5	6	5	5	5	4	4	4	4

NOTE: If a length of grade falls on a boundary condition, the equivalent from the longer grade category is used. For any grade steeper than the percent shown, use the next higher grade category.

Table 7.8 PCEs for Buses on Specific Freeway Upgrades

GRADE (%)	PASSENGER-CAR EQUIVALENT, E_B
0–3	1.6
4[a]	1.6
5[a]	3.0
6[a]	5.5

[a] Use generally restricted to grades more than 1/4 mi long.

Table 7.9 Adjustment Factors for the Character of the Traffic on Basic Freeway Segments

TRAFFIC STREAM TYPE	FACTORS, f_p
Weekday or Commuter	1.0
Other	0.75–0.90[a]

[a] Engineering judgment and/or local data must be used in selecting an exact value.

Table 7.10 Adjustment Factors for Directional Distribution on General Terrain Segments of 2-Lane Highways

Directional Distribution	100/0	90/10	80/20	70/30	60/40	50/50
Adjustment Factor, f_d	0.71	0.75	0.83	0.89	0.94	1.00

Table 7.11 Level of Service Criteria for General 2-Lane Highways

			LEVEL TERRAIN								ROLLING TERRAIN								MOUNTAINOUS TERRAIN						
				v/c RATIO[a]																				
	PERCENT TIME DELAY	AVG[b] SPEED	PERCENT NO PASSING ZONES						AVG[b] SPEED	PERCENT NO PASSING ZONES						AVG[b] SPEED	PERCENT NO PASSING ZONES					
LOS			0	20	40	60	80	100		0	20	40	60	80	100		0	20	40	60	80	100
A	≤ 30	≥ 58	0.15	0.12	0.09	0.07	0.05	0.04	≥ 57	0.15	0.10	0.07	0.05	0.04	0.03	≥ 56	0.14	0.09	0.07	0.04	0.02	0.01
B	≤ 45	≥ 55	0.27	0.24	0.21	0.19	0.17	0.16	≥ 54	0.26	0.23	0.19	0.17	0.15	0.13	≥ 54	0.25	0.20	0.16	0.13	0.12	0.10
C	≤ 60	≥ 52	0.43	0.39	0.36	0.34	0.33	0.32	≥ 51	0.42	0.39	0.35	0.32	0.30	0.28	≥ 49	0.39	0.33	0.28	0.23	0.20	0.16
D	≤ 75	≥ 50	0.64	0.62	0.60	0.59	0.58	0.57	≥ 49	0.62	0.57	0.52	0.48	0.46	0.43	≥ 45	0.58	0.50	0.45	0.40	0.37	0.33
E	> 75	≥ 45	1.00	1.00	1.00	1.00	1.00	1.00	≥ 40	0.97	0.94	0.92	0.91	0.90	0.90	≥ 35	0.91	0.87	0.84	0.82	0.80	0.78
F	100	< 45	—	—	—	—	—	—	< 40	—	—	—	—	—	—	< 35	—	—	—	—	—	—

[a] Ratio of flow rate to an ideal capacity of 2,800 pcph in both directions.
[b] Average travel speed of all vehicles (in mph) for highways with design speed ≥ 60 mph; for highways with lower design speeds, reduce speed by 4 mph for each 10-mph reduction in design speed below 60 mph; assumes that speed is not restricted to lower values by regulation.

Table 7.12 Level of Service Criteria for Specific Grades on 2-Lane Highways

LEVEL OF SERVICE	AVERAGE UPGRADE SPEED (MPH)
A	≥ 55
B	≥ 50
C	≥ 45
D	≥ 40
E	≥ 25–40[a]
F	< 25–40[a]

[a] The exact speed at which capacity occurs varies with the percentage and length of grade, traffic compositions, and volume; computational procedures are provided to find this value.

Table 7.13 PHF for 2-Lane Highways Based on Random Flow

A. LEVEL-OF-SERVICE DETERMINATIONS

TOTAL 2-WAY HOURLY VOLUME (VPH)	PEAK HOUR FACTOR (PHF)	TOTAL 2-WAY HOURLY VOLUME (VPH)	PEAK HOUR FACTOR (PHF)
100	0.83	1,000	0.93
200	0.87	1,100	0.94
300	0.90	1,200	0.94
400	0.91	1,300	0.94
500	0.91	1,400	0.94
600	0.92	1,500	0.95
700	0.92	1,600	0.95
800	0.93	1,700	0.95
900	0.93	1,800	0.95
		≥ 1,900	0.96

B. SERVICE FLOW-RATE DETERMINATIONS

Level of Service	A	B	C	D	E
Peak Hour Factor	0.91	0.92	0.94	0.95	1.00

Table 7.14 Adjustment Factors (f_w) for the Combined Effect of Narrow Lanes and Restricted Shoulder Widths of 2-Lane Highways

USABLE[a] SHOULDER WIDTH (FT)	12-FT LANES		11-FT LANES		10-FT LANES		9-FT LANES	
	LOS A–D	LOS[b] E	LOS A–D	LOS[b] E	LOS A–D	LOS[b] E	LOS A–D	LOS[b] E
≥ 6	1.00	1.00	0.93	0.94	0.84	0.87	0.70	0.76
4	0.92	0.97	0.85	0.92	0.77	0.85	0.65	0.74
2	0.81	0.93	0.75	0.88	0.68	0.81	0.57	0.70
0	0.70	0.88	0.65	0.82	0.58	0.75	0.49	0.66

[a] Where shoulder width is different on each side of the roadway, use the average shoulder width.
[b] Factor applies for all speeds less than 45 mph.

Table 7.15 Average PCEs for Trucks, Recreational Vehicles, and Buses on 2-Lane Highways over General Terrain

VEHICLE TYPE	LEVEL OF SERVICE	TYPE OF TERRAIN		
		LEVEL	ROLLING	MOUNTAINOUS
Trucks, E_T	A	2.0	4.0	7.0
	B and C	2.2	5.0	10.0
	D and E	2.0	5.0	12.0
RV's E_R	A	2.2	3.2	5.0
	B and C	2.5	3.9	5.2
	D and E	1.6	3.3	5.2
Buses, E_B	A	1.8	3.0	5.7
	B and C	2.0	3.4	6.0
	D and E	1.6	2.9	6.5

Table 7.16 Adjustment Factors for Directional Distribution on Specific Grades of 2-Lane Highways

PERCENT OF TRAFFIC ON UPGRADE	ADJUSTMENT FACTOR
100	0.58
90	0.64
80	0.70
70	0.78
60	0.87
50	1.00
40	1.20
30	1.50

Table 7.17 Values of v/c Ratio* Versus Speed, Percent Grade, and Percent No-passing Zones for Specific Grades

PERCENT GRADE	AVERAGE UPGRADE SPEED (MPH)	PERCENT NO PASSING ZONES					
		0	20	40	60	80	100
3	55	0.27	0.23	0.19	0.17	0.14	0.12
	52.5	0.42	0.38	0.33	0.31	0.29	0.27
	50	0.64	0.59	0.55	0.52	0.49	0.47
	45	1.00	0.95	0.91	0.88	0.86	0.84
	42.5	1.00	0.98	0.97	0.96	0.95	0.94
	40	1.00	1.00	1.00	1.00	1.00	1.00
4	55	0.25	0.21	0.18	0.16	0.13	0.11
	52.5	0.40	0.36	0.31	0.29	0.27	0.25
	50	0.61	0.56	0.52	0.49	0.47	0.45
	45	0.97	0.92	0.88	0.85	0.83	0.81
	42.5	0.99	0.96	0.95	0.94	0.93	0.92
	40	1.00	1.00	1.00	1.00	1.00	1.00
5	55	0.21	0.17	0.14	0.12	0.10	0.08
	52.5	0.36	0.31	0.27	0.24	0.22	0.20
	50	0.57	0.49	0.45	0.41	0.39	0.37
	45	0.93	0.84	0.79	0.75	0.72	0.70
	42.5	0.97	0.90	0.87	0.85	0.83	0.82
	40	0.98	0.96	0.95	0.94	0.93	0.92
	35	1.00	1.00	1.00	1.00	1.00	1.00
6	55	0.12	0.10	0.08	0.06	0.05	0.04
	52.5	0.27	0.22	0.18	0.16	0.14	0.13
	50	0.48	0.40	0.35	0.31	0.28	0.26
	45	0.49	0.76	0.68	0.63	0.59	0.55
	42.5	0.93	0.84	0.78	0.74	0.70	0.67
	40	0.97	0.91	0.87	0.83	0.81	0.78
	35	1.00	0.96	0.95	0.93	0.91	0.90
	30	1.00	0.99	0.99	0.98	0.98	0.98
7	55	0.00	0.00	0.00	0.00	0.00	0.00
	52.5	0.13	0.10	0.08	0.07	0.05	0.04
	50	0.34	0.27	0.22	0.18	0.15	0.12
	45	0.77	0.65	0.55	0.46	0.40	0.35
	42.5	0.86	0.75	0.67	0.60	0.54	0.48
	40	0.93	0.82	0.75	0.69	0.64	0.59
	35	1.00	0.91	0.87	0.82	0.79	0.76
	30	1.00	0.95	0.92	0.90	0.88	0.86

* Ratio of flow rate to ideal capacity of 2,800 pcph, assuming passenger-car operation is unaffected by grade.

NOTE: Interpolate for intermediate values of "Percent No Passing Zone"; round "Percent Grade" to the next higher integer value.

Table 7.18 PCEs for Specific Grades on 2-Lane Highways

GRADE (%)	LENGTH OF GRADE (MI)	AVERAGE UPGRADE SPEED (MPH)					
		55.0	52.5	50.0	45.0	40.0	30.0
0	All	2.1	1.8	1.6	1.4	1.3	1.3
3	¼	2.9	2.3	2.0	1.7	1.6	1.5
	½	3.7	2.9	2.4	2.0	1.8	1.7
	¾	4.8	3.6	2.9	2.3	2.0	1.9
	1	6.5	4.6	3.5	2.6	2.3	2.1
	1½	11.2	6.6	5.1	3.4	2.9	2.5
	2	19.8	9.3	6.7	4.6	3.7	2.9
	3	71.0	21.0	10.8	7.3	5.6	3.8
	4	a	48.0	20.5	11.3	7.7	4.9
4	¼	3.2	2.5	2.2	1.8	1.7	1.6
	½	4.4	3.4	2.8	2.2	2.0	1.9
	¾	6.3	4.4	3.5	2.7	2.3	2.1
	1	9.6	6.3	4.5	3.2	2.7	2.4
	1½	19.5	10.3	7.4	4.7	3.8	3.1
	2	43.0	16.1	10.8	6.9	5.3	3.8
	3	a	48.0	20.0	12.5	9.0	5.5
	4	a	a	51.0	22.8	13.8	7.4
5	¼	3.6	2.8	2.3	2.0	1.8	1.7
	½	5.4	3.9	3.2	2.5	2.2	2.0
	¾	8.3	5.7	4.3	3.1	2.7	2.4
	1	14.1	8.4	5.9	4.0	3.3	2.8
	1½	34.0	16.0	10.8	6.3	4.9	3.8
	2	91.0	28.3	17.4	10.2	7.5	4.8
	3	a	a	37.0	22.0	14.6	7.8
	4	a	a	a	55.0	25.0	11.5

Table 7.18—Continued

GRADE (%)	LENGTH OF GRADE (MI)	AVERAGE UPGRADE SPEED (MPH)					
		55.0	52.5	50.0	45.0	40.0	30.0
6	¼	4.0	3.1	2.5	2.1	1.9	1.8
	½	6.5	4.8	3.7	2.8	2.4	2.2
	¾	11.0	7.2	5.2	3.7	3.1	2.7
	1	20.4	11.7	7.8	4.9	4.0	3.3
	1½	60.0	25.2	16.0	8.5	6.4	4.7
	2	a	50.0	28.2	15.3	10.7	6.3
	3	a	a	70.0	38.0	23.9	11.3
	4	a	a	a	90.0	45.0	18.1
7	¼	4.5	3.4	2.7	2.2	2.0	1.9
	½	7.9	5.7	4.2	3.2	2.7	2.4
	¾	14.5	9.1	6.3	4.3	3.6	3.0
	1	31.4	16.0	10.0	6.1	4.8	3.8
	1½	a	39.5	23.5	11.5	8.4	5.8
	2	a	88.0	46.0	22.8	15.4	8.2
	3	a	a	a	66.0	38.5	16.1
	4	a	a	a	a	a	28.0

[a] Speed not attainable on grade specified.
NOTE: Round "Percent Grade" to next higher integer value.

Table 7.19 Maximum AADTs Versus Level of Service and Type of Terrain for 2-Lane Rural Highways

K-FACTOR	LEVEL OF SERVICE				
	A	B	C	D	E
LEVEL TERRAIN					
0.10	2,400	4,800	7,900	13,500	22,900
0.11	2,200	4,400	7,200	12,200	20,800
0.12	2,000	4,000	6,600	11,200	19,000
0.13	1,900	3,700	6,100	10,400	17,600
0.14	1,700	3,400	5,700	9,600	16,300
0.15	1,600	3,200	5,300	9,000	15,200
ROLLING TERRAIN					
0.10	1,100	2,800	5,200	8,000	14,800
0.11	1,000	2,500	4,700	7,200	13,500
0.12	900	2,300	4,400	6,600	12,300
0.13	900	2,100	4,000	6,100	11,400
0.14	800	2,000	3,700	5,700	10,600
0.15	700	1,800	3,500	5,300	9,900
MOUNTAINOUS TERRAIN					
0.10	500	1,300	2,400	3,700	8,100
0.11	400	1,200	2,200	3,400	7,300
0.12	400	1,100	2,000	3,100	6,700
0.13	400	1,000	1,800	2,900	6,200
0.14	300	900	1,700	2,700	5,800
0.15	300	900	1,600	2,500	5,400

Note: All values rounded to the nearest 100 vpd. Assumed conditions include 60/40 directional split, 14 percent trucks, 4 percent RVs, no buses, and PHF values from Table 7.13. For level terrain, 20 percent no-passing zones were assumed; for rolling terrain, 40 percent no-passing zones; for mountainous terrain, 60 percent no-passing zones.

CHAPTER 8

Capacity and Level of Service at Signalized Intersections

The level of service at any intersection on a highway has a significant effect on the overall operating performance of that highway. Thus, improvement of the level of service at each intersection usually results in an improvement of the overall operating performance of the highway. An analysis procedure that provides for the determination of capacity or level of service at intersections is therefore an important tool for designers, operation personnel, and policy makers. Factors that affect the level of service at intersections include the flow and distribution of traffic, the geometric characteristics, and the signalization system.

A major difference between consideration of flow on highway segments and that at intersections is that only through flows are used in computing the levels of service at highway segments (see Chapter 7), whereas turning flows are of significant importance when computing the levels of service at signalized intersections. The signalization system, which includes the allocation of time among the conflicting movements of traffic and pedestrians at the intersection, is also an important factor. For example, the distribution of green times among these conflicting flows significantly affects both the capacity and operation of the intersection. Other factors such as lane widths, traffic composition, grade, and speed also affect the level of service at intersections in a similar manner as for highway segments.

Until recently, the procedures given in the 1965 *Highway Capacity Manual* (HCM), of the Transportation Research Board, have served as the principal means of determining levels of service. Research results since 1965 have indicated the need to revise the procedures given in that manual. This chapter presents procedures for determining the capacity and level of service at signalized intersections, based on results of some recent research projects and field observations.

DEFINITIONS OF SOME COMMON TERMS

Most of the terms commonly used in capacity and level of service analyses of signalized intersections were defined in Chapter 6. However, some additional terms need to be defined and others need to be redefined to understand their use in this chapter.

Permitted turning movements are those made within gaps of an opposing traffic stream or through a conflicting pedestrian flow. For example, when a right turn is made while pedestrians are crossing a conflicting crosswalk, the right turn is a permitted turning movement. Similarly, when a left turn is made between two consecutive vehicles of the opposing traffic stream, the left turn is a permitted turn. The suitability of permitted turns at a given intersection depends on the geometric characteristics of the intersection, the turning volume, and the opposing volume.

Protected turns are those turns protected from any conflicts with vehicles in an opposing stream or pedestrians on a conflicting crosswalk. A permitted turn takes more time than a similar protected turn and will use more of the available green time.

Traffic conditions is the term used to describe the details of the traffic characteristics at an approach. They include the distribution of the total volume on the approach into left-, through and, right-turn movements; the percentage distribution by vehicle type of the total flow on each approach; whether bus stops and/or parking bays are located within the intersection area; and the flow of pedestrians crossing the intersections.

Roadway conditions is a term used to describe the geometric characteristics of the approach. They include the number and width of lanes, grades, and the allocation of the lanes for different uses, including the designation of a parking lane.

Signalization conditions is a term used to describe the details of the signal operation. They include the type of signal control, the phasing sequence, the timing, and an evaluation of signal progression on each approach.

Flow ratio (v/s) is the ratio of the actual flow rate v on an approach or lane group to the saturation flow rate s.

CAPACITY AT SIGNALIZED INTERSECTIONS

The capacity at a signalized intersection is given for each approach and is defined as the maximum rate of flow on the approach that can go through the intersection under prevailing traffic, roadway, and signalized conditions. Capacity is given in vehicles per hour (vph), but it is based on the flow during a peak 15 min period. The capacity of the intersection as a whole is not considered, rather emphasis is placed on providing suitable facilities for the major movements of the intersections. Capacity is therefore meaningfully applied only to major movements or approaches of the intersection. Note also that, in comparison with other locations such as freeway segments, the capacity of an intersection approach is not as strongly correlated with the level of service. It is therefore necessary that both the level of service and the capacity be analyzed separately when signalized intersections are being evaluated.

In the capacity analysis, volume/capacity (v/c) ratios are computed for each movement and for the sum of the critical movements of lane groups. The (v/c) ratios are determined by dividing the peak 15 min rate of flow on an approach or lane group by the capacity of the approach or lane group. A lane group is a set of lanes on an approach carrying a set of traffic streams. Both the traffic streams and the geometric characteristics are taken into consideration when the lane groups are formulated. The basic principle, however, is to obtain the least number of lane groups that adequately describes the intersection operation.

The concept of a saturation flow or saturation flow rate s is used to determine the capacity of a lane group. The saturation flow rate is the maximum flow rate on the approach or lane group that can go through the intersection under prevailing traffic and roadway conditions when 100 percent effective green time is available. The saturation flow rate is given in units of vph of effective green time (vphg).

The capacity of an approach or lane group is given as

$$c_i = s_i(g_i/C) \tag{8.1}$$

where

c_i = capacity of lane group i in vph
s_i = saturation flow rate for lane group or approach i (vph of green, vphg)
(g_i/C) = green ratio for lane group or approach i
g_i = effective green for lane group i or approach i
C = cycle length

The ratio of flow to capacity (v/c) is usually referred to as the *degree of saturation* and can be expressed as

$$(v/c)_i = X_i = \frac{v_i}{s_i(g_i/C)} \tag{8.2}$$

where

$X_i = (v/c)$ ratio for lane group or approach i
v_i = actual flow rate for lane group or approach i (vph)
s_i = saturation flow for lane group or approach i (vphg)
g_i = effective green time for lane group i or approach i (sec)

It can be seen that when the flow rate equals capacity, X_i equals 1.00, and when flow rate equals zero, X_i equals zero.

When the overall intersection is to be evaluated with respect to its geometry and the total cycle time, the concept of critical (v/c) ratio (X_c) is used. The critical (v/c) ratio is usually obtained for the overall intersection but considers only the critical lane groups or approaches, which are those lane groups or approaches that have the maximum flow ratio (v/s) for each signal phase. For example, in a two-phase signalized intersection, if the north approach has a higher (v/s) ratio than the south approach, more time will be required for vehicles on the north approach to go through the intersection during the north–south green phase, and the phase length will be based on the green time requirements for the north approach. The north approach will therefore be the critical approach for the north–south phase. The critical v/c ratio for the whole intersection is given as

$$X_c = \sum_i (v/s)_{ci} \frac{C}{C - L} \qquad (8.3)$$

where

X_c = critical v/c ratio for the intersection

$\sum_i (v/s)_{ci}$ = summation of the ratios for all critical lanes, groups, or approaches

C = cycle length (sec)

L = total lost time per cycle computed as the sum of "start up" and change interval lost time, minus the portion of change interval used by vehicles for each critical signal phase

Eq. 8.3 can be used to estimate the signal timing for the intersection if this is unknown and a critical (v/c) ratio is specified for the intersection. When the critical (v/c) ratio is less than 1.00, the cycle length provided is adequate for all critical movements to go through the intersection if the green time is proportionately distributed to the different phases. That is, for the assumed phase sequence, all movements in the intersection will be provided with adequate green times if the total green time is proportionately divided among all phases. If the total green time is not properly allocated to the different phases, it is possible to have a critical (v/c) ratio of less than 1.00, but with one or more individual oversaturated movements within a cycle.

LEVEL OF SERVICE AT SIGNALIZED INTERSECTIONS

Several methods have been developed for determining the capacity and level of service at signalized intersections since the publication of the 1965 HCM. The procedures presented here are those given in the 1985 HCM. These procedures deal with the computation of the level of service at the intersection approaches and the level of service at the intersection as a whole. See Additional Readings at the end of this chapter for other methods of determining capacity and level of service at signalized intersections.

Delay is used to define the level of service at signalized intersections, since delay not only indicates the amount of lost travel time and fuel consumption, it is also a measure of the frustration and discomfort of motorists. Delay, however, depends on the red time, which in turn depends on the length of the cycle.

Reasonable levels of service can therefore be obtained for short cycle lengths, even though the (v/c) ratio is as high as 0.9. To the extent that signal coordination reduces delay, different levels of service may also be obtained for the same (v/c) ratio when the effect of signal coordination changes.

The procedures can be used for either a detailed or operational evaluation of a given intersection or a general planning estimate of the overall performance of an existing or planned signalized intersection. At the design level of analysis, more input data are required and a direct estimate of the level of service can be made. It is also possible at this level of analysis to determine the effect of changing signal timing.

OPERATION ANALYSIS

The procedure at the operation level of analysis can be used to determine the capacity or level of service at the approaches of an existing signalized intersection or the overall level of service at an existing intersection. The procedure can also be used in the detailed design of a given intersection. In using the procedure to analyze an existing signal, operational data such as phasing sequence, signal timing, and geometric details (lane widths, number of lanes) are known. The procedure is used to determine the level of service at which the intersection is performing in terms of delay, which is then related to the level of service at the intersection. In using the procedure for detailed design, the operational data usually are not known and therefore have to be computed or assumed. The delay and level of service are then determined.

The LOS criteria are given in terms of the average stopped delay per vehicle during an analysis period of 15 min. Six levels of service are prescribed and the criteria are shown in Table 8.1.*

Level of service A describes that level of operation at which the average delay per vehicle is 5.0 sec or less. At level of service A, vehicles arrive mainly during the green phases, resulting in only a few vehicles stopping at the intersection. Short cycle lengths may help in obtaining low delays.

Level of service B describes that level of operation at which delay per vehicle ranges from 5.1 to 15.0 sec. At level of service B, the number of vehicles stopped at the intersection is higher than that for level of service A, but progression is still good and the cycle length may also be short.

Level of service C describes that level of operation at which delay per vehicle ranges from 15.1 to 25 sec. At level of service C, many vehicles go through the intersection without stopping, but a significant number of vehicles are stopped. In addition, not all vehicles at an approach clear the intersection during a few cycles (cycle failure). The higher delay may be due to the significant number of vehicles arriving during the red phase (poor progression) and/or relatively long cycle lengths.

---

*For convenience in looking up values, all tables referenced in this chapter are gathered in an appendix to the chapter beginning on page 357.

Level of service D describes that level of operation at which the delay per vehicle ranges from 25.1 to 40 sec. At level of service D, vehicles are stopped at the intersection, resulting in the longer delay. The number of individual cycles failing is now noticeable. The longer delay at this level of service is due to a combination of two or more of several factors that include long cycle lengths, high (v/c) ratios, and unfavorable progression.

Level of service E describes that level of operation at which the delay per vehicle ranges from 40.1 to 60 sec. At level of service E, individual cycles frequently fail. This long delay, which is usually taken as the limit of acceptable delay, generally indicates high (v/c) ratios, long cycle lengths, and poor progression.

Level of service F describes that level of operation at which the delay per vehicle is greater than 60.0 sec. This long delay is usually unacceptable to most motorists. At level of service F, the phenomenon known as oversaturation usually occurs—that is, arrival flow rates are greater than the capacity at the intersection. Long delay can also occur as a result of poor progression and long cycle lengths. Note that this level of service can occur when approaches have high (v/c) ratios, which are less than 1.00, but also have many individual cycles failing.

It should be emphasized once more that, in contrast to other locations, the level of service at a signalized intersection does not have a simple one-to-one relationship with capacity. For example, at freeway segments, the (v/c) ratio is 1.00 at the upper limits of level of service E. At the signalized intersection, however, it is possible for the delay to be unacceptable at level of service F while the (v/c) ratio is less than 1.00, even as low as 0.75. When long delays occur at such (v/c) ratios, it may be due to a combination of two or more of the following conditions:

- Long cycle lengths
- Green time is not properly distributed, resulting in a longer red time for one or more lane groups—that is, there is one or more disadvantaged lane groups
- A poor signal progression, which results in a large percentage of the vehicles on the approach arriving during the red phase

It is also possible to have short delays at an approach when the (v/c) ratio equals 1.00—that is, saturated approach—which can occur if the following conditions exist:

- Short cycle length
- Favorable signal progression, resulting in a high percentage of vehicles arriving during the green phase

Clearly, level of service F does not necessarily indicate that the intersection, approach, or lane group is oversaturated, nor can it be automatically assumed that the demand flow is less than capacity for a level of service range of A to E. It is therefore imperative that both the capacity and level of service analyses be carried out when a signalized intersection is to be evaluated fully.

Methodology of Operation Analysis Procedure

In order to simplify the analysis procedure, the tasks involved are presented in the flow chart shown in Figure 8.1. The tasks have been divided into five modules: (1) input module; (2) volume adjustment module; (3) saturation flow rate module; (4) capacity analysis module; and (5) level of service module. Each of these modules will be discussed in turn, including a detailed description of each task involved.

Input Module

All tasks under this module involve the collection and presentation of the data that will be required for the analysis. The tasks involved are

- Identifying and recording the geometric characteristics
- Identifying and specifying the traffic conditions
- Specifying the signalized conditions

Figure 8.1   Flow Chart for Operation Analysis Procedure

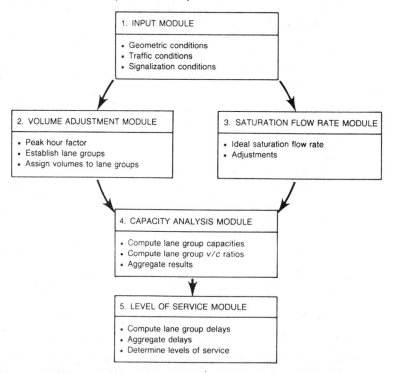

Source: Reproduced from *Highway Capacity Manual*, Special Report 209, Transportation Research Board, Washington, D.C., 1985.

Specifying Geometric Conditions. The physical configuration of the inter-section is obtained in terms of the number of lanes, lane width, movement by lane, parking locations, lengths of storage bays, and so forth and is recorded on the appropriate form, shown in Figure 8.2. In cases where the physical con-figuration of the intersection is unknown, the planning level of analysis may be used to determine a suitable configuration or state and local policies and/or guidelines can be used. If no guidelines are available, the following may be used.

1. Provide an exclusive left-turn lane when a fully protected left-turn phase is to be provided.

2. When left-turn volume is higher than 100 vph, an exclusive left-turn lane should be considered if space is available. Note however that an exclusive left-turn lane may be provided at volumes lower than 100 vph if state or local guidelines require it.

3. Consider the provision of double exclusive left-turn lanes when left-turn volumes are higher than 300 vph.

4. Length of storage bays should be adequate for the turning volume, so that the safety or capacity of the approach is not affected negatively. The length of the turning bay can be determined from Figure 8.3 and Table 8.2.*

5. When right-turn volumes are higher than 300 vph and adjacent mainline volume is also higher than 300 vph, provide an exclusive right-turn lane. Double exclusive right-turn lanes may be provided when right-turn volume is higher than 350 vph. This volume is higher than that for exclusive left-turn lanes because right-turning vehicles usually do not conflict with an opposing vehicle flow but face only a conflicting pedestrian flow. Right turns are therefore more efficient

6. Maximum volume on any through or through plus right-turn lane should not be greater than 450 vph. Note that, although this represents a reason-able starting point, major approaches with the necessary allocated green time will accommodate higher volumes.

7. Lane widths should be 12 ft, unless this is prevented by known restrictions.

8. Use the current local practice for parking at intersections to determine whether parking should be permitted.

These guidelines are presented as suggestions and not as specific design stan-dards. Local and state guidelines, if available, should take precedence over these suggestions.

---

*See the appendix to this chapter.

Figure 8.2 Input Worksheet for Operation Level of Analysis

| | | | **INPUT WORKSHEET** | | | |

Intersection:_____ Date:_____

Analyst:_____ Time Period Analyzed:_____ Area Type: ☐ CBD ☐ Other

Project No.:_____ City/State:_____

VOLUME AND GEOMETRICS

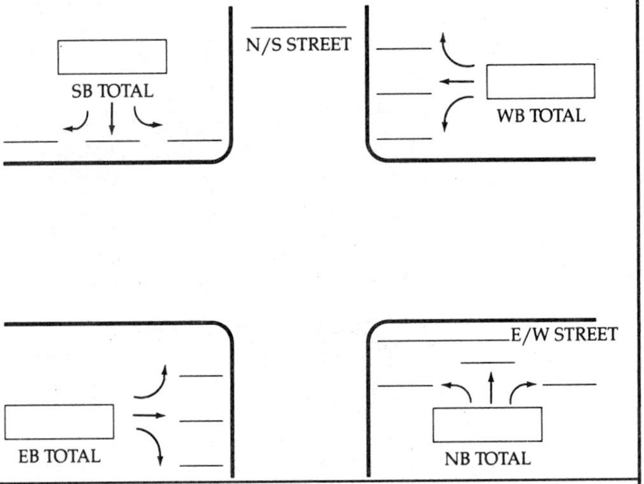

NORTH

SB TOTAL

N/S STREET

WB TOTAL

IDENTIFY IN DIAGRAM:

1. Volumes
2. Lanes, lane widths
3. Movements by lane
4. Parking (PKG) locations
5. Bay storage lengths
6. Islands (physical or painted)
7. Bus stops

EB TOTAL

E/W STREET

NB TOTAL

TRAFFIC AND ROADWAY CONDITIONS

Approach	Grade (%)	% HV	Adj. Pkg. Lane Y or N	Adj. Pkg. Lane N_m	Buses (N_B)	PHF	Conf. Peds. (peds./hr)	Pedestrian Button Y or N	Pedestrian Button Min. Timing	Arr. Type
EB										
WB										
NB										
SB										

Grade: + up, − down
HV: veh. with more than 4 wheels
N_m: pkg. maneuvers/hr

N_B: buses stopping/hr
PHF: peak-hour factor
Conf. Peds: Conflicting peds./hr

Min. Timing: min. green for pedestrian crossing
Arr. Type: Type 1-5

PHASING

DIAGRAM

Timing	G = Y + R =	G = Y + R =	G = Y + R =	G = Y + R =	G = Y + R =	G = Y + R =	G = Y + R =	G = Y + R =
Pretimed or Actuated								

Protected turns   Permitted turns   - - - - - - Pedestrian   Cycle Length_____Sec

Source: Reproduced from *Highway Capacity Manual*, Special Report 209, Transportation Research Board, Washington, D.C., 1985.

Figure 8.3 Left-Turn Bay Length Versus Turning Volume

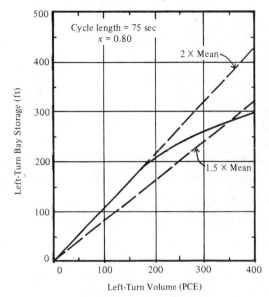

Notes: 1. Relationship is based on random arrivals and 5 percent probability of storage bay overflow. 2. Convert total mixed volume to PCE. Use PCE factors given in Table 8.2b. 3. Value obtained from this graph is for a cycle length of 75 sec and (v/c) ratio of 0.8. To correct for other cycle lengths and (v/c) ratios, use multiplication factors given in Table 8.2a.

Source: Reproduced from *Highway Capacity Manual*, Special Report 209, Transportation Research Board, Washington, D.C., 1985.

Specifying Traffic Conditions. This phase involves the recording of both pedestrian and vehicular hourly volumes on the appropriate arrows of the form shown on Figure 8.2. Pedestrian volumes are recorded such that those that conflict with a given stream of right-turning vehicles are in the same direction as the conflicting right-turning vehicles. For example, pedestrians on the north crosswalk will conflict with the westbound (WB) right-turning vehicles and should be recorded in the WB row of the form. Similarly, pedestrians on the east crosswalk will conflict with the northbound (NB) right-turning vehicles and should be recorded in the NB row of the form. Details of traffic volume should include the percentage of heavy vehicles (%HV) in each movement, where heavy vehicles are defined as all vehicles having six or more wheels on the pavement. In recording the number of buses, only buses that stop to pick up or discharge passengers on either side of the intersection are included. Buses that go through the intersection without stopping to pick up or discharge passengers are considered heavy vehicles.

The level of coordination between the lights at the intersection being studied and those at adjacent intersections is a critical characteristic, and it is determined in terms of the type of vehicle arrival at the intersection. Five types of arrivals have been identified.

- **Type 1**, which represents the worst condition of arrival, is a dense platoon arriving at the beginning of the red phase.
- **Type 2**, which, while better than Type 1, is still considered unfavorable, is a dense platoon arriving in the middle of the red phase or a dispersed platoon arriving throughout the red phase.
- **Type 3**, which represents an average condition and usually occurs at isolated intersections or at intersections that have different cycle lengths from those of adjacent intersections, entails the random arrival of vehicles.
- **Type 4**, which is usually considered as a moderately favorable platoon condition, is a dense platoon arriving in the middle of a green phase or a dispersed platoon arriving throughout the green phase.
- **Type 5**, which represents the best condition of arrival, is a dense platoon arriving at the start of the green phase.

It is necessary to determine, as accurately as possible, the type of arrival for the intersection being considered, since both the estimate of delay and the determination of the level of service will be significantly affected by the arrival type used in the analysis. Field observation is the best way to determine the arrival type, although time–space diagrams for the street being considered could be used for an approximate estimation. In using field observations, the percentage of vehicles arriving during the green phase is determined and the arrival type is then obtained for the platoon ratio for the approach,[2] which is given as

$$R_p = \frac{PVG}{PTG} \tag{8.4}$$

where

R_p = platoon ratio
PVG = percentage of all vehicles in the movement arriving during the green phase
PTG = percentage of the cycle that is green for the movement = $(G/C) \times 100$

The arrival type is obtained from Table 8.3, which gives a range of platoon ratios for each arrival type.

The number of parking maneuvers at the approach is another factor that influences capacity and level of service of the approach. A parking maneuver is when a vehicle enters or leaves a parking space. The parking capacity adjacent to analysis lane groups is therefore given as the number of parking maneuvers per hour (N_m) that occur within 250 ft of the intersection.

Specifying Signalized Conditions. Details of the signal system should be specified, including a phase diagram, the green, amber, and cycle lengths. The phasing scheme at an intersection determines which traffic stream or streams are

Figure 8.4   Phase Plans for Pretimed and Actuated Signals

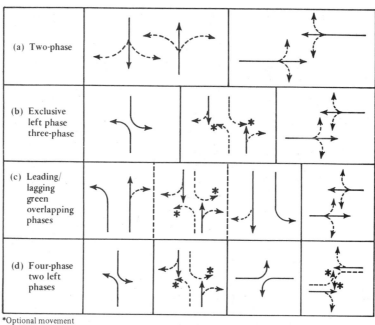

*Optional movement

Source: Reproduced from *Highway Capacity Manual*, Special Report 209, Transportation Research Board, Washington, D.C., 1985.

given the right of way at the intersection and therefore has a significant effect on the level of service. A poorly designed phasing scheme may result in unnecessary delay. Figure 8.4 shows different phase plans. The two-phase system (a) is generally used when no left-turn phases are required, and the three-phase system (b) is used when only one approach requires a separate left-turn phase. It is sometimes necessary to provide leading and/or lagging green overlapping phases to provide for both protected and permitted left turns. Phasing system (c) can be used for this provision. Phasing system (d) is a four-phase system, that is typically used when left-turn phases are required on all approaches of a four-legged intersection. Figure 8.5 shows two optional phasing schemes that may be used for actuated signals. It is necessary to differentiate permitted turns from protected turns. This is usually done by using dotted lines for permitted turns and solid lines for protected turns. It is also necessary to identify those phases that are actuated and whether pedestrian-actuated phases will be used. In cases where pedestrian-actuated phases will not be used, it is necessary to specify the minimum green time for each phase that will allow pedestrians to safely cross the appropriate approach. This minimum green time can be estimated from Eq. 8.5 as[3]

$$G_p = 7.0 + (W/U_p) - Y \qquad (8.5)$$

Figure 8.5 Optional Phase Plans for Actuated Signals

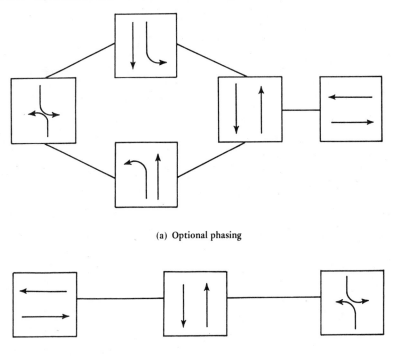

(a) Optional phasing

(b) Three-phase without options

Source: Reproduced from *Highway Capacity Manual*, Special Report 209, Transportation Research Board, Washington, D.C., 1985.

where

G_p = minimum green time (sec)

W = distance from curb to center of the farthest travel lane on the street being crossed or to the nearest pedestrian refuge island (ft)

Y = change interval: yellow and all red time (sec)

U_p = walking speed of pedestrians

A value of 4 ft/sec has been suggested for U_p,[4] which is assumed to represent the 15th percentile pedestrian walking speed. This value of 4 ft/sec is generally lower than the observed average walking speed of pedestrians but will provide safer crossing conditions for those pedestrians who walk below the average speed. When available, local guidelines should be used in preference to Eq. 8.5. An alternative way to compute the minimum green time is that presented in Chapter 6. The difference between the minimum green times obtained from the two methods depends only on the value of Y. When Y is 2, the same minimum green time will be obtained. As Y increases, the difference between the two minimum green times will also increase by the amount Y is greater than 2. Also, as Y increases, the value of the minimum green time obtained from Eq. 8.5 decreases. This is because high values of Y indicate an all red phase, which may be used by the pedestrian to cross the appropriate approach. The expression given in Chapter 6, will normally give a more conservative value for the minimum green time than will Eq. 8.5.

In cases where actuated signal phases are used, the green times and cycle lengths are not constant since they depend on the demand volumes at any given time. It is therefore necessary to obtain these values from the field at the same time that traffic volumes are being obtained for the analysis.

Table 8.4 summarizes the parameters required for the input module. If all the information shown in Table 8.4 is not readily available, default values shown in Table 8.5 can be used.

Volume Adjustment Module

Three main tasks are involved in this module: adjustment of hourly volumes to peak 15 min flow rates, establishing lane groups, and adjustment of lane group flows, taking lane utilization into consideration. Figure 8.6 shows a worksheet that can be used for this module.

Adjustment of Hourly Volumes. Earlier, we saw that the analysis for level of service requires demand flow rates that are based on the peak 15 min flow rate. It is therefore necessary to convert hourly volumes to 15 min flow rates by dividing the hourly volumes by the peak hour factor (PHF). Note that, although not all movements of an approach may peak at the same time, dividing all hourly volumes by the PHF assumes that the peaking occurs for all movements at the same time, which is a conservative assumption.

Lane Group Determination. The lane groups for each approach are established using the following guidelines.

1. Separate lane groups should be established for exclusive left-turn lane(s) and for exclusive right-turn lane(s).
2. When exclusive left-turn lane(s) and/or exclusive right-turn lane(s) are provided on an approach, all through lanes are generally established as a single lane group.
3. When an approach with more than one lane also has a shared left-turn lane, the operation of the shared left-turn lane should be evaluated as shown below to determine whether it is effectively operating as an exclusive left-turn lane because of the high volume of left-turn vehicles on it.

Figure 8.7 shows typical lane groups used for analysis. Note that, when two or more lanes have been established as a single lane group for analysis, all subsequent computations must consider these lanes as a single entity.

Particular attention should be paid to shared left-turn and through (LT/TH) lanes to determine the effective mode of operation—that is, whether the shared LT/TH lane is effectively operating as a left-turn lane or a shared lane. The

Figure 8.6   Volume Adjustment Worksheet

VOLUME ADJUSTMENT WORKSHEET

① Appr.	② Mvt.	③ Mvt. Volume (vph)	④ Peak Hour Factor PHF	⑤ Flow Rate v_p (vph) ③ ÷ ④	⑥ Lane Group	⑦ Flow rate in Lane Group v_g (vph)	⑧ Number of Lanes N	⑨ Lane Utilization Factor U	⑩ Adj. Flow v (vph) ⑦ × ⑨	⑪ Prop. of LT or RT P_{LT} or P_{RT}
	LT									
EB	TH									
	RT									
	LT									
WB	TH									
	RT									
	LT									
NB	TH									
	RT									
	LT									
SB	TH									
	RT									

Source: Reproduced from *Highway Capacity Manual*, Special Report 209, Transportation Research Board, Washington, D.C., 1985.

Figure 8.7 Typical Lane Groups for Analysis

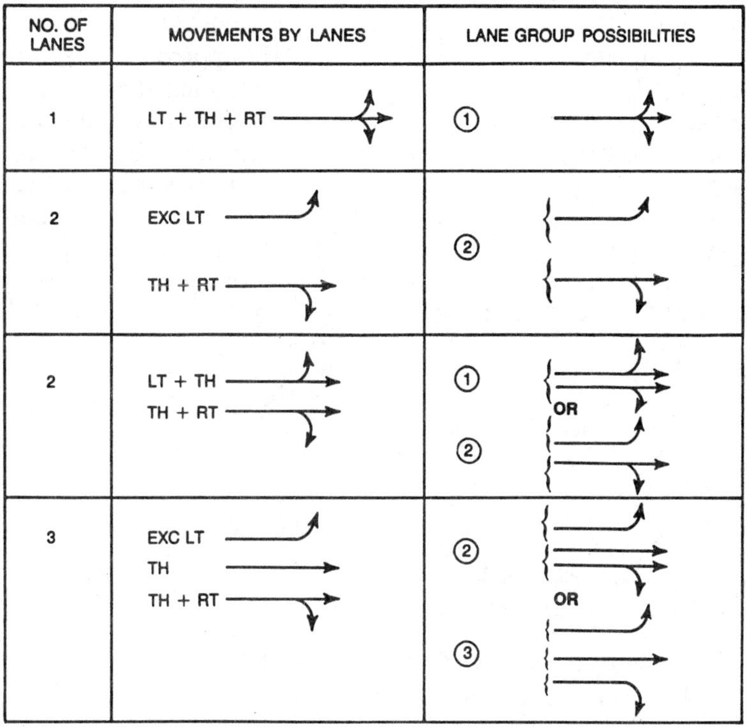

NO. OF LANES	MOVEMENTS BY LANES	LANE GROUP POSSIBILITIES
1	LT + TH + RT	①
2	EXC LT TH + RT	②
2	LT + TH TH + RT	① OR ②
3	EXC LT TH TH + RT	② OR ③

Source: Reproduced from *Highway Capacity Manual*, Special Report 209, Transportation Research Board, Washington, D.C., 1985.

evaluation of the shared left-turn lane is carried out by converting the left-turn flow to an equivalent flow of through vehicles using Eq. 8.6.

$$v_{LE} = v_L \frac{1800}{1400 - v_o} \tag{8.6}$$

where

v_{LE} = approximate equivalent left-turn flow rate (vph)

v_L = actual left-turn flow rate (vph)

v_o = total opposing flow rate (vph)

It can be seen from Eq. 8.6 that, when the opposing flow is equal to or greater than 1400, v_{LE} is meaningless because it is not feasible to have left-turn movements through opposing flows when the opposing volume is greater than 1400 vph. It is therefore necessary to consider the provision of a protected left-turn phase in such cases.

The decision on whether the shared left-turn lane is operating basically as a left-turn lane is based on whether or not the flow rate carried by the leftmost lane is greater than the average flow rate on the other lanes of the approach. The assumption is made that the equivalent left-turn flow (v_{LE}) will completely occupy the leftmost lane during the critical conditions, and the remaining flow will be carried equally by the remaining lanes of the approach. When the flow rate on the leftmost lane due to left-turn vehicles is less than the average flow rate on the other lanes, it is assumed that through vehicles will share the leftmost lane to obtain equilibrium, and the whole approach can be assumed to be a single lane group. When the flow rate on the leftmost lane due to left-turn vehicles is greater than the average flow rate on the remaining lanes of the approach, equilibrium is not achieved, and the leftmost lane initially operates as an exclusive left-turn lane, and should be designed as a separate lane group. Eq. 8.7 and Eq. 8.8, respectively, present these criteria in mathematical form.

$$v_{LE} < \frac{v_a - v_L}{N - 1} \tag{8.7}$$

$$v_{LE} \geq \frac{v_a - v_L}{N - 1} \tag{8.8}$$

where

v_{LE} = actual equivalent left-turn flow rate (vph)

v_a = total flow on the approach (vph)

N = total number of lanes on the approach

In Eq. 8.7, equilibrium is achieved and the leftmost lane is shared and a single lane group is used. In Eq. 8.8 equilibrium is not achieved and the leftmost lane acts as an exclusive left-turn lane and should be established as a separate lane group.

The demand flow rate (v_g) for each lane group is also determined as this is required for the next task in this module.

Adjustment of Lane Group Flows. It is generally known that, when two or more traffic lanes carry the same movement of traffic, the volume is usually not divided equally among the lanes. This results in one lane carrying a higher volume of traffc than any other lane, that is, unequal utilization of the lanes. This phenomenon is accounted for by adjusting the peak 15 min flow rate for each lane group by multiplying this flow rate by the appropriate lane utilization factor shown in Table 8.6. The lane utilization factors were obtained by assuming that 52.5 percent of the demand flow is carried by the more heavily loaded lane in a group of two lanes and 36.7 percent is carried by the most heavily loaded lane in a group of three. The adjusted demand flow rate for any lane group is therefore given as

$$v_i = v_{gi} U_i \tag{8.9}$$

where

v_i = adjusted demand flow rate for lane group i (vph)

v_{gi} = unadjusted demand flow rate for group i (vph)

U_i = lane utilization factor for lane group i

When the demand volume of any lane group approaches the capacity of that lane group—that is, (v/c) approaches 1.0—the utilization of the lanes tends to be more equally spread, and the lane utilization factor may be set to 1.0. The use of the lane utilization factor provides the opportunity for the worst lane in a lane group to be analyzed but is not required if the average situation is being analyzed. In this case the lane utilization factor is set as 1.0.

Saturation Flow Module

This module provides for the computation of a saturation flow rate for each lane group. The saturation flow rate is defined as the flow rate in vph that the lane group can carry if it has the green indication continuously, that is, $g/C = 1$. The saturation flow rate (s) depends on an ideal saturation flow (s_o), which is usually taken as 1800 vph of hour of green time per lane. This ideal saturation flow is then adjusted for the prevailing conditions to obtain the saturation flow for the lane group being considered. The adjustment is made by introducing factors that correct for the number of lanes, lane width, the percent of heavy vehicles in the traffic, approach grade, parking activity, local buses stopping within the intersection, area type, and right and left turns. The saturation flow is given as

$$s = (s_0)(N)(f_w)(f_{HV})(f_g)(f_p)(f_{bb})(f_a)(f_{RT})(f_{LT}) \qquad \textbf{(8.10)}$$

where

s = saturation flow rate for the subject lane group, expressed as a total for all lanes in lane group under prevailing conditions (vphg)

s_o = ideal saturation flow rate per lane usually taken as 1800 (vphg/ln)

N = number of lanes in lane group

f_w = adjustment factor for lane width (see Table 8.7)

f_{HV} = adjustment factor for heavy vehicles in the traffic stream (see Table 8.8)

f_g = adjustment factor for approach grade (see Table 8.9)

f_p = adjustment factor for the existence of parking lane adjacent to the lane group and the parking activity on that lane (see Table 8.10)

f_{bb} = adjustment factor for the blocking effect of local buses stopping within the intersection area (see Table 8.11)

f_a = adjustment factor for area type

f_{RT} = adjustment factor for right turns in the lane group (see Table 8.12)

f_{LT} = adjustment factor for left turns in the lane group (see Table 8.13)

Although the necessity for using some of these adjustment factors was presented in Chapter 7, the basis for using each of them is given here again to facilitate easy comprehension of the material.

Lane Width Adjustment Factor, f_w. This factor depends on the average width of the lanes in a lane group. It is used to account for the reduction in saturation flow rates when lane widths are less than 12 ft and the increase in saturation flow rates when lane widths are greater than 12 ft. The adjustment factors are obtained from Table 8.7. When lane widths are 16 ft or greater, such lanes may be divided into two narrow lanes of 8 ft each.

Heavy Vehicle Adjustment Factor, f_{HV}. The heavy vehicle adjustment factor is related to the percentage of heavy vehicles in the lane group. This factor corrects for the additional delay and reduction in saturation flow due to the presence of heavy vehicles in the traffic stream. The additional delay and reduction in saturation flow are mainly due to the difference between the operational capabilities of the heavy vehicles and passenger cars and the additional space taken up by heavy vehicles. The appropriate factor is selected from Table 8.8.

Grade Adjustment Factor, f_g. This factor is related to the slope of the approach being considered. It is used to correct for the effect of slopes on the speed of vehicles, both passenger cars and heavy vehicles, since passenger cars are also affected by grade. This effect is different for up-slope and down-slope conditions, therefore, the direction of the slope should be taken into consideration as shown in Table 8.9.

Parking Adjustment Factor, f_p. On-street parking close to an intersection causes friction between parking and nonparking vehicles, which results in a reduction of the maximum flow rate that the approach can handle. This effect is corrected for by using a parking adjustment factor (see Table 8.10) on the base saturation flow. This factor depends on the number of lanes in the lane group and the number of parking maneuvers per hour. The values given for the parking adjustment factor indicate that the higher the number of lanes in a given lane group, the less effect parking has on the saturation flow, and the higher the number of parking maneuvers, the greater the effect. This adjustment factor should be applied only to the lane group immediately adjacent to the parking lane.

Area Type Adjustment Factor, f_a. The general types of activities in the area at which the intersection is located have a significant effect on speed and therefore on saturation volume at an approach. For example, because of the complexity of intersections located in business areas, these intersections operate less efficiently than intersections at other areas. This is corrected for by using the area type adjustment factor f_a and is given as

$f_a = 0.90$ for central business district (CBD)
$f_a = 1.0$ for all other areas

Bus Blockage Adjustment Factor, f_{bb}. When buses have to stop on a travel lane to discharge or pick up passengers, some of the vehicles immediately behind the bus will also have to stop. This results in a decrease in the maximum volume that can be handled by that lane. This effect is corrected for by using the bus blockage adjustment factor, which is related to the number of buses in an hour that stop on the travel lane to pick up or discharge passengers and the number of lanes in the lane group that are affected. Table 8.11 gives values for these factors.

Right-Turn Adjustment Factor, f_{RT}. This factor accounts for the effect of right-turning vehicles on the saturation flow. It depends on the phasing system—that is, protected, permitted, or protected plus permitted—the conflicting pedestrian volume, and the proportion of right-turning vehicles that use the protected portion of a protected plus permitted phase. This proportion can be determined from a field study or, alternatively, can be estimated from the signal timing by assuming that the proportion of the right-turning phase that is protected and the proportion of the right turns that use the protected phase are approximately equal. The right-turning volume may also be reduced if right-turn-on-red is allowed by subtracting the number of vehicles that turn during the red phase from the total right-turn volume. This reduction is, however, done on the hourly volume before it is converted to the equivalent peak 15 min flow rate. Table 8.12 gives the right-turn adjustment factors.

Left-Turn Adjustment Factor, f_{LT}. This adjustment factor is used to account for the fact that left-turn movements take more time than through movements. The values of this factor also depend on the type of phasing (protected, permitted, or protected plus permitted), the type of lane used for left turns (that is, exclusive or shared lane), the proportion of left-turn vehicles using a shared lane, and the opposing flow rate when there is a permitted left-turn phase. The adjustment factors are given in Table 8.13. It can be seen from Table 8.13 that special computations are required to determine the left-turn adjustment factors when the left turns are made during a permitted left-turn phase. These special computations are required to account for equilibrium flows due to the conflicting effect of left-turning vehicles, through vehicles, and opposing vehicles. The chart shown in Figure 8.8 can be used to facilitate the computation. These computations take into account the three component flows during any green phase:

- The shared lane through flow at the start of the green until the lane is blocked by a left-turn vehicle waiting to turn
- The flow from a shared lane or left-turn lane during the period when the opposing flow is unsaturated
- The left turns accomplished during the amber phase by vehicles already waiting to turn left

The appropriate left-turn adjustment factor is determined through the following computations.

Figure 8.8 Supplemental Worksheet for Computing Left-Turn Adjustment Factors for Permitted Left Turns

SUPPLEMENTAL WORKSHEET FOR LEFT-TURN ADJUSTMENT FACTOR, f_{LT}				
INPUT VARIABLES	EB	WB	NB	SB
Cycle Length, C (sec)				
Effective Green, g (sec)				
Number of Lanes, N				
Total Approach Flow Rate, v_a (vph)				
Mainline Flow Rate, v_M (vph)				
Left-Turn Flow Rate, v_{LT} (vph)				
Proportion of LT, P_{LT}				
Opposing Lanes, N_o				
Opposing Flow Rate, v_o (vph)				
Prop. of LT in Opp. Vol., P_{LTO}				
COMPUTATIONS	EB	WB	NB	SB
$S_{op} = \dfrac{1800\, N_o}{1 + P_{LTO}\left[\dfrac{400 + v_M}{1400 - v_M}\right]}$				
$Y_o = v_o / S_{op}$				
$g_u = (g - CY_o) / (1 - Y_o)$				
$f_s = (875 - 0.625\, v_o) / 1000$				
$P_l = P_{LT}\left[1 + \dfrac{(N-1)\,g}{f_s g_u + 4.5}\right]$				
$g_q = g - g_u$				
$P_T = 1 - P_l$				
$g_f = 2\dfrac{P_T}{P_l}\left[1 - P_T^{\,0.5\, g_q}\right]$				
$E_L = 1800 / (1400 - v_o)$				
$f_m = \dfrac{g_f}{g} + \dfrac{g_u}{g}\left[\dfrac{1}{1 + P_l\,(E_L - 1)}\right] + \dfrac{2}{g}(1 + P_l)$				
$f_{LT} = (f_m + N - 1) / N$				

Source: Reproduced from *Highway Capacity Manual*, Special Report 209, Transportation Research Board, Washington, D.C., 1985.

Step 1. It is first necessary to estimate the saturation flow rate for the opposing approach from Eq. 8.11 as[5]

$$S_{op} = \frac{1800N_o}{1 + P_{LTo}(400 + v_m)/(1400 - v_m)} \tag{8.11}$$

where

S_{op} = saturation flow rate for opposing approach (vphg)

N_o = number of mainline lanes opposing the permitted left turns, not including left-turn or right-turn lanes in the opposing flow

P_{LTo} = proportion of left turns in opposing flow

v_m = mainline approach flow rate, not including left turns from an exclusive lane or on a one-lane approach, in vph; the maximum value of v_m is 1399. This value is used for all $v_m > 1399$.

It can be seen that the denominator of Eq. 8.11 is a representation of the weighted average through-vehicle equivalent for the opposing flow.

Step 2. The flow ratio for the opposing flow is then computed as

$$Y_o = \frac{v_o}{S_{op}} \tag{8.12}$$

where

Y_o = flow ratio for opposing approach

v_o = opposing flow rate, discounting left turns from an exclusive lane or one-lane approach (the maximum value of v_o is 1399), which is used for all $v_o > 1399$.

Step 3. Next we compute the fraction of the green phase that is not blocked by an opposing queue.

$$g_u = \frac{C(g/C - Y_o)}{1 - Y_o} = \frac{g - CY_o}{1 - Y_o} \tag{8.13}$$

$$g_u = 0 \quad \text{if } Y_o > g/C$$

where

g_u = portion of green phase not blocked by the clearing of an opposing queue of vehicles in seconds

Y_o = flow ratio of opposing flow

C = cycle length in seconds

It can be seen from Eq. 8.13, that g_u should be greater than zero—that is, g/C should be greater than Y_o if there should be any left-turn capacity apart from that provided during the amber phase. Thus, the opposing green ratio (g/C) should be greater than the opposing flow ratio $(Y_o = v_o/s_{op})$. It is useful to consider this requirement, when cycle length is to be assumed, for example, in the design of a signal system.

Step 4. The saturation factor for left turns is computed from Eq. 8.14 using the opposing flow.

$$f_s = \frac{875 - 0.625v_o}{1000} \tag{8.14}$$

where

f_s = left turn saturation factor

v_o = opposing flow rate (including left-turning vehicles when they are made from a shared lane on a multilane approach during a permitted phase, but not including left turns made from a single-lane approach or from an exclusive left through)

Step 5. It is also necessary to determine the proportion of left-turn flow in a shared lane when the left turn is made from a shared lane. This is given as

$$P_L = P_{LT}\frac{1 + (N-1)g}{f_s g_u + 4.5} \tag{8.15}$$

where

P_L = proportion of left turns in shared median or left-turn lane

P_{LT} = proportion of left turns in lane group flow

N = number of lanes in lane group or approach

g = effective green time (sec)

f_s = left-turn saturation factor

g_u = portion of green not blocked by the clearing of an opposed queue of vehicles (sec)

Note that $P_L = 1.0$ when the left turns are made from an exclusive left-turn lane.

Step 6. The next computation is the determination of that portion of the green phase during which the vehicles on a shared lane will continue to move until left-turning vehicles block the flow. This portion of the green time is estimated as

$$g_f = \frac{2P_T}{P_L}(1 - P_T^{0.5g_q}) \tag{8.16}$$

where

g_f = initial portion of green phase, during which through vehicles may move in a shared LT/TH lane with movement continuing until arrival of first LT vehicle, which waits until the opposing queue clears, thereby blocking the lane from the remaining portion g_q (sec)

g_q = portion of green phase blocked to left-turning vehicles by the clearing of an opposing queue of vehicles $g_q = g - g_u$ (sec)

g_u = portion of green phase not blocked by the clearing of an opposing queue of vehicles (sec)

P_T = proportion of through vehicles in shared median or left lane

P_L = proportion of left turns in shared median or left-turn lane

When a separate left-turn lane is being considered, $P_T = 0$ and $g_f = 0$.

Step 7. It is also necessary to determine the through-vehicle equivalent for the opposed left-turn vehicles during the portion of the green phase when the flow is less than the saturated flow. This value is approximately estimated as

$$E_L = \frac{1800}{1400 - v_o} \tag{8.17}$$

where

E_L = through-vehicle equivalent for opposed left turns

v_o = opposing flow rate

Step 8. These parameters are then used to compute the left-turn factor for a shared LT/TH lane or an exclusive LT lane as

$$f_m = \frac{g_f}{g} + \frac{g_u}{g} \frac{1}{1 + P_L(E_L - 1)} + \frac{2}{g}(1 + P_L) \tag{8.18}$$

(Note: $f_m \leq 1.00$)

The factor f_m is applicable only to the one lane used for left turns. The factor f_m is, therefore, equal to f_{LT} where there is only a single lane on the approach or a left-turn lane is being considered. When shared lanes are used on a multilane approach, the left turn factor (f_{LT}) for the lane group is computed as

$$f_{LT} = \frac{f_m + N - 1}{N} \tag{8.19}$$

where

f_{LT} = left-turn factor for lane group

f_m = left-turn factor for single lane

N = number of lanes in lane group

To carry out the computations required to determine f_{LT}, the cycle length and green times must be known. When these are unknown, it is suggested that a cycle length of 60 to 90 sec be assumed and green times allotted proportionately to the average lane flows.[6] This assumption can then be revised when detailed information is available, although this is seldom necessary since the effect of cycle length and green time on the final factor is negligible. An alternative approach is to use one of the methods presented in Chapter 6 to determine appropriate green times. Several of the equations given for this special computation of the left-turn factors are based on expressions empirically developed during studies listed in Additional Readings at the end of this chapter.

If the user of the procedure does not wish to determine the adjustment factors used in the computation of the saturation flow, a suggested value for saturation flow is 1600 vphg.[7]

Field Determination of Saturation Flow.   An alternative to the use of adjustment factors is to determine directly the saturation flow, in the field, using the following procedure.

Two people are needed to carry out the procedure, with one being the timer, equipped with a stop watch, and the other the recorder. The form shown in Figure 8.9 is used to record the data. An observation point is selected at the intersection such that a clear vision of the traffic signals and the stop line or crosswalk is maintained. A reference point is selected to indicate when a vehicle has entered the intersection. This reference point is usually the stop line so that all vehicles that cross the stop line are considered as having entered the intersection. The following steps are then carried out for each cycle and for each lane.

Step 1.   The timer starts the stop watch at the beginning of the green phase and shouts this out to the recorder.

Step 2.   The recorder immediately notes the last vehicle in the stopped queue and describes it to the timer and also notes which vehicles are heavy vehicles.

Step 3.   The timer then counts aloud each vehicle in the queue as its rear axle crosses the reference point (that is, "one," "two," "three," and so on). The recorder also notes which vehicles turn left or right.

Step 4.   The timer calls out the times that the fourth, tenth, and last vehicles in the queue cross the reference points, and these are recorded by the recorder.

Step 5.   In cases where queued vehicles are still entering the intersection at the end of the green phase, the number of the last vehicle at the end of the green should be identified by the timer and shouted to the recorder so that number can be recorded.

Step 6.   The width of the lane and the slope of the approach are then measured and recorded, together with any unusual occurrences that might have affected the saturation flow.

Step 7.   Since the flow just after the start of the green phase is less than saturation flow, the time considered for calculating the saturation flow is that between

Figure 8.9   Field Sheet for Direct Observation of Prevailing Saturation Flow Ratio

FIELD SHEET – SATURATION FLOW STUDY

Location:_____

Date:___/___/___ Time:_____ City:_____

_____Bound Traffic; Approaching From the_____

Observers:_____ Weather:_____

Movements Allowed
- ☐ Thru
- ☐ Right Turn
- ☐ Left Turn

Identify all Lane Movements
& The Lane Studied

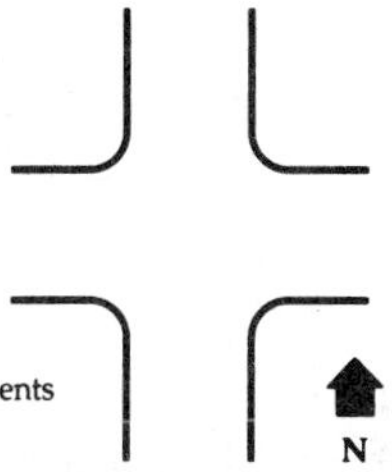

N

Veh. in Queue	Cycle 1			Cycle 2			Cycle 3			Cycle 4			Cycle 5			Cycle 6		
	Time	HV	T	Time	HV	T	Time	HV	T	Time	HV	T	Time	HV	T	Time	HV	T
1																		
2																		
3																		
4																		
5																		
6																		
7																		
8																		
9																		
10																		
11																		
12																		
13																		
14																		
15																		
16																		
17																		
18																		
19																		
20																		
End of Saturation																		
End of Green																		
No. Veh. > 20																		
No. Veh. on Yellow																		

HV = Heavy Vehicles (Vehicles with more than 4 tires)
T = Turning Vehicles (L = Left, R = Right)
Pedestrians and buses which block vehicles should be noted with the time that they block traffic, i.e.,
P12 = pedestrians blocked traffic for 12 sec
B15 = bus blocked traffic for 15 sec

Grade_____ Area Type_____

Source: Reproduced from *Highway Capacity Manual*, Special Report 209, Transportation Research Board, Washington, D.C., 1985.

the time the rear axle of the fourth car crosses the reference point (t_4) and the time the rear axle of the last vehicle queued at the beginning of the green crosses the same reference point (t_n). The saturation flow is then determined from Eq. 8.20.

$$\text{saturation flow} = \frac{3600}{(t_4 - t_n)/(n - 4)} \qquad (8.20)$$

where n is the number of vehicles queued at the beginning of the green.

The data recorded on heavy vehicles, turning vehicles, and approach geometrics can be used in the future if adjustment factors are to be applied.

Capacity Analysis Module

In this module, results of the computations carried out in the previous modules are used to determine the important capacity variable, which include

- Flow ratios for the different lane groups
- Capacities for the different lane groups
- (v/c) ratios for the different lane groups
- The critical (v/c) ratio for the overall intersection

The adjusted demand volume obtained for each lane group in the volume adjustment module is divided by the saturation flow for the appropriate lane group determined in the saturation flow module to obtain the flow ratio (v_i/s_i) for that lane group. The capacity of each lane group is then determined by using Eq. 8.1 as

$$c_i = s_i(g_i/C)$$

Using Eq. 8.2, the (v/c) ratio is then computed for each lane group as

$$X_i = (v_i/c_i)$$

Similarly, using Eq. 8.3, the critical (v/c) ratio X_c is then computed for the intersection as

$$X_c = \sum_i (v/s)_{ci} \frac{C}{C - L}$$

The identification of the critical lane group for each green phase is necessary before the critical (v/c) ratio (X_c) can be determined for the intersection. This identification is relatively simple when there are no overlapping phases, as the lane group with the maximum flow ratio (v_i/s_i) during each green phase is the critical lane group for that phase. When the phases overlap, however, identification of the critical lane group is not as simple. The basic principle used in this case is that the critical (v/c) ratio for the intersection is based on the combinations of lane groups that will use up the largest amount of the capacity available. This is demonstrated by Figure 8.10, which shows a phasing system that provides for exclusive left-turn lanes in the south and north approaches and overlapping phases.

There are only two lane groups during phase A—that is, EBLT/TH/RT and WBLT/TH/RT. The critical lane group is simply selected as the lane group with the greater (v_i/s_i) ratio. The three other phases, however, include overlapping movements, and the critical lane group is not as straightforward to identify. It can

Figure 8.10 Illustrative Example for Determining Critical Lane Group

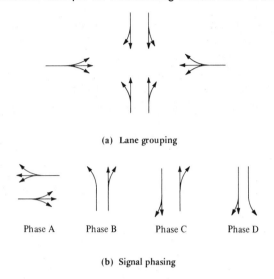

(a) Lane grouping

Phase A   Phase B   Phase C   Phase D

(b) Signal phasing

be seen that the NBTH/RT lane group moves during phases B and C and therefore overlaps with the SBTH/RT lane group, which moves during phases C and D, while the NBLT lane group moves only during phase B and the SBLT lane group moves only during phase D. The NBTH/RT lane group can therefore be critical for the sum of phases B and C, whereas the SBLT lane group can be critical for phase D. Two possibilities therefore exist for the (v/s) ratios for the overlapping phases B, C, and D:

(1) NBTH/RT + SBLT
(2) SBTH/RT + NBLT

The critical lane flow ratio for the intersection is therefore the maximum flow ratio of the following:

EBLT/TH/RT + NBTH/RT + SBLT
EBLT/TH/RT + SBTH/RT + NBLT
WBLT/TH/RT + NBTH/RT + SBLT
WBLT/TH/RT + SBTH/RT + NBLT

Any one phase or portion of a phase can have only one critical lane group. If a critical lane group has been determined for the sum of phases i and j, no other lane group can be critical for either phase i or j, or any combination of phases that includes phase i or j. Noted also that in determining the signal timing for any intersection—that is, design of the intersection—the critical lane group is used.

When exclusive left-turn lane groups with a protected permitted phase are used, these lane groups are usually separated in this module, and it is assumed that all the flow takes place during the protected phase. This assumption could lead to a protected-phase (v/c) ratio or critical (v/c) ratio that is too high. When this occurs, it is useful at this stage to reassign some of the flow to the permitted portion of the phase rather than changing the signal timing or roadway geometrics. The amount of left-turn flow that can be reassigned to the permitted portion is restricted to the capacity of the permitted phase, which is given as

$$C_{LT} = (1400 - v_o)(g/C)_{P_{LT}} \qquad\qquad (8.21)$$

or

$$C_{LT} = 2 \text{ vehicles per signal cycle}$$

where

C_{LT} = capacity of permitted left-turn phase (vph)

v_o = opposing through and right-turn flow rate (vph)

$(g/C)_{P_{LT}}$ = effective green ratio for the permitted left-turn phase

This flow rate can then be subtracted from the protected flow rate and all the necessary computations carried out again.

The computation of the critical (v/c) ratio (X_c) completes the definition of the capacity characteristics of the intersection. As stated earlier, these characteristics must be evaluated separately and in conjunction with the delay and levels of service that are determined in the next module. Following are some key points that should be kept in mind when the capacity characteristics are being evaluated.

1. When critical (v/c) ratio is greater than 1.00, the geometric and signal characteristics are inadequate for the critical demand flows at the intersection. The operating characteristics at the intersection may be improved by increasing the cycle length, changing the cycle phase, and/or changing the roadway geometrics.

2. When there is a large variation in the (v/c) ratio for the different critical lane groups but the critical (v/c) ratio is acceptable, the green time is not proportionately distributed and reallocation of the green time should be considered.

3. A protected left-turn phase should be considered when the use of permitted left turns results in drastic reductions of the saturation flow rate for the appropriate lane group.

4. When the critical flow rate is greater than 0.9, it is quite likely that the existing signal and geometric characteristics will not be adequate for the demand flow rate. Consideration should therefore be given to changing either the signal timing and/or the roadway geometrics.

5. If the (v/c) ratios are above acceptable limits and protected turning phases have been included for the turning movements with high flows, then changes in roadway geometrics will be required to reduce the (v/c) ratios.

Figure 8.11 Capacity Analysis Worksheet

CAPACITY ANALYSIS WORKSHEET								
LANE GROUP		③ Adj. Flow Rate v (vph)	④ Adj. Sat. Flow Rate s (vphg)	⑤ Flow Ratio v/s ③ ÷ ④	⑥ Green Ratio g/C	⑦ Lane Group Capacity c (vph) ④ × ⑥	⑧ v/c Ratio X ③ ÷ ⑦	⑨ Critical ? Lane Group
① Appr.	② Lane Group Movements							
EB								
WB								
NB								
SB								

Cycle Length, C _____ sec

Lost Time Per Cycle, L _____ sec

$$\sum_i (v/s)_{ci} = \underline{\hspace{2cm}}$$

$$X_c = \frac{\sum (v/s)_{ci} \times C}{C - L} = \underline{\hspace{2cm}}$$

Source: Reproduced from *Highway Capacity Manual*, Special Report 209, Transportation Research Board, Washington, D.C., 1985.

The computation required for this module may be carried out in the format shown in Figure 8.11.

Level of Service Module

The results obtained from the volume adjustment, saturation flow rate, and capacity analysis modules are now used in this module to determine the average stopped time delay per vehicle in each lane group and hence the level of service for each approach and the intersection as a whole. The computation first involves the determination of the uniform and incremental delays. The uniform delay is that delay that will occur in a lane group if vehicles arrive with a uniform distribution and saturation does not occur during any cycle. Uniform delay is determined as[8]

$$d_{1i} = 0.38C \frac{(1 - g_i/C)^2}{1 - (g_i/C)(X_i)} \tag{8.22}$$

where

d_{1i} = uniform delay (sec/vehicle) for lane group i
C = cycle length (sec)
g_i = effective green time for lane group i (sec)
X_i = (v/c) ratio for lane group i

The incremental delay takes into consideration that the arrivals are not uniform but random and that some cycles will overflow. It is given as

$$d_{2i} = 173x_i^2[(X_i - 1) + \sqrt{(X_i - 1)^2 + (16X_i/c_i)}] \tag{8.23}$$

where

d_{2i} = incremental delay (sec/vehicle)
c_i = capacity of lane group i

The total delay for lane group i is given as

$$d_i = d_{1i} + d_{2i}$$

$$= 0.38C \frac{(1 - g_i/C)^2}{1 - (g_i/C)(X_i)}$$

$$+ 173X_i^2[(X_i - 1) + \sqrt{(X_i - 1)^2 + (16X_i/c_i)}] \tag{8.24}$$

The delay obtained from Eq. 8.24 is based on the assumption that arrivals are random. The arrival distribution is, however, affected by the operation of the adjacent intersection, with the result that, in most cases, arrivals are not truly random but are platooned. It is therefore necessary to adjust the delay computed from Eq. 8.24 by multiplying it by a progression factor (PF) given in Table 8.14, which is based on the arrival type. It can be seen from Table 8.14 that when the arrival type is favorable, for example, arrival type 5, the PF is low in comparison with an unfavorable arrival type, for example, type 1 when the PF is high. The adjusted delay is then given as

$$d_{ia} = d_i \text{PF}_i \tag{8.25}$$

where

d_{ia} = adjusted delay for lane group i (sec/vehicle)
d_i = random delay for lane group i (sec/vehicle)
PF_i = progression factor for lane group i

The level of service for each lane group can then be determined from Table 8.1 using the computed adjusted delays.

Approach Delay. Having determined the average stopped delay for each lane group, we can now determine the average stopped delay for any approach as the weighted average of the stopped delays of all lane groups on that approach. The approach delay is given as

$$d_A = \frac{\sum\limits_{i=1}^{n_A} (d_{ia}v_i)}{\sum\limits_{i=1}^{n_A} v_i} \tag{8.26}$$

where

d_A = delay for approach A (sec/vehicle)
d_{ia} = adjusted delay for lane group i on approach A (sec/vehicle)
v_i = adjusted flow rate for lane group i (vph)
n_A = number of lane groups on approach A

The level of service of approach A can then be determined from Table 8.1.

Intersection Delay. The average intersection stopped delay is found in a manner similar to the approach delay. In this case, the weighted average of the delays at all approaches is the average stopped delay at the intersection. The average intersection delay is therefore given as

$$d_I = \sum\limits_{A=1}^{T} \frac{d_A v_A}{\sum\limits_{A}^{T} v_A} \tag{8.27}$$

where

d_I = average stopped delay for the intersection (sec/vehicle)
d_A = adjusted delay for approach A (sec/vehicle)
v_A = adjusted flow rate for approach A (vph)
T = number of approaches at the intersection

The level of service for the intersection is then determined from Table 8.1, using the average stopped delay for the intersection. The computation required to determine the different levels of service may be organized in the format shown in Figure 8.12.

It should be emphasized again that short or acceptable delays do not automatically indicate adequate capacity. Both the capacity and the delay should be

Figure 8.12   Level of Service Worksheet

LEVEL-OF-SERVICE WORKSHEET												
Lane Group	First Term Delay					Second Term Delay				Total Delay & LOS		
① Appr.	② Lane Group Move-ments	③ v/c Ratio X	④ Green Ratio g/C	⑤ Cycle Length C (sec)	⑥ Delay d_1 (sec/veh)	⑦ Lane Group Capacity c (vph)	⑧ Delay d_2 (sec/veh)	⑨ Progression Factor PF	⑩ Lane Group Delay (sec/veh) (⑥+⑧)×⑨	⑪ Lane Group LOS	⑫ Approach Delay (sec/veh)	⑬ Appr. LOS
EB												
WB												
NB												
SB												

Intersection Delay _____ sec/veh   Intersection LOS _____

Source: Reproduced from *Highway Capacity Manual*, Special Report 209, Transportation Research Board, Washington, D.C., 1985.

considered in the evaluation of any intersection. Where long and unacceptable delays are determined, it is necessary to find the cause of the delay. For example, if (v/c) ratios are low and delay is long, the most probable cause for the long delay is that the cycle length is too long and/or the progression (arrival type) is unfavorable. Delay can therefore be reduced by improving the arrival type, by coordinating the intersection signal with the signals at adjacent intersections, and/or by reducing the cycle length at the intersection. When delay is long but arrival types are favorable, the most probable cause is that the intersection geometrics are inadequate and/or the signal timing is improperly designed.

Example 8–1   Computing Level of Service at a Signalized Intersection Using the Operation Level of Analysis

Figure 8.13 shows peak hour volumes, pedestrian volumes, geometric layout, traffic mix, and signal timings at an isolated pretimed signalized intersection. If the intersection is located outside the business district and studies have shown that the PHF at the intersection is 0.85, what is the expected level of service if curb parking is allowed on each approach? Use the operation evaluation level of analysis. There is no all red phase and the amber time is 3 sec. To demonstrate the application of the operational level of analysis procedure, the problem will be solved by systematically going through each of the modules discussed.

■ **Input Module**

The tasks in this module are to identify the geometric, traffic, and signalization conditions.

Geometric Conditions.   Since this is an existing intersection, the actual geometrics are shown in Figure 8.13. Parking is permitted on each approach, and the number of parking maneuvers is recorded for each approach as shown.

Traffic Conditions.   Figure 8.13 shows volumes at the intersection as well as other traffic conditions, including the PHF, which is 0.85, the %HV, and so forth.

Signalization Condition.   There is no pedestrian push button since the intersection is pretimed. "N" is therefore recorded in the pedestrian push button column. It is also necessary to estimate the minimum time required for pedestrians to safely cross each street. Using Eq. 8.5, the minimum green times required for pedestrian crossings are

$$\text{EB and WB approaches} = 7 + \frac{36}{4} - 3 = 13 \text{ sec}$$

$$\text{NB and SB approaches} = 7 + \frac{52}{4} - 3 = 17 \text{ sec}$$

These results allow for crossing the whole width of road in one attempt. There is also no coordination since the lights are pretimed. Arrivals will therefore be assumed to be random, and arrival type 3 is recorded in Figure 8.13.

Figure 8.13 Traffic Volume, Geometric Layout, and Signal Phasing for Example 8–1

INPUT WORKSHEET

Intersection:_____ Date:_____

Analyst:_____ Time Period Analyzed:_____ Area Type: ☐ CBD ☐ Other

Project No.:_____ City/State:_____

VOLUME AND GEOMETRICS

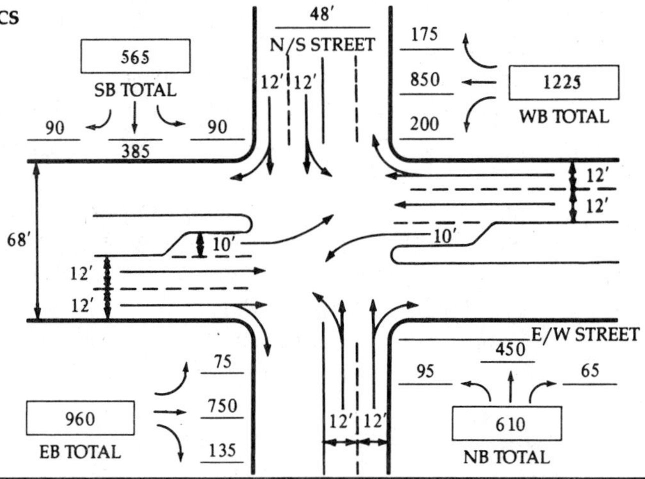

NORTH

48'
N/S STREET
12' 12'

565
SB TOTAL
90 385 90

175
850
200
1225
WB TOTAL

12'
12'

68'
12'
12'
10'
10'

IDENTIFY IN DIAGRAM:
1. Volumes
2. Lanes, lane widths
3. Movements by lane
4. Parking (PKG) locations
5. Bay storage lengths
6. Islands (physical or painted)
7. Bus stops

E/W STREET
450
95 65

75
750
135

960
EB TOTAL

12' 12'

610
NB TOTAL

TRAFFIC AND ROADWAY CONDITIONS

Approach	Grade (%)	% HV	Adj. Pkg. Lane Y or N	Adj. Pkg. Lane N_m	Buses (N_B)	PHF	Conf. Peds. (peds./hr)	Pedestrian Button Y or N	Pedestrian Button Min. Timing	Arr. Type
EB	−4	4	Y	10	10	0.85	60	N	24	3
WB	+4	4	Y	10	10	0.85	60	N	24	3
NB	+2	3	Y	10	10	0.85	60	N	19	3
SB	−2	3	Y	10	10	0.85	60	N	19	3

Grade: + up, − down $\qquad$ N_B: buses stopping/hr $\qquad$ Min. Timing: min. green for
HV: veh. with more than 4 wheels $\quad$ PHF: peak-hour factor $\qquad\qquad$ pedestrian crossing
N_m: pkg. maneuvers/hr $\qquad$ Conf. Peds: Conflicting peds./hr $\quad$ Arr. Type: Type 1-5

PHASING

DIAGRAM

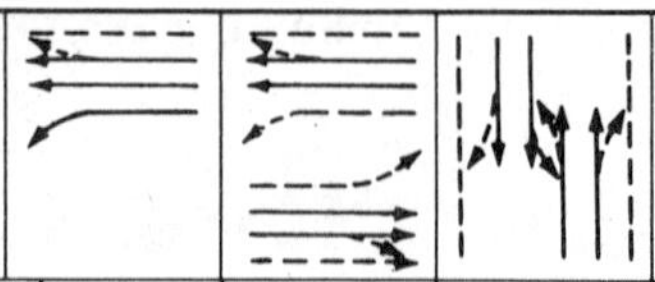

Timing	G = 13 Y + R = 87	G = 40 Y + R = 60	G = 35 Y + R = 65	G = Y + R =	G = Y + R =	G = Y + R =	G = Y + R =	G = Y + R =	G = Y + R =

Pretimed or Actuated

↱ Protected turns $\qquad$ ↱ Permitted turns $\qquad$ ------ Pedestrian $\qquad$ Cycle Length 100 Sec

Source: Adapted from *Highway Capacity Manual*, Special Report 209, Transportation Research Board, Washington, D.C., 1985.

The existing signal phases are shown in Figure 8.13, under Phasing Diagram, and the phase lengths are recorded as shown in the diagram.

■ Volume Adjustment Module

The computations required for the volume adjustment module are carried out using the volume adjustment worksheet shown in Figure 8.14.

Flow Rate. The volumes are adjusted to the peak 15 min flow, using the PHF, and are recorded in column 5 of Figure 8.14. For example,

$$\text{EBLT flow rate} = \frac{75}{0.85} = 88 \text{ vph}$$

Lane Groups. The lane groups selected are shown in Figure 8.14. The EB and WB left lanes are exclusive left-turn lanes and are therefore made a single lane group, and the through and right lanes are made into a single group. Since the lanes on the north and south approaches are all served by the same green phase and there are no exclusive left- or right-turn lanes, all lanes in the north approach are made into a single lane group and all lanes in the south approach are made into a single lane group. Note that WB left turns are protected/permitted, whereas EB left turns are only permitted.

Adjust Volume. The total volume in each lane group is determined first and recorded in column 7 of Figure 8.14. For example, the total WB through and right-lane group has $(1000 + 206) = 1206$ vph. The number of lanes and total volume in each lane group are recorded in the appropriate columns in Figure 8.14. For example, the NB approach has 2 lanes, whereas for the WB approach, the left-turn lane group has 1 lane and the through and right-turn lane group has 2 lanes. The volumes are then adjusted using the appropriate lane utilization factors obtained from Table 8.6. The flow rate for each lane group is then adjusted for lane utilization. For example, the adjusted flow rate for the EB lane group $= 1041 \times 1.05 = 1093$. The proportion of left turns and right turns are also computed for each lane group. For example, the proportion of EB through and right-turn flow rate that turns right is $= 159/1041 = 0.15$.

■ Saturation Flow Rate Module

The computations carried out in this module are shown in Figure 8.15. The ideal saturation flow is taken as 1800 passenger cars per hour of green per lane and recorded in column 3. The adjustment factors are then obtained from the respective tables and recorded as shown.

Lane Width Adjustment Factor, f_w. The lane width adjustment factors are obtained from Table 8.7

For 12 ft lanes: $f_w = 1.00$
For 10 ft lanes: $f_w = 0.93$

Figure 8.14   Volume Adjustment Worksheet for Example 8–1

① Appr.	② Mvt.	③ Mvt. Volume (vph)	④ Peak Hour Factor PHF	⑤ Flow Rate v_p (vph) ③ ÷ ④	⑥ Lane Group	⑦ Flow rate in Lane Group v_g (vph)	⑧ Number of Lanes N	⑨ Lane Utilization Factor U Table 8-6	⑩ Adj. Flow v (vph) ⑦ × ⑨	⑪ Prop. of LT or RT P_{LT} or P_{RT}
	LT	75	0.85	88		88	1	1.00	88	1.0_{LT}
EB	TH	750	0.85	882		1041	2	1.05	1093	0.15_{RT}
	RT	135	0.85	159						
	LT	200	0.85	235		235	1	1.00	235	1.0_{LT}
WB	TH	850	0.85	1000		1206	2	1.05	1266	0.17_{RT}
	RT	175	0.85	206						
	LT	95	0.85	112						
NB	TH	450	0.85	529		717	2	1.05	753	0.16_{LT} 0.11_{RT}
	RT	65	0.85	76						
	LT	90	0.85	106						
SB	TH	385	0.85	453		665	2	1.05	698	0.16_{LT} 0.16_{RT}
	RT	90	0.85	106						

VOLUME ADJUSTMENT WORKSHEET

Source: Adapted from *Highway Capacity Manual*, Special Report 209, Transportation Research Board, Washington, D.C., 1985.

Figure 8.15 Worksheet for the Saturation Flow Rate Module for Example 8–1

SATURATION FLOW ADJUSTMENT WORKSHEET												
LANE GROUPS				**ADJUSTMENT FACTORS**								⑬
① Appr.	② Lane Group Movements	③ Ideal Sat. Flow (pcphgpl)	④ No. of Lanes N	⑤ Lane Width f_w Table 8-7	⑥ Heavy Veh f_{HV} Table 8-8	⑦ Grade f_g Table 8-9	⑧ Pkg. f_p Table 8-10	⑨ Bus Blockage f_{bb} Table 8-11	⑩ Area Type f_a	⑪ Right Turn f_{RT} Table 8-12	⑫ Left Turn f_{LT} Table 8-13	Adj. Sat. Flow Rate s (vphg)
	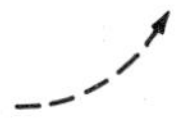	1800	1	0.93	0.98	1.02	1.0	1.0	1.0	1.0	0.13	218
EB		1800	2	1.0	0.98	1.02	0.92	0.98	1.0	0.97	1.0	3147
	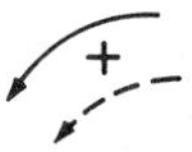	1800	1	0.93	0.98	0.98	1.0	1.0	1.0	1.0	0.95	1527
WB	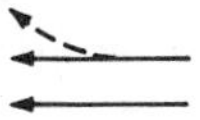	1800	2	1.0	0.98	0.98	0.92	0.98	1.0	0.97	1.0	3024
NB		1800	2	1.0	.985	0.99	0.92	0.98	1.0	0.98	0.68	2109
SB		1800	2	1.0	.985	1.01	0.92	0.98	1.0	0.97	0.65	2036

Source: Adapted from *Highway Capacity Manual*, Special Report 209, Transportation Research Board, Washington, D.C., 1985.

Heavy Vehicle Adjustment Factor, f_{HV}.   These factors are based on the percentage of heavy vehicles in the traffic stream on each approach and are obtained directly from Table 8.8.

In the NB and SB approaches, the percentage of heavy vehicles is 3, giving $f_{HV} = 0.985$. In the EB and WB approaches, the percentage of heavy vehicles is 4, giving $f_{HV} = 0.98$.

Grade Adjustment Factor, f_g.   These factors are obtained directly from Table 8.9, based on the approach grade.

EB approach − 4 percent: $f_g = 1.02$
WB approach + 4 percent: $f_g = 0.98$
NB approach + 2 percent: $f_g = 0.99$
SB approach − 2 percent: $f_g = 1.01$

Parking Adjustment Factor, f_p.   The parking adjustment factors are obtained from Table 8.10. In this case, there are 10 parking maneuvers per hour at each approach. Each of the lane groups affected by the parking maneuvers has two lanes. The parking adjustment factor for each lane group affected by parking is therefore 0.92. Note that only the right lane group on each approach is affected by parking.

Bus Blockage Adjustment Factor, f_{bb}.   These factors depend on the number of bus blockages per hour and the number of lanes in the lane group affected. The factors are obtained from Table 8.11 and recorded in Figure 8.15. Since there are 10 buses stopping at each approach per hour and each of the lane groups affected consists of 2 lanes, $f_{bb} = 0.98$. Note that the left-turn lane groups are not affected by stopping buses.

Area Type Adjustment Factor, f_a.   The intersection is not located within a business district; therefore, the area type adjustment factor is 1.00 for all approaches.

Right-Turn Lane Adjustment Factor, f_{RT}.   Adjustment is required only for lane groups with right-turning vehicles. All other lane groups will have an f_{RT} of 1.0. The factors are obtained from Table 8.12.

For example the EBTH/RT lane group is considered a shared permitted right-turn phase. The factor can be obtained either directly from Table 8.12 or from the expression

$$f_{RT} = 1.0 - P_{RT}\left(0.15 + \frac{Pds}{2100}\right)$$

where P_{RT} is the proportion of right turns in the lane group and Pds is the volume of conflicting pedestrians. For example, the proportion of right turns in the EB through and right lane is 0.15 and the number of conflicting pedestrians is 60, which gives $f_{RT} = 1.0 - 0.15[0.15 + (60/2100)] = 0.97$.

Left-Turn Lane Adjustment Factor, f_{LT}. The left-turn adjustment factors for the WB approach are obtained directly from Table 8.13 as 0.95, since the left turns on this approach are made from exclusive left-turn lanes with protected plus permitted phasing. It is, however, necessary to use the special procedure for the other approaches, as the left turns at all of these are made during a permitted phase. The computation required for each of these is shown in Figure 8.16. This involves the solution of each of Eqs. 8.11 through 8.19 as shown in the figure.

The adjusted saturated flow rate is then determined using Eq. 8.10 and the factors listed in Figure 8.15. For example, the adjusted saturation flow rate for the EB left lane group is

$$(1800)(1)(0.93)(0.98)(1.02)(1.0)(1.0)(1.0)(1.0)(0.13) = 218 \text{ vphg}$$

- **Capacity Analysis Module**

In this module, we compute the capacity and the (v/c) ratio for each lane group and the aggregate results. The computations required are carried out in Figure 8.17. The adjusted flow rates and adjusted saturation flow rates are extracted from Figures 8.14 and 8.15, respectively, and recorded in Figure 8.17, together with the appropriate lane group movement as shown. The flow ratios are then computed and recorded in column 5 of the figure. Since this is an analysis problem, the signal times are known and the g/C ratios are determined, assuming a lost time of 3 sec per phase. The amber time is given as 3 sec per phase.

$$\text{NB:} \quad (g/C) = \frac{35 - 3 + 3}{100} = 0.35$$

$$\text{EB:} \quad (g/C) = \frac{40 - 3 + 3}{100} = 0.40$$

The critical lane groups are then determined from the flow ratios. With the phase plan shown in Figure 8.13, it can be seen that during phase 3 only the NBTH/RT/LT and the SBTH/RT/LT lane groups move, and they do not move during any other phase. The NBTH/RT/LT lane group having the higher (v/s) ratio of 0.36 is therefore the critical lane group. Phases 1 and 2, however, overlap. The WBLT and WBTH/RT traffic moves during phases 1 and 2, and the EBLT and EBTH/RT traffic moves only during phase 2. The critical combination for phases 1 and 2 is therefore obtained from

WBLT + EBTH/RT: $(v/s) = 0.15 + 0.35 = 0.50$
WBTH/RT + EBLT: $(v/s) = 0.42 + 0.40 = 0.82$

The critical lane combination phases 1 and 2 is therefore WBTH/RT + EBLT. The sum of the (v/s) ratios for all the critical lane groups is then determined as $0.42 + 0.40 + 0.36 = 1.18$. The critical (v/c) ratio for the intersection is then computed as 1.30 as shown.

Figure 8.16   Supplemental Worksheet for Computation of Left-Turn Adjustment Factors for Permitted Left Turns of Example 8–1

SUPPLEMENTAL WORKSHEET FOR LEFT-TURN ADJUSTMENT FACTOR, f_{LT}				
INPUT VARIABLES	EB	WB	NB	SB
Cycle Length, C (sec)	100		100	100
Effective Green, g (sec)	40		35	35
Number of Lanes, N	1		2	2
Total Approach Flow Rate, v_a (vph)	1129		717	665
Mainline Flow Rate, v_M (vph)	1041		717	665
Left-Turn Flow Rate, v_{LT} (vph)	88		112	106
Proportion of LT, P_{LT}	1.0		0.16	0.16
Opposing Lanes, N_o	2		2	2
Opposing Flow Rate, v_o (vph)	1206		665	717
Prop. of LT in Opp. Vol., P_{LTO}	0.0		0.16	0.16
COMPUTATIONS	EB	WB	NB	SB
$S_{op} = \dfrac{1800\, N_o}{1 + P_{LTO} \left[\dfrac{400 + v_M}{1400 - v_M} \right]}$	3600		2853	2922
$Y_o = v_o / S_{op}$	0.335		0.23	0.25
$g_u = (g - CY_o) / (1 - Y_o)$	9.77		15.58	13.33
$f_s = (875 - 0.625\, v_o) / 1000$	—		0.46	0.43
$P_l = P_{LT} \left[1 + \dfrac{(N - 1)g}{f_s g_u + 4.5} \right]$	1.0		0.64	0.71
$g_q = g - g_u$	30.23		19.42	21.67
$P_T = 1 - P_L$	0.0		0.36	0.29
$g_f = 2 \dfrac{P_T}{P_l} \left[1 - P_T^{\,0.5\, g_q} \right]$	0.0		1.125	0.82
$E_L = 1800 / (1400 - v_o)$	9.28		2.45	2.64
$f_m = \dfrac{g_f}{g} + \dfrac{g_u}{g} \left[\dfrac{1}{1 + P_l (E_l - 1)} \right] + \dfrac{2}{g}(1 + P_L)$	0.13		0.36	0.30
$f_{LT} = (f_m + N - 1) / N$	0.13		0.68	0.65

Source: Adapted from *Highway Capacity Manual*, Special Report 209, Transportation Research Board, Washington, D.C., 1985.

Figure 8.17 Worksheet for the Capacity Analysis Module for Example 8–1

CAPACITY ANALYSIS WORKSHEET								
LANE GROUP		③ Adj. Flow Rate v (vph)	④ Adj. Sat. Flow Rate s (vphg)	⑤ Flow Ratio v/s ③ ÷ ④	⑥ Green Ratio g/C	⑦ Lane Group Capacity c (vph) ④ × ⑥	⑧ v/c Ratio X ③ ÷ ⑦	⑨ Critical ? Lane Group
① Appr.	② Lane Group Movements							
EB		88	218	0.40	0.40	87	1.01	✓
	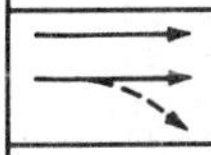	1093	3147	0.35	0.40	1259	0.87	
WB		235	1527	0.15	0.53	809	0.29	
		1266	3024	0.42	0.53	1603	0.79	✓
NB		753	2109	0.36	0.35	738	1.02	✓
SB		698	2036	0.34	0.35	713	0.98	

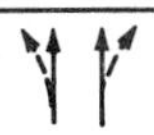

Cycle Length, C $\underline{\quad 100 \quad}$ sec

Lost Time Per Cycle, L $\underline{\quad 9 \quad}$ sec

$$\Sigma\,(v/s)_{ci} = \underline{\quad 1.18 \quad}$$

$$X_c = \frac{\Sigma\,(v/s)_{ci} \times C}{C - L} = \underline{\quad 1.30 \quad}$$

Source: Adapted from *Highway Capacity Manual*, Special Report 209, Transportation Research Board, Washington, D.C., 1985.

■ Level of Service Module

Figure 8.18 is used to carry out the required computations. The uniform delay (d_1) and the incremental delay d_2 are calculated for each phase using the appropriate equations

$$d_1 = 0.38C \frac{(1 - g/C)^2}{1 - (g/C)x}$$

and

$$d_2 = 173x^2[(x - 1) + \sqrt{(x - 1)^2 + (16x/c)}]$$

Since the intersection is isolated, type 3 arrival type is assumed, which gives a PF of 1.0 from Table 8.14. The total delay for each lane group is obtained by summing the uniform and incremental delays and the level of service obtained from Table 8.1. The average delay for each approach is found by determining the weighted average of the delays of the lane groups in that approach. Similarly, the average delay for the intersection is found by determining the weighted delay for all approaches. For example,

$$\text{EB approach delay} = \frac{(88)(100.79) + (1093)(25.85)}{88 + 1093} = 31.43 \text{ sec}$$

intersection delay

$$= \frac{31.43(88 + 1093) + 15.38(235 + 1266) + (753)(55.57) + (698)(45.9)}{88 + 1093 + 235 + 1266 + 753 + 698}$$

$$= 32.46$$

The results of the analysis indicate that the critical v/c ratio is higher than 1.0. This suggests that both the geometric and signal characteristics are inadequate for the critical demand flows at the intersection. In addition, the (v/c) ratios for the EB left-turn lane group and the NB and SB approaches are much higher than those for the EB through and right-turn lane group and the WB approach. In fact, the EB left-turn lane group and the NB approach are oversaturated, indicating that the green time is not proportionally distributed. Also, the NB and SB approaches operate at a level of service E, which normally would be unacceptable, although the overall level of service at the intersection is D. To improve the LOS at this intersection, consideration should be given to improving the geometrics and the signal timing. For example, consideration should be given to increasing the cycle length and/or adding a left-turn lane to the east, north, and south approaches. Operation at this level of service may occur only during the peak hour. If there is significant volume reduction during non-peak hours, the LOS during those times will significantly improve.

The computations required to solve Example 8–1 were carried out manually to facilitate a detailed description of each step of the procedure. Note, however,

Figure 8.18   Worksheet for the Level of Service Module for Example 8–1

LEVEL-OF-SERVICE WORKSHEET												
Lane Group		First Term Delay				Second Term Delay				Total Delay & LOS		
① Appr.	② Lane Group Move-ments	③ v/c Ratio X	④ Green Ratio g/C	⑤ Cycle Length C (sec)	⑥ Delay d_1 (sec/veh)	⑦ Lane Group Capacity c (vph)	⑧ Delay d_2 (sec/veh)	⑨ Progression Factor PF Table 8-14	⑩ Lane Group Delay (sec/veh) (⑥+⑧)×⑨	⑪ Lane Group LOS Table 8-1	⑫ Approach Delay (sec/veh)	⑬ Appr. LOS Table 8-1
EB		1.01	0.4	100	22.95	87	77.84	1	100.79	F	31.43	D
		0.87	0.4	100	20.98	1259	4.87	1	25.85	D		
WB		0.29	0.53	100	9.92	809	0.06	1	9.98	B	15.38	C
		0.79	0.53	100	14.44	1603	1.94	1	16.38	C		
NB		1.02	0.35	100	24.96	738	30.61	1	55.57	E	55.57	E
SB		0.98	0.35	100	24.44	713	21.54	1	45.98	E	45.98	E

Intersection Delay ___32.46___ sec/veh   Intersection LOS ___D___ (Table 8-1)

Source: Adapted from *Highway Capacity Manual*, Special Report 209, Transportation Research Board, Washington, D.C., 1985.

that computer programs are now available that can be used to carry out these computations. For example, CAPCALC 85, produced by Roger Creighton Associates, Inc., is a microcomputer program that can be used to determine the level of service of a signalized intersection at both the "planning" and the "operation and design" levels of analysis. The Federal Highway Administration (FHWA) has also produced microcomputer software packages for use with the 1985 *Highway Capacity Manual*. These software packages can be used in solving this example or any similar problem.

PLANNING ANALYSIS

The planning level of analysis can be used to determine the required geometrics of an intersection for a given demand flow or to determine whether a given demand flow will be less or greater than the capacity for a given set of geometric characteristics. Delay is not assessed at this level of analysis because details of the signalized system are usually not known at the planning stage. The data required for this analysis consist mainly of the intersection geometric details and demand volumes. The intersection geometric details for this level of analysis consist only of the number of lanes on each approach and the use of each lane. The demand volumes are given in vph and could be the estimated peak hour volume or could be based on the maximum flow rate during a 15 min period of the peak hour. When the maximum 15 min flow rate is used, the demand volume is obtained by dividing the peak hour volume by the assumed PHF. The volumes are given in numbers of mixed vehicles, consisting of passenger cars, trucks, buses, and recreational vehicles. At this level of analysis, factors such as lane widths, parking conditions, and traffic mix are not considered, since these are not usually available during the planning stage. However, to determine the appropriate demand at the intersection, the volume on each lane is required. The demand volume on each approach is therefore assigned uniformly to each lane, as far as this is possible, using the projected flow conditions. The primary reason for assigning uniform distribution is that traffic at any approach of an intersection will tend to redistribute among the lanes to obtain an equal number of passenger cars (PCEs) on each approach lane. The assignment is therefore carried out, using the following guidelines.

1. When there is an exclusive right-turn or left-turn lane, total right-turn volume is assigned to the exclusive right-turn lane, and total left-turn volume is assigned to the exclusive left-turn lane.

2. When there are two or more lanes available for the same movement, uniformly distribute the volume for that movement among the available lanes.

3. When there is an exclusive left-turn lane but no exclusive right-turn lane, the through and right-turn vehicles for the approach are uniformly distributed among the available through lanes, but with the provision that the total right-turn volume should be assigned to the rightmost travel lane. The PCE volume for right-turning volume is 1.00 for the planning level of analysis.

4. When there is a shared left-turn/through lane, distribute the number of vehicles to obtain the same number of PCEs in each lane, with the provision that all left-turn vehicles must be assigned to the leftmost lane. PCEs for shared left-turn volumes are obtained from Table 8.15. It is necessary to use the equivalency factor in this case because the left-turn vehicles on the shared lanes will impede the through vehicles using the same shared lane. It is useful to identify by an asterisk (*) those median lanes that carry left-turn and through traffic and are impeded by opposing through traffic. Figure 8.19 is a worksheet that can be used to compute the volume distribution for shared left/through lanes.

Figure 8.19 Worksheet for Lane Distribution for Shared Left/Through Lanes

LANE DISTRIBUTION FOR SHARED LEFT/THRU LANES ON A MULTILANE APPROACH WITH PERMITTED LEFT TURN LANES (OPTIONAL WORKSHEET)

①	②	③	④	⑤	⑥	⑦	⑧	⑨	⑩	⑪
V_O Opposing Volume (vph)	PCE_{LT}	V_{LT}	LT Equiv. PCE's	Total Volume (TH+RT)	Total	No. of Lanes on Approach	Equiv. Volume Per Lane	Thru Vehicles in LT+TH Lane	Vol. in LT+TH Lane	Vol. in ea. of the Remaining Lanes
$\begin{aligned} 0 - 199 &= 1.1 \\ 200 - 599 &= 2.0 \\ 600 - 799 &= 3.0 \\ 800 - 999 &= 4.0 \\ \geq 1{,}000 &= 5.0 \end{aligned}$			② × ③		④ + ⑤		⑥ ÷ ⑦	⑧ − ④	③ + ⑨	$\dfrac{(③+⑤)-⑩}{⑦-1.0}$
APPR.										
EB LT										
WB LT										
NB LT										
SB LT										

Source: Reproduced from *Highway Capacity Manual*, Special Report 209, Transportation Research Board, Washington, D.C., 1985.

Figure 8.20   Planning Application Worksheet

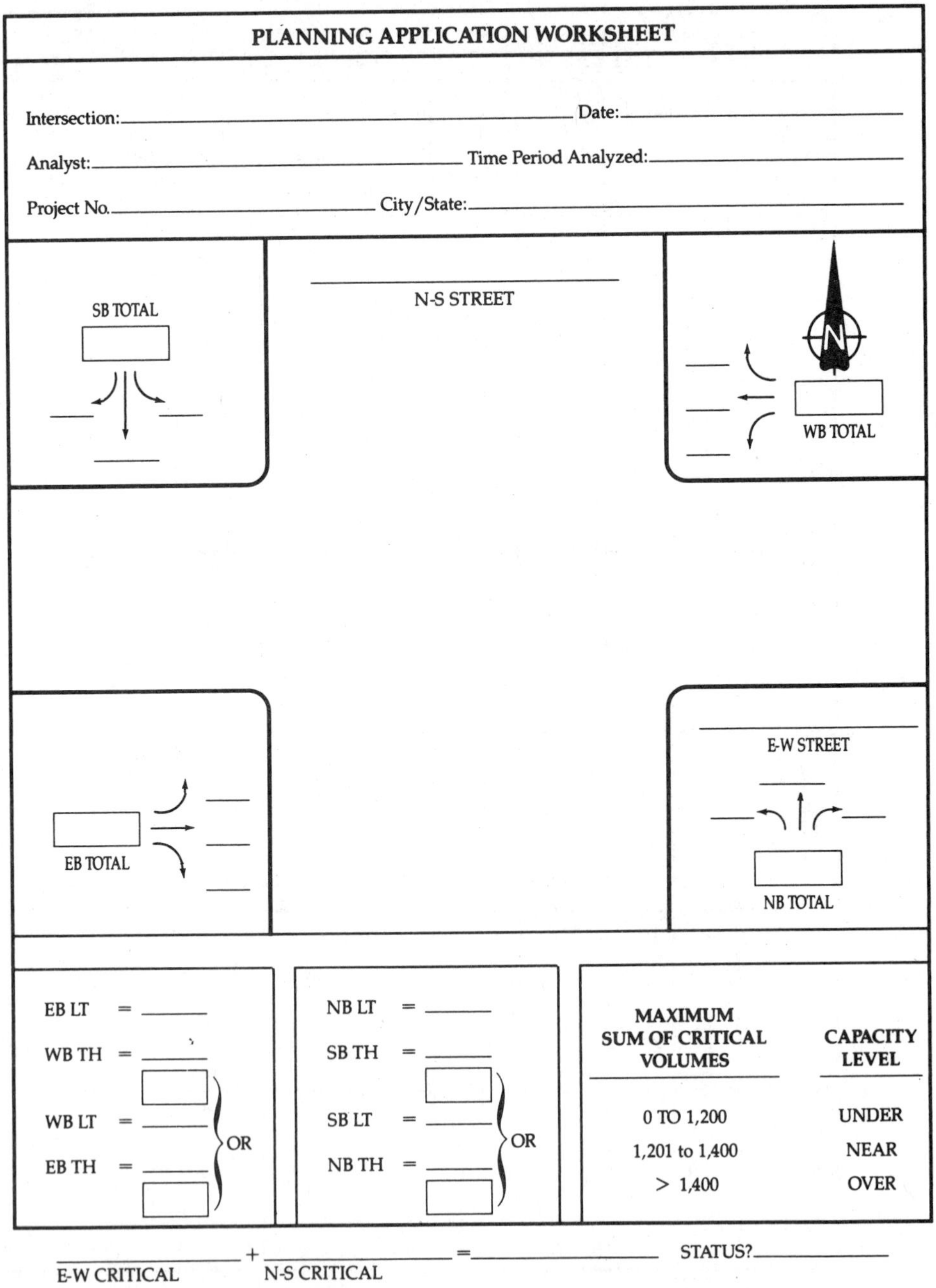

PLANNING APPLICATION WORKSHEET

Intersection:_____ Date:_____

Analyst:_____ Time Period Analyzed:_____

Project No._____ City/State:_____

SB TOTAL

N-S STREET

WB TOTAL

E-W STREET

EB TOTAL

NB TOTAL

		MAXIMUM SUM OF CRITICAL VOLUMES	CAPACITY LEVEL
EB LT = _____	NB LT = _____		
WB TH = _____	SB TH = _____		
WB LT = _____ OR	SB LT = _____ OR	0 TO 1,200	UNDER
EB TH = _____	NB TH = _____	1,201 to 1,400	NEAR
		> 1,400	OVER

_____ + _____ = _____   STATUS?_____
E-W CRITICAL   N-S CRITICAL

Source: Reproduced from *Highway Capacity Manual*, Special Report 209, Transportation Research Board, Washington, D.C., 1985.

To use these guidelines effectively, it helps to make a sketch of the intersection since this will facilitate the correct assignment of the demand volume to the appropriate lanes. Figure 8.20 shows a basic worksheet for this purpose.

Capacity Analysis

At the planning analysis level, details of the signal operations are usually unknown. The demand volume used for analysis is the sum of the critical volumes for the intersection. The critical volume for each street at the intersection is determined by considering the critical conflicting movements for that street. For example, the critical conflicts for the east–west street on a four-legged intersection are (1) the west approach left-turn movement with the east approach through movement and right-turn movement, and (2) the east approach left-turn movement with the west approach through movement and right-turn movement. Similarly, the critical movements for the north–south street are (1) the north approach left-turn movement with the south approach through movement and south approach right-turn movement, and (2) the south approach left-turn movement with the north approach through movement and north approach right-turn movement.

The critical volume for the east–west street is therefore the higher of the following:

- The sum of the east approach left-turn volume, the maximum west approach single-lane through volume, and the west approach right-turn volume
- The sum of the west approach left-turn volume, the maximum east approach single-lane through volume, and the east approach right-turn volume

The critical volume for the north–south street is the higher of the following:

- The sum of the north approach left-turn volume, the maximum south approach single-lane through volume, and the south approach right-turn volume
- The sum of the south approach left-turn volume, the maximum north approach single-lane through volume, and the north approach right-turn volume

In computing the critical volumes, only those vehicles that impede the left-turn vehicles of the opposing movement should be added. Use the following guidelines to ensure that only those vehicles that create impedance are considered.

1. Left-turning vehicles in a separate left-turn lane are not included as part of the opposing mainline traffic.
2. When through traffic is impeded by left-turn vehicles sharing the same lane, these left-turn vehicles are included as part of the opposing vehicles

and should be converted to PCEs as described in Example 8–3, using left-turn factors given in Table 8.15.

3. Right-turn vehicles on a separate right-turn lane are not included as part of the opposing mainline traffic.

The total critical volume used in the analysis is the sum of all critical volumes. The total critical volume for the intersection is then compared with the critical volumes given in Table 8.16 to determine whether the intersection will be operating under capacity, near capacity, or over capacity for the projected conditions.

If it is found that the intersection will be operating under capacity, it can be concluded that excessive delays are not anticipated. When operating conditions are such that the intersection will be operating, near capacity, it is quite probable that the actual demand may be higher or lower than the capacity of the intersection. It is therefore possible to have unstable flows that could result in a wide range of delays. In cases where it is determined that the intersection will be operating over capacity, it can be concluded that excessive delays will occur.

Example 8–2   Computing Capacity at a Signalized Intersection Using the Planning Level of Analysis

Figure 8.21 shows the design volumes and the layout of a new signalized intersection being planned for projected traffic growth in 10 yr at the intersection of 10th and Main streets. Using the planning level of analysis, determine whether the capacity of the proposed design will be adequate.

- Assign the approach volumes given in the boxes to the appropriate lanes as shown in the top portion of Figure 8.21. For example, for the SB total, there is an exclusive left-turn lane with a total left-turn volume of 175 vehicles assigned to it. The total through and right-turn vehicles are then distributed equally to the remaining lanes, giving a volume of 313 on each of the remaining lanes. This exercise is repeated for each approach as shown in Figure 8.21.

- Find the critical volume for each street, as shown in the lower portion of Figure 8.21. For example, the critical volume for 10th Street is found by considering the N–S direction. That is,

$$SBLT + NBTH = 175 + 475 = 650$$

$$SBTH + NBLT = 313 + 110 = 423$$

The critical volume for 10th Street is therefore 650. The critical volume for Main Street is similarly found as 663. Thus,

$$\text{total critical volume} = 650 + 663 = 1313$$

Figure 8.21 Data and Solution for Example 8–2

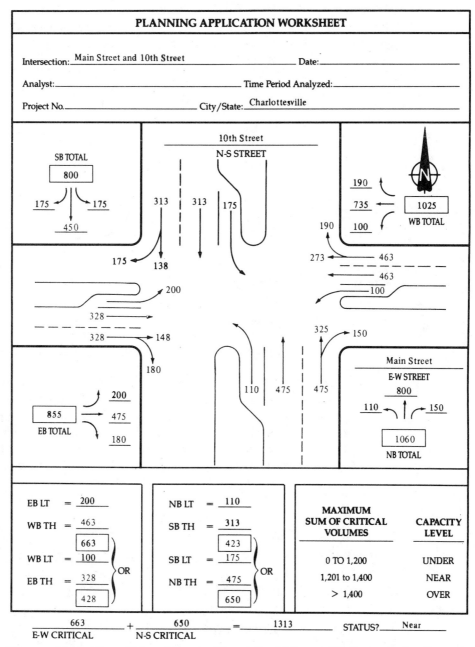

Source: Adapted from *Highway Capacity Manual*, Special Report 209, Transportation Research Board, Washington, D.C., 1985.

- Use Table 8.16 to determine the capacity level. The criteria indicate that the intersection will operate near capacity and a wide range of delays can be expected.

- Test whether operating conditions may be improved by increasing by one lane the number of WB and NB through lanes. Accordingly, the critical volume for 10th Street is

$$175 + \left(\frac{475 + 475}{3}\right) = 492$$

and the critical volume for Main Street is

$$200 + \left(\frac{463 + 463}{3}\right) = 509$$

Thus, the total critical volume is 1001, which will provide for an operation under capacity and excessive delay will not be anticipated.

Example 8–3   Determining a Suitable Layout for a Signalized Intersection Using the Planning Level of Analysis

Figure 8.22 shows the estimated maximum flows at a signalized intersection planned for an urban area. Using this data, we have to determine a suitable layout for the intersection that will provide a condition of under capacity. Due to buildings adjacent to Riverbend Drive, only two lanes can be provided on each approach of this road.

- Determine the number of lanes on each approach. Exclusive left-turn lanes should be provided when left-turn volumes are higher than 100 vph. Exclusive left-turn lanes are therefore provided at both approaches of Route 250. Since only two lanes can be provided on Riverbend Drive, exclusive left-turn lanes are not provided to accommodate the through and right-turn vehicles.

North approach (SB):   2 lanes (due to restriction) (all vehicles)
South approach (NB):   2 lanes (due to restriction) (all vehicles)
East approach (WB):   1 exclusive left-turn lane (left-turn vehicles only)
West approach (EB):   1 exclusive left-turn lane (left-turn vehicles only)

For the east approach (through and right-turn vehicles), assuming 450 vehicles per lane,

$$\text{WBTH} + \text{WBRT} = \frac{775 + 180}{450} = 2.12 \quad \text{(use 2 lanes)}$$

Figure 8.22 Data and Solution for Example 8–3

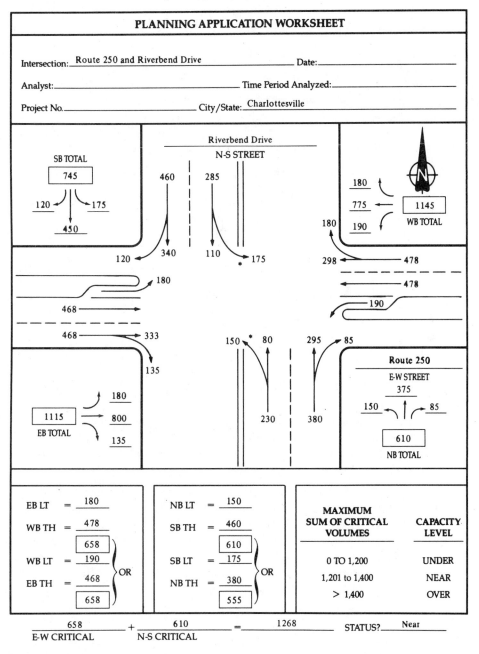

Source: Adapted from *Highway Capacity Manual*, Special Report 209, Transportation Research Board, Washington, D.C., 1985.

For the west approach, through and right-turn vehicles.

$$\text{EBTH} + \text{EBRT} = \frac{800 + 135}{450} = 2.07 \quad (\text{use 2 lanes})$$

A sketch of the assumed geometrics is shown in Figure 8.22.

■ Determine the demand volumes. Since it is not possible to provide an exclusive left-turn phase for the N–S approaches, the left-turning vehicles on Riverbend Drive will be on shared lanes. Left turns that cause impedance—that is, left turns on shared lanes that conflict with an opposing flow—are marked with an asterisk as shown in Fig. 8.22.

For the *east approach*,

$$\text{WBLT} = 190 \text{ vph}$$

$$\text{remaining lanes} = \frac{775 + 180}{2}$$

$$= 477.5 \quad (\text{say, 478 vehicles per lane per hour})$$

For the *west approach*,

$$\text{EBLT} = 180 \text{ vph}$$

$$\text{remaining lanes} = \frac{800 + 135}{2}$$

$$= 467.5 \quad (\text{say, 468 vehicles per lane per hour})$$

The through lanes of the east and west approaches have volumes slightly greater than 450 vph, which is acceptable since approaches with the necessary allotted green times will accommodate volumes greater than 450 vplph.

For the *south approach*, which has a shared left-turn/through lane, it is first necessary to find the PCE of the left-turning vehicles. The procedure for assigning the volumes is shown in Figure 8.23.

The opposing volume to the NBLT is 570, which is the sum of the SB right-turn and straight through volumes. Figure 8.22 shows the PCE for left turns (PCE_{LT}) is 2.0, which is recorded in column 2. The NBLT volume is recorded in column 3, and the PCE of the left-turn volume, which is found by multiplying the left-turn volume by the PCE_{LT} of 2, is recorded in column 4. The through and right-turn volume is recorded in column 5.

The PCE volume of the left turns in column 4 is then added to the through and right-turn volume and recorded in column 6. Using the number of lanes on the approach, the total equivalent volume is equally distributed and the equivalent volume per lane is recorded in column 8. It is,

Figure 8.23 Volume Distribution Involving a Shared Left-Turn/Through Lane for Example 8–3

LANE DISTRIBUTION FOR SHARED LEFT/THRU LANES ON A MULTILANE APPROACH WITH PERMITTED LEFT TURN LANES (OPTIONAL WORKSHEET)										
①	②	③	④	⑤	⑥	⑦	⑧	⑨	⑩	⑪
V_O Opposing Volume (vph)	PCE_{LT}	V_{LT}	LT Equiv. PCE's	Total Volume (TH+RT)	Total	No. of Lanes on Approach	Equiv. Volume Per Lane	Thru Vehicles in LT+TH Lane	Vol. in LT+TH Lane	Vol. in ea. of the Remaining Lanes
$0 - 199 = 1.1$ $200 - 599 = 2.0$ $600 - 799 = 3.0$ $800 - 999 = 4.0$ $\geq 1{,}000 = 5.0$ APPR.	②×③		④+⑤		⑥÷⑦		⑧−④		③+⑨	$\dfrac{(③+⑤)-⑩}{⑦-1.0}$
EB LT										
WB LT										
NB LT	570	2.0	150	300	460	760	2	380	80	230	380
SB LT	460	2.0	175	350	570	920	2	460	110	285	460

Source: Adapted from *Highway Capacity Manual*, Special Report 209, Transportation Research Board, Washington, D.C., 1985.

Figure 8.24   Revised Solution for Example 8–3

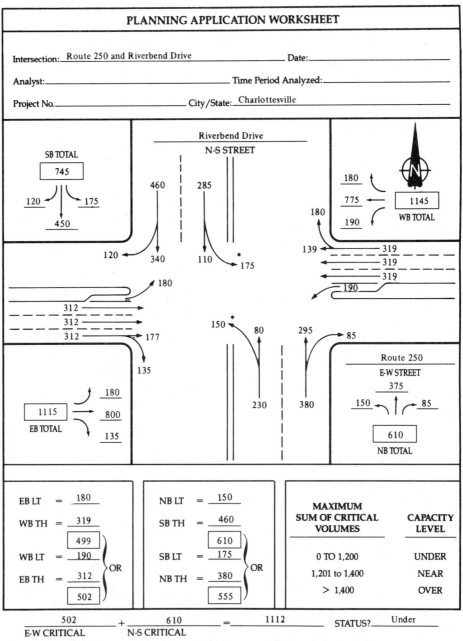

Source: Adapted from *Highway Capacity Manual*, Special Report 209, Transportation Research Board, Washington, D.C., 1985.

however, necessary to convert the total PCE volumes on each lane back to actual volumes in terms of vph, since at this level of analysis only actual number of mixed vph is used. This conversion is carried out under columns 9, 10, and 11, and the actual volume on each lane is then recorded in columns 10 and 11. The procedure is checked by finding the sum of the volumes under columns 10 and 11 for the approach. This sum should be equal to the demand volume at that approach.

For the *north approach*, the procedure used for the south approach is also used. The volumes assigned are shown in Figure 8.22. The critical volumes on Riverbend Drive and Route 250 are as shown in Figure 8.22. The sum of the critical volumes is 1268, as shown on Figure 8.22. From Table 8.16, we find that the condition will be near capacity. Since the condition of under capacity is required at this intersection, some changes will have to be made to the geometrics of the intersection. Let us increase by one each the lanes in the E–W direction. (Note that this is not necessarily the best change; it is used to demonstrate the procedure). The computation should be repeated from the assignment of the demand volume.

- Recalculate the demand volume. (The volumes on the north and south approaches remain the same.)

For the *east approach*,

$$WBLT = 190 \text{ vph}$$

$$\text{remaining lanes} = \frac{(775 + 180)}{3} = 319 \text{ vehicles per hour per lane}$$

For the *west approach*,

$$EBLT = 180 \text{ vph}$$

$$\text{remaining lanes} = \frac{(800 + 135)}{3} = 312 \text{ vehicles per hour per lane}$$

These revised volumes are shown in Figure 8.24 and the condition is now under capacity. The additional lanes have improved the performance of the intersection.

SUMMARY

The complexity of a signalized intersection requires that several factors be considered when its quality of operation is being evaluated. In particular, the prevailing traffic conditions, geometric characteristics, and signal characteristics must be used in determining the level of service at the intersection. This requirement makes the determination and improvement of level of service at a signalized intersection much more complex than at other locations of a highway. For example, in Chapter 7, the primary characteristics used in determining capacity and level of service at segments of highways are those of geometry of the highway and traffic composition. Since the geometry is usually fixed, the capacity of a highway segment can be improved by improving the highway geometry if consideration is given to some variations over time in traffic composition. This, however, cannot be done easily at signalized intersections because of the added factor of the green time allocation to the different traffic streams, which has a significant impact on the operation of the intersection.

The procedures presented in this chapter for the determination of delay and level of service at a signalized intersection take into consideration the most recent results of research in this area. In particular, the procedures not only take into consideration such factors as traffic mix, lane width, and grades but also factors inherent in the signal system itself, such as whether the signal is pretimed, semi-actuated, or fully actuated.

In using the procedures for design, remember that the cycle length has a significant effect on the delay at an intersection and therefore on the level of service. Thus, it is useful first to determine a suitable cycle and suitable phase lengths, using one of the methods discussed in Chapter 6 or the method presented in this chapter, and then to use this information to determine the level of service.

PROBLEMS

8–1 Figure 8.25 shows peak hour volumes and other traffic and geometric characteristics of an intersection located in the CBD of a city. Determine the overall level of service at the intersection. Use operation analysis.

8–2 Figure 8.26 shows an isolated intersection at Third Street and Ells Avenue outside the CBD of a city. Third Street is one way in the NB direction, with parking allowed on either side. Ells Avenue is a major arterial with a separate left-turn lane on the EB approach. Traffic volume counts at the intersection have shown that the growth rate over the past 3 yr is 4 percent per annum, and it is predicted that this rate will continue for the next 5 yr. The existing peak hour volumes and traffic characteristics are shown in Figure 8.26. Determine a suitable timing of the existing three-phase signal that will be suitable for the traffic volumes in 5 yr. Assume that, except for the traffic volumes, characteristics will remain the same. Also determine the LOS at which the intersection will operate. A critical (v/c) ratio of 0.85 or lower is required at the intersection.

8–3 Figure 8.27 shows projected peak hour volumes in 5 years and other traffic and geometric characteristics for the intersection of 3rd and K streets in the CBD of an urban area. Determine whether the existing signal timing will be suitable for the projected demand if the level of service at each approach must be D or better. If the existing system

FIGURE 8.25

INPUT WORKSHEET

Intersection:_____ Date:_____

Analyst:_____ Time Period Analyzed:_____ Area Type: ☐ CBD ☐ Other

Project No.:_____ City/State:_____

VOLUME AND GEOMETRICS

NORTH

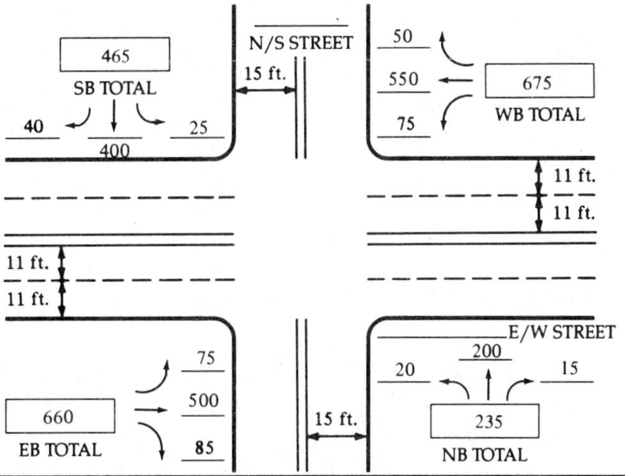

N/S STREET
15 ft.

465
SB TOTAL
40 ↵ ↓ ↘ 25
400

50
550 ←
75
675
WB TOTAL

11 ft.
11 ft.

11 ft.
11 ft.

IDENTIFY IN DIAGRAM:

1. Volumes
2. Lanes, lane widths
3. Movements by lane
4. Parking (PKG) locations
5. Bay storage lengths
6. Islands (physical or painted)
7. Bus stops

75
500
660
85
EB TOTAL

15 ft.

E/W STREET
200
20 ← ↑ 15
235
NB TOTAL

TRAFFIC AND ROADWAY CONDITIONS

Approach	Grade (%)	% HV	Adj. Pkg. Lane Y or N	Adj. Pkg. Lane N_m	Buses (N_B)	PHF	Conf. Peds. (peds./hr)	Pedestrian Button Y or N	Pedestrian Button Min. Timing	Arr. Type
EB	0	5	N	–	0	0.85	100	N	–	4
WB	0	5	N	–	0	0.85	100	N	–	2
NB	0	8	N	–	0	0.85	100	N	–	3
SB	0	8	N	–	0	0.85	100	N	–	3

Grade: + up, − down N_B: buses stopping/hr   Min. Timing: min. green for
HV: veh. with more than 4 wheels   PHF: peak-hour factor   pedestrian crossing
N_m: pkg. maneuvers/hr   Conf. Peds: Conflicting peds./hr   Arr. Type: Type 1-5

PHASING

DIAGRAM

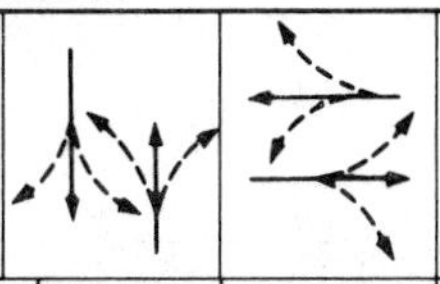

Timing	G = 39 Y + R = 61	G = 55 Y + R = 45	G = Y + R =	G = Y + R =	G = Y + R =	G = Y + R =	G = Y + R =	G = Y + R =
Pretimed or Actuated	P	P						

↗ Protected turns   ⌁ Permitted turns   - - - - Pedestrian   Cycle Length 100 Sec

Source: Adapted from *Highway Capacity Manual*, Special Report 209, Transportation Research Board, Washington, D.C., 1985.

FIGURE 8.26

Source: Adapted from *Highway Capacity Manual*, Special Report 209, Transportation Research Board, Washington, D.C., 1985.

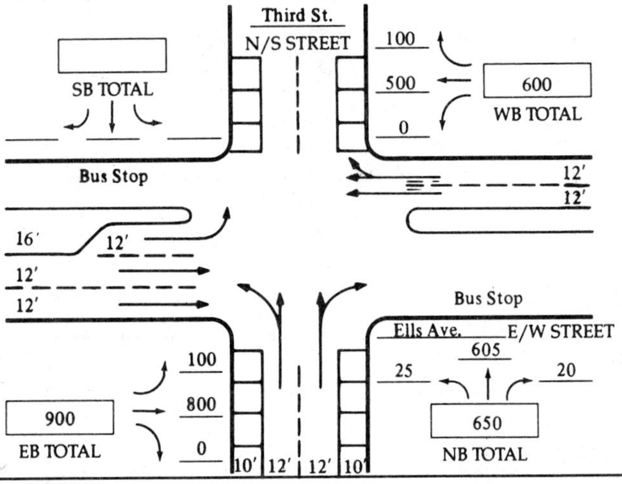

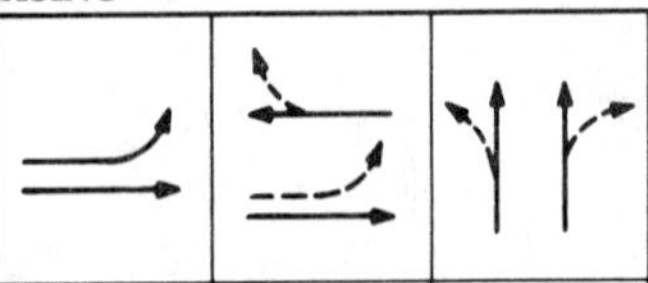

FIGURE 8.27

INPUT WORKSHEET

Intersection:_____   Date:_____

Analyst:_____ Time Period Analyzed:_____ Area Type: ☐ CBD ☐ Other

Project No.:_____ City/State:_____

VOLUME AND GEOMETRICS

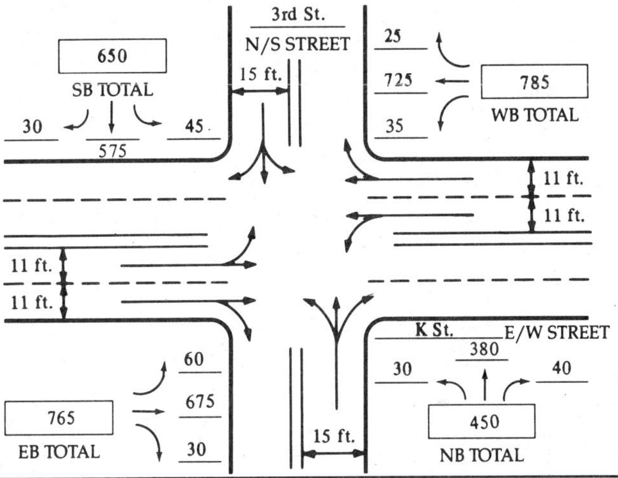

3rd St.
N/S STREET

650
SB TOTAL
575
30 / ↓ ↘ 45

15 ft.

25
725 ←
35
785
WB TOTAL

11 ft.
11 ft.

NORTH

11 ft.
11 ft.

IDENTIFY IN DIAGRAM:

1. Volumes
2. Lanes, lane widths
3. Movements by lane
4. Parking (PKG) locations
5. Bay storage lengths
6. Islands (physical or painted)
7. Bus stops

60
675
765
EB TOTAL
30

K St.   E/W STREET
380
30 / ↓ ↘ 40
450
NB TOTAL

15 ft.

TRAFFIC AND ROADWAY CONDITIONS

Approach	Grade (%)	% HV	Adj. Pkg. Lane Y or N	Adj. Pkg. Lane N_m	Buses (N_B)	PHF	Conf. Peds. (peds./hr)	Pedestrian Button Y or N	Pedestrian Button Min. Timing	Arr. Type
EB	−1	4	N	—	0	0.95	60	N		3
WB	−1	4	N	—	0	0.95	60	N		3
NB	0	6	N	—	0	0.95	60	N		3
SB	0	6	N	—	0	0.95	60	N		3

Grade: + up, − down
HV: veh. with more than 4 wheels
N_m: pkg. maneuvers/hr

N_B: buses stopping/hr
PHF: peak-hour factor
Conf. Peds: Conflicting peds./hr

Min. Timing: min. green for pedestrian crossing
Arr. Type: Type 1-5

PHASING

DIAGRAM	↖↑↗→	⋏↑⋎							
Timing	G = 30 Y + R = 40	G = 34 Y + R = 36	G = Y + R =	G = Y + R =	G = Y + R =	G = Y + R =	G = Y + R =	G = Y + R =	G = Y + R =
Pretimed or Actuated	P	P							

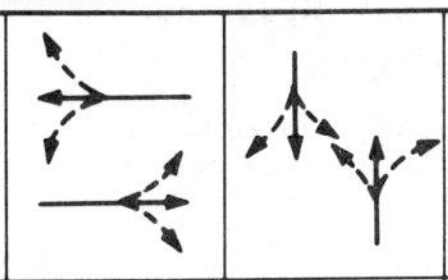

⟋ Protected turns	_ ⟋ Permitted turns	- - - - - Pedestrian	Cycle Length_____Sec

Source: Adapted from *Highway Capacity Manual*, Special Report 209, Transportation Research Board, Washington, D.C., 1985.

will not satisfy this requirement, make suitable changes to the phasing and/or signal timing that will achieve the level of service requirement of D. Determine the critical volume and the intersection overall level of service, using your phasing and/or signal timing. A critical (v/c) ratio of 0.85 or lower is required at the intersection.

8–4 Repeat Problem 8–3, assuming that the intersection is located outside the CBD. There is adequate right of way available for significant geometric improvement and the level of service on each approach should be C or better.

8–5 Figure 8.28 shows the estimated maximum flows at an isolated signalized intersection planned for an urban area. Using the planning level of analysis, determine a suitable intersection layout and phasing sequence that will provide a condition of under capacity.

FIGURE 8.28

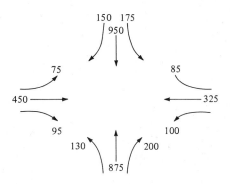

8–6 It is estimated that traffic will grow at a rate of 4 percent per annum at the intersection described in Problem 8–5. Using the planning level of analysis, determine under what conditions the intersection will be operating after 10 yr of operation if your recommended geometric layout is adopted.

8–7 Using your solution for Problem 8–5 and the operation analysis procedure, determine the overall level of service at the intersection if it is located within the CBD. All approaches have 0 percent grades and 3 percent heavy vehicles in the traffic stream. Parking is not allowed and there is no bus stop on any approach. The PHF at each approach is 0.90. There are 60 conflicting pedestrians at each approach, and there are no pedestrian buttons. A critical (v/c) ratio of 0.85 or lower is desired. Use a four-phase signal (one phase for each approach).

8–8 Figure 8.28 shows existing peak hour volumes at an isolated four-legged intersection, with each approach consisting of two 10 ft wide lanes. A signal system is to be installed at the intersection in 3 yr. Traffic growth is projected at 3 percent per annum. Using the planning level of analysis, determine whether it will be necessary to improve the intersection geometry before the installation of the signals.

8–9 Using the peak hour volumes given for Problem 8–8, determine a suitable intersection layout if the growth rate is 4 percent per annum, the signal system is to be installed in 5 years, and a condition of under capacity is required.

REFERENCES

1. *Highway Capacity Manual*, Special Report 209, National Research Council, Transportation Research Board, Washington, D.C., 1985.
2. Ibid.
3. Ibid.
4. Ibid.
5. Ibid.
6. Ibid.
7. Ibid.
8. Ibid.

ADDITIONAL READINGS

Akcelik, R. (ed.), *Signalized Intersection Capacity and Timing Guide*, Signalized Intersection Capacity Workshop Papers—Internal Report AIR 1094–1, Australian Road Research Board, Nunawading, Victoria, Australia, 1979.

Akcelik, R. *Traffic Signals: Capacity and Timing Analysis*, Research Report Number 123, Australian Road Research Board, Nunawading, Victoria, Australia, 1982.

Interim Materials on Highway Capacity, Transportation Research Circular 212, National Research Council, Transportation Research Board, Washington, D.C., January 1980.

JHK & Associates, in cooperation with the Traffic Institute, Northwestern University, *NCHRP Signalized Intersections Capacity Method*, National Research Council, Transportation Research Board, Washington, D.C., May 1982.

JHK & Associates and the Traffic Institute, Northwestern University, *Development of an Improved Highway Capacity Manual*, NCHRP 3–28(2), National Research Council, Transportation Research Board, Washington, D.C., 1979.

McInerney, H. B., and S. G. Peterson, "Intersection Capacity Measurement Through Critical Movement Summations: A Planning Tool," *Traffic Engineering* 41(June 1971).

Miller, A. J., "Australian Road Capacity Guide," *Australian Road Research Bulletin* 3(1981).

Reilly, W. R., C. C. Gardner, and J. H. Kell, *A Technique for Measurement of Delay at Intersections*, Federal Highway Administration Report No. RD–76–135/137, U.S. Department of Transportation, Federal Highway Administration, Washington, D.C., November 1976.

Shenk, E., W. R. McShane, J. M. Ulerio, and R. P. Roess, *Highway Capacity Software User's Manual*, U.S. Department of Transportation, Federal Highway Administration, Washington, D.C., 1986.

CHAPTER 8 APPENDIX: TABLES

All tables in this appendix are reprinted from *Highway Capacity Manual*, Special Report 209, Transportation Research Board, Washington, D.C., 1985. Part (a) of Table 8.2 is from C. J. Messer, "Guidelines for Signalized Left-Turn Treatments," *Implementation* Package *FHWA-IP-81-4*, U.S. Department of Transportation, Federal Highway Administration, Washington, D.C., 1981.

Table 8.1 Level of Service Criteria for Signalized Intersections (Operation Analysis Level)

LEVEL OF SERVICE	STOPPED DELAY PER VEHICLE (SEC)
A	≤ 5.0
B	5.1 to 15.0
C	15.1 to 25.0
D	25.1 to 40.0
E	40.1 to 60.0
F	> 60.0

Table 8.2 Left-Turn Bay Length Adjustment Factors

(a) Multiplication Factors to Correct for Cycle Length and (v/c) Ratio

v/c RATIO, X	CYCLE LENGTH, C (SEC)				
	60	70	80	90	100
0.50	0.70	0.76	0.84	0.89	0.94
0.55	0.71	0.77	0.85	0.90	0.95
0.60	0.73	0.79	0.87	0.92	0.97
0.65	0.75	0.81	0.89	0.94	1.00
0.70	0.77	0.84	0.92	0.98	1.03
0.75	0.82	0.88	0.98	1.03	1.09
0.80	0.88	0.95	1.05	1.11	1.17
0.85	0.99	1.06	1.18	1.24	1.31
0.90	1.17	1.26	1.40	1.48	1.56
0.95	1.61	1.74	1.92	2.03	2.14

(b) Passenger Car Equivalents

TYPE OF TURN	OPPOSING VOLUME (VPH)	PASSENGER CAR EQUIVALENT (PCE)
Protected	—	1.05
Permitted	0 to 199	1.1
	200 to 599	2.0
	600 to 799	3.0
	800 to 999	4.0
	$\geq 1,000$	5.0

Table 8.3 Relationship Between Arrival Type and Platoon Ratio

ARRIVAL TYPE	RANGE OF PLATOON RATIO, R_p
1	0.00 to 0.50
2	0.51 to 0.85
3	0.86 to 1.15
4	1.16 to 1.50
5	≥ 1.51

Table 8.4   Input Data Needs for Each Analysis Lane Group

TYPE OF CONDITION	PARAMETER	SYMBOL
Geometric Conditions	Area Type	CBD or other
	Number of Lanes	N
	Lane Widths, ft	W
	Grades, %	+ (Upgrade)
	Existence of Exclusive LT or RT Lanes	− (Downgrade)
	Length of Storage Bay, LT or RT Lanes	L_s
	Parking Conditions	Y or N
Traffic Conditions	Volumes by Movement, vph	V_i
	Peak-Hour Factor	PHF
	Percent Heavy Vehicles	%HV
	Conflicting Pedestrian Flow Rate, peds/hr	$PEDS$
	Number of Local Buses Stopping in Intersection	N_B
	Parking Activity, pkg maneuvers/hr	N_m
	Arrival Type	
Signalization Conditions	Cycle Length, sec	C
	Green Times, sec	G_i
	Actuated vs Pretimed Operation	A or P
	Pedestrian Push-Button?	Y or N
	Minimum Pedestrian Green	G_p
	Phase Plan	

Table 8.5   Default Values for Use in Operational Analysis

PARAMETER	DEFAULT VALUE	
Conflicting Pedestrian Flow Rate, peds/hr	Low Ped. Flow	50 peds/hr
	Moderate Ped. Flow	200 peds/hr
	High Ped. Flow	400 peds/hr
Percent Heavy Vehicles, %HV	2%	
Peak-Hour Factor, PHF	0.90	
Grade	0%	
Number of Buses, N_B	0 buses/hr	
Number of Parking Maneuvers, N_m	20 maneuvers/hr (where parking exists)	
Arrival Type	3	

Table 8.6 Lane Utilization Factors

NO. OF THROUGH LANES IN GROUP (EXCLUDING LANES USED BY LEFT-TURNING VEHICLES)	LANE UTILIZATION FACTOR, U
1	1.00
2	1.05
≥ 3	1.10

Table 8.7 Adjustment Factors for Lane Widths

Lane Width (ft)	Lane Width Factor, f_w
8	0.87
9	0.90
10	0.93
11	0.97
12	1.00
13	1.03
14	1.07
15	1.10
≥ 16	Use 2 lanes

Table 8.8 Adjustment Factors for Heavy Vehicles

Percent Heavy Vehicles	Heavy Vehicles Factor, f_{HV}
0	1.00
2	0.99
4	0.98
6	0.97
8	0.96
10	0.95
15	0.93
20	0.91
25	0.89
30	0.87

Table 8.9 Adjustment Factors for Approach Grade

	Downhill			Level	Uphill		
Grade	−6	−4	−2	0	+2	+4	+6
Grade factor, f_g	1.03	1.02	1.01	1.00	0.99	0.98	0.97

Table 8.10 Adjustment Factors for Parking

No. of Lanes in Lane Group	No Parking	Number of Parking Maneuvers per Hour, N_m				
		0	10	20	30	40
1	1.00	0.90	0.85	0.80	0.75	0.70
2	1.00	0.95	0.92	0.89	0.87	0.85
3	1.00	0.97	0.95	0.93	0.91	0.89

Table 8.11 Adjustment Factors for Bus Blockage (f_{bb})

No of Lanes in Lane Group	Number of Buses Stopping per Hour, N_B				
	0	10	20	30	40
1	1.00	0.96	0.92	0.88	0.83
2	1.00	0.98	0.96	0.94	0.92
3	1.00	0.99	0.97	0.96	0.94

Table 8.12 Adjustment Factors for Right Turns

CASE	TYPE OF LANE GROUP	RIGHT-TURN FACTORS, f_{RT}						
1	EXCLUSIVE RT LANE; PROTECTED RT PHASING	0.85						
2	EXCLUSIVE RT LANE; PERMITTED RT PHASING	$f_{RT} = 0.85 - (\text{peds}/2{,}100)$ peds $\leq$ 1,700; $f_{RT} = 0.05$ peds $>$ 1,700						
		No. of Conf. Pedestrians (peds)	0	50 (Low)	100	200 (Mod)	300	400 (High)	500
		Factor	0.85	0.83	0.80	0.75	0.71	0.66	0.61
		No. of Conf. Pedestrians (peds)	600	800	1,000	1,200	1,400	1,600	≥ 1,700
		Factor	0.56	0.47	0.37	0.28	0.18	0.05	0.05
3	EXCLUSIVE RT LANE; PROTECTED PLUS PERMITTED PHASING	$f_{RT} = 0.85 - (1 - P_{RTA})(\text{peds}/2{,}100)$; $f_{RT} = 0.05$ (minimum)						
		No. of Conf. Pedestrians (peds)	Prop. of RT Using Prot. Phase, P_{RTA}					
			0.00	0.20	0.40	0.60	0.80	1.00
		0	0.85	0.85	0.85	0.85	0.85	0.85
		50 (Low)	0.83	0.83	0.84	0.84	0.85	0.85
		100	0.80	0.81	0.82	0.83	0.84	0.85
		200 (Mod)	0.75	0.77	0.79	0.81	0.83	0.85
		300	0.71	0.74	0.76	0.79	0.82	0.85
		400 (High)	0.66	0.70	0.74	0.77	0.81	0.85
		600	0.56	0.62	0.68	0.74	0.79	0.85
		800	0.47	0.55	0.62	0.70	0.77	0.85
		1,000	0.37	0.47	0.56	0.66	0.75	0.85
		1,400	0.18	0.32	0.45	0.58	0.72	0.85
		≥ 1,700	0.05	0.20	0.36	0.53	0.69	0.85
4	SHARED RT LANE; PROTECTED PHASING	$f_{RT} = 1.0 - 0.15\,P_{RT}$						
		Prop. of RT in Lane, P_{RT}	0.00	0.20	0.40	0.60	0.80	1.00
		Factor	1.00	0.97	0.94	0.91	0.88	0.85
5	SHARED RT LANE; PERMITTED PHASING	$f_{RT} = 1.0 - P_{RT}\,[0.15 + (\text{peds}/2{,}100)]$; $f_{RT} = 0.05$ (minimum)						
		No. of Conf. Pedestrians (peds)	Prop. of RT in Lane Group, P_{RT}					
			0.00	0.20	0.40	0.60	0.80	1.00
		0	1.00	0.97	0.94	0.91	0.88	0.85
		50 (Low)	1.00	0.97	0.93	0.90	0.86	0.83
		100	1.00	0.96	0.92	0.88	0.84	0.80
		200 (Mod)	1.00	0.95	0.90	0.85	0.80	0.75
		400 (High)	1.00	0.93	0.86	0.80	0.73	0.66
		600	1.00	0.91	0.83	0.74	0.65	0.56
		800	1.00	0.89	0.79	0.68	0.58	0.47
		1,000	1.00	0.87	0.75	0.62	0.50	0.37
		1,400	1.00	0.84	0.67	0.51	0.35	0.18
		≥ 1,700	1.00	0.81	0.62	0.42	0.23	0.05

Continued

Table 8.12—_Continued_

CASE	TYPE OF LANE GROUP	RIGHT-TURN FACTORS, f_{RT}						
6	SHARED RT LANE; PROTECTED PLUS PERMITTED PHASING	$f_{RT} = 1.0 - P_{RT} [0.15 + (\text{peds}/2,100)(1 - P_{RTA})]$ $f_{RT} = 0.05$ (minimum)						

CASE	TYPE OF LANE GROUP	RIGHT-TURN FACTORS, f_{RT}
6	SHARED RT LANE; PROTECTED PLUS PERMITTED PHASING	$f_{RT} = 1.0 - P_{RT} [0.15 + (\text{peds}/2,100)(1 - P_{RTA})]$ $f_{RT} = 0.05$ (minimum)

Prop. RT's Using Prot. Phase P_{RTA}	No. of Conf. Peds. (peds)	Prop. of RT's in Lane Group P_{RT}					
		0.00	0.20	0.40	0.60	0.80	1.00
0.00	All	Same as Case 5					
0.20	0	1.00	0.97	0.94	0.91	0.88	0.85
	50	1.00	0.97	0.93	0.90	0.86	0.83
	200	1.00	0.95	0.91	0.86	0.82	0.77
	400	1.00	0.94	0.88	0.82	0.76	0.70
	600	1.00	0.92	0.85	0.77	0.70	0.62
	1,000	1.00	0.89	0.79	0.68	0.58	0.47
	1,400	1.00	0.86	0.73	0.59	0.45	0.32
	≥ 1,700	1.00	0.81	0.62	0.42	0.23	0.20
0.40	0	1.00	0.97	0.94	0.91	0.88	0.85
	50	1.00	0.97	0.94	0.91	0.87	0.84
	200	1.00	0.96	0.92	0.88	0.83	0.79
	400	1.00	0.95	0.89	0.84	0.79	0.74
	600	1.00	0.94	0.87	0.81	0.74	0.68
	1,000	1.00	0.91	0.83	0.74	0.65	0.56
	1,400	1.00	0.89	0.78	0.67	0.56	0.45
	≥ 1,700	1.00	0.87	0.75	0.62	0.49	0.36
0.60	0	1.00	0.97	0.94	0.91	0.88	0.85
	50	1.00	0.97	0.94	0.90	0.87	0.84
	200	1.00	0.96	0.92	0.89	0.85	0.81
	400	1.00	0.95	0.91	0.86	0.82	0.77
	600	1.00	0.94	0.89	0.84	0.79	0.74
	1,000	1.00	0.93	0.86	0.80	0.73	0.66
	1,400	1.00	0.92	0.83	0.75	0.67	0.58
	≥ 1,700	1.00	0.91	0.81	0.72	0.62	0.53
0.80	0	1.00	0.97	0.94	0.91	0.88	0.85
	50	1.00	0.97	0.94	0.91	0.88	0.85
	200	1.00	0.97	0.93	0.90	0.86	0.83
	400	1.00	0.96	0.92	0.89	0.85	0.81
	600	1.00	0.96	0.92	0.88	0.83	0.79
	1,000	1.00	0.95	0.90	0.85	0.80	0.75
	1,400	1.00	0.94	0.89	0.83	0.77	0.72
	≥ 1,700	1.00	0.94	0.88	0.81	0.75	0.69
1.00	All	Same as Case 4					

CASE	TYPE OF LANE GROUP	RIGHT-TURN FACTORS, f_{RT}
7	SINGLE LANE APPROACH	$f_{RT} = 0.90 - P_{RT} [0.135 + (\text{peds}/2,100)]$ $f_{RT} = 0.05$ (minimum)

No. of Conf. Peds. (peds)		Prop. of RT's in Single Lane P_{RT}					
		0.00	0.20	0.40	0.60	0.80	1.00
0		1.00	0.87	0.85	0.82	0.79	0.77
50	(Low)	1.00	0.87	0.84	0.81	0.77	0.74
100		1.00	0.86	0.83	0.79	0.76	0.72
200	(Mod)	1.00	0.86	0.81	0.77	0.72	0.68
300		1.00	0.85	0.79	0.74	0.69	0.64
400	(High)	1.00	0.84	0.78	0.72	0.65	0.59
600		1.00	0.82	0.74	0.66	0.59	0.51
800		1.00	0.80	0.71	0.61	0.52	0.42
1,000		1.00	0.79	0.67	0.56	0.45	0.34
1,200		1.00	0.77	0.64	0.51	0.38	0.25
1,400		1.00	0.75	0.61	0.46	0.31	0.16
≥ 1,700		1.00	0.73	0.55	0.38	0.21	0.05

CASE	TYPE OF LANE GROUP	RIGHT-TURN FACTORS, f_{RT}
8	DOUBLE EXCLUSIVE RT LANE; PROTECTED PHASING	0.75

Table 8.13   Adjustment Factors for Left Turns

CASE	TYPE OF LANE GROUP	LEFT-TURN FACTOR, f_{LT}					
1	EXCLUSIVE LT LANE; PROTECTED PHASING	0.95					
2	EXCLUSIVE LT LANE; PERMITTED PHASING	Special Procedure					
3	EXCLUSIVE LT LANE; PROTECTED PLUS PERMITTED PHASING	0.95[a]					
4	SHARED LT LANE; PROTECTED PHASING	$f_{LT} = 1.0/(1.0 + 0.05\ P_{LT})$					
		Prop. of LT's in Lane, P_{LT}	0.00	0.20	0.40	0.60	0.80	1.00
		Factor	1.00	0.99	0.98	0.97	0.96	0.95
5	SHARED LT LANE; PERMITTED PHASING	Special Procedure					
6	SHARED LT LANE; PROTECTED PLUS PERMITTED PHASING	$f_{LT} = (1{,}400 - V_o)/[(1{,}400 - V_o) + (235 + 0.435\ V_o)P_{LT}]\quad V_o \leq 1{,}220$ vph					

(Case 6, second equation): $f_{LT} = 1/[1 + 4.525\ P_{LT}]\quad V_o > 1{,}220$ vph

Opposing Volume V_o	Prop. of Left Turns, P_{LT}					
	0.00	0.20	0.40	0.60	0.80	1.00
0	1.00	0.97	0.94	0.91	0.88	0.86
200	1.00	0.95	0.90	0.86	0.82	0.78
400	1.00	0.92	0.85	0.80	0.75	0.70
600	1.00	0.88	0.79	0.72	0.66	0.61
800	1.00	0.83	0.71	0.62	0.55	0.49
1,000	1.00	0.74	0.58	0.48	0.41	0.36
1,200	1.00	0.55	0.38	0.29	0.24	0.20
≥ 1,220	1.00	0.52	0.36	0.27	0.22	0.18

CASE	TYPE OF LANE GROUP	LEFT-TURN FACTOR, f_{LT}
7	SINGLE LANE APPROACH	Special Procedures
8	DOUBLE EXCLUSIVE LT LANE; PROTECTED PHASING	0.92

[a] This value is a starting estimate. Solutions are iterated for this case.

Table 8.14 Progression Adjustment Factors

TYPE OF SIGNAL	LANE GROUP TYPES	v/c RATIO, X	ARRIVAL TYPE[a]				
			1	2	3	4	5
Pretimed	TH, RT	≤ 0.6	1.85	1.35	1.00	0.72	0.53
		0.8	1.50	1.22	1.00	0.82	0.67
		1.0	1.40	1.18	1.00	0.90	0.82
Actuated	TH, RT	≤ 0.6	1.54	1.08	0.85	0.62	0.40
		0.8	1.25	0.98	0.85	0.71	0.50
		1.0	1.16	0.94	0.85	0.78	0.61
Semiactuated	Main St. TH, RT[b]	≤ 0.6	1.85	1.35	1.00	0.72	0.42
		0.8	1.50	1.22	1.00	0.82	0.53
		1.0	1.40	1.18	1.00	0.90	0.65
Semiactuated	Side St. TH, RT[b]	≤ 0.6	1.48	1.18	1.00	0.86	0.70
		0.8	1.20	1.07	1.00	0.98	0.89
		1.0	1.12	1.04	1.00	1.00	1.00
	All LT[c]	all	1.00	1.00	1.00	1.00	1.00

[a] See Table 8.3.

[b] Semiactuated signals are typically timed to give all extra green time to the main street. This effect should be taken into account in the allocation of green times.

[c] This category refers to exclusive LT lane groups with protected phasing only. When LT's are included in a lane group encompassing an entire approach, use factor for the overall lane group type. Where heavy LT's are intentionally coordinated, apply factors for the appropriate through movement.

Table 8.15 PCEs for Shared Left-Turn Lanes
(Planning Level of Analysis)

Opposing Through and Right-Turn Volume, V_o (vph)	Passenger Car Equivalent (PCE)
0 to 199	1.1
200 to 599	2.0
600 to 799	3.0
800 to 999	4.0
≤ 1,000	5.0

Table 8.16 Capacity Criteria for Planning Analysis of Signalized
Intersections

CRITICAL VOLUME FOR INTERSECTION, VPH	RELATIONSHIP TO PROBABLE CAPACITY
0 to 1,200	Under Capacity
1,201 to 1,400	Near Capacity
≥ 1,401	Over Capacity

Highway and Traffic Planning

Transportation planning involves a process that includes the elements of situation and problem definition, search for solutions and performance analysis, as well as evaluation and choice of project. The process is useful for describing the effects of a proposed transportation alternative and for explaining not only how a new transportation system will benefit the traveler but what its impacts will be on the community. The highway and traffic engineer is responsible for developing forecasts of travel demand; conducting evaluations based on economic and noneconomic factors; and identifying alternatives for short-, medium-, and long-range purposes.

CHAPTER 9

The Transportation Planning Process

This chapter explains how decisions to build transportation facilities are reached and highlights the major elements of the process. Transportation planning has become institutionalized and federal guidelines, regulations, and requirements for local planning are often a driving force behind existing planning methods.

The formation of the nation's transportation system has been evolutionary and not part of a grand plan. The system now in place is the product of many individual decisions to build or improve its various parts such as bridges, highways, tunnels, harbors, railway stations, and airport runways. Most of these transportation facilities were selected for construction or improvement because it was the conclusion of those involved that the project would result in an overall improvement in the situation as it existed at the time.

Among the factors usually considered to justify a transportation project are improvements in traffic flow and safety, savings in energy consumption and travel time, economic growth, and accessibility. Some transportation projects, however, may be selected for political reasons, for example, to stimulate employment in a particular region, to compete with other cities or states for prestige, to attract industry, to respond to pressures from a political constituency, or to gain personal benefit from a particular route location or construction project. In some instances, transportation projects may *not* be selected for construction because of opposition by those who would be adversely affected. For example, a new highway may require the taking of residential property, or the construction of an airport may introduce noise due to low-flying planes. Whatever the reason for selecting or rejecting a transportation project, a process was followed that led to the conclusion to build or not to build.

The process for planning transportation systems is a rational one that seeks to furnish unbiased information about the effects that the proposed transportation project will have on the community and its expected users. For example, if noise or air pollution is a concern, then the process will examine and estimate how much additional noise or air pollution will occur if the transportation facility is built. Usually cost is a major factor and the process will include an estimate of the construction, maintenance, and operating costs.

The process is applicable to any transportation project or system, because the kinds of problems that transportation engineers work on will vary over time. Considerable change in emphasis in transportation has occurred over a 200 year period, with such modes as canals, railroads, highways, air, and public transit being dominant at one time or another. Thus, the activities of transportation engineers have varied considerably during this period, depending on society's needs and concerns. Examples of societal concern include energy conservation, traffic congestion, environmental impacts, safety, security, efficiency, productivity, and community preservation.

The transportation planning process is not intended to furnish a decision or to give a single result that must be followed, although it can do so in relatively simple situations. Rather, the process is intended to give the appropriate information to those who will be responsible for deciding if the transportation project should go forward.

BASIC ELEMENTS OF TRANSPORTATION PLANNING

The transportation planning process comprises seven basic elements which are interrelated and not necessarily carried out sequentially. The information learned in one phase of the process may be helpful in some earlier or later phase so that there is a continuity of effort, which results finally in a decision. The elements in the process are as follows:

- Situation definition
- Problem definition
- Search for solutions
- Analysis of performance
- Evaluation of alternatives
- Choice of project
- Specification and construction

These elements are described and illustrated, using a scenario in which a community is investigating the feasibility of constructing a new bridge.

Situation Definition

The first step in the planning process is situation definition, which involves all of the activities required to understand the present situation that gave rise to the perceived need for a transportation improvement. In this phase, the basic factors that created the present situation are described, and the scope of the system to be studied is delineated. The present system is analyzed, and its characteristics are described. Information about the surrounding area, its people, and their travel habits may be obtained. Previous reports and studies that may be relevant to the present situation are reviewed and summarized. The scope of the study and the domain of the system to be investigated are delineated.

In our hypothetical community in which a new bridge is being considered, situation definition involves developing a description of the present highway and transportation services in the region, measuring present travel patterns and highway traffic volumes, reviewing prior studies, geological maps, and soil conditions, and delineating the scope of the study and the area affected.

Problem Definition

The purpose of this step is to describe the problem in terms of the objectives to be accomplished by the project and to translate the objectives into criteria that can be quantified. Objectives are statements of purpose such as: to reduce traffic congestion, to improve safety, or to maximize net highway user benefits. Criteria are the measures of effectiveness that can be used to tell how effective a proposed transportation project will be in meeting the stated objectives. For example, the objective "to reduce traffic congestion" might use "travel time" as the measure of effectiveness. There may be other considerations that the planner must know about that will set constraints on what is proposed and these should be described. The characteristics of an acceptable system are identified and specific limitations and requirements should be noted. Also, any pertinent standards and restrictions that the proposed transportation project must conform to should be understood.

An objective for the bridge project might be to reduce travel congestion on other roads or to reduce travel time between certain areas. Criteria to measure how well these objectives are achieved are average delay or average travel distance. Constraints placed on the project could be physical limitations such as the presence of other structures, topography, or historic buildings. Design standards for bridge width, clearances, loadings, and capacity would also be noted.

Search for Solutions

In this phase of the planning process, consideration is given to a variety of ideas, designs, locations, and system configurations that might be a solution to the problem. This is the *brainstorming* stage where many options may be proposed for later testing and evaluation. The alternatives generated can be suggested by any group or organization. In fact, the planning study may have been originated to determine the feasibility of a particular project or idea. The transportation engineer has a variety of options available in any particular situation, and any or all might be considered in this idea-generating phase. Among the options that could be used are different types of transportation technology or vehicles, various system or network arrangements, and different methods of operation. This phase will also include preliminary feasibility studies that might narrow the range of choices to those that appear most promising. Some data gathering, field testing, and cost estimating may be necessary at this stage to determine the practicality and financial feasibility of the alternatives being proposed.

In the case of our hypothetical bridge construction, a variety of options might be considered, including different locations and bridge types. The study should also include the option of not building the bridge and might also consider what alternatives are available other than a new bridge. Operating policies that might be considered include various toll charges and methods of collection.

Analysis of Performance

The purpose of performance analysis is to estimate how each of the proposed alternatives would perform under present and future conditions. The criteria identified in the previous steps are calculated for each transportation option. Included in this step is a determination of the investment cost to build the transportation project as well as annual costs for maintenance and operation. This element would also involve the use of mathematical models for estimating travel demand. The number of persons or vehicles that will use the system would be determined, and these results, expressed in vehicles or persons per hour, serve as the basis for project design. Other information about the usage of the system, such as trip length, travel by time of day, and vehicle occupancy, would also be determined and used in calculating user benefits for various criteria or measures of effectiveness. Other effects of the transportation project, such as noise and air pollution levels and acres of land required, would be estimated if required. These nonuser impacts would be calculated in situations where the transportation project could have significant impacts on the community or adjacent environment.

This task is sometimes referred to as the transportation planning process, but it is really the systems analysis process that integrates system supply on a network with travel demand forecasts to show equilibrium travel flows.[1] The forecasting model system and related network simulation are discussed in Chapter 10.

To analyze the performance of the new bridge we are contemplating, we would prepare preliminary cost estimates for each location being considered. Then we would compute an estimate of the traffic that would use the bridge for various toll levels and bridge widths. The average trip length and average travel time for bridge users would be determined and compared with existing or no-build conditions. Other impacts, such as land required, visual effects, noise levels, and air or water quality changes, would also be computed.

Evaluation of Alternatives

The purpose of the evaluation phase is to determine how well each alternative will achieve the objectives of the project as defined by the criteria. The performance data produced in the analysis phase are used to compute the benefits and costs that will occur if the project is selected. In cases where the results cannot be reduced to a single monetary value, a weighted ranking for each alternative might be produced and compared with other proposed projects. The benefit-cost ratio (described in Chapter 11) for each project might be calculated to show the extent to which the project would be a sound economic investment. Other economic tests might be applied, such as the net present worth of benefits and costs. In complex situations where there are many criteria expressed in both monetary and nonmonetary terms, the results might simply be shown in a cost-effectiveness matrix that allows the person or group making the decision to see exactly how each alternative performs for each of the criteria and at what cost. The results might also be plotted on a chart that allows a visual comparison of each project and its performance to be made.

In our evaluation of the bridge project, we would analyze the benefits and costs to determine if the project meets the criteria of profitability. If the result is positive, then the evaluation of alternative sites would require comparison of many factors, for both engineering and economic feasibility and environmental impact. A cost-effectiveness matrix that compares the cost of each alternative with how effective that alternative is in achieving goals would be produced to assist in the evaluation.

Choice of Project

The final project selection is made after considering all the factors involved. In a simple situation, for example, where the project has been authorized and is in the design phase, a single criterion such as cost might be used and the project chosen would be the one with the lowest cost. With a more complex project, however, many factors will have to be considered and selection will be based on how the results are perceived by those involved in decision making. If the project involves the community, it may be necessary to hold public hearings and it is possible that a bond issue or referendum may be required. Perhaps none of the alternatives will meet the criteria or standards and additional investigations may be necessary. The transportation engineer will furnish a recommendation, and he or she may have developed a strong opinion concerning which alternative to select. If the engineer is not careful, such bias could result in the early elimination of contenders or the presentation to decision makers of inferior projects for comparison. If the engineer is acting professionally and ethically, he or she will perform the task such that the information necessary to make a choice is available and that every alternative has been considered.

In choosing whether or not to build our hypothetical bridge, decision makers will look carefully at the revenue–cost forecasts and will probably select the project that appears to be most financially sound. The site location will be selected based on a careful study of the factors involved. The information gathered in the earlier phases will be used, together with judgmental and political considerations, to arrive at a final project selection.

Specification and Construction

Once the transportation project has been selected, a detailed design phase is begun in which each of the components of the facility is specified. For a transportation facility, this involves its physical location, geometric dimensions, and structural configuration. Design plans are produced that can be used by contractors to estimate the cost to build the project. When a construction firm is selected, these plans will be the basis on which the project will be built.

For our bridge project, with a decision made to proceed, we will produce a design that includes the type of superstructure, piers and foundations, roadway widths and approaches, as well as appurtenances such as toll booths, traffic signals, and lighting. These plans will be submitted to contractors who will prepare a bid for the construction. If the bids are below the amount of funds available and the contractor is acceptable to the client, the project will proceed to the construction phase. Upon completion, the new bridge will be turned over to the local transportation authority for operation and maintenance.

PLANNING THE RELOCATION OF A RURAL ROAD: AN EXAMPLE

To illustrate how the transportation planning process works, an example is described for a rural road relocation project. Each of the activities involved in the project is discussed in terms of the seven-step planning process previously described.[2,3]

Step 1: Situation Definition.   The project is a proposed relocation or reconstruction of 3.3 mi of US 1A located in the coastal town of Harrington, Maine. The town center, the focal point of the project, is located near the intersection of US 1 and US 1A, on the banks of the Harrington River, an estuary of the Gulf of Maine. (See Figure 9.1.) The town of Harrington has 553 residents, of which 420 reside within the study area and 350 in the town center. The population has been declining in recent years and many young people have left because of lack of employment opportunities. Most of the town's industry is agriculture or fishing and a realignment of the road that damaged the environment would also affect the town's livelihood. There are 10 business establishments within the study area and 20 percent of the town's retail sales are tourist related. The average daily

Figure 9.1   Location Map for Highway US 1 and Highway US 1A

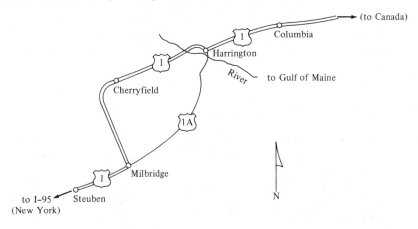

traffic is 2620 vehicles per day, of which 69 percent represent through traffic and 31 percent represent local traffic.

Step 2: Problem Definition. The state Department of Transportation wishes to improve US 1A, primarily to reduce the high accident rate on this road in the vicinity of the town center. The problem is caused by a narrow bridge that carries the traffic on route 1A into the town center, the poor horizontal and vertical alignment of the road within the town center, and a dangerous intersection where US 1A and US 1 meet. The accident rate on US 1A in the vicinity of the town center is four times greater than the statewide rate. A secondary purpose of the proposed relocation is to improve the level of service for through traffic by increasing the average speed on the relocated highway.

The measures of effectiveness for the project will be the accident rate, travel time, and construction cost. Other aspects that will be considered relate to the effects that each alternative will have on the number of businesses and residences that would be displaced, the noise and air quality, and the changes in natural ecology. The criteria that will be used to measure these nonuser effects will be the number of businesses and homes displaced, and the air quality, noise levels, and acres of salt marsh and trees affected.

Step 3: Search for Solutions. The Department of Transportation has identified four alternative routes, as illustrated in Figure 9.2, in addition to the present route—alternative 0—referred to as the "do-nothing" alternative. All routes begin at the same location—the town of West Harrington, which is 3 mi southwest of the center of Harrington—and end at a common point northeast of the center. The alternatives are as follows.

Alternative 1: This road bypasses the town to the south on a new location across the Harrington River. The road will be 2 lanes, each 12 ft wide with 8 ft shoulders. The minimum right-of-way width is 120 ft. A new bridge would be constructed about 1/2 mi downstream from the old bridge.

Alternative 2: This alternative would use the existing route 1A into town but with improvements in the horizontal and vertical alignment throughout its length and the construction of a new bridge. The geometric specifications would be the same as for alternative 1.

Alternative 3: This new road would merge with US 1, west of Harrington and then continue through town. It would use the Route 1 bridge, which was recently constructed. Geometric specifications are as for the other alternatives.

Alternative 4: This road merges with US 1 and uses the Route 1 bridge, as does alternative 3. However, it bypasses the town on a new alignment.

Step 4: Analysis of Performance. The measures of effectiveness are calculated for each of the alternatives. The results of these calculations are shown in Table 9.1 for alternatives 1 through 4, as well as the null or do-nothing alternative. The relative ranking of each alternative based on criteria results is presented in Table 9.2. For example, the average speed on the existing road is 25 mph, whereas for alternatives 1 and 4, the speed is 55 mph, and for alternatives 2 and 3, the speed is 30 mph. Similarly, the accident rate, which now is four times the statewide

Figure 9.2   Alternative Routes for Highway Relocation

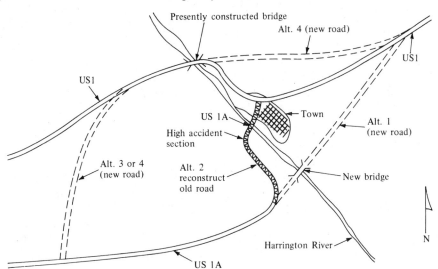

average, would be reduced to 0.6 for alternative 4 and 1.2 for alternative 1. The project cost ranges from $1.18 million for alternative 3 to $1.58 million for alternative 2. Other items that are calculated include the number of residences displaced, the volume of traffic within the town both now and in the future, air quality, noise, lost taxes, and acreage of trees removed.

Table 9.1   Measures of Effectiveness for Rural Road Alternatives

	Alternatives				
Criteria	0	1	2	3	4
Speed (mph)	25	55	30	30	55
Distance (m)	3.7	3.2	3.8	3.8	3.7
Travel time (min)	8.9	3.5	7.6	7.6	4.0
Accident factor[a]	4	1.2	3.5	2.5	0.6
Cost ($ million)	0	1.50	1.58	1.18	1.54
Residence displaced	0	0	7	3	0
City traffic					
Present	2620	1400	2620	2520	1250
Future	4350	2325	4350	4180	2075
Air quality (μg/m^3 CO)	825	306	825	536	386
Noise (dBA)	73	70	73	73	70
Tax loss	None	Slight	High	Moderate	Slight
Tree acreage (acres)	None	Slight	Slight	25	28
Runoff	None	Some	Some	Much	Much

[a]Relative to statewide average for this type of facility.

Table 9.2 Ranking of Alternatives

Criteria/Alternative	Alternatives				
	0	*1*	*2*	*3*	*4*
Travel time	4	1	3	3	2
Accident factor[a]	5	2	4	3	1
Cost ($ millions)	1	3	5	2	4
Residences displaced	1	1	3	2	1
Air quality	4	1	4	3	2
Noise	2	1	2	2	1
Tax loss	1	2	4	3	2
Tree acreage	1	2	2	3	4
Increased runoff	1	2	2	3	3

Note: 1 = highest; 5 = lowest.

[a]Relative to statewide average for this type of facility.

Step 5: Evaluation of Alternatives. Each of the alternatives is compared with each of the others to assess the improvement that would occur for each of the criteria. In this example, we consider the following measures of effectiveness and their relationship to project cost.

Travel time: Every alternative improves the travel time over the existing conditions. The best is alternative 1, followed by alternative 4. Alternatives 2 and 3 are equal but not a great improvement over present conditions (Figure 9.3).

Accident factor: The best accident record will occur with alternative 4, followed by alternatives 1, 3, and 2 (Figure 9.4).

Figure 9.3 Travel Time Between West Harrington and US 1 Versus Cost

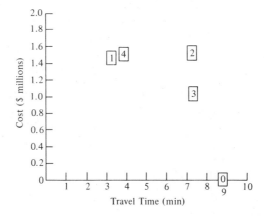

Figure 9.4 Accident Factor (Relative to Statewide Average) Versus Cost

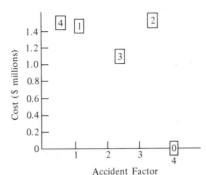

Cost: The least cost project is simply to do nothing, but the dramatic improvements in travel time and safety indicate that the proposed project should probably be constructed. Alternative 3 is lowest in cost at $1.18 million. Alternative 2 is highest in cost but would not be as safe as alternative 3 and produces the same travel time. Thus, alternative 2 would be eliminated. Alternative 1 will cost $0.32 million more than alternative 3 but would reduce the accident factor by 1.3 and travel time by 4.1 min. Alternative 4 would cost $0.04 million more than alternative 1 and would increase travel time while decreasing the accident factor. These cost-effectiveness values are shown in Figures 9.3 and 9.4. They indicate that alternatives 1 and 4 are both more attractive than alternatives 2 and 3 because they will produce significant improvements in travel time and accidents.

Displaced residences: Three residences will be displaced if alternative 3 is selected and seven residences will be displaced if alternative 2 is selected. No residences will have to be removed if either alternative 1 or 4 are built.

Air quality in the town: Alternative 1 will produce the highest air quality, followed by alternatives 4, 3, and 2. The air quality improvement results from removing a significant amount of the slow-moving through traffic from the center of the city to a high-speed road where most of the pollution will be disbursed.

Noise levels: Noise levels are lower for alternatives 1 and 4.

Tax Losses: Tax losses will be slight for alternatives 1 and 4, moderate for alternative 3, and high for alternative 2.

Tree acreage: Alternatives 3 and 4 would eliminate 25 and 28 acres, respectively. Alternative 1 would result in slight losses and alternative 2 no loss.

Runoff: There would be no runoff for alternative 0, some for alternatives 1 and 2, and a considerable amount for alternatives 3 and 4.

Step 6: Choice of Project.   From strictly a cost point of view the Department of Transportation would select alternative 3 since it results in travel time and safety improvements at the lowest cost. However, if additional funds are available, then alternative 1 or 4 would be considered. Since alternative 1 is lower in cost than alternative 4 and is equal or better than alternative 3 for each criterion related to community impacts, this alternative would be the one most likely to be selected.

In the selection process, each alternative would be reviewed and comments received from citizens and elected officials. These would be used to assist in the design process so that environmental and community effects would be minimized.

Step 7: Specifications and Construction. The choice has been made, and alternative 1 has been ranked of sufficiently high priority so that it will be constructed. This alternative involves building a new bridge across the Harrington River and a new road connecting Route 1A with US 1. The designs for the bridge and road will be prepared. Detailed estimates of the cost to construct will be made, and the project will be announced for bid. The construction company that produces the lowest bid and can meet other qualifications will be awarded the contract and the road will be built. Upon completion, the road will be turned over to the Department of Transportation, who will be responsible for its maintenance and operation. Follow-up studies will be conducted to determine how successful the road was in meeting its objectives and, where necessary, modifications will be made to improve its performance.

Institutionalization of Transportation Planning

The transportation planning process that has been introduced is based on the systems approach to problem solving and is quite general in its structure. The process can be applied to many cases for transportation decision making, such as intercity high-speed rail feasibility studies, airport location, port and harbor development, and urban transportation systems. The most common application is in urban areas, where it has been mandated by law since 1962 when the Federal Aid Highway Act required that all transportation projects in urbanized areas of 50,000 or more population be based on a transportation planning process that was continuous, comprehensive, and coordinated with other modes.

Because the urban transportation planning process provides an institutionalized and formalized planning structure, it is next introduced to identify the environment in which the transportation planner works. The forecasting modeling process that has evolved is presented in Chapter 10 to provide an illustration of the methodology. The planning process used at other problem levels is a variation of this basic approach.

URBAN TRANSPORTATION PLANNING

Urban transportation planning involves the evaluation and selection of highway or transit facilities to serve present and future land uses. For example, the construction of a new shopping center, airport, or convention center will require additional transportation services. Also, new residential development, office space, and industrial parks will generate additional traffic, requiring that roads and transit services be created or expanded.

The planning process must also consider other proposed developments and improvements that will occur within the planning period. Furthermore, the urban transportation planning process has been improved through the efforts of the Federal Highway Administration and the Urban Mass Transportation Administration of the U.S. Department of Transportation by the preparation of manuals and computer programs that assist in organizing data and forecasting travel flows.[4,5]

Urban transportation planning is concerned with two separate time horizons. The first is a short-term activity intended to select projects that can be implemented within a one- to three-year period. These projects are designed to provide better management of existing facilities by making them as efficient as possible. The second time horizon deals with the long-range transportation needs of an area and identifies the projects to be constructed over a 20 yr period.

Short-term projects involve programs such as traffic signal timing to improve flow, car and van pooling to reduce congestion, park-and-ride fringe parking lots to increase transit ridership, and transit improvements.

Long-term projects involve programs such as adding new highway sections, additional bus lines or freeway lanes, rapid transit systems and extensions, or access roads to airports or shopping malls.

The urban transportation planning process can be carried out in terms of the seven steps outlined previously and is usually described as follows.

Inventory of Existing Travel and Facilities

This is the data-gathering activity in which urban travel characteristics are described for each defined geographic unit or traffic zone within the study area. Inventories and surveys are made to determine traffic volumes, land uses, origins and destinations of travelers, population, employment and economic activity. Inventories are made of existing transportation facilities, both highway and transit, and their capacity, speed, travel time, and traffic volume are determined. The information gathered is summarized for each traffic zone and for the existing highway and transit system.

Establish Goals and Objectives

The urban transportation study is carried out to develop a program of highway and transit projects that should be completed in the future. Thus, a statement of goals, objectives, and standards is prepared that identifies deficiencies in the existing system, desired improvements, and what is to be achieved by the transportation improvements.

For example, if a transit authority is considering the possibility of extending an existing rail line into a newly developed area of the city, its objectives for the new service might be to maximize its revenue from operations, maximize ridership, promote development, and attract the largest number of auto users so as to relieve traffic congestion.

Generation of Alternatives

In this phase of the urban transportation planning process, the alternatives to be analyzed will be identified. It may also be necessary to analyze the travel effects of different land-use plans and consider various life-style scenarios. The transportation options available to the urban transportation planner include various technologies, network configurations, links, nodes, vehicles, operating policies, and organizational arrangements.

In the case of a transit line extension, the technologies could be rail rapid transit or bus. The network configuration could be defined by a single line, two branches, or a geometric configuration such as a radial or grid pattern. The links, which represent a homogeneous section of the transportation system, could be varied in length, speed, waiting time, capacity, and direction. The nodes, which represent the end points of a link, could be a transit station or the line terminus. The vehicles could be singly driven buses or multicar trains. The operating policy could involve 10 min headways during peak hours and 30 min headways in off-peak hours or other combinations. The organizational arrangements could be private or public. These and other alternatives would be considered in this phase of the planning process.

Estimation of Project Cost and Travel Demand

This activity in the urban transportation planning process involves two separate tasks. The first is to determine the project cost and the second is to estimate the amount of traffic expected in the future. The estimation of facility cost is relatively straightforward, whereas the estimation of future traffic flows is a complex undertaking requiring the use of mathematical models and computers.

Future traffic is determined by forecasting future land use in terms of economic activity and population and then relating this to the future traffic that this land use will produce. With the land-use forecasts established, in terms of number of jobs, residents, auto ownership and income, and so forth, then the traffic that this land use will produce can be determined. This is carried out in a four-step process that includes the determination of the number of trips generated, the origin and destination of trips, the mode of transportation used by each trip (for example, auto, bus, rail), and the route taken by each trip. The urban traffic forecasting process thus involves four distinct activities that are called trip generation, trip distribution, modal split, and network assignment.

When the travel forecasting process is completed, the highway and transit volumes on each link of the system will be estimated. The actual amount of traffic is not known until it occurs. These results can be compared with the present capacity of the system to determine the operating level of service.

Evaluation of Alternatives

This phase of the process is similar in concept to what was described earlier but can be complex in practice because of the conflicting objectives and diverse groups that will be affected by an urban transportation project.

Among the groups that could be affected are the traveling public (user), the highway or transit agency (operator), and the nontraveling public (community). Each of these groups will have different objectives and viewpoints concerning how well the system performs. The traveling public wants to improve speed, safety, and comfort; the transportation agency wishes to minimize cost; and the community wants to preserve its life style and minimize any adverse impacts.

The purpose of the evaluation process is to identify feasible alternatives in terms of cost and traffic capacity, to estimate the effects of each alternative in terms of the objectives expressed, and to assist in identifying those alternatives that will serve the traveling public and be acceptable to the community.

Choice of Project

Selection of a project will be based on a process that will ultimately involve elected officials and the public. Quite often funds to build an urban transportation project, such as a subway system, may involve a public referendum. In other cases, a vote by a state legislature may be required before funds are committed. A multiyear program will then be produced that outlines the projects to be carried out over the next 20 years. With approval in hand, the project can proceed to the specification and construction phase.

Transportation Planning Organization

In carrying out the urban transportation planning process, it is quite common to organize several committees that represent various community interests and viewpoints. These committees are the policy committee, the technical committee, and the citizens' advisory committee.

Policy Committee

The policy committee comprises elected or appointed officials, such as the mayor and director of public works, who represent the governing bodies or agencies that will be affected by the results. This committee makes the basic policy decisions and acts as a board of directors for the study. They will decide on management aspects of the study as well as key issues of financial and political nature.

Technical Committee

The technical committee is composed of the engineering and planning staffs who are responsible for carrying out the work and evaluating the technical aspects of the project. This group will make the necessary evaluations and cost comparisons for each project alternative and supervise the technical details of the entire process. Typically, the technical committee will include highway, transit, and traffic engineers, as well as other specialists in land-use planning, economics, and computers.

Citizens' Advisory Committee

The citizens' advisory committee is composed of a cross section of the community and may include representatives from labor, business, League of Women Voters, interested citizens, and other community interest groups. This committee's function is to express community goals and objectives, suggest alternatives, and to react to proposed alternatives. Through this committee structure, an open dialogue is produced between the policy makers, technical staff, and the community. It is the expectation that when a selection is made and recommendations are produced by the study, a consensus of all interested parties has been reached. Although this is not always possible, the role of a citizens' advisory committee should be to increase communications and, it is hoped, result in plans that reflect community interests.

METHODOLOGY AND ANALYSIS

To accomplish the objectives and tasks of the urban transportation planning process, a technical effort referred to as the *urban transportation forecasting process* is carried out to analyze the performance of various alternatives. There are four basic elements and related tasks in the process, as illustrated in Figure 9.5. These are (1) data collection (or inventories), (2) analysis of existing conditions and calibration of forecasting techniques, (3) forecasts of future travel demand, and (4) analysis of the results. These elements and related tasks are described in the following sections.

Defining the Study Area

Prior to collecting and summarizing the data, it is usually necessary to delineate the study area boundaries and to further subdivide the area into traffic zones for data tabulation and analysis. The size of the zone will depend on the nature of the transportation study, and it is important that the number of zones be adequate for the type of problem being investigated. Often census tracts or census enumeration districts are used as traffic zones because population data are easily available by this geographic designation.

The selection of these zones is based on the following criteria.

1. Socioeconomic characteristics should be homogeneous.
2. Intrazonal trips should be minimized.
3. Physical, political, and historical boundaries should be utilized where possible.
4. Zones should not be created within other zones.
5. The zone system should generate and attract approximately equal trips, households, population, or area.
6. Zones should use census tract boundaries where possible.

An illustration of analysis zones for a transportation study is shown in Figure 9.6.

Figure 9.5   Urban Travel Forecasting Process

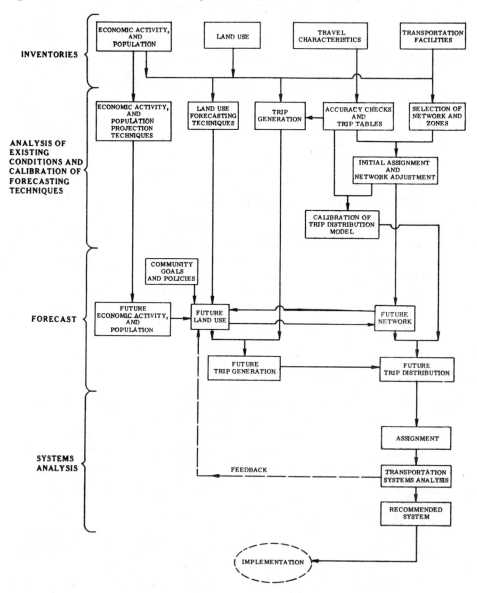

Source: Reproduced from *Computer Programs for Urban Transportation Planning*, U.S. Department of Transportation, Washington, D.C., 1977.

Figure 9.6　Analysis Zones for Transportation Study

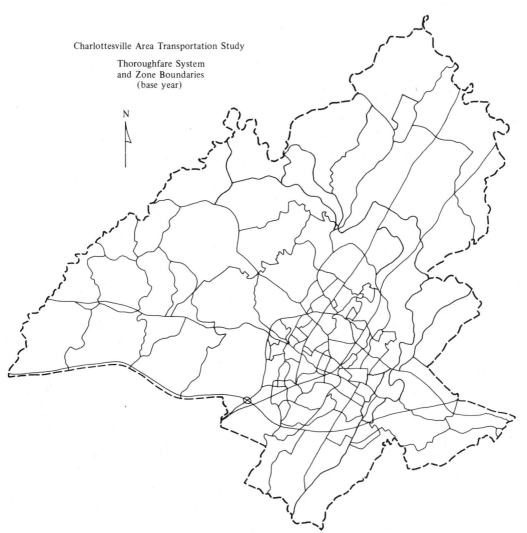

Charlottesville Area Transportation Study

Thoroughfare System
and Zone Boundaries
(base year)

N

Source:　Adapted from Virginia Department of Transportation, Richmond, Va.

Data Collection

The data collection phase provides information about the city and its people that will serve as the basis for developing travel demand estimates. The data include information about economic activity (employment, sales volume, income, and so forth), land use (type, intensity), travel characteristics (trip and traveler profile), and transportation facilities (capacity, travel speed, and so forth). This phase may involve surveys or it can be based on previously collected data. Urban travel data have been collected periodically since the transportation studies of the 1950s and 1960s, which has created extensive travel data bases in many cities. For this reason, less travel data are usually required now or the existing data base is updated for a new study, thus reducing time and cost.

Population and Economic Data

Once a zone system for the study area is established, population and socio-economic forecasts prepared at a regional or statewide level are used. These are allocated to the study area and then the totals are distributed to each zone. This process can be accomplished by using either a ratio technique or small area land-use allocation models.[6]

The population and economic data will usually be furnished by the agencies responsible for planning and economic development, whereas travel and transportation data are the responsibility of the traffic engineer. For this reason, we will focus our attention on the data required to describe travel characteristics and the transportation system.

Transportation Inventories

Transportation system inventories involve a description of the existing transportation services provided, the facilities available and their condition, location of routes and schedules, maintenance and operating costs, system capacity and existing traffic, volumes, speed and delay, and property and equipment. The types of data collected about the current system will depend on the specifics of the problem.

For a highway planning study, the system would be classified functionally into categories that reflect their principal use. There would be the major arterial system, minor arterials, collector roads, and local service. Physical features of the road system would include number of lanes, pavement and approach width, traffic signals, and traffic control devices. Street and highway capacity would be determined, including capacity of intersections. Traffic volume data would be determined for intersections and highway links. Travel times along the arterial highway system would also be determined.

A computerized network of the existing street and highway system is produced. The network consists of a series of links, centroids, and nodes. A *link* is a portion of the highway system that can be described by its capacity, lane width, and speed. A *node* is the end point of a link and represents an intersection or location where a link changes direction, capacity, width, or speed. A *centroid* is the location within a zone where trips are considered to begin and end. Coding of the network requires information from the highway inventory in terms of link speeds, length, and capacities. The network is then coded to locate zone centroids, nodes, and the street system. A portion of a link–node map is illustrated in Figure 9.7.

For a transit planning study, the present routes and schedules would be observed, including headways, transfer points, location of bus stops, terminals, and parking facilities. Information about the bus fleet, such as its number, size,

Figure 9.7 Link–Node Map for Highway System

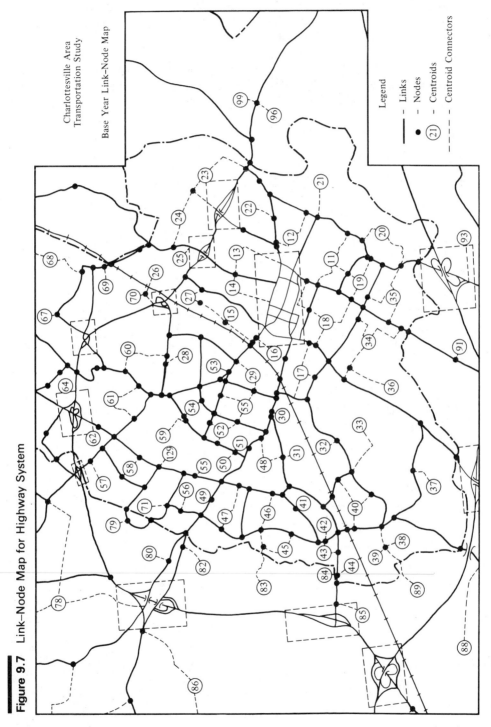

Charlottesville Area
Transportation Study

Base Year Link–Node Map

Legend

Links —
Nodes –
Centroids — ㉑
Centroid Connectors – – –

Source: Reproduced from Virginia Department of Transportation, Richmond, Va.

and age, would be identified. Maintenance facilities and maintenance schedules would be determined as would be the organization and financial condition of the transit companies furnishing service in the area. Other data would include revenue and operating expenses.

The transportation facility inventories provide the basis for establishing the networks that will be studied to determine present and future traffic flows. Data needs can include the following items:

- Public streets and highways
  - Rights of way
  - Roadway and shoulder widths
  - Location of curbed sections
  - Location of structures, such as bridges, overpasses, underpasses, and major culverts
  - Overhead structure clearances
  - Railroad crossings
  - Location of critical curves or grades
  - Identification of routes by governmental unit having maintenance jurisdiction
  - Functional classification
  - Street lighting
- Land use and zoning controls
- Traffic generators
  - Schools
  - Parks
  - Stadiums
  - Shopping centers
  - Office complexes
- Laws, ordinances, and regulations
- Traffic control devices
  - Traffic signs
  - Signals
  - Pavement markings
- Transit system
  - Routes by street
  - Location and lengths of stops and bus layover spaces
  - Location of off-street terminals
  - Change of mode facilities

- Parking facilities
- Traffic volumes
- Travel times
- Intersection and roadway capacity

In many instances, the data will already have been collected and is available in the files of city, county, or state offices. In other instances, some data may be more essential than others. A careful evaluation of the data needs should be undertaken prior to the study.

Travel Surveys

Travel surveys are obtained to establish a complete understanding of the travel patterns within the study area. For single projects, it may be sufficient to use traffic counts on existing roads or counts of passengers riding the present system. However, to understand why people travel and where they wish to go, origin–destination (O–D) survey data can be useful. The O–D survey asks questions about each trip that was made on a specific day, such as where the trip began and ended, the purpose of the trip, time of day, vehicle used (auto or transit), and about the person making the trip, age, sex, income, vehicle owner, and so forth. The O–D survey may be completed as a home interview, or people may be asked questions while riding the bus or stopped at a roadside interview station. Sometimes the information is requested by telephone or by return postcard. O–D surveys are rarely made in communities where these data have been collected previously. Due to the high cost of O–D surveys, prior data are updated and U.S. Census travel-to-work data are used.

O–D data are compared with other sources to assure the accuracy and consistency of the results. Among the tests used are crosschecks that compare the number of dwelling units or trips per dwelling unit observed in the survey with published data. Screenline checks can be made to compare the number of reported trips that cross a defined boundary, such as a bridge or two parts of a city, with the number actually observed. For example, the number of cars observed crossing one or more bridges might be compared with the number estimated from the surveys. It is also possible to assign trips to the existing network to compare how well the data actually replicates actual travel. If the screenline crossings are significantly different from those produced by the data, it is possible to make adjustments in the O–D results so that conformance with actual conditions is assured.

Following the O–D checking procedure, a set of *trip tables* is prepared that shows the number of trips between each zone in the study area. These tables can be subdivided by trip purpose, truck trips, and taxi trips. Tables are also prepared that list the socioeconomic characteristics for each zone and the travel time between zones.

Calibration and Forecasting

Calibration of forecasting techniques are concerned with establishing mathematical relationships that can be used in the forecasting process itself. Usually, analysis of the data will reveal the effect on travel demand of factors such as land use, socioeconomic characteristics, or transportation system factors. For example, a multiple regression analysis may show that the following relationship exists for the number of trips per household in a residential zone:

$$T = 0.82 + 1.3P + 2.1A$$

where

T = number of trips per household
P = number of persons per household
A = number of autos per household

Other mathematical formulas will be used to establish the relationships for trip length, percentage of trips by auto or transit, or the particular travel route selected.

Developing the forecasts of travel is carried out by using the relationships developed in the calibration process. These formulas use estimates of future land use, socioeconomic characteristics, and transportation conditions.

Forecasting can be summarized in a simplified way by indicating the task that each step in the process is intended to perform. These are as follows.

1. *Population and economic analysis* determines the magnitude and extent of activity in the urban area.
2. *Land use analysis* determines where the activities will be located.
3. *Trip generation* determines how many trips each activity will produce or attract.
4. *Trip distribution* determines the origin or destination of trips that are generated at a given activity.
5. *Modal split* determines which mode of transportation will be used to make the trip.
6. *Traffic assignment* determines which route on the transportation network will be used when making the trip.

Computers are used extensively in the urban transportation planning process. Over the years a package of programs has been developed by the Federal Highway Administration (FHWA) and the Urban Mass Transportation Administration (UMTA). These are called the Urban Transportation Planning System (UTPS).

A microcomputer version of the UTPS strategy was developed by the FHWA and is referred to as the Quick Response System (QRS) because it is not data intensive and uses a set of general models that are transferable. The techniques have been computerized,[7] and various versions of the original UTPS program have been developed for microcomputers under such acronyms as MINUTP, TRANPLAN, and MICROTRIPS.[8]

SUMMARY

Transportation projects are selected for construction based on a variety of factors and considerations. The transportation planning process is useful when it can assist decision makers and others in the community to select a course of action for improving transportation services.

The seven-step planning process is a useful guide for organizing the work necessary to develop a plan. Although the process does not produce a single answer, it assists the transportation planner or engineer in carrying out a logical procedure that will result in a solution to the problem. The process also is valuable as a means of describing the effects of each course of action and for explaining to those involved how the new transportation system will benefit the traveler and what its impacts will be on the community.

The urban transportation planning process is an institutionalized version of the basic seven-step approach to planning for any new transportation project. An understanding of the elements of urban transportation planning is essential to place in perspective the analytical processes for estimating travel demand. It is this so-called forecasting model system to which the next chapter is devoted.

PROBLEMS

9–1 Explain why the transportation planning process is not intended to furnish a decision or give a single result. Do you agree or disagree with this approach?

9–2 Describe the steps that an engineer should follow if he or she were asked to determine the need for a grade-separated railroad grade crossing that would replace an at-grade crossing of a 2-lane highway with a rail line.

9–3 Describe the basic steps in the transportation planning process. Show these in a flow diagram and explain each step.

9–4 Select a current transportation problem in your community or state with which you are familiar or interested. Briefly describe the situation and the problem involved. Indicate the options available and the major impacts of each option on the community.

9–5 You have been asked to evaluate a proposal that tolls be increased on roads and bridges within your state. Describe the general planning and analysis process that you would use in carrying out this task.

9–6 You have been asked to prepare a study that will result in a decision regarding improved transportation between an airport and the city it serves. The city has a population of 500,000 but the airport also serves a region of approximately 3 million people. It is anticipated that the region will grow to approximately double its size in 15 yr and that air travel will increase by 150 percent. Briefly outline the elements necessary to undertake the study and show these in a flow diagram.

9–7 Using the information furnished in the example in this chapter regarding the relocation of a rural road, prepare a short report for the Highway Planning Commission explaining the alternatives and the issues involved. Explain which alternative you would recommend and why.

9–8 List four ways of obtaining origin–destination information. Which would produce the most accurate results?

9–9 Define the following terms: (a) link, (b) node, (c) centroid, and (d) network.

9–10 Draw a link-node map of the streets and highways within the boundaries of your university campus. For each link, show travel times and distance (to nearest 0.1 mi).

REFERENCES

1. Marvin, L. Manheim, *Fundamentals of Transportation Systems Analysis*, MIT Press, Cambridge, Mass., 1979.

2. Marvin, L. Manheim et al., *Transportation Decision Making: A Guide to Social and Environmental Considerations*, NCHRP Report 156, Transportation Research Board, National Research Council, Washington, D.C., 1975.

3. Maine Department of Transportation, "U.S. Route 1A, Harrington, Maine, Final Environmental Impact Statement." Edwards and Kelsey (December 1972).

4. U.S. Department of Transportation, Federal Highway Administration, *Computer Programs for Urban Transportation*, Government Printing Office, Washington, D.C., April 1977.

5. U.S. Department of Transportation, *Microcomputers in Transportation*, Software Source Book, Government Printing Office, Washington, D.C., February 1986.

6. *Forecasting Inputs to Transportation Planning*, NCHRP Report 266, Transportation Research Board, National Research Council, Washington, D.C., 1983.

7. *Quick Response Urban Travel Estimation Techniques and Transferrable Parameters*, Users Guide, NCHRP Report 187, Transportation Research Board, National Research Council, Washington, D.C., 1978.

8. U.S. Department of Transportation, *Microcomputers in Transportation*.

ADDITIONAL READINGS

Blanchard, Benjamin, S., and Walter J. Fabrycky, *Systems Engineering and Analysis*, Prentice-Hall, Englewood Cliffs, N.J., 1981.

Chadwick, George, *A Systems View of Planning*, Pergamon Press, Oxford, 1974.

deNeufville, Richard, and Joseph H. Stafford, *Systems Analysis for Engineers and Managers*, McGraw-Hill Book Company, New York, 1971.

Dickey, John W., *Metropolitan Transportation Planning*, Scripta Book Company, Washington, D.C., 1975.

Hall, A. D., *A Methodology for Systems Engineering*, Van Nostrand, Princeton, N.J., 1962.

Homburger, Wolfgang, S., Ed., *Transportation and Traffic Engineering Handbook*, Institute of Transportation Engineers, Washington, D.C., 1982, Chapters 11 and 12.

Hutchinson, B. G., *Principles of Urban Transport Systems Planning*, Scripta Book Company, Washington, D.C., 1974.

Yu, Jason C., *Transportation Engineering: Introduction to Planning, Design, and Operations*, Elsevier, New York, 1982.

Stopher, Peter R., and Arnim H. Meyburg, *Urban Transportation Modeling and Planning*, Lexington Books, D. C. Heath, Lexington, Mass., 1975.

10

Forecasting
Travel Demand

Travel demand is expressed as the number of persons or vehicles per unit time that can be expected to travel on a given segment of a transportation system, under a set of given land-use, socioeconomic, and environmental conditions. Forecasts of travel demand are used to establish the loads on future or modified transportation system alternatives. The methods for forecasting travel demand can range from a simple extrapolation of observed trends to a sophisticated computerized process involving extensive data gathering and mathematical modeling. The travel demand forecasting process is as much an art as it is a science because judgments are required concerning the various parameters—that is, population, car ownership, and so forth—that provide the basis for a travel forecast. The methods used in forecasting demand will depend on the availability of data as described in Chapter 9 and specific constraints on the project, such as availability of funds and project schedules. Sources and techniques for establishing the data base for transportation studies are given in *Data Requirements for Metropolitan Transportation Planning.*[1] Survey techniques and in-depth descriptions of available data can be found in the Additional Readings at the end of this chapter.

DEMAND FORECASTING APPROACHES

There are two basic demand forecasting situations in transportation planning. The first involves travel demand studies for urban areas and the second deals with intercity travel demand. Urban travel demand forecasts, when first developed in the 1950s and 1960s, required that extensive data bases be prepared using home interview and/or roadside interview surveys. The information gathered provided useful insight concerning the characteristics of the trip maker—that is, age, sex, income, auto ownership, and so forth—the land use at each end of the trip, and the mode of travel. Travel data could then be aggregated by zone and/or be used at a more disaggregated level—that is, household or individual—to formulate relationships between variables and to calibrate models.

In the intercity case, data generally are aggregated to a greater extent than for urban travel forecasting, such as city population, average city income, and travel time or travel cost between city pairs. In this chapter we describe the urban travel demand forecasting process, for introductory purposes, recognizing that the underlying concepts also apply to intercity travel demand.

The data bases that were established in urban transportation studies during the 1955–1970 period were used for the calibration and testing of models for trip generation, distribution, modal choice, and traffic assignment. These data collection and calibration efforts involved a significant investment of money and personnel resources and consequent studies were based on updating of the existing data base and the use of models that had been developed previously.[2,3]

Factors Influencing Travel Demand

Three factors affect the demand for urban travel: the location and intensity of land use; the socioeconomic characteristics of people living in the area; and the extent, cost, and quality of available transportation services. These factors are incorporated in most travel forecasting procedures.

Land-use characteristics are a primary determinant of travel demand. The amount of traffic generated by a parcel of land depends on how the land is used. For example, shopping centers, residential complexes, and office buildings produce different traffic generation patterns.[4]

Socioeconomic characteristics of the people living within the city also influence the demand for transportation. Life styles and values affect how people decide to use their resources for transportation. For example, a residential area consisting of high-income, white-collar workers will generate more trips by automobile per person than a residential area populated primarily by retirees.[5,6]

The availability of transportation facilities and services, referred to as the supply, also affects the demand for travel. Travelers are sensitive to the level of service provided by alternative transportation modes, and in deciding whether to travel at all or which mode to use, they consider attributes such as travel time, cost, convenience, comfort, and safety.[7,8]

Preliminary Tasks for Travel Forecasting

Prior to the technical task of travel forecasting, the study area must be delineated into a set of traffic zones that form the basis for analysis of travel movements within, into, and out of the urban area. (See Figure 9.6 and the discussion in Chapter 9 for further details.) The set of zones can be aggregated into larger units, called districts, for certain analytical techniques or analyses that work at such levels. Land-use estimates are also developed.

Travel forecasting is solely within the domain of the transportation planner and is an integral part of site development and traffic engineering studies as well as areawide transportation planning. Techniques that represent the state-of-the-practice of each task are described in this chapter to introduce the topic and to illustrate how demand forecast can be determined. Variations of each forecasting technique can be readily found in the literature. Much of this material is covered in advanced courses.

The travel demand forecasting approach illustrated here is the sequential steps of trip generation, trip distribution, modal choice, and traffic assignment. Simultaneous model structures have also been used in practice, particularly to forecast intercity travel.[9]

Trip Tables

Transportation demand data are ultimately summarized on a *trip table*, which is a matrix listing the number of trips from zone to zone.[10] Two basic types of trip tables used in transportation planning are origin–destination (O–D) and production–attraction (P–A). Trip tables are developed by origin zone for an O–D matrix or by production zone for a P–A matrix. The O–D trip table consists of a matrix of trips from each zone (the origin) to each other zone (the destination).

Nonhome-based trips are separated from home-based trips and sorted by residence zones. The residence zone of the home-based trip becomes the production zone and the attraction zone is either the nonresidence origin zone or destination zone.

A P–A table consists of a zone-to-zone matrix of trips, which is used in calibration of mathematical models to generate and distribute future trips in an urban area when the gravity model (to be described later) is used for trip distribution. Each trip has two ends, one a production and the other an attraction. For trips that have an end at home, the home end of the trip is considered to be the production and the other end the attraction. For example, an employee usually makes two home-based work trips each day, one to work and one from work. There would then be two productions (at the home end) and two attractions (at the work end).

The P–A tables are used in the development of trip generation and trip distribution models, for each trip purpose using O–D survey data. A separate P–A trip table is developed for each trip purpose, either as auto trips or person trips. When forecasts of future trip making are completed, they are converted to O–D tables, which list the number of autos or persons that are estimated to travel between zones. The O–D table (which represents person trips made within the study area) is combined with taxi, truck, and external trips (made from outside the study area) to produce a total vehicle O–D table. The final trip matrix is used in the traffic assignment step where each trip is *assigned* to the highway network, thus producing a forecast of traffic volumes on each highway link.

TRIP GENERATION

Trip generation is the process of determining the number of trips that will begin or end in each traffic zone within a study area. Since the trips are determined without regard to destination, they are referred to as *trip ends*. Each trip has two ends, and these are described in terms of trip purpose, or whether the trips are either *produced* by a traffic zone or *attracted* to a traffic zone. For example, a trip home to work would be considered to have a trip end produced in the home zone and attracted to the work zone. Trip generation analysis has two functions: (1) to develop a relationship between trip end production or attraction and land use, and (2) to use the relationship developed to estimate the number of trips generated at some future date under a new set of land-use conditions. To illustrate the process we will consider two methods: cross classification and rates based on activity units. Another commonly used method is regression analysis, which has been applied to estimate both productions and attractions.[11,12]

Cross Classification

Cross classification is a technique developed by the Federal Highway Administration (FHWA) to determine the number of trips that begin or end at the home. Home-based trip generation is a very useful number because it can represent up to 80 percent of all trips. The first step is to develop a relationship between socioeconomic measures and trip production. The two variables most commonly used are average income and auto ownership. Other variables that could be considered are household size and stage in the household life cycle. The relationships are developed based on results of income data and O–D surveys. Figure 10.1 illustrates how income within a zone varies with average income of the zone.

To illustrate the trip generation process we will consider a set of hypothetical data shown in Table 10.1, which lists income, car ownership, and trip data for 20 households, as obtained from an O–D survey.[13] From the information in Table 10.1, a matrix is produced that shows the number and percentage of households as a function of auto ownership and income grouping (see Table 10.2).

Figure 10.1 Average Zonal Income Versus Households in Income Category

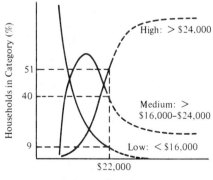

Source: Reproduced from *Computer Programs for Urban Transportation Planning*, U.S. Department of Transportation, Washington, D.C., April 1977.

Table 10.1 Survey Data Showing Trips per Household, Income, and Auto Ownership

Household Number	Trips Produced/ Household	Household Income	Autos/ Household
1	2	$ 8,000	0
2	4	12,000	0
3	10	34,000	2
4	5	22,000	0
5	5	9,000	1
6	15	34,000	3
7	7	19,000	1
8	4	18,000	0
9	6	14,000	1
10	13	38,000	3
11	8	36,000	1
12	6	16,000	1
13	9	14,000	2
14	11	22,000	2
15	10	22,000	2
16	11	26,000	2
17	12	30,000	2
18	8	22,000	1
19	8	26,000	1
20	6	14,000	1

The numerical values in each cell represent the number of households observed in each combination of income–auto ownership category. The value in parentheses is the percentage observed at each income level. In actual practice, the sample size would be at least 25 data points per cell to assure statistical accuracy. The data shown in Table 10.2 can be used to develop relationships between the percent of households in each auto ownership category by household income, as illustrated in Figure 10.2.

Table 10.2 Percent of Households in Each Income Category Versus Car Ownership

Income ($000)	Autos Owned			Total
	0	1	2	
≤ 12	2(67)	1(33)	0(0)	3(100)
12–18	1(25)	3(50)	1(25)	5(100)
18–24	1(20)	2(40)	2(40)	5(100)
24–30		1(33)	2(67)	3(100)
> 30		1(25)	3(75)	4(100)
Total	4	8	8	20

Figure 10.2   Households by Auto Ownership and Income Category

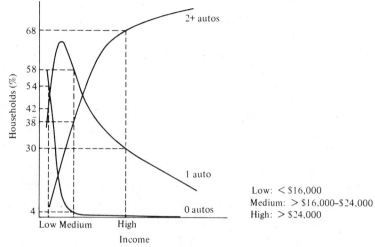

Source: Reproduced from *Computer Programs for Urban Transportation Planning*, U.S. Department of Transportation, Washington, D.C., April 1977.

A second table produced from the data in Table 10.1 shows the number of trips per household versus income and cars owned. The results, shown in Table 10.3, are illustrated in Figure 10.3, which depicts the relationship between trips per household by income and auto ownership. The table indicates that for a given income, trip generation increases with the number of cars owned. Similarly, for a given car ownership, trip generation increases with the rise in income.

As a further refinement, the O–D data can be used to determine the percentage of trips by each trip purpose for each income category. These results are

Figure 10.3   Trips per Household by Auto Ownership and Income Category

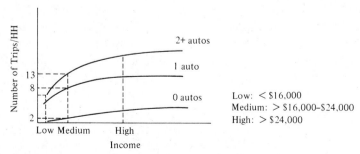

Source: Reproduced from *Computer Programs for Urban Transportation Planning*, U.S. Department of Transportation, Washington, D.C., April 1977.

Table 10.3 Average Trips per Household Versus Income and Car Ownership

Income ($000)	Autos Owned		
	0	1	2
≤ 12	3	5	—
12–18	4	6	9
18–24	5	7.5	10.5
24–30	—	8.5	11.5
> 30	—	8.5	13.3

shown in Figure 10.4, wherein three trip purposes are used: Home-based work (HBW), Home-based other (HBO), and nonhome based (NHB). The terminology refers to the origination of a trip as either at the home or not at the home. The trip generation model that has been developed based on survey data can now be used to estimate the number of home- and nonhome-based trips for each trip purpose. This is illustrated in the following example.

Example 10–1 Computing Trips Generated in a Suburban Zone

Consider a zone that is located in a suburban area of a city. The population and employment data for the zone are as follows.

Number of dwelling units: 60
Average income per dwelling unit: $22,000
Retail employment: 220 persons
Nonretail employment: 650 persons

We want to determine the number of trips generated in this zone for each trip purpose, assuming that the characteristics depicted in Figures 10.1–10.4 apply in this situation. The problem is solved in four basic steps.

Step 1. Determine the percentage of households in each economic category. These results can be determined by analysis of census data for the area. A typical plot of average zonal income versus income distribution is shown in Figure 10.1. For an average zonal income of $22,000, the following distribution is observed:

Income ($)	Households (%)
Low (under 16,000)	9
Medium (16,000–24,000)	40
High (over 24,000)	51

Figure 10.4 Trips by Purpose and Income Category

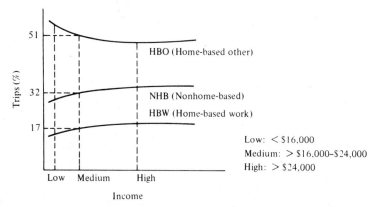

Source: Reproduced from *Computer Programs for Urban Transportation Planning*, U.S. Department of Transportation, Washington, D.C., April 1977.

Step 2. Determine the distribution of auto ownership per household for each income category. A typical curve showing percent of households, at each income level, that own zero, one, or two + autos is shown in Figure 10.2, and the results are listed in Table 10.4.

Table 10.4 Percentage of Households in Each Income Category Versus Auto Ownership

	Cars/Household		
Income	0	1	2+
Low	54	42	4
Medium	4	58	38
High	2	30	68

Table 10.4 shows that 58 percent of medium income families own one auto per household. Also, from the previous step we know that our zone, with an average income of $22,000, contains 40 percent of households in the medium income category. Thus, we can calculate that of the 60 households in that zone, there will be $60 \times 0.40 \times 0.58 = 14$ medium income households that own one auto.

Step 3. Determine the number of trips per household for each income–auto ownership category. A typical curve showing the relationship between trips per household, household income, and auto ownership is shown as Figure 10.3. The results are listed in Table 10.5.

Table 10.5   Number of Trips per Household

	Autos/Household		
Income	0	1	2+
Low	1	6	7
Medium	2	8	13
High	3	11	15

The table shows that a medium income household owning one auto will generate eight trips per day.

Step 4.   Calculate the total number of trips generated in the zone. This is done by computing the number of households in each income–auto ownership category multiplying this result by the number of trips per household, as determined in step 3, and summing the result. Thus,

$$P_{ij} = HH \times I_i(\%) \times A_{ij}(\%) \times (P_{hh})_{ij}$$

where

HH = number of households in the zone

$I_i(\%)$ = percentage of households in zone with income level i (low, medium, or high)

$A_{ij}(\%)$ = percentage of households in income level i with j autos per household ($j = 0$, 1, or 2+)

$(P_{hh})_{ij}$ = number of trips produced in a household at income level i and auto ownership j

P_{ij} = number of trips generated in the zone by householders with income level i and auto ownership j

Table 10.6   Number of Trips per Day Generated by Sixty Households

	Income/ Auto Ownership	Total Trips by Income Group
$60 \times 0.09 \times 0.54 \times \ 1 = \ \ 3$ trips	L, 0	
$60 \times 0.09 \times 0.42 \times \ 6 = \ 14$ trips	L, 1	
$60 \times 0.09 \times 0.04 \times \ 7 = \ \ 2$ trips	L, 2+	19
$60 \times 0.40 \times 0.04 + \ 2 = \ \ 2$ trips	M, 0	
$60 \times 0.40 \times 0.58 \times \ 8 = 111$ trips	M, 1	
$60 \times 0.40 \times 0.38 \times 13 = 119$ trips	M, 2+	232
$60 \times 0.51 \times 0.02 \times \ 3 = \ \ 2$ trips	H, 0	
$60 \times 0.51 \times 0.30 \times 11 = 101$ trips	H, 1	
$60 \times 0.51 \times 0.68 \times 15 = 312$ trips	H, 2+	415
Total $= 666$ trips		666

The actual calculations are shown in Table 10.6. For a zone with 60 households and an average income of $22,000, the number of trips generated is 666 auto trips per day.

Step 5. Determine the percentage of trips by trip purpose. As a final step, we can calculate the number of trips that are HBW, HBO, and NHB. If these percentages are 17, 51, and 32, respectively (see Figure 10.4), for the medium income category, then the number of trips from the zone for the three trip purposes are $232 \times 0.17 = 40$ HBW, $232 \times 0.51 = 118$ HBO, and $232 \times 0.32 = 74$ NHB. Similar calculations would be made for other income groups. The final result is obtained using the following percentage: low income, 15, 55, and 30; high income, 18, 48, and 34, to yield 118 HBW, 327 HBO, and 221 NHB productions.

We have illustrated how trip generation values are determined and used to calculate the number of trips in each zone. Values for each income or auto ownership category can be developed using survey data or published statistics compiled for other cities. Table 10.7 lists detailed trip generation characteristics for urbanized areas in the 50,000–100,000 and 100,000–250,000 population range. An extensive amount of useful trip generation data is available in *Quick Response Urban Travel Estimation Techniques and Transferrable Parameters*.[14]

Rates Based on Activity Units

The preceding section illustrated how trip generation is determined for residential zones where the basic unit is the household. Trips generated at the household end are referred to as *productions*, and they are *attracted* to zones for purposes such as work, shopping, visiting friends, and medical trips. Thus, an activity unit can be described by measures such as square feet of floor space or number of employees. Trip generation rates for attraction zones can be determined from survey data or are tabulated in some of the sources listed in the References and Additional Readings at the end of this chapter. Trip attraction rates are illustrated in Table 10.8.

The data for the zone, as given earlier, can be applied to determine the trip attractions. (Recall there are 220 retail employees and 650 nonretail employees in the zone.) For HBW,

$$
\begin{array}{rll}
60\,\text{HH} & \times\,0.0\,\text{T/HH} & = \quad 0 \\
220\,\text{RE} & \times\,1.7\,\text{T/RE} & = \quad 374 \\
650\,\text{NRE} & \times\,1.7\,\text{T/NRE} & = \underline{1105} \\
& & \quad 1479
\end{array}
$$

where

$\quad$ RE = number of retail employees

$\quad$ NRE = number of nonretail employees

Table 10.7 Trip Generation Characteristics

URBANIZED AREA POPULATION: 50,000-100,000

Income Range 1970 $ (000's)	Avg Autos Per HH	Average Daily Person Trips Per HH	% HH by Autos Owned				Average Daily Person Trips Per HH by No. of Autos/HH				% Average Daily Person Trips by Purpose		
			0	1	2	3+	0	1	2	3+	HBW	HBNW	NHB
0-3	0.56	4.5	53	39	7	1	2.0	6.5	11.5	12.5	21	57	22
3-4	0.81	6.8	32	58	10	1	2.2	8.0	13.0	15.0	21	57	22
4-5	0.88	8.4	26	61	12	1	2.6	9.5	14.5	16.5	21	57	22
5-6	0.99	10.2	20	62	17	1	3.0	11.0	15.5	18.0	18	59	23
6-7	1.07	11.9	15	64	20	1	3.0	12.5	16.5	19.5	18	59	23
7-8	1.17	13.2	11	64	23	2	3.5	13.3	17.0	21.5	16	61	23
8-9	1.25	14.4	8	62	28	2	4.8	14.0	17.5	22.5	16	61	23
9-10	1.31	15.1	6	60	32	2	5.5	14.3	17.5	24.0	16	61	23
10-12.5	1.47	16.4	3	49	44	3	6.2	15.0	18.5	25.5	15	62	23
12.5-15	1.69	17.7	2	38	52	8	6.1	15.0	19.0	25.5	14	62	24
15-20	1.85	18.0	2	28	57	13	6.0	13.5	19.5	23.0	13	62	25
20-25	2.03	19.0	1	21	58	20	6.0	13.0	20.0	23.0	13	62	25
25+	2.07	19.2	1	19	59	21	6.0	12.5	20.0	23.0	13	62	25
Weighted Average	1.55	14.1	12	47	35	6	4.6	12.6	17.2	21.4	16	61	23

URBANIZED AREA POPULATION: 100,000-250,000

Income Range 1970 $ (000's)	Avg Autos Per HH	Average Daily Person Trips Per HH	% HH by Autos Owned				Average Daily Person Trips Per HH by No. of Autos/HH				% Average Daily Person Trips by Purpose		
			0	1	2	3+	0	1	2	3+	HBW	HBNW	NHB
0-3	0.49	4.0	57	37	6	0	1.0	7.5	10.5	13.8	20	63	17
3-4	0.72	6.8	36	56	8	0	1.7	9.2	13.3	16.4	22	60	18
4-5	0.81	8.4	29	61	10	0	2.5	10.2	14.5	17.6	22	58	20
5-6	0.94	10.2	21	65	13	1	3.5	11.4	14.5	19.0	22	58	20
6-7	1.01	11.7	17	66	16	1	4.5	12.5	15.6	20.5	20	58	22
7-8	1.14	13.6	12	65	21	2	5.4	13.8	17.0	22.2	20	57	23
8-9	1.25	15.3	9	61	28	2	5.8	15.0	17.5	23.0	20	57	23
9-10	1.34	16.2	6	58	33	3	6.3	15.8	18.0	23.5	19	57	24
10-12.5	1.50	17.3	4	50	40	6	6.8	16.0	19.0	24.5	19	57	24
12.5-15	1.65	18.7	2	40	51	7	7.0	16.0	20.4	25.0	19	56	25
15-20	1.85	19.6	2	28	57	13	7.2	15.0	21.0	25.5	18	56	26
20-25	2.01	20.4	1	20	61	18	7.5	15.0	21.0	25.5	18	55	27
25+	2.07	20.6	1	19	59	21	7.5	15.0	21.0	25.2	18	55	27
Wt. Avg.	1.55	14.5	14	48	33	6	5.4	13.7	18.4	22.4	20	57	23

Source: Adapted from *Quick-Response Urban Travel Estimation Techniques and Transferrable Parameters, User's Guide*, NCHRP Report No. 187, Transportation Research Board, Washington, D.C., 1978.

Table 10.8 Trip Generation Rates by Trip Purpose and Employee Category

	Attractions per Household	Attractions per Nonretail Employee	Attractions per Downtown Retail Employee	Attractions per Other Retail Employee
HBW	—	1.7	1.7	1.7
HBO	1.0	2.0	5.0	10.0
NHB	1.0	1.0	3.0	5.0

Source: Adapted from *Computer Programs for Urban Transportation Planning*, U.S. Department of Transportation, Washington, D.C., April 1977.

TRIP DISTRIBUTION

Trip distribution analysis is a process by which the trips generated in one zone are allocated to other zones in the study area. For example, if the trip generation analysis results in an estimate of 200 HBW trips in zone 10, then the trip distribution analysis would determine how many of these trips would be made between zone 10 and zone 6, and so forth.

Several basic methods are used for trip distribution. Among these are the gravity model, growth factor models, and intervening opportunities. The gravity model is preferred because it uses the attributes of the transportation system and land-use characteristics and has been calibrated extensively for many urban areas. Growth factor models, which were more widely used in the 1950s and 1960s, require that the origin–destination matrix be known for the base (or current) year, as well as an estimate of the number of future trip ends in each zone. The intervening opportunities model and other models are available but not widely used in practice.

Gravity Model

The most widely used and documented trip distribution model is the so-called gravity model, which states that the number of trips between two zones is directly proportional to the number of trip attractions generated by the zone of destination and inversely proportional to a function of time of travel between the two zones.

Mathematically, the gravity model is expressed as follows.

$$T_{ij} = P_i \left[\frac{A_j F_{ij} K_{ij}}{\sum_j A_j F_{ij} K_{ij}} \right]$$

where

T_{ij} = number of trips that are produced in zone i and attracted to zone j

P_i = total number of trips produced in zone i

A_j = number of trips attracted to zone j

F_{ij} = a value which is an inverse function of travel time

K_{ij} = socioeconomic adjustment factor for interchange ij

The values of P_i and A_j have been determined in the trip generation process. The sum of P_i's for all zones must equal the sum of A_j's for all zones. K_{ij} values are used when the estimated trip interchange must be adjusted to assure that it agrees with the observed trip interchange.

The values for F_{ij} are determined by a calibrating process in which trip generation values as measured in the O–D survey are distributed using the gravity model.[15] After each distribution process is completed, the percentage of trips in each trip length category produced by the gravity model is compared with the percentage of trips recorded in the O–D survey. If the percentages do not agree, then the F_{ij} factors that were used in the distribution process are adjusted and another gravity model trip distribution is performed. The calibration process is

Figure 10.5   Trip Length Distribution: Actual Versus Gravity Model

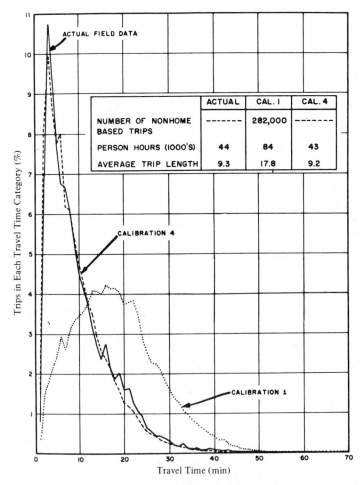

Source: Reproduced from *Calibrating and Testing a Gravity Model for Any Size Urban Area*, U.S. Department of Commerce, Washington, D.C., October 1965.

Figure 10.6 Calibration of F Factors

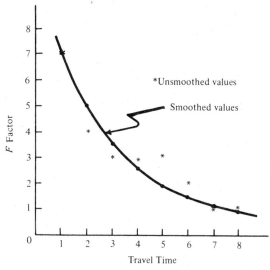

*Unsmoothed values

Smoothed values

F Factor (y-axis)

Travel Time (x-axis)

Source: Reproduced from *Computer Programs for Urban Transportation Planning*, U.S. Department of Transportation, Washington, D.C., April 1977.

continued until the trip length percentages are in agreement. Figure 10.5 illustrates the results of four calibrations and shows the comparison with travel time frequency. Note that calibration 4 is very close to the observed values. Thus the F values used in this calibration are presumed to be the correct values. Also note that the values for F decrease as travel time increases. Figure 10.6 illustrates F values for calibrations of a gravity model. (Normally this curve is a semi-log plot.) F values can also be determined using travel time values and an inverse relationship between F and t. For example, the relationship for F might be in the form t^{-1}, t^{-2}, e^{-t}, and so forth.

Example 10–2 Use of Calibrated F Factors and Iteration

To illustrate the application of the gravity model, consider a study area consisting of three zones. The data have been determined as follows. The number of productions and attractions have been computed for each zone by methods described in the section on trip generation, and the average travel times between each zone have been determined. Both are shown in Tables 10.9 and 10.10. Assume K_{ij} is the same unit value for all zones. Finally, the F values have been calibrated as previously described and are shown in Table 10.11 for each travel time increment. Note that the intrazonal travel time for zone 1 is larger than most of the other interzone times, because of the geographical characteristics of the zone and the lack of access within the area. This zone could represent conditions in a congested downtown area.

Table 10.9 Trip Productions and Attractions for a Three-Zone Study Area

Zone	1	2	3	Total
Trip Productions	140	330	280	750
Trip Attractions	300	270	180	750

Table 10.10 Travel Times Between Zones (min)

Zone	1	2	3
1	5	2	3
2	2	6	6
3	3	6	5

Table 10.11 Travel Time Versus Friction Factor

Time	F
1	82
2	52
3	50
4	41
5	39
6	26
7	20
8	13

Note: Obtained from the calibration process.

- Determine the number of trips between each zone using the gravity model formula and the given data. (Note: F_{ij} is obtained by using the travel times in Table 10.10, and selecting the correct F value from Table 10.11. For example, travel time is 2 min between zones 1 and 2. The corresponding F value is 52.)

$$T_{ij} = P_i \left(\frac{A_j F_{ij} K_{ij}}{\sum\limits_{j=1}^{n} A_j F_{ij} K_{ij}} \right) \qquad K_{ij} = 1 \text{ for all zones}$$

$$T_{1-1} = 140 \times \frac{300 \times 39}{300 \times 39 + 270 \times 52 + 180 \times 50} = 47$$

$$T_{1-2} = 140 \times \frac{270 \times 52}{300 \times 39 + 270 \times 52 + 180 \times 50} = 57$$

$$T_{1-3} = 140 \times \frac{180 \times 50}{300 \times 39 + 270 \times 52 \times 180 \times 50} = 36$$

$$P_1 = 140$$

- Make similar calculations for zones 2 and 3.

$$T_{2-1} = 188; \quad T_{2-2} = 85; \quad T_{2-3} = 57; \quad P_2 = 330$$

$$T_{3-1} = 144; \quad T_{3-2} = 68; \quad T_{3-3} = 68; \quad P_3 = 280$$

- Summarize the results as shown in Table 10.12. Note that the sum of the productions in each zone is equal to the number of productions given in the problem statement. However, the number of attractions estimated in the trip distribution phase differs from the number of attractions given. For zone 1 the correct number is 300 whereas the computed value is 379. Values for zone 2 are 270 versus 210 and for zone 3 are 180 versus 161.
- Calculate the adjusted attraction factors according to the following formula.[16]

$$A_{jk} = \frac{A_j}{C_{j(k-1)}} A_{j(k-1)}$$

where

A_{jk} = adjusted attraction factor for attraction zone (column) j, iteration k.

$A_{jk} = A_j$ when $k = 1$

C_{jk} = actual attraction (column) total for zone j, iteration k

A_j = desired attraction total for attraction zone (column) j

j = attraction zone number, $j = 1, 2, \ldots n$

n = number of zones

k = iteration number, $k = 1, 2, \ldots m$

m = number of iterations

Table 10.12 Zone to Zone Trips: First Iteration

Zone	1	2	3	P
1	47	57	36	140
2	188	85	57	330
3	144	68	68	280
Computed A'	379	210	161	750
Given A	300	270	180	

- To produce a mathematically correct result, repeat the trip distribution computations using modified attraction values, so that the numbers attracted will be increased or reduced as required. For zone 1, the estimated attractions were too great. Therefore, the new attraction factors are adjusted downward by multiplying the original attraction value by the ratio of the original to estimated attraction values.

$$\text{Zone 1} \quad A'_1 = 300 \times \frac{300}{379} = 237$$

Similarly,

$$\text{Zone 2} \quad A'_2 = 270 \times \frac{270}{210} = 347$$

and

$$\text{Zone 3} \quad A'_3 = 180 \times \frac{180}{161} = 201$$

- Apply the gravity model formula in each iteration to calculate zonal trip interchanges using the adjusted attraction factors obtained from the preceding iteration. In practice, the gravity model formula thus becomes

$$\left(T_{ijk} = \frac{P_i A_{jk} F_{ij} K_{ij}}{\displaystyle\sum_{j=1}^{n} A_{jk} F_{ij} K_{ij}} \right)_p$$

where T_{ijk} is the trip interchange between i and j for iteration k and $A_{jk} = A_j$ when $K = 1$. Subscript j goes through one complete cycle every time k changes, and i goes through one complete cycle every time j changes. The above formula is enclosed in brackets, which are subscripted to indicate that the complete process is performed for each trip purpose.

- Perform a second iteration using the adjusted attraction values.

$$T_{1-1} = 140 \times \frac{237 \times 39}{237 \times 39 + 347 \times 52 + 201 \times 50} = 34$$

$$T_{1-2} = 140 \times \frac{347 \times 52}{237 \times 39 + 347 \times 52 + 201 \times 50} = 68$$

$$T_{1-3} = 140 \times \frac{201 \times 50}{237 \times 39 + 347 \times 52 + 201 \times 50} = 37$$

$$P_1 = 140$$

- Make similar calculations for zones 2 and 3.

$$T_{2-1} = 153; \quad T_{2-2} = 112; \quad T_{2-3} = 65; \quad P_2 = 330$$

$$T_{3-1} = 116; \quad T_{3-2} = 88; \quad T_{3-3} = 76; \quad P_3 = 280$$

Table 10.13 Zone to Zone Trips: Second Iteration

Zone		1	2	3	P
	1	34	68	38	140
	2	153	112	65	330
	3	116	88	76	280
Computed A''		303	268	179	750
Given	A	300	270	180	750

- List the results as shown in Table 10.13. Note that, in each case, the sum of the attractions is now much closer to the given value. The process will be continued until there is reasonable agreement (within 5 percent) between the A that is estimated using the gravity model and the values that are furnished in the trip generation phase.

Growth Factor Models

Trip distribution can also be computed when the only data available are the origins and destinations between each zone for the current or base year and the trip generation values for each zone for the future year. This method was widely used when O–D data were available but the gravity model and calibrations for F factors had not yet become operational. Growth factor models are used primarily to distribute trips between zones in the study area and zones in cities external to the study area. The most popular growth factor model is the Fratar method, which is a mathematical formula that proportions future trip generation estimates to each zone as a function of the product of the current trips between the two zones T_{ij} and the growth factor of the attracting zone G_j. Thus,

$$T_{ij} = (t_i G_i) \frac{t_{ij} G_j}{\sum_x t_{ix} G_x}$$

where

T_{ij} = number of trips estimated from zone i to zone j

t_i = present trip generation in zone i

G_x = growth factor of zone x

$T_i = t_i G_i$ = future trip generation in zone i

t_{ix} = number of trips between zone i and other zones x

To illustrate the application of the growth factor method consider the following example.

Example 10–3   Forcasting Trips Using Growth Factor Model

A study area consists of four zones (A, B, C, and D). An O–D survey indicates that the number of trips between each zone is as shown in Table 10.14. Planning estimates for the area indicate that in 5 years the number of trips in each zone will increase by the growth factor shown in Table 10.15 and that trip generation will be increased to the amounts shown in the last column of the table. Determinine the number of trips between each zone for future conditions.

Table 10.14   Present Trips Between Zones

	A	B	C	D
A	—	400	100	100
B	400	—	300	—
C	100	300	—	300
D	100	—	300	—
Total	600	700	700	400

Table 10.15   Present Trip Generation and Growth Factors

Zone	Present Trip Generation (trips/day)	Growth Factor	Trip Generation in 5 Years
A	600	1.2	720
B	700	1.1	770
C	700	1.4	980
D	400	1.3	520

Using the Fratar formula, we can calculate the number of trips between zones A and B, A and C, A and D, and so forth. Note that we obtain two values for each zone pair (that is, T_{AB} and T_{BA}). These values are averaged, yielding a value $\bar{T}_{AB} = (T_{AB} + T_{BA})/2$.

The calculations are as follows:

$$T_{AB} = 600 \times 1.2 \frac{400 \times 1.1}{(400 \times 1.1) + (100 \times 1.4) + (100 \times 1.3)} = 446$$

$$T_{BA} = 700 \times 1.1 \frac{400 \times 1.2}{(400 \times 1.2) + (300 \times 1.4)} = 411$$

$$\bar{T}_{AB} = \frac{446 + 411}{2} = 428$$

Table 10.16 First Estimate of Trips Between Zones

	A	B	C	D	Estimated Total Trip Generation	Actual Trip Generation
A	—	428	141	124	693	720
B	428	—	372	—	800	770
C	141	372	—	430	943	980
D	124	—	430	—	554	520
Totals	693	800	943	554		

Similar calculations yield

$$\bar{T}_{AC} = 141; \qquad \bar{T}_{AD} = 124; \qquad \bar{T}_{BC} = 372; \qquad \bar{T}_{CD} = 430$$

The results of the preceding calculations have produced the first estimate (or iteration) of future trip distribution and are shown in Table 10.16. Note, however, that the totals for each zone do not equal the values of future trip generation as stated earlier. For example, the trip generation in zone A is estimated as 693 trips, whereas the correct value is 720 trips. Similarly, the estimate for zone B is 800 trips, whereas the desired value is 770 trips.

We now proceed with a second iteration in which the input data is the number of trips between zones as previously calculated. Also, new growth factors are computed as the ratio of the trip generation expected to occur in 5 years and the trip generation estimated in the preceding calculation. The values are given in Table 10.17.

Table 10.17 Growth Factors for Second Iteration

Zone	Estimated Trip Generation	Actual Trip Generation	Growth Factor
A	693	720	1.04
B	800	770	0.96
C	943	980	1.04
D	554	520	0.94

The calculations for the second iteration are left to the reader and can be repeated as many times as needed until the estimated and actual trip generation values are close in agreement.

MODAL SPLIT

Modal split is that aspect of the demand analysis process that determines the number (or percentage) of trips between zones that are made by automobile and by transit. The selection of one mode or another is a complex process that depends on factors such as the traveler's income, the availability of transit service or auto ownership, and the relative advantages of each mode in terms of travel time, cost, comfort, convenience, and safety. Modal split models attempt to replicate the relevant characteristics of the traveler, the transportation system, and the trip itself, such that a realistic estimate of the number of trips by each mode for each zonal pair is obtained. A discussion of the many modal split models is beyond the scope of this chapter, and the interested student should refer to other primary sources cited.[17–21]

Types of Modal Split Models

Since public transportation is an important factor, primarily in larger cities, modal split calculations may only involve distinguishing trip interchanges as either auto or transit. Depending on the level of detail required, three types of transit estimating procedures are used: (1) direct generation of transit trips, (2) use of trip end models, and (3) trip interchange modal split models.

For *direct generation* methods, transit trips are generated directly, either by estimating total person trips or auto driver trips. Figure 10.7 is a graph that illustrates the relationship between transit trips per day per 1000 population and persons per acre versus auto ownership. As density of population increases, it can be expected that transit riding will also increase for a given level of auto ownership.

Trip end models determine the percentage of total person trip generation productions that will use transit. The estimates are made prior to the trip distribution phase and are based on land-use or socioeconomic characteristics

Figure 10.7   Number of Transit Trips by Population Density and Auto Ownership per Household

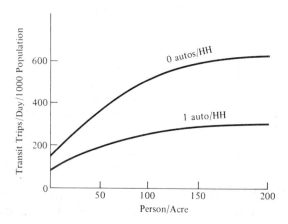

of the zone. They do not incorporate the quality of service. The procedure follows.

1. Generate total person trip productions and attractions by trip purpose.
2. Compute the urban travel factor.
3. Determine the percentage of these trips by transit using a modal split curve (see Figure 10.8).
4. Apply auto occupancy factors.
5. Distribute transit and auto trips separately.

The modal split model shown in Figure 10.8 is based on two factors, households per auto and persons per square mile. The product of these variables is called the urban travel factor (UTF). Percentage of travel by transit will increase but in an *S* curve fashion as the UTF increases.

The third traditional estimating procedure for modal split is the use of *trip interchange models*. These incorporate system level of service variables such as relative travel time, relative travel cost, economic status of the trip maker, and relative travel service. The following is a description of one such model developed for the National Cooperative Highway Research Program, referred to as the QRS method.[22] This model illustrates the basic elements that are considered in estimating modal split.

Figure 10.8 Transit Mode Split Versus Urban Travel Factor

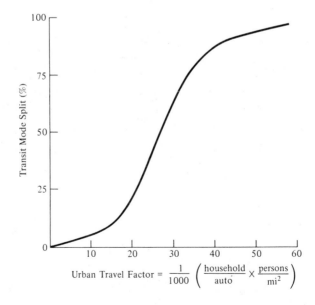

The QRS method is based on the following relationship:

$$MS_t = \frac{I_{ijt}^{-b}}{I_{ijt}^{-b} + I_{ija}^{-b}} \times 100 \quad \text{or} \quad \frac{I_{ija}^{b}}{I_{ijt}^{b} + I_{ija}^{b}} \times 100$$

$$MS_a = (1 - MS_t) \times 100$$

where

MS_t = proportion of trips between zone i and j using transit

MS_a = proportion of trips betwen zone i and j using auto

I_{ijm} = a value referred to as the *impedance* of travel of mode m, between i and j which is a measure of the total cost of the trip. [Impedance = (in-vehicle time min) + (2.5 × excess time min) + (3 × trip cost in $/income per min).]

b = an exponent, which depends on trip purpose

(Note that in-vehicle time is time spent traveling in the vehicle and excess time is time spent traveling but not in the vehicle (waiting, walking, and so forth.)

The impedance value is determined for each zone pair and represents a measure of the expenditure required to make the trip by either auto or transit. The data required for estimating modal split include (1) distance between zones by auto and transit, (2) transit fare, (3) out-of-pocket auto cost, (4) parking cost, (5) highway and transit speed, (6) exponent values, b, (7) median income, and (8) excess time, which includes the time required to walk to a transit vehicle and time waiting or transferring. Assume 120,000 min per year.

Example 10–4 Computing Modal Split Using the QRS Model

To illustrate the application of this method, assume that the data shown in Table 10.18 have been developed for travel between a suburban zone S and a downtown zone D. Determine the percent of work trips by auto and transit. An exponent value of 2.0 is used for work travel. Median income is $12,000 per year.

Table 10.18 Travel Data Between Two Zones, S and D

	Auto	Transit
Distance	10 mi	8 mi
Cost per mile	$0.15	$0.10
Excess time	5 min	8 min
Parking cost	$1.50 (or 0.75/trip)	—
Speed	30 mph	20 mph

Then,

$$I_{SDA} = \left(\frac{10}{30} \times 60\right) + (2.5 \times 5) + \left\{\frac{3 \times [(1.50/2) + 0.15 \times 10]}{12,000/120,000}\right\} = 20 + 12.5 + 67.5$$

$$= 100.0 \text{ equivalent min}$$

$$I_{SDT} = \left(\frac{8}{20} \times 60\right) + (2.5 \times 8) + \left[\frac{3 \times (8 \times 0.10)}{12,000/120,000}\right] = 24 + 20 + 24$$

$$= 68 \text{ equivalent min}$$

$$MS_a = \frac{(68)^2}{(68)^2 + (100)^2} \times 100 = 31.6 \text{ percent}$$

$$MS_t = (1 - 0.316) \times 100 = 68.4 \text{ percent.}$$

Thus the modal split of travel by transit between zones S and D is 68.4 percent and by highway the value is 31.6 percent. These percentages are applied to the estimated trip distribution values to determine the number of trips by each mode. If, for example, the number of work trips between zones S and D were computed to be 500, then the number by auto would be $500 \times 0.316 = 158$, and by transit the number of trips would be $500 \times 0.684 = 342$.

Choice Models

An alternative approach used in transportation demand analysis is to consider the relative utility of each mode as a summation of each modal attribute. Then the choice of a mode is expressed as a probability distribution. For example, if the utility of each mode is

$$U_x = \sum_{i=1}^{n} a_i X_i$$

where

U_x = utility of mode x

n = number of attributes

X_i = attribute value (time, cost, and so forth)

a_i = coefficient value for attribute i

Then if 2 modes (A and B) are being considered, the probability of selecting mode A can be written as

$$P(A) = \frac{e^{U_A}}{e^{U_A} + e^{U_B}}$$

This form is called the logit model and provides a convenient way to compute mode choice. Choice models are utilized within the urban transportation planning process primarily in the modal split phase but are also used in transit marketing studies and in directly estimating travel demand.[23,24]

TRAFFIC ASSIGNMENT

The final step in the transportation forecasting process is to determine the actual street and highway routes that will be used and the number of automobiles and buses that can be expected on each highway segment. The procedure used to determine the expected traffic volumes is known as *traffic assignment*.[25-27] Up to this point, we know the number of trips by transit and auto that will travel between zones. We now *assign* these trips to a logical highway route and sum up the results for each highway segment. Our result is what we have been seeking: a forecast of the average daily or peak hour traffic volumes that will occur on the urban transportation system that serves the study area.

To carry out a trip assignment, the following data are required. First, we need to know how many trips will be made from one zone to another (this information was determined in the trip distribution phase). Second, we need to know the available highway or transit routes between zones and how long it will take to travel on each route, and third, we need a decision rule (or algorithim) that states the criteria by which motorists or transit users will select a route.

Basic Approaches

Three basic approaches can be used for traffic assignment purposes: (1) diversion curves, (2) minimum time path (all-or-nothing) assignment, and (3) minimum time path with capacity restraint. The *diversion curve* method is similar in approach to a modal split curve. The traffic between two routes is determined as a function of relative travel time or cost. Figure 10.9 illustrates a diversion curve based on travel time ratio.

The *minimum time path* method assigns all trips to those links that comprise the shortest time path between the two zones. *Capacity restraint* is a refinement of the minimum path method in that, after all traffic has been assigned to a link, the travel times on each link are adjusted, based on the capacity of the link and the number of trips on each link. The capacity restraint method requires repeated assignments and adjustments of travel time until a balance is achieved.

Minimum Path Algorithm

The traffic assignment process is illustrated using the minimum path algorithm. This method is selected because it is commonly used, generally produces accurate results, and adequately demonstrates the basic principles involved.

The minimum path assignment is based on the theory that a motorist or transit user will select the quickest route between any O–D pair. In other words, the traveler will always select the route that represents minimum travel time. Thus, to determine which route that will be, it is necessary to find the shortest

Figure 10.9 Travel Time Ratio Versus Percentage of Travel on Route B

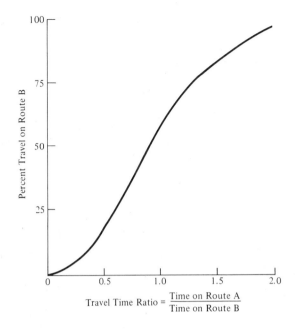

Percent Travel on Route B

$$\text{Travel Time Ratio} = \frac{\text{Time on Route A}}{\text{Time on Route B}}$$

route from the zone of origin to all other destination zones. The results can be depicted as a tree, referred to as a *skim tree*. All trips from that zone are assigned to links on the skim tree. To determine the minimum path, a procedure is used that finds the shortest path without having to test all possible combinations.

The algorithm that will be used in the example that follows is to connect all nodes from the home (originating) node and keep all paths as contenders until one path to the same node is a faster route than others, at which juncture those links on the slower path are eliminated.

The general mathematical algorithm that describes the process is that we wish to select paths that minimize the expression

$$\sum_{\text{all } ij} V_{ij} T_{ij}$$

where
V_{ij} = volume on link i, j
T_{ij} = travel time on link i, j
i, j = adjacent nodes

Example 10–5 Finding Minimum Paths in a Network

To illustrate the process of path building, consider the following 16-node network with travel times on each link shown for each zone pair.

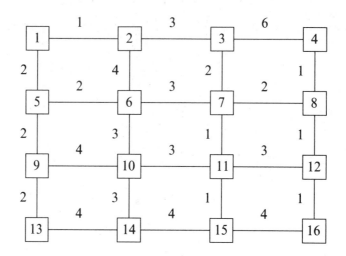

The link and node network is a representative of the road and street system. We wish to determine the shortest travel path from zone 1 (home node) to all other zones. To determine minimum time paths from node 1 to all other nodes, proceed as follows.

Step 1. Determine the time to nodes connected to node 1. Time to node 2 is 1 min. Time to node 5 is 2 min. Times are noted near zone in diagram.

Step 2. From the node closest to the home node (node 2 is closest to home node 1), make connections to nearest nodes. These are nodes 3 and 6. Write the cumulative travel time at each node.

Step 3. From the node closest to the home node (node 5), make connections to the nearest node (nodes 6 and 9).

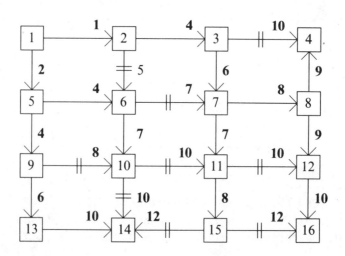

Step 4. Time to node 6 via node 5 is shorter than via zone 2. Therefore link 2–6 is deleted.

Step 5. Three nodes are equally close to the home node (nodes 3, 6, and 9). Select lowest numbered node (3) and add corresponding links to nodes 4 and 7.

Step 6. Of the three equally close nodes, node 6 is the next closest to the home node. Connect to zone 7 and 10. Eliminate link 6–7.

Step 7. Building proceeds from node 9 to nodes 10 and 13. Eliminate link 9–10.

Step 8. Build from zone 7.

Step 9. Build from zone 13.

Step 10. Build from zone 10, eliminate node 10–11.

Step 11. Build from zone 11.

Step 12. Build from zone 8, eliminate node 11–12.

Step 13. Build from zone 15, eliminate node 14–15.

Step 14. Build from zone 12, eliminate node 15–16.

To find the minimum path from any node to node 1, following the path backwards. Thus, for example, the links on the minimum path from zone 1 to zone 11 are 7–11, 3–7, 2–3, and 1–2. This process is then repeated for the other 15 zones, to produce the *skim trees* for each of the zones in the study area. Figure 10.10 illustrates the skim tree produced for zone 1.

Figure 10.10   Minimum Path Tree for Zone 1

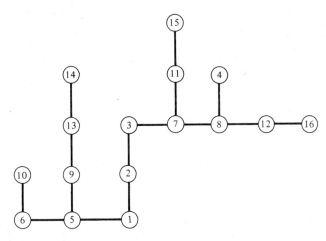

Example 10–6   Network Loading Using Minimum Path Method

The links that are on the minimum path for each of the nodes connecting node 1 are shown in Table 10.19. Also shown are the number of auto trips between zone 1 and all other zones. From these results, the number of trips on each link is determined. To illustrate, link 1–2 is used by trips from zone 1 and zones 2, 3, 4, 7, 8, 11, 12, 15, and 16. Thus, the trip between these zone pairs are assigned to link 1–2 as illustrated in Table 10.19. The volumes are 50, 75, 80, 60, 30, 80, 25, 20, and 85 for a total of 505 trips on link 1–2 from zone 1.

Table 10.19   Links on Minimum Path for Trips from Zone 1

From	To	Trips	Links on the Minimum Path
1	2	50	1–2
	3	75	1–2, 2–3
	4	80	1–2, 2–3, 3–7, 7–8, 4–8
	5	100	1–5
	6	125	1–5, 5–6
	7	60	1–2, 2–3, 3–7
	8	30	1–2, 2–3, 3–7, 7–8
	9	90	1–5, 5–9
	10	40	1–5, 5–6, 6–10
	11	80	1–2, 2–3, 3–7, 7–11
	12	25	1–2, 2–3, 3–7, 7–8, 8–12
	13	70	1–5, 5–9, 9–13
	14	60	1–5, 5–9, 9–13, 13–14
	15	20	1–2, 2–3, 3–7, 7–11, 11–15
	16	85	1–2, 2–3, 3–7, 7–8, 8–12, 12–16

Table 10.20   Assignment of Trips from Zone 1 to Links on Highway Network

Link	Trips on Link	
1–2	50, 75, 80, 60, 30, 80, 25, 20, 85 =	505
2–3	75, 80, 60, 30, 80, 25, 20, 85 =	455
1–5	100, 125, 90, 40, 70, 60 =	485
5–6	125, 40 =	165
7–8	80, 30, 25, 85 =	220
4–8	80 =	80
5–9	90, 70, 60 =	220
6–10	40 =	40
7–11	80, 20 =	100
8–12	25, 85 =	110
9–13	70, 60 =	130
11–15	20 =	20
12–16	85 =	85
13–14	60 =	60

We can now calculate the number of trips that should be assigned to each link of those that have been generated in zone 1 and distributed to zones 2 through 16 (Table 10.20). A similar process of network loading would be completed for all other zone pairs.

A modification of the process just described is known as capacity restraint. The number of trips assigned to each link is compared with the capacity of the link to determine the extent to which link travel times have been reduced. Using relationships between volume and travel time (or speed) similar to those derived in Chapter 5, it is possible to recalculate the new link travel time. A reassignment is then made based on these new values. The iteration process continues until an equilibrium balance is achieved.

The process of calculating the travel demand for an urban transportation system is now completed. The results of this work will be used to determine where improvements will be needed in the system, for making economic evaluations of project priority, and in the geometric and pavement design phases. In actual practice the calculations are carried out by computer as the process is computational intensive as the number of zones increases.

OTHER METHODS FOR FORECASTING DEMAND

This chapter has described how travel demand is forecasted by illustrating the process, using the four-step procedure of trip generation, distribution, modal split, and traffic assignment. There are many variations within each of these steps and the interested reader should refer to the references cited for additional details. Furthermore, there are other methods that can be used to forecast demand. Some of these are listed below but are not described in detail since the topic of transportation demand forecasting could be the subject of a textbook or an entire course.

Trend Analysis

This approach to demand estimation is based simply on the extrapolation of past trends into the future. For example, to forecast the amount of traffic on a rural road, traffic count data from previous years are plotted versus time. Then to compute the volume of traffic at some future date, the *trend line* is extrapolated forward or the average growth rate is used. Often a mathematical expression is developed using statistical techniques, and quite often a semi-log relationship is used. Although simple in application, trend line analysis has the disadvantage that future demand estimates are based on extrapolations of the past, with no allowance made for changes that may be time dependent.

Demand Elasticity

Travel demand can also be determined if the relationship between demand and a key service variable (such as travel cost) is known. If V = volume (demand) at a given service level X, then the elasticity of demand, $E(V)$, is the percent change in volume divided by the percent change in service level or

$$E(V) = \frac{\% \; \Delta \; \text{in} \; V}{\% \; \Delta \; \text{in} \; X}$$

$$E(V) = \frac{\Delta V/V}{\Delta X/X} = \frac{X}{V} \frac{\Delta V}{\Delta X}$$

Example 10–7   Forecasting Transit Ridership Reduction Due to Increase in Fares

To illustrate the use of demand elasticity, a rule of thumb in the transit industry states that for each 1 percent increase in fares, there will be one-third of 1 percent reduction in ridership. If current ridership is 2000/day at a fare of 30¢, what will the ridership be if the fare is increased to 40¢?

In this case $E(V) = 1/3$, and

$$\frac{1}{3} = \frac{X}{V} \frac{\Delta V}{\Delta X} = \frac{30}{2000} \times \frac{\Delta V}{10}$$

$$\Delta V = 222$$

or new ridership is $2000 - 222 = 1778/\text{day}$.

SUMMARY

The process of forecasting travel demand is necessary to determine the number of persons or vehicles that will use a new transportation system or component. The methods used to forecast demand include extrapolation of past trends, elasticity of demand, and relating travel demand to socioeconomic variables.

Urban travel demand forecasting is a complex process because demand for urban travel is influenced by the location and intensity of land use, the socioeconomic characteristics of the population, and the extent, cost, and quality of transportation services.

Forecasting urban travel demand involves a series of tasks. These are population and economic analysis, land-use forecasts, trip generation, trip distribution, modal split, and traffic assignment. The development of computer programs to calculate the elements within each task has greatly simplified implementation of the demand forecasting process.

Travel demand forecasts are also required for completing an economic evaluation of various system alternatives. This topic is described in the next chapter.

Travel Time	F_{ij}
1	2.0
4	1.6
6	1.0
9	0.9
10	0.86
11	0.82
12	0.80
15	0.68
20	0.49

10–6 The following table shows the productions and attractions used in the first iteration of a trip distribution procedure and the productions and attractions that resulted. Determine the number of productions and attractions that should be used for each zone in the second iteration.

	1	2	3	4
P	100	200	400	600
A	300	100	200	700
P^1	100	200	400	600
A^1	250	150	300	600

10–7 The Jeffersonville Transportation Study area has been divided into four large districts (traffic zones). The following data have been compiled.

District	Production	Attractions
1	1000	1000
2	2000	700
3	3000	6000
4	2200	500

District	Travel Time (min)			
	1	2	3	4
1	5	8	12	15
2	8	5	10	8
3	12	10	5	7
4	15	8	7	5

Travel Time	F_{ij}
1	2.00
5	1.30
6	1.10
7	1.00
8	0.95
10	0.85
12	0.80
15	0.65

After the first iteration the trip table was

District	1	2	3	4	
1	183	94	677	46	1000
2	256	244	1372	128	2000
3	250	186	2404	160	3000
4	180	183	1657	180	2200
A's	869	707	6110	514	8200

Complete the second iteration.

10–8 For the travel pattern in Figure 10.11 develop a Fratar distribution for two iterations.

FIGURE 10.11

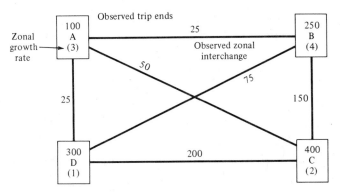

10–9 What data are required in order to use (**a**) the gravity model and (**b**) the Fratar model?

10–10 Assign the vehicle trips shown in the O–D trip table to the network, using the all-or-nothing assignment technique shown in Figure 10.12. Make a list of the links in the network and indicate the volume assigned to each. Calculate the total vehicle minutes of travel. Show the minimum path and assign traffic for each of the five nodes.

FIGURE 10.12

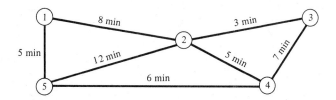

From/To	1	2	3	4	5
1	0	100	100	200	150
2	400	0	200	100	500
3	200	100	0	100	150
4	250	150	300	0	400
5	200	100	50	350	0

10–11 Figure 10.13 represents travel times on the links connecting six zonal centroids. Determine the minimum path from each zone to each other zone.

FIGURE 10.13

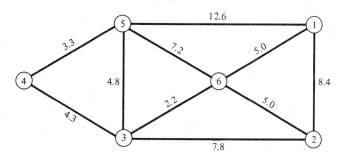

Use the all-or-nothing trip assignment method to determine the total trips for each link after all of the trips from the following two-way trip table have been loaded onto the network.

Zones	Trips Between Zones					
	1	2	3	4	5	6
1	0	1000	1100	400	1000	1300
2	—	0	1050	700	1100	1200
3	—	—	0	1200	1150	1600
4	—	—	—	0	800	400
5	—	—	—	—	0	700
6	—	—	—	—	—	0

REFERENCES

1. *Data Requirements for Metropolitan Transportation Planning*, NCHRP Report 120, Highway Research Board, Washington, D.C., 1971.

2. *Quick Response Urban Travel Estimation Techniques and Transferable Parameters, Users Guide*, NCHRP Report No. 187, National Research Council, Transportation Research Board, Washington, D.C., 1978.

3. *Guidelines for Trip Generation Analysis*, U.S. Department of Transportation, Federal Highway Administration, Washington, D.C., April 1973.

4. *Quick Response Urban Travel Estimation Techniques.*

5. Ibid.

6. *Guidelines for Trip Generation Analysis.*

7. Ibid.

8. *Application of New Travel Demand Forecasting Techniques to Transportation Planning: A Study of Individual Choice Modes*, U.S. Department of Transportation, Federal Highway Administration, Washington, D.C., 1977.

9. Adib Kanafani, *Transportation Demand Analysis*, McGraw-Hill Book Company, New York, 1983.

10. *Computer Programs for Urban Transportation Planning*, U.S. Department of Transportation, Federal Highway Administration, Washington, D.C., 1977.

11. Michael D. Meyer and Eric J. Miller, *Urban Transportation Planning: A Decision Oriented Approach*, McGraw-Hill Book Company, New York, 1984.

12. Lawrence C. Caldwell III and M. J. Demetsky, *An Evaluation of the Transferability of Cross Classification Trip Generation Models*, VHTRC 78-R39, Virginia Highway and Transportation Research Council, Charlottesville, Va., 1978.

13. *Computer Programs for Urban Transportation Planning.*

14. *Quick Response Urban Travel Estimation Techniques.*

15. *Computer Programs for Urban Transportation Planning.*

16. Ibid.

17. Peter R. Stopher and Arnim H. Meyberg, *Urban Transportation Modeling and Planning*, Lexington Books, D.C. Heath, Lexington, Mass., 1975.

18. *Mode Choice Forecasting Methodology, Development and Calibration of the Washington Mode Choice Models*, Technical Report 8, R. H. Pratt Associates, Inc., Washington, D.C., June 1973, Appendix II.

19. Marvin L. Manheim, *Fundamentals of Transport Systems Analysis*, MIT Press, Cambridge, Mass., 1979.

20. *Application of Disaggregate Travel Demand Models*, NCHRP Report 253, Transportation Research Board, National Research Council, Washington, D.C., 1982.

21. *New Approaches to Understanding Travel Behavior*, NCHRP Report 250, Transportation Research Board, National Research Council, Washington, D.C., 1982.

22. *Quick Response Urban Travel Estimation Techniques.*

23. Manheim, *Fundamentals of Transport Systems Analysis.*

24. *Application of Disaggregate Travel Demand Models.*

25. *Traffic Assignment and Distribution for Small Urban Areas*, U.S. Department of Commerce, Bureau of Public Roads, Washington, D.C., September 1965.

26. Cosmis Corporation, *Traffic Assignment*, U.S. Department of Transportation, Federal Highway Administration, Washington, D.C., 1974.

27. *UTPS Network Development Manual*, U.S. Department of Transportation, Washington, D.C., 1974.

ADDITIONAL READINGS

Black, John, *Urban Transport Planning*, Johns Hopkins University Press, Baltimore, Md., 1981.

Calibrating and Testing a Gravity Model for Any Size Urban Area, U.S. Department of Transportation, Federal Highway Administration, Washington, D.C., March 1975.

Census Data and Urban Transportation Planning, Special Report 145, Transportation Research Board, National Research Council, Washington, D.C., 1974.

Consequences of Small Sample O–D Data Collection in the Transportation Planning Process, U.S. Department of Transportation, Federal Highway Administration, Washington, D.C., 1976.

Forecasting Impacts to Transportation Planning, NCHRP Report No. 266, Transportation Research Board, National Research Council, Washington, D.C., 1983.

Gooding, D. I., and Robert C. Knighton, *Forecasting Independent Variables: Responsibilities and Techniques*, New York Department of Transportation, Planning Research Unit, Albany, N.Y., July 1976.

Guide for Local Area Population Projections, Technical Paper 39, U.S. Department of Commerce, Bureau of Census, Washington, D.C., July 1977.

Guidelines for Designing Travel Surveys for Statewide Transportation Planning, U.S. Department of Transportation, Federal Highway Administration, Washington, D.C., 1976.

Hutchinson, B. G., *Principles of Urban Transport Systems Planning*, Scripta Book Company, Washington, D.C., 1974.

An Introduction to Urban Development Models and Guidelines for Their Use in Urban Transportation Planning, U.S. Department of Transportation, Federal Highway Administration, Washington, D.C., 1975.

Keller, C. Richard, and Joe Mehra, *Site Impact Traffic Evaluation Handbook*, U.S. Department of Transportation, Federal Highway Administration, Washington, D.C., January 1985.

Mehra, Joe, and C. Richard Keller, *Development of Application of Trip Generation Rates*, FHWA/PL/85/003, U.S. Department of Transportation, Federal Highway Administration, Washington, D.C., January 1985.

Microcomputers in Transportation, Software and Source Book, U.S. Department of Transportation, Washington, D.C., 1986.

Parking Generation (interim report), Institute of Transportation Engineers, Washington, D.C., 1985.

Population Forecasting Methods, U.S. Department of Transportation, Federal Highway Administration, Washington, D.C., 1964.

Putman, Steven H., *Laboratory Testing of Predictive Land Use Models: Some Comparisons*, U.S. Department of Transportation, Washington, D.C., 1976.

Putman, Steven H., *Integrated Policy Analysis of Metropolitan Transportation and Location*, U.S. Department of Transportation, Washington, D.C., 1980.

Report of the Conference on Economic and Demographic Methods for Projections of Population, American Statistical Association, Washington, D.C., 1977.

Trip Generation, 3rd ed. (with updates), Institute of Transportation Engineers, Washington, D.C., 1983.

Urban Development Models, TRB Special Report 97, Transportation Research Board, National Research Council, Washington, D.C., 1968.

Urban Mass Transportation Surveys, U.S. Department of Transportation, Federal Highway Administration and Urban Mass Transit Administration, Washington, D.C., 1972.

Urban Origin-Destination Surveys, U.S. Department of Transportation, Federal Highway Administration, Washington, D.C., 1973.

CHAPTER 11

Evaluating Transportation Alternatives

In the previous chapter we described methods and techniques for establishing the demand for transportation services under a given set of conditions. The results of this process furnish the necessary input data to prepare an evaluation of the relative worth of alternative projects. In this chapter we describe various ways in which transportation project evaluations are carried out.

BASIC ISSUES IN EVALUATION

The basic concept of an evaluation is simple and straightforward, but the actual process itself can be complex and involved. A transportation project is usually proposed because of a perceived problem or need. For example, a project to eliminate a railroad grade crossing may be based on citizen complaints about accidents or time delays at the crossing site. In most instances, there are many ways to solve the problem, and each solution or alternative will result in a unique outcome in terms of project cost and results. In the railroad grade crossing example, one solution would be to install gates and flashing lights; another solution would be to construct a grade-separated overpass. These two solutions are quite different in terms of their costs and effectiveness. The first solution will be less costly than the second, but it will also be less effective in reducing accidents and delays.

A transportation system or project can be viewed as a mechanism for producing a result desired by society *at a price*. The question is, Will the benefits of the project be worth the cost? In some instances, the results may be confined to the users of the system (as in the case of the grade crossing), whereas in other instances, those affected may include persons in the community who do not use the system.

Objectives of Evaluation

The objective of an evaluation is to furnish the appropriate information about the outcome of each alternative so that a selection can be made. The evaluation process should be viewed as an activity in which information relevant to the selection is available to the person or group who will make a decision. An essential input in the process is to know what information will be important in making a project selection. In some instances, a single criterion may be paramount (such as cost); in other cases, there may be many objectives to be achieved. The decision maker may wish to have the relative outcome of each alternative expressed as a single number, whereas at other times, it may be more helpful to see the results individually for each criteria and each alternative.

There are many methods and approaches for preparing a transportation project evaluation, and each one can be useful when correctly applied. In this chapter we first describe the considerations in selecting an evaluation method and discuss issues that are raised in the evaluation process. We then describe two classes of evaluation methods that are based on a single measure of effectiveness: the first reduces all outcomes to a monetary value and the second reduces all outcomes to a single number. Finally, we discuss evaluation as a fact-finding process in which all outcomes are reported separately in a matrix format so that the decision maker has complete information about the project outcome. We also discuss how this information can be used in public forums for citizen input and how the decision process can be extended to include public participation.

Evaluations can also be made after a project is completed to determine if the outcomes for the project were as had been anticipated. Post facto evaluation can be very helpful in formulating information useful for similar decisions made elsewhere or in making modifications in original designs. In the final section of this chapter, we discuss the issues in post facto evaluations and illustrate the results of evaluations for completed projects.

Selecting an Evaluation Procedure

Prior to beginning an evaluation of a transportation alternative, the engineer or planner should consider a number of basic questions and issues. These will assist in determining the proper approach to be taken, what data will be needed, and what analytical techniques should be used. These issues are discussed in the following paragraphs.

Who will use the information and what is their viewpoint? A transportation project can affect a variety of groups in different ways. In some instances, only one or a few groups may be involved; in other cases, many factions might have an interest. Examples of groups that could be affected by a transportation project include the system users, transportation management, labor, citizens in the community, business, and local, state and, national governments. Each of these groups will be concerned with something different, and the viewpoint that each represents will influence the evaluation process itself.

For smaller, self-contained projects, those groups with something to gain or lose by the project—the *stakeholders*—will usually be limited to the system users and transportation management. For larger, regional-scale transportation projects, the number and variety of stakeholders will increase because the project will affect many people beyond the users and management. For example, a major project could increase business in the downtown area, or expanded construction activity could trigger an economic boom in the area. The project might also require the taking of land, or it could create negative environmental effects. Thus, if the viewpoint is that of an individual traveler or business, the analysis can be made on narrow economic grounds. If the viewpoint is the community at large, then the analysis must consider a wider spectrum of concerns.

If the viewpoint is that of a local community, then transfer funds by grants from the state or federal government would not be considered as a cost, but increases in land values within the area would be considered as a benefit. However, if the viewpoint were expanded to a regional or state level, these grants and land value increases would be viewed as costs to the region or as transfers of benefits from one area to another. Thus, a clear definition of whose viewpoint is being considered in the evaluation is necessary if proper consideration is to be given to how these groups are either positively or negatively affected by each proposed alternative.

Selecting and Measuring Evaluation Criteria

What are the relevant criteria and how should these be measured? A transportation project is intended to accomplish one or more goals and objectives. These are expressed as criteria, and the numerical or relative results for each criteria are called measures of effectiveness. For example, in a railroad grade crossing problem, if the goal is to reduce accidents, the criteria can be measured as the number of accidents expected to occur for each of the alternatives considered. If another goal is to reduce waiting time, the criteria could be the number of minutes per vehicle consumed at the grade crossing. Nonquantifiable criteria can also be used and expressed in a relative scale such as high, medium, low, and so forth.

Criteria selection is a basic element of the evaluation process because it becomes the basis on which each project is measured. Thus, it is important that the criteria be related as closely as possible to the stated objective. To use a nontransportation example for illustration, if the objective of a course is to learn traffic and highway engineering, then a relevant criterion to measure results is exam grades, whereas a less relevant criterion is the number of class lectures attended. Both are measures of class performance but the first is more relevant in measuring how well one achieved the stated objective.

Measures of Effectiveness

Another related issue is the use of measures of effectiveness in the evaluation process itself. One approach is to convert each measure of effectiveness to a common unit, and then for each alternative, compute the summation for all measures. A common unit is money, and it may be possible to make a transformation of the relevant criteria to equivalent dollars and then compare each alternative from an economic point of view. For example, if the cost of an accident is known and the value of travel time can be determined, then, for the railroad grade crossing problem, it would be possible to compute a single number that would represent the total cost involved for each alternative, since construction, maintenance, and operating costs are already known in dollar terms, and the accident and time costs can be computed using conversion rates.

Another common-unit approach is to convert each measure of effectiveness to a numerical score. For example, if a project alternative does well in one criterion, it is given a high score, and if it does poorly in another criterion, it is given a low score. A single number can be calculated that represents the weighted average score of all the measures of effectiveness that were considered. This approach is similar to calculating grades in a course. The instructor establishes a set of criteria to measure a student's performance (for example, homework, midterms, class attendance, final, participation, and term paper) and a weight for each. The overall measure of the student's performance is the weighted sum of the outcome for each measure of effectiveness. Measures of effectiveness should be independent of each other if a summation procedure (such as adding grades) is to be used in the evaluation. If the criteria are correlated, then adding up the weighted scores will bias the outcome.

Finally, the measures of effectiveness can simply be reported for each alternative in a matrix form with no attempt made to combine them. This approach furnishes the maximum amount of information without prejudging as to how the measures of effectiveness should be combined or their relative importance.

Criteria not only must be relevant to the problem but must have other attributes as well. They should be easy to measure and be sensitive to changes made in each alternative. Also, it is advisable to limit the number of criteria to those that will be most helpful in reaching a decision in order to keep the problem manageable for both the engineer who is doing the work and the person(s) who will act on the result. Too much information can be confusing and counterproductive and, rather than being helpful, could create uncertainty and division or encourage a decision on political or other nonquantitative basis. Some examples of criteria used in transportation evaluation are listed in Table 11.1.

Table 11.1   Criteria for Evaluating Transportation Alternatives

- Capital Costs
  Construction
  Right of way
  Vehicles
- Maintenance Costs
- Facility Operating Costs
- Travel Time Cost
  Total hours and cost of system travel
  Average door-to-door speed
  Distribution of door-to-door speeds
- Vehicle Operating Costs
- Accident Costs

Evaluation Procedures and Decision Making

A final question is how well the evaluation process will assist in making a decision. The decision maker will, of course, want to know what the costs of the project will be, and in many instances, this alone will determine the outcome. Another question may be, do the benefits justify the expenditure of funds for transportation or would the money be best spent elsewhere? The decision maker will also want to know if the proposed project is likely to produce the stated results, that is, how confident we can be of the predicted outcomes.

It may be necessary to carry out a sensitivity analysis that shows a range of values, rather than a single number. Also, evaluations of similar projects elsewhere may provide clues as to the probable success of the proposed venture. The decision maker may also wish to know if all the alternatives have been considered and how they compare with the one being recommended. Are there better ways of accomplishing the objective, using management and traffic control strategies, that would eliminate the need for costly construction projects? For example, providing separate bus and carpool lanes could result in significantly increasing the passenger-carrying capacity of a freeway without the need to build additional highway lanes.

Another factor is the cost to highway users as the result of travel delays during construction. Also of interest is the length of time necessary to finish the project, since people in public office are often interested in seeing work completed during their administration. The source of funds for the project and other matters dealing with its implementation will also be of concern. Thus, in addition to the fairly straightforward problem of evaluation based on a selected set of measurable criteria, the transportation engineer must really be prepared to answer any and all questions about the project and its implications.

The evaluation process requires that the engineer have all the appropriate facts about a proposed project and be able to convey these in a clear and logical manner so that decision making is facilitated. In addition to the formal numerical summaries of each project (which will be described in the remainder of the chapter), the engineer must also be prepared to answer other questions about the project that relate to its political and financial feasibility. In the final analysis, the selection itself will be based on a variety of factors and considerations that reflect all of the inputs that a decision maker receives from the appropriate source.

EVALUATION BASED ON ECONOMIC CRITERIA

To begin the discussion of economic evaluation, it is helpful to consider the relationship between the supply and demand for transportation services. Consider a particular transportation project, such as a section of roadway or a bridge. Further, assume that we can calculate the cost involved for a motorist to travel on the facility. (These costs would include fuel, tolls, travel time, maintenance, and other actual or perceived out-of-pocket expenses.) Using methods described in Chapter 10, it should be possible to calculate the traffic volumes (or demand) for various values of user cost. We would intuitively expect that as the cost of using the facility decreased, the number of vehicles per day would increase. This relationship is shown schematically in Figure 11.1, and represents the demand curve for the facility for a particular group of motorists. A demand curve could shift upward or downward and have a different slope for users with different incomes or for various trip purposes. If the curve moved upward, it would indicate a greater willingness to pay, reflecting perhaps a group with a higher income. If the slope approached horizontal, it would indicate that demand is elastic, that is, a small change in price would result in a large change in volume. If the slope approached vertical, it would indicate that the demand is inelastic, that is, a large change in price has little effect on demand. As an example, the price of gasoline is said to be inelastic because people continue to drive when gas prices increase.

Consider that the cost to travel on this facility is P_1. Then the number of trips per unit time will be V_1, and the total cost for all users over a given period per hour or day will be $(P_1) \times (V_1)$. This amount can be shown graphically as the area $0P_1D_1V_1$. As can be seen from the demand curve, all but the last user would have been willing to pay more than the actual price. For example, V_0 users would have been willing to pay P_0 to use the facility, whereas they paid the lesser amount P_1. The area under the demand curve $0DD_1V_1$ is the amount that the V_1 users would be willing to pay. If we subtract the area $0P_1D_1V_1$, which is the amount the users actually paid, we are left with area P_1DD_1. This triangular area is referred to as a *consumer surplus* and represents the value of the economic benefit for the current users of the facility.

Suppose the price for using the facility is reduced. to P_2 because of various improvements that have been made to the facility. The total user cost is now equal to $(P_2) \times (V_2)$, and the consumer surplus is the triangular area between the demand curve and P_2D_2 or P_2DD_2. The net benefit of the project is the net increase in consumer surplus, or area P_2DD_2 minus area P_1DD_1 and is represented by the shaded trapezoidal area $P_2P_1D_1D_2$. That area is made up of two parts. The first is the reduction in total cost paid by the original travelers, V_1, and is represented

Figure 11.1   Demand Curve for Travel on a Given Facility

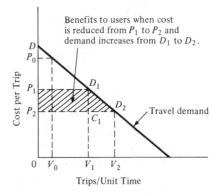

Trips/Unit Time

by the rectangular area $P_2 P_1 D_1 C_1$. The second is the consumer surplus earned by the new users $V_2 - V_1$ and is represented by the triangular area $C_1 D_1 D_2$.

We now have a theoretical basis to calculate the net benefits to users of an improved transportation facility, which can then be compared with the improvement cost. The formula for user benefits is

$$B_{2,1} = \frac{1}{2}(P_1 - P_2)(V_1 + V_2) \tag{11.1}$$

where

$B_{2,1}$ = net benefits to transport users
P_1 = user cost of unimproved facility
P_2 = user cost of improved facility
V_1 = volume of travel on unimproved facility
V_2 = volume of travel on improved facility

In many situations it is not practical to develop a demand curve. In these instances, the value for the volume that is used in economic calculations is taken to be the number of trips that will occur on the improved facility. Eq. 11.2, which replaces the term $\frac{1}{2}(V_1 + V_2)$ of Eq. 11.1 with V_2, has been commonly used in highway engineering studies. This formula will overstate benefits unless demand is inelastic—that is, the demand curve is vertical.

$$B_{2,1} = (P_1 - P_2)(V_2) \tag{11.2}$$

To consider the economic worth of improving this transportation facility, we calculate the cost of the improvement and compare it with the cost of maintaining the facility in its present condition (the do-nothing alternative). One approach is to consider the difference in costs and compare this with the difference in benefits and then to select the project if the net increase in benefits exceeds the net increase in costs. Another approach is to consider the total costs of each alternative, including user and facility costs, and then to select the project that has the lowest total cost. To carry out this type of economic evaluation, it is necessary to develop the elements of cost for both the facility and the users. These include costs for construction, maintenance, and operation and user costs for travel time, accidents, vehicle maintenance, and operation. The elements of cost are discussed in the following section.

Elements of Cost

The cost of a transportation facility improvement includes two components: *first cost* and *continuing costs* for annual maintenance, operation, and administration. Since we are concerned with cost differences, those costs that are common to both projects can be excluded. The first cost for a highway or transit project may include engineering design, right of way, and construction. Each transportation project is unique and the specifics of the design will dictate what items will be required and at what cost. Maintenance and operating costs of the facility must also be determined. These are recurring costs that will be incurred over the life of the facility and are usually based on historical data for similar projects. For example, if one alternative involves the purchase of buses, then the first (or capital) cost is the price of the bus and the operating and maintenance cost will be known from manufacturer data or experience.

Expenses for administration or other overhead charges are usually excluded in an economic evaluation because they will be incurred regardless of whether or not the project is selected. Other excluded costs are those that have already been incurred. These are known as *sunk costs* and as such are not relevant to the decision of what to do in the future since these expenditures have already been made. For most capital projects, a service life must be determined and a salvage value estimated. (Salvage value is the worth of an asset at the end of its service life.) For example, a transit bus may be considered to have a service life of 12 years and a salvage value of $2,000, and a concrete pavement may have a service life of 15 years and no salvage value. Suggested service lives for various facilities can be obtained from various trade or transportation organizations, such as the American Bus Association, the American Association of State Highway and Transportation Officials, or the American Public Transit Association.

Three commonly used measures of user costs are included in a transportation project evaluation: costs for vehicle operation, travel time costs, and costs of accidents. These costs are sometimes referred to as *benefits*, the implication being that the improvements to a transportation facility will reduce the cost for the users—that is, lower the perceived price, as shown on the demand curve— and result in a user benefit. It is simpler to consider these items in terms of cost because the data are needed in this format for purposes of an economic evaluation.

User costs for motor vehicle operation are significant items in a highway project evaluation. For example, a road improvement that eliminates grades, curves, and traffic signals as well as shortening the route can result in major cost reductions to the motorist. A procedure for computing vehicle running costs is furnished by AASHTO.[1] Other agencies, such as the U.S. Department of Transportation and various vehicle manufacturers, furnish data about vehicle costs on highways or at intersections.

One of the most important reasons for making transportation improvements is to increase speed or to reduce travel time. In the world of trade and commerce, time is money, and in recent years new business ventures that furnish overnight delivery of small packages have grown and flourished. Transoceanic airline service replaced steamships because airplanes reduced the time to cross the ocean from 6–9 days to 6–9 hours.

The method of handling travel time savings in an economic analysis has stirred considerable debate over the years. No one disagrees that time savings have an economic value, but the question is how should these be converted to dollar amounts (if at all)? One problem is that a typical stream of traffic contains both private and commercial vehicles, each of which values time savings quite differently. Time savings for a trucking firm can be translated directly into savings in labor cost by using an hourly rate for personnel and equipment. Personal travel, on the other hand, is made for a variety of reasons; some are work related but many are not (shopping, school, social, recreation). Time saved in traveling to and from work can be related to wages earned but time saved in other pursuits has little if any economic basis for conversion.

The value of time saved also depends on the length of trip and family income. If time savings are small—less than 5 minutes—they will not be perceived as significant and therefore have little value. If time savings are above a threshold level where they will make a noticeable difference in total travel time—over 15 minutes—then they could have significant economic value. For example, the American Association of State Highway and Transportation Officials (AASHTO) recommended that for an average income of $20,000, time savings for trips less than 5 minutes are worth only $0.27 per hour, whereas the value increases by a factor of 18, to $2.32 per hour for trips greater than 15 minutes. The problem in using this type of time value data is it is often difficult and costly to develop trip length and income distributions at the level of detail required.

The apparent monetary savings from even small travel time reductions can be quite large. For example, if a highway project that will carry average daily traffic (ADT) of 50,000 autos saves only 2 minutes per traveler and the value of time for the average motorist is estimated conservatively at $5.00 per hour, the total minimum annual saving is $50,000 \times (2/60) \times 365 \times 5 = \$3,041,667$. At 10 percent interest, these savings could justify spending a total of $25.895 million for a 20 year project life. Clearly, this result is an exaggeration of the actual benefits received. Although travel time does represent an economic benefit, the conversion to a dollar value is always open to question. AASHTO has used the average value approach ($3.00 per hour), whereas others would argue that time savings should be credited only for commercial uses or stated simply in terms of actual value of number of hours saved.

Loss of life, injury, and property damage incurred in a transportation accident is a continuing national concern. Following every major air tragedy is an extensive investigation and often expenditures of funds are then authorized to improve the nation's air navigation system. Similarly, the 55 mph speed limit imposed by Congress following the oil crisis of 1973–1974 was retained long after the crisis was ended because it is credited with saving lives on the nation's highways. (As of 1987, states may increase the speed limit to 65 mph on certain highways.) It has been well established that the accident rate per million vehicle miles is substantially lower on limited access highways than on four-lane undivided roads. The total number of accidents has been estimated by AASHTO as 1.07 per 100 million vehicle miles on freeways versus 6.65 on four-lane undivided highways. Reflecting the economic costs of accidents requires an estimate of the number and type of accidents that are likely to occur over the life of the facility and an estimate of the value of each occurrence. Property damage and injury-related accidents can be valued using insurance data. The value of a human life is "priceless" but in economic terms, measures such as future earnings have been used. There is no simple numerical answer to the question, what is the value of a human life or the cost of an accident, although everyone would agree that economic value does exist. Published data vary widely and the most prudent course, if an economic value is desired, is to select a value that appears most appropriate for the given situation.

Economic Evaluation Methods

Four basic methods are used in an economic evaluation of a transportation project: net present worth (NPW), equivalent uniform annual cost (EUAC), benefit-cost ratio (BCR), and internal rate of return (ROR). Each method will produce the same results, and the reason for selecting one over the other is simply a matter of convenience or preference for how the results will be presented. Since transportation projects are usually built to serve traffic over a long period of time, it is necessary to consider the change in the value of money over the life of a project.

Present worth (PW) is perhaps the simplest and most straightforward of the methods, since it represents the current value of all the costs that will be incurred over the lifetime of the project. The general expression for PW of a project is

$$\text{PW} = \sum_{n=0}^{N} \frac{C_n}{(1+i)^n} \tag{11.3}$$

given

$$P/F = \frac{1}{(1+i)^n}$$

where

C_n = facility and user costs incurred in year n

N = service life of the facility

i = rate of interest

P = present worth factor of a future amount F

F = future amount

Eq. 11.3 can be written as

$$P = (P/F)F$$

where P/F is a factor used to determine P given F. Expressing the PW relationship as shown is useful in that the computation becomes a problem of dimensional analysis. The F's on the right hand side of the equation cancel, leaving P as a result.

Net present worth (NPW) is the present worth of a given cash flow that has both receipts and disbursements. The use of an interest rate in an economic evaluation is common practice as it represents the cost of capital. Money spent on a transportation project is no longer available for investment. Therefore, a minimal value of interest rate is the rate that would have been earned if the money were invested elsewhere. For example, if $1000 is deposited in a bank at 8 percent interest, its value in 5 years will be $1000(1 + 0.08)^5 = 1469.33$. Thus the PW of having $1469.33 in 5 years at 8 percent interest is equal to $1000, and the opportunity cost is 8 percent. Discount rates can be higher or lower, depending on risk of investment and economic conditions.

It is helpful to use a cash flow diagram to depict the costs and revenues that will occur over the lifetime of a project. Time is plotted as the horizontal axis and money as the vertical axis, as illustrated in Figure 11.2. Using Eq. 11.4, we can calculate the NPW of the project, which is

$$\text{NPW} = \sum_{n=0}^{N} \frac{R_n}{(1+i)^n} + \frac{S}{(1+i)^N} - \sum_{n=0}^{N} \frac{M_n + O_n + U_n}{(1+i)^n} - C_o \qquad \textbf{(11.4)}$$

where

C_o = initial construction cost
M = maintenance cost
O = operating cost
U = user cost
S = salvage
R = revenues

In this manner we have converted a time stream of costs and revenues into a single number: its PW. The term $1/(1+i)^n$ is known as the PW factor of a single payment and is written as $P/F - i - N$.

Equivalent uniform annual worth (EUAW) is a conversion of a given cash flow to a series of equal annual amounts. If the amounts are considered to occur at the

Figure 11.2 Typical Cash Flow Diagram for a Transportation Alternative and Equivalence as Net Present Worth or Annual Cost

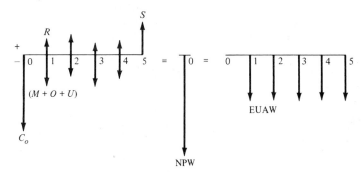

end of the interest period, then the formula is

$$\text{EUAW} = \text{NPW}\left[\frac{i(1+i)^N}{(1+i)^N - 1}\right] = \text{NPW}(A/P - i - N) \tag{11.5}$$

Similarly,

$$\text{NPW} = \text{EUAW}\left[\frac{(1+i)^N - 1}{i(1+i)^N}\right] = \text{EUAW}(P/A - i - N)$$

where

A = uniform payment, an annuity

The term in the brackets in Eq. 11.5 is referred to as the capital recovery factor and represents the amount necessary to repay \$1 if N equal payments are made at interest rate i. For example, if a loan is made for \$5000 to be repaid in equal monthly payments over a 5 year period at 1 percent per month, then the amount is

$$\text{EUAW} = 5000\left[\frac{0.01(1 + 0.01)^{60}}{(1 + 0.01)^{60} - 1}\right] = 5000(0.02225) = 111.25$$

Thus 60 payments of \$111.25 would repay a \$5000 debt, including both principal and interest. We can convert the NPW of a cash flow to an EUAW by multiplying the PW by the capital recovery factor. The inverse of the capital recovery factor is the PW for a uniform series.

$$\begin{aligned}
\text{EUAW} &= \text{NPW}(A/P - i - N) \\
\text{NPW} &= \text{EUAW}(P/A - i - N)
\end{aligned} \tag{11.6}$$

Values of the basic formulas required to convert a monetary value from a future to present time period $(P/F - i - N)$ and from a present time period to equal end-of-period payments $(A/P - i - N)$ are tabulated in textbooks on engineering economics. Table 11.2 lists values of PW factors and capital recovery factors for a selected range of interest rates and time periods.

The *benefit cost ratio* (BCR) is a ratio of the present worth of net project benefits and net project costs. This method is used in situations where it is desired to show the extent to which an investment in a transportation project will result in a benefit to society. To do this, it is necessary to make project comparisons to determine how the added investment compares with the added benefits. The formula for BCR is

$$\text{BCR}_{2/1} = \frac{B_{2/1}}{C_{2/1}}$$

Table 11.2 Present Worth and Capital Recovery Factors

N	i = 5		i = 10		i = 15		i = 20	
	(P/F − 5 − N)	(A/P − 5 − N)	P/F	A/P	P/F	A/P	P/F	A/P
1	0.9524	1.0500	0.9091	1.1000	0.8696	1.1500	0.8333	1.2000
2	0.9070	0.5378	0.8265	0.5762	0.7561	0.6151	0.6945	0.6546
3	0.8638	0.3672	0.7513	0.4021	0.6575	0.4380	0.5787	0.4747
4	0.8227	0.2820	0.6830	0.3155	0.5718	0.3503	0.4823	0.3863
5	0.7835	0.2310	0.6209	0.2638	0.4972	0.2983	0.4019	0.3344
10	0.6139	0.1295	0.3856	0.1628	0.2472	0.1993	0.1615	0.2385
15	0.4810	0.0963	0.2394	0.1315	0.1229	0.1710	0.0649	0.2129
20	0.3769	0.0802	0.1487	0.1175	0.0611	0.1598	0.0261	0.2054

where

$B_{2/1}$ = reduction in user and operation costs between higher cost alternative 2 and lower cost alternative 1, expressed in PW or EUAW

$C_{2/1}$ = increase in facility costs expressed as PW or EUAW

If the BCR is 1 or greater, then the higher cost alternative is economically attractive. If the BCR is less than 1, this alternative is discarded.

Correct application of the BCR method requires first that costs for each alternative be converted to PW or EUAW values. The proposals must be ranked in ascending order of capital cost, including the do-nothing alternative, which usually has little if any initial cost. Incremental BCRs are calculated for pairs of projects, beginning with the lowest cost alternative. If the higher cost alternative yields a BCR less than 1, it is eliminated and the next higher cost alternative is compared with the lower cost alternative. If the higher cost alternative yields a BCR equal to or greater than 1, it is retained and the lower cost alternative is eliminated. This process continues until every alternative has been compared. The alternative selected is the one with the highest initial cost and a BCR of 1 or more with respect to lower cost alternatives and a BCR less than 1 for higher cost projects.

The *internal rate-of-return* (ROR) method determines the interest rate at which the PW of reductions in user and operation costs $B_{2/1}$ equals the PW of increases in facility costs $C_{2/1}$. If the ROR exceeds the interest rate (referred to as minimum attractive rate of return), the higher cost project is retained. If the ROR is less than the interest rate, the higher priced project is eliminated. The procedure for comparison is similar to that used in the BCR method.

Example 11–1 Illustration of Economic Analysis Methods

The Department of Traffic is considering three improvement plans for a heavily traveled intersection within the city. The intersection improvement is expected to achieve three goals: improve travel speeds, increase safety, and reduce operating expenses for motorists. The annual dollar value of savings compared with existing conditions for each criteria as well as additional construction and maintenance costs shown in Table 11.3. If the economic life of the road is considered to be 50 years and the discount rate is 3 percent, which alternative should be selected?

Table 11.3 Cost and Benefits for Alternative Improvement Plans

Alternative	Construction Cost	Annual Saving in Accidents*	Annual Travel Time Benefits*	Annual Operating Savings*	Annual Additional Maintenance Cost*
I	$185,000	$5,000	$3,000	$ 500	$1,500
II	220,000	5,000	6,500	500	2,500
III	310,000	7,000	6,000	2,800	3,000

*Compared with existing (do-nothing) conditions.

- Compute the NPW of each project.

$$(P/A - 3 - 50) = \frac{(1+i)^N - 1}{i(1+i)^N} = \frac{(1+0.03)^{50} - 1}{0.03(1+0.03)^{50}} = 25.729$$

$$\begin{aligned}
\text{NPW}_I &= -185{,}000 + (-1500 + 5000 + 3000 + 500)(P/A - 3 - 50) \\
&= -185{,}000 + (7000)(25.729) = -185{,}000 + 180{,}103 \\
&= -4897
\end{aligned}$$

$$\begin{aligned}
\text{NPW}_{II} &= -220{,}000 + (-2500 + 5000 + 6500 + 500)(P/A - 3 - 50) \\
&= -220{,}000 + (9500)(25.729) = -220{,}000 + 244{,}425 \\
&= +24{,}465
\end{aligned}$$

$$\begin{aligned}
\text{NPW}_{III} &= -310{,}000 + (-3000 + 7000 + 6000 + 2800)(P/A - 3 - 50) \\
&= -310{,}000 + (12{,}800)(25.729) = -310{,}000 + 329{,}331 \\
&= +19{,}331
\end{aligned}$$

The project with the highest NPW, is alternative II.

- Solve by the EUAW method. Note $(A/P - 3 - 50) = 1/25.729$.

$$\begin{aligned}
\text{EUAW}_I &= -185{,}000(A/P - 3 - 50) - 1500 + 5000 + 3000 + 500 \\
&= -185{,}000(0.03887) + 7000 \\
&= -7190 + 7000 = -190
\end{aligned}$$

$$\begin{aligned}
\text{EUAW}_{II} &= -220{,}000(A/P - 3 - 50) - 2500 + 5000 + 6500 + 500 \\
&= 220{,}000(0.03887) + 9500 \\
&= -8551 + 9500 = +949
\end{aligned}$$

$$\begin{aligned}
\text{EUAW}_{III} &= -310{,}000(0.03887) - 3000 + 7000 + 6000 + 2800 \\
&= 12{,}050 + 12{,}800 = +750
\end{aligned}$$

The project with the highest EUAW is alternative II, which is as expected since EUAW = NPW(0.03887).

- Solve by the BCR method.
  1. Compare the BCR of alternative I with respect to do nothing (DN).

$$\text{BCR}_{I/DN} = \frac{180{,}103}{185{,}000} = 0.97$$

Since $\text{BCR}_{I/DN}$ is less than 1, we would not build alternative I.

2. Compare BCR of alternative II with respect to DN.

$$\text{BCR}_{\text{II/DN}} = \frac{244,425}{220,000} = 1.11$$

Since BCR > 1, we would select alternative II over DN.

3. Compare BCR of alternative III with respect to alternative II.

$$\text{BCR} = \frac{(329,331) - (244,425)}{(310,000) - (220,000)} = \frac{84,906}{90,000} = 0.94$$

Since BCR is less than 1, we would not select alternative III, and we reach the same conclusion as previously, which is to select alternate II.

■ Solve by the ROR method. In this situation, we solve for the value of interest rate for which NPW = 0.

1. Compute ROR for alternative I versus DN. (Recall that all values are with respect to existing conditions.)

$$\text{NPW} = 0 = -185,000 + (-1500 + 5000 + 3000 + 500)$$
$$\times (P/A - i - 50)$$
$$(P/A - i - 50) = 185,000/7000$$
$$(P/A - i - 50) = 26.428$$
$$i = 2.6 \text{ percent}$$

Since the ROR is lower than 3 percent we discard alternative I.

2. Compute ROR for alternative II versus DN.

$$\text{NPW} = 0 = -220,000 + (-2500 + 5000 + 6500 + 500)$$
$$\times (P/A - i - 50)$$
$$(P/A - i - 50) = 220,000/9500$$
$$(P/A - i - 50) = 23.16$$
$$i = 3.6 \text{ percent}$$

Since ROR is greater than 3 percent, select alternative II over DN.

3. Compute ROR for alternative III versus alternative II.

$$NPW = 0 = -(310,000 - 220,000) + (12,800 - 9500)$$
$$\times (P/A - i - 50)$$
$$(P/A - i - 50) = 90,000/3300$$
$$(P/A - i - 50) = 27.27$$
$$i = 2.7 \text{ percent}$$

Since the increased investment in alternative III yields an ROR less than 3 percent, we do not select it but again pick alternative II.

The preceding example illustrates the basic procedures used in an economic evaluation. Four separate methods were used, each producing the same result. The PW method is simplest to understand and apply and is recommended for most purposes when the economic lives of each alternative are equal. The BCR gives less information than meets the eye and must be carefully applied if it is to produce the correct answer. (For example, the alternative with the highest BCR with respect to the do-nothing case is not necessarily the best.) The ROR method requires more calculations but does provide additional information. For example, the highest ROR in the preceding problem was 3.6 percent. This says that if the minimum attractive ROR were greater than 3.6 percent (say 5 percent), none of the projects would be economically attractive.

EVALUATION BASED ON MULTIPLE CRITERIA

The many problems associated with economic methods limit their usefulness. Among these are

- Converting criteria directly into dollar amounts (a questionable practice)
- Choosing the appropriate value of interest rate and service life, which can influence the result
- Distinguishing between the user groups that benefit from a project and those who pay
- Failing to distinguish between user and other groups who benefit and those who lose by the project
- Including all costs, even external costs

For these reasons, economic evaluation methods should be used either in narrowly focused projects or as one of many inputs in larger projects. The next section discusses evaluation methods that seek to include measurable criteria that are not translated only into monetary terms.

Rating and Ranking

In cases where criteria values cannot be transformed into monetary amounts, using numerical scores is helpful in comparing the relative worth of alternatives. This approach is commonly used in many applications. The basic equation states:

$$S_i = \sum_{j=1}^{N} K_j V_{ij} \qquad\qquad (11.7)$$

where

S_i = total value of score of alternative i

K_j = weight placed on criteria j

V_{ij} = relative value achieved by criteria j for alternative i

The application of this method is illustrated by the following example.[2]

Example 11–2 Evaluating Light-Rail Transit Alternatives Using the Rating and Ranking Method

A transportation agency is considering the construction of a light-rail transit line from the center of town to a growing suburban region. The transit agency wishes to examine five alternative alignments, each of which has advantages and disadvantages in terms of cost, ridership, and service provided. The alternatives differ in length of the line, location, types of vehicles used, seating arrangements, operating speeds, and numbers of stops. The agency wants to evaluate each alternative using a six-step ranking process.

Step 1. Identify the goals and objectives of the project. The transit agency has determined that five major objectives should be achieved by the new transit line.

1. Net revenue generated by fares should be as large as possible with respect to the capital investment.
2. Ridership on the transit line should be maximized.
3. Service on the system should be comfortable and convenient.
4. The transit line should extend as far as possible to promote development and accessibility.
5. The transit line should divert as many auto users as possible during the peak hour in order to reduce highway congestion.

Step 2. Develop the alternatives that will be tested. In this case five alternatives have been identified as feasible candidates. These vary in length from 5 to 8 miles and will cost between $4.5 and $6.0 million per year over the lifetime of the project, at an interest rate of 8 percent. The alignment, the amount of the system below, at, and above grade, vehicle size, headways, number of trains, and other physical and operational features of the line are determined in this step.

Step 3. Define an appropriate measure of effectiveness for each objective. For the objectives listed in step 1, the following measures of effectiveness are selected.

Objective	Measure of Effectiveness
1	Net annual revenue divided by annual capital cost
2	Total daily ridership
3	Percent of riders seated during the peak hour
4	Miles of extension into the corridor
5	Number of auto drivers diverted to transit

Step 4. Determine the relative weight for each objective. This step requires a subjective judgment on the part of the group making the evaluation and will vary among individuals and vested interests. One approach is to allocate the weights on a 100 point scale (just as would be done in developing final grade averages for a course). Another approach is to rank each objective in order of importance and then use a formula of proportionality to obtain relative weights. In this example, the objectives are ranked as shown in Table 11.4. The weighting factor is determined by assigning the value n to the highest ranked alternative, $n - 1$ to the next highest, and so forth and computing a relative weight as

$$K_j = \frac{W_j}{\sum_{j=1}^{n} W_j}$$

where

K_j = weighting factor of objective j
W_j = relative weight for objective j

The resulting values for each objective are shown in Table 11.4. Objective 1, which is to generate revenue, will be worth 30 points, whereas objective 5, which is to divert auto drivers, is weighted 12 points. Other weighting methods, such as by ballot or group consensus, could be used. It is not necessary to use weights that total 100 as any range of values can be selected, and the final results normalized to 100 at the end of the process.

Table 11.4 Ranking and Weights for Each Objective

Objective	Ranking	Relative Weight	Weighting Factor ($\times 100$)
1	1	5	30
2	2	4	24
3	3	3	17
4	3	3	17
5	4	2	12
Total		17	100

Step 5. Determine the value of each measure of effectiveness. In this step the measures of effectiveness are calculated for each alternative. Techniques for demand estimation, as described in Chapter 10, are used to obtain daily and hourly ridership on the line. Cost estimates are developed based on the length of line, number of vehicles and stations, right of way costs, electrification, and so forth. Revenues are computed and ridership volumes during the peak hour are estimated. In some instances, forecasts are difficult to make and a best or most likely estimate is produced. Since it is the relative performance of each alternative that is of interest, relative values of performance measures can be used. The estimated values for each measure of effectiveness and each alternative are shown in Table 11.5.

Table 11.5 Estimated Values for Measures of Effectiveness

Measure of Effectiveness	Alternatives				
	I	II	III	IV	V
Annual return on investment (%)	13.0	14.0	11.0	13.5	15
Daily ridership (000)	25	23	20	18	17
Passengers seated in peak hour (%)	25	35	40	50	50
Length of line (mi)	8	7	6	5	5
Auto drivers diverted (000)	3.5	3.0	2.0	1.5	1.5

Step 6. Compute a score and ranking for each alternative. The score for each alternative is computed by considering each measure of effectiveness and awarding the maximum score to the alternative with the highest value and a proportionate amount to the other alternatives. Consider the first criterion, return on

Table 11.6 Point Score for Candidate Transit Lines

Measure of Effectiveness	Alternatives				
	I	II	III	IV	V
1	26.0	28.0	22.0	27.0	30.0
2	24.0	22.1	19.2	17.3	16.3
3	8.5	11.9	13.6	17.0	17.0
4	17.0	14.9	12.8	10.6	10.6
5	12.0	10.3	6.9	5.1	5.1
Total	87.5	87.2	74.5	77.0	79.0

investment. Alternative V achieves the highest value and is awarded 30 points. The value for alternative I is calculated as $(13/15)(30) = 26$. The results are shown in Table 11.6. (An alternative approach is to award the maximum points to the highest valued alternative and zero points to the lowest.)

The ranking of the alternatives in order of preference is I, II, V, IV, and III. Alternatives I and II are clearly superior to the others and are very similar in ranking. These two will bear further investigation prior to making a decision.

Ranking and rating evaluation is an attractive approach because it can accommodate a wide variety of criteria and can incorporate various viewpoints. Reducing all inputs to a single number is a convenient way to rate the alternatives. The principal disadvantage is that the dependence on a numerical outcome masks the major issues underlying the selection and the tradeoffs involved.

Another problem with ranking methods is that the mathematical form for the rating value (Eq. 11.7) is a summation of the products of the criteria weight and the relative value. For this mathematical operation to be correct, the scale of measurement must be a constant interval (for example, temperature). If the ranking values are ordinal (such as numbering of a sports team), the ranking formula cannot be used.

Also, the rank order could be reversed by changing the ranking of the objectives. For example, if the short line were ranked highest (since it cost less), then the total rating would be reversed with $S_I = 81.8$ and $S_V = 85.4$.

Finally, there is the problem of communicating the final results to decision makers. Given the numbers, the interpretation is often difficult to visualize. People think in concrete terms and are only able to judge alternatives when they are presented realistically.

In the next section we describe a more general and comprehensive approach to evaluation that furnishes information for decision making but stops short of computing numerical values for each alternative.

Cost Effectiveness

Cost effectiveness attempts to be comprehensive in its approach while using the best attributes of economic evaluation. In this method, the criteria that reflect the goals of the project are listed separately from project costs. Thus, the project criteria are considered to be measures of its effectiveness and the costs are considered as the investment required if that effectiveness value is to be achieved. This approach uses data from economic analysis but permits other intangible effects, such as environmental consequences, to be measured as well. The following example illustrates the use of the cost-effectiveness method.[3]

Example 11–3 Evaluating Metropolitan Transportation Plans Using Cost Effectiveness

Five alternative system plans are being considered for a major metropolitan area. They are intended to provide added capacity, improved levels of service, and

Table 11.7 Benefit-Cost Comparisons for Highway and Transit Alternatives

Plan Comparisons	Annual Cost Difference ($ million)	Annual Savings ($ million)	BCR
A versus B	28.58	21.26	0.74
A versus C	104.14	116.15	1.12
C versus D	22.66	17.16	0.76
C versus E	16.73	19.75	1.18

Source: Adapted from *Alternative Multimodal Passenger Transportation Systems*, NCHRP Report 146, Transportation Research Board, Washington, D.C., 1973.

reductions in travel time during peak hours. Plan A retains the status quo with no major improvements; Plan B is an all-rail system; Plan C is all highways; Plan D is a mix of rail transit and highways; and Plan E is a mix of express buses and highways. An economic evaluation has been completed for the project with the results shown in Table 11.7.

Plan B, the all-rail system, and Plan D, the combination rail and highway system, have BCRs of less than 1, whereas Plan C, all highways, and Plan E, highways and express buses, have BCRs greater than 1. These results would suggest that the highway–bus alternative is preferable to the highway–rail transit alternatives. To examine these options more fully, noneconomic impacts have been determined for each and are displayed as an evaluation matrix in Table 11.8.

Table 11.8 Measure of Effectiveness Data for Alternative Highway–Transit Plans

Measure of Effectiveness	Plan A Null	Plan B All Rail	Plan C All Highway	Plan D Rail and Highway	Plan E Bus and Highway
Persons displaced	0	660	8000	8000	8000
Businesses displaced	0	15	183	183	183
Fatal accidents	159	158	137	136	134
Personal injuries	6767	6714	5596	5544	5517
Emission of CO (tons)	2396	2383	2233	2222	2215
Emissions of HCD (tons)	204	203	190	189	188
Average door-to-door trip speed (mph)	15.9	16.2	21.0	21.2	21.5
Average transit trip speed (mph)	6.8	7.6	6.8	7.6	7.8
Annual transit passengers (millions)	154.2	161.7	154.2	161.7	165.2
Annual cost ($ millions)	2.58	31.16	106.72	129.38	123.44
Interest rate	8	8	8	8	8

Source: Adapted from *Alternative Multimodal Passenger Transportation Systems*, NCHRP Report 146, Transportation Research Board, Washington, D.C., 1973.

Figure 11.3 Relationship Between Annual Cost and Passengers Carried

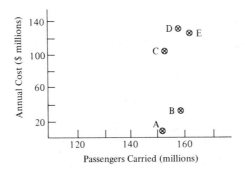

Among the measures of interest are numbers of persons and businesses displaced, number of fatal and personal injury accidents, emissions of carbon monoxide and hydrocarbons, and average travel speeds by highway and transit.

An examination of Table 11.8 yields several observations. In terms of number of transit passengers carried, Plan E ranks highest, followed by Plan B. The relationship between annual cost and transit passengers carried is illustrated in Figure 11.3. This cost-effectiveness analysis indicates that Plan B produces a significant increase in transit passengers over Plan A, Although plans C, D, and E are much more costly, they do not produce many more transit riders for the added investment.

Community impacts are reflected in the number of homes and businesses displaced and the extent of environmental pollution. Figure 11.4 illustrates the results for number of businesses displaced, and Figure 11.5 depicts the results for emissions by hydrocarbons.

In terms of businesses displaced versus transit passengers carried, Plans C, D, and E require considerable displacement of property owners with very little increase in transit patronage. Plan B is clearly preferred if the impact on the community is to be minimized. On the other hand, Plan C, which is considerably more costly than Plan B, results in a significant reduction in pollution levels, whereas

Figure 11.4 Relationship Between Passengers Carried and Businesses Displaced

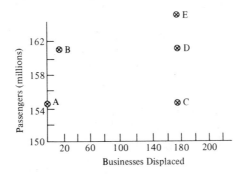

Figure 11.5   Annual Cost Versus Hydrocarbon Emissions

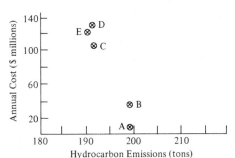

the other two plans, D and E, although more expensive than C, have little further impact on pollution levels.

The items described are but a few of the many relationships that could be examined. They do, however, illustrate the cost-effectiveness procedure and the various conflicting tradeoffs that can result. One conclusion that seems evident is that, although the BCR for Plan B is less than 1, this plan bears further investigation since it produces several environmentally and socially beneficial effects at a relatively low cost. A sensitivity analysis of the benefit-cost study would show that if the interest rate were reduced to 4 percent or the value of travel time were increased by $0.30 per hour, the rail transit plan, B, would have a BCR greater than 1.

The cost-effectiveness approach does not yield a recommended result as do economic methods or ranking schemes. However, it is a valuable tool because it defines more fully the impacts of each course of action and helps to clarify the issues. With more complete information should come a better decision. Rather than closing out the analysis, the approach opens it up and permits a wide variety of factors to be considered.

Evaluation as a Fact-Finding Process

The preceding discussion of economic and rating methods for evaluation has illustrated the technique and application of these approaches. These so-called rational methods are inadequate when the transportation alternatives create a large number of impacts on a wide variety of individuals and groups. Under these conditions, the evaluation process is primarily one of fact finding to provide the essential information from which a decision can be made. The evaluation procedure for complex projects is illustrated in Figure 11.6 and should include four activities that follow the development and organization of basic data and the identification of the major problems or issues that must be addressed in the evaluation process.[4] The activities are as follows.

Figure 11.6   Evaluation Procedure for Complex Projects

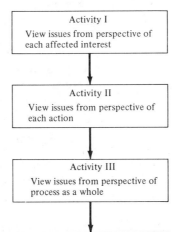

Source: Redrawn from *Transportation Decision Making*, NCHRP Report 156, Transportation Research Board, Washington, D.C., 1975.

Activity 1.   View the issues from the perspective of each affected interest group. The thrust of this activity is to view the consequences of proposed alternatives as they affect particular groups and individuals and as those groups perceive them. For each interest group, the information should be examined and a statement prepared that indicates how that group will react to each alternative. This step can be considered as a means of understanding where each of the stakeholders are coming from, what their biases, likes, and dislikes are, what problems they represent, and so forth.

Activity 2.   View the issues from the perspective of each alternative. The purpose of this activity is to describe each alternative in terms of its advantages and disadvantages. Each proposed project is discussed from the point of view of community concern, feasibility, equity, and potential acceptability.

Activity 3.   View the issues from the perspective of the process as a whole. In this step, all of the alternatives and each of the issues (criteria) are examined together to see if patterns develop from which general statements can be made. For example, there may be one or more alternatives that prove to have so many disadvantages that they can be eliminated. There may be several groups who share the same viewpoint or who are in fierce opposition. Compromise solutions may emerge that will reconcile conflicts, or there may be popular alternatives that generate little controversy.

Activity 4.   Summarize the results of the previous activities. In this activity, the result of the evaluation is documented for use by decision makers and other interested individuals. The report should include a description of the alternatives, the major advantages and disadvantages of each, major areas of conflict and agreement, and the alternatives that have the greatest potential for success in accomplishing major objectives and achieving public acceptance.

The above process is comprehensive and goes beyond a simple listing of criteria and alternatives. One essential feature of the approach is that it requires the analyst to furnish values of each measure of effectiveness for each alternative, without attempting to reduce the results to a single numerical value. The information is then used to make judgments about the relative merits of each alternative.

Trade-Off and Balance Sheet Approaches

According to Lockwood and Wagner, in any evaluation process there are several important conditions that must be met if the difference among alternative projects is to be adequately considered.[5]

1. All alternatives should be evaluated in a framework of common objectives. Measures of effectiveness should be derived from the objectives covering all impact areas.
2. The incidence and timing of impacts on groups and areas should be identified for all impact categories.
3. Standards or accepted impact significance thresholds for measures of effectiveness should be indicated where accepted or required by law.
4. All measures of effectiveness should be treated at an equal level of detail and appropriate scale.
5. Uncertainties or probabilities or both should be expressed for each impact category.
6. A sensitivity analysis of key impacts to variations in major dimensions of alternatives is important for understanding impacts.

Balance-sheet or trade-off approaches satisfy these criteria in contrast to economic or weighting schemes, which provide little in the way of comparative information. These approaches display the impacts of plan alternatives to various groups. The method is based on the viewpoint that individuals or groups that review the data will introduce their own sets of values and weights, and then reach a judgment based on the merits of each alternative using all the data in a disaggregated fashion. Each impact category or goal may have more than one measure of effectiveness. To determine the cost effectiveness of each impact category in a balance-sheet framework, it is useful to compare proposals with the do-nothing alternative. In this way, the positive and negative impacts of not constructing a new project are fully understood.

Evaluation of Completed Projects

The material on evaluation discussed thus far has dealt with the evaluation of plans, and the focus has been to answer the "what if" questions in sufficient detail so that a good decision will be made. Another form of evaluation is to examine the results of a project after it has been implemented to determine (1) how effective it has been in accomplishing its objectives, (2) what can be learned that is useful for other project decisions, (3) what changes should be made to improve the current situation, or (4) if the project should be continued or abandoned.

The subject of postevaluation of transportation projects is closely related to the more general topic of experimental design. If, for example, the physical effects of a particular medical treatment are to be determined, the typical procedure is to identify two different groups with similar characteristics, one serves as a control group and the other as the experimental group. Then the treatment is applied to the experimental group but not to the control group. Differences are measured before and after treatment for both groups, and the net effect is considered to be the result of the treatment itself.

Example 11–4   Evaluating a Completed Transit Project

A transit authority wishes to evaluate the effectiveness of new bus shelters on transit ridership as well as acceptance by the community. A series of new shelters were built along one bus route but not on the other lines. Bus ridership was measured before and after the shelters had been installed on the test line and on a control line where nothing new had been added. The ridership results are shown in Table 11.9. The line with new shelters increased ridership by 13.3 percent, whereas the line without shelters increased by only 2.5 percent. In the absence of any other factors, we can conclude that the effect of the new shelters was to increase ridership by $(13.3 - 2.5) = 10.8$ percent.

Table 11.9   Transit Ridership

	Before	After	Change (%)
Line A: new shelters	1500	1700	13.3
Line B: no shelters	1950	2000	2.5

Another tool for evaluation of completed projects is to conduct a survey of users of the new facility. The questionnaire can probe in greater depth why the person is using the facility and to what extent the project improvement influenced the choice to use the system. In the bus shelter example, a survey of bus riders would inquire if passengers were long time bus riders or are new riders. If new riders, the survey would ask the reasons for riding to find out how many of the new riders considered the new shelters a factor. (Other reasons for riding could be new in the neighborhood, car being repaired, gasoline prices have just gone up, and so forth.) The survey would also ask the old riders to comment on the new shelters. Thus, the survey would corroborate the before and after ridership data as well as furnish additional information about the riders themselves and how they reacted to the new project. This set of complete information would be very valuable in deciding whether or not to implement a bus shelter program for the entire city.

In the transportation field, it is difficult to achieve a "neat" experimental design with a well-defined control and experimental group because (1) transportation projects influence a wide range of outcomes, (2) implementation times are very long and therefore funds are not available to gather before data, (3) a control group is difficult to identify, and (4) changes that occur over a long time period are difficult to connect with a single event, such as a new transportation system. An example of a postevaluation for a major transportation project is the Bay Area Rapid Transit impact study. The approach to postevaluation in this instance was to predict the effect on the region of measures such as air pollution, noise, travel time, and so forth without the rail transit system and then measure the actual amounts with the rapid transit system in place. In this approach, a control group was impossible to obtain but in its place a forecast was made of conditions in the region if a rapid transit system had not been constructed. The obvious difficulty with this method is that it depends on the accuracy of the forecast and must take into account all the changes that have occurred in the region that might have an effect on the impact measures of interest.

Another type of transportation project postevaluation is to make comparisons between different systems or technologies that serve similar travel markets. These comparisons can be helpful to decision makers in other localities because they furnish useful information about what happened in an actual situation. A postevaluation study can be useful because it can consider the actual results for many variables, whereas a preproject evaluation of a mode or technology tends to focus primarily on cost factors or is based on hypothetical situations that require many questionable assumptions.

An example of this type of postproject evaluation is a comparison of a rail transit line serving downtown Philadelphia and a suburb of New Jersey with an express bus line connecting downtown Washington, D.C. with the Virginia suburbs.[6] The rail line, known as the Lindenwold Line, serves 12 stations with 24 hour service per day, whereas the busway, known as the Shirley Highway, extends for 11 miles, with no stations along the way and with bus service provided on exclusive lanes only during the peak hour. Both systems serve relatively low-density, auto-oriented residential areas with heavy travel during the peak hours. A comparative analysis of each project was made after they had been in operation for several years. Measures of effectiveness were considered from the viewpoint of the passenger, the operator, and the community. Data were collected for each system and for each measure of effectiveness. A detailed evaluation for each parameter was prepared that described how each system performed and discussed its advantages and disadvantages. To illustrate, consider the evaluation of one service parameter—*reliability*—expressed as schedule adherence. The variance from scheduled travel times may result from traffic delays, vehicle breakdowns, or adverse weather conditions. It depends mostly on the control that exists over the system. By far the most significant factor for reliability is operation on private rights of way.

Table 11.10 Comparative Evaluation of Completed Rail and Bus Transit

Measure of Effectiveness	Lindenwold (Rail)	Shirley (Bus)	Higher Rated System
Availability	Good	Poor	L
Absolute travel time	Very good	Good	L
Reliability	Very good	Poor	L
Comfort	Good	Poor	L
Convenience	Good	Fair	L
Safety and security	Very good	Good	L
Area coverage	Good	Very good	S
Frequency	Very good	Very poor	L
Investment cost	Very poor	Fair	S
Operating cost	Good	Fair	L
Capacity	Good	Poor	L
Passenger attraction	Very good	Good	L
System impact	Very good	Good	L

Source: Adapted from V. R. Vuchic and R. M. Stanger, "Lindenwold Rail Line and Shirley Busway: A Comparison," *Highway Research Record*, HRR No. 459, Highway Research Board, Washington, D.C., 1973.

- *Lindenwold:* That year 99.15 percent of all trains ran less than 5 min late, and the following year the figure was 97 percent.
- *Shirley:* Surveys conducted over a 4 day period indicated that 22 percent arrived before schedule time, 32 percent were more than 6 minutes late and only 46 percent arrived at the scheduled time within a 5 minute period.
- *Comparison:* The Lindenwold Line (fixed rail) is superior to the Shirley busway (bus on freeway) with respect to reliability.

A summary of the comparative evaluations of the two completed systems is shown in Table 11.10.

A detailed analysis of the results would indicate that each system has advantages and disadvantages. The principal reasons why the rail system appears more attractive than the bus is because it provides all day service, is simpler to understand and use, and produces a higher quality of service.

SUMMARY

The evaluation process for selecting a transportation project has been described. Various methods have been presented that, when used in the proper context, can assist a decision maker in making a selection. The most important attribute of an evaluation method is its ability to correctly describe the outcomes of a given alternative. The evaluation process begins with a statement of the goals and objectives of the proposed project and these are converted into measures of effectiveness. Evaluation methods differ by the way in which measures of effectiveness are considered.

Economic evaluation methods require that each measure of effectiveness be converted into dollar units. Numerical ranking methods require that each measure of effectiveness be translated to an equivalent score. Both methods produce a single number to indicate the total worth of the project. Cost-effectiveness methods require only that each measure of effectiveness be displayed in matrix form, and it is the analysts' task to develop relationships between various impacts and the costs involved. For projects with many impacts that will influence a wide variety of individuals and groups, the evaluation process is essentially one of fact finding, and the projects must be considered from the viewpoint of the stakeholders and community. The reasons for selecting a project will include many factors in addition to simply how the project performs. A decision maker must consider issues such as implementation, schedules, financing, and legal and political matters.

When a project has been completed, a postevaluation can be a useful means to examine the effectiveness of the results. To conduct a postevaluation, it is necessary to separate the effect of the project on each measure of effectiveness from other effects. A standard procedure is the use of a control group for comparative purposes, but this is usually not possible for most transportation projects. A typical procedure is to compare the results with a forecast of the region without the project in place. Postevaluations can also be used to compare alternative modes and technologies, using a wide range of measures of effectiveness.

The bottom line in an evaluation procedure is its effectiveness in assisting decision makers to arrive at a solution that will best accomplish the intended goals.

PROBLEMS

11-1 Describe four basic issues that should be considered prior to selection of an evaluation procedure.

11-2 List the basic criteria used for evaluating transportation alternatives. What units are used for measurement?

11-3 Average demand on a rural roadway ranges from zero to 500 vehicles per day when the cost per trip goes from $1.50 to zero.

(a) Calculate the net user benefits per year if the cost decreases from $1.00 to $0.75 per trip (assume a linear demand function).

(b) Compare the value calculated in (a) with the benefits as calculated in typical highway studies.

11-4 Estimate the average unit costs for (a) operating a standard vehicle on a level roadway, (b) travel time for a truck company, (c) single vehicle property damage, (d) personal injury, and (e) fatality.

11-5 Derive the equation to compute the equivalent annual cost given the capital cost of a highway, such that $A = (A/P) \times P$, where A/P is the capital recovery factor. Compute the equivalent annual cost if the capital cost of a transportation project is $100,000, interest = 10 percent, and $n = 15$ years.

11–6 A highway project is expected to cost $1,500,000 initially. The annual operating and maintenance cost after the first year is $2,000 and will increase by $250 each year for the next 10 years. At the end of the fifth year the project must be resurfaced at a cost of $300,000.

(a) Calculate the present worth of costs for this project if the interest rate is 8 percent.

(b) Convert the value obtained in **(a)** to equivalent uniform annual costs.

11–7 Three designs have been proposed to improve traffic flow at a major intersection in a heavily traveled suburban area. The first alternative involves improved traffic signaling. The second alternative includes traffic signal improvements and intersection widening for exclusive left turns. The third alternative includes extensive reconstruction, including a grade separation structure. The construction costs are listed in the following table for each alternative as well as annual maintenance and user costs. Determine which alternative is preferred, based on economic criteria if the analysis period is 20 years and the interest rate is 15 percent. Show that the result is the same using the present worth, equivalent annual cost, benefit-cost ratio, and rate of return methods.

Alternative	Capital Cost ($)	Annual Maintenance ($)	Annual User Cost ($)	Salvage Value ($)
Present condition	—	15,000	500,000	—
Traffic signals	440,000	10,000	401,000	15,000
Intersection widening	790,000	9,000	350,000	11,000
Grade separation	1,250,000	8,000	301,000	—

11–8 The light-rail transit line described in this chapter is being evaluated by another group of stakeholders. In examining the objectives, they place the following rankings on each as follows.

Objective	Ranking
1	5
2	3
3	1
4	2
5	4

Using this revised information, determine the weighted score for each alternative and comment on your result.

11–9 You have been hired as a consultant to a medium-sized city to develop and implement a procedure for evaluating whether or not to build a highway bypass around the CBD. Write a short report describing your proposal and recommendation as to how the city should proceed with this process.

11–10 The following data have been developed for four alternative transportation plans for a high-speed transit line that will connect a major airport with the downtown area of a large city. Prepare an evaluation report for these proposals by considering the cost effectiveness of each attribute. Show your results in graphical form and comment on each proposal.

		Rail Alternatives			
Measure of Effectiveness	Existing Service	Plan A	Plan B	Plan C	Plan D
Persons displaced	0	264	3200	3200	3200
Businesses displaced	0	23	275	275	275
Average door-to-door trip speed (mph)	10.2	38	45	46	48
Annual passengers (millions)	118.6	124.4	118.6	124.4	127.0
Annual cost (millions)	—	16.4	20.2	23.8	22.7

11–11 A new carpool lane has replaced one lane of an existing 6-lane highway. During peak hours the lane is restricted to cars carrying three or more passengers. After five months of operation, the carpool lane handles 800 autos per hour, whereas the existing lanes are operating at capacity levels of 1500 vph/ln at occupancy rate of 1.2. How would you determine if the new carpool lane is successful or if the lane should be open to all traffic.

REFERENCES

1. *A Manual on User Benefit Analysis of Highway and Bus Transit Improvements*, American Association of State Highway and Transportation Officials, Washington, D.C., 1978.

2. William Jessiman et al., "A Rational Decision Making Technique for Transportation Planning," *Highway Research Record* 180(1967):71–80.

3. Frederick R. Frye, *Alternative Multimodal Passenger Transportation Systems*, NCHRP Report 146, Transportation Research Board, National Research Council, Washington, D.C., 1973.

4. Marvin L. Manheim et al., *Transportation Decision Making*, NCHRP Report 156, Transportation Research Board, National Research Council, Washington, D.C., 1975.

5. Stephen C. Lockwood and Fredrick A. Wagner, *Methodological Framework for the TSM Planning Process*, TRB Special Report 172, Transportation Research Board, National Research Council, Washington, D.C., 1973.

6. V. R. Vuchic and R. M. Stanger, "Lindenwold Rail Line and Shirley Busway: A Comparison," *Highway Research Record* 459(1973):13–28.

ADDITIONAL READINGS

Dickey, John W., *Metropolitan Transportation Planning*, 2nd ed., McGraw-Hill Book Company, New York, 1983.

Hay, William W., *An Introduction to Transportation Engineering*, John Wiley & Sons, New York, 1977.

Hutchinson, B. G., *Principles of Urban Transport Systems Planning*, Scripta Book Co., Washington, D.C., 1974.

Manheim, Marvin L., *Fundamentals of Transportation Systems Analysis*, MIT Press, Cambridge, Mass., 1979.

Meyer, Michael D., and Eric J. Miller, *Urban Transportation Planning: A Decision Oriented Approach*, McGraw-Hill Book Company, New York, 1984.

Morlok, Edward K., *Introduction to Transportation Engineering and Planning*, McGraw-Hill Book Company, New York, 1978.

Oglesby, Clarkson H., and R. Gary Hicks, *Highway Engineering*, 4th ed., John Wiley & Sons, New York, 1982.

Paquette, Radnor J., Norman J. Ashford, and Paul M. Wright, *Transportation Engineering*, John Wiley & Sons, New York, 1982.

Thomas, Edwin N., and Joseph L. Schofer, *Strategies for the Evaluation of Alternative Transportation Plans*, NCHRP Report 96, Transportation Research Board, National Research Council, Washington, D.C., 1970.

Wohl, Martin, and Brian V. Martin, *Traffic Systems Analysis*, McGraw-Hill Book Company, New York, 1967.

Transportation Systems Management

In the preceding chapters we described the process by which transportation plans are developed and evaluated. The context for this planning process is the need to establish long-range plans for the construction of transportation facilities, such as highways and mass transit systems, as well as to evaluate short-range improvements. In many situations today, the major transportation facilities are already built and, as described in the preceding chapters, the problem faced by transportation engineers is to operate these systems at their most productive and efficient levels. The term used to describe this operational planning process is *transportation systems management* (TSM). This chapter describes various TSM strategies that can be used to improve transportation performance and illustrates how these alternatives are evaluated.

The basic objective of TSM is to create more efficient use of existing facilities through improved management and operation of vehicles and the roadway. For example, a highway that carries 1500 autos per day with only one person per vehicle could increase its carrying capacity to 6000 persons per day with four persons per car. Thus, ridesharing represents an effective TSM action. Other TSM actions include park-and-ride, separate lanes for high occupancy vehicles, bicycle and pedestrian facilities, and traffic signal improvements to increase capacity and reduce delay at intersections.

The planning process for a TSM study is illustrated in Figure 12.1. The process entails defining the problem, establishing system objectives and criteria, deriving measures of effectiveness, identifying TSM actions, estimating costs and impacts, selecting TSM projects, and post facto monitoring of project results. The process is similar to that used for long-range planning but is less data intensive and often relies on field tests or before and after studies for evaluation. Since most TSM projects are relatively less costly than major construction and are easily implemented, it is often possible to try out an idea as a demonstration project and, if not successful, discontinue the service or method when the trial period is over. If the idea proves successful, modifications might improve it further, and it may also be implemented in other areas of the city or state.

Often, TSM projects are initiated to accomplish a nontransportation objective. For example, the Environmental Protection Agency has been mandated to

Figure 12.1 TSM Planning Process

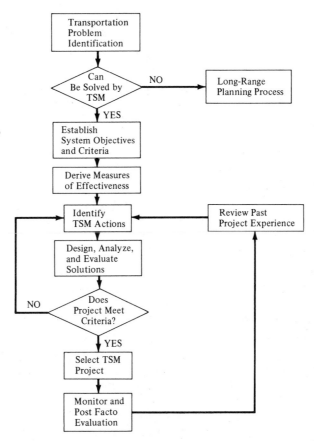

Source: Redrawn from *Transportation System Management*, Special Report 172, Transportation Research Board, Washington, D.C., 1977.

assure that air quality standards are maintained. The U.S. Department of Transportation is concerned with highway safety, energy conservation, cost effectiveness, reduction in peak hour travel, and minimizing transit operating expenses. These and other objectives can be addressed through TSM by examining many diverse ways to move people.

TSM STRATEGIES

According to the Institute of Transportation Engineers, TSM strategies can be classified into three basic categories: creating efficient use of road space, reducing vehicle use in congested areas, and improving transit service.[1] Within each category are a variety of strategies and tactics that can be used. These are described in the following sections.

TSM Actions to Create Efficient Use of Road Space

The actions in this group are intended to improve the flow of traffic, usually without altering the total number of vehicles that use the roadway during an average day. These TSM project improvements are intended to reduce travel time and delay for motorists, pedestrians, and transit users. Techniques used include traffic operations and traffic signalization, improvements for vehicle travel, bicycle and pedestrian projects, priorities in roadway assignments for high occupancy vehicles, traffic restrictions in residential and congested areas, parking management, altering work schedules, and improving the coordination between modes.

Traffic Operations Improvements

TSM strategies in this category are similar to those a traffic engineer carries out in normal operations. These include (1) widening intersections, (2) creating one-way streets, (3) installing separate lanes for right and left turns, and (4) restricting turning movements (especially left turns).

Traffic Signalization Improvements

Many tools and techniques can be used to improve traffic signal performance. Usually improvements are made that will benefit all traffic flow, but signals also can be used to favor high occupancy vehicles, buses, and carpools, or to create a safe environment in residential areas. Traffic signals can be improved by

- Physically improving local intersections
- Coordinating signal timing for arterial roadways
- Computerizing areawide signal coordination in downtown gird networks
- Controlling traffic for freeways, including changeable message signs
- Metering ramps
- Television monitoring of traffic

Improvements for Pedestrians and Bicycles

These TSM strategies encourage nonmotorized travel by making walking and bicycling safer and more pleasant. Among the most effective means are widening sidewalks and providing lighting, benches and pedestrian malls, grade separations, such as underpasses or overpasses to avoid conflicts at intersections, bikeways, and pedestrian controls at intersections.

Special Roadway Designations

This TSM strategy is based on the idea that if one or more lanes of traffic are reserved for high-occupancy vehicles, then the per lane person-carrying capacity will be increased, use of public transportation will be stimulated, energy will be conserved and service reliability will be improved. Roadways can be specially designated in a variety of ways such as the following.

- Reserving the curb lane of an arterial for buses only
- Reserving an entire street usually in the CBD for the exclusive use of buses
- Reserving a lane in the opposite direction of traffic (contraflow lane) when traffic in one direction is heavy and in the other direction is light
- Setting up reversible freeway lanes for heavy inbound morning traffic and outbound evening traffic
- Providing bypass lanes on freeway ramps for use by buses or high-occupancy vehicles
- Providing bus or carpool lanes on freeways (referred to as diamond lanes)
- Permitting use of highway shoulders during peak hours
- Using the center lane for left turns only, usually for all vehicles in off-peak hours and for buses only in peak hours

Vehicle Restrictions in Pedestrian Areas

In instances when traffic volume is excessive or pedestrian–vehicular conflicts exist, it may be necessary to restrict the area for use by autos and truck traffic. This strategy can also improve air quality and safety. As with any TSM strategy that restricts or limits use, its success will often depend on whether or not the restrictions cause severe hardship on those no longer permitted to use the facility. For this reason, traffic restricting strategies must also be accompanied by other improvements such as transit, parking, traffic operations, and marketing. Types of vehicle restriction include

- Peak hour pricing in which tolls or charges are placed on vehicles that use the area during congested hours
- Auto-restricted areas that limit the use of autos within the CBD. (One technique is to divide the CBD into zones and restrict travel across zone boundaries. Transit is permitted into the CBD using street rights of way that define the zones.)
- Pedestrian malls that restrict streets to pedestrian traffic only or sometimes pedestrian and transit traffic
- Vehicle restrictions in residential areas, including stop signs, street closings, speed bumps, and diagonal diverters

Parking Management

Parking availability is a means of regulating the flow of traffic because when parking becomes more scarce and expensive, the number of vehicles entering the area will be reduced. Some people will shift to transit or carpools; others will park at the periphery of the city, where space is available at more reasonable rates. The TSM strategies for parking include

- Curb parking restrictions to reduce the amount of on-street parking
- Off-street parking restrictions such as pricing differentials to discourage all day parking, elimination of free parking, and parking subsidies

- Preferential parking for cars and vanpools that serves as an incentive for rideshare
- Parking rate charges designed to encourage ridesharing or to limit vehicular traffic

Work-Schedule Management

Traffic congestion can also be reduced if peak period flows can be spread out over a longer period of time. If work traffic is normally concentrated between the hours of 7:30 a.m. and 8:30 a.m. then a reduction in traffic density would be possible if it were spread from 7:00 a.m. to 9:00 a.m. This TSM strategy is controlled by employers and will depend on work schedules within the organizations. The approaches used are

- Staggered work hours that require an adjustment in the beginning and ending work times
- Flex-time that permits an employee to begin and end the work day within a flexible time range
- A four-day work week
- Substitution of communications for transportation (which may become more prevalent as computers are introduced into the home)

Intermodal Coordination

Most TSM strategies require that the traveler change modes at some point. For this reason, coordination between modes must be considered, including improvements between transit, carpools, autos, walking, and bicycles. Many possibilities exist for intermodal coordination, such as park-and-ride facilities to assist the transfer between autos and rail or bus transit and transit interface improvements that simplify the transfer between modes by eliminating fare collection or locating bus stops to minimize walking.

TSM Actions to Reduce Vehicle Volume

In the previous section, we listed a variety of actions intended primarily to improve the efficiency of road space. For example, better traffic signal timing should result in fewer delays and higher average speeds. These actions may have secondary effects by encouraging more or less vehicular travel as motorists adjust to the changed level of service. Thus, if a road has been improved by traffic signalization, more people may decide to use it or if a freeway lane is dedicated to exclusive use by buses, then some people may shift from auto to transit. The TSM actions described in this section are primarily intended to reduce the volume of vehicles and in this way save energy and reduce congestion. These actions will also result in more efficient use of road space. Thus, all TSM actions are interrelated and should result in both efficient use of road space and less congestion.

Techniques available to reduce vehicle volume on streets and highways include increasing vehicle occupancy through ridesharing, imposing economic disincentives on auto users, encouraging the use of travel by other modes, and reduction of truck traffic.

Increasing Vehicle Occupancy Through Ridesharing

Every day, throughout the nation, millions of people drive to work alone. Average car occupancy in most cities is less than 1.5 persons per vehicle. If many of these motorists would rideshare, then fewer autos would be driven resulting in less traffic congestion. Four methods of ridesharing are

- Carpools
- Vanpools
- Subscription bus service
- Shared-ride taxi service

Carpools involve 2 to 6 persons who share driving to work, usually on a rotating basis. Vanpools are prearranged, usually by the employer, and involve 8 to 15 persons who each pay a prorated share of vehicle and operating costs. Subscription bus services are usually provided by a transit company to prearranged groups of 30 to 40 persons who commute to work from the same general area. The success of ridesharing depends on how well the proposed strategy compares with the auto in terms of travel time, cost, and convenience. Other TSM strategies, such as reserved parking for ridesharing and separate lanes for high occupancy vehicles, provide additional incentives. Carpooling is the strategy that has the greatest potential because it is easily implemented, requiring very little in the way of organization or new vehicles. Many transportation agencies provide carpool matching and information services that assist people in communities to find rides.

Discouraging Auto Use by Economic Means

The use of tolls and other user charges was mentioned earlier as a means of restricting autos in congested pedestrian areas. This strategy can be used as well to reduce or eliminate traffic on congested streets or highways. Among the approaches that could be used are

- Increased parking fees in downtown areas
- Special licenses permitting parking in restricted areas
- Charges for each trip based on vehicle metering of travel
- Tolls on highways and bridges

The two most common methods are parking charges and tolls. Other economic disincentives, such as congestion toll pricing or drastic increases in parking charges, have not been widely accepted in the United States because they are politically unpopular and are usually opposed by the motoring public as well as the business community.

Encouraging Travel by Means Other Than Auto

If motorists would leave the car at home and travel to work by bus, train, bike, or on foot, then traffic congestion on streets and highways would be reduced. For this to occur, the service provided by the other modes must be perceived as comparable or better than driving alone. Several available methods intended to make these modes look more attractive are

- Restrictions in auto travel such that transit becomes more competitive, for example, tolls, auto-restricted zones, freeway ramp metering, parking elimination
- Provision of fringe parking lots in outlying areas connected with high-speed rapid transit or express bus service
- Provision of special facilities such as downtown distributors or bikeways in areas where autos are prohibited

Park-and-ride facilities have proven to be an effective means of encouraging transit use. The provision of parking lots at outlying locations is relatively easy to implement, and they can be built at low cost. For a park-and-ride facility to be successful, a high level of transit service must be provided, including preferential treatment on some highways, and the total cost to the motorist must be lower than the cost by auto. The parking lot should be constructed so as to be compatible with the neighborhood and located in an area where travel demand and traffic congestion are sufficient to warrant service.

Reduction of Truck Traffic in Congested Areas

The conflict between auto and truck traffic on city streets can create major traffic congestion and delay problems. Curbside loading and unloading, double parking, and truck sizes can result in significant interference to the smooth flow of traffic. Among solutions proposed are to

- Arrange deliveries during off-peak hours
- Limit the size of trucks permitted for downtown delivery
- Create separate truck-only streets
- Streamline truck deliveries and scheduling operations

Attempts to restrict truck operations either in time or space usually are difficult to achieve because of the large number of trucking firms involved, difficulties in arranging deliveries and pickups during nonworking hours, and resistance by local businesses and unions. Trucks to a great extent already schedule themselves outside of peak hours so they can avoid slow speeds during hours of congestion. The most effective strategy is to accommodate trucking needs by building off-street loading areas and special truck lanes.

Improving Transit Service

One of the goals of TSM is to reduce traffic congestion. If economical, reliable, and fast transit service is available, it is likely that some auto users will switch to transit service thus reducing vehicle volumes. Any of the TSM actions that improve the flow of traffic should also improve transit service, but if service is to be truly competitive with the automobile, then special treatment is required. If transit is to be effective it should travel on its own right of way, have flexible routing patterns, be "hassle free" and easy to use. Techniques used to improve transit include providing express bus services, shuttle services from fringe parking areas to downtown, internal circulation in low density areas, improved flexibility in route scheduling and dispatching, simplified fare collection procedures, park-and-ride facilities, shelters, bus stop signs, bus fleet modernization, and improved passenger information services. Carrying out a transit improvement program requires an integrated approach that incorporates each of these actions where appropriate so that the net effect is a system that operates as effectively and efficiently as possible.

TSM ACTION PROFILES

A convenient way to examine various TSM strategies is to describe the problem that each action addresses, the conditions necessary for application of the action, potential problems of implementation, and potential evaluation factors. Table 12.1 lists a variety of TSM actions that could be considered for use in solving a particular transportation problem.

To illustrate how each action can be described, the attributes of three TSM strategies are profiled in this section. These are staggered work hours, travel on freeway shoulders during peak periods, and two-way left-turn lanes. Similar profiles of each TSM strategy listed in Table 12.1 are described in the report *Simplified Procedures for Evaluating Low-Cost TSM Projects.*[2]

Staggered Work Hours

Problems Addressed

- Traffic congestion consistently develops during peak commuting periods in or near an employment center.
- Congestion is expected to increase if business expansion or plans are approved and implemented.
- Transit vehicles serving an employment center are consistently over-crowded during parts or all of the peak commuting periods.

Conditions of Application

- *Slack capacity:* There should be periods of low-traffic within the hours that work trips will be staggered. If traffic congestion is uniform, this strategy will not be effective.

Table 12.1 TSM Actions Available for Solving Transportation Problems

Staggered work hours
Flexible work hours
Increased peak period tolls
Toll discounts for carpools during peak hours
Residential parking permits
Neighborhood traffic barriers
Park-and-ride lots along transit routes
On-street parking bans during peak hours
Parking reserved for short-term use
Increased parking rates
Parking rate fines and time-limit adjustments
Expanded off-street parking
Freeway ramp control
Travel on freeway shoulders during peak periods
One-way streets
Reversible lanes
Two-way left-turn lanes
New street segments
Signal phases for left turns
Employer-based carpool matching programs
Employer vanpool programs
Freeway lanes reserved for buses and carpools
Priority freeway access/egress for buses and carpools
Arterial street lanes reserved for express buses or carpools
Shuttle buses or vans
Bus transfer stations
Expanded regular-route bus service
Pedestrian-only streets
Community transit services

- *Employment concentration:* The industrial center should contain about 40 percent of employment in a 1–2 mi radius. Large firms with 1000 or more employees are more effective than many small firms.
- *Transit availability:* Transit routes should serve the center with service convenient to changes in work schedules.

Implementation Problems

- Some businesses and labor organizations may not cooperate.
- A reduction may occur in van- or carpools when work times are shifted.

Evaluation Factors

- Changes in vehicle volumes entering and leaving the employment center
- Changes in peak operating speeds or levels of service
- Changes in average loads on transit routes
- Changes in ridership and revenue

Travel on Freeway Shoulders

Problems Addressed

- Traffic congestion consistently develops on a freeway or in a freeway corridor during peak commuting periods.
- Localized traffic congestion develops on a freeway as a result of a lane drop.

Conditions for Application

- *Geometric conditions:* There should be wide shoulders (minimum 10 ft) horizontal and vertical alignment for safe speed on shoulder lanes.
- *Location:* This strategy should be implemented on freeways where most drivers are commuters and thus familiar with use of shoulders for moving traffic during peak periods.
- *Traffic composition:* If more than 3 percent of traffic stream is comprised of trucks, this strategy may cause traffic flow disruptions and accidents due to lane-change movements.

Implementation Problems

- The shoulder is not available for breakdowns and traffic enforcement.
- Use of the shoulder will attract travelers who would normally use other routes, thus limiting the effectiveness of this strategy.
- This action could simply shift the problem to another location such as the ramps or points where the shoulder terminates.

Evaluation Factors

- Increases in traffic volumes
- Changes in peak period speeds or levels of service
- Changes in traffic volumes on parallel facilities
- Increased cost of freeway operations for ramp modifications, maintenance, and enforcement
- Changes in accident rate

Two-Way Left-Turn Lanes

Problems Addressed

- Left turns on arterial streets at midblock locations interfere with through traffic and cause delays.
- Left turns are also a major contributor to accidents, especially rear-end collisions.

Conditions for Application

- Since a separate lane is designated solely for left turns, sufficient roadway width is required. Streets with 4 lanes, each 12 ft wide, can be restriped with lanes $9\frac{1}{2}$ ft wide, thus creating an additional turning lane 10 ft wide. It should be understood, however, that narrowing lanes could result in some compromise in safety, as 12 ft lanes are preferred. Lane narrowing should be permitted only when sufficient right of way is unavailable. Speed limit reductions should be posted along the stretch where lanes have been reduced in width.
- Left-turn movements should be distributed along the arterials to avoid long queues. This strategy is most appropriate where strip development exists. At major shopping centers separate left-turn signals are recommended.
- Distances between traffic signals should be sufficient to allow left-turn storage lanes at intersections. If signals are placed closer than 1000 ft apart, less than half of the lane may be available for two-way operation.

Evaluation Factors

- Change in peak operating speed along arterial street
- Change in number of accidents involving left turns
- Cost to modify intersections, maintain pavement markings, and traffic signs

Each of the TSM strategies described involve both advantages and disadvantages. Although a strategy may solve one problem, it may create others. Also, the project may be accepted by some groups but opposed by others. It is helpful to understand the likely effects as well as the limitations of each TSM strategy before the project is selected. Thus, profiles of each TSM action are helpful in deciding if a given action should be used in a given situation.

TSM PROJECT EVALUATION

An Example Using Travel Demand Analysis Methods

The following example illustrates how the planning process for a TSM project is used in preparing an evaluation. The situation involves an urban arterial corridor in a large metropolitan area. For additional details, see *Measures of Effectiveness for TSM Strategies.*[3]

Existing Conditions

The arterial corridor studied (Lee Street) is 2 mi in length and connects two freeways, as illustrated in Figure 12.2. Traffic volumes increase toward the CBD, going north. Railroad tracks, which parallel the road on the east side for its entire length, limit traffic access from that direction. The width of the arterial is 60 ft and is striped for 4 lanes throughout its length. Bus service consists of five routes that provide headways between 6 and 30 min. No special or preferential bus facilities exist in the corridor and all routes terminate in the downtown area. At the peak load point along the arterial there are 25 buses per hour.

Objectives

Improvements to the corridor are intended to accomplish the following objectives:

- Improve travel time for transit users
- Improve comfort and convenience
- Increase capacity
- Minimize capital and operating costs

Figure 12.2 Urban Arterial Corridor

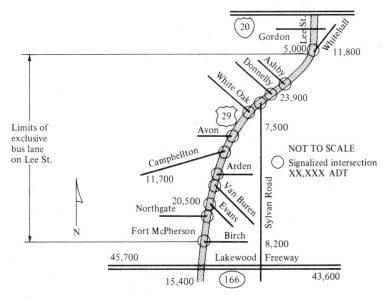

Source: Redrawn from Charles M. Abrams et al., *Measures of Effectiveness for TSM Strategies*, U.S. Department of Transportation, Washington, D.C., 1979.

Proposed TSM Actions

The TSM program being evaluated consists of the following elements.

1. *Exclusive Bus Lanes in Peak Hours.* The arterial will be striped as 5 lanes. During the AM peak, one northbound lane will be reserved for buses only, and during the PM peak, one southbound lane will be reserved for buses only. During off-peak hours, the center lane will be used for two-way left turns. See Figure 12.3.

2. *Express Bus Service.* Two of the bus routes will operate express bus service.

3. *Park-and-Ride Facilities.* Park-and-ride lots will be built at two locations, one at Campbellton Road and the other near the Lakewood Freeway (Figure 12.2).

4. *Reduced Headways on Existing Bus Routes.* The headways on two bus routes will be reduced from 20 to 12 min. Three new bus routes will be added and will stop at park-and-ride lots every 8 min.

Selection of Measures of Effectiveness

For each objective, a measure of effectiveness is used that best represents the improvement. These are

- Point-to-point travel time
- Frequency of bus service
- Volume to capacity ratio
- Equivalent annual cost of each action

Estimating Travel Demand

Since this is a short-term planning study, the data and modeling techniques available from the local planning agency will be used. The Urban Transportation Planning System (UTPS) is selected because of its ability to perform modal split

Figure 12.3   Bus Headways and Lane Designations

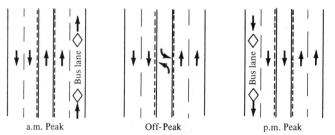

Source: Redrawn from Charles M. Abrams et al., *Measures of Effectiveness for TSM Strategies,* U.S. Department of Transportation, Washington, D.C., 1979.

as well as transit and highway assignment. The key procedure used to evaluate shifts in transit use is the modal split model, which is of the form

$$P_t(x) = \frac{U_t(x_i b)^B}{U_A(x_i b)^B + U_t(x_i b)^B}$$

$$P_A(x) = 1 - P_t(x)$$

where

$P_t(x)$ = probability of an individual using transit for a trip

$P_A(x)$ = probability of an individual using auto for a trip

$U_t(x_i b)$ = utility function incorporating appropriate costs, x_i, of using transit and the coefficients b_i of the various costs

$U_A(x_i b)$ = similar function for auto

B = constant that varies by trip purpose

The modal split is calibrated by three trip purposes: home-based work, home-based other, and nonhome based. The variables used are transit in-vehicle time, $2.5 \times$ (walk + wait time), driving time to transit stop, out-of-pocket cost, transit fare, and parking cost.

For the proposed TSM actions, the highway network was modified to simulate planned conditions, such that where lanes were added, link capacities and speeds were increased. Trip table data were available, but it was necessary to update these to present conditions. A Fratar model was used for this purpose because growth in the region was generally stable over the period and zonal population and employment data were available to develop growth factors.

Impact Evaluation

The measures of effectiveness were calculated by determining the differences between the existing conditions and those with the TSM actions in place. The results of the impact evaluation are as follows.

Travel Time: Overall the proposed project is estimated to reduce peak period person hours of travel by about 2 percent in the corridor. The reduction is due mainly to the increase in express bus service and improved highway travel. Door-to-door transit travel times will be substantially reduced from an average of 48.3 min to 40.8 min, whereas auto travel is reduced only from 29.2 min to 28.9 min.

Comfort and Convenience: Frequency of bus service was selected as the measure of effectiveness, and this would increase substantially if the proposed actions were placed in effect.

Capacity of the Transportation System: The volume-capacity ratios at selected locations decreased slightly with the proposed TSM actions in place. This is due mainly to a modest (2 percent) reduction in highway travel. There was a significant demand for park-and-ride facilities, which is estimated to serve 1000 vehicles at maximum accumulation.

Equivalent Annual Cost: The capital cost of the proposed action is $3.9 million, which includes 33 new buses, 1000 parking spaces, and 2.5 mi of bus lane. At an interest rate of 8 percent and a 10 yr life, the equivalent annual cost is $0.58 million. Operating and maintenance costs are estimated to increase by $1.1 million. In addition, the operating deficits for the transit authority will increase by $0.95 million.

Project Analysis and Selection

The proposed TSM actions in this corridor contribute to the achievement of the objectives; however, the effectiveness of the strategy is quite modest. Improvements in auto traffic were about 2 percent, whereas the increases in transit patronage and levels of service were more substantial. On the other hand, the annual cost of the project is estimated to be approximately $2.63 million. Whether the project will be selected will depend on the values for each measure of effectiveness developed in the analysis, as well as the relative importance of each objective, availability of funds, and other projects that have been proposed for the region.

An Example Using Simplified Procedures for Low-Cost Projects

The preceding example illustrated how the transportation planning process, which is typically used for long-range studies, can be applied to the evaluation of TSM actions. The procedure is still somewhat cumbersome, however, since it depends on an extensive data base, the use of calibrated models for trip distribution and modal split, as well as a link-node network for traffic assignment. Many TSM projects are low-cost improvements and to evaluate them by using such complex procedures is both impractical and costly. The following examples illustrates an evaluation procedure intended for low-cost projects. Additional details of this approach are described in *Simplified Procedures for Evaluating Low-Cost TSM Projects.*[4] The procedure follows four basic steps: (1) analyze problems and their setting, (2) identify and screen candidate solutions, (3) design, analyze, and evaluate solutions, and (4) recommend an action plan. The following problem involves a travel corridor in a small city and for this reason transit does not play a major role.

Analyze Problems and Setting

A 2-mi corridor connects the CBD of a small town with a residential district and a freeway interchange. The artery, Lisbon Street, is a 2-lane road that carries 5000–8000 vpd. See Figure 12.4. Strip development borders the road and also serves as frontage for residential development.

The residents and merchants in the community have complained about delays and unsafe driving conditions. In response, the county engineering office conducted field observations, including traffic movements at intersections, travel time runs, speed and delay studies, and accident data. In addition, the physical condition of the road was noted, including surface, curbs, signals, markings, and signs.

Figure 12.4 Lisbon Street Corridor

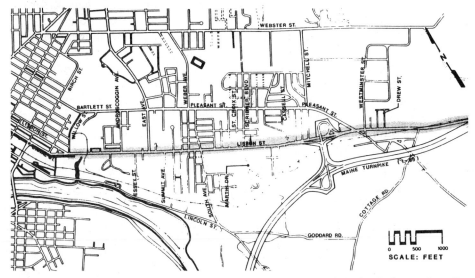

Source: Reproduced from J. H. Batchelder et al., *Simplified Procedures for Evaluating Low-Cost TSM Projects: User's Manual*, NCHRP Report 263, Transportation Research Board, Washington, D.C., 1983.

Table 12.2 Criteria to Establish Deficiency of Roadway Segments

- Supply Capacity
  Lane width under 11 ft
  Signs and pavement marking not in conformance with *Manual of Uniform Traffic Control Devices* (MUTCD)
  Traffic control not in conformance with MUTCD
  Curb radii less than 20 ft
  Sight distance not in conformance with ITE/AASHTO standards

- Service Quality
  Intersection level of service lower than C or critical lane volumes greater than 1200 cars/hr
  Road segment level of service lower than C or average lane volume greater than 1200 cars/hr
  Peak average travel speed less than 20 mph

- Accidents
  Three-year average accident rate exceeds station average with 95 percent confidence
  Average rates
    Signalized intersections: 1.22 per million entering vehicles
    Nonsignalized intersections: 0.39 per million entering vehicles
    Road segments: 3.74 per million vehicles

Source: Reproduced from J. H. Batchelder et al., *Simplified Procedures for Evaluating Low-Cost TSM Projects: User's Manual*, NCHRP Report 263, Transportation Research Board, Washington, D.C., 1983.

The results of this data-gathering phase were then compared with a set of preestablished deficiency criteria, shown in Table 12.2. The performance measures noted in the field were then plotted, showing the observed value at each intersection or along the road itself. These results are shown in Figure 12.5 for three performance measures: peak hour volume/capacity ratio; average peak travel speed, and accident rate factor. The profiles show locations along the corridor where problems exist and also suggest possible relationships among the problems. Notes on the margin of the existing condition profile indicate the probable causes of each deficiency. For example, low travel speeds between Summit Street and South Street are due to faulty traffic signal operation and curb parking near intersections.

Figure 12.5 Observed Conditions Along Lisbon Street Corridor

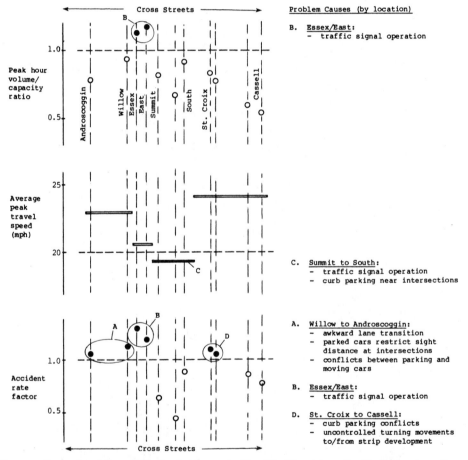

Source: Reproduced from J. H. Batchelder et al., *Simplified Procedures for Evaluating Low-Cost TSM Projects: User's Manual*, NCHRP Report 263, Transportation Research Board, Washington, D.C., 1983.

Figure 12.6 Problem Summary for Lisbon Street Corridor

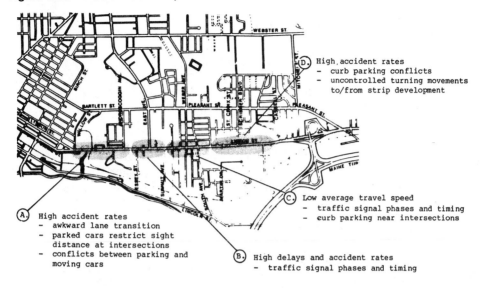

(D.) High accident rates
- curb parking conflicts
- uncontrolled turning movements to/from strip development

(C.) Low average travel speed
- traffic signal phases and timing
- curb parking near intersections

(A.) High accident rates
- awkward lane transition
- parked cars restrict sight distance at intersections
- conflicts between parking and moving cars

(B.) High delays and accident rates
- traffic signal phases and timing

- Bartlett/Pleasant Sts. do not provide access to commercial establishments; any upgrading likely to be opposed by residents.

- No other feasible, parallel routes exist; Lincoln and Webster Sts. serve different corridors

- No land use changes expected in near-term that would change travel volumes or patterns in corridor.

- Lisbon St. right-of-way sufficient to accomodate near-term traffic given expected low growth in region.

- Moderate traffic improvements are not expected to change regional or corridor travel, so improvement analysis can be limited to narrow corridor along Lisbon St.

- Air quality in corridor meets EPA standards.

Source: Reproduced from J. H. Batchelder et al., *Simplified Procedures for Evaluating Low-Cost TSM Projects: User's Manual*, NCHRP Report 263, Transportation Research Board, Washington, D.C., 1983.

The problems are summarized on a site map, Figure 12.6, by road segment. In this case, there are four problem areas along the corridor. Other constraints are noted, for example, no other feasible parallel routes exist, so improvements must be limited to the narrow corridor along Lisbon Street.

Identify and Screen Candidate Solutions

As is typically the case in many communities, funds to correct street and highway problems are limited. In this instance, solutions must be restricted to those not requiring the taking of right of way or the purchase of new traffic control equipment. Also, any solution that is likely to create opposition and delay in implementation would be eliminated. Thus, solutions are limited to traffic and parking controls. Ideas such as ridesharing, transit, or physical improvements are discarded as being impractical, ineffective, or too costly.

Possible solutions are identified with the aid of a set of TSM Action Identification Tables and TSM Action Profiles. The TSM Action Identification Tables are intended to assist the engineer in identifying general approaches and specific types of TSM actions that may be applicable in solving short-term transportation problems. Tables have been prepared in the *Simplified Procedures Manual* for six problem types: (1) isolated intersections or street segments, (2) corridors, (3) residential communities, (4) employment centers, (5) commercial centers, and (6) regional, state, and national problems.[5] For each problem type, a subset of problems is identified. The tables are organized around five factors: (1) location and scale of the problem, (2) nature of the problem, (3) underlying deficiency in transportation system, (4) strategy or approach to the solution, and (5) location or scale of action.

In the Lisbon Street example, the problem falls under the category of "Isolated Intersections or Street Segments" and the subset of this problem includes (1) vehicle flow conflicts and accidents and (2) traffic congestion. Table 12.3 shows the Action Identification Table for these two problems. Other problem solution ideas would be obtained from the Action Identification Table for corridors.

The Action Identification Tables and Action Profiles (described earlier) assist the engineer in identifying candidate solutions. A list of possible actions is prepared, together with notes or comments pertaining to their relevance to the problem being studied. For the Lisbon Street example, a list of 20 possible action improvements was developed. Comments concerning their appropriateness were prepared and a recommendation made as to whether or not the action should be pursued (based on a judgment assessment of feasibility and effectiveness). The action screening worksheet is shown in Figure 12.7. The process resulted in a recommendation to pursue 12 of the 20 candidate actions. These were combined into two initial action packages. The first set of actions included minimum improvements required to solve the problem at the lowest cost. The second set included roadway widening along one segment of the corridor. The initial actions are shown in Figure 12.8. Essentially, the improvement program involves removal of parking to open up lanes for travel or turning movements and to modify signal operations.

Design, Analyze, and Evaluate Solutions

An analysis plan is developed that will permit an evaluation of each criteria considered to be important for the project. The criteria selected are as follows:

- Travel lanes should be at least 11 ft wide
- Signs, signals, and markings conform to MUTCD
- Turning radii at least 20 ft
- Adequate separation between turning and through traffic
- Adequate buffer between curb travel lanes and sidewalk flows
- Safe crossing time during each signal cycle
- Level of service C or higher
- Peak travel times 20 mph or higher
- Parking provided within 300 ft of stores
- Budget limited to $200,000

Table 12.3 Action Identification: Isolated Problems at Intersections or on Street Segments

Underlying Transportations Deficiencies	Corrective Strategies	Actions Applied at Intersections	Actions Applied on Street Segments
Problem 1: *Vehicle Flow Conflicts and Accidents*			
Turning and parking movements inhibit traffic flow	⟶ 1A. *Reduce delays and conflicts by separating flows.*	—New signals —Signal phasing and timing changes —Turn lanes —Striping —Traffic police —Islands	—2-way left-turn lanes —Expanded off-street parking or loading areas —On-street parking restrictions or removal[1]
	1B. *Reduce delays and conflicts by diverting movements.*	—Left turn prohibitions —Jug-handles —On-street parking restrictions near intersections	—Medians —Side street and curb cut closures —On-street parking restrictions or removal —Expanded off-street parking or loading areas
Inadequate sight or stopping distances	⟶ 1C. *Increase time available for driver reaction.*	—New signals or stop signs —Signal phasing and timing changes —Warning devices	—Speed restriction —On-street parking restrictions or removal —Warning devices
Problem 2: *Traffic Congestion*			
Turning and parking movements inhibit traffic flow	⟶ 2A&B. Same as strategies 1A and 1B listed for "Vehicle Flow Conflicts and Accidents"		
Inadequate capacity to handle peak traffic volumes	⟶ 2C. *Reduce delays by adding capacity.*	—New lanes —Signal timing	

Source: Adapted from J. H. Batchelder et al., *Simplified Procedures for Evaluating Low-Cost TSM Projects: User's Manual*, NCHRP Report 263, Transportation Research Board, Washington, D.C., 1983.

Figure 12.7   Candidate TSM Actions for Lisbon Street Corridor

CANDIDATE STRATEGIES/ACTIONS	NOTES/COMMENTS	PURSUE?
• Separate traffic flows at intersections to reduce accidents and delays		
– new traffic signals	– major intersections already signalized	No
– separate turn phases	– significant turning movements at 5 intersections	Yes
– left-turn lanes	– ditto; adequate right-of-way, but need to remove parking	Yes
– striping/islands	– probably useful at some locations	Yes
• Separate traffic flows on street segments to reduce delays		
– two-way left turn lanes	– inadequate right-of-way	No
• Divert turning movements to reduce accidents and delays		
– left-turn prohibitions	– added traffic on side streets and driver confusion rule this out	No
– street and curb-cut closures	– no alternative entrances to many abutting properties; entrances to some side streets could be closed but this would hinder access to off-street parking and residences; must be done selectively	Maybe
– medians	– added travel distances and increased turns at intersections (including illegal U-turns) would disrupt traffic more than existing midblock turns	No
• Move parking from street to reduce accidents and delays		
– remove parking near intersections	– adequate vacancy rate in nearby spaces along most of corridor, so little opposition expected from merchants	Yes
– remove parking in midblock	– in most locations, adequate off-street or side street parking exists near stores; may need to retain parking in some blocks of Lisbon St.	Yes
– add off-street parking	– not needed, vacant spaces can absorb cars currently parking on Lisbon St.	No
• Increase time for driver reaction to avoid accidents		
– new signals, signs or warning devices	– most intersections meet signal and signing standards, but some signing could be improved	Yes
– parking removal near intersections	– should be effective in improving sight distance at problem	Yes
• Add capacity at intersections to reduce delays		
– new lanes	– can use width of parking lanes to add turning or through lanes if curb radii are increased	Yes
– signal timing	– retiming signals will add effective capacity to at least two intersections	Yes
• Add capacity on street segments to reduce delays		
– new lane	– existing parking lanes can be safely used as travel lanes if curb radii are increased	Yes
• Add street capacity to corridor to reduce delays		
– new segments or lanes on parallel streets	– not feasibile in residential neighborhoods	No
• Improve use of existing corridor capacity to reduce delays		
– signal coordination	– physically interconnecting Essex/East St. signals should be considered; manual coordination probably adequate elsewhere	Yes
– reversible lanes	– inadequate right-of-way; balanced traffic	No
– one-way pairs	– too much local traffic; no good route for other half of pair	No

Source: Reproduced from J. H. Batchelder et al., *Simplified Procedures for Evaluating Low-Cost TSM Projects: User's Manual*, NCHRP Report 263, Transportation Research Board, Washington, D.C., 1983.

Figure 12.8 Action Packages for Further Evaluation

Package #1: improvements requiring only
a small amount of new traffic control
equipment and no new right-of-way

- All sections:
  - increase curb radii
  - remove parking near major intersections

- Section A (Willow to Androscoggin):
  - add turn and travel lanes through restriping

- Section B (Essex to East):
  - connect traffic signals
  - add turn lanes and phases

- Section C (Summit to South):
  - restripe for 2 approach lanes at intersections

- Section D (St. Croix to Cassell):
  - remove parking
  - restripe for 2 lanes in each direction

Package #2: improvements requiring only
a small amount of new traffic control
equipment and/or right-of-way

- All actions included in Package #1

- Section A:
  - widen to add travel lanes

Source: Reproduced from J. H. Batchelder et al., *Simplified Procedures for Evaluating Low-Cost TSM Projects: User's Manual*, NCHRP Report 263, Transportation Research Board, Washington, D.C., 1983.

The analysis plan for the project is shown in Figure 12.9. It addresses each of the criteria posed. Since this is a short-term action, factors such as traffic growth and changing travel patterns need not be considered. The techniques used to carry out the analysis plan are listed in Figure 12.10. Many of the traffic engineering analysis methods required to solve this problem have been described in earlier chapters of this textbook. Table 12.4 lists the methods suggested for carrying out travel time estimation procedures. Method selection aids for computing other variables are contained in *Simplified Procedures for Evaluating Low-Cost TSM 4 Projects.*[6]

Figure 12.9 Analysis Plan for Corridor Evaluation

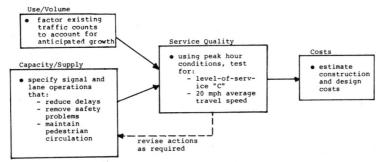

Source: Reproduced from J. H. Batchelder et al., *Simplified Procedures for Evaluating Low-Cost TSM Projects: User's Manual*, NCHRP Report 263, Transportation Research Board, Washington, D.C., 1983.

Figure 12.10   Techniques Required to Complete Analysis

● Service Quality

   - Traffic delays at intersections - Use
     critical lane method to test specific
     lane and signal options for achievement
     of level-of-service C during peak hour.

   - Average travel speed - Intersection
     improvements and parking restrictions
     should be sufficient to meet 20 mph
     standard. Proposed phasing sequences
     should be checked for interference.

   - Parking availability - Field survey showed
     vacancy rates along side streets and in
     adjacents lots are adequate to accomodate
     any reductions in on-street spaces.

   - Pedestrian access and circulation -
     Existing sidewalks and crosswalks are
     sufficient to accommodate increased flows.
     Anticipated pedestrian and turning traffic
     volumes do not warrant phase separations.
     Signal timing must continue to allow
     adequate crossing time for pedestrians.

● Safety

   Examine accident records to
   identify correctible safety problems.

● Supply/Capacity

   Specify different signal phasings and lane
   configurations and operations for analysis.

● Financial

   Prepare cost estimates for engineering and
   construction/implementation of recommended
   actions. Changes in on-going street
   maintenance costs are expected to be
   minimal.

● Public Costs

   - Accidents - Intersection safety improve-
     ments identified in safety analysis
     should sufficiently reduce accident rates.

   - Air quality - No problems exist or are
     anticipated in the corridor.

Source: Reproduced from J. H. Batchelder et al., *Simplified Procedures for Evaluating Low-Cost TSM Projects: User's Manual*, NCHRP Report 263, Transportation Research Board, Washington, D.C., 1983.

The principal elements of this problem required the analysis and design of seven intersections and three short segments of roadway. The engineering analyses required to solve the problem were examination of accident records and safety conditions and the use of critical lane analysis and signal timing equations to test for operation at level of service C or better. Similar techniques were applied to the remainder of the corridor and final checks were made of signal phasing and timing to assure that no conflicts existed. Finally, the results were summarized and used in estimating costs.

Recommend an Action Plan

The major constraints on the project are the requirement that costs not exceed $200,000 and that the project be implemented over a 3 to 5 yr period. Accordingly, the recommended improvements were grouped in order of priority such that projects selected first addressed the most significant problems and second, could be completed within a 2 to 3 yr period. Other projects were identified that should be implemented but are deferred subject to availability of funds. The recommended action plan, shown in Figure 12.11, consists of the following projects. Detailed diagrams of each improvement are also prepared as illustrated in Figure 12.12.

Priority 1 ($55,000)

- Section A
  - Restripe for 1 eastbound and 2 westbound lanes
  - Paint transition zone to start of curb parking
  - Increase curb radii

Figure 12.11 Recommended Action Plan

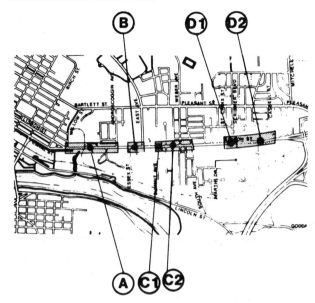

Source: Reproduced from J. H. Batchelder et al., *Simplified Procedures for Evaluating Low-Cost TSM Projects: User's Manual*, NCHRP Report 263, Transportation Research Board, Washington, D.C., 1983.

- Section B
    - Change lane-use designations (left-turn lanes)
    - Change traffic signal phases and timing
    - Increase curb radii

Priority 2 ($60,000)

- Section C
    - Relocate traffic signal heads
    - Prohibit parking near intersections
    - Restripe for 2 approach lanes
    - Increase curb radii
- Section D
    - Remove parking and restripe for 2 lanes in each direction
    - Upgrade traffic signal hardware
    - Increase curb radii

Priority 3 ($75,000)

- Section C2
    - Prohibit parking near intersections

Figure 12.12 Example of a Project Improvement Diagram for Lisbon Street Corridor

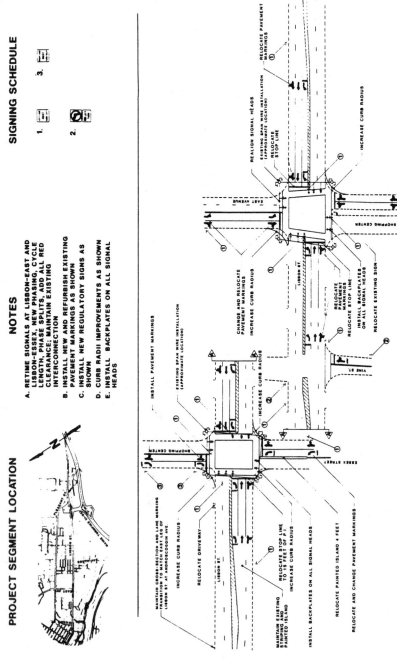

Source: Reproduced from J. H. Batchelder et al., *Simplified Procedures for Evaluating Low-Cost TSM Projects: User's Manual*, NCHRP Report 263, Transportation Research Board, Washington, D.C., 1983.

Table 12.4 Methods Selection Table Travel Time Estimation

Spot or Segment Analysis	Appropriate Techniques
■ Vehicular delays at intersections	
■ Check for acceptable level of service.	Use of capacity adequacy techniques cited in the Methods Selection Table for supply/capacity estimation.
Estimate average travel time, delay, or queue duration.	Apply analytical techniques and worksheets. Apply simulation models of signalized intersection operations (mainly applicable to complex, high-volume intersections).
Set signal timing to minimize delays.	Apply Webster's equation or similar procedure.
■ Vehicular travel time along roadway segment.	
Check for acceptable level of service.	Use capacity adequacy techniques cited in the Methods Selection Table for supply/capacity estimation.
Estimate average travel time.	Use curves relating speed to volume/capacity ratio for different types of roadways.
■ Travel time along transit route segment.	Apply an analytical procedure to calculate route segment running times. Use average operating speeds observed under similar conditions.

Source: Adapted from J. H. Batchelder et al., *Simplified Procedures for Evaluating Low-Cost TSM Projects: User's Manual*, NCHRP Report 263, Transportation Research Board, Washington, D.C., 1983.

- ▪ Restripe for 2 approach lanes
- ▪ Increase curb radii
- ■ Section D2
  - ▪ Remove parking and restripe for 2 lanes in each direction
  - ▪ Increase curb radii

SUMMARY

Transportation systems management (TSM) is that aspect of transportation engineering that seeks to improve the existing street and highway system by making changes in its operations. A wide variety of changes can be made both to the system itself (supply) and how the system is used (demand).

This chapter has described the many TSM techniques that are available for consideration in solving a given transportation problem. The actions that can be selected will improve the use of road space, reduce vehicle use, and improve transit service. Each action has its advantages and disadvantages and these should be fully understood before it is recommended.

A TSM planning process should be followed when evaluating any action. The process includes identification of the problem, establishing a set of objectives and criteria, identification of candidate TSM actions, design and evaluation of solutions, preparation of a recommended action plan, and after the project is completed, an evaluation of its actual performance.

To carry out a TSM analysis and design requires techniques and skills of traffic engineering and highway planning. Among these are signal timing, capacity analysis, speed and accident studies, pavement striping and signing, and parking regulations. The application of these topics has been described in previous chapters, and the techniques can be applied in the design and testing of TSM alternatives.

PROBLEMS

12–1 For the following situations, indicate if a TSM project is an appropriate solution. For each situation where TSM would be selected, describe two ways in which the problem could be solved.

(a) Traffic is congested on downtown city streets.

(b) A surburban employment center lacks in parking facilities.

(c) A freeway during peak a.m. and p.m. periods has a level of service D.

(d) Traffic accidents increase along an arterial street.

12–2 Describe two situations with which you are familiar in which TSM actions have been undertaken to accomplish the following objectives:

> Create efficient use of road space
> Reduce vehicle volume
> Improve transit service

12–3 Prepare a TSM action profile for the following actions: (a) increased peak period tolls, (b) one-way streets, and (c) employer vanpool programs. In your profile consider the problem addressed, implementation problems, and evaluation factors.

12–4 Select a transportation corridor in your community for which traffic congestion is a problem. Prepare a report discussing how you would proceed to develop an improvement program. In your report include the following items.

> Description of existing conditions
> Objectives to be accomplished
> Proposed TSM actions to be investigated
> Measures of effectiveness that best represent the improvement
> Travel demand data required
> Impact analysis and project selection

12–5 What TSM strategies could you suggest that would improve the parking situation at your college or university? What is the likelihood that these actions would be implemented by the university administration?

12–6 Contact the traffic engineer in your community and ask what TSM projects have been completed in the past 3 to 4 years. Prepare a case study of one of these projects. In your report consider the following items.

>   The problem and its setting
>   Candidate solutions that were considered
>   The design, analysis, and evaluation of solutions
>   Recommended action plan

12–7 A proposal has been made to convert one lane (in each direction) of a 6-lane freeway for use by carpools only. The lane would extend for 10 mi. Currently the lanes operate at capacity, in a.m. and p.m. peak hours with an average of 1.2 persons per car. It is expected that about 20 percent of drivers will carpool with an average of 3 persons per auto. If the jam density is 124 vpm and mean free speed is 58 mph, what will be the travel time savings under these conditions? (Assume linear relationships between speed and density.) Peak hours are 7–9 a.m. and 4–6 p.m.

12–8 A separate bus lane is proposed that will replace a single lane of a freeway with a capacity of 2000 vph. If auto occupancy is 1.5 persons per vehicle, what is the fewest bus trips per hour required, if each bus carried 50 passengers? What would be the average headway for each bus? Comment on the advantages and disadvantages of this proposal.

12–9 A large corporation located in a surburban area wishes to limit the need for parking by encouraging its employees to vanpool to work. The vans will be driven by an employee who rides free, and each van will accommodate 10 riders who live an average of 10 mi from the work site. A van costs $18,000, and maintenance and operating costs are $0.50 per mile.

(a) Determine the monthly charge for each passenger, assuming the life of a van is 8 years and interest rates are 10 percent.

(b) If 50 vans are in use, how many fewer parking spaces are needed?

(c) If a parking space costs $4000 to construct, how much could the company spend for the vanpool program and still break even?

REFERENCES

1. ITE Committee 6Y–17, *Systems Management Optimization*, Technical Council Information Report, Institute of Transportation Engineers, Washington, D.C., March 1980.

2. J. H. Batchelder et al., *Simplified Procedures for Evaluating Low-Cost TSM Projects: User's Manual*, NCHRP Report 263, Transportation Research Board, National Research Council, Washington, D.C., 1983.

3. Charles M. Abrams et al., *Measures of Effectiveness for TSM Strategies*, U.S. Department of Transportation, Federal Highway Administration, Washington, D.C., 1979.

4. Batchelder et al., *Simplified Procedures*.

5. Ibid.

6. Ibid.

ADDITIONAL READINGS

Interplan Corporation, *Transportation System Management: State of the Art*, Urban Mass Transportation Administration, Washington, D.C., 1976.

Morlok, Edward K., *Introduction to Transportation Engineering and Planning*, McGraw-Hill Book Company, New York, 1978.

New York State Department of Transportation, *Energy Impacts of Transportation Systems Management Actions*, Urban Mass Transportation Administration, Washington, D.C., 1981.

Oram, Richard L., *Transportation System Management: A Bibliography of Technical Reports*, Urban Mass Transportation Administration, Washington, D.C., 1976.

Paquette, Radnor J., Norman J. Ashford, and Paul M. Wright, *Transportation Engineering*, John Wiley and Sons, New York, 1982.

Transportation System Management, Special Report 172, National Academy of Sciences, Transportation Research Board, Washington, D.C., 1977.

Urbitran Associates, *Transportation System Management: Implementation and Impacts*, Urban Mass Transportation Administration, Washington, D.C., 1982.

Voorhees, Alan M., Inc., *TSM: An Assessment of Impacts*, Urban Mass Transportation Administration, Washington, D.C., 1978.

Location and Geometric Design

Highway location involves the acquisition of data concerning the terrain upon which the road will traverse and the economical siting of an alignment. To be considered are factors of earthwork, geologic conditions, and land use. Geometric design principles are used to establish the horizontal and vertical alignment, including consideration of the driver, the vehicle, and roadway characteristics. Design of parking and terminal facilities must be considered as they form an integral part of the total system. Since the new highway will alter existing patterns of surface and subsurface flow—and be influenced by it—careful attention to the design of drainage facilities is required.

Highway Surveys and Location

Selecting the location of a proposed highway is an important initial step in its design. The decision to select a particular location is usually based on topography, soil characteristics, environmental factors such as noise and air pollution, and economic factors. The data required for the decision process are usually obtained from different types of surveys, depending on the factors being considered. Most engineering consultants and state agencies presently involved in highway locations use computerized techniques to process the vast amounts of data that are generally handled in the decision process. These techniques include remote sensing, which uses aerial photographs for the preparation of maps, and computer graphics, which is a combination of the analysis of computer-generated data with a display on a computer monitor. In this chapter we present a brief description of the current techniques used in highway surveys to collect and analyze the required data and the steps involved in the procedure for locating highways. We also cover earthwork computations and mass diagram, since an estimate of the amount of earthwork associated with any given location is required for an economic evaluation of the highway at that location. The result of the economic evaluation aids in the decision to accept or reject that location.

TECHNIQUES FOR HIGHWAY SURVEYS

Highway surveys usually involve measuring and computing horizontal and vertical angles, vertical heights (elevations), and horizontal distances. The surveys are then used to prepare base maps with contour lines and longitudinal cross sections, as may be required. Highway surveying techniques have been revolutionized during the past decade due to the rapid development of electronic equipment and computers. These techniques can be grouped into three general categories:

- Conventional ground surveys
- Remote sensing
- Computer graphics

Conventional Ground Surveys

Conventional ground surveys were the only location techniques available to highway engineers until the relatively recent developments in electronics. The most important equipment used were the *transit* for measuring angles in both vertical and horizontal planes, the *level* for measuring changes in elevation, and the *tape* for measuring horizontal distances. However, *electronic distance measuring* (*EDM*) *devices* are now widely used.

The Transit

The transit consists of a telescope, with vertical and horizontal cross hairs, a graduated arc or vernier for reading vertical angles, and a graduated circular plate for reading horizontal angles. The telescope is mounted so it can rotate vertically about a horizontal axis, and the amount of rotation can be directly obtained from the graduated arc. Two vertical arms support the telescope on its horizontal axis, with the graduated arc attached to one of the arms. The arms are attached to a circular plate, which can rotate horizontally with reference to the graduated circular plate, thereby providing a means for measuring horizontal angles. The correct name for the transit is *theodolite*, although the term *transit* is commonly used in the United States for the conventional double-center instrument. A typical engineer's transit is shown in Figure 13.1(a).

Measurement of Horizontal Angles.   A horizontal angle, such as angle *XOY* in Figure 13.1(b), can be measured by setting the engineer's transit directly over the point *O*. (Details for setting up a transit can be found in any surveying textbook.) The telescope is then turned toward point *X*, and viewed first over its top and then by sighting through the telescope to obtain a sightline of *OX*, using the horizontal tangent screws provided. The graduated horizontal scale is then set to zero and locked into position, using the clamp screws provided. The telescope is then rotated in the horizontal plane until the line of sight is *OY*. The horizontal angle (α) is then read from the graduated scale.

Measurement of Vertical Angles.   The transit measures vertical angles by using the fixed graduated arc or vertical vernier and a spirit level attached to the telescope. When a transit is used to determine the vertical angle to a point, the angle obtained is the angle of elevation or the angle of depression from the horizontal. The transit is set up in a way similar to that for measuring horizontal angles, ensuring that the longitudinal axis is in the horizontal position. The telescope is then approximately sighted at the point. The tangent screw of the telescope is then used to set the horizontal cross hair at the point. The vertical angle is obtained from the graduated arc or the vertical vernier.

The Level

The essential parts of a level are the telescope, with vertical and horizontal cross hairs, a level bar, a spindle, and a leveling head. The level bar on which the telescope is mounted is rigidly fixed to the spindle. The level tube is attached to the telescope or the level bar so that it is parallel to the telescope. The spindle is

Figure 13.1 Typical Surveying Equipment

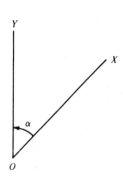

(a) Typical engineer's transit

(b) Measurement of a horizontal angle using a transit

(c) Dumpy level

(d) Wooden surveying rod

Source: Photos by Ed Deasy, Virginia Transportation Research Council, Charlottesville, Va.

fitted into the leveling head in such a way that allows the level to rotate about the spindle as an axis, with the leveling head attached to a tripod. The level also carries a bubble that indicates whether the level is properly centered. The centering of the bubble is done by using the leveling screws provided.

The most common types of levels are the dumpy level, shown in Figure 13.1(c), and the self-leveling or automatic levels. The specific features of these levels are that the telescope of the dumpy level is permanently fixed to the level bar and the self-leveling level automatically levels itself.

A leveling rod is required to measure changes in elevation. These are wooden rods of rectangular cross sections and are graduated in feet or meters so that differences in heights can be measured. Figure 13.1(d) shows a typical leveling rod, graduated in feet.

The difference in elevation between two points, A and B, is obtained by setting and leveling the level at a third point. The rod is then placed at the first point, say, point A, and the level sighted at the rod. The reading indicated by the cross hair is then recorded. The level rod is then transferred to point B, the level is again sighted at the rod, and the reading indicated by the cross hair recorded. The difference between these readings gives the difference in elevation between points A and B. If the elevation of one of the points is known, the elevation of the other point can then be determined.

Measuring Tapes

Tapes are used for direct measurement of horizontal distances. They are available in several materials, but the types used for engineering work usually are made of steel or a woven nonmetallic or metallic material. They are available in lengths of 50, 100, and 150 ft, when graduated in feet. Tapes graduated in both U.S. and metric units are also available.

Electronic Distance Measuring Devices

Several EDM devices with slightly different advantages are available, but they all work on the same general principle. They consist mainly of a transmitter located at one end and a reflector at the other end of the distance to be measured. The transmitter transmits either a light beam, a low-power laser beam, or a high-frequency radio beam in the form of very short waves, which are picked up and reflected back by the reflector to the transmitter. The difference in phase between the transmitted and reflected waves is measured electronically and used to determine the distance between the transmitter and the reflector. Figure 13.2(a) shows the TOPCON DM-A2. This equipment can measure up to about 3300 ft in average atmospheric conditions. When used in conjunction with the slope reduction calculator, it can also measure slope and height differences. Special features permit the operator to change automatically the display from slope to horizontal distance, and so forth. Units can also be changed from meters to feet. More recent EDM devices can be mounted on the framework of a theodolite, which permits the determination of vertical and horizontal angles. The result is that distances and directions can be determined from a single instrument setup. Figure 13.2(b)

Figure 13.2 Typical Electronic Surveying Instruments

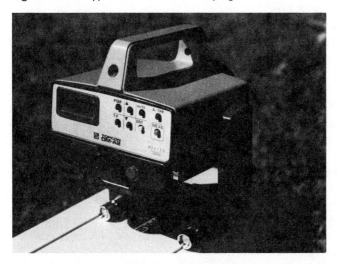

(a) TOPCON DM-A2 electronic distance measuring device

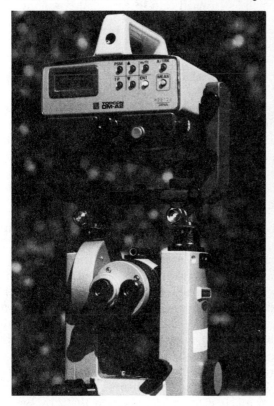

(b) TOPCON DM-A2 mounted on a theodolite for angular and linear measurements

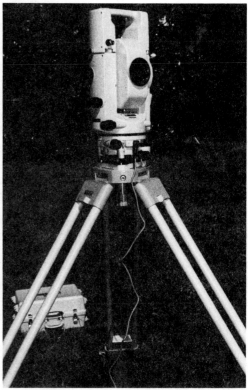

(c) TOPCON GTS-2S geodetic total station

Source: Photos by Ed Deasy, Virginia Transportation Research Council, Charlottesville, Va. with permission of TOPCON, Paramus, N.J.

shows a set up of TOPCON DM-A2 on a theodolite for angular and linear measurement. Figure 13.2(c) shows a geodetic total station, consisting of an EDM and a theodolite in a single instrument.

Remote Sensing

The most commonly used remote sensing method is aerial photography, usually referred to a photogrammetry. Photogrammetry is the process by which distances between objects and other information are obtained indirectly, by taking measurements from aerial photographs of the objects. This process is fast and cheap for large projects but can be very expensive for small projects. The break-even size for which photogrammetry can be used varies between 30 and 100 acres, depending on the circumstances of the specific project. The successful use of the method depends on the type of terrain. Difficulties will arise when it is used for terrain with the following characteristics:

- Areas of thick forest, such as tropical rain forests, that cover the ground surface
- Areas that contain deep canyons or tall buildings, which may conceal the ground surface on the photographs
- Areas that photograph as uniform shades, such as plains and some deserts.

The most common uses of photogrammetry in highway engineering are identification of suitable locations for highways and preparation of base maps with contours of 2 or 5 ft intervals. In both of these uses, the first task is to obtain the aerial photographs of the area if none are available. A brief description of the methodology used to obtain and interpret the photographs is therefore presented. Readers interested in a more detailed treatment of the subject should refer to sources listed in Additional Readings at the end of this chapter.

Obtaining and Interpreting Aerial Photographs.   The photographs are taken from airplanes, with the axis of the camera at a near vertical position. The axis should be exactly vertical, but this position is usually difficult to obtain, because the motion of the aircraft may cause some tilting of the camera up to a maximum of about 5°, although this value on the average is about 1°. Photographs taken this way are vertical aerial photographs and are most useful for highway mapping. In some cases, however, the axis of the camera may be intentionally tilted so that a greater area will be covered by a single photograph. Photographs of this type are known as oblique photographs.

Vertical aerial photographs are taken in a square format, usually 9 in. × 9 in., so that the area covered by each photograph is also a square. The airplane flies over the area to be photographed in parallel runs such that any two adjacent photographs overlap both in the direction of flight and in the direction perpendicular to flight. The overlap in the direction of flight is the forward or end overlap and provides an overlap of about 60 percent for any two consecutive photographs. The overlap in the direction perpendicular to flight is the side

Figure 13.3 End and Side Overlaps in Aerial Photography

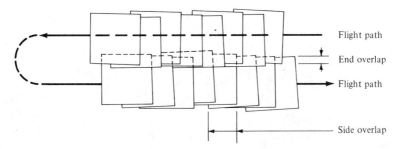

overlap and provides an overlap of about 25 percent between consecutive flight lines, as shown in Figure 13.3. This ensures that each point on the ground is photographed at least twice, which is necessary for obtaining a three-dimensional view of the area. Figure 13.4 shows a set of consecutive vertical aerial photographs, usually called stereopairs. Certain features of the photographs are worth pointing out. The four marks A, B, C, and D that appear in the middle of each side of the photographs are known as fiducial marks. The intersection of the lines joining opposite fiducial marks gives the geometric center of the photograph, which is also called the principal point. It is also necessary to select a set of marked points on the ground that can be easily identified on the photographs as control points. These control points are used to set elevations on the maps developed from the aerial photographs.

Figure 13.4 Set of Stereopairs

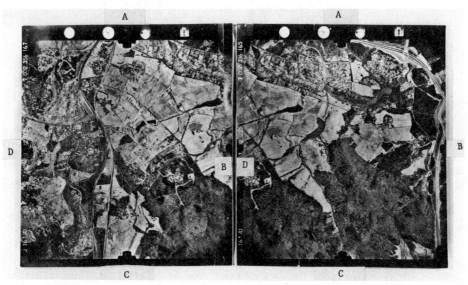

Source: Photos by Ed Deasy, Virginia Transportation Research Council, Charlottesville, Va.

Figure 13.5 Mirror Stereoscope

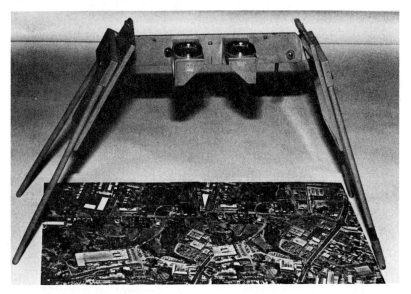

Source: Photo by Nicholas J. Garber.

Figure 13.6 Planicomp CIVO-Analytical Stereo-Plotter

Source: Photo by Sarath C. Joshua with permission of ZEISS, Inc., Thornwood, N.Y.

The information on the aerial photographs is then used to convert these photographs into maps. The instruments used for this process are known as stereoscopes or stereo-plotters and vary from a simple mirror stereoscope, shown in Figure 13.5, to more complex types such as the stereo-plotter shown in Figure 13.6. All of these stereoscopes use the principle of stereoscopy, which is the ability of humans to see objects in three dimension when these objects are viewed by both eyes. When a set of stereopairs is properly placed under a stereoscope, so that an object on the left photograph is viewed by the left eye and the same object is viewed by the right eye, but on the right photograph, the observer perceives the object in three dimensions, and therefore sees the area in the photograph in three dimensions. This technique can be used to produce maps showing features on the ground and the topography of the area.

Scale at a Point on a Vertical Photograph

Figure 13.7 shows a schematic of a single photograph taken of points M and N. The camera is located at O, and it is assumed that the axis is vertical. H is the flying height above the datum XX, and h_m and h_n are the elevations of M and N. If the images are recorded on plane FF, the scale at any point such as M'' on the photograph can be found. For example, the scale at M'' is given as OM''/OM.

Since triangles OMK and $OM''N''$ are similar,

$$\frac{OM''}{OM} = \frac{ON''}{OK}$$

but $ON'' = f = $ focal length of camera and $OK = H - h_m$. The scale at M'' is $f/(H - h_m)$. Similarly, scale at P'' is $f/(H - h_p)$ and scale at N'' is $f/(H - h_n)$.

Figure 13.7 Schematic of a Single Aerial Photograph

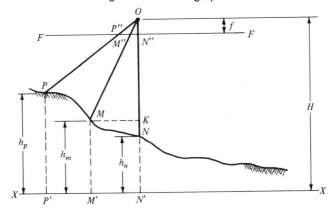

In general, the scale at any point is given as

$$S = \frac{f}{H - h}$$ (13.1)

where

   S = scale at a given point on the photograph
   f = focal length of camera lens
   H = flying height
   h = elevation of point at which the scale is being determined

   Thus, even when the axis of the camera is perfectly vertical, the scale on the photograph varies from point to point and depends on the elevation of the point at which the scale is to be determined. One cannot therefore refer to the scale of an aerial photograph but can refer to the scale *at a point* on an aerial photograph. An approximate overall scale may, however, be obtained by determining the average height (h_{av}) of different points of the photograph and substituting h_{av} for h in Eq. 13.1.

Example 13–1   Computing Elevation of a Point from an Aerial Photograph

The elevations of two points A and B on an aerial photograph are 500 ft and 565 ft, respectively. The scales at these points on the photograph are $1:10{,}000$ and $1:9{,}870$, respectively. Determine the elevation at point C, if the scale at C is $1:8{,}000$.

$$S = \frac{f}{H - h}$$

$$\frac{1}{10{,}000} = \frac{f}{H - 500}$$

$$\frac{1}{9870} = \frac{f}{H - 565}$$

$$H - 500 = 10{,}000f$$

$$H - 565 = 9870f$$

$$65 = 130f$$

where $f = 0.5$ ft $= 6$ in., and $H - 500 = 10{,}000 \times 0.5$, giving $H = 5500$ ft.
   At C,

$$\frac{1}{8000} = \frac{0.5}{5500 - h_C}$$

$$h_C = 5500 - 4000 = 1500 \text{ ft}$$

Example 13–2 Computing Flying Height of an Airplane Used in Aerial Photography

Vertical photographs are to be taken of an area whose mean ground elevation is 2000 ft. If the scale of the photographs should be approximately 1 : 15,000, determine the flying height if the focal length of the camera lens is 7.5 in.

$$f = 7.5 = 0.625 \text{ ft}$$

$$S = \frac{1}{15,000} = \frac{0.625}{H - 2000}$$

$$H = 11,375 \text{ ft}$$

Distances and Elevations from Aerial Photographs

Figure 13.8 is a schematic of a stereopair of vertical photographs taken on a flight. D is the orthogonal projection to the datum plane of a point A located on the ground. The height AD is therefore the elevation of point A. O and O' are the camera positions for the two phtotographs, with OO' being the air base. Images of A on the photographs are a' and a'', whereas d' and d'' are the images of D. The lines $a'd'$ and $a''d''$ represent the radial shift, which is caused by the elevation of point A above point D on the datum plane. This radial shift is known as the parallax, and it is an important parameter when elevations and distances are to be obtained from aerial photographs. The parallax at any point of an aerial photograph can be determined by the use of an instrument called the parallax bar, shown in Figure 13.9. Two identical dots on the bar known as half-marks appear

Figure 13.8 Schematic of a Stereopair of Vertical Photographs

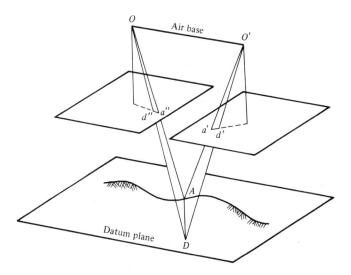

Figure 13.9 Parallax Bar

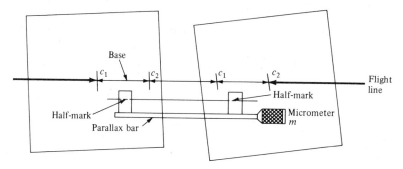

as a single floating dot above the three dimensional view (terrain model) when stereoscopic viewing is attained. The distance between the half-marks can be changed by turning the micrometer, which results in the single dot moving up and down over the terrain model. The micrometer is moved until the dot appears to be just touching the point on the model at which the parallax is required. The distance between the half-marks as read by the micrometer is the parallax at that point. It can be shown that

$$\frac{H-h}{f}=\frac{B}{p}$$ **(13.2)**

where

H = flying height
h = elevation of a given point
p = parallax at the given point
B = air base
f = focal length of the camera

The X and Y coordinates of any point can also be found from the parallax by using the expression for scale as given in Eq. 13.1.

$$S=\frac{f}{H-h}$$

From Eq. 13.2,

$$H=\frac{fB}{p}+h$$

Substituting for H in Eq. 13.1 gives

$$S=\frac{p}{B}$$ **(13.3)**

By establishing a coordinate system in the terrain with the x axis being parallel to the photographic axis of flight, the coordinates of any point are given as

$$X = \frac{x}{S} = \frac{xB}{p} \tag{13.4}$$

and

$$Y = \frac{y}{S} = \frac{yB}{p} \tag{13.5}$$

where X and Y are ground coordinates of the point, and x and y are coordinates of the image of the point on the left photograph.

The horizontal distance between any two points A and B on the ground can therefore be obtained from Eq. 13.6.

$$D = \sqrt{(X_A - X_B)^2 + (Y_A - Y_B)^2} \tag{13.6}$$

Example 13–3   Computing the Distance Between Two Points
on an Aerial Photograph

The scale at the image of a control point on an aerial photograph is $1:15,000$ and the elevation of the point is 1300 ft. The focal length of the camera lens is 6 in. Determine the elevations of points B and C and the horizontal ground distance between them for the following data. Also determine the parallax at B if the air base is 200 ft.

	B	C
x coordinate on axis of left photograph	6 in.	7 in.
y coordinate on axis of left photograph	3 in.	4 in.
Scale on photograph	$1:17,500$	$1:17,400$

For the control point,

$$\frac{1}{15,000} = \frac{0.5}{H - 1300} \qquad \text{or} \qquad H = 8800 \text{ ft}$$

Thus, the elevation at B is found

$$\frac{1}{17,500} = \frac{0.5}{8800 - h_B} \qquad \text{or} \qquad h_B = 50 \text{ ft}$$

and elevation at C is found

$$\frac{1}{17,400} = \frac{0.5}{8800 - h_C} \quad \text{or} \quad h_C = 100 \text{ ft}$$

$X_B = 0.5/(1/17,500) = 8750 \text{ ft}$
$Y_B = 0.25/(1/17,500) = 4375 \text{ ft}$
$X_C = (7.0/12.0)/(1/17,400) = 10,150 \text{ ft}$
$Y_C = (4.0/12)/(1/17,400) = 5800 \text{ ft}$

The distance between B and C_1 using Eq. 13.6 is

$$D = \sqrt{[(8750 - 10,150)^2 + (4375 - 5800)^2]}$$

$$= \sqrt{(1,960,000 + 2,030,625)}$$

$$= 1997.65 \text{ ft}$$

The parallax at B is

$$p_B = \frac{0.5 \times 200}{8750} = 0.01143 \text{ ft (from Eq. 13.4)},$$

$$p_B = \frac{0.25 \times 200}{4375} = 0.01143 \text{ ft (from Eq. 13.5)}$$

Computer Graphics

Computer graphics, when used for highway location, is usually the combination of photogrammetry and computer techniques. The information obtained from photogrammetry is stored in a computer, which is linked with a stereo-plotter and computer monitor. With input of the appropriate command, the terrain model or the horizontal and vertical alignment are obtained and displayed on the monitor. It is therefore easy to change some control points and obtain a new alignment of the highway on the screen, permitting the designer to see immediately the effect of any changes he or she makes.

Figure 13.10 shows the setup of a Galileo stereo digitizer and the Intergraph/Intermap work station, which has good computer graphics capabilities. A typical work station is controlled by system software that covers four main areas of design work:

- Preparatory work
- Restitution and plotting
- Data transfer
- Systems calibration and diagnosis

Figure 13.10   Galileo Stereo Digitizer Intergraph/Intermap Work Station

Source: Photo by Sarath C. Joshua with permission of Galileo Corporation of America, East Chester, N.Y.

The software for preparatory work is used for the input of control data, the determination of the calibration of camera characteristics, and manuscript preparation. The restitution and plotting software programs are used for measuring data from photographs, automatically locating fiducial marks, digital plotting, and on-line computing of horizontal and vertical angles, vertical and slope distances, and radii and centers of circles. The data transfer programs store and check all data and either print them out or put them in a form suitable for use by other programs. The systems calibration and diagnosis programs are used for calibration of the measuring system and interactive checking of all the important functions of the optomechanical units, the operating elements, control electronics, and the digital plotting table.

PRINCIPLES OF HIGHWAY LOCATION

The basic principle for locating highways is that roadway elements such as curvature and grade must blend with each other to produce a system that provides for the easy flow of traffic at the design capacity, while meeting design criteria and safety standards. The highway should also cause a minimal disruption to historic and archeological sites and to other land-use activities. Environmental impact studies are therefore required in most cases before a highway location is finally agreed upon.

The highway location process involves four phases:

- Office study of existing information
- Reconnaissance survey

- Preliminary location survey
- Final location survey

Office Study of Existing Information

The first phase in any highway location study is the examination of all available data of the area in which the road is to be constructed. This phase is usually carried out in the office prior to any field or photogrammetric investigation. All the available data are collected and examined. These data can be obtained from existing engineering reports, maps, aerial photographs and charts, which are usually available at one or more of the state's departments of transportation, agriculture, geology, hydrology, or mining. The type and amount of data collected and examined depend on the type of highway being considered, but in general data should be obtained on the following characteristics of the area:

- Engineering, including, topography, geology, climate and traffic volumes
- Social and demographic, including land use and zoning patterns
- Environmental, including types of wild life, location of recreational, historic, and archeological sites and the possible effects of air, noise, and water pollution
- Economic, including unit costs for construction and the trend of agricultural, commercial, and industrial activities

Preliminary analysis of the data obtained will indicate whether any of the specific sites should be excluded from further consideration because of one or more of the above characteristics. For example, if it is found that a site of historic and archeological importance is located within an area being considered for possible route location, it may be immediately decided that any route that traverses that site should be excluded from further consideration. At the completion of this phase of the study, the engineer will be able to select general areas through which the highway can traverse.

Reconnaissance Survey

The object of this phase of the study is to identify several feasible routes, each within a band of a limited width of a few hundred feet. When rural roads are being considered, there is often very little or no information available on maps or photographs. Aerial photography is therefore widely used to obtain the required information. Feasible routes are identified by a stereoscopic examination of the aerial photographs, taking into consideration factors such as:

- Terrain and soil conditions
- Serviceability of route to industrial and population areas

- Crossing of other transportation facilities, such as rivers, railroads, and other highways
- Directness of route

Control points between the two terminals are determined for each feasible route. For example, a unique bridge site with no alternative may be taken as a primary control point. The feasible routes identified are then plotted on photographic base maps.

Preliminary Location Survey

During this phase of the study, the positions of the feasible routes are set as closely as possible by establishing all the control points and determining preliminary vertical and horizontal alignments for each. Preliminary alignments are used to evaluate the economic and environmental feasibility of the alternative routes.

Economic Evaluation

Economic evaluation of each alternative route is carried out to determine the future effect of investing the resources necessary to construct the highway.

The benefit-cost ratio method described in Chapter 11 is generally used for this evaluation. Factors usually taken into consideration include road user costs, construction costs, maintenance costs, road user benefits, and any disbenefits, which may include adverse impacts due to dislocation of families, businesses, and so forth. The results obtained from the economic evaluation of the feasible routes provide valuable information to the decision maker. For example, these results will provide information on the economic resources that will be gained or lost if a particular location is selected. This information is also used to aid the policy maker in determining whether the highway should be built, and if so, what type of highway it should be.

Environmental Evaluation

Construction of a highway at any location will have a significant impact on its surroundings. A highway is therefore an integral part of the local environment and must be considered as such. This environment includes plant, animal, and human communities and encompasses social, physical, natural, and man-made variables. These variables are interrelated in a manner that maintains equilibrium and sustains the life style of the different communities. The construction of a highway at a given location may result in significant changes in one or more variables, which in turn may offset the equilibrium and result in a significant adverse effect on the environment. This may lead to a reduction of the quality of life of the animals and/or human communities. It is therefore essential that the environmental impact of any alignment selected be fully evaluated.

Federal legislation has been enacted that sets forth the requirements of the environmental evaluation required for different types of projects. In general, the requirements call for the submission of environmental impact statements for many projects. These statements should include:

- A detailed description of alternatives
- The probable environmental impact including the assessment of positive and negative effects
- An analysis of short-term impact as differentiated from long-term impact
- Any secondary effects, which may be in the form of changes in the patterns of social and economic activities
- Probable adverse environmental effects that cannot be avoided if the project is constructed
- Any irreversible and irretrievable resources that have been committed

In cases where an environmental impact study is required, it is conducted at this stage to determine the environmental impact of each alternative route. Such a study will determine the negative and/or positive effects the highway facility will have on the environment. For example, the construction of a freeway at grade through an urban area may result in an unacceptable noise level for the residents of the area, which is a negative impact, or the highway facility may be located so that it provides better access to jobs and recreation centers, which will be considered a positive impact. Public hearings are also usually held at this stage to provide an opportunity for constituents to give their views on the positive and negative impacts of the proposed alternatives.

The best alternative, based on all the factors considered, is then selected as the preliminary alignment of the highway.

Final Location Survey

The final location survey is the detailed layout of the selected route, during which time final horizontal and vertical alignments are determined and the final positions of structures and drainage channels are also determined. The conventional method used is first to set out the points of intersections (PI) of the straight portions of the highway and then to fit a suitable horizontal curve between these. This is usually a trial-and-error-process until, in the designer's opinion, the best alignment is obtained, taking both engineering and aesthetic factors into consideration. Splines and curve templates are available that can be used in this process. The spline is a flexible plastic guide that can be bent into different positions and is used to lay out different curvilinear alignments, from which the most suitable is selected. Curve templates are transparencies giving circular curves, three-center compound curves, and spiral curves of different radii and different standard scales. Figure 13.11 shows circular curve templates, and Figure 13.12 shows three-centered curve templates. The spline is first used to obtain a hand-fitted smooth curve, which fits in with the requirements of grade, cross sections, curvature, and drainage. The hand-fitted curve is then changed to a more defined curve by using the standard templates.

Figure 13.11   Circular Curve Template

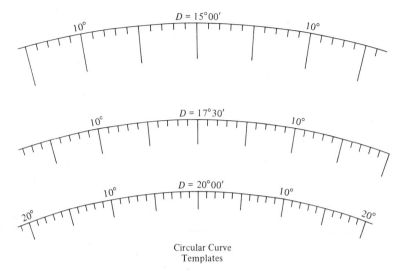

Circular Curve
Templates

The recent availability of computer-based techniques has significantly en-
hanced this process since a proposed highway can be displayed on a monitor,
which enables the designer to have a driver's eye view of both the horizontal and
vertical alignments of the road. The designer can therefore change either or both
alignments until the best alignment is achieved.

Detailed design of the vertical and horizontal alignments are then carried out
to obtain deflection angles for horizontal curves and cuts or fills for vertical
curves and straight sections of the highway. The design of horizontal and vertical
curves is presented in Chapter 14.

Location of Recreational and Scenic Routes

The location process for recreational and scenic routes follow the same steps as
discussed earlier, but the designer of these types of roads must be aware that their
primary purpose is different. For example, although it is essential for freeways
and arterial routes to be as direct as possible from one terminal to the other, a
circuitous alignment may be desirable for recreational and scenic routes to
provide access to recreational sites such as lakes or camp sites or to provide
special scenic views. The designer must realize, however, the importance of adopt-
ing adequate design standards, as given in Chapter 14. Three additional factors
should be considered in the location of recreational and scenic routes.

1.   Design speeds are usually low, therefore special provisions should be made
     to discourage fast driving, for example, by providing a narrow road bed.

Figure 13.12 Centered Curve Template

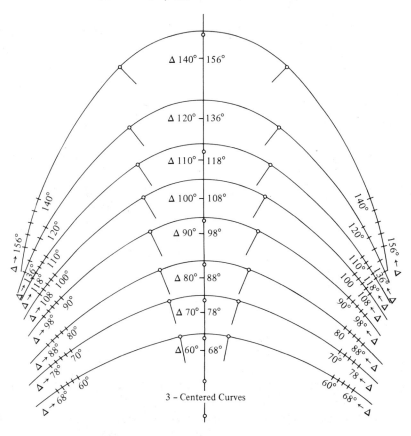

3 – Centered Curves

2. Location should be such that the conflict between the driver's attention on the road and the need to enjoy the scenic view is minimized. This can be achieved by providing turn-outs with wide shoulders and adequate turning space at regular intervals or providing only straight alignments when the view is spectacular.

3. Location should be such that minimum disruption is caused to the area.

Special guides are available for the location and design of recreational and scenic routes. The reader is referred to *A Policy on Geometric Design of Highway and Streets*, published by the American Association of State Highway and Transportation Officials, for more information on this topic.

Location of Highways in Urban Areas

Urban areas usually present complex conditions that must be considered in the highway location process. In addition to factors discussed under office study and reconnaissance survey, other factors that significantly influence the location of highways in urban areas include:

- Connection to local city streets
- Right-of-way acquisition
- Coordination of the highway system and other transportation systems
- Adequate provisions for pedestrians

Connection to Local Streets

When the location of an expressway or urban freeway is being planned, it is important that adequate thought be given to which local streets should connect with on and off ramps to the expressway or freeway. The main factor to consider is the existing travel pattern in the area. The location should enhance the flow of traffic on the local streets. The techniques of traffic assignment, as discussed in Chapter 10, can be used to determine the effect of the proposed highway on the volume and traffic flow on the existing streets. The location should provide for adequate site distances at all ramps. Ramps should not be placed at intervals that will cause confusion or increase the accident potential on the freeway or expressway.

Right-of-Way Acquisition

One factor that significantly affects the location of highways in urban areas is the cost of acquiring right of way. This cost is largely dependent on the predominant land use on the proposed location of the highway. Costs tend to be much higher in commercial areas, and landowners in these areas are often unwilling to give up their land for highway construction. Thus, freeways and expressways in urban areas have been placed on continuous elevated structures in order to avoid the acquisition of rights of way and the disruption of commercial and residential activities. This method of design has the advantage of minimal interference with existing land-use activities, but it is usually objected to by occupiers of adjacent land, because of noise or for aesthetic reasons. The elevated structures are also very expensive to construct and therefore do not eliminate the problem of high costs.

Coordination of the Highway System with Other Transportation Systems

Many urban planners now realize the importance of a balanced transportation system and strive toward a fully integrated highway and public transportation system. This factor should therefore be taken into consideration during the location process of an urban highway. Several approaches have been considered, but the main objective is to provide new facilities that will increase the overall level of service of the transportation system in the urban area. In Washington, D.C., for example, park-and-ride facilities have been provided at transportation terminals to facilitate the use of the subway system, and exclusive bus lanes have been used to reduce the travel time of express buses during the peak hour.

Figure 13.13   Washington Metropolitan Area Transit Authority Rail System in the
Median of Route 66, Northern Virginia

Source: Photo by Ed Deasy, Virginia Transportation Research Council, Charlottesville, Va.

Another form of transportation system integration is the multiple use of
rights of way by both highway and the transit agencies. In this case the right of
way is shared between the two agencies and bus or rail facilities are constructed
either in the median or alongside the freeway. Examples of this include the
WMATA rail system in the median of Interstate 66 in Northern Virginia (see
Figure 13.13), Congress Street and Dan Ryan Expressway in Chicago, and some
sections of Bay Area Rapid Transit in San Francisco.

Adequate Provisions for Pedestrians

Providing adequate facilities for pedestrians should be an important factor in
deciding the location of highways, particularly for highways in urban areas.
Pedestrians are an integral part of any highway system but are more common in
urban areas than in rural areas. Therefore, special attention must be given to the
provision of adequate pedestrian facilities in planning and designing urban highways. Facilities that should be provided include sidewalks, crosswalks, traffic
control features, curb cuts, and ramps for the handicapped. In heavily congested
urban areas, the need for grade-separated facilities, such as overhead bridges
and/or subways, may have a significant effect on the final location of the highway. Although vehicular traffic demands in urban areas are of primary concern
in deciding the location of highways in these areas, the provision of adequate
pedestrian facilities must also be of concern because pedestrians are an indis-
pensable vital component of the urban area.

Principles of Bridge Location

The basic principle for locating highway bridges is that the highway location should determine the bridge location, not the reverse. In most cases when the bridge is located first, the resulting highway alignment is not the best. The general procedure for most highways is, therefore, first to determine the best highway location and this determines the bridge site. In some cases, this may result in skewed bridges, which are more expensive to construct, or locations where foundation problems exist. When serious problems of this nature occur, all factors such as highway alignments, construction costs of bridge deck and foundation, and construction cost of bridge approaches should be considered in order to determine a compromise route alignment that will give a suitable bridge site. This will include the economic analysis of the benefits and costs as discussed in Chapters 9 and 11.

A detailed report should be obtained for the bridge site selected to determine whether there are any factors that make the site unacceptable. This report should include accurate data on soil stratification, the engineering properties of each soil stratum at the location, the crushing strength of bedrock, and water levels in the channel or waterway.

When the waterway to be crossed requires a major bridge structure, however, it is necessary to identify first a narrow section of the waterway with suitable foundation conditions for the location of the bridge and then determine acceptable highway alignments that cross the waterway at that section. This will significantly reduce the cost of bridge construction.

Effect of Terrain on Route Location

One factor that significantly influences the selection of a highway location is the terrain of the land, which in turn affects the laying of the grade line. The primary factor that the designer considers on laying the grade line is the amount of earthwork that will be necessary for the selected grade line. One method to reduce the amount of earthwork is to set the grade line as closely as possible to the natural ground level. This is not always possible, especially in undulating or hilly terrain. The least overall cost may also be obtained if the grade line is set such that there is a balance between the excavated volume and the volume of embankment. Another factor that should be considered in laying the grade line is the existence of fixed points, such as railway crossings, intersections with other highways, and in some cases existing bridges, which require that the grade be set to meet them. When the route traverses flat or swampy areas, the grade line must be set high enough above the water level to facilitate proper drainage and to provide adequate cover to the natural soil. The height of the grade line is usually dictated by the expected flood water level. Grade lines should also be set such that the minimum sight distance requirements as discussed in Chapter 3 are obtained. The criteria for selecting maximum and minimum grade lines are presented in Chapter 14. In addition to these guidelines, the amount of earthwork associated with any grade line influences the decision on whether the grade line should be accepted or rejected.

Volumes of Earthwork

One of the major objectives in selecting a particular location for a highway is to minimize the amount of earthwork required for the project. The estimation of the amount of earthwork involved for each alternative location is therefore required both at the preliminary and final stages.

To determine the amount of earthwork involved for a given grade line, cross sections are taken at regular intervals along the grade line. The cross sections are usually spaced 100 ft apart, although this distance is sometimes increased for preliminary engineering. These cross sections are obtained by plotting the natural ground levels and proposed grade profile of the highway along a line perpendicular to the grade line to indicate areas of excavation and areas of fill. Figure 13.14 shows three types of cross sections. When the computation is done manually, the cross sections are plotted on standard cross-section paper, usually to a scale of 1 in. to 10 ft for both the horizontal and vertical directions. The areas of cuts and fills at each cross section are then determined by the use of a planimeter or any other suitable method. Surveying books document the different methods for area computation. The volume of earthwork is then computed from the cross-sectional areas.

Figure 13.14   Types of Cross Sections

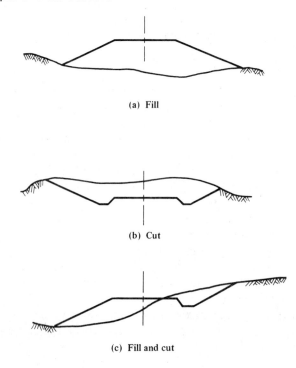

(a) Fill

(b) Cut

(c) Fill and cut

A common method of determining the volume is that of average end areas. The procedure is based on the assumption that the volume between two consecutive cross sections is the average of their areas multiplied by the distance between them as given in Eq. 13.7.

$$V = \frac{L}{54}(A_1 + A_2)$$ (13.7)

where

V = volume (yd³)
A_1 and A_2 = end areas (ft²)
L = distance between cross sections (ft)

Note, however, that Eq. 13.7 is accurate only when $A_1 = A_2$ but is only approximate when $A_1 \neq A_2$. In cases where there is a significant difference between A_1 and A_2 (for example, when one of the end areas approaches zero), it is possible to obtain an error of about 50 percent. This condition usually exists when the grade line moves from a cut section to a fill section as shown in Figure 13.14(c). When such a condition exists it is advisable to calculate the volume as a pyramid as shown in Eq. 13.8.

$$V = \frac{1}{81}(\text{area of base})(\text{length})$$ (13.8)

where V is volume in cubic yards. The area of the base is given in square feet and the length is given in feet.

The average end area method has been found to be sufficiently accurate for most ordinary earthwork computations, since cross sections are usually taken at some distance apart with no consideration given to any minor inequalities on the natural ground between cross sections.

The prismoidal formula is sometimes used when a more accurate volume is desired. It is an application of Simpson's rule and is given as

$$V = \frac{L}{6}(A_1 + 4A_m + A_2)$$ (13.9)

where

V = volume (ft³)
A_1 and A_2 = end areas (ft²)
A_m = middle area determined by averaging corresponding linear dimensions of the end sections (ft²) (note: not by averaging the end areas)

It can be shown that the difference between the volume given by the end area formula and that given by the prismoidal formula, commonly referred to as the prismoidal correction, can be obtained from Eq. 13.10 as

$$C_v = \frac{L}{12}(C_1 - C_2)(d_1 - d_2) \tag{13.10}$$

where

C_v = difference in volume between average area method and prismoidal method (prismoidal correction) in cubic feet

C_1 and C_2 = center heights at end sections

d_1 and d_2 = distance between slope stakes at corresponding end sections

The prismoidal correction is algebraically subtracted from the volume obtained by using the average-end-area formula (Eq. 13.7) to obtain a more accurate volume.

It is common practice in earthwork construction to move suitable materials from cut sections to fill sections to reduce to a minimum, the amount of material borrowed from borrow pits. When the materials excavated from cut sections are compacted at the fill sections, they fill less volume than was originally occupied. This phenomenon is referred to as *shrinkage* and should be accounted for when excavated material is to be reused as fill material. The amount of shrinkage depends on the type of material. Shrinkages of up to 50 percent have been observed for some soils. However, shrinkage factors used are generally between 1.10 and 1.25 for high fills and between 1.20 and 1.25 for low fills. These factors are applied to the fill volume in order to determine the required quantity of fill material. Table 13.1 is a typical volume sheet for computing fill and cut quantities. Column 1 gives the station numbers at which the cross sections are taken. In this case, the cross sections are taken at 100 ft intervals. Values in columns 2 and 3 are the areas of the cross sections.

These areas are assumed to have been computed and are given. The volumes in cubic yards, given in columns 4 and 5, are computed from Eq. 13.7. For example, between station 0 and station 1,

$$\text{cut volume} = \frac{100}{54}(3 + 2)$$

$$= 9.26 \text{ ft}^3$$

$$\approx 9 \text{ ft}^3$$

$$\text{fill volume} = \frac{100}{54}(18 + 50)$$

$$= 125.93 \text{ ft}^3$$

$$\approx 126 \text{ ft}^3$$

Table 13.1 Computation of Fill and Cut Volumes

1	End Area (ft²)		Volume (yd³)				Net Volume (4–7)		10
	2	3	4	5	6	7	8	9	
Station	Total Cut	Total Fill	Total Cut	Fill	Shrinkage 10 percent	Total Fill (5+6)	Fill	Cut	Mass Diagram Ordinate
0	3	18							
			9	126	13	139	130		
1	2	50							−130
			7	272	27	299	292		
2	2	97							−422
			11	420	42	462	451		
3	4	130							−873
			22	335	34	369	347		
4	8	51							−1220
			89	178	18	196	107		
5	40	45							−1327
			157	120	12	132		25	
6	45	20							−1302
			231	46	5	51		180	
7	80	5							−1122
			374	13	1	14		360	
8	122	2							−762
			467	4	0	4		463	
9	130	0							−299
			500	0	0	0		500	
10	140	0							201
			444	6	1	7		437	
11	100	3							638
			333	61	6	67		266	
12	80	30							904
			287	93	9	102		185	
13	75	20							1089
			231	130	13	143		88	
14	50	50							1177
			130	241	24	265	135		
15	20	80							1042
			56	333	33	366	310		
16	10	100							732
			19	407	41	448	429		
17	0	120							303
			6	444	44	488	482		
18	3	120							−179
			80	315	31	346	266		
19	40	50							−445
			130	148	15	163	33		
20	30	30							−478

Figure 13.15   Mass Haul Diagram for Computation Shown in Table 13.1

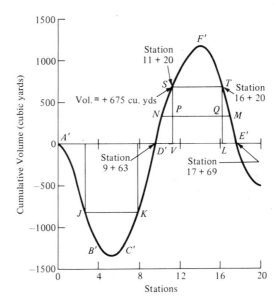

Since Eq. 13.7 is used to compute the cut and fill volumes, the values obtained are only approximate. They may, however, be used to assess the suitability of the alignment. Column 6 is 10 percent of column 5, which is the additional fill material required to form the embankment because of shrinkage. Column 7 is the sum of columns 5 and 6, which is the total amount of fill required. The net volume at each section is given in column 8 or 9. It is obtained by subtracting the fill volume (column 7) from the cut volume (column 4). When this value is negative, the net volume is a fill and when it is positive, the net volume is a cut. The negative sign indicates an amount of material that should be obtained to form the embankment, and the positive sign indicates an excess amount of material excavated. The algebraic cumulative values are given in column 10. These values are used to draw the mass diagram shown in Figure 13.15.

Mass Diagram.   The main reasons for calculating earthwork quantities are to enable the engineer to estimate construction costs and to enable prospective contractors to estimate the project's contract price. The unit price for excavating and transporting the excavated material is usually given for a maximum distance known as the *free-haul* distance. The length of this distance may vary from 500 ft to 3000 ft, depending on the state. When it is necessary to transport excavated material greater than the free-haul distance, the added transportation is paid for under a bid item known as *overhaul*. The overhaul distance is the difference between the total distance the material is transported and the free-haul distance. Some states do not consider overhaul, and thus require a unit price for excavation that allows for long hauls. When overhaul is considered, the quantity is given either in units of station-yards or cubic-yard-mile. The station yard is the overhaul distance in stations multiplied by the volume in cubic yards overhauled. Stations are usually located 100 ft apart. Cubic-yard-mile is the volume in cubic yards multiplied by the overhaul distance in miles.

The mass diagram presents in graphical form the net amount of cut and fill along the length of the project and also indicates the direction of haul. This provides a convenient means for computing payment and provides a valuable aid to engineers supervising work in the field. Figure 13.15 shows a mass haul diagram drawn for the data shown in Table 13.1. It is useful to understand the following characteristics of a mass diagram.

1. Sections of the curve with steep slopes indicate areas where large amounts of earthwork are to be done, whereas areas of small amounts of earthwork are indicated by flat slopes. For example, in Figure 13.15, the amount of earthwork between points A^1 and B^1 (station 0 to station 4) is much greater than that between points B^1 and C^1 (station 4 to station 6).

2. Raising sections of the mass curve indicate areas where excavation exceeds fill, (C^1 to D^1), whereas falling sections indicate where fill exceeds excavation (A^1 to B^1).

3. The net difference of earthwork between any two points is the difference between the ordinates at these points.

4. Points at which the earthwork changes from cut to fill or from fill to cut are generally indicated by zero slopes. The points of zero slope may not always coincide exactly with the point at which the grade line changes from cut to fill or fill to cut, since in some cases, the irregularity of the cross slope at this point may result in a net excess of excavation or fill.

5. Any horizontal line drawn to intersect two points within the same curve indicates a balance of excavation and fill quantities between the two points. Balance points on the base line are therefore A^1, D^1, and E^1.

6. In cases where there is a balance of fill and excavated quantities, both ends of the mass curve will lie on the base line. This, however, does not always occur, as shown in Figure 13.15, where there is a net fill of 478 yd³. This material will have to be borrowed from a borrow pit. When an excess of material is indicated by the mass curve ending above the base line, that amount of excess material must be wasted.

The procedure for computing the overhaul is illustrated using the upper part of the curve between balanced points D^1 and E^1 and assuming that the free-haul distance is 500 ft. Since the curve rises between D^1 and F^1 and falls between F^1 and E^1, it follows that the net excavated material obtained between stations $(9 + 63)$ and $(11 + 20)$ should be moved to the area between stations $(16 + 20)$ and $(17 + 69)$. The procedure is illustrated by the following steps:

Step 1. Since the horizontal distance is greater than 500 ft, it is first necessary to find the free-haul section. This is done by drawing a horizontal line 500 ft in length, within the curve, with it ends on the curve. This is line ST in Figure

13.15. Note that if the distance between balance points on the base line is less than the free-haul distance, there is no overhaul. For example, if the free-haul distance in this example was 1000 ft, there is no overhaul between D^1 and E^1 because the distance between these two points is only about 800 ft. For the free-haul distance of 500 ft, however, the ends of the balanced line are located at stations $(11 + 20)$ and $(16 + 20)$. This means that all earthworks above ST are not subject to additional overhaul payment. The total volume of 675 ft³ excavated between stations $(9 + 63)$ and $(11 + 20)$ should be moved and placed on the embankment between stations $(16 + 20)$ and $(17 + 69)$.

Step 2.   The distance through which the excavated material is moved is next determined. This distance is taken as that between the center of mass of the excavation and the center of mass of the fill. The three common manual methods used to determine the center of mass of the embankment and the fill are the graphical method, the planimeter method, and the method of moments.

The *graphical method* is approximate and suitable only for mass curves with fairly uniform shapes. A horizontal line is drawn, bisecting the ordinate representing the volume that is to be overhauled. When this line is extended to the curve, its length is approximately the distance between the centers of mass of the excavation and the fill. In Figure 13.15, line PQ bisects the ordinates SV and TL, which represent the volume to be overhauled. The line PQ is then extended to points N and M on the curve. Line NM represents the approximate distance between the centers of mass. Length of NM is approximately 6.4 stations.

The *planimeter method* uses Eq. 13.11 to determine the location of the center of mass from the line representing the volume to be overhauled as

$$D = \frac{\text{area} \times \text{horizontal scale} \times \text{vertical scale}}{\text{volume ordinate}} \qquad \textbf{(13.11)}$$

where

$D =$ distance of center of mass from the line representing the overhaul
    volume in stations

area $=$ planimetered area under volume curve (in.²)

horizontal scale $=$ horizontal scale of mass diagrams in stations per inch

vertical scale $=$ vertical scale of mass diagrams in stations per inch

The center of mass of the excavation in the example is found as follows. In locating the center of mass for the cut section, the area under the volume curve between stations $(9 + 63)$ and $(11 + 20)$ was measured using a planimeter and determined to be 0.284 in.²

Horizontal scale $= 4$ stations per inch
Vertical scale $= 500$ ft³ per inch
Volume ordinate $= 675$

Using Eq. 13.11, distance of the center of mass of the excavation from line SV—that is, station $(11 + 20)$ is

$$D = \frac{0.284 \times 4 \times 500}{675} = 0.841 \text{ stations}$$

$$= 84 \text{ ft}$$

The center of mass of the excavation is therefore located at station $(11 + 20) - 84 \text{ ft} = \text{station} (10 + 36)$.

Similarly, for the embankment,

$$\text{planimeter area} = 0.257 \text{ in.}^2$$

$$D = \frac{0.257 \times 4 \times 500}{675} = 0.76 \text{ station}$$

The center of mass of the embankment is located at station $(16 + 20) + 76 \text{ ft} = \text{station} (16 + 96)$.

The haul distance is the distance between the two centers of mass.

$$\text{station} (16 + 96) - \text{station} (10 + 36) = 6.60 \text{ stations}$$

In the *method of moments*, the sum of the moments about a point of each volume in the fill or embankment section of the mass curve is divided by the total overhaul volume to determine the center of mass from the point about which the moments are taken. The volume between consecutive stations is obtained from column 8 or 9 of Table 13.1.

In this case, the distance between center of mass of the excavation and station $(11 + 20)$ is found by taking moments about station $(11 + 20)$.

Station	Volume (yd³)	Distance (ft)	Moment
$(9 + 63)$ to $(10 + 00)$	185	138.5	25,622
$(10 + 00)$ to $(11 + 00)$	437	70.0	30,590
$(11 + 00)$ to $(11 + 20)$	53	10.0	530
Total	675		56,742

Note that the distance (that is, lever arm) is the distance between the midpoint of the stations being considered and the station at which the moment is being taken. For example, the midpoint between stations $(9 + 63)$ and $(10 + 00)$ is station $(9 + 81.5)$. Distance from station $(11 + 20)$ is station $(11 + 20) - $ station $(9 + 81.5) = 1.385$ stations $= 138.5$ ft.

Since the distance of the center of mass from station $11 + 20 = 56,742/675 = 84$ ft, the center of mass is therefore located at station $(11 + 20) - 0.84$ stations, which is $10 + 36$. The same station is obtained in the planimeter method.

Similarly the distance between the center of mass of the embankment and station $(16 + 20)$ is found by taking moments about station $(16 + 20)$.

Station	Volume (yd^3)	Distance (ft)	Moment
$16 + 20$ to $17 + 00$	343	40.0	13,720
$17 + 00$ to $17 + 69$	332	114.5	38,014
Total	675		51,734

Since the distance of the center of mass from station $(16 + 20) = 51,734/675 \approx 76.6$ ft, the center of mass of embankment is located at station $(16 + 96.6)$, which is approximately the same as the planimeter method.

Step 3.   Having determined the total haul, the overhaul is determined by subtracting the free-haul from the total haul—that is, $(6.60 - 5.00) = 1.60$ stations—using a total haul of 6.60 stations.

Step 4.   The overhaul is then computed by multiplying the overhaul volume by the overhaul distance—that is, $675 \times 1.60 = 1080$ yd^3 stations.

It should be noted that the same procedure is used for the lower part of the curve between A^1 and D^1. In this case, the free-haul balance points are J and K and the overhaul volume is about 810 yd^3.

Computer programs are now available that can be used to compute cross-sectional areas and volumes directly from the elevations given at the cross sections. Some programs will also compute the ordinate values for a mass diagram and determine the overhaul, if this is necessary.

PREPARATION OF HIGHWAY PLANS

Once the final location of the highway system is determined, it is then necessary to provide the plans and specifications for the facility. The plans and specifications of a highway are the instructions under which the highway is constructed. They are also used for the preparation of engineer's estimates and contractor's bids. When a contract is let out for the construction of a highway, the plans and specifications are part of the contract documents and are therefore considered legal documents. The plans are drawings that contain all details necessary for proper construction, whereas the specifications give written instructions on quality and type of materials, Figure 13.16 (top) shows an example of a highway plan. Figure 13.16 (bottom) shows the vertical alignment, sometimes referred to as the *profile*, indicating the natural ground surface and the center line of the road, with details of vertical curves. The horizontal alignment is usually drawn to a scale of 1 in. to 100 ft, although in some cases, the scale of 1 in. to 50 ft is used to provide greater detail. In drawing the vertical alignment, the horizontal scale used is the same as that of the horizontal alignment, but the vertical scale is exaggerated five

Figure 13.16 Highway Plan (top: horizontal alignment; bottom: vertical alignment)

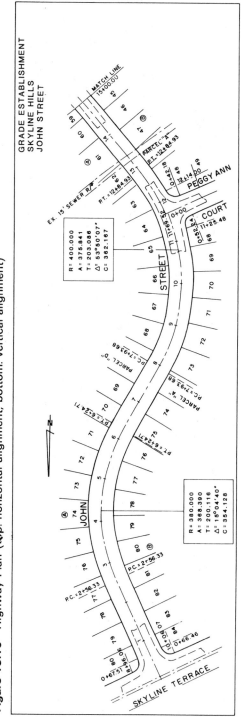

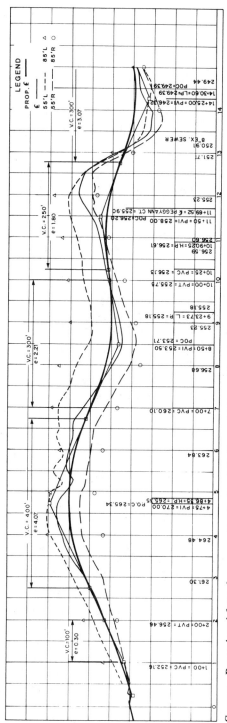

Source: Reproduced from forms supplied by Byrd, Tallamy, McDonald and Lewis, Fairfax, Va.

to ten times. The vertical alignment may also give estimated earthwork quantities at regular intervals, usually at 100 ft stations.

Most state agencies require consultants to prepare final design drawings on standard sheets 36 in. × 22 in. These drawings are then usually reduced to facilitate easy handling in the field during construction. Other drawings showing typical cross sections and specific features such as pipe culverts and concrete box culverts are also provided. Standard drawings of some of these features that occur frequently in highway construction have been provided by some states and can be obtained directly from the highway agencies. Consultants may not have to produce them as part of their scope of work.

SUMMARY

The selection of a suitable location for a new highway requires information obtained from highway surveys. These surveys can be carried out by either conventional ground methods or use of electronic equipment and computers. A brief description of some of the more commonly used methods of surveys has been presented to introduce the reader to these techniques.

A detailed discussion of the four phases of the highway location process has also been presented to provide the reader with the information required and the tasks involved in selecting the location of a highway. The computation of earthwork volumes is also presented since the amount of earthwork required for any particular location may significantly influence the decision to either reject or select that location. Note, however, that the final selection of a highway location, particularly in an urban area, is not now purely in the hands of the engineers. The reason is that citizen groups, with interest in the environment and historical preservation, have been formed and they can be extremely vocal in opposing highway locations that, in their opinion, conflict with their objectives. Thus, in selecting a highway location, the engineeer must take into consideration the environmental impact of the road on its surroundings. Federal regulations also require environmental impact statements for many types of projects, which require a detailed analysis of environmental impacts.

PROBLEMS

13–1 Briefly compare the factors that should be considered in locating an urban freeway with those for a rural freeway.

13–2 The slope distance between two points A and B on a highway is 150 ft. If the difference in elevation between these is 15 ft, what is the horizontal distance between these points?

13–3 Describe how each of the following could be used in highway survey and location: (a) aerial photogrammetry, (b) computer graphics, and (c) conventional survey.

13–4 A photograph is to be obtained at a scale of 1 : 10,000 by aerial photogrammetry. If the focal length of the camera to be used is 6.5 in., determine the height at which the aircraft should be flown if the average elevation of the terrain is 950 ft.

13–5 The distance in the x direction between two control points on a vertical aerial photograph is 4.5 in. If the distance between these same two points is 3.6 in. on another photograph having a scale of 1 : 24,000, determine the scale of the first vertical aerial photograph. If the focal length of the camera is 6.0 in. and the average elevation at these points is 100 ft, determine the flying height from which each photograph was taken.

13–6 The scale at the image of a well-defined object on an aerial photograph is 1 : 24,000 and the elevation of the object is 1500 ft. The focal length of the camera lens is 6.5 in. If the air base (B) is 250 ft, determine the elevation of two points A and C and the distance between them if the coordinates of A and C are as given below.

	A	C
x coordinate	5.5 in.	6.5 in.
y coordinate	3.5 in.	5.0 in.
Scale of photograph	1 : 13,000	1 : 17,400

13–7 Using an appropriate diagram, discuss the importance of the side overlap and forward overlap in aerial photography.

13–8 Discuss briefly the factors that are of specific importance in the location of scenic routes.

13–9 Under what conditions will you prefer borrowing new material from a borrow pit for a highway embankment to using material excavated from an adjacent section of the road?

13–10 Using the data given in Table 13.1, determine the total overhaul cost if the free haul is 700 ft and the overhaul cost is $7.50 per cubic yard station.

ADDITIONAL READINGS

American Association of State Highway and Transportation Officials, *A Policy on Geometric Design of Highways and Streets*, Washington, D.C., 1984.

Davis, R. E., F. S. Foote, J. M. Anderson, and E. M. Makhail, *Surveying Theory and Practice*, McGraw-Hill Book Company, New York, 1981.

Mayfield, G. M., "Engineering Surveying," *Journal of the Surveying and Mapping Division ASCE* 102(SU1). Proceedings paper 12648, December 7, 1976.

Mence, C. F., and D. W. Gibson, *Route Surveying*, 5th ed., Harper & Row, New York, 1980.

Sanyaolu, A., "New Analytic Formulas and Automatic Checks for Earth Works," *The Canada Surveyor* 32(June 1978).

CHAPTER 14

Geometric Design of Highway Facilities

The geometric design of highway facilities deals with the proportioning of the physical elements of highways, such as vertical and horizontal curves, lane widths, cross sections, and parking bays. The characteristics of driver, vehicle, and road, as discussed in Chapter 3, serve as the basis for determining the physical dimensions of these elements. For example, lengths of vertical curves or radii of circular curves are determined such that the minimum stopping sight distance is provided on the curve for the design speed of the highway. Apart from federal requirements, which all states have to satisfy on roads built with federal funds, the laws that limit size and weight of motor vehicles may vary from state to state, resulting in variation of design standards. The basic objective in geometric design of highways, however, is to produce a smooth-flowing, accident-free facility. This can be achieved by having a consistent design standard along the highway that satisfies the characteristics of the drivers and vehicles.

The American Association of State Highway and Transportation Officials (AASHTO) plays a very important role in the development of guidelines and standards used in highway geometric design. The membership of this association consists of representatives from all state highway and transportation departments and the Federal Highway Administration (FHWA). The association has several technical committees that consider suggested standards from individual states. When a standard is approved by the required majority, it is declared as adopted and is accepted by all members of the association. *A Policy on Geometric Design of Highways and Streets*, published by AASHTO, gives all of these standards for geometric design of highways.[1] However, the guidelines given in that publication are based on technical data collected prior to the enactment of the Surface Transportation Assistance Act of 1982, which permitted the increase of the dimensions of truck, tractor-trailer combinations. Care should therefore be taken in using the guidelines when highways are being designed to carry these larger truck combinations.

HIGHWAY FUNCTIONAL CLASSIFICATION

Highways are classified according to their respective functions in terms of the character of the service they are providing. This classification system facilitates the systematic development of highways and the logical assignment of highway responsibilities among different jurisdictions. Highways and streets are primarily described as rural or urban roads, depending on the area in which they are located. This primary classification is essential since urban and rural areas have fundamentally different characteristics, particularly those related to type of land use and population density, which significantly influence travel patterns. Following the primary classification, highways are then classified separately for urban and rural areas under the following categories:

- Principal arterials
- Minor arterials
- Collectors
- Local roads and streets

Freeways are not listed as a separate functional class since they are generally classified as part of the principal arterial system. Note, however, that freeways have unique geometric criteria that require special consideration during design.

Functional System of Urban Roads

Urban roads are all highway facilities within urban areas. Urban areas are usually designated by state and local officials and have populations of 5,000 or more. They are further subdivided into urbanized areas with populations of 50,000 or more and small urban areas with populations between 5,000 and 50,000. Urban roads are functionally classified into principal arterials, minor arterials, collectors, and locals. Figure 14.1 is a schematic of the different classes.

Urban Principal Arterial System

This system of highways serves the major activity centers of the urban area and consists mainly of the highest traffic volume corridors. It carries a high proportion of the total vehicle miles of travel within the urban area and carries most trips with origin or destination within the urban area. The system also serves as a through system for trips that by-pass the central business districts (CBDs) of urbanized areas. All controlled-access facilities are within this system, although controlled access is not necessarily a condition for a highway to be classified as an urban principal arterial. Highways within this system are further divided into three subclasses based mainly on the type of access to the facility: (1) interstate, with fully controlled access and grade-separated interchanges, (2) expressways that have controlled access but may also include at-grade intersections (particularly expressways) and (3) other principal arterials (with partial or no controlled access).

Figure 14.1   Schematic of the Functional Classes of Suburban Roads

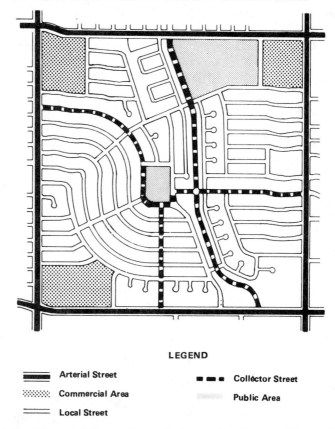

LEGEND

▬▬▬ **Arterial Street**   ■ ■ ■ **Collector Street**

:::::::: **Commercial Area**   **Public Area**

═══ **Local Street**

Source: Reproduced from *A Policy on Geometric Design of Highways and Streets*, Washington, D.C.: The American Association of State Highway and Transportation Officials, copyright 1984. Used by permission.

Urban Minor Arterial System

Streets and highways that interconnect with and augment the urban primary arterials are classified as urban minor arterials. This system serves trips of moderate length and puts more emphasis on land access than the primary arterial system. All arterials not classified as primary are included in this class. Although highways within this system may serve as local bus routes and connect communities within the urban areas, they do not normally go through identifiable neighborhoods.

Urban Collector Street System

The main purpose of streets within this system is to collect traffic from local streets in residential areas or in CBDs and convey it to the arterial system. Thus, collector streets usually go through residential areas and facilitate traffic circulation within residential, commercial, and industrial areas.

Urban Local Street System

This system consists of all other streets within the urban area that are not included in the three systems described earlier. The primary purposes of these streets are to provide access to abutting land and connection to the collector streets. Through traffic is deliberately discouraged on these streets.

Functional System of Rural Roads

Highway facilities outside urban areas form the rural road system. These highways are also divided into the categories of principal arterials, minor arterials, major collectors, minor collectors, and locals. Figure 14.2 is a schematic of the system.

Rural Principal Arterial System

This system consists of a network of highways that serves most of the interstate trips and a substantial amount of intrastate trips. Virtually all highway trips between urbanized areas and a high percentage of the trips between small urban areas with populations of 25,000 or more are made on this system. The system is

Figure 14.2 Schematic of the Functional Classes of Rural Roads

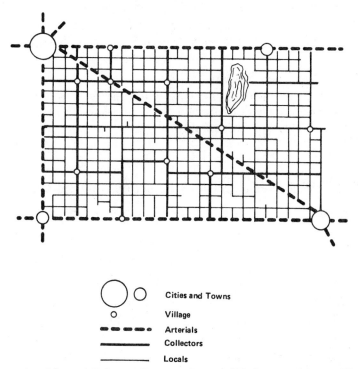

○ ◯   Cities and Towns

○   Village

━ ━ ━ ━ ○   Arterials

━━━━━   Collectors

─────   Locals

Source: Reproduced from *A Policy on Geometric Design of Highways and Streets*, Washington, D.C.: The American Association of State Highway and Transportation Officials, copyright 1984. Used by permission.

further divided into freeways—divided highways with fully controlled access, none at grade intersections—and other principal arterials consisting of all principal arterials not classified as freeways.

Rural Minor Arterial System

This system of roads augments the principal arterial system in the formation of a network of roads that connect cities, large towns, and other traffic generators, such as large resorts. Travel speeds on these roads are usually as high as those on the principal arterial system.

Rural Collector System

Highways within this system primarily carry traffic within individual counties, and trip distances are usually shorter than those on the arterial roads. This system of roads is subdivided into major collector roads and minor collector roads.

Rural Major Collector System.   Routes under this system primarily carry traffic to and from county seats and large cities that are not directly served by the arterial system. The system also carries the main intracounty traffic.

Rural Minor Collector System.   This system consists of routes that collect traffic from local roads and convey it to other facilities. One important function of minor collector roads is that they provide linkage between rural hinterland and locally important traffic generators such as small communities.

Rural Local Road System

This system consists of all roads within the rural area not classified within the other systems. These roads serve trips of relatively short distances and connect adjacent lands with the collector roads.

FACTORS INFLUENCING HIGHWAY DESIGN

Highway design is based on several design standards and controls, which in turn depend on:

- Functional classification of the highway being designed
- Expected traffic volume and vehicle mix
- Design speed
- Topography of the area in which the highway will be located
- Level of service to be provided
- Available funds
- Safety
- Social and environmental factors

These factors are, however, not independent. For example, the design speed depends on the functional classification of the highway, and to a certain extent, the functional classification depends on the expected traffic volume. The design speed may also depend on the topography, particularly in cases where limited funds are available. In general, however, the principal factors used to determine the standards to which a particular highway will be designed are the level of service to be provided, the expected traffic volume, the design speed, and the design vehicle for the highway. These factors coupled with the basic characteristics of drivers, vehicles, and road, as discussed in Chapter 3, are used to determine standards for the geometric characteristics of the highway, such as cross section and horizontal and vertical alignments. For example, appropriate geometric standards should be selected to maintain a desired level of service for a known proportional distribution of different types of vehicles.

HIGHWAY DESIGN STANDARDS

Selection of geometric design standards is the first step in the design of any highway. This is essential as no single set of geometric standards can be used for all highways. For example, geometric standards that may be suitable for a scenic mountain road with low average daily traffic (ADT) will be inadequate for a freeway carrying heavy traffic. The characteristics of the highway should therefore be considered in selecting the geometric standards.

Design Hourly Volume

The design hourly volume (DHV) is the projected hourly volume that is used for design. This volume is usually taken as a percentage of the expected ADT on the highway. Figure 14.3 shows the relationship between traffic hourly volumes as a percentage of ADT and the number of hours in one year with higher volumes. This relationship was computed from the analysis of traffic count data over a wide range of volumes and geographic conditions. Figure 14.3 shows, for example, that an hourly volume equal to 12 percent of the ADT is exceeded at 85 percent of locations during 20 hours in the entire year. A close examination of this curve also shows that between 0 and about 25 highest hours, a small increase in the number of hours results in a significant reduction in the percentage of ADT, whereas a relatively large increase in number of hours at the right of the 30th highest hour results in only a slight decrease in the percentage of ADT. This characteristic of the curve has led to the conclusion that it will be uneconomical to select DHV greater than that which will be exceeded during only 29 hours in a year. The 30th highest hourly volume is therefore usually selected as the DHV. Experience has also shown that the 30th highest hourly volume as a percentage of ADT varies only slightly from year to year, even when significant changes of ADT occur. It has also been shown that, excluding rural highways with unusually high or low fluctuation in traffic volume, the 30th highest hourly volume for rural highways is usually between 12 percent and 18 percent of the ADT, with the average being 15 percent.[2]

Figure 14.3 Relationship Between Hourly Volume (Two Way) and Annual Average Daily Traffic on Rural Roads

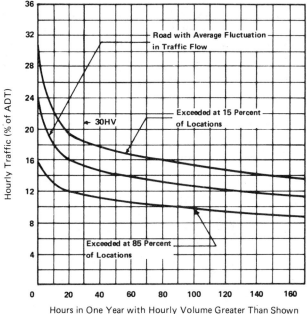

Source: Reproduced from *A Policy on Geometric Design of Highways and Streets*, Washington, D.C.: The American Association of State Highway and Transportation Officials, copyright 1984. Used by permission.

Note, however, that the 30th highest hourly volume should not be indiscriminately used as the DHV, particularly on highways with unusual or high seasonal fluctuation in the traffic flow. Although the percentage of annual average daily traffic (AADT) represented by the 30th highest hourly volume on such highways may not be significantly different from those on most rural roads, this criterion may not be applicable, since the seasonal fluctuation results in a high percentage of high-volume hours and low percentage of low-volume hours. For example, economic consideration may not permit the design to be carried out for the 30th highest hourly volume, but at the same time, the design should not be such that severe congestion occurs during peak hours. A compromise is to select a DHV that will result in traffic operating at a somewhat slightly lower level of service than that which normally exists on rural roads with normal fluctuations. It is therefore suggested that for this type of rural road, it is desirable to select 50 percent of the volume that occurs for only a few peak hours during the design year as the DHV, even though this may not be equal to the 30th highest hourly volume. This may result in some congestion during the peak hour, but the capacity of the highway will normally not be exceeded.

The 30th highest hourly volume may also be used as the DHV for urban highways. It is usually determined by applying between 8 percent and 12 percent to the ADT. Other relations may, however, be used for highways with seasonal fluctuation in traffic flow much different from that on rural roads. One alternative is to use the average of the highest afternoon peak hour volume for each week in the year as the DHV.

Design Speed

Design speed is defined as the "maximum safe speed that can be maintained over a specified section of highway when conditions are favorable such that the design features of the highway govern."[3] Design speed depends on the type of highway, the topography of the area in which the highway is located, and the land use of the adjacent area. It is important that the design speed selected not be significantly different from the speed at which motorists will expect to drive. For example, a low design speed should not automatically be selected for a rural collector road, because when such a road is located in an area of flat topography, motorists will tend to drive at high speeds. The average trip length on the highway is another factor that should be considered in selecting the design speed. In general, highways with longer average trips should be designed for higher speeds.

Design speeds range from 20 mph to 70 mph, with intermediate values at 10 mph increments. Design elements show no significant difference when increments are less than 10 mph but do show very large differences with increments of 15 mph or higher. In general, however, freeways are designed for 60–70 mph, whereas design speeds for other arterial roads range from 30 mph to 60 mph. Tables 14.1 and 14.2 give values for minimum design speeds for different classes of highway.

Note that a design speed is selected to achieve a desired level of operation and safety on the highway. It is one of the first parameters selected in the design process because several other design variables are determined from it.

Table 14.1   Minimum Design Speeds for Rural Collector Roads

Type of Terrain	Minimum Design Speeds (mph) for Design Volumes				
	Current ADT 0–400	Current ADT over 400	DHV 100–200	DHV 200–400	DHV over 400
Level	40	50	50	60	60
Rolling	30	40	40	50	50
Mountainous	20	30	30	40	40

Source: Adapted from *A Policy on Geometric Design of Highways and Streets*, Washington, D.C.: The American Association of State Highway and Transportation Officials, copyright 1984. Used by permission.

Table 14.2   Minimum Design Speeds for Freeways and Arterials

	Design Speed	
Facility	Urban (mph)	Rural (mph)
Freeways	50–60 preferred	70, but 60 in mountains
Arterials	40–60, but 30 in built-up areas	50–70

Source: Adapted from *A Policy on Geometric Design of Highways and Streets*, Washington, D.C.: The American Association of State Highway and Transportation Officials, copyright 1984. Used by permission.

Design Vehicle

The design vehicle is that vehicle selected to represent all vehicles on the highway. Its weight, dimensions, and operating characteristics will be used to establish the geometric standards of the highway, such as radii at intersections and radii of turning roadways. The different classes of vehicles and their dimensions were discussed in Chapter 3. The vehicle type selected as the design vehicle is the largest that is likely to use the highway with considerable frequency. The selected design vehicle is then used to determine such critical design features as radii at intersections and radii of turning roadways.

Cross Section Elements

The principal elements of a highway cross section consist of the travel lanes, shoulders, and medians (for some multilane highways). Marginal elements include median and roadside barriers, curbs, gutters, guardrail, sidewalks, and side slopes. Figure 14.4 shows the cross section for a two-lane highway and Figure 14.5 shows that for a multilane highway.

Figure 14.4   Cross Section for Two-Lane Highways

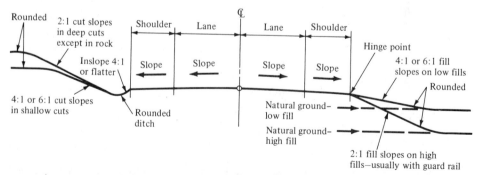

Source: Redrawn from *A Policy on Geometric Design of Highways and Streets*, Washington, D.C.: The American Association of State Highway and Transportation Officials, copyright 1984. Used by permission.

Figure 14.5 Cross Section for Multilane Highways (Half Section)

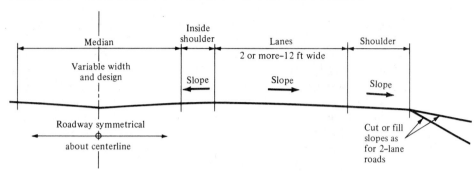

Source: Redrawn from *A Policy on Geometric Design of Highways and Streets*, Washington, D.C.: The American Association of State Highway and Transportation Officials, copyright 1984. Used by permission.

Width of Travel Lanes

Travel lane widths usually vary from 10 ft to 12 ft. Most arterials have 12 ft travel lanes since the extra cost for constructing 12 ft lanes over 10 ft lanes is usually offset by the lower maintenance cost for shoulders and pavement surface, resulting in the reduction of wheel concentrations at the pavement edges. On 2-lane, two-way rural roads, lane widths of 10 ft or 11 ft may be used, but two factors must be considered when selecting a lane width less than 12 ft wide. The first is that it has been shown that when pavement surfaces are less than 22 ft, the clearances between opposing vehicles and between vehicles and the edge of the pavement tend to be inadequate.[4] Secondly, as discussed in Chapters 7 and 8, the capacity of a highway significantly decreases as the lane width is reduced from 12 ft. Lane widths of 10 ft are therefore used only on low-speed facilities. Lanes of 9 ft are occasionally used in urban areas if traffic volume is low and there are extreme right-of-way constraints.

Shoulders

Figures 14.4 and 14.5 show the sections of the highway cross section designated as shoulders. The shoulder is contiguous with the traveled lane and provides an area along the highway for vehicles to stop, particularly during an emergency. Shoulders are also used to laterally support the pavement structure. The shoulder width is known as either *graded* or *usable*, depending on the section of the shoulder being considered. The graded shoulder width is the whole width of the shoulder measured from the edge of the traveled pavement to the intersection of the shoulder slope and the plane of the side slope. The usable shoulder width is that part of the graded shoulder that can be used to accommodate parked vehicles. The usable width is the same as the graded width when the side slope is equal to or flatter than 4 : 1, as the shoulder break is usually rounded to a width between 4 ft and 6 ft, thereby increasing the usable width.

When a vehicle stops on the shoulder, it is desirable for it to be at least 1 ft and preferably 2 ft from the edge of the pavement. Based on this, AASHTO recommends that usable shoulder widths of at least 10 ft and preferably 12 ft be used on highways having a large number of trucks and on highways with high traffic volumes and high speeds. However, it may not always be feasible to provide this minimum width, particularly when the terrain is difficult or when traffic volume is low. A minimum shoulder width of 2 ft may therefore be used on the lowest type of highways but 6 to 8 ft widths should preferably be used. The width for usable shoulders within the median for divided arterials having two lanes in each direction may, however, be reduced to 3 ft, since drivers rarely use the median shoulder for stopping on these roads. The usable median shoulder width for divided arterials with three or more lanes in each direction should be at least 8 ft, since drivers in difficulty on the lane next to the median often find it difficult to maneuver to the outside shoulder.

It is essential that all shoulders be flush with the edge of the traveled lane and be sloped to facilitate the drainage of surface water on the traveled lanes. Recommended slopes are 2 percent to 6 percent for bituminous and concrete-surfaced shoulders and 4 percent to 6 percent for gravel or crushed-rock shoulders.

Medians

A median is the section of a divided highway that separates the lanes in opposing directions. The width is the distance between the edges of the inside lanes, including the median shoulders. The functions of a median include:

- Providing a recovery area for out-of-control vehicles
- Separating opposing traffic
- Providing stopping areas during emergencies
- Providing storage areas for left-turning and U-turning vehicles
- Providing refuge for pedestrians
- Reducing the effect of headlight glare
- Providing temporary lanes and cross-overs during maintenance operations

Medians can either be raised, flushed, or depressed. Raised medians are frequently used in urban arterial streets; they facilitate the control of left-turn traffic at intersections by using part of the median width for left-turn-only lanes. Some disadvantages associated with raised medians include possible loss of control of the vehicle by the driver if the median is accidentally struck and the casting of a shadow from oncoming headlights, which results in drivers having difficulty seeing the curb.

Flushed medians are commonly used on urban arterials. They can also be used on freeways, but with a median barrier. To facilitate drainage of surface water, the flushed median should be crowned. The practice in urban areas of converting flushed medians into two-way left-turn lanes is rather popular, as this helps to increase the capacity of the urban highway while providing some of the features of a median.

Depressed medians are generally used on freeways and are more effective in draining surface water. A side slope of 6 : 1 is suggested for depressed medians, although a slope of 4 : 1 may be adequate.

Median widths vary from a minimum of 2 ft to 80 or more ft. Median widths should be as wide as possible but should be balanced with the other elements of the cross section and the cost involved. In general, the wider the median, the more effective it is in providing safe operating conditions. AASHTO recommends a minimum width of 10 ft for four-lane urban freeways, which is adequate for two 4 ft shoulders, and a 2 ft median barrier.[5] A minimum of 22 ft, preferably 26 ft, is recommended for six or more lanes of freeway.

Median widths for urban collector streets, however, vary from 2 ft to 40 ft, depending on the median treatment. For example, when the median is a paint-striped separation, 2 to 4 ft wide medians are required. For narrow raised or curbed areas, 2 to 6 ft wide medians are required, and for curbed sections, 16 to 40 ft wide medians are required. The larger width is necessary for curbed sections because it provides space for protecting vehicles crossing an intersection. It can also be used for landscape treatment.

Median and Roadside Barriers

AASHTO defines a median barrier "as a longitudinal system used to prevent an errant vehicle from crossing the portion of a divided highway separating the traveled ways for traffic in opposite directions."[6] Roadside barriers, on the other hand, protect vehicles from hazards on the roadside. They may also be used to shield pedestrians and property from the traffic stream. The provision of median barriers must be considered when traffic volumes are high and when access to multilane highways and other highways is only partially controlled. However, when the median of a divided highway has physical characteristics that may create unsafe conditions, such as a sudden lateral drop off or obstacles, the provision of a median barrier should be considered regardless of the traffic volume or the median width. Roadside barriers should be provided whenever conditions exist on the side of the road that warrant protection of vehicles. For example, when the slope of an embankment is high or when there is a roadside object such as a view of an overhead bridge, the provision of a roadside barrier is warranted. Figures 14.6 and 14.7 show typical roadside and median barriers currently in use. For additional information on selecting, locating, and designing median and roadside barriers, see reference 6 and the Michie and Bronstad publications listed in Additional Readings at the end of this chapter.

Curbs and Gutters

Curbs are raised structures made of either Portland cement concrete or bituminous concrete (rolled asphalt curbs) that are used mainly on urban highways to delineate pavement edges and pedestrian walkways. Curbs are also used to control drainage, improve aesthetics, and reduce right of way. Curbs can be generally classified as barriers or mountables. Barrier curbs are relatively high, ranging from 6 to 8 in., with steep sides and are designed to prevent vehicles from leaving the highway. Mountable curbs are designed so that vehicles can cross them if

Figure 14.6 Typical Roadside Barriers

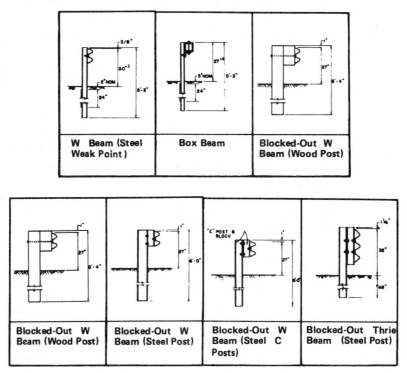

Source: Reproduced from *A Policy on Geometric Design of Highways and Streets*, Washington, D.C.: The American Association of State Highway and Transportation Officials, copyright 1984. Used by permission.

necessary. Figure 14.8 shows some typical highway curbs. Both barrier and mountable curbs may be designed separately or as an integral part of the pavement. In general, barrier curbs should not be used in conjunction with traffic barriers such as bridge railings or median and roadside barriers because they could contribute to vehicles rolling over the traffic barriers. Barrier curbs should also be avoided on highways with design speeds greater than 40 mph because at such speeds, it is usually difficult for drivers to retain control of the vehicle after an impact with the curb.

Gutters or drainage ditches are usually located on the pavement side of a curb to provide the principal drainage facility for the highway. They are sloped to prevent any hazard to traffic, and usually have cross slopes of 5 percent to 8 percent and are 1 to 6 ft wide. Gutters can be designed as V-type sections or as broad, flat, rounded sections.

Guard Rails

Guard rails are longitudinal barriers placed on the outside of sharp curves and at sections with high fills. Their main function is to prevent vehicles from leaving the

Figure 14.7 Typical Median Barriers

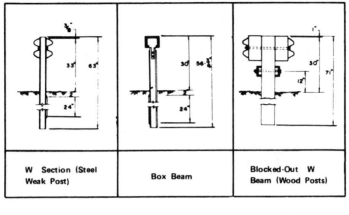

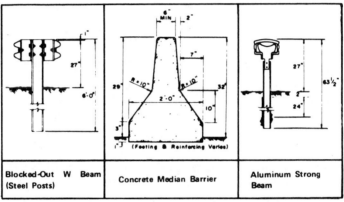

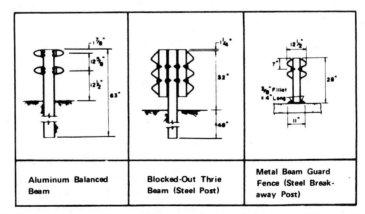

Figure 14.8   Typical Highway Curbs

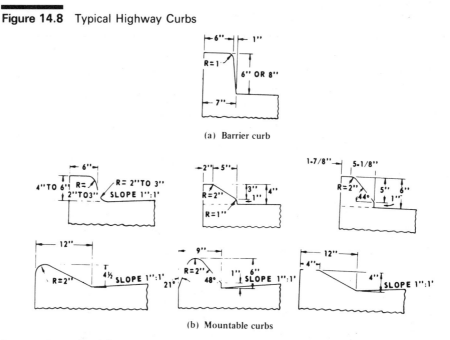

(a) Barrier curb

(b) Mountable curbs

Source: Reproduced from *A Policy on Geometric Design of Highways and Streets*, Washington, D.C.: The American Association of State Highway and Transportation Officials, copyright 1984. Used by permission.

roadbed. They are installed at embankments higher than 8 ft and when shoulder slopes are greater than 4 : 1. Research has been done on the design of guard rails, particularly in the treatment of the end sections. The main objective of these research activities has been to select the best materials and shapes for guard rails. Shapes commonly used include the W beam and the box beam. Research has also led to the development of the weak post system, which provides for the post to collapse on impact, with the rail deflecting and absorbing the energy due to impact. A summary of some of the research results can be found in the Michie and Bronstad publications listed in Additional Readings at the end of this chapter.

Sidewalks

Sidewalks are usually provided on roads in urban areas but very seldom are they provided in rural areas. The provision of sidewalks in rural areas should, however, be evaluated in the planning of rural roads to determine if any sections of the road require them. For example, rural high-speed highways may require sidewalks at areas with high pedestrian concentrations, such as areas adjacent to schools, industrial plants, and local businesses. Generally, sidewalks should be provided when pedestrian traffic is high along main or high-speed roads in either rural or urban areas. When no shoulders are provided on arterials, sidewalks are necessary even when pedestrian traffic is low. In urban areas, sidewalks should also be provided along both sides of collector streets that serve as pedestrian access to schools, parks, shopping centers, and transit stops, and along collector streets in commercial areas. Sidewalks should have a minimum clear width of 4 ft in residential areas and a range of 4 to 8 ft in commercial areas.

To encourage pedestrian use of sidewalks during all weather conditions, they should have all-weather surfaces; otherwise pedestrians will tend to use the traffic lanes.

Cross Slopes

Pavements on straight sections of two-lane and multilane highways without medians are sloped from the middle downward to both sides of the highway. This provides a cross slope, whose cross section can be either curved or plane or a combination of the two. The parabola is generally used for the curved cross section. In this case, the highest point of the pavement (the crown) is slightly rounded, with the cross slope increasing toward the pavement edge. Plane cross slopes consist of uniform slopes at both sides of the crown. The curved cross section has one advantage in that the slope increases outward to the pavement edge, thereby enhancing the flow of surface water away from the pavement. One major disadvantage is that they are difficult to construct.

The cross slopes on divided highways are provided by either crowning the pavement in each direction, as shown in Figure 14.9(a), or by sloping the whole pavement in one direction as shown in Figure 14.9(b). The advantage of draining the pavement in each direction separately is that surface water is quickly drained away from the traveled roadway during heavy rain storms, whereas the disadvantage is that more drainage facilities, such as inlets and underground drains, are required. This method is therefore mainly used at areas with heavy rain and snow falls.

Figure 14.9 Basic Cross Slope Arrangements for Divided Highways

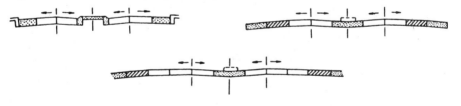

(a) Each pavement slopes two ways.

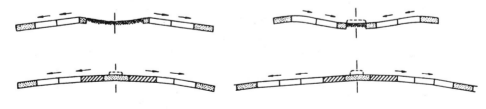

(b) Each pavement slopes one way.

Source: Reproduced from *A Policy on Geometric Design of Highways and Streets*, Washington, D.C.: The American Association of State Highway and Transportation Officials, copyright 1984. Used by permission.

In determining the rate of cross slope for design, two conflicting factors should be considered. Although a steep cross slope is required for drainage purposes, it may be undesirable in that vehicles will tend to drift to the edge of the pavement, particularly under icy conditions. Recommended rates of cross slopes are 1.5 percent to 2 percent for high-type pavements and 1.5 percent to 3 percent for intermediate-type pavements. High-type pavements have wearing surfaces that can adequately support the expected traffic load without visible distress due to fatigue and are not susceptible to weather conditions. Intermediate-type pavements have wearing surfaces that range from surface treated to those with qualities just below that of high-type pavements. Low-type pavements are used mainly for low-cost roads and have wearing surfaces ranging from untreated loose material to surface-treated earth.

Side Slopes

Side slopes are provided on embankments and fills to provide stability for earthworks. They also serve as a safety feature by providing a recovery area for out-of-control vehicles. When being considered as a safety feature, the important sections of the cross slope are the hinge point, the foreslope, and the toe of the slope as shown in Figure 14.10. The hinge point is potentially hazardous because it may cause vehicles to jump into the air while crossing it, resulting in loss of control by the driver. Rounding the hinge point enhances the control of the vehicle by the driver. The fore slope is the area that serves principally as a recovery area, where vehicle speeds can be reduced and other recovery maneuvers taken to regain control of the vehicle. The gradient of the foreslope should therefore not be high. Slopes of 3 to 1 or flatter are generally used for high embankments. This can be increased only when conditions at the site dictate it. The recommended values for foreslopes and backslopes are given in Table 14.3.

Figure 14.10   Regions of a Cross Slope

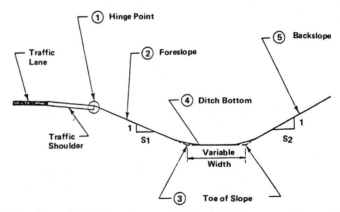

Source: Reproduced from *A Policy on Geometric Design of Highways and Streets*, Washington, D.C.: The American Association of State Highway and Transportation Officials, copyright 1984. Used by permission.

Table 14.3 Guide for Earth Slope Design

Height of Cut or Fill (ft)	Earth Slope, Horizontal to Vertical, for Type of Terrain		
	Flat or Rolling	Moderately Steep	Steep
0–4	6 : 1	6 : 1	4 : 1
4–10	4 : 1	4 : 1	2 : 1*
10–15	4 : 1	2.50 : 1	1.75 : 1*
15–20	2 : 1*	2 : 1*	1.75 : 1*
Over 20	2 : 1*	2 : 1*	1.75 : 1*

*Slopes 2 : 1 or steeper should be subject to a soil stability analysis and should be reviewed for safety.

Source: Adapted from *A Policy on Geometric Design of Highways and Streets*, Washington, D.C.: The American Association of State Highway and Transportation Officials, copyright 1984. Used by permission.

To facilitate the safe movement of vehicles from the fore slope to the back slope, the toe of slope is rounded up as shown in Figure 14.10.

Right of Way

The right of way is the total land area acquired for the construction of the highway. Its width should be enough to accommodate all the elements of the highway cross section, any planned widening of the highway, and any public utility facilities that will be installed along the highway. In some cases, however, the side slopes may be located outside the right of way on easement areas. The right of way for 2-lane urban collector streets should be between 40 and 60 ft, whereas the desirable minimum for 2-lane arterials is 84 ft. Right-of-way widths for divided arterials range from about 92 to 292 ft when frontage roads are not included. A minimum desirable width is 134 ft, which allows for a median at least

Table 14.4 Recommended Minimum Right-of-Way Widths for Ground Level Freeways

Number of Lanes	Typical		Restricted	
	With One-Way Frontage Roads on Either Side	Without Frontage Roads	With Two-Way Frontage Roads on One Side	Without Frontage Roads
4	315*	225*	140	100
6	315	225	175	135
8	340	230	200	160

*Provides for wider medians.

Source: Adapted from *A Policy on Geometric Design of Highways and Streets*, Washington, D.C.: The American Association of State Highway and Transportation Officials, copyright 1984. Used by permission.

16 ft wide. The minimum right-of-way widths for freeways depend on the number of lanes and the existence of a frontage road. Table 14.4 shows the recommended minimum widths. The widths for typical freeway sections include 12 ft lane widths, a 56 ft median, 50 ft outer separation, 30 ft frontage roads, and 15 ft borders.[7] For the restricted cross section, the widths include a 24 ft outer separation, 12 ft lanes, a 10 ft median for 4-lane highways and 22 ft medians for 6- or 8-lane highways, 30 ft two-way frontage road (where this exists), with a 10 ft border.

DESIGN OF THE ALIGNMENT

The vertical and horizontal layouts of the highway make up the alignment. The design of the alignment primarily depends on the design speed selected for the highway. The least costly alignment is one that generally takes the form of the natural topography. This is not often possible, however, as the designer has to adhere to certain standards that may not exist on the natural topography. It is important that the alignment of a given section has consistent standards to avoid sudden changes in the vertical and horizontal layout of the highway. It is also important that both horizontal and vertical alignments be designed to complement each other, as this will result in a safer and more attractive highway. Factors that should be considered to achieve this include the proper balancing of grades of tangents with curvatures of horizontal curves and the location of horizontal and vertical curves with respect to each other. For example, a design that achieves horizontal curves with large radii at the expense of steep or long grades is a poor design. Similarly, if sharp horizontal curves are placed at or near the top of pronounced crest vertical curves or at or near the bottom of a pronounced sag vertical curve, this will create hazardous sections of the highway. It is important that this coordination of the vertical and horizontal alignments be considered at the early stages of preliminary design.

Vertical Alignment

The vertical alignment of a highway consists of straight sections of the highway known as grades, or tangents, connected by vertical curves. The design of the vertical alignment therefore involves the selection of suitable grades for the tangent sections and the design of the vertical curves. The topography of the area through which the road traverses has a significant impact on the design of the vertical alignment. For highway design, topography is generally classified into three groups: level terrain, rolling terrain, and mountainous terrain.

Level terrain is relatively flat, and horizontal and vertical sight distances are generally long or can be achieved without much construction difficulty or major expense.

Rolling terrain has natural slopes that often rise above and fall below the highway grade, with occasional steep slopes that restrict the normal vertical and horizontal alignments.

Mountainous terrain has sudden changes in ground elevation in both the longitudinal and transverse directions, thereby, requiring frequent hillside excavations to achieve acceptable horizontal and vertical alignments. The criteria for describing different topographies were presented in Chapter 7.

Grades

The effect of grade on the performance of heavy vehicles was discussed in Chapter 3, where it was shown that the speed of a heavy vehicle can be significantly reduced if the grade is steep and/or long. In Chapter 7, we noted that steep grades not only affect the performance of heavy vehicles, but also affect the performance of passenger cars. This fact is therefore taken into consideration when the level of service at specific grades of 2-lane, two-way rural highways is being computed. In order to limit the effect of grades on vehicular operation, the maximum grade on any highway should be selected judiciously.

The selection of maximum grades for a highway depends on the design speed and the design vehicle. It is generally accepted that grades of 4 percent to 5 percent have little or no effect on passenger cars, except for those with high weight/horsepower ratios as in compact and subcompact cars. As the grade increases above 5 percent, however, speeds of passenger cars decrease on upgrades and increase on downgrades.

The grade has much more impact on trucks than it does on passenger cars. Extensive studies have been conducted and results (some of which were presented in Chapter 7) have shown that truck speed may increase up to 5 percent on downgrades, and reduce by 7 percent on upgrades, depending on the percent and length of the grade.

The impact of grades on recreational vehicles is more significant than that for passenger cars, but not as critical as that for trucks. It is, however, very difficult to establish maximum grades for recreational routes, and it may be necessary to provide climbing lanes on steep grades when the percentage of recreational vehicles is high.

Maximum grades have been established, based on the operating characteristics of the design vehicle on the highway. These vary from 5 percent for a design speed of 70 mph to between 7 percent and 12 percent for 30 mph, depending on the type of highway. Table 14.5 gives recommended values of maximum grades. Note, however, that these recommended maximum grades should not be used frequently, particularly when grades are long and the traffic includes a high percentage of trucks. On the other hand, when grade lengths are less than 500 ft and roads are one-way in the downgrade direction, maximum grades may be increased by 1 percent to 2 percent, particularly on low volume rural highways.

Minimum grades depend on the drainage conditions of the highway. Zero percent grades may be used on uncurbed pavements with adequate cross slopes to laterally drain the surface water. When pavements are curbed, however, a longitudinal grade should be provided to facilitate the longitudinal flow of the surface water. It is customary to use a minimum of 0.5 percent in such cases, although this may be reduced to 0.3 percent on high-type pavement constructed on suitably crowned, firm ground.

Table 14.5 Recommended Maximum Grades

Rural Collectors[a]

Type of Terrain	Design Speed (mph)					
	20	30	40	50	60	70
	Grades (%)					
Level	7	7	7	6	5	4
Rolling	10	9	8	7	6	5
Mountainous	12	10	10	9	8	6

Urban Collectors[a]

Type of Terrain	Design Speed (mph)					
	20	30	40	50	60	70
	Grades (%)					
Level	9	9	9	7	6	5
Rolling	12	11	10	8	7	6
Mountainous	14	12	12	10	9	7

Rural Arterials

Type of Terrain	Design Speed (mph)		
	50	60	70
	Grades (%)		
Level	4	3	3
Rolling	5	4	4
Mountainous	7	6	5

Freeways[b]

Type of Terrain	Design Speed (mph)		
	50	60	70
	Grades (%)		
Level	4	3	3
Rolling	5	4	4
Mountainous	6	6	5

[a]Maximum grades shown for rural and urban conditions of short lengths (less than 500 ft) and on one-way downgrades may be 1 percent steeper.

[b]Grades 1 percent steeper than the value shown may be used for extreme cases in urban areas where development precludes the use of flatter grades and for one-way downgrades except in mountainous terrain.

Source: Adapted from *A Policy on Geometric Design of Highways and Streets*, Washington, D.C.: The American Association of State Highway and Transportation Officials, copyright 1984. Used by permission.

Figure 14.11   Types of Vertical Curves

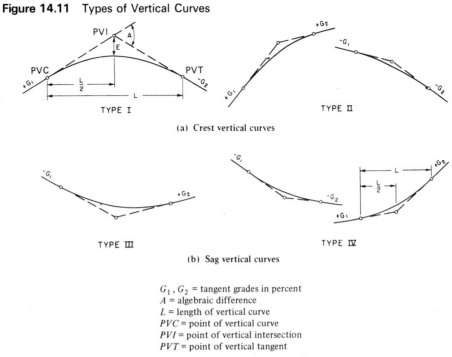

(a) Crest vertical curves

(b) Sag vertical curves

G_1, G_2 = tangent grades in percent
A = algebraic difference
L = length of vertical curve
PVC = point of vertical curve
PVI = point of vertical intersection
PVT = point of vertical tangent

Source: Reproduced from *A Policy on Geometric Design of Highways and Streets*, Washington, D.C.: The American Association of State Highway and Transportation Officials, copyright 1984. Used by permission.

Vertical Curves

Vertical curves are used to provide a gradual change from one tangent grade to another, so that vehicles may run smoothly as they traverse the highway. These curves are usually parabolic in shape because the vertical offsets of parabolas can be computed easily. They are classified as crest vertical curves or sag vertical curves. The different types of vertical curves are shown in Figure 14.11.

The main criteria used for designing vertical curves are

- Provision of minimum stopping sight distance
- Adequate drainage
- Comfortable in operation
- Pleasant appearance

Only the first criterion is associated with crest vertical curves, whereas all four criteria are associated with sag vertical curves.

Crest Vertical Curves.   The design of a crest vertical curve involves the determination of the minimum length that will satisfy the sight distance requirements and then the computation of the elevation on the curve at regular intervals. It was shown in Chapter 3 that to certify the stopping sight distance requirements, the following expressions for the minimum length of the curve apply.

For $S < L$

$$L = \frac{AS^2}{200(\sqrt{H_1} + \sqrt{H_2})^2}$$

and for $S > L$

$$L = 2S - \frac{200(\sqrt{H_1} + \sqrt{H_2})^2}{A}$$

where

  L = minimum length of vertical curve
  S = minimum stopping sight distance
  A = algebraic difference in grades (percent)
  H_1 = height of driver's eye above roadway surface (ft)
  H_2 = height of object above road surface (ft)

When H_1 is assumed to be 3.5 ft and H_2 0.5 ft, then

$$L = \frac{AS^2}{1329} \qquad \text{(for } S < L\text{)} \qquad\qquad \textbf{(14.1)}$$

and

$$L = 2S - \frac{1329}{A} \qquad \text{(for } S > L\text{)} \qquad\qquad \textbf{(14.2)}$$

Equation 14.1 can be written as

$$L = KA \qquad\qquad \textbf{(14.3)}$$

where K is the length of the vertical curve per percent change in A. K can therefore be used as a simple and convenient way to establish design control for crest vertical curves.

Table 14.6 gives values for K based on stopping sight distance requirements. The use of K as a design control is convenient, since the value for any design speed will represent all combinations of A and L for that speed. Similarly, K values can be computed for the case where the sight distance is greater than the vertical curve. A graph of A versus the minimum length of crest vertical curves for different speeds is shown in Figure 14.12. The K values plotted are the upper rounded values, which are higher than the computed values. The lower left section of Figure 14.12 gives values for L when S is greater than L. This section of the curve gives zero values for L for low values of A, because the sight line of the

Table 14.6   Design Controls for Crest Vertical Curves Based on Stopping Sight Distance

Design Speed (mph)	Assumed Speed for Condition (mph)	Coefficient of Friction, f	Stopping Sight Distance, Rounded for Design (ft)	Rate of Vertical Curvature, K[a] (length in feet per % of A)	
				Computed[b]	Rounded for Design
20	20–20	0.40	125–125	8.6–8.6	10–10
25	24–25	0.38	150–150	14.4–16.1	20–20
30	28–30	0.35	200–200	23.7–28.8	30–30
35	32–35	0.34	225–250	35.7–46.4	40–50
40	36–40	0.32	275–325	53.6–73.9	60–80
45	40–45	0.31	325–400	76.4–110.2	80–120
50	44–50	0.30	400–475	106.6–160.0	110–160
55	48–55	0.30	450–550	140.4–217.6	150–220
60	52–60	0.29	525–650	189.2–302.2	190–310
65	55–65	0.29	550–725	227.1–394.3	230–400
70	58–70	0.28	625–850	282.8–530.9	290–540

[a]Different K values for the same speed result from using unequal coefficients of friction.

[b]Using computed values of stopping sight distance.

Source: Adapted from *A Policy on Geometric Design of Highways and Streets*, Washington, D.C.: The American Association of State Highway and Transportation Officials, copyright 1984. Used by permission.

driver goes over the apex of the curve. The minimum lengths obtained for the case of S greater than L are therefore not practical design values and are generally not used. The common practice for this condition is for individual states to set minimum limits, which range from 100 ft to 300 ft. Alternatively, minimum lengths can be set at 3 times the design speed, which gives values that are directly dependent on the design speed.

Having determined the minimum length of the crest vertical curve, the elevations of the curve at regular intervals can then be determined. This is done by considering the properties of the parabola. Consider the symmetrical crest vertical curve in the form of a parabola shown in Figure 14.13.

From the properties of a parabola, $y = ax^2$, where a is a constant. Rate of change of slope is

$$\frac{d^2Y}{dx^2} = 2a$$

but

$$T_1 = T_2 = T$$

$$L = 2T$$

Figure 14.12   Minimum Lengths for Crest Vertical Curves

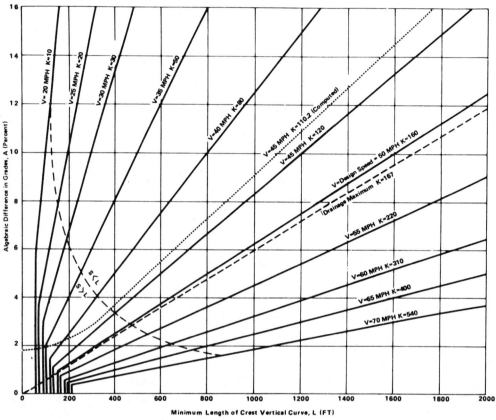

If the total change in slope is A, then

$$2a = \frac{A}{100L}$$

and

$$a = \frac{A}{200L}$$

The equation of the curve can therefore be written as

$$Y = \frac{A}{200L} x^2 \qquad\qquad \textbf{(14.4)}$$

Figure 14.13 Layout of a Crest Vertical Curve for Design

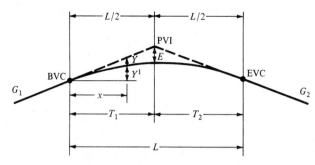

PVI = point of vertical intersection
BVC = beginning of vertical curve (same point as PVC)
EVC = end of vertical curve (same point as PVT)
E = external distance
G_1, G_2 = grades of tangents (%)
L = length of curve
A = algebraic difference of grades, $G_1 - G_2$

when $x = L/2$. The external distance E from the point of vertical intersection (PVI) to the curve is given as

$$E = \frac{A}{200L} \left(\frac{L}{2}\right)^2 = \frac{AL}{800} \qquad (14.5)$$

If stations are given in 100 ft intervals, E can be given as

$$E = \frac{AN}{8} \qquad (14.6)$$

where N is the length of the curve in stations. The vertical offset Y at any point on the curve can also be given in terms of E. Substituting $800E/L$ for A in Eq. 14.4 will give

$$Y = \left(\frac{x}{L/2}\right)^2 E \qquad (14.7)$$

The location of the high point of the crest vertical curve is frequently of interest to the designer because of drainage requirements. The distance between the beginning of the vertical curve (BVC) and the high point can be determined by considering the expression for Y^1 (see Figure 14.13).

$$Y^1 = \frac{G_1 x}{100} - Y$$

$$= \frac{G_1 x}{100} - \frac{A}{200L} x^2$$

$$= \frac{G_1 x}{100} - \left(\frac{G_1 - G_2}{200L}\right) x^2 \qquad (14.8)$$

Differentiating Eq. 14.8 and equating it to zero will give the value of x for the high point on the curve.

$$\frac{dY^1}{dx} = \frac{G_1}{100} - \left(\frac{G_1 - G_2}{100L}\right)x = 0 \qquad (14.9)$$

Therefore,

$$x_{high} = \frac{100L}{(G_1 - G_2)} \frac{G_1}{100} = \frac{LG_1}{(G_1 - G_2)} \text{ (ft)}$$

where x_{high} = distance from BVC to the turning point—that is, the point with the highest elevation on the curve (ft).

Similarly, the difference in elevation between the BVC and the turning point can be obtained by substituting the expression x_{high} for x in Eq. 14.8 and obtaining

$$Y^1_{high} = \frac{LG_1^2}{200(G_1 - G_2)} \text{ (ft)} \qquad (14.10)$$

The design of a crest vertical curve involves the use of some of the properties of a parabola discussed in this section and will generally proceed in the following manner.

Step 1. Determine the minimum length of curve to satisfy sight distance requirements.

Step 2. Determine from the layout plans the station and elevation of the PVI— that is, the point where the grades intersect.

Step 3. Compute the elevations of the BVC and end of vertical curve (EVC).

Step 4. Compute offsets y from the tangent to the curve at equal distances, usually 100 ft apart.

Step 5. Compute elevations on curve.

Example 14–1 Design of Crest Vertical Curve

A crest vertical curve joining a +2 percent and a −3 percent grade is to be designed for 60 mph. If the tangents intersect at station (335 + 35.00) at an elevation of 200 ft, determine the stations and elevations of the BVC and EVC. Also calculate the elevations of intermediate points on the curve spaced 100 ft apart.

A sketch of the curve is shown in Figure 14.14.

For a design speed of 60 mph, $K = 310$, using the higher rounded value in Table 14.6.

$$\text{minimum length} = 310 \times [2 - (-3)] = 1550 \text{ ft}$$

Figure 14.14 Layout of a Vertical Curve for Example 14–1

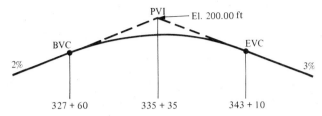

Note that the minimum length can also be obtained directly from Figure 14.12.

$$\text{station of BVC} = (335 + 35) - \left(\frac{15 + 50}{2}\right) = 327 + 60$$

$$\text{station of EVC} = (327 + 60) + (15 + 50) = 343 + 10$$

$$\text{elevation of BVC} = 200 - 0.02 \times \frac{1550}{2} = 184.5 \text{ ft}$$

The remainder of the computation is efficiently done using the format shown in Table 14.7.

Sag Vertical Curves. The design of sag vertical curves takes the same format as for crest vertical curves. The minimum length requirements were discussed in detail in Chapter 3. The headlight sight distance requirement is generally used for design purposes since stopping sight distances based on this requirement are generally within the limits of practical accepted values. The expressions developed in Chapter 3 for minimum length of sag vertical curves based on this requirement are

$$L = 2S - \frac{400 + 3.5S}{A} \qquad \text{(for } S > L) \qquad \textbf{(14.11)}$$

and

$$L = \frac{AS^2}{400 + 3.5S} \qquad \text{(for } S < L) \qquad \textbf{(14.12)}$$

The design control for sag vertical curves is also conveniently expressed in terms of K rate for all values of A. The computed values of K are shown in Table 14.8, and Figure 14.15 shows the required minimum lengths.

Computation of the elevations at different points on the curve takes the same form as that for the crest vertical curve. In this case, however, the offset Y is added to the appropriate tangent elevation to obtain the curve elevation since the formation elevation of the curve is higher than the tangent.

Table 14.7 Elevation Computations for Example 14–1

Station	Distance from BVC (x) (ft)	Tangent Elevation (ft)	Offset $\left(y=\dfrac{Ax^2}{200L}\right)$ (ft)	Curve Elevation (Tangent Elevation – Offset) (ft)
BVC 327 + 60	0	200 − 15.5 = 184.50	0.00	184.50
328 + 60	100	184.5 + 2 = 186.50	0.16	186.34
329 + 60	200	188.50	0.65	187.85
330 + 60	300	190.50	1.45	189.05
331 + 60	400	192.50	2.58	189.92
332 + 60	500	194.50	4.03	190.47
333 + 60	600	196.50	5.81	190.69
334 + 60	700	198.50	7.90	190.60
335 + 60	800	200.50	10.32	190.18
336 + 60	900	202.50	13.06	189.44
337 + 60	1000	204.50	16.13	188.37
338 + 60	1100	206.50	19.52	186.98
339 + 60	1200	208.50	23.23	185.27
340 + 60	1300	210.50	27.26	183.24
341 + 60	1400	212.50	31.61	180.89
342 + 60	1500	214.50	36.29	178.21
EVC 343 + 10	1550	215.50	38.75	176.75

Table 14.8 Design Controls for Sag Vertical Curves Based on Stopping Sight Distance

Design Speed (mph)	Assumed Speed for Condition (mph)	Coefficient of Friction, f	Stopping Sight Distance, Rounded for Design (ft)	Rate of Vertical Curvature, K [a] (length in feet per % of A)	
				Computed [b]	Rounded for Design
20	20–20	0.40	125–125	14.7–14.7	20–20
25	24–25	0.38	150–150	21.7–23.5	30–30
30	28–30	0.35	200–200	30.8–35.3	40–40
35	32–35	0.34	225–250	40.8–48.6	50–50
40	36–40	0.32	275–325	53.4–65.6	60–70
45	40–45	0.31	325–400	67.0–84.2	70–90
50	44–50	0.30	400–475	82.5–105.6	90–110
55	48–55	0.30	450–550	97.6–126.7	100–130
60	52–60	0.29	525–650	116.7–153.4	120–160
65	55–65	0.29	550–725	129.9–178.6	130–180
70	58–70	0.28	625–850	147.7–211.3	150–220

[a]Different K values for the same speed result from using unequal coefficients of friction.

[b]Using computed values of stopping sight distance.

Source: Adapted from *A Policy on Geometric Design of Highways and Streets*, Washington, D.C.: The American Association of State Highway and Transportation Officials, copyright 1984. Used by permission.

Figure 14.15   Minimum Lengths for Sag Vertical Curves

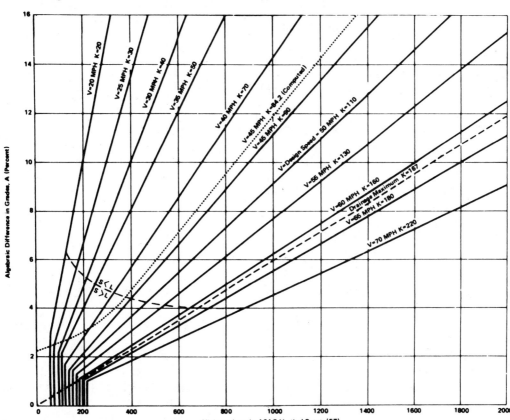

Figure 14.16   Layout of a Sag Vertical Curve for Example 14–2

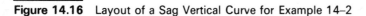
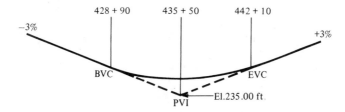

Table 14.9 Elevation Computations for Example 14–2

Station	Distance from BVC (x) (ft)	Tangent Elevation (ft)	Offset $\left(y = \dfrac{Ax^2}{200L}\right)$ (ft)	Curve Elevation (Tangent Elevation + Offset) (ft)
BVC 428 + 90	0	254.80	0	254.80
429 + 90	100	251.80	0.23	252.03
430 + 90	200	248.80	0.91	249.79
431 + 90	300	245.80	2.05	247.85
432 + 90	400	242.80	3.64	246.44
433 + 90	500	239.80	5.68	245.48
434 + 90	600	236.80	8.18	244.98
435 + 90	700	233.80	11.14	244.94
436 + 90	800	230.80	14.55	245.35
437 + 90	900	227.80	18.41	246.21
438 + 90	1000	224.80	22.73	247.53
439 + 90	1100	221.80	27.50	249.30
440 + 90	1200	218.80	32.73	251.53
441 + 90	1300	215.80	38.41	254.21
EVC 442 + 10	1320	215.20	39.60	254.80

Example 14–2   Design of Sag Vertical Curve

A sag vertical curve joins a -3 percent grade and a $+3$ percent grade. If the PVI of the grades is at station $(435 + 50)$ and has an elevation of 235 ft, determine the station and elevation of the BVC and EVC for a design speed of 70 mph. Also compute the elevation on the curve at 100 ft intervals. Figure 14.16 shows a layout of the curve.

For a design speed of 70 mph, $K = 220$, using the higher rounded value in Table 14.8.

$$\text{length of curve} = 220 \times 6 = 1320 \text{ ft}$$

$$\text{station of BVC} = (435 + 50) - (6 + 60) = 428 + 90$$

$$\text{station of EVC} = (435 + 50) + (6 + 60) = 442 + 10$$

$$\text{elevation of BVC} = 235 + 0.03 \times 660 = 254.8$$

$$\text{elevation of EVC} = 235 + 0.03 \times 660 = 254.8$$

The computation of the elevations is shown in Table 14.9.

Horizontal Alignment

The horizontal alignment consists of straight sections of the road, known as tangents, connected by horizontal curves. The curves are usually segments of circles, which have radii that will provide for a smooth flow of traffic along the curve. It was shown in Chapter 3 that the minimum radius of a horizontal curve depends on the design speed u of the highway, the superelevation e, and the coefficient of side friction f_s. This relationship was shown to be

$$R = \frac{u^2}{15(e + f_s)} \qquad e + f = \frac{v^2}{15R} \qquad f_{max} = \frac{v^2}{15R}$$

The design of the horizontal alignment therefore entails the determination of the minimum radius, the determination of the length of the curve, and the computation of the horizontal offsets from the tangents to the curve to facilitate the setting out of the curve. In some cases, to avoid a sudden change from a tangent with infinite radius to a curve of finite radius, a curve with radii varying from infinite value to the radius of the circular curve is placed between the circular curve and the tangent. Such a curve is known as a spiral or transition curve.

Horizontal Curves

There are four types of horizontal curves: simple, compound, reversed, and spiral. Computations required for each of them is presented in turn.

Simple curves. Figure 14.17 shows a layout sketch of a simple horizontal curve. As stated earlier, the curve is a segment of a circle with radius R. The point at which the curve begins is known as the point of curve (PC), and the point at which it ends is known as the point of tangent (PT). The point at which the two tangents intersect is known as the point of intersection (PI) or vertex (V). The simple circular curve is described either by its radius—for example, 200 ft radius

Figure 14.17 Layout of a Simple Horizontal Curve

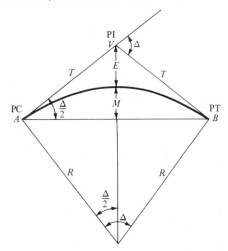

R = radius of circular curve PC = point of curve
T = tangent length PT = point of tangent
Δ = deflection angle PI = point of intersection
M = middle ordinate

Figure 14.18   Arc and Chord Definitions for a Circular Curve

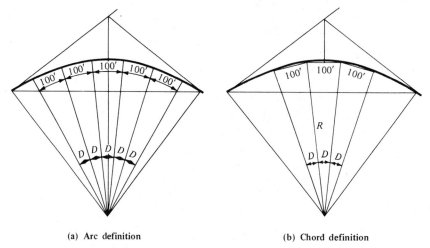

(a) Arc definition   (b) Chord definition

curve—or by the degree of the curve. Two definitions exist for the degree of the curve; the arc and the chord.

The arc defines the curve in terms of the angle subtended at the center by a circular arc 100 ft in length [Figure 14.18(a)]. This means that for a 2° curve, for example, an arc of 100 ft will be subtended by an angle of 2° at the center. If D is the degree of the curve, then

$$\theta = \frac{\pi D}{180} \quad \text{(rad)}$$

$$R\theta = L$$

$$\frac{R\pi D}{180} = 100$$

where L is the length of an arc subtending an angle of θ in radians at the center. Therefore,

$$R = \frac{180 \times 100}{\pi D} \tag{14.13}$$

giving

$$R = \frac{5729.6}{D}$$

The radius of the curve can therefore be determined if the degree of curve is known.

The chord defines the curve in terms of the angle subtended at the center by a chord of 100 ft [see Figure 14.18(b)] and the radius is given as

$$R = \frac{50}{\sin D/2} \tag{14.14}$$

where D is the angle in degrees subtended at the center by a 100 ft chord. The arc definition is commonly used for highway work and the chord definition is commonly used for railway work.

Formulas for Simple Circular Curves. Referring to Figure 14.17 and using the properties of a circle, the two tangent lengths AV and BV are equal and designated as T. The angle Δ formed by the two tangents is known as the deflection angle. We can therefore obtain the tangent length as

$$T = R \tan \frac{\Delta}{2} \tag{14.15}$$

The length C of the chord AB which is known as the long chord, is given as

$$C = 2R \sin \frac{\Delta}{2} \tag{14.16}$$

The external distance E is the distance from the point of intersection to the curve on a radial line and is given as

$$E = R \sec \frac{\Delta}{2} - R$$

$$= R \left(\frac{1}{\cos \frac{\Delta}{2}} - 1 \right) \tag{14.17}$$

The middle ordinate M is the distance between the midpoint of the long chord and the midpoint of the curve and is given as

$$M = R - R \cos \frac{\Delta}{2}$$

$$= R \left(1 - \cos \frac{\Delta}{2} \right) \tag{14.18}$$

The length of the curve L is given as

$$L = \frac{R \Delta \pi}{180} \tag{14.19}$$

Setting Out of a Simple Horizontal Curve. Simple horizontal curves are usually set out in the field by staking out points on the curve using deflection angles measured from the tangent at the point of curve and the lengths of the chords joining consecutive whole stations. Figure 14.19 is a schematic of the procedure involved. The first deflection angle VAp determined to the first whole station on the curve, which is usually less than a station away from the PC, is equal to $\delta_1/2$ (properties of a circle).

Figure 14.19   Deflection Angles on a Simple Circular Curve

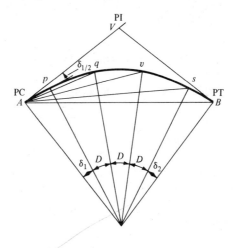

The next deflection angle VAq is

$$\frac{\delta_1}{2} + \frac{D}{2}$$

and the next angle VAv is

$$\frac{\delta_1}{2} + \frac{D}{2} + \frac{D}{2} = \frac{\delta_1}{2} + D$$

and the next VAs angle is

$$\frac{\delta_1}{2} + \frac{D}{2} + \frac{D}{2} + \frac{D}{2} = \frac{\delta}{2} + \frac{3D}{2}$$

and the last angle VAB is

$$\frac{\delta_1}{2} + \frac{D}{2} + \frac{D}{2} + \frac{D}{2} + \frac{\delta_2}{2} = \frac{\delta_1}{2} + \frac{\delta_2}{2} + \frac{3D}{2} = \frac{\Delta}{2}$$

To set out the horizontal curve, it is necessary to determine δ_1 and δ_2. The length of the first arc l_1 is related to δ_1 as

$$l_1 = \frac{R\pi}{180} \delta_1 \qquad\qquad \textbf{(14.20)}$$

giving

$$R = \frac{l_1 \times 180}{\delta_1 \pi}$$

From Eq. 14.19

$$R = \frac{180L}{\Delta\pi}.$$

Therefore,

$$\frac{l_1}{\delta_1} = \frac{L}{\Delta} = \frac{l_2}{\delta_2}$$

In setting out a simple horizontal curve in the field, the PC and PT are first located and staked. Deflection angles from the PC to each whole station are then computed. A transit is mounted over the PC and each whole station is located on the ground using the appropriate deflection angle, and the chord distance is measured from the preceding station.

Note that the lengths l_1 and l_2 are measured along the curve and the corresponding chord lengths should be calculated, particularly when circular curves are sharp. These chord lengths can be calculated from

$$C_1 = 2R \sin \frac{\delta_1}{2} \tag{14.21}$$

$$C_D = 2R \sin \frac{D}{2}$$

$$C_2 = 2R \sin \frac{\delta_2}{2}$$

where C_1, C_D, and C_2 are the first, intermediate, and last chords, respectively.

Example 14–3 Design of a Horizontal Curve

The deflection angle of a $3°$ curve is $50°25'$. If the PC is located at station $(331 + 38.75)$, determine the length of the curve and the station of PT. Also determine the deflection angles for setting out the curve at whole stations from the PC. Figure 14.20 shows a layout of the curve.

$$\text{radius of curve} = \frac{5729.6}{D} = \frac{5729.6}{3}$$

$$= 1909.9 \quad (\text{say, } 1910 \text{ ft})$$

$$\text{length of curve} = \frac{R\Delta\pi}{180} = \frac{1910 \times 50.4167\pi}{180}$$

$$= 1680.68 \text{ ft}$$

Therefore, station at PT is equal to station $(331 + 38.75) + (16 + 80.68)$

Figure 14.20 Layout of Curve for Example 14–3

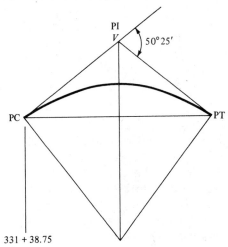

331 + 38.75

stations $= 348 + 19.43$. The distance between PC and the first station is $332 - (331 + 38.75)$ stations $= 61.25$ ft.

$$\frac{\delta_1}{\Delta} = \frac{l_1}{L}$$

$$\frac{\delta_1}{50.4167} = \frac{61.25}{1680.68}$$

Therefore,

$$\delta_1 = 1.837°$$

$$C_1 = 2 \times 1910 \sin\left(\frac{1.837}{2}\right) = 61.24 \text{ ft}$$

The first deflection angle to station 332 is $\delta_1/2 = 0.9185° = 55'7''$.
    Similarly,

$$l_2 = (348 + 19.43) - 348 \text{ stations} = 19.43 \text{ ft}$$

$$\frac{\delta_2}{2} = \frac{19.43}{1680.65} \times \frac{50.4167}{2} = 0.29°$$

$$= 17'24''$$

$$C_2 = 2 \times 1910 \sin(0.29°)$$

$$= 19.33 \text{ ft}$$

Table 14.10 Computations of Deflection Angles and Chord Lengths for Example 14–3

Station	Deflection Angle	Chord Length (ft)
PC 331 + 38.75	0	0
332	0°55′7″	61.24
333	2°25′7″	100.00
334	3°55′7″	100.00
335	5°25′7″	100.00
336	6°55′7″	100.00
337	8°25′7″	100.00
338	9°55′7″	100.00
339	11°25′7″	100.00
340	12°55′7″	100.00
341	14°25′7″	100.00
342	15°55′7″	100.00
343	17°25′7″	100.00
344	18°55′7″	100.00
345	20°25′7″	100.00
346	21°55′7″	100.00
347	23°25′7″	100.00
348	24°55′7″	100.00
PT 348 + 19.40	25°12′30″	19.33

The other deflection angles are computed in Table 14.10

$$D = 3°$$

$$C_D = 2 \times 1910 \sin\left(\frac{3}{2}\right)$$

$$= 99.996 \approx 100 \text{ ft}$$

Note that the deflection angle to PT is half the deflection angle of the tangents. This serves as a check of the computation. Also, because the curve is flat, the chord lengths are approximately equal to the arc lengths.

Compound Curves. Compound curves consist of two or more curves in succession, turning in the same direction, with any two successive curves having a common tangent point. Figure 14.21 shows a typical layout of a compound curve, consisting of two simple curves. These curves are used mainly in obtaining desirable shapes of the horizontal alignment, particularly at at-grade intersections, ramps of interchanges, and highway sections in difficult topographic conditions. To avoid abrupt changes in the alignment, the radii of any two consecutive simple curves that form a compound curve should not be widely different. AASHTO recommends that the ratio of the flatter radius to the sharper radius in open

Figure 14.21 Layout of a Compound Curve

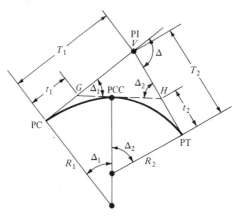

R_1, R_2 = radii of simple curves forming compound curve
Δ_1, Δ_2 = deflection angles of simple curves
Δ = deflection angle of compound curve
t_1, t_2 = tangent lengths of simple curves
T_1, T_2 = tangent lengths of compound curve
PCC = point of compound curve

highways should not be greater than 1.5 : 1. This ratio may be increased to 2 : 1 at intersections, where drivers can adjust to sudden changes in curvature and speed. The maximum desirable ratio recommended by AASHTO, however, is 1.75 : 1. In cases where ratios higher than 2 : 1 are to be used, a spiral curve should be placed between the simple curves.

To provide smooth transition from a flat curve to a sharp curve, and to facilitate a reasonable deceleration rate on a series of curves of decreasing radii, the length of each curve should not be too short. Minimum lengths based on the radius of each curve as recommended by AASHTO are given in Table 14.11.

Figure 14.21 shows seven variables: R_1, R_2, Δ_1, Δ_2, Δ, T_1, T_2, six of which are independent, since $\Delta = \Delta_1 + \Delta_2$. Several solutions can be devloped for the com-

Table 14.11 Lengths of Circular Arc for a Compound Intersection Curve When Followed by a Curve of One-Half Radius or Preceded by a Curve of Double Radius

Length of Circular Arc (ft)	Radius (ft)						
	100	*150*	*200*	*250*	*300*	*400*	*500 or more*
Minimum	40	50	60	80	100	120	140
Desirable	60	70	90	120	140	180	200

Source: Adapted from *A Policy on Geometric Design of Highways and Streets*, Washington, D.C.: The American Association of State Highway and Transportation Officials, copyright 1984. Used by permission.

pound curve, but only the vertex triangle method is presented here, since this is the method frequently used in highway design.

In Figure 14.21, R_1 and R_2 are usually known.

$$\Delta = \Delta_1 + \Delta_2 \tag{14.22}$$

$$t_1 = R_1 \tan \frac{\Delta_1}{2} \tag{14.23}$$

$$t_2 = R_2 \tan \frac{\Delta_2}{2} \tag{14.24}$$

$$\frac{\overline{VG}}{\sin \Delta_2} = \frac{\overline{VH}}{\sin \Delta_1} = \frac{t_1 + t_2}{\sin(180 - \Delta)} = \frac{t_1 + t_2}{\sin \Delta} \tag{14.25}$$

$$T_1 = VG + t_1 \tag{14.26}$$

$$T_2 = VH + t_2 \tag{14.27}$$

where

R_1 and R_2 = radii of simple curves forming the compound curve
Δ_1 and Δ_2 = deflection angles of simple curves
t_1 and t_2 = tangent lengths of simple curves
T_1 and T_2 = tangent lengths of compound curves
Δ = deflection angle of compound curve

To be able to lay out the curve, it is necessary to know the value of at least one other variable. Usually, Δ_1 or Δ_2 can be obtained from the layout plans. Eqs. 14.22 through 14.27 can then be used to solve for Δ_1 or Δ_2, VG, VH, t_1, t_2, T_1, and T_2. Deflection angles can then be determined for each simple curve in turn.

Example 14–4   Design of a Compound Curve

Figure 14.22 shows a compound curve that is to be set out at an intersection. If the point of compound curve (PCC) is located at station $(565 + 35)$, determine the deflection angles for setting out the curve.

$$t_1 = 500 \tan \frac{34}{2} = 152.87 \text{ ft}$$

$$t_2 = 350 \tan \frac{26}{2} = 80.80 \text{ ft}$$

For length of horizontal curve of 500 ft radius,

$$L = R\Delta_1 \frac{\pi}{180} = 500 \times \frac{34\pi}{180}$$

$$= 296.71 \text{ ft}$$

Figure 14.22 Compound Curve for Example 14–4

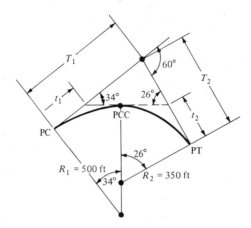

For length of horizontal curve of 350 ft radius,

$$L = 350 \frac{26\pi}{180}$$

$$= 158.82 \text{ (ft)}$$

Therefore, station of PC is equal to $(565 + 35.00) - (2 + 96.71) = 562 + 38.29$ and station of PT is equal to $(565 + 35.00) + (1 + 58.82) = 566 + 93.82$.
    For curve of 500 ft radius,

$$\frac{D}{2} = \frac{5729.6}{2 \times 500} = 5°43'47'' \qquad \text{(from Eq. 14.13)}$$

$$l_1 = (563 + 00) - (562 + 38.29) = 61.71 \text{ (ft)}$$

$$\frac{\delta_1}{l_1} = \frac{\Delta}{L}$$

$$\frac{\delta_1}{2} = \frac{61.71 \times 34}{2 \times 296.71} = 3°32'8''$$

$$l_2 = (565 + 35.00) - (565 + 00) = 35 \text{ (ft)}$$

$$\frac{\delta_2}{2} = \frac{35 \times 34}{2 \times 296.71} = 2°0'19''$$

Table 14.12 Computations for Example 14–4

| | **500 ft Radius Curve** | |
Station	Deflection Angle	Chord Length (ft)
PC 562 + 35.00	0	0
563	3°32'8"	61.66'
564	9°15'55"	99.84'
565	14°59'42"	99.84'
565 + 35.00	17°00'01"	35.00'

| | **350 ft Radius Curve** | |
Station	Deflection Angle	Chord Length (ft)
PC 565 + 35.00	0	0
566	5°19'16"	64.9'
566 + 93.80	13°00'00"	93.5'

For curve of 350 ft radius,

$$\frac{D}{2} = \frac{5729.6}{2 \times 350} = 8°11'7''$$

$$l_1 = (566 + 00) - (565 + 35.00) = 65 \text{ ft}$$

$$\frac{\delta_1}{2} = \frac{65 \times 26}{2 \times 158.82} = 5°19'16''$$

$$l_2 = (566 + 93.82) - (566 + 00) = 93.82 \text{ ft}$$

$$\frac{\delta_2}{2} = \frac{93.82 \times 26}{2 \times 158.8} = 7°40'44''$$

For computation of the deflection angles, see Table 14.12.

Note that deflection angles for the 350 ft radius curve are turned from the common tangent with the transit located at PCC. Also note that, with the relatively flat simple curves, the calculated lengths of the chords are almost equal to the corresponding arc lengths.

Reverse Curves.   Reverse curves usually consist of two simple curves with equal radii turning in opposite directions with a common tangent. They are generally

Figure 14.23 Geometry of a Reverse Curve with Parallel Tangents

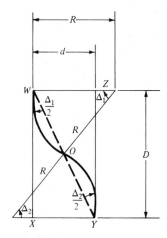

R = radius of simple curves
Δ_1, Δ_2 = deflection angles of simple curves
d = distance between parallel tangents
D = distance between tangent points

used to change the alignment of a highway. Figure 14.23 shows a reverse curve with parallel tangents. If d and D are known, it is necessary to determine Δ_1, Δ_2, and R to set out the curve

$$\Delta = \Delta_1 = \Delta_2$$

$$\text{angle } OWX = \frac{\Delta_1}{2} = \frac{\Delta}{2}$$

$$\text{angle } OYZ = \frac{\Delta_1}{2} = \frac{\Delta_2}{2}$$

Line WOY is therefore a straight line, hence

$$\tan\frac{\Delta}{2} = \frac{d}{D}$$

$$d = R - R\cos\Delta_1 + R - R\cos\Delta_2$$

$$= 2R(1 - \cos\Delta)$$

$$R = \frac{d}{2(1 - \cos\Delta)}$$

If d and R are known,

$$\cos\Delta = 1 - \frac{d}{2R}$$

$$D = d\cot\frac{\Delta}{2}$$

Reverse curves are usually not recommended because sudden changes to the alignment may result in drivers finding it difficult to keep on their lanes. When it is necessary to reverse the alignment, a preferable design consists of two simple horizontal curves, separated by a sufficient length of tangent between them, to achieve full superelevation. Alternatively, the simple curves may be separated by an equivalent length of spiral, which is is described in the next section.

Transition (Spiral) Curves. Transition curves are placed between tangents and circular curves or between two adjacent circular curves having substantially different radii. The use of transition curves provides a vehicle path that gradually increases or decreases the radial force as the vehicle enters or leaves a circular curve. The degree of a transition curve placed between a tangent and a circular curve varies from zero at the tangent end to the degree of the circular curve at the curve end. When placed between two circular curves, the degree varies from that of the first circular curve to that of the second circular curve.

The minimum length of a transition curve for highways is given as[8]

$$L = \frac{3.15u^3}{RC} \qquad\qquad (14.28)$$

where

L = minimum length of curve (ft)

u = Speed (mph)

R = radius of curve (ft)

C = rate of increase of radial acceleration

C is an empirical factor that indicates the level of comfort and safety involved. Values used for C in highway engineering vary from 1 to 3.

The computation for setting out the transition curve is beyond the scope of this book. In fact, many highway agencies do not use transition curves, as drivers will usually guide their vehicles into circular curves gradually. For a more comprehensive treatment of this topic, interested readers may refer to *Surveying Theory and Practice*, listed in the Additional Readings at the end of this chapter. A practical alternative for determining the minimum length of a spiral is to use the length required for superelevation run-off (see following section).

Superelevation Run-Off

In Chapter 3 we showed that highway pavements on circular curves should be superelevated to counteract the effect of centrifugal force. The length of highway section required to achieve a full superelevated section from a section with adverse crown removed or vice versa is known as the superelevation run-off. This length depends on the design speed, the rate of superelevation, and the pavement width. Table 14.13 gives values of the required lengths for 2-lane highways.

Table 14.13   Length Required for Superelevation Run-Off, 2-Lane Pavements

Superelevation Rate, e	Length of Run-Off (ft) for Design Speed (mph) of						
	20	*30*	*40*	*50*	*60*	*65*	*70*
	12 ft lanes						
.02	30	35	40	50	55	60	60
.04	60	70	85	95	110	115	120
.06	95	110	125	145	160	170	180
.08	125	145	170	190	215	230	240
.10	160	180	210	240	270	290	300
	10 ft lanes						
.02	25	30	35	40	45	50	50
.04	50	60	70	80	90	95	100
.06	80	90	105	120	135	145	150
.08	105	120	140	160	180	190	200
.10	130	150	175	200	225	240	250
Design minimum length regardless of superelevation	50	100	125	150	175	190	200

Source: Adapted from *A Policy on Geometric Design of Highways and Streets*, Washington, D.C.: The American Association of State Highway and Transportation Officials, copyright 1984. Used by permission.

AASHTO recommends that, for pavements with more than two lanes, the lengths given in Table 14.13 should be adjusted as follows:

- Three-lane pavements: 1.2 times the corresponding length for two-lane highways
- Four-lane, undivided pavements: 1.5 times the corresponding length for two-lane highways
- Six-lane, undivided pavement: 2.0 times the corresponding length for two-lane highways[9]

To avoid sudden changes in pavement profiles, several highway agencies have adopted minimum values ranging from 100 to 250 ft, regardless of width and superelevation rate.

In cases where transition curves are used, it is common practice for the fully superelevated section to be achieved over the whole length of the spiral. The use of Eq. 14.28, however, sometimes gives values for the length of the spiral curve, that are not exactly the same as the values given in Table 14.13. When this occurs, it is appropriate to use the superelevation run-off for the length of the spiral curve.

Attainment of Superelevation

It is essential that the change from a crowned cross section to a superelevated one be achieved without causing any discomfort to motorists or creating unsafe conditions. One of four methods can be used to achieve this change on undivided highways.

1. A crowned pavement is rotated about the profile of the center line.
2. A crowned pavement is rotated about the profile of the inside edge.
3. A crowned pavement is rotated about the profile of the outside edge.
4. A straight cross slope pavement is rotated about the profile of the outside edge.

Figure 14.24(a) is a schematic of method 1. This is the most commonly used method since the distortion obtained is less than those for the other methods. The procedure used is first to raise the outside edge of the pavement relative to the centerline, until the outer half of the cross section is horizontal (point B). The outer edge is then raised by an additional amount to obtain a straight cross section. Note that at this point the inside edge is still at its original elevation as indicated at point C. The whole cross section is then rotated as a unit about the centerline profile until the full superelevation is achieved at point E. Figure 14.24(b) illustrates method 2, where the centerline profile is raised with respect to the inside pavement edge to obtain half the required change, while the remaining half is achieved by raising the outside pavement edge with respect to the profile of the centerline. Method 3, demonstrated by Figure 14.24(c), is similar to method 2. The only difference is that the change is effected below the outside edge profile. Figure 14.24(d) illustrates method 4, which is used for sections of straight cross slopes.

Note, however, that straight lines have been conveniently used to illustrate the different methods, but in practice, the angular breaks are appropriately rounded by using short vertical curves, as shown in Figure 14.24(e).

Superelevation is achieved on divided highways by using one of three methods. Method 1 involves superelevating the whole cross section, including the median, as a plane section. The rotation in most cases is done about the centerline of the median. This method is used only for highways with narrow medians and moderate superelevation rates, since large differences in elevation can occur between the extreme pavement edges if the median is wide. Method 2 involves rotating each pavement separately around the median edges, while keeping the median in a horizontal plane. This method is used mainly for pavements with median widths of 30 ft or less, although it can be used for any median, because by keeping the median in the horizontal plane, the difference in elevation between the extreme pavement edges does not exceed the pavement superelevation. In method 3, the two pavements are treated separately, resulting in variable elevation differences between the median edges. This method generally is used on pavements with median widths of 40 ft or greater. The large difference in elevation between the extreme pavement edges is avoided by providing a compensatory slope across the median.

Figure 14.24 Diagrammatic Profiles Showing Methods of Attaining Superelevation for a Curve to the Right

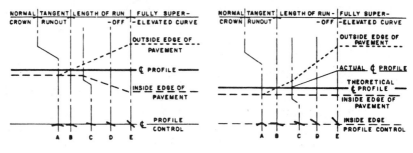

(a) Crowned pavement revolved about centerline   (b) Crowned pavement revolved about inside edge

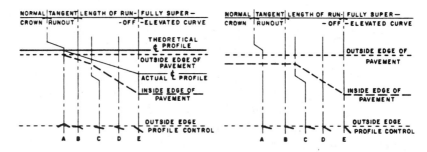

(c) Crowned pavement revolved about outside edge   (d) Straight cross slope pavement revolved about outside edge

(e) Angular breaks appropriately rounded (dotted lines)

Source: Reproduced from *A Policy on Geometric Design of Highways and Streets*, Washington, D.C.: The American Association of State Highway and Transportation Officials, copyright 1984. Used by permission.

SPECIAL FACILITIES FOR HEAVY VEHICLES ON STEEP GRADES

Recent statistics indicate a continual increase in the annual vehicle miles of large trucks on the nations highways over the past few years. We noted in both Chapter 3 and Chapter 7 that large trucks have different operating characteristics from those of passenger cars, and that this difference increases as the grade of a section of highway increases. These factors make it necessary to consider the provision of special facilities on sections of highways with steep grades carrying high volumes of heavy vehicles. The two facilities normally considered are climbing lanes and emergency escape ramps.

Climbing Lanes

A climbing lane is an extra lane in the upgrade direction for use by heavy vehicles whose speeds are significantly reduced by the grade. A climbing lane eliminates the need for drivers of other vehicles using the normal upgrade lane to reduce their speed, which would be necessary if they encounter a slow-moving heavy vehicle just ahead on the same lane. In the past, the provision of climbing lanes was not common because of the added cost of construction. The increasing rate of accidents directly associated with the reduction in speed of heavy vehicles on steep sections of 2-lane highways and the significant reduction of the capacity of these sections when heavy vehicles are present have made it necessary to seriously consider the provision of climbing lanes on these highways. The basic condition that suggests the use of a climbing lane is when a grade is longer than its critical length. The critical length is that length that will cause a speed reduction of the heavy vehicle by 10 mph or more. The climbing lane is provided only if, in addition to the critical length requirement, the volume of traffic is sufficiently high, and there is a high enough percentage of heavy vehicles in the traffic stream. This reduction in speed is used since the difference in speeds between adjacent levels of service is usually about 10 mph. For example, in Chapter 7 it was shown that the operating speed on freeway segments is about 54 mph for level of service C and about 46 mph for level of service D. The length of the climbing lane will depend on the physical characteristics of the grade, but a general guideline is that the climbing lane should be long enough to facilitate the heavy vehicle rejoining the main traffic stream without causing a hazardous condition.

Climbing lanes are not frequently used on multilane highways, since relatively faster moving vehicles can pass the slower moving vehicles by using one of the other lanes. In addition, it is usually difficult to justify the provision of climbing lanes based on capacity since the multilane highways usually have adequate capacity to carry the traffic demand, including the normal percentage of slow-moving vehicles without becoming too congested.

Emergency Escape Ramps

An emergency escape ramp is one provided on the downgrade of a highway for the use of a driver who has lost control of the vehicle because of brake failure. The objective is to provide a lane that diverges away from the main traffic stream while the uncontrolled vehicle's speed is gradually reduced and the vehicle is eventually brought to rest. The four basic types of design commonly used are shown in Figure 14.25. The sandpile shown in Figure 14.25(a) has a pile of loose dry sand at the end of the escape ramp on which the vehicle is brought to rest. The sandpile provides an increased rolling resistance, and it is placed with an upgrade to use the influence of gravity in bringing the vehicle to a stop. Sandpiles are usually not longer than 400 ft. The descending and horizontal grades shown in Figures 14.25(b) and 14.25(c), respectively, do not use this influence of gravity in stopping the vehicle, but rely mainly on the increased rolling resistance provided by the arresting bed, which is made up of loose aggregates. These escape ramps

Figure 14.25   Basic Types of Emergency Escape Ramps

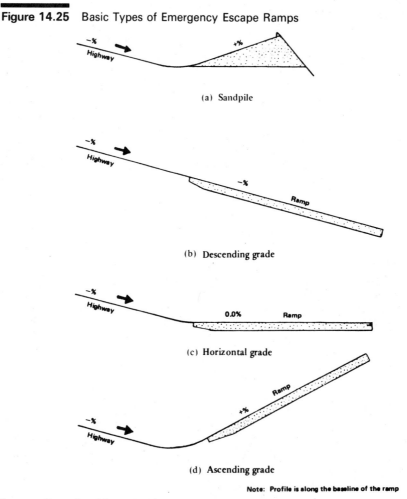

(a) Sandpile

(b) Descending grade

(c) Horizontal grade

(d) Ascending grade

Note: Profile is along the baseline of the ramp

Source: Reproduced from *A Policy on Geometric Design of Highways and Streets*, Washington, D.C.: The American Association of State Highway and Transportation Officials, copyright 1984. Used by permission.

can therefore be very long. The ascending-grade ramp shown in Figure 14.25(d) combines the effect of gravity and the increased rolling resistance by providing both the upgrade and the arresting bed. Thus, these ramps are relatively shorter than the descending or horizontal grade ramps.

BICYCLE FACILITIES

Use of the bicycle as a viable mode of transportation is now recognized, particularly in urban areas. The bicycle has therefore become an important factor that is considered in the design of urban highways with the result that a wide variety of bicycle-related projects have been implemented.

Although most of the existing highway and street systems can be used to carry bicycles, it is often necessary to carry out improvements to these facilities to guarantee the safety of bicycle riders. Guides for the development of new bicycle facilities have been developed by AASHTO.[10]

Following is a brief description of the design for bicycle lanes and bicycle paths.

Bicycle Lanes

A bicycle (bike) lane is that part of the street or highway specifically reserved for the exclusive or preferential use of bicycle riders. Bicycle lanes can be delineated by striping, signing, or pavement markings. These lanes should always be one-way with traffic. Figure 14.26 shows typical bicycle lane cross sections. The minimum width under ideal conditions is 4 ft but should be increased when conditions demand it.[11] For example, when bike lanes are located on a curbed urban street, with parking allowed at both sides, as shown in Figure 14.26(a), a minimum width of 5 ft is required. Note that when parking is provided, the bicycle lane should always be located between the parking lane and the traveled lanes for automobiles, as shown in Figure 14.26(a). In cases where parking is not allowed, as in Figure 14.26(b), a minimum bike lane of 5 ft should be provided, because cyclists tend to ride away from the curb to avoid hitting it. In cases where

Figure 14.26   Typical Bicycle Lane Cross Sections

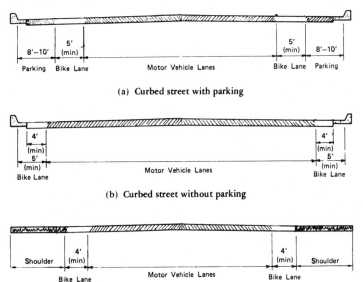

(a) Curbed street with parking

(b) Curbed street without parking

(Not to Scale)   (Metric Conversion: 1 Ft. = 0.3m.)

(c) Street or highway without curb or gutter

Source: Reproduced from *Guide for Development of New Bicycle Facilities, 1981*, Washington, D.C.: The American Association of State Highway and Transportation Officials, copyright 1981. Used by permission.

the highway has no curb or gutter, the bicycle lane should be located between the shoulders and the travel lanes of the motor vehicles. A minimum bicycle lane of 4 ft could be provided if additional maneuvering width is provided by the shoulders.

Special consideration should be given to pavement markings at intersections, since through movement of bicycles will tend to conflict with right-turning movement of motor vehicles, and left-turning movement of bicycles will tend to conflict with through movement of motor vehicles. Pavements should therefore be striped to encourage both cyclists and motorists to commence their turning maneuvers in

Figure 14.27   Channelization of Bicycle Lanes Approaching Motor Vehicle Right-Turn-Only Lanes

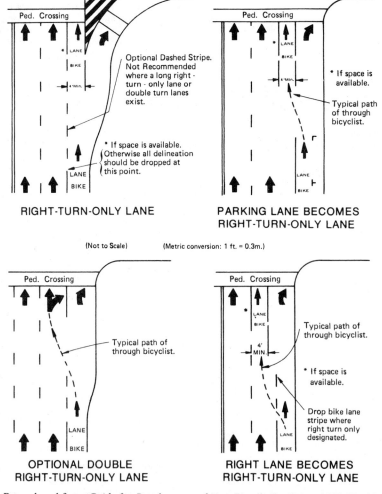

a merging configuration some distance away from the intersection. Figure 14.27 shows alternate ways for achieving this, as suggested by AASHTO.[12]

Bicycle Paths

Bicycle (bike) paths are physically separated from automobile traffic, and are used exclusively by cyclists. The design criteria for bicycle paths are somewhat similar to those for highways, but some of these criteria are governed by bicycle operating characteristics, which are significantly different from those of automobiles. Designers should be aware of those bicycle operating characteristics that are similar to and different from automobile characteristics. Important design considerations for a safe bicycle path include the width, the design speed, the horizontal alignment, and the vertical alignment.

Width of Bicycle Paths

A typical layout of a bicycle path on a separate right of way is shown in Figure 14.28. The minimum width recommended by AASHTO is 10 ft for two-way paths and 5 ft for one-way paths.[13] Under certain conditions, the width for the two-way path may be reduced to 8 ft. These conditions include:

- Low bicycle volumes at all times
- Vertical and horizontal alignments properly designed to provide frequent safe passing conditions
- Only occasional use of path by pedestrians
- No loaded maintenance vehicles on path

Just as shoulders are provided on highways, it is necessary to provide a uniform graded area at least 2 ft wide at both sides of the path.

When bicycle paths are located adjacent to a highway, a wide area must be provided between the highway and the bicycle path to facilitate the independent use of the bicycle path. In cases where this area is less than 5 ft wide, physical barriers, such as dense shrubs or a fence, should be used to separate the two facilities.

Design Speed

AASHTO recommends that on paved paths, a minimum design speed of 20 mph be used, but this should be increased to 30 mph for grades greater than 4 percent or where strong prevailing tail winds exist. The minimum design speed can also be reduced to 15 mph on unpaved bicycle paths, as riding speeds tend to be lower on these paths.[14]

Horizontal Alignment

Similar criteria exist for bicycle paths as for highways, in that the minimum radius of horizontal curves depends on the velocity and the superelevation and is given as

$$R = \frac{u^2}{15(e + f_s)}$$

Figure 14.28   Bicycle Path on Separated Right of Way

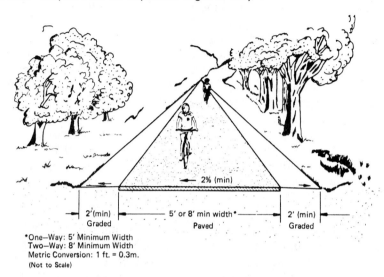

*One—Way: 5' Minimum Width
Two—Way: 8' Minimum Width
Metric Conversion: 1 ft. = 0.3m.
(Not to Scale)

Source: Reproduced from *Guide for Development of New Bicycle Facilities, 1981*, Washington, D.C.: The American Association of State Highway and Transportation Officials, copyright 1981. Used by permission.

Figure 14.29   Stopping Sight Distances for Bicycles

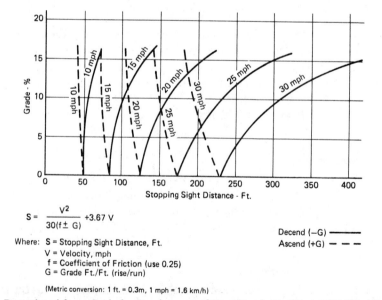

$$S = \frac{V^2}{30(f \pm G)} + 3.67 V$$

Where: S = Stopping Sight Distance, Ft.
          V = Velocity, mph
          f = Coefficient of Friction (use 0.25)
          G = Grade Ft./Ft. (rise/run)

Decend (−G) ———
Ascend (+G) − − −

(Metric conversion: 1 ft. = 0.3m, 1 mph = 1.6 km/h)

Source: Reproduced from *Guide for Development of New Bicycle Facilities, 1981*, Washington, D.C.: The American Association of State Highway and Transportation Officials, copyright 1981. Used by permission.

where

R = minimum radius of the curve (ft)

u = design speed (mph)

e = rate of superelevation

f_s = coefficient of side friction

The factors to be considered here are the superelevation rate and the coefficient of side friction. Superelevation rates for bicycle paths vary from a minimum of 2 percent, which will facilitate surface drainage, to a maximum of 5 percent. Higher superelevation rates may result in some cyclists finding it difficult to maneuver around the curve. Coefficents of side friction used for design vary from 0.3 to 0.22. These values correspond with those used for highways at design speeds of 15 and 30 mph, respectively. It is suggested that for unpaved bicycle paths, values varying from 0.15 to 0.11 should be used.[15]

Figure 14.30   Minimum Lengths of Vertical Curves for Bicycles

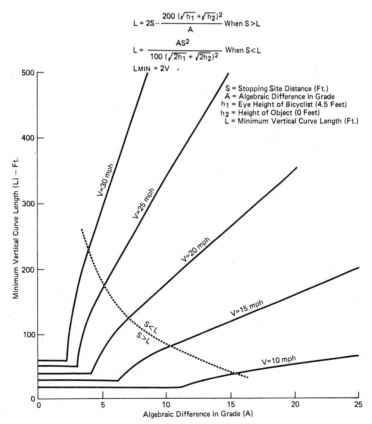

Source: Reproduced from *Guide for Development of New Bicycle Facilities, 1981,* Washington, D.C.: The American Association of State Highway and Transportation Officials, copyright 1981. Used by permission.

Vertical Alignment

The design of the vertical alignment for a bicycle path is similar to that for a highway, in that the selection of suitable grades and the design of appropriate vertical curves are required.

Grades should not exceed 5 percent and should be lower when possible, particularly on long inclines. This is necessary to enable cyclists to easily ascend the grade and to prevent excessive speed while descending the grade.

The vertical curves are designed to provide the minimum length as dictated by the stopping sight distance requirement. Figure 14.29 shows minimum stopping sight distances for different design speeds, and Figure 14.30 gives minimum lengths for vertical curves. The minimum stopping sight distances given in Figure 14.29 are based on a reaction time of 2.5 sec, which is the value normally used for bicyclists.

PARKING FACILITIES

The geometric design of parking facilities mainly involves the dimensioning and arranging of parking bays to provide safe and easy access without seriously restricting the flow of traffic on the adjacent traveling lanes. Guidelines for the design of on-street and off-street parking facilities are presented.

Design of On-Street Parking Facilities

On-street parking facilities may be designed with the parking bays parallel to the curb or inclined to the curb. Figure 14.31 shows the angles of inclination commonly used for curb parking and the associated dimensions for automobile parking. It can be seen that the number of parking bays that can be fitted along a given length of curb increases as the angle of inclination increases up to 90°. The higher the inclination angle, however, the greater is the encroachment of the parking bays on the traveling pavement of the highway. Parking bays inclined at angles to the curb severely interfere with the movement of traffic on the highway, with the result that accident rates tend to be higher on sections of roads with angle parking than on those with parallel parking. It should be noted that the dimensions given in Figure 14.31 are for automobiles. When parking bays are to be provided for trucks and other types of vehicles, the dimensions should be determined on the basis of the dimensions of the specific vehicle being considered.

Design of Off-Street Parking Facilities—Surface Car Parks

The primary aim in designing off-street parking facilities is to obtain the maximum possible number of spaces. Figures 14.32 and 14.33 show different types of layouts that can be used in a surface lot. The most important consideration is that the layout should be such that parking a vehicle involves only one distinct maneuver, without necessity to reverse. The different layouts also indicate that the

Figure 14.31 Sheet Space Used for Various Parking Configurations

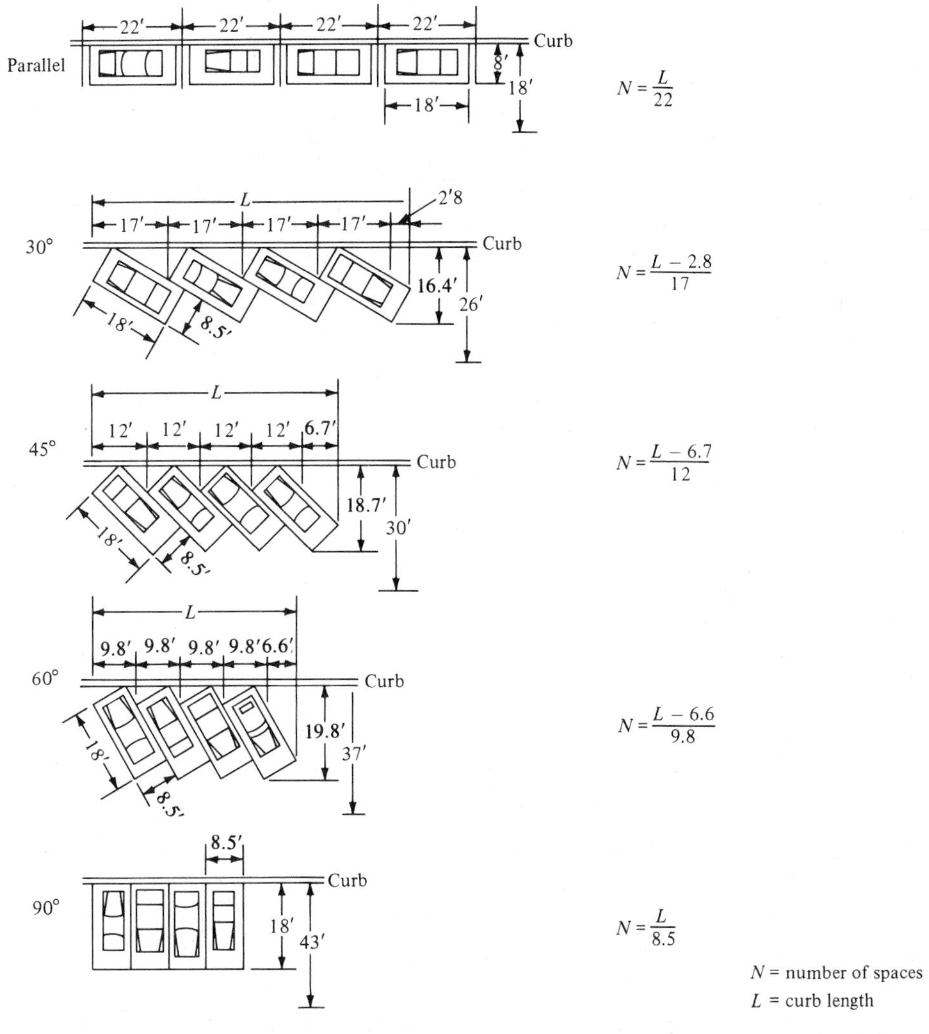

$$N = \frac{L}{22}$$

$$N = \frac{L - 2.8}{17}$$

$$N = \frac{L - 6.7}{12}$$

$$N = \frac{L - 6.6}{9.8}$$

$$N = \frac{L}{8.5}$$

N = number of spaces
L = curb length

Source: Redrawn from R. H. Burrage and E. G. Mogren, *Parking*, ENO Foundation for Highway Traffic Control, Inc., Saugatuck, Conn. 1957.

parking space is most efficiently used when the parking bays are inclined at 90° to the direction of traffic flow. The use of the herringbone layout (Figure 14.33) facilitates traffic circulation since it provides for one-way flow of traffic on each aisle.

Design of Off-Street Parking Facilities—Garages

Parking garages basically consist of several platforms, supported by columns, that are placed in such a way that facilitates an efficient arrangement of parking

Figure 14.32   Parking Stall Layout

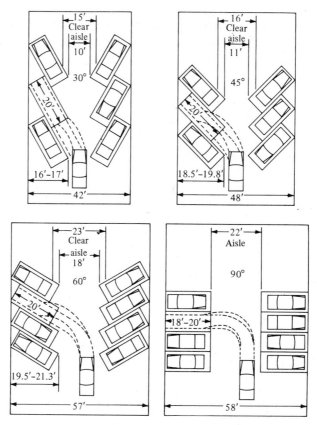

bays and aisles. Access ramps connect each level with the one above. The gradient of these ramps is usually not greater than 1 in 10 on straight ramps and 1 in 12 on the centerline of curved ramps. It is recommended that the radius of curved ramps measured to the end of the outer curve should not be less than 70 ft and that the maximum superelevation should be 0.15 ft/ft. It is also recommended that the width should not be less than 16 ft for curved ramps and 9 ft for straight ramps. These ramps can be one-way or two-way, although one-way ramps are preferable. When two-way ramps are used, the lanes must be clearly marked and perhaps physically divided, particularly at curves and turning points. This helps to avoid head-on collisions, that may occur as drivers cut corners or swing wide at bends.

In some cases, the platforms are connected by elevators into which cars are driven or mechanically placed. These elevators then lift the car to the appropriate level for parking. The vehicle is then either removed by an attendant who eventually parks it or it is mechanically removed and then parked by that attendant.

The size of the receiving area is an important factor in garage design and depends on whether the cars are owner-parked (self-parking) or attendant-

Figure 14.33 Herringbone Layout of Parking Stalls in an On-Surface Lot

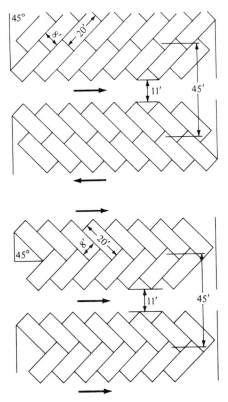

parked. When cars are self-parked, very little or no reservoir space is required, since drivers need pause only for a short time to pick up a ticket from a machine or to be given one by an attendant. When cars are parked by attendants, the driver must stop and leave the vehicle. The attendant then enters and drives the vehicle to the parking bay. Thus, a reservoir space must be provided that will accommodate temporary storage for entering vehicles. The size of this reservoir space depends on the ratio of the rate of storage of the vehicles to the rate of arrival of the vehicles. The rate of storage must take into consideration the time required to transfer the vehicle from its driver to the attendant. The number of temporary storage bays can be determined by applying the queuing theory presented in Chapter 5. Figure 14.34 shows the size of reservoir space required for different ratios of storage rate to arrival rate, with overloading occurring only 1 percent of the time.

COMPUTER USE IN GEOMETRIC DESIGN

Computer programs are now available to carry out all of the designs discussed in this chapter. Most highway agencies have developed programs suitable for their

Figure 14.34   Reservoir Space Required If Facility Is Overloaded Less Than 1 Percent of Time

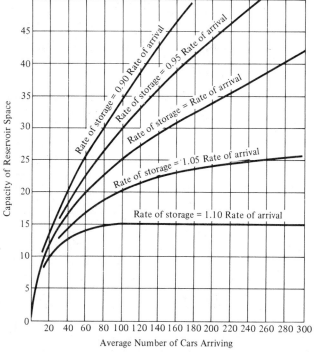

Source: Redrawn from E. R. Ricker, *Traffic Design of Parking Garages*, ENO Foundation for Highway Traffic Control, Inc., Saugatuck, Conn. 1957.

individual hardware systems, which may not easily be used by another agency with a noncompatible system. AASHTO regularly publishes an index of available software packages.[16] The province of Manitoba, Canada, for example, has developed a software package known as the Roadway Design System for the IBM Model 3033 computer. This package will undertake the complete design process from preliminary highway location to the detailed design of vertical and geometric alignments. A similar program, known as ICES ROADS, is currently used by several consultants and design agencies. This package was developed for the IBM Model 3081. This is a comprehensive system used in all phases of highway design. A sister program is the ICES COGO System, which is used for roadway surveying and geometric design. Several programs have also been developed for use with microcomputers, offering even small consulting firms the opportunity to use computer-aided methods in their design work.

SUMMARY

The fundamental principles of highway geometric design are presented in this chapter together with the necessary formulas required for the design of the vertical and horizontal alignments of the highway. The basic characteristics of the driver, vehicle, and road are used to determine the geometric characteristics of any highway or bikeway. The fundamental characteristic on which several design standards are based is the design speed, which depends on the type of facility being considered.

The material presented in this chapter provides the mathematical tools required by an engineer engaged in the geometric design of highways. Since any motor vehicle on a highway has to be parked at one time or another, the design of parking bays and garages is therefore included.

A brief introduction to available computer programs for geometric design of highways is also presented. The intention is to let the reader know of the opportunities available in computer-aided design in highway engineering, rather than the presentations of different computer programs, since many of those currently in use may soon become outdated as new programs become available.

PROBLEMS

14–1 A rural arterial highway located in a mountainous terrain is to be designed to carry a DHV of 800 vph, 98 percent of which are passenger cars. Determine the following: **(a)** a suitable design speed, **(b)** lane and usable shoulder widths, **(c)** maximum desirable grade, **(d)** minimum radius of horizontal curve, **(e)** minimum length of crest vertical curves, and, **(f)** minimum length of sag vertical curves, for the maximum grade.

14–2 Repeat problem 14–1 for a freeway in rolling terrain with a DHV of 1250 vph, 25 percent of which are trucks.

14–3 With the aid of an appropriate diagram, describe a suitable method you will use to achieve superelevation on a 6-lane rural freeway (3 lanes in each direction) with 50 ft wide median.

14–4 Discuss briefly the factors you will consider in locating and designing a vertical curve on an urban freeway.

14–5 A +2 percent grade on an arterial highway intersects with a −1 percent grade at station (535 + 24.25) at an elevation of 300 ft. If the design speed of the highway is 65 mph, determine the stations and elevations of the BVC and EVC and the elevation of each 100 ft station.

14–6 A $+g_1$ grade is connected by a vertical curve to a $-g_2$ grade (g_2 less than g_1). If the rate of change of gradient of the curve is constant, show that the height y of any point of the curve above the BVC is given as

$$100y = g_1 x - \frac{(g_1 - g_2)x^2}{2L}$$

14–7 A sag vertical curve connects a −1.5 percent grade with a +2.5 percent grade on a rural arterial highway. If the criterion selected for design is the minimum stopping sight distance and the design speed of the highway is 70 mph, compute the elevation of the curve at 100 ft stations if the grades intersect at station 475 + 00 at an elevation of 300 ft.

14–8 A circular curve connects two tangents that deflect at an angle of 48°. The point of intersection of the tangents is located at station (948 + 67.32). If the design speed of the highway is 60 mph, determine the point of curve, the point of the tangent, and the deflection angles to whole stations for laying out the curve. (Select appropriate values for e and f.)

14–9 To determine the radius of an existing curve, two chords xy and yz of 135 ft each were marked out on the curve. The distance between y and the chord xz was found to be 15 ft. Determine the radius of the curve and the angle you will set out from xz to get a line to the center of the curve.

14–10 A compound circular curve having radii of 600 ft and 450 ft is to be designed to connect two tangents deflecting by 75°. If the deflection angle of the first curve is 45° and the PCC is located at station (675 + 35.25), determine the deflection angles and the corresponding chord lengths for setting out the curve.

14–11 An arterial road is to be connected to a frontage road by a reverse curve with parallel tangents. The distance between the centerline of the two roads is 60 ft. If the PC of the curve is located at station (38 + 25.31) and the deflection angle is 25°, determine the station of PT.

14–12 A parking lot is to be designed for a shopping center. If the area available is 400 ft × 500 ft, design a suitable layout for achieving each of the following objectives—that is, one layout for each objective.

(a) Provide the maximum number of spaces.
(b) Provide the maximum number of spaces while facilitating traffic circulation by providing one-way flow on each aisle.

REFERENCES

1. *A Policy on Geometric Design of Highways and Streets*, American Association of State Highway and Transportation Officials, Washington, D.C., 1984.

2. Ibid.

3. Ibid.

4. Highway Research Board, *Relationship of Highway Geometry to Traffic Accidents*, Highway Research Record 312, National Research Council, Transportation Research Board, Washington, D.C., 1970.

5. *Policy on Geometric Design*.

6. *Guide for Selecting, Locating and Designing Traffic Barriers*, American Association of State Highway and Transportation Officials, Washington, D.C., 1977.

7. *Policy on Geometric Design*.

8. Ibid.

9. Ibid.

10. *Guide for Development of New Bicycle Facilities*, American Association of State Highway and Transportation Officials, Washington, D.C., 1981.

11. Ibid.

12. Ibid.

13. Ibid.

14. Ibid.

15. Ibid.

16. *Computer System Index*, American Association of State Highway and Transportation Officials, Washington, D.C., 1983.

ADDITIONAL READINGS

Davis, R. E., F. S. Foote, J. M. Anderson, and E. M. Mikhail, *Surveying Theory and Practice*, McGraw-Hill Book Company, New York, 1981.

Michie, J. D., and M. E. Bronstad, *Location and Selection and Maintenance of Highway Traffic Barriers*, NCHRP Report 118, National Research Council, Transportation Research Board, Washington, D.C., 1971.

Michie, J. D., and M. E. Bronstad, *Guardrail Crash Test Elevation—New Concepts and Designs*, NCHRP Report 129, National Research Council, Transportation Research Board, Washington, D.C., 1982.

CHAPTER 15

Drainage and Drainage Systems

Provision of adequate drainage is an important factor in the location and geometric design of highways. Drainage facilities on any highway or street should adequately provide for the flow of water away from the surface of the pavement to properly designed channels. Inadequate drainage will eventually result in serious damage to the highway structure. In addition, traffic may be slowed by accumulated water on the pavement, and accidents may occur as a result of hydroplaning and loss of visibility from splash and spray. The importance of adequate drainage is recognized in the amount of highway construction dollars allocated to drainage facilities. About 20 percent to 25 percent of highway construction dollars are spent for erosion control and drainage structures, such as culverts, bridges, channels, and ditches.

The highway engineer is primarily concerned with two sources of water. The first source, surface water, is that which occurs as rain or snow. Some of this is absorbed into the soil, and the remainder remains on the surface of the ground and should be removed from the highway pavement. Drainage for this source of water is referred to as *surface drainage*. The second source, ground water, is that which flows in underground streams. This may become important in highway cuts or at locations where a high water table exists near the pavement structure. Drainage for this source of water is referred to as *subsurface drainage*.

In this chapter, we present the fundamental design principles for surface and subsurface drainage facilities. The principles of hydrology necessary for understanding the design concepts are also included, together with a brief discussion on erosion prevention.

SURFACE DRAINAGE

Surface drainage encompasses all means by which surface water is removed from the pavement and right of way of the highway or street. A properly designed highway surface drainage system should effectively intercept all surface and watershed run-off and direct this water into adequately designed channels and gutters for eventual discharge into the natural waterways. Water seeping through cracks in the highway riding surface and shoulder areas into underlying layers of the pavement may result in serious damage to the highway pavement. The major source of water for this type of intrusion is surface run-off. An adequately designed surface drainage system will therefore minimize this type of damage. The surface drainage system for rural highways should include adequate transverse and longitudinal slopes on both the pavement and shoulder to ensure positive run-off, and longitudinal channels (ditches), culverts, and bridges to provide for the discharge of the surface water to the natural waterways. Storm drains and inlets are also provided on the median of divided highways in rural areas. In urban areas, the surface drainage system also includes adequate longitudinal and transverse slopes, but the longitudinal drains are usually underground pipe drains, designed to carry both surface run-off and ground water. Curbs and gutters may also be used in urban and rural areas to control street run-off, although they are more frequently used in urban areas.

Transverse Slopes

The main objective for providing slopes in the transverse direction is to facilitate the removal of surface water from the pavement surface in the shortest possible time. This is achieved by crowning the surface at the center of the pavement, thereby providing cross slopes on either side of the centerline or providing a slope in one direction across the pavement width. The different ways in which this is achieved were discussed in Chapter 14. Shoulders, however, are usually sloped to drain away from the pavement, except on highways with raised narrow medians. The need for high cross slopes to facilitate drainage is somewhat in conflict with the need for relatively flat cross slopes for driver comfort. Selection of a suitable cross slope is, therefore, usually a compromise between the two requirements. It has been determined that cross slopes of 2 percent or less do not significantly affect driver comfort, particularly with respect to the driver's effort in steering. Recommended values of cross slopes for different shoulders and types of highway pavement were presented in Chapter 14.

Longitudinal Slopes

A minimum gradient in the longitudinal direction of the highway is required to obtain adequate slope in the longitudinal channels, particularly at cut sections. Slopes in longitudinal channels should generally not be less than 0.2 percent for highways in very flat terrain. Although zero percent grades may be used on uncurbed pavements with adequate cross slopes, a minimum of 0.5 percent is recommended for curbed pavements. This may be reduced to 0.3 percent on suitably crowned high type pavements constructed on firm ground.

Longitudinal Channels

Longitudinal channels (ditches) are constructed along the sides of the highway to collect the surface water that runs off from the pavement surface, subsurface drains, and other areas of the highway right of way. When the highway pavement is located at a lower level than the adjacent ground, such as in cuts, water is prevented from flowing onto the pavement by constructing a longitudinal drain (intercepting drain) at the top of the cut to intercept the water. The water collected by the longitudinal ditches is then transported to a natural or artificial drainage channel and eventually to a natural waterway.

Curbs and Gutters

As pointed out in Chapter 14, curbs and gutters can be used to control drainage in addition to other functions, which include preventing the encroachment of vehicles on adjacent areas and delineating pavement edges. Some typical shapes for curbs and gutters were presented in Chapter 14. Curbs and gutters are used more frequently in urban areas, particularly in residential areas, where they are used in conjunction with storm sewer systems to control street run-off. When it is necessary to provide relatively long continuous sections of curbs in urban areas, the inlets to the storm sewers must be adequately designed for both size and spacing so that the impounding of large amounts of water on the pavement surface is prevented.

HIGHWAY DRAINAGE STRUCTURES

Drainage structures are constructed to carry traffic over natural waterways that flow below the right of way of the highway. These structures also provide for the flow of water below the highway, along the natural channel, without significant alteration or disturbance to its normal course. One of the main concerns of the highway engineer is to provide an adequate size structure, such that the waterway opening is sufficiently large to discharge the expected flow of water. Inadequately sized structures can result in water impounding, which may lead to failure of the adjacent sections of the highway due to embankments being submerged in water for long periods.

The two general categories of drainage structures are major and minor. Major structures are those with clear spans greater than 20 feet, whereas minor structures are those with clear spans of 20 feet or less. Major structures are usually large bridges, although multiple-span culverts may also be included in this class. Minor structures include small bridges and culverts.

Major Structures

It is beyond the scope of this book to discuss the different types of bridges. Emphasis is therefore placed on selecting the span and vertical clearance requirements for such structures. The bridge deck should be located above the high water mark. The clearance above the high water mark depends on whether the waterway is navigable. If the waterway is navigable, the clearance above the high water mark should allow the largest ship using the channel to pass underneath the bridge without colliding with the bridge deck. The clearance height, type, and spacing of piers also depend on the probability of ice jams and the extent to which floating logs and debris appear on the waterway during high water.

An examination of the banks on either side of the waterway will indicate the location of the high water mark, since this is usually associated with signs of erosion and debris deposits. Local residents, who have lived and observed the waterway during flood stages over a number of years, can also give reliable information on the location of the high water mark. Stream gauges that have been installed in the waterway for many years can also provide data that can be used to locate the high water mark.

Minor Structures

Minor structures, consisting of short-span bridges and culverts, are the predominant type of drainage structures on highways. Although openings for these structures are not designed to be adequate for the worst flood conditions, they should be large enough to accommodate the flow conditions that might occur during the normal life expectancy of the structure. Provision should also be made for preventing clogging of the structure due to floating debris and large boulders rolling from the banks of steep channels.

Culverts are made of different materials and in different shapes. Materials used to construct culverts include concrete (reinforced and unreinforced), corrugated steel, and corrugated aluminum. Other materials may also be used to line the interior of the culvert to prevent corrosion and abrasion or to reduce hydraulic resistance. For example, asphaltic concrete may be used to line corrugated metal culverts. The different shapes normally used in culvert construction include circular, rectangular (box), elliptical, pipe arch, metal box, and arch. Figure 15.1 shows a corrugated metal arch culvert and a rectangular (box) culvert.

EROSION CONTROL

Continuous flow of surface water over shoulders, side slopes, and unlined channels often results in soil eroding from the adjacent areas of the pavement. Erosion can lead to conditions that are detrimental to the pavement structure and other adjacent facilities. For example, soil erosion of shoulders and side slopes can result in failure of embankment and cut sections, and soil erosion of highway channels often results in the pollution of nearby lakes and streams. Prevention of erosion is an important factor when highway drainage is being considered, and the methods used to prevent erosion are now briefly discussed.

Intercepting Drains

Provision of an intercepting drain at the top of a cut helps to prevent erosion of the side slopes of cut sections, since the water is intercepted and prevented from flowing freely down the side slopes. The water is collected and transported in the intercepting drain to paved spillways that are placed at strategic locations on the

Figure 15.1   Different Types of Culverts

(a) Corrugated metal arch culvert (ARMCO)

Source: Reproduced from J. M. Normann, R. J. Houghtalen, and W. J. Johnston, *Hydraulic Design of Highway Culverts*, Report No. FHWA–IP–85–15, U.S. Department of Transportation, Office of Implementation, McLean, Va., September 1985.

(b) Rectangular (box) culvert

Source: Photo by Lewis Woodson, Virginia Transportation Research Council, Charlottesville, Va.

side of the cut. The water is then transported through these protected spillways to the longitudinal ditches alongside the highway.

Curbs and Gutters

Curbs and gutters can be used to protect unsurfaced shoulders on rural highways from eroding. They are placed along the edge of the pavement such that surface water is prevented from flowing over and eroding the unpaved shoulders. Curbs and gutters can also be used to protect embankment slopes from erosion when paved shoulders are used. In this case, the curbs and gutters are placed on the outside edge of the paved shoulders and the surface water is then directed to paved spillways located at strategic positions and transported to the longitudinal drain at the bottom of the embankment.

Turf Cover

Using a firm turf cover on unpaved shoulders, ditches, embankments, and cut slopes is an efficient and economic method of preventing erosion when slopes are flatter than 3 to 1. The turf cover is commonly developed by sowing suitable grasses immediately after grading. The two main disadvantages of using turf cover on unpaved shoulders are that turf cover cannot resist continued traffic and it loses its firmness under conditions of heavy rains.

Slope and Channel Linings

When the highway is subjected to extensive erosion, more effective preventive action than any of those already described is necessary. For example, when cut and embankment side slopes are steep and are located in mountainous areas subjected to heavy rain and snow falls, additional measures should be taken to prevent erosion. A commonly used method is to line the slope surface with rip-rap or hand-placed rock.

Channel linings are also used to protect longitudinal channels from eroding. The lining is placed along the sides and in the bottom of the drain. Protective linings can be categorized into two general groups: flexible and rigid. Flexible linings include dense-graded bituminous mixtures and rock rip-rap, whereas rigid linings include Portland Cement concrete and soil cement. Rigid linings are much more effective in preventing erosion under severe conditions, but they are more expensive and, because of their smoothness, tend to create high unacceptable velocities at the end of the linings. When the use of rigid lining results in high velocities, a suitable energy dissipator must be placed at the lower end of the channel to prevent excessive erosion. The energy dissipator is not required if the water discharges into a rocky stream or into a deep pool. Detailed design procedures for linings are presented under Channel Design.

HYDROLOGIC CONSIDERATIONS

Hydrology is the science that deals with the characteristics and distribution of water in the atmosphere, on the earth's surface, and in the ground. The basic phenomenon in hydrology is the cycle that consists of precipitation occurring on the ground in the form of water, snow, hail, and so forth and returning to the atmosphere in the form of vapor. It is customary in hydrology to refer to all forms of precipitation as rainfall, with precipitation usually measured in terms of the equivalent depth of water that is accumulated on the ground surface.

Highway engineers are primarily concerned with three properties of rainfall: the rate of fall, known as *intensity*; the length of time for a given intensity, known as *duration*; and the probable number of years that will elapse before a given combination of intensity and duration will be repeated, known as *frequency*. The U.S. Weather Bureau has a network of automatic rainfall instruments that collect data on intensity and duration over the entire country. These data are used to draw rainfall intensity curves from which rainfall intensity for a given return period and duration can be obtained. Figure 15.2 is an example of a set of rainfall intensity curves, and Figure 15.3 shows rainfall intensities for different parts of the United States for a storm of 1 hr duration and 1 yr frequency. The use of the rainfall intensity curves is illustrated in Figure 15.2. An intensity of 1.57 in./hr is obtained for a 10 year storm having a duration of 1.30 hours.

Note that any estimate of rainfall intensity, duration, or frequency made from these data is based on probability laws. For example, if a culvert is designed to carry a "100 year" flood, then the probability is 1 in 100 that the culvert will flow full in any one year. This does not mean that a precipitation of the designed intensity and duration will occur exactly once every 100 years. In fact, it is likely that precipitations of higher intensities could occur one or more times before a time lapse of 100 years, although the probability of this happening is low. This suggests that drainage facilities should be designed for very rare storms to reduce the chance of overflowing to a minimum. Designing for this condition, however, results in very large facilities that cause the cost of the drainage facility to be very high. The decision on what frequency should be selected for design purposes must therefore be based on a comparison of the capital cost for the drainage facility and the cost to the public in case the highway is severely damaged by storm run-off. Factors usually considered in making this decision include the importance of the highway, the volume of traffic on the highway and the population density of the area. Recommended storm frequencies referred to as return periods are shown in Table 15.1.

Other hydrologic variables that the engineer uses to determine surface run-off rates are the drainage area, the run-off coefficient, and the time of concentration.

The *drainage area* is that area of land that contributes to the run-off at the point where the channel capacity is to be determined. This area is normally determined from a topographic map.

The *run-off coefficient*, C, is the ratio of run-off to rainfall for the drainage area. The run-off coefficient depends on the type of ground cover, the slope of the

Figure 15.2 Rainfall Intensity Curves

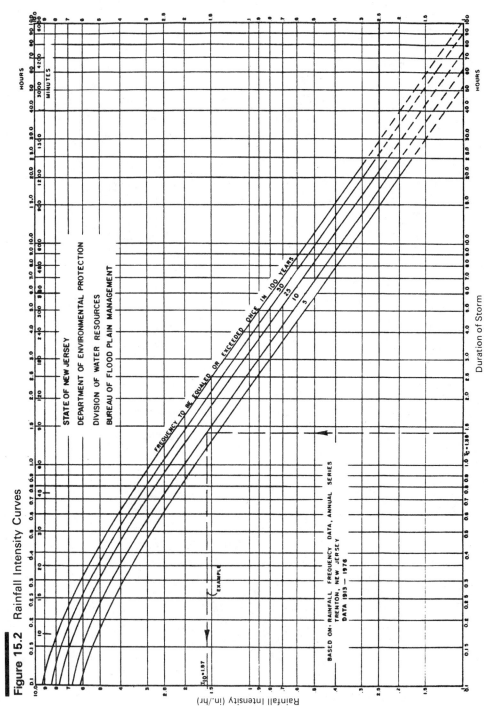

Source: Reproduced from George S. Koslov, *Road Surface Drainage Design, Construction and Maintenance Guide for Pavements*, Report No. FHWA/NJ–82–004, State of New Jersey, Department of Transportation, Trenton, N.J., June 1981.

Figure 15.3 One-Hour/One-Year Frequency Precipitation Rates for the United States

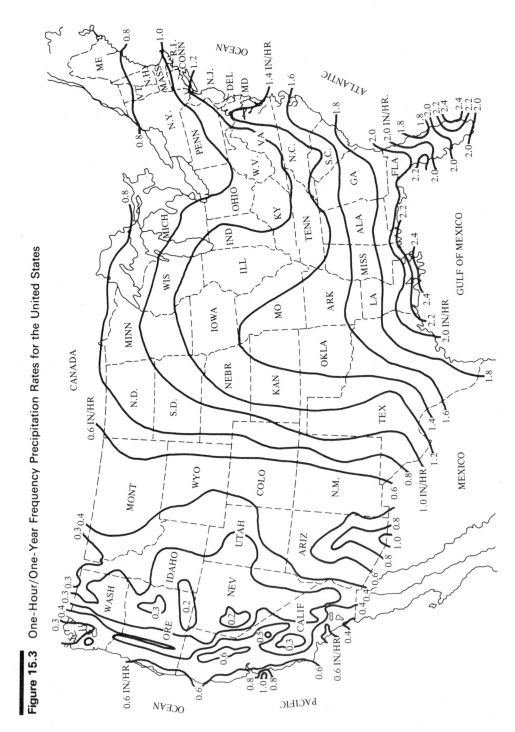

Source: Redrawn from *Rainfall Frequency Atlas of the United States*, U.S. Department of Commerce, Washington, D.C., 1981.

Table 15.1 Recommended Storm Return Periods for Design

Locations and Conditions	Periods (years) Proposed for New York State	
	Ditches	Culverts
Low-volume roads in rural areas	2–5	5–10
High-volume secondary roads in rural areas	5–10	5–10
Primary roads in rural or suburban areas	5–10	10–25
Important roads or where backwater may cause excessive property damage or danger to life	25–50	50–100
Town and rural county roads	25	
Suburban county roads, state highways, and municipal streets	50–100	

	Presently in Use in the State of New Jersey	
	Interstate Design	State Design
For the entire system	15	10
At low points	25	15
For cross drains	50	25

	FHWA
Bridges on important highways or where backwater may cause excessive property damage or result in loss of the bridge	50 to 100 years
Bridges on less important roads or culverts on important roads	25 years
Culverts on secondary roads, storm sewers, side ditches	5 to 10 years
Storm-water inlets, gutter flow	1 to 2 years

Source: Adapted from George S. Koslov, *Road Surface Drainage Design, Construction and Maintenance Guide for Pavements*, Report No. FHWA/NJ-82-004, State of New Jersey, Department of Transportation, Trenton, N.J., June 1981.

drainage area, storm duration, prior wetting, and the slope of the ground. Several suggestions have been made to adjust C for storm duration, prior wetting, and other factors. However, these adjustments are probably not necessary for small drainage areas. Average values assumed to be constant for any given storm are therefore used. Representative values for C for different run-off surfaces are given in Table 15.2.

Table 15.2 Values of Run-Off Coefficients, C

Type of Surface	Coefficient, C*
Rural Areas	
Concrete sheet asphalt pavement	0.8–0.9
Asphalt macadam pavement	0.6–0.8
Gravel roadways or shoulders	0.4–0.6
Bare earth	0.2–0.9
Steep grassed areas (2 : 1)	0.5–0.7
Turf meadows	0.1–0.4
Forested areas	0.1–0.3
Cultivated fields	0.2–0.4
Urban Areas	
Flat residential, with about 30 percent of area impervious	0.40
Flat residential, with about 60 percent of area impervious	0.55
Moderately steep residential, with about 50 percent of area impervious	0.65
Moderately steep built-up area, with about 70 percent of area impervious	0.80
Flat commercial, with about 90 percent of area impervious	0.80

*For flat slopes or permeable soil, use the lower values. For steep slopes or impermeable soil, use the higher values.

Source: Adapted from George S. Koslov, *Road Surface Drainage Design, Construction and Maintenance Guide for Pavements*, Report No. FHWA/NJ–82–004, State of New Jersey, Department of Transportation, Trenton, N.J., June 1981.

In cases where the drainage area consists of different ground characteristics with different run-off coefficients, a representative value C_w is computed by determining the weighted coefficient using Eq. 15.1.

$$C_w = \frac{\sum_{i=1}^{n} C_i A_i}{\sum_{i=1}^{n} A_i} \qquad (15.1)$$

where

C_w = weighted run-off coefficient for the whole drainage area

C_i = run-off coefficient for watershed i

A_i = area of watershed i

n = number of different watersheds in the drainage area

The *time of concentration,* t_c, is the time required for the run-off to flow from the farthest point of the drainage area to the outlet. The time of concentration for a drainage area must be determined for the design of the associated drainage channels. Time of concentration depends on several factors, including the size and shape of the drainage area, the type of surface, the slope of the drainage area, the rainfall intensity, and whether the flow is entirely over land or partly channelized. The time of concentration generally consists of one or more of three components of travel times, depending on the location of the drainage systems. These are the times for overland flow, gutter or storm sewer flow (mainly in urban areas), and channel flow.

Figure 15.4 Average Velocities for Estimating Travel Time for Overland Flow

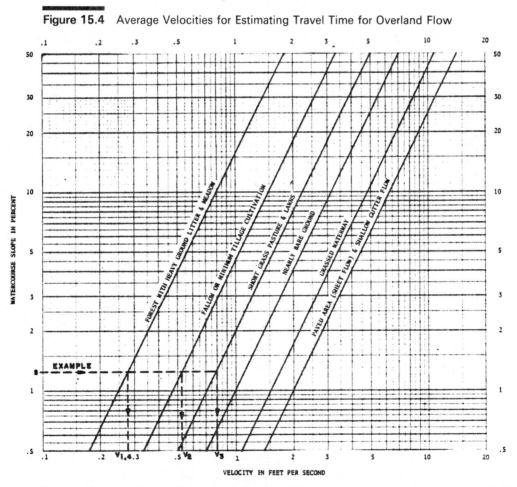

Source: Reproduced from George S. Koslov, *Road Surface Drainage Design, Construction and Maintenance Guide for Pavements,* Report No. FHWA/NJ–82–004, State of New Jersey, Department of Transportation, Trenton, N.J., June 1981.

Travel time for overland flow, t_o, is computed by dividing the total length of the overland flow—that is, the distance between the farthest point of the drainage area and the channel or storm sewer inlet—by the average velocity. The overland flow velocity depends on the slope and type of ground of the drainage area. Figure 15.4 can be used to determine average overland flow velocities for different slopes and types of ground.

The travel time for flow in the gutter or storm sewer is the sum of the travel times in each component of the gutter and/or storm sewer system between the farthest inlet and the outlet. Although velocities in the different components may be different, the use of the average velocity for the whole system does not usually result in large errors. When gutters are shallow, the curve for overland flow in paved areas in Figure 15.4 can be used to determine the average velocity.

The travel time in the open channel is determined in a similar way as that for the flow in gutter or storm sewer. In this case, however, the velocity of flow in the open channel must be determined first by using an appropriate equation such as Manning's formula. This is discussed in more detail under Channel Design. The total time of concentration is then determined as the sum of all travel time components.

Determination of Run-Offs

The amount of run-off for any combination of intensity and duration depends on the type of surface. For example, run-off will be much higher on rocky or bare impervious slopes, roofs, and pavements than on plowed land or heavy forest. The highway engineer is therefore interested in determining the proportion of rainfall that remains as run-off. This determination is not easy since the run-off rate for any given area during a single rainfall is not usually constant. Several methods for estimating run-off are available, but only two commonly used methods are presented.

Rational Method

The rational formula is based on the premise that the rate of run-off for any storm depends on the average intensity of the storm, the size of the drainage area, and the type of drainage area surface. Note that, for any given storm, the rainfall intensity is not usually constant over a large area, nor during the entire duration of the storm. The rational formula therefore uses the theory that, for a rainfall of average intensity (I) falling over an impervious area of size A, the maximum rate of run-off at the outlet to the drainage area occurs when the whole drainage area is contributing to the run-off and this run-off rate is constant. This requires that the storm duration be at least equal to the time of concentration, which is the time required for the run-off to flow from the farthest point of the drainage area to the outlet. This condition is not always satisfied in practice, particularly in large drainage areas. It is therefore customary for the rational formula to be used for relatively small drainage areas not greater than 200 acres. The rational formula is given as

$$Q = CIA \qquad (15.2)$$

where

Q = peak rate of run-off in cubic feet per second

A = drainage area in acres

I = average intensity in inches per hour for a selected frequency and duration equal to at least the time of concentration

C = a coefficient representing the fraction of rainfall that remains on the surface of the ground (run-off coefficient)

Although the formula is not dimensionally correct, results of studies indicate that a rainfall of 1 in./hr falling over an area of 1 acre produces a run-off of 1.008 ft³/sec if there are no losses. The run-off value Q is therefore almost exactly equal to the product of I and A. The losses due to infiltration and evaporation are accounted for by C. Values for C can be obtained from Table 15.2.

Example 15–1 Computing Rate of Run-Off Using the Rational Formula

A 175 acre rural drainage area consists of three different watershed areas as follows.

Steep grassed = 50 percent
Forested area = 30 percent
Cultivated area = 20 percent

If the time of concentration for the drainage area is 1.5 hr, determine the run-off rate for a storm of 50 yr frequency. Assume that the rainfall intensity curves in Figure 15.2 are applicable to this drainage area.

The weighted run-off coefficient should first be determined for the whole drainage area. From Table 15.2, mean coefficients for the different watersheds are

Steep grassed = 0.6
Forested area = 0.2
Cultivated area = 0.3

$$C_w = \frac{175(0.5 \times 0.6 + 0.3 \times 0.2 + 0.2 \times 0.3)}{175} = 0.42$$

The storm intensity for a duration of at least 1.5 hr (time of concentration) and frequency of 50 yr is obtained from Figure 15.2 as approximately 2 in./hr. From Eq. 15.2

$$Q = 0.42 \times 2 \times 175 = 147 \text{ (ft}^3\text{/sec)}$$

U.S. Soil Conservation Service Methods

Two methods developed by the U.S. Soil Conservation Service (SCS) are commonly used in determining surface run-offs in highway engineering. These methods are usually referred to as TR–55[1] and TR–20.[2] The TR–55 method can be used to estimate run-off volumes and peak rate of discharge, whereas the TR–20 method is a computerized method for formulating and developing run-off hydrographs and route hydrographs through both channel reaches and reservoirs. It is beyond the scope of this book to give details of the TR–20 method, but interested readers can refer to the SCS publication for details.[3]

TR–55 Method. This method consists of two parts. The first part determines the depth of run-off h in inches, and the second part estimates the peak discharge using the value of h obtained and a graph that relates the time of concentration (hours) with the unit peak discharge ($ft^3/sec/mi^2/in.$). This procedure is usually referred to as the graphical method.

The fundamental premise used in developing this method is that the depth of run-off h in inches depends on the precipitation P in inches and the water retained in the soil F, where F is the difference between the volume of precipitation and the volume of run-off. Also, some of the precipitation occurring at the early stage of the storm and known as initial abstraction (I_a) will not be part of the run-off. The basic relationship assumed is

$$\frac{F}{S} = \frac{h}{P - I_a}$$

where S is the potential maximum retention.

However, taking into consideration the initial abstraction

$$F = (P - I_a) - h$$

substituting for F in the basic equation gives

$$\frac{(P - I_a) - h}{S} = \frac{h}{P - I_a}$$

which gives

$$h = \frac{(P - I_a)^2}{(P - I_a) + S} \tag{15.3}$$

The initial abstraction depends on several factors, including land use, treatment, and condition; interception; and infiltration and the soil moisture just prior to the storm. An empirical relationship was developed by SCS for I_a, which is given as

$$I_a = 0.2S$$

Substituting for I_a in Eq. 15.3 gives

$$h = \frac{(P - 0.2S)^2}{P + 0.8S}$$ (15.4)

where

$S = (1000/CN) - 10$ (obtained from empirical studies)

CN = run-off curve number

The run-off curve number accounts for the watershed characteristics, which include the soil type, land use, hydrologic condition of the cover, and soil moisture just prior to the storm (antecedent soil moisture). In order to determine values for CN, SCS has developed classification systems for soil type, cover, and antecedent soil moisture.

Soils are divided into four groups, A, B, C, and D, as follows:

Group A: deep sand, deep loess, aggregated silts

Group B: shallow loess, sandy loam

Group C: clay loams, shallow sandy loam, soils low in organic content, and soils usually high in clay

Group D: soils that swell significantly when wet, heavy plastic clays, and certain saline soils

Land use and hydrologic conditions of the cover are incorporated in a complex classification as shown in Table 15.3. Table 15.3 shows 14 different classifications for land use. In addition, agricultural land uses are usually subdivided into specific treatment or practices, to account for the differences in hydrologic run-off potential among these treatments or practices. The hydrologic condition is considered in terms of the level of land management, which is given in three classes: poor, fair, and good.

The effect of antecedent moisture on rate of run-off is taken into consideration by classifying soil conditions into three categories: I, II, and III as follows:

Condition I: soils are dry but not to wilting point; satisfactory cultivation has taken place

Condition II: average condition

Condition III: heavy rainfall, or light rainfall and low temperatures have occurred within the last five days; saturated soils.

Table 15.4 gives seasonal rainfall limits as guidelines for determining the antecedent moisture condition (AMC).

Table 15.3 Run-Off Curve Numbers for Condition II

Land-use Description/Treatment/Hydrologic Condition		Hydrologic Soil Group			
		A	B	C	D
Residential[1]					
Average lot size:	Average Percent Impervious:[2]				
1/8 acre or less	65	77	85	90	92
1/4 acre	38	61	75	83	87
1/3 acre	30	57	72	81	86
1/2 acre	25	54	70	80	85
1 acre	20	51	68	79	84
Paved parking lots, roofs, driveways, and so forth[3]		98	98	98	98
Streets and roads					
Paved with curbs and storm sewers[3]		98	98	98	98
Gravel		76	85	89	91
Dirt		72	82	87	89
Commercial and business areas (85 percent impervious)		89	92	94	95
Industrial districts (72 percent impervious)		81	88	91	93
Open spaces, lawns, parks, golf courses, cemeteries, and so forth					
Good condition: grass cover on 75 percent or more of the area		39	61	74	80
Fair condition: grass cover on 50 percent to 75 percent of the area		49	69	79	84
Fallow					
Straight row	—	77	86	91	94
Row crops					
Straight row	Poor	72	81	88	91
Straight row	Good	67	78	85	89
Contoured	Poor	70	79	84	88
Contoured	Good	65	75	82	86
Contoured and terraced	Poor	66	74	80	82
Contoured and terraced	Good	62	71	78	81
Small grain					
Straight row	Poor	65	76	84	88
Straight row	Good	63	75	83	87
Contoured	Poor	63	74	82	85
Contoured	Good	61	73	81	84
Contoured and terraced	Poor	61	72	79	82
Contoured and terraced	Good	59	70	78	81
Close-seeded legumes[4] or rotation meadow					
Straight row	Poor	66	77	85	89
Straight row	Good	58	72	81	85

Continued

Table 15.3—*Continued*

Land-Use Description/Treatment/Hydrologic Condition		Hydrologic Soil Group			
		A	B	C	D
Contoured	Poor	64	75	83	85
Contoured	Good	55	69	78	83
Contoured and terraced	Poor	63	73	80	83
Contoured and terraced	Good	51	67	76	80
Pasture or range					
	Poor	68	79	86	89
	Fair	49	69	79	84
	Good	39	61	74	80
Contoured	Poor	47	67	81	88
Contoured	Fair	25	59	75	83
Contoured	Good	6	35	70	79
Meadow	Good	30	58	71	78
Woods or forest land	Poor	45	66	77	83
	Fair	36	60	73	79
	Good	25	55	70	77
Farmsteads	—	59	74	82	86

[1]Curve numbers are computed assuming the run-off from the house and driveway is directed toward the street with a minimum of roof water directed to lawns where additional infiltration could occur.

[2]The remaining pervious areas (lawn) are considered to be in good pasture condition for these curve numbers.

[3]In some warmer climates of the country, a curve number of 95 may be used.

[4]Close-drilled or broadcast.

Source: Adapted from Richard H. McCuen, *A Guide to Hydrologic Analysis Using SCS Methods,* copyright © 1982. Reprinted by permission of Prentice-Hall, Inc., Englewood Cliffs, N.J.

Table 15.3 gives *CN* values for antecedent moisture condition II. When antecedent moisture condition I or III exists, Table 15.3 is first used to determine the condition II *CN* value, and Table 15.5 is then used to obtain the appropriate value for either condition I or III.

The first task in the second part of the procedure is to determine the unit peak discharge (q'_p) (ft³/sec/mi²/in.) from Figure 15.5, which relates the time of

Table 15.4 Total Five-Day Antecedent Rainfall (in.)

AMC	Dormant Season	Growing Season
I	≤ 0.5	< 1.4
II	0.5 to 1.1	1.4 to 2.1
III	> 1.1	> 2.1

Table 15.5 Corresponding Run-Off Curve Numbers for Conditions I and III

CN for Condition II	Corresponding CN for Condition	
	I	III
100	100	100
95	87	99
90	78	98
85	70	97
80	63	94
75	57	91
70	51	87
65	45	83
60	40	79
55	35	75
50	31	70
45	27	65
40	23	60
35	19	55
30	15	50
25	12	45
20	9	39
15	7	33
10	4	26
5	2	17
0	0	0

Source: Adapted from Richard H. McCuen, *A Guide to Hydrologic Analysis Using SCS Methods*, copyright © 1982. Reprinted by permission of Prentice-Hall, Inc., Englewood Cliffs, N.J.

Figure 15.5 Peak Discharge in ft^3/sec/mi^2 per Inch of Run-Off Versus Time of Concentration, T_c, for 24 hr, Type II Storm Distribution

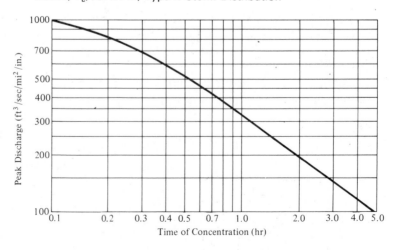

concentration (hours) and the unit peak discharge for a 24 hr type II storm. The peak discharge is then given as

$$q_p = q'_p A h \tag{15.5}$$

where A is the drainage area (mi²).

Example 15–2   Computing Peak Discharge Using the TR–55 Method

A 0.5 mi² drainage area consists of 20 percent residential (1/2 acre lots), 30 percent row crops, with straight row treatment and good hydrologic condition, and 50 percent wooded area with good hydrologic condition. If the soil is classified as group C, with an AMC III, determine the peak discharge if the 24 hr precipitation is 6 in. and the time of concentration is 2 hr.

A weighted CN value should first be determined as follows.

Land Use/Treatment/Hydrologic Condition	CN for AMC II	CN for AMC III
Residential (1/2 acre lots) (20 percent)	80	94
Row crops/straight row/good (30 percent)	85	97
Wooded/—/good (50 percent)	70	87

weighted $CN = 0.2 \times 94 + 0.3 \times 97 + 0.5 \times 87 = 91.4$

$$S = \frac{1000}{CN} - 10 = \frac{1000}{91.4} - 10 = 0.94$$

From Eq. 15-4,

$$h = \frac{(P - 0.2S)^2}{P + 0.8S} = \frac{(6 - 0.2 \times 0.94)^2}{6 + 0.8 \times 0.94} = \frac{33.78}{6.75}$$

$$= 5.0 \text{ in.}$$

The peak unit discharge is determined from Figure 15.5 for a time of concentration of 2 hr, $q'_p = 190 \text{ ft}^3/\text{sec/mi}^2/\text{in.}$

$$q_p = 190 \times 0.5 \times 5.0 = 475 \text{ ft}^3/\text{sec}$$

Unit Hydrographs

A hydrograph is a plot that shows the relationship between stream flow (ordinate scale) and time (abscissa scale). A unit hydrograph is the hydrograph obtained for a run-off due to water of 1 in. in depth, uniformly distributed over the whole drainage area, for a given storm with a specified duration. The unit hydrograph represents all the effects of the different characteristics of the drainage area, such as ground cover, slope, and soil characteristics, acting concurrently. The unit hydrograph therefore gives the run-off characteristics of the drainage area.

It has been shown that similar storms having the same duration, but not necessarily the same intensity, will produce unit hydrographs that are similar in shape. This means that for a given type of rainfall, the time scale of the unit hydrograph is constant for a given drainage area and its ordinates are approximately proportional to the run-off volumes. Thus, the unit hydrograph can be used to estimate the expected run-offs from similar storms having different intensities.

A unit hydrograph can be used to determine effectively peak discharges from drainage areas as large as 2000 mi^2.

Computer Models for Generating Flood Hydrographs

The use of computer models to generate flood hydrographs is becoming increasingly common. Some of these models merely solve existing empirical models, whereas others use simulation techniques. Most simulation models provide for the drainage area to be divided into smaller subareas with similar characteristics. A design storm is then synthesized for each subarea and the volume due to losses such as infiltration and interception deducted. The flow of the remaining water is then simulated using an overland flow routine. The overland flows from the subareas are collected by adjacent channels, which are eventually linked together to obtain the total response of the drainage area to the design storm.

The validity of any simulation model is increased by using measured historical data to calibrate the parameters of the model. One major disadvantage of simulation models is that they usually require a large amount of input data and extensive user experience to obtain reliable results.

HYDRAULIC DESIGN OF HIGHWAY DRAINAGE STRUCTURES

The ultimate objective in determining the hydraulic requirements for any highway drainage structure is to provide a suitable structure size that will economically and efficiently dispose of the expected run-off. Certain hydraulic requirements should also be met to avoid erosion and/or sedimentation in the system.

Design of Open Channels

An important design consideration is that the flow velocity in the channel should not be so low as to cause deposits of transported material nor too high as to cause erosion of the channel. The velocity that will satisfy this condition usually depends on the shape and size of the channel, the type of lining in the channel, the quantity of water being transported, and the type of material suspended in the water. The most appropriate channel gradient range to produce the required velocity is between 1 percent and 5 percent. For most types of linings, sedimentation is usually a problem when slopes are less than 1 percent, and excessive erosion of the lining will occur when slopes are higher than 5 percent. Tables 15.6, 15.7, and 15.8 give recommended maximum velocities for preventing erosion.

Table 15.6 Maximum Permissible Velocities in Erodible Channels, Based on Uniform Flow in Continuously Wet, Aged Channels

	Velocities		
Material	Clear Water (ft/sec)	Water Carrying Fine Silts (ft/sec)	Water Carrying Sand and Gravel (ft/sec)
Fine sand (noncolloidal)	1.5	2.5	1.5
Sandy loam (noncolloidal)	1.7	2.5	2.0
Silt loam (noncolloidal)	2.0	3.0	2.0
Ordinary firm loam	2.5	3.5	2.2
Volcanic ash	2.5	3.5	2.0
Fine gravel	2.5	5.0	3.7
Stiff clay (very colloidal)	3.7	5.0	3.0
Graded, loam to cobbles (noncolloidal)	3.7	5.0	5.0
Graded, silt to cobbles (colloidal)	4.0	5.5	5.0
Alluvial silts (noncolloidal)	2.0	3.5	2.0
Alluvial silts (colloidal)	3.7	5.0	3.0
Coarse gravel (noncolloidal)	4.0	6.0	6.5
Cobbles and shingles	5.0	5.5	6.5
Shales and hard pans	6.0	6.0	5.0

Note: For sinuous (winding) channels, multiply allowable velocity by 0.95 for slightly sinuous channels; by 0.9 for moderately sinuous channels; and by 0.8 for highly sinuous channels.

Source: Adapted from George S. Koslov, *Road Surface Drainage Design, Construction and Maintenance Guide for Pavements*, Report No. FHWA/NJ–82–004, State of New Jersey, Department of Transportation, Trenton, N.J., June 1981.

Attention should also be paid to the point at which the channel discharges into the natural waterway. For example, if the drainage channel at the point of discharge is at a much higher elevation than the natural waterway, the water should be discharged through a spillway or chute to prevent erosion.

Design Principles

The hydraulic design of a drainage ditch for a given storm entails the determination of the minimum cross-sectional area of the ditch that will accommodate the flow due to that storm and prevent water from overflowing the sides of the ditch. The most commonly used formula for this purpose is Manning's formula, which assumes uniform steady flow in the channel and gives the mean velocity in the channel as

$$v = \frac{1.486}{n} R^{2/3} S^{1/2} \qquad (15.6)$$

Table 15.7 Maximum Allowable Water Velocities for Different Types of Ditch Linings

	Maximum Velocity (ft/sec)
Natural Soil Linings	
Bedrock or rip-rap sides and bottoms	15–18
Gravel bottom, rip-rap sides	8–10
Clean gravel	6–7
Silty gravel	2–5
Clayey gravel	5–7
Clean sand	1–2
Silty sand	2–3
Clayey sand	3–4
Silt	3–4
Light clay	2–3
Heavy clay	2–3
Vegetative Linings	
Average turf, erosion-resistant soil	4–5
Average turf, easily eroded soil	3–4
Dense turf, erosion-resistant soil	6–8
Gravel bottom, brushy sides	4–5
Dense weeds	5–6
Dense brush	4–5
Dense willows	8–9
Paved Linings	
Gravel bottom, concrete sides	10
Mortared rip-rap	8–10
Concrete or asphalt	18–20

Source: Adapted from *Drainage Design for New York State*, U.S. Department of Commerce, Washington, D.C., November 1974.

where

v = average discharge velocity (ft/sec)

R = mean hydraulic radius of flow in the channel (ft)

$\quad = \dfrac{a}{P}$

a = channel cross-sectional area (ft^2)

P = wetted perimeter (ft)

S = longitudinal slope in channel (ft/ft)

n = Manning's roughness coefficient

Manning's roughness depends on the type of material used to line the surface of the ditch. Table 15.9 gives recommended values for the roughness coefficients

Table 15.8 Maximum Permissible Velocities in Channels Lined with Uniform Strands of Various Well-Maintained Grass Covers

Cover	Slope Range (%)	Maximum Permissible Velocity on[a]	
		Erosion-Resistant Soils (ft/sec)	Easily Eroded Soils (ft/sec)
Bermuda grass	0–5	8	6
	5–10	7	5
	Over 10	6	4
Buffalo grass Kentucky bluegrass Smooth brome Blue grama	0–5	7	5
	5–10	6	4
	Over 10	5	3
Grass mixture	0–5[b]	5	4
	5–10[b]	4	3
Lespedeza sericea Weeping lovegrass Yellow bluestem Kudzu Alfalfa Crabgrass	0–5[c]	3.5	2.5
Common lespedeza[d] Sudangrass[d]	0–5[c]	3.5	2.5

[a]Use velocities over 5 ft/sec only where good covers and proper maintenance can be obtained.

[b]Do not use on slopes steeper than 10 percent.

[c]Not recommended for use on slopes steeper than 5 percent.

[d]Annuals, used on mild slopes or as temporary protection until permanent covers are established.

Source: Adapted from *Drainage Design for New York State*, U.S. Department of Commerce, Washington, D.C., Novermber 1974.

for different lining materials. The flow in the channel is then given as

$$Q = va = \frac{1.486}{n} aR^{2/3}S^{1/2} \qquad (15.7)$$

where Q is the discharge (ft³/sec).

The Federal Highway Administration (FHWA) has published a series of charts for channels of different cross sections that can be used to solve Eq. 15.7.[4] Figures 15.6 and 15.7 are two examples of these charts.

Since Manning's formula assumes uniform steady flow in the channel, it is now necessary to discuss the concepts of steady, unsteady, uniform, and non-uniform flows.

Table 15.9 Manning's Roughness Coefficients

	Manning's n range [2]
I. Closed conduits:	
A. Concrete pipe	0.011–0.013
B. Corrugated-metal pipe or pipe-arch:	
1. 2⅔ by ½-in. corrugation (riveted pipe): [3]	
a. Plain or fully coated	0.024
b. Paved invert (range values are for 25 and 50 percent of circumference paved):	
(1) Flow full depth	0.021–0.018
(2) Flow 0.8 depth	0.021–0.016
(3) Flow 0.6 depth	0.019–0.013
2. 6 by 2-in. corrugation (field bolted)	0.03
C. Vitrified clay pipe	0.012–0.014
D. Cast-iron pipe, uncoated	0.013
E. Steel pipe	0.009–0.011
F. Brick	0.014–0.017
G. Monolithic concrete:	
1. Wood forms, rough	0.015–0.017
2. Wood forms, smooth	0.012–0.014
3. Steel forms	0.012–0.013
H. Cemented rubble masonry walls:	
1. Concrete floor and top	0.017–0.022
2. Natural floor	0.019–0.025
I. Laminated treated wood	0.015–0.017
J. Vitrified clay liner plates	0.015
II. Open channels, lined [4] (straight alinement): [5]	
A. Concrete, with surfaces as indicated:	
1. Formed, no finish	0.013–0.017
2. Trowel finish	0.012–0.014
3. Float finish	0.013–0.015
4. Float finish, some gravel on bottom	0.015–0.017
5. Gunite, good section	0.016–0.019
6. Gunite, wavy section	0.018–0.022
B. Concrete, bottom float finished, sides as indicated:	
1. Dressed stone in mortar	0.015–0.017
2. Random stone in mortar	0.017–0.020
3. Cement rubble masonry	0.020–0.025
4. Cement rubble masonry, plastered	0.016–0.020
5. Dry rubble (riprap)	0.020–0.030
C. Gravel bottom, sides as indicated:	
1. Formed concrete	0.017–0.020
2. Random stone in mortar	0.020–0.023
3. Dry rubble (riprap)	0.023–0.033
D. Brick	0.014–0.017
E. Asphalt:	
1. Smooth	0.013
2. Rough	0.016
F. Wood, planed, clean	0.011–0.013
G. Concrete-lined excavated rock:	
1. Good section	0.017–0.020
2. Irregular section	0.022–0.027
III. Open channels, excavated [4] (straight alinement, [5] natural lining):	
A. Earth, uniform section:	
1. Clean, recently completed	0.016–0.018
2. Clean, after weathering	0.018–0.020
3. With short grass, few weeds	0.022–0.027
4. In gravelly soil, uniform section, clean	0.022–0.025
B. Earth, fairly uniform section:	
1. No vegetation	0.022–0.025
2. Grass, some weeds	0.025–0.030
3. Dense weeds or aquatic plants in deep channels	0.030–0.035
4. Sides clean, gravel bottom	0.025–0.030
5. Sides clean, cobble bottom	0.030–0.040
C. Dragline excavated or dredged:	
1. No vegetation	0.028–0.033
2. Light brush on banks	0.035–0.050
D. Rock:	
1. Based on design section	0.035
2. Based on actual mean section:	
a. Smooth and uniform	0.035–0.040
b. Jagged and irregular	0.040–0.045
E. Channels not maintained, weeds and brush uncut:	
1. Dense weeds, high as flow depth	0.08–0.12
2. Clean bottom, brush on sides	0.05–0.08
3. Clean bottom, brush on sides, highest stage of flow	0.07–0.11
4. Dense brush, high stage	0.10–0.14

	Manning's n range [2]
IV. Highway channels and swales with maintained vegetation [6] [7] (values shown are for velocities of 2 and 6 f.p.s.):	
A. Depth of flow up to 0.7 foot:	
1. Bermudagrass, Kentucky bluegrass, buffalograss:	
a. Mowed to 2 inches	0.07–0.045
b. Length 4–6 inches	0.09–0.05
2. Good stand, any grass:	
a. Length about 12 inches	0.18–0.09
b. Length about 24 inches	0.30–0.15
3. Fair stand, any grass:	
a. Length about 12 inches	0.14–0.08
b. Length about 24 inches	0.25–0.13
B. Depth of flow 0.7–1.5 feet:	
1. Bermudagrass, Kentucky bluegrass, buffalograss:	
a. Mowed to 2 inches	0.05–0.035
b. Length 4 to 6 inches	0.06–0.04
2. Good stand, any grass:	
a. Length about 12 inches	0.12–0.07
b. Length about 24 inches	0.20–0.10
3. Fair stand, any grass:	
a. Length about 12 inches	0.10–0.06
b. Length about 24 inches	0.17–0.09
V. Street and expressway gutters:	
A. Concrete gutter, troweled finish	0.012
B. Asphalt pavement:	
1. Smooth texture	0.013
2. Rough texture	0.016
C. Concrete gutter with asphalt pavement:	
1. Smooth	0.013
2. Rough	0.015
D. Concrete pavement:	
1. Float finish	0.014
2. Broom finish	0.016
E. For gutters with small slope, where sediment may accumulate, increase above values of n by	0.002
VI. Natural stream channels: [8]	
A. Minor streams [9] (surface width at flood stage less than 100 ft.):	
1. Fairly regular section:	
a. Some grass and weeds, little or no brush	0.030–0.035
b. Dense growth of weeds, depth of flow materially greater than weed height	0.035–0.05
c. Some weeds, light brush on banks	0.035–0.05
d. Some weeds, heavy brush on banks	0.05–0.07
e. Some weeds, dense willows on banks	0.06–0.08
f. For trees within channel, with branches submerged at high stage, increase all above values by	0.01–0.02
2. Irregular sections, with pools, slight channel meander; increase values given in 1a–e about	0.01–0.02
3. Mountain streams, no vegetation in channel, banks usually steep, trees and brush along banks submerged at high stage:	
a. Bottom of gravel, cobbles, and few boulders	0.04–0.05
b. Bottom of cobbles, with large boulders	0.05–0.07
B. Flood plains (adjacent to natural streams):	
1. Pasture, no brush:	
a. Short grass	0.030–0.035
b. High grass	0.035–0.05
2. Cultivated areas:	
a. No crop	0.03–0.04
b. Mature row crops	0.035–0.045
c. Mature field crops	0.04–0.05
3. Heavy weeds, scattered brush	0.05–0.07
4. Light brush and trees: [10]	
a. Winter	0.05–0.06
b. Summer	0.06–0.08
5. Medium to dense brush: [10]	
a. Winter	0.07–0.11
b. Summer	0.10–0.16
6. Dense willows, summer, not bent over by current	0.15–0.20
7. Cleared land with tree stumps, 100–150 per acre:	
a. No sprouts	0.04–0.05
b. With heavy growth of sprouts	0.06–0.08
8. Heavy stand of timber, a few down trees, little undergrowth:	
a. Flood depth below branches	0.10–0.12
b. Flood depth reaches branches	0.12–0.16
C. Major streams (surface width at flood stage more than 100 ft.): Roughness coefficient is usually less than for minor streams of similar description on account of less effective resistance offered by irregular banks or vegetation on banks. Values of n may be somewhat reduced. Follow recommendation in publication cited [8] if possible. The value of n for larger streams of most regular section, with no boulders or brush, may be in the range of	0.028–0.033

Source: Reproduced from *Design Charts for Open Channels*, U.S. Department of Transportation, Washington, D.C., 1980.

Figure 15.6 Graphical Solution of Manning's Equation for a 2:1 Side Slope Trapezoidal Channel

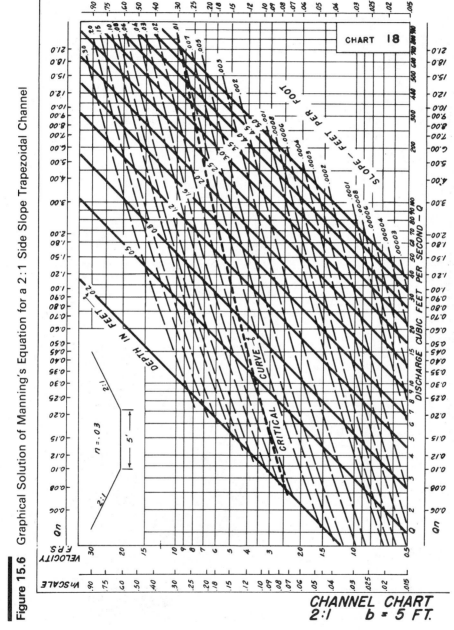

Source: Reproduced from *Design Charts for Open Channels*, U.S. Department of Transportation, Washington, D.C., 1980.

Figure 15.7 Graphical Solution of Manning's Equation for a Rectangular Channel, *b* = 6 ft

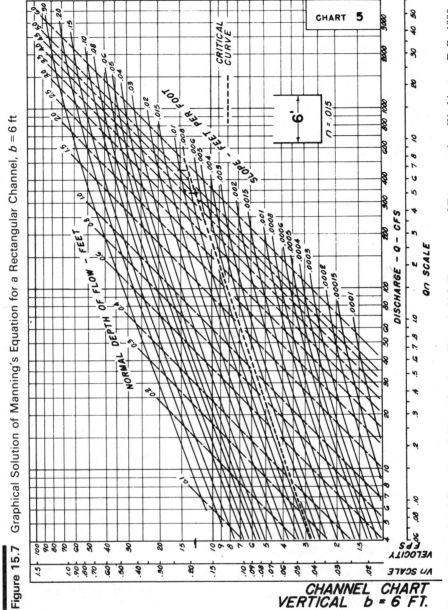

Source: Reproduced from *Design Charts for Open Channels*, U.S. Department of Transportation, Washington, D.C., 1980.

Open channel flows can be grouped into two general categories: steady and unsteady. When the rate of discharge does not vary with time, the flow is steady; conversely, the flow is unsteady when the rate of discharge varies with time. Steady flows are further grouped into uniform and nonuniform, depending on the channel characteristics. Uniform flows are obtained when the channel properties, such as slope, roughness, and cross section, are constant along the length of the channel, whereas nonuniform flows are obtained when these properties vary along the length of the channel. When uniform flow is achieved in the channel, the depth d and velocity v_n are taken as normal and the slope of the water surface is parallel to the slope of the channel. Since it is extremely difficult to obtain the exact same channel properties along the length of the channel, it is very difficult to obtain uniform flow conditions in practice. Nevertheless, Manning's equation can be used to obtain practical solutions to stream flow problems in highway engineering since, in most cases, the error involved is small.

Flows in channels can also be tranquil or rapid. Tranquil flow is similar to the flow of water in an open channel with a relatively flat longitudinal slope, whereas rapid flow is similar to water tumbling down a steep slope. The depth at which the flow in any channel changes from tranquil to rapid is known as the critical depth. When the flow depth is greater than the critical depth, the flow is known as subcritical. This type of flow often occurs in streams in the plains and broad valley regions. When the flow depth is less than the critical depth, the flow is known as supercritical. This type of flow is prevalent in steep flumes and mountain streams. The critical depth may also be defined as the depth of flow at which the specific energy is minimum. The critical depth depends only on the shape of the channel and the discharge. This implies that for any given channel cross section, there is only one critical depth for a given discharge.

The velocity and channel slope corresponding to uniform flow at critical depth are known as critical velocity and critical slope, respectively. The flow velocity and channel slope are therefore higher than the critical values when flow is supercritical, and lower when flow is subcritical.

Consider a channel consisting of four sections, each with a different grade, as shown in Figure 15.8. The slopes of the first two sections are less than the critical slope (although the slope of the second section is less than that of the first section), resulting in a subcritical flow in both sections. The slope of the third section is higher than the critical slope, resulting in a supercritical flow. Flow in the fourth section is subcritical, as the slope is less than the critical slope. Let us now consider how the depth of water changes as the water flows along all four sections of the channel.

As the water flows through section A of the channel, the depth of flow is greater than the critical flow h_C because the flow is subcritical. The slope of section B reduces, resulting in a lower velocity and a higher flow depth ($h_B > h_A$). This increase in depth takes place gradually and begins somewhere upstream in section A. The change from subcritical to supercritical flow in section C with the steep grade also takes place smoothly over some distance. The reduction in flow depth occurs gradually and begins somewhere upstream in the subcritical section. However, when the slope changes again to a value less than the critical slope, the

Figure 15.8 Schematic of the Effect of Critical Depth on Flow in Open Prismatic Channels

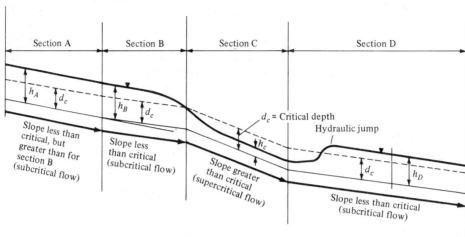

flow changes abruptly from supercritical to subcritical, resulting in a *hydraulic jump*, in which some of the energy is absorbed by the resulting turbulence. This shows that downstream conditions may change the depth upstream of a subcritical flow, which means that *control is downstream.* Thus, when flow is subcritical, any downstream changes in slope, cross section, or intersection with another stream will result in the gradual change of depth upstream, known as *backwater curve.* Conversely, supercritical flow is normally not affected by downstream conditions and *control is upstream.*

It should be noted that the FHWA curves for the solution of Manning's formula are only suitable for flows not affected by backwater.

Design Procedure

The design of a highway drainage channel is accomplished in two steps. The first step is to determine the channel cross section that will economically and effectively transport the expected run-off to a natural waterway. The second step is to determine whether the channel requires any erosion protection and if so, what type of ining is to be used.

Determination of Cross Section. The Manning solution is solved for an assumed channel cross section to determine whether the channel is large enough for the run-off of the design storm. This solution may be carried out manually or by using the appropriate FHWA chart. Both methods are demonstrated in the following examples.

Example 15–3 Manual Design of an Open Channel

Determine a suitable cross section for a channel to carry an estimated run-off of 340 ft³/sec if the slope of the channel is 1 percent. Take *n* in Manning's formula as 0.015.

The solution is carried out by selecting a channel section and then using Manning's formula to determine the flow depth required for the estimated run-off. Let us assume a rectangular channel 6 ft wide and the following data.

d = flow depth

Cross sectional area = $6d$

Wetted perimeter = $6 + 2d$

Hydraulic radius $R = 6d/(6 + 2d)$

$$Q = 340 = \frac{1.486}{0.015}(6d)\left(\frac{6d}{6 + 2d}\right)^{2/3}(0.01)^{1/2}$$

This equation is solved by trial and error to obtain $d \approx 4$ ft.

Alternatively, the FHWA chart shown in Figure 15.7 can be used. Enter the chart at $Q = 340$ ft^3/sec, move vertically to intersect the channel slope of 1 percent (0.01), and normal depth of 4 ft is read from the normal depth lines.

The critical depth can also be obtained directly from the chart by entering the chart at 340 ft^3/sec and moving vertically to intersect the critical curve. This gives a critical depth of about 4.5 ft, which means that the flow is supercritical.

The critical and flow velocities can be obtained similarly. The chart is entered at 340 ft^3/sec, we move vertically to the 0.01 slope, and then to the vertical velocity scale to read the flow velocity as approximately 14 ft/sec. The critical velocity is obtained in a similar manner by moving vertically to the critical curve. The critical velocity is about 13 ft/sec. The critical slope is about 0.007.

The solution indicates that if a rectangular channel 6 ft wide is to be used to carry the run-off of 340 ft^3, the channel must be at least 4 ft deep. However, it is necessary to provide a freeboard of at least 1 ft, which makes the depth for this channel 5 ft.

The same chart can be used for different values of n, when all other conditions are the same. For example, if n is 0.02 in Example 15–3, Qn is first determined as $340 \times 0.02 = 6.8$. Then enter the chart at $Qn = 6.8$ on the Qn scale, move vertically to the 0.01 slope, and read the flow depth as approximately 5 ft. The velocity is also read from the Vn scale as $Vn = 0.225$, which gives V as $0.225/0.02 = 11.25$ ft/sec.

Example 15–4 Computing Discharge Flow from a Trapezoidal Channel

A trapezoidal channel has $2:1$ side slopes, a 5 ft bottom width, and a depth of 4 ft. If the channel is on a slope of 2 percent and $n = 0.030$, determine the discharge flow, velocity, and type of flow. The chart shown in Figure 15.6 is used to solve this problem.

The intersection of the 2 percent slope (0.02) and the 4 ft depth line is located. Move vertically to the horizontal discharge scale to determine the discharge,

$Q = 600 \, \text{ft}^3/\text{sec}$. Similarly, the velocity is 12.5 ft/sec. The intersection of the 2 percent slope and 4 ft depth lies above the critical curve, thus, the flow is supercritical.

Determination of Suitable Lining.   The traditional method for determining suitable lining material for a given channel is to select a lining material such that the flow velocity is less than the permissible velocity to prevent erosion of the lining. However, research results have shown that the maximum permissible depth of flow (d_{max}) for flexible linings should be the main criterion for selecting channel lining.[5] Rigid channels, such as those made of concrete or soil cement, do not usually erode in normal highway work, therefore have no maximum permissible depth of flow to prevent erosion. Thus, maximum flow depth for rigid channels depends only on the freeboard required over the water surface.

The maximum permissible depth of flow for different flexible lining materials can be obtained from charts given in *Design of Stable Channels with Flexible Linings*.[6] Figures 15.9 through 15.11 show some of these charts.

Figure 15.9   Maximum Permissible Depth of Flow, d_{max}, for Channels Lined with Jute Mesh

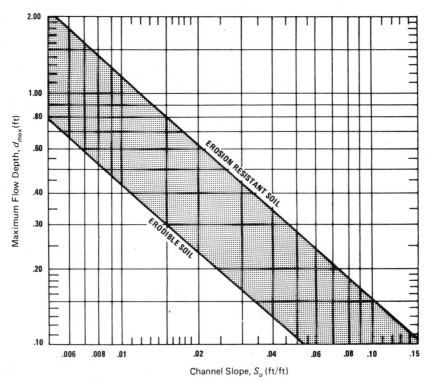

Source: Reproduced from *Design of Stable Channels with Flexible Linings*, Hydraulic Engineering Circular No. 15, U.S. Department of Transportation, Washington, D.C., October 1975.

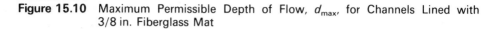

Figure 15.10 Maximum Permissible Depth of Flow, d_{max}, for Channels Lined with 3/8 in. Fiberglass Mat

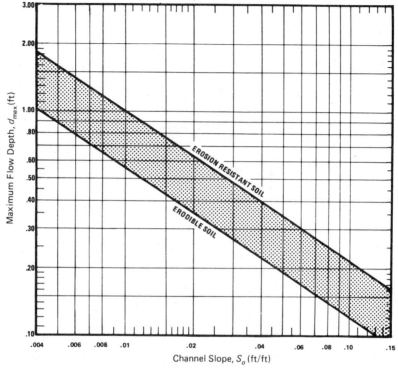

Source: Reproduced from *Design of Stable Channels with Flexible Linings*, Hydraulic Engineering Circular No. 15, U.S. Department of Transportation, Washington, D.C., October 1975.

The design procedure for flexible linings is described in the following steps.

Step 1. Determine channel cross section as in Example 15–3 or Example 15–4.

Step 2. Determine the maximum top width T of the selected cross section and select a suitable lining.

Step 3. Using the appropriate d_{max} for the lining selected (obtained from the appropriate chart), determine the hydraulic radius R and the cross sectional area a for the selected cross section. This can be done by calculation or by using the chart shown in Figure 15.12.

Step 4. Determine the flow velocity V using the channel slope S and the hydraulic radius R determined in step 3.

Step 5. Determine the allowable flow in the channel using the velocity determined in step 4 and the cross sectional area a in step 3—that is, $Q = aV$.

Step 6. Compare the flow determined in step 5 with the designed flow. If there is a difference between the two flows, then the selected lining material is not suitable. For example, if the flow obtained in step 5 is less than the design flow,

Figure 15.11 Maximum Permissible Depth of Flow, d_{max}, for Channels Lined with Bermuda Grass. Good Stand, Cut to Various Lengths

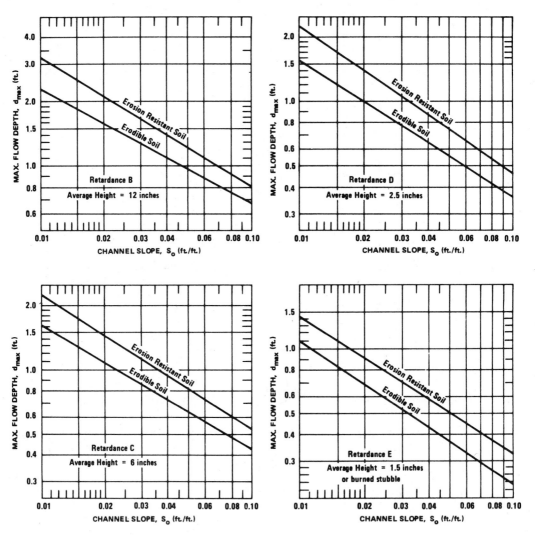

Note: Use on slopes steeper than 10 percent is not recommended

Source: Reproduced from *Design of Stable Channels with Flexible Pavements*, Hydraulic Engineering Circular No. 15, U.S. Department of Transportation, Washington, D.C., October 1975.

Figure 15.12 Trapezoidal Channel Geometry

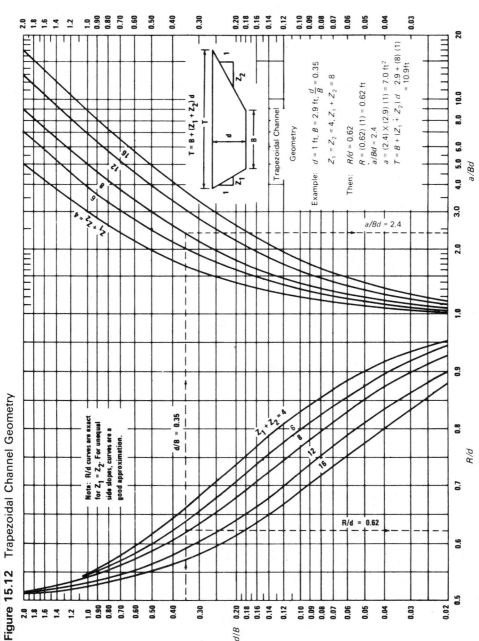

Source: Reproduced from *Design of Stable Channels with Flexible Pavements*, Hydraulic Engineering Circular No. 15, U.S. Department of Transportation, Washington, D.C., October 1975.

then the lining is inadequate, and if it is much greater, the channel is overdesigned. When the flows are much different, then another lining material should be selected and steps 3 through 6 repeated.

Note that Manning's formula can be used to determine the allowable velocity in step 4, if the Manning coefficient of the material selected is known. Since it is not usually easy to determine these coefficients, curves obtained from test results and provided in *Design of Stable Channels with Flexible Linings* can be used.[7] Figures 15.13 and 15.14 are flow velocity curves for jute mesh and fiberglass mat, for which maximum permissible depth charts are shown in Figures 15.9 and 15.10.

Example 15–5   Checking for Lining Suitability

Determine whether jute mesh is suitable for the flow and channel section in Example 15–3. Assume erosion resistant soil.

$$d_{max} = 1.20 \qquad \text{(Figure 15.9)}$$

$$\text{wetted perimeter} = 2 \times 1.20 + 6 = 8.40 \text{ (ft)}$$

$$\text{cross sectional area} = 1.20 \times 6 = 7.20 \text{ (ft}^2\text{)}$$

$$R = \frac{7.20}{8.40} = 0.86$$

Selecting jute mesh as the lining and a channel slope of 0.01, the flow velocity for the lining is obtained from Figure 15.13 as

$$V = 61.53(0.86)^{1.028}(0.01)^{0.431} = 7.24 \text{ ft/sec}$$

and the maximum allowable flow is

$$Q = 7.24 \times 7.20 = 52.13 \text{ (ft}^3\text{)}$$

Jute mesh is not suitable. A rigid channel is probably appropriate for this flow.

Design of Culverts

Several complex hydraulic phenomena occur when water flows through highway culverts. The hydraulic design of culverts is therefore more complex than open channel design. The main factors considered in culvert design are the location of the culvert, the hydrologic characteristics of the watershed being served by the culvert, economy, and type of flow control.

Culvert Location

The most appropriate location of a culvert is in the existing channel bed, with the center line and slope of the culvert coinciding with that of the channel. At this

Figure 15.13 Flow Velocity for Channels Lined with Jute Mesh

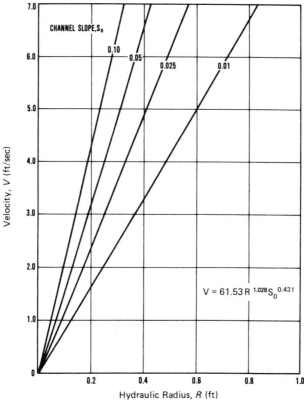

Source: Reproduced from *Design of Stable Channels with Flexible Linings*, Hydraulic Engineering Circular No. 15, U.S. Department of Transportation, Washington, D.C., October 1975.

location, the minimum cost associated with earth and channel work is achieved and stream flow disturbance is minimized. However, other locations may have to be selected in some cases; for example, relocation of a stream channel may be necessary to avoid an extremely long culvert. The basic principle used in locating culverts is that abrupt stream changes at the inlet and outlet of the culvert should be avoided.

Special consideration should be given to culverts located in mountainous areas. When culverts are located in the natural channel in these areas, high fills and long channels are usually required, which result in high construction costs. Culverts are therefore sometimes located on the side of the steep valley. When this is done, adequate measures must be taken to prevent erosion.

Hydrologic and Economic Considerations

The hydrologic and economic considerations are similar to those for open channel design, in that the design flow rate is based on the storm with an acceptable

Figure 15.14 Flow Velocity for Channels Lined with 3/8 in. Fiberglass Mat

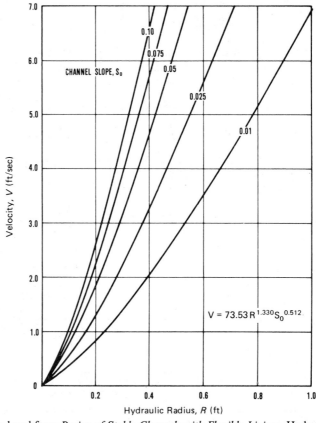

Source: Reproduced from *Design of Stable Channels with Flexible Linings*, Hydraulic Engineering Circular No. 15, U.S. Department of Transportation, Washington, D.C., October 1975.

return period (frequency). This return period is selected such that construction and maintenance costs balance the probable cost of damage to adjacent properties if the storm should occur. Since the occurrence of any given storm within a given period is a stochastic phenomenon, risk analysis is used to determine the appropriate return period. Risk analysis directly relates the culvert design to economic theory and identifies the probable financial consequences of both underdesign and overdesign.

Culverts are designed for the peak flow rate of the design storm. The peak flow rate is obtained from a unit hydrograph at the culvert site, developed from stream flow and rainfall records for a number of storm events.

Other Factors

Other factors that should be considered in culvert design are tailwater and upstream storage conditions.

Tailwater. *Tailwater* is defined as the depth of water above the culvert outlet invert, as the water flows out of the culvert. Designing the culvert capacity must take tailwater into consideration, particularly when the design is based on the outlet conditions. High tailwater elevations may occur due to the hydraulic resistance of the channel or during flood events if the flow downstream is obstructed. Field observation and maps should be used to identify conditions that facilitate high tailwater elevations. These conditions include channel constrictions, intersections with other water courses, downstream impounds, channel obstructions, and tidal effects. If these conditions do not exist, the tailwater elevation is based on the elevation of the water surface in the natural channel.

Upstream Storage. The ability of the channel to store large quantities of water upstream from the culvert may have some effect on the design of the culvert capacity. The storage capacity upstream should therefore be checked using large-scale contour maps, from which topographic information is obtained.

Hydraulic Design of Culverts

As stated earlier, it is extremely difficult to carry out an exact theoretical analysis of culvert flow, because of the many complex hydraulic phenomena that occur. For example, different flow types may exist at different times in the same culvert, depending on the tailwater elevation.

The design procedure presented here is that developed by FHWA and published in *Hydraulic Design of Highway Culverts*.[8] The control section of the culvert is used to classify different culvert flows, which are then analyzed. The location at which a unique relationship exists between the flow rate and the depth of flow upstream from the culvert is the control section. When the flow is dictated by the inlet geometry, then the control section is the culvert inlet, that is, the upstream end of the culvert, and the flow is *inlet controlled*. When the flow is governed by a combination of the tailwater, culvert inlet, and the characteristics of the culvert barrel, the flow is *outlet controlled*. Although it is possible for the flow in a culvert to change from one control to the other and back, the design is based on the *minimum performance* concept, which provides for the culvert to perform at a level that is never lower than the designed level. This means that the culvert may perform at a more efficient level, that is, a higher flow may be obtained for a given headwater level. The design procedure uses several design charts and nomographs, developed from a combination of theory and numerous hydraulic test results.

Inlet Control. The flow in culverts operating under inlet control conditions is supercritical with high velocities and low depths. Four different flows under inlet control are shown in Figure 15.15. The flow type depends on whether the inlet and/or outlet of the culvert are submerged. In Figure 15.15(a), both the inlet and outlet are above the water surface. In this case, the flow within the culvert is supercritical, the culvert is partly full throughout its length, and the flow depth approaches normal at the outlet end. In Figure 15.15(b), only the downstream end (outlet) of the culvert is submerged, but this submergence does not result in outlet control. The flow in the culvert just passed the culvert entrance (inlet) is

Figure 15.15   Types of Inlet Control

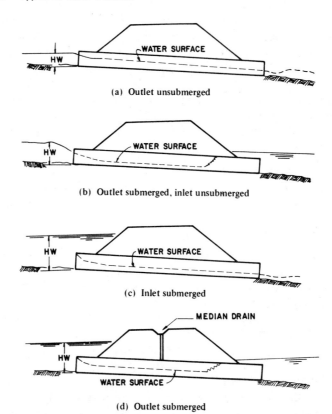

(a) Outlet unsubmerged

(b) Outlet submerged, inlet unsubmerged

(c) Inlet submerged

(d) Outlet submerged

Source: Reproduced from J. M. Normann, R. J. Houghtalen, and W. J. Johnston, *Hydraulic Design of Highway Culverts*, Report No. FHWA–IP–85–15, U.S. Department of Transportation, Office of Implementation, McLean, Va., September 1985.

supercritical, and a hydraulic jump occurs within the culvert. Figure 15.15(c) shows the inlet end of the culvert submerged, with the water flowing freely at the outlet. The culvert is partly full along its length and the flow is supercritical within the culvert, as the critical depth is located just past the culvert inlet. Also, the flow depth at the culvert outlet approaches normal. This example of inlet control is more typical of design conditions. Figure 15.15(d) shows both the inlet and outlet of the culvert submerged, but the culvert is only partly full over part of its length. A hydraulic jump occurs within the culvert resulting in a full culvert along its remaining length. Under these conditions, pressures less than atmospheric may develop, creating an unstable situation, with the culvert alternating between partly full flow and full flow. This is prevented by providing the median inlet shown.

Several factors affect the performance of a culvert under inlet control conditions. These include the inlet area, the inlet shape, the inlet configuration, and the headwater depth. Several methods are available to increase the performance of culverts under inlet control. These include the use of special configurations for inlet edges and beveled edges at the culvert inlet. A detailed description of these methods is given in *Hydraulic Design of Highway Culverts*[9] and are briefly discussed later.

Model tests have been used to determine flow relationships between headwater (depth of water above the culvert inlet invert) and flow for culverts operating under inlet control conditions. The basic condition used to develop these equations is whether or not the inlet is submerged. The inlet performs as an orifice when it is submerged and as a weir when it is not submerged. Two equations were developed for the unsubmerged condition. The first equation (Eq. 15.8) is based on the specific head at critical depth, and the second equation (Eq. 15.9) is exponential and similar to a weir equation. Eq. 15.8 has more theoretical support, but Eq. 15.9 is simpler to use. Both equations will give adequate results.[10] Eq. 15.10 gives the relationship for the submerged condition.

For the *unsubmerged condition*,

$$\frac{HW_i}{D} = \frac{H_c}{D} + K\left[\frac{Q}{(A)(D)^{0.5}}\right]^M - 0.5S \tag{15.8}$$

$$\frac{HW_i}{D} = K\left[\frac{Q}{(A)(D)^{0.5}}\right]^M \tag{15.9}$$

For the *submerged condition*,

$$\frac{HW_i}{D} = c\left[\frac{Q}{(A)(D)^{0.5}}\right]^2 + Y - 0.5S \tag{15.10}$$

where
HW_i = required headwater depth above inlet control section invert (ft)
D = interior height of culvert barrel (ft)
H_c = specific head at critical depth—that is, $d_c + (V_c^2/2g)$ (ft)
d_c = critical depth
V = flow velocity
Q = discharge (ft³/sec)
A = full cross-sectional area of culvert barrel (ft²)
S = culvert barrel slope (ft/ft)
K, M, c, Y = constants from Table 15.10

Note that the last term ($-0.5S$) in Eqs. 15.8 and 15.10 should be replaced by ($+0.7S$) when mitered corners are used. Eqs. 15.8 and 15.9 apply up to about $Q/(A)(D)^{0.5} = 3.5$. Eq. 15.10 applies above about $Q/(A)(D)^{0.5} = 4.0$.

Several charts for different culvert shapes have been developed based on these equations and can be found in *Hydraulic Design for Highway Culverts*.[11] Figure 15.16 is the chart for rectangular box culverts under inlet control, with flared wingwalls and beveled edge at top of inlet, and Figure 15.17 shows the chart for a circular pipe culvert under inlet control.

These charts are used to determine the depth of the headwater required to accommodate the design flow through the selected culvert configuration under

Table 15.10 Constants for Inlet Control Design Equations

Shape and Material	Inlet Edge Design	Unsubmerged			Submerged	
		Equation No.	K	M	c	Y
Circular Concrete	Square edge with headwall	15.8	0.0098	2.0	0.0398	0.67
	Groove end with headwall		.0078	2.0	0.292	.74
	Groove end projecting		.0045	2.0	.0317	.69
Circular CMP	Headwall	15.8	.0078	2.0	.0379	.69
	Mitered to slope		.0210	1.33	.0463	.75
	Projecting		.0340	1.50	.0553	.54
Circular	Beveled ring, 45° bevels	15.8	.0018	2.50	.0300	.74
	Beveled ring, 33.7° bevels		.0018	2.50	.0243	.83
Rectangular Box	30° to 75° wingwall flares	15.8	.026	1.0	.0385	.81
	90° and 15° wingwall flares		.061	0.75	.0400	.80
	0° wingwall flares		.061	0.75	.0423	.82
Rectangular Box	45° wingwall flare, $d = .043D$	15.9	.510	.667	.0309	.80
	18° to 33.7° wingwall flare, $d = .083D$		.486	.667	.0249	.83
Rectangular Box	90° headwall with 3/4 in. chamfers	15.9	.515	.667	.0375	.79
	90° headwall with 45° bevels		.495	.667	.0314	.82
	90° headwall with 33.7° bevels		.486	.667	.0252	.865
Rectangular Box	3/4 in. chamfers; 45° skewed headwall	15.9	.522	.667	.0402	.73
	3/4 in. chamfers; 30° skewed headwall		.533	.667	.0425	.705
	3/4 in. chamfers; 15° skewed headwall		.545	.667	.04505	.68
	45° bevels; 10°–45° skewed headwall		.498	.667	.0327	.75
Rectangular Box 3/4 in. chamfers	45° nonoffset wingwall flares	15.9	.497	.667	.0339	.803
	18.4° nonoffset wingwall flares		.493	.667	.0361	.806

Continued

Table 15.10—*Continued*

Shape and Material	Inlet Edge Design	Unsubmerged			Submerged	
		Equation No.	K	M	c	Y
Rectangular Box Top Bevels	18.4° nonoffset wingwall flares 30° skewed barrel		.495	.667	.0386	.71
	45° offset wingwall flares	15.9	.497	.667	.0302	.835
	33.7° offset wingwall flares		.495	.667	.0252	.881
	18.4° offset wingwall flares		.493	.667	.0227	.887
C M Boxes	90° headwall	15.8	.0083	2.0	.0379	.69
	Thick wall projecting		.0145	1.75	.0419	.64
	Thin wall projecting		.0340	1.5	.0496	.57

Source: Adapted from J. M. Normann, R. J. Houghtalen, and W. J. Johnston, *Hydraulic Design of Highway Culverts*, Report No. FHWA–IP–85–15, U.S. Department of Transportation, Office of Implementation, McLean, Va., September 1985.

inlet control conditions. Alternatively, the iteration required to solve any of the equations may be carried out by using a computer. The use of the chart is demonstrated in Example 15–6.

Example 15–6 Computing Inlet Invert for a Box Culvert

Determine the required inlet invert for a 5 ft × 5 ft box culvert under inlet control with 45° flared wingwalls and beveled edge for the following flow conditions.

Peak flow = 250 ft^3/sec
Design headwater elevation (EL_{hd}) = 230.5 ft (based on adjacent structures)
Stream bed elevation at face of inlet = 224.0 ft

The chart shown in Figure 15.16 is applicable, and the solution is carried out to demonstrate the consecutive steps required.

Step 1. Select the size of the culvert and locate the design flow rate on the appropriate scales (points A and B, respectively). Note that for rectangular box culverts, the flow rate per width of barrel width is used.

$$\frac{Q}{NB} = \frac{250}{5} = 50 \text{ ft}^3/\text{sec/ft for point B}$$

Figure 15.16 Headwater Depth for Inlet Control, Rectangular Box Culverts, Flared Wingwalls 18° to 33.7° and 45° with Beveled Edge at Top of Inlet

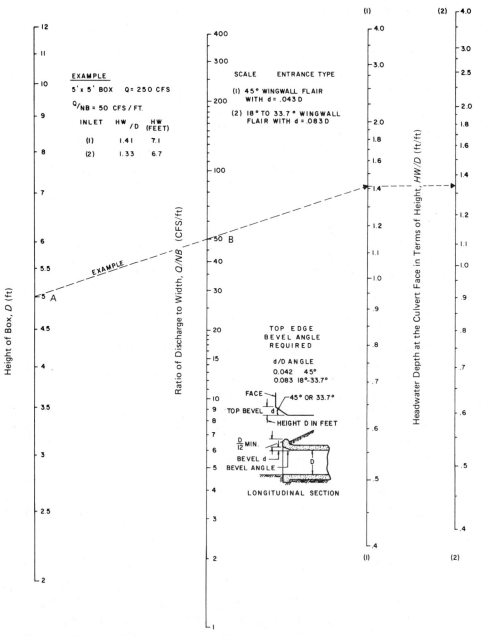

Source: Reproduced from J. M. Normann, R. J. Houghtalen, and W. J. Johnston, *Hydraulic Design of Highway Culverts*, Report No. FHWA–IP–85–15, U.S. Department of Transportation, Office of Implementation, McLean, Va., September 1985.

Figure 15.17 Headwater Depth for Concrete Pipe Culverts with Inlet Control

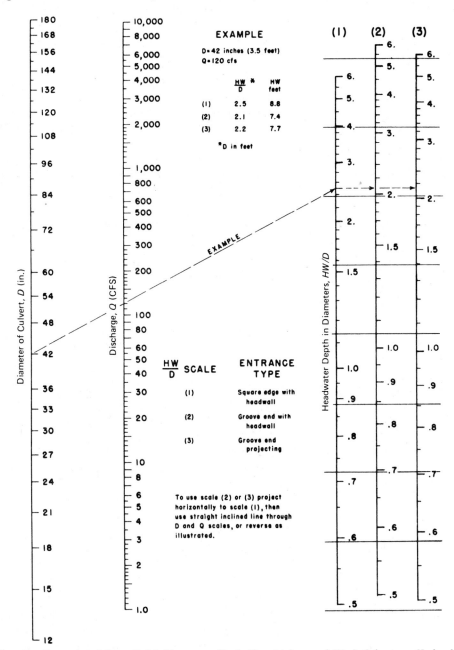

Source: Reproduced from J. M. Normann, R. J. Houghtalen, and W. J. Johnston, *Hydraulic Design of Highway Culverts*, Report No. FHWA–IP–85–15, U.S. Department of Transportation, Office of Implementation, McLean, Va., September 1985.

Step 2. Draw a straight line through points A and B and extend this line to the first headwater/culvert height (HW/D) scale. Read the value on this scale. (Note that the first line is a turning line and alternate values of (HW/D) can be obtained by drawing a horizontal line from this point to the other scales as shown.) Using the first line in this example (HW/D) = 1.41.

Step 3. The required headwater is determined by multiplying the reading obtained in step 2, that is, value for (HW/D) by the culvert depth, $HW = 1.41 \times 5 \simeq 7.1$. This value is used for HW_i (required headwater depth above inlet control invert, ft) if the approach velocity head is neglected. When the approach velocity head is not neglected, then

$$HW_i = HW - \frac{V^2}{2g}$$

Neglecting the approach velocity head in this problem, $HW_i = 7.1$ ft.

Step 4. The required depression (fall)—that is, the depth below the stream bed at which the invert should be located is obtained as follows.

$$HW_d = EL_{hd} - EL_{sf} \qquad \textbf{(15.11)}$$

and

$$\text{fall} = HW_i - HW_d \qquad \textbf{(15.12)}$$

where

HW_d = design headwater depth (ft)
EL_{hd} = design headwater elevation (ft)
EL_{sf} = elevation at the stream bed at the face (ft)

In this case,

$$HW_d = 230.5 - 224.0 = 6.5 \text{ ft}$$

but required depth is 7.1 ft.

$$\text{fall} = 7.1 - 6.5 = 0.6 \text{ ft} \approx 7 \text{ in.}$$

The invert elevation is therefore $224.0 - 0.6 = 223.4$ ft. Note that the value obtained for the fall may be either negative, zero, or positive. When a negative or zero value is obtained, use zero. When a positive value, that is, regarded as being too large, is obtained, another culvert configuration must be selected and the procedure repeated. In this case, a fall of 7 in. is acceptable and the culvert is located with its inlet invert at 223.4 ft.

Outlet Control. A culvert flows under outlet control when the barrel is incapable of transporting as much flow as the inlet opening will receive. Figure 15.18 shows different types of flows under outlet control conditions, where the control section is located at the downstream end of the culvert or beyond. In Figure 15.18(a), both the inlet and outlet of the culvert are submerged and the water flows under pressure along the whole length of the culvert with the culvert completely full. This is a common design assumption although it does not often occur in practice. Figure 15.18(b) shows the inlet unsubmerged and the outlet submerged. This usually occurs when the headwater depth is low, resulting in the top of the culvert being above the surface of the water as the water contracts into the culvert. In Figure 15.18(c), the outlet is unsubmerged and the culvert flows full along its whole length because of the high depth of the headwater. This condition

Figure 15.18 Types of Outlet Control

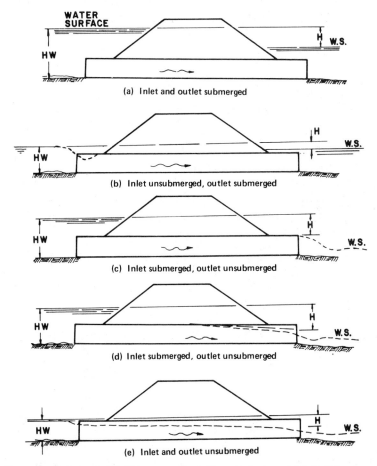

(a) Inlet and outlet submerged

(b) Inlet unsubmerged, outlet submerged

(c) Inlet submerged, outlet unsubmerged

(d) Inlet submerged, outlet unsubmerged

(e) Inlet and outlet unsubmerged

Source: Reproduced from J. M. Normann, R. J. Houghtalen, and W. J. Johnston, *Hydraulic Design of Highway Culverts*, Report No. FHWA–IP–85–15, U.S. Department of Transportation, Office of Implementation, McLean, Va., September 1985.

does not occur often, as it requires very high inlet heads. High outlet velocities are obtained under this condition. In Figure 15.18(d), the culvert inlet is submerged and the outlet unsubmerged, with a low tailwater depth. The flow in the culvert is therefore partly full along part of its length. The flow is also subcritical along part of the culvert's length, but the critical depth is passed just upstream of the outlet. Figure 15.18(e) shows both the culvert's inlet and outlet unsubmerged with the culvert partly full along its entire length, with the flow being subcritical.

In addition to the factors that affect the performance of culverts under inlet control, the performance of culverts under outlet control is also affected by the tailwater depth and certain culvert characteristics, which include the roughness, area, shape, slope, and length.

The hydraulic analysis of culverts flowing under outlet control is based on energy balance. The total energy loss through the culvert is given as

$$H_L = H_e + H_f + H_o + H_b + H_j + H_g \tag{15.13}$$

where

H_L = total energy required
H_e = energy loss at entrance
H_f = friction loss
H_o = energy loss at exit
H_b = bend loss
H_j = energy loss at junction
H_g = energy loss at safety grates

Losses due to bend, junction, and grates occur only when these features are incorporated in the culvert. For culverts without these features, the total head loss is given as[12]

$$H_L = \left(1 + k_e + \frac{29n^2 L}{R^{1.33}}\right) \frac{V^2}{2g} \tag{15.14}$$

where

k_e = factor based on various inlet configurations (see Table 15.11)
n = Manning's coefficients for culverts (see Table 15.12)
R = hydraulic radius of the full culvert barrel $= \dfrac{a}{p}$ (ft)
L = length of culvert barrel (ft)
V = velocity in the barrel

When special features such as grates, bends, and junctions are incorporated in the culvert, the appropriate additional losses may be determined from one or more of the following equations.

Table 15.11 Entrance Loss Coefficients

Type of Structure and Design of Entrance	Coefficient, k_e
Pipe, Concrete	
Projecting from fill, socket end (groove-end)	0.2
Projecting from fill, sq. cut end	0.5
Headwall or headwall and wingwalls	
Socket end of pipe (groove-end)	0.2
Square-edge	0.5
Rounded (radius = $1/12D$)	0.2
Mitered to conform to fill slope	0.7
End-section conforming to fill slope	0.5
Beveled edges, 33.7° or 45° bevels	0.2
Side- or slope-tapered inlet	0.2
Pipe or Pipe-Arch, Corrugated Metal	
Projecting from fill (no headwall)	0.9
Headwall or headwall and wingwalls square-edge	0.5
Mitered to conform to fill slope, paved or unpaved slope	0.7
End-section conforming to fill slope	0.5
Beveled edges, 33.7° or 45° bevels	0.2
Side- or slope-tapered inlet	0.2
Box, Reinforced Concrete	
Headwall parallel to embankment (no wingwalls)	
Square-edged on 3 edges	0.5
Rounded on 3 edges to radius of 1/12 barrel	
dimension, or beveled edges on 3 sides	0.2
Wingwalls at 30° to 75° to barrel	
Square-edged at crown	0.4
Crown edge rounded to radius of 1/12 barrel	
dimension, or beveled top edge	0.2
Wingwall at 10° to 25° to barrel	
Square-edged at crown	0.5
Wingwalls parallel (extension of sides)	
Square-edged at crown	0.7
Side- or slope-tapered inlet	0.2

Source: Adapted from J. M. Normann, R. J. Houghtalen, and W. J. Johnston, *Hydraulic Design of Highway Culverts*, Report No. FHWA–IP–85–15, U.S. Department of Transportation, Office of Implementation, McLean, Va., September 1985.

The bend loss H_b is given as

$$H_b = k_b \frac{v^2}{2g} \qquad (15.15)$$

where

k_b = bend loss coefficient (see Table 15.13)

v = flow velocity in the culvert barrel (ft/sec)

Table 15.12 Manning's Coefficients for Culverts

Type of Conduit	Wall and Joint Description	Manning n
Concrete pipe	Good joints, smooth walls	0.011–0.013
	Good joints, rough walls	0.014–0.016
	Poor joints, rough walls	0.016–0.017
Concrete box	Good joints, smooth finished walls	0.012–0.015
	Poor joints, rough, unfinished walls	0.014–0.018
Corrugated metal pipes and boxes, annular corrugations (Manning n varies with barrel size)	$2\frac{3}{2}$ by $\frac{1}{2}$ in. corrugations	0.027–0.022
	6 by 1 in. corrugations	0.025–0.022
	5 by 1 in. corrugations	0.026–0.025
	3 by 1 in. corrugations	0.028–0.027
	6 by 2 in. structural plate corrugations	0.035–0.033
	9 by $2\frac{1}{2}$ in. structural plate corrugations	0.037–0.033
Corrugated metal pipes, helical corrugations, full circular flow	$2\frac{2}{3}$ by $\frac{1}{2}$ in. corrugations 24 in. plate width	0.012–0.024
Spiral rib metal pipe	$\frac{3}{4}$ by $\frac{3}{4}$ in. recesses at 12 in. spacing, good joints	0.012–0.013

Source: Adapted from J. M. Normann, R. J. Houghtalen, and W. J. Johnston, *Hydraulic Design of Highway Culverts*, Report No. FHWA–IP–85–15, U.S. Department of Transportation, Office of Implementation, McLean, Va., September 1985.

The junction loss H_j is given as

$$H_j = y' + H_{v1} - H_{v2} \qquad (15.16)$$

where

   y' = change in hydraulic grade line through the junction
      $= (Q_2 v_2 - Q_1 v_1 - Q_3 v_3 \cos \theta_j)/[0.5(a_1 + a_2)g]$
   Q_i = flow rate in barrel i (see Figure 15.19)
   v_i = velocity in barrel i (ft/sec)
   a_i = cross-sectional area of barrel i
   θ_j = angle of the lateral with respect to the outlet conduit
   H_{v1} = velocity head in the upstream conduit (ft)
   H_{v2} = velocity head in the downstream conduit (ft)

Table 15.13 Loss Coefficients for Bends

Radius of Bend	Angle of Bend (°)		
Equivalent Diameter	90°	45°	22.5°
1	0.50	0.37	0.25
2	0.30	0.22	0.15
4	0.25	0.19	0.12
6	0.15	0.11	0.08
8	0.15	0.11	0.08

Source: Adapted from Ray F. Linsley and Joseph B. Franzini, *Water Resources Engineering*, copyright © 1979, McGraw-Hill Book Company.

The head loss due to bar grate (H_g) is given as[13]

$$H_g = k_g \frac{W}{x} \frac{v_u^2}{2g} \sin \theta_g \qquad (15.17)$$

where

x = minimum clear spacing between bars (ft)
W = maximum cross-sectional width of the bars facing the flow (ft)
θ_g = angle of grates with respect to the horizontal. (degrees)
k_g = dimensionless bar shape factor
  = 2.42 for sharp-edged rectangular bars
  = 1.83 for rectangular bars with semicircular upstream face
  = 1.79 for circular bars
  = 1.67 for rectangular bars with semicircular upstream and downstream faces

Note that Eqs. 15.16 and 15.17 are both empirical, and caution must be exercised in using them.

Figure 15.20 is a schematic of the energy grade lines for a culvert flowing full. If the total energies at the inlet and outlet are equated, then

$$HW_o + \frac{v_u^2}{2g} = TW + \frac{v_d^2}{2g} + H_L \qquad (15.18)$$

where

HW_o = headwater depth above the outlet invert (ft)
v_u = approach velocity
TW = tailwater depth above the outlet invert (ft)
v_d = downstream velocity (ft/sec)
H_L = sum of all losses

Figure 15.19 Culvert Junction

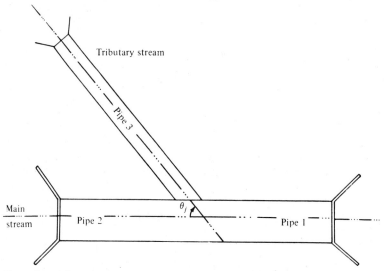

Source: Reproduced from J. M. Normann, R. J. Houghtalen, and W. J. Johnston, *Hydraulic Design of Highway Culverts*, Report No. FHWA–IP–85–15, U.S. Department of Transportation, Office of Implementation, McLean, Va., September 1985.

When the approach and downstream velocity heads are both neglected, we obtain

$$HW_o = TW + H_L \tag{15.19}$$

Note that Eqs. 15.13, 15.14, 15.18, and 15.19 were developed for the culvert flowing full and therefore apply to the conditions shown in Figure 15.18(a), (b), and (c). Additional calculations may be required for the conditions shown in Figure 15.18(d) and (e). These additional calculations are beyond the scope of

Figure 15.20 Full Flow Energy Grade Line (EGL) and Hydraulic Grade Line (HGL)

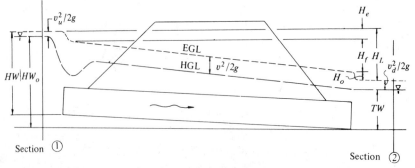

Source: Reproduced from J. M. Normann, R. J. Houghtalen, and W. J. Johnston, *Hydraulic Design of Highway Culverts*, Report No. FHWA–IP–85–15, U.S. Department of Transportation, Office of Implementation, McLean, Va., September 1985.

Figure 15.21 Headwater Depth for Concrete Box Culverts Flowing Full ($n = 0.012$)

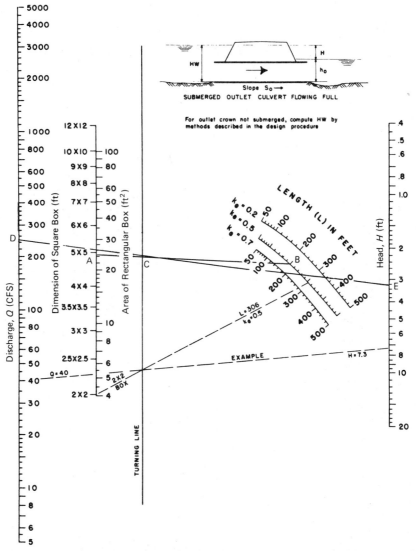

this book but are discussed in detail in *Hydraulic Design of Highway Culverts* for those interested readers.[14]

Nomographs have also been developed for solving Eq. 15.18, for different configurations of culverts flowing full and performing under outlet control. Only entrance, friction, and exit losses are considered in the nomographs. Figures 15.21 and 15.22 show examples of these nomographs for a concrete box culvert and

Figure 15.22   Headwater Depth for Concrete Pipe Flowing Full ($n = 0.012$)

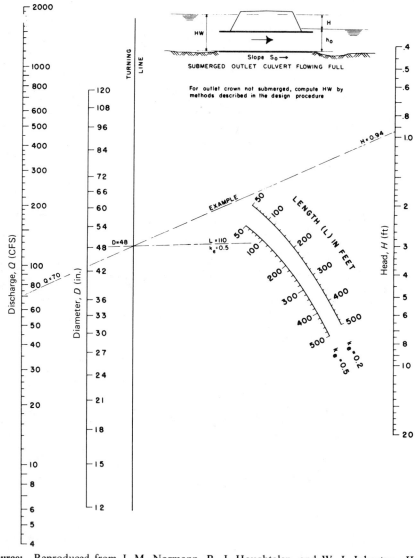

Source: Reproduced from J. M. Normann, R. J. Houghtalen, and W. J. Johnston, *Hydraulic Design of Highway Culverts*, Report No. FHWA–IP–85–15, U.S. Department of Transportation, Office of Implementation, McLean, Va., September 1985.

circular concrete pipe culvert. Figures 15.23 and 15.24 show the critical depth charts for these culverts, which we also use in the design.

The nomographs for outlet control conditions are also used to determine the depth of the headwater required to accommodate the design flow through the selected culvert configuration under outlet control. The procedure is demonstrated in Example 15–7.

Figure 15.23 Critical Depth for Rectangular Sections

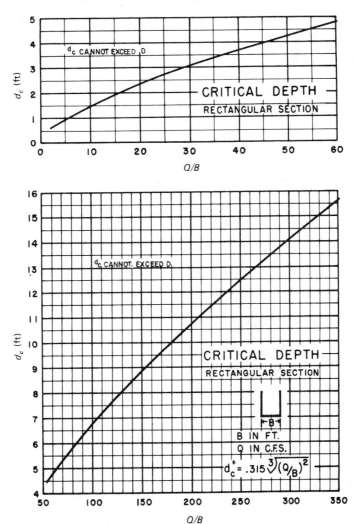

Source: Reproduced from J. M. Normann, R. J. Houghtalen, and W. J. Johnston, *Hydraulic Design of Highway Culverts*, Report No. FHWA–IP–85–15, U.S. Department of Transportation, Office of Implementation, McLean, Va., September 1985.

Example 15–7 Computing Required Headwater Elevation for a Culvert Flowing Full Under Outlet Control

Determine the headwater elevation for Example 15–6 if the culvert is flowing full under outlet control, the tailwater depth above the outlet invert at the design flow rate is 6.5 ft, the length of culvert is 200 ft, and the natural stream slopes at 2 percent. The tailwater depth is determined using normal depth or back water depth calculations or from on-site inspection. Assume $n = 0.012$.

Figure 15.24 Critical Depth for Circular Pipes

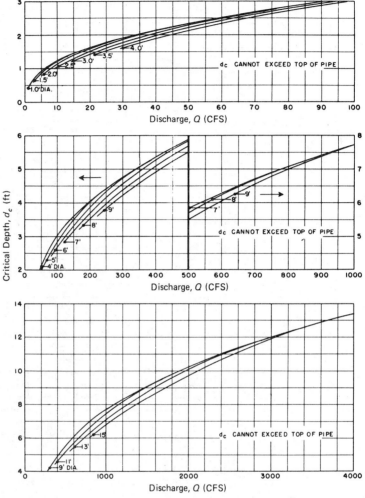

Source: Reproduced from J. M. Normann, R. J. Houghtalen, and W. J. Johnston, *Hydraulic Design of Highway Culverts*, Report No. FHWA–IP–85–15, U.S. Department of Transportation, Office of Implementation, McLean, Va., September 1985.

The charts shown in Figures 15.21 and 15.23 are used in the following steps.

Step 1. From Figure 15.23, determine the critical depth

$$Q/B = 50 \qquad d_c = 4.3 \text{ ft}$$

Alternatively, d_c may be obtained from the equation

$$d_c = 0.315 \sqrt[3]{(Q/B)^2}$$

Step 2. The depth (h_o) from the outlet invert to the hydraulic grade line is then determined. This is taken as ($d_c + D)/2$ or the tailwater depth TW, whichever is greater. In this case,

$$TW = 6.5 \text{ ft}$$

$$(d_c + D)/2 = \frac{4.3 + 5}{2} \simeq 4.7 \text{ ft}$$

$$h_o = 6.5 \text{ ft}$$

Step 3. The inlet coefficient k_e is obtained from Table 15.11 as 0.5.

Step 4. Locate the size, length, and k_e of the culvert as shown at A and B in Figure 15.21. Draw a straight line through A and B and locate the intersection C of this line with the turning line. (Note that when the culvert material has a different n value than that given in the nomograph, an adjusted length L_1 of the culvert is determined as

$$L_1 = L\left(\frac{n_1}{n}\right)^2 \tag{15.20}$$

where

n_1 = desired Manning coefficient

n = Manning n value for outlet control chart

Step 5. Locate the discharge D on the discharge scale and draw a straight line joining C and D. Extend this line to the head loss scale at E and determine the energy loss through the culvert. The total head (H) loss is obtained as 3.3 ft.

Step 6. The required outlet headwater elevation EL_{ho} is computed as

$$EL_{ho} = E_{Lo} + H + h_o \tag{15.21}$$

where E_{Lo} is the invert elevation at outlet. In this case,

$$E_{Lo} = 223.4 \text{ (inlet invert from Example 15–6)} - 0.02 \times 200 = 219.4 \text{ ft}$$

and

$$EL_{ho} = 219.4 + 3.3 + 6.5 = 229.2$$

If this computed outlet control headwater elevation is higher than the design headwater elevation, another culvert should be selected and the procedure repeated.

The design headwater elevation is 230.5 ft (from Example 15–6). The 5 ft × 5 ft culvert is therefore acceptable.

In the design of a culvert, the headwater elevations are computed for inlet and outlet controls and the control with the higher headwater elevation is selected as the controlling condition. In the above case for example, the headwater elevation for inlet control $= 223.4 + 7.1 = 230.5$ ft, and the headwater elevation for outlet control is 229.2 ft, which means that the inlet control governs.

The outlet velocity is then determined for the governing control. When the inlet control governs, the normal depth velocity is taken as the outlet velocity. When the outlet control governs, the outlet velocity is determined by using the area of flow at the outlet, which is based on the culvert's geometry and the following conditions.

1. If the tailwater depth is less than critical depth, use the critical depth.
2. Use the tailwater depth if the level of the tailwater is between the critical depth and the top of the culvert barrel.
3. Use the height of the culvert barrel if the tailwater is above the top of the barrel.

In this example since the inlet control governs, the outlet velocity should be determined, based on the normal depth.

The whole design procedure can be carried out, using a table similar to the one in Figure 15.25, which facilitates the trial of different pipe configurations.

Computer Programs for Culvert Design and Analysis

Examples 15–6 and 15–7 indicate how tedious the analysis or design of a culvert can be, even with the use of available charts. Several computer and hand calculator programs are presently available, however, that can be used to increase the accuracy of the results and to reduce significantly the time it takes for design or analysis.

Programs available for hand calculators include the Calculator Design Series (CDS) 1,2,3 and the Calculator Design Series (CDS) 4, developed by FHWA.[15,16] The programs in Calculator Design Series 1,2,3 are suitable for the Compucorp 325 Scientist, the HP-65, and the TI-59 programmable calculators, whereas the CDS 4 is suitable only for the TX-54. The CDS 1,2,3 consists of a set of subroutines that are run sequentially, some of which provide inputs for subsequent ones. The program can be used to analyze culvert sizes with different inlet configurations. The outputs include the dimensions of the barrel, the performance data, and the outlet velocities.

The CDS 4 program can be used to analyze corrugated metal and concrete culverts. This program also consists of a series of subroutines.

FHWA has also produced various software packages, some of which are suitable for use on mainframe computers and others on personal computers. The program H4-2 can be used to design pipe-arch culverts on a mainframe computer, whereas H4-6, which is also run on a mainframe, can be used to obtain a list of optional circular and box culverts for the specified site and hydrologic conditions.

A program written in basic for use on an IBM/PC has also been developed by FHWA. The program analyzes a culvert specified by the user for a given set of hydrologic data and site conditions. It is suitable for circular, rectangular, elliptical, arch, and other geometrics defined by the user and will compute the hydraulic characteristics of the culvert.

Note, however, that both computer hardware and software in all fields of engineering are rapidly being improved, which makes it imperative that designers

Figure 15.25 Work Sheet for Culvert Design

PROJECT : ______________________

STATION : ______________________

SHEET ______ OF ______

CULVERT DESIGN FORM

DESIGNER / DATE : ________ / ________

REVIEWER / DATE : ________ / ________

HYDROLOGICAL DATA

SEE ADD'L. SHTS.

☐ METHOD : ______________________

☐ DRAINAGE AREA : ________ ☐ STREAM SLOPE : ________

☐ CHANNEL SHAPE : ______________________

☐ ROUTING : ________ ☐ OTHER : ________

DESIGN FLOWS/TAILWATER

R. I. (YEARS)	FLOW (cfs)	TW (ft)

EL_{hd} : ________ (ft)

ROADWAY ELEVATION : ________ (ft)

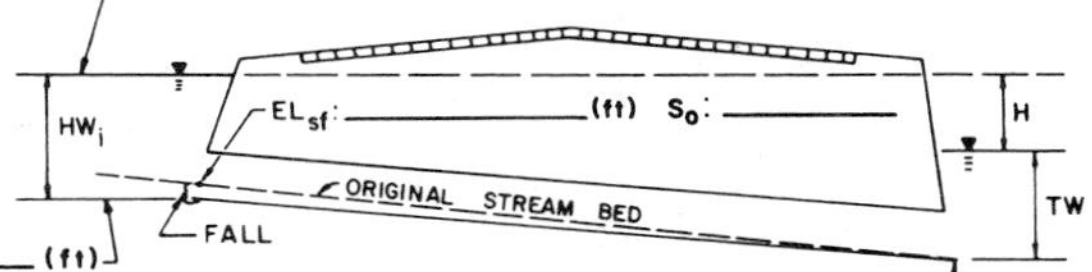

EL_i ________ (ft)

EL_o : ________ (ft)

$S \approx S_o - FALL / L_a$

S = ________

L_a = ________

CULVERT DESCRIPTION: MATERIAL - SHAPE - SIZE - ENTRANCE	TOTAL FLOW Q (cfs)	FLOW PER BARREL Q/N (1)	HEADWATER CALCULATIONS: INLET CONTROL HW_i/D (2)	HW_i	FALL (3)	EL_{hi} (4)	OUTLET CONTROL TW (5)	d_c	$\frac{d_c+D}{2}$	h_o (6)	k_e	H (7)	EL_{ho} (8)	CONTROL HEADWATER ELEVATION	OUTLET VELOCITY	COMMENTS

TECHNICAL FOOTNOTES:

(1) USE Q/NB FOR BOX CULVERTS

(2) $HW_i/D = HW/D$ OR HW_f/D FROM DESIGN CHARTS

(3) $FALL = HW_i - (EL_{hd} - EL_{sf})$; FALL IS ZERO FOR CULVERTS ON GRADE

(4) $EL_{hi} = HW_i + EL_i$ (INVERT OF INLET CONTROL SECTION)

(5) TW BASED ON DOWN STREAM CONTROL OR FLOW DEPTH IN CHANNEL.

(6) $h_o = TW$ or $(d_c + D/2)$ (WHICHEVER IS GREATER)

(7) $H = \left[1 + k_e + (29n^2 L)/R^{1.33}\right] V^2/2g$

(8) $EL_{ho} = EL_o + H + h_o$

SUBSCRIPT DEFINITIONS :

a. APPROXIMATE
f. CULVERT FACE
hd. DESIGN HEADWATER
hi. HEADWATER IN INLET CONTROL
ho. HEADWATER IN OUTLET CONTROL
i. INLET CONTROL SECTION
o. OUTLET
sf. STREAMBED AT CULVERT FACE
tw. TAILWATER

COMMENTS / DISCUSSION :

CULVERT BARREL SELECTED :

SIZE : ______________________

SHAPE : ______________________

MATERIAL : ____________ n: ________

ENTRANCE : ______________________

Source: Reproduced from J. M. Normann, R. J. Houghtalen, and W. J. Johnston, *Hydraulic Design of Highway Culverts*, Report No. FHWA–IP–85–15, U.S. Department of Transportation, Office of Implementation, McLean, Va., September 1985.

keep abreast of the development of new software packages and hardware equipment.

Inlet Configuration

It was stated earlier that a culvert's performance is affected by the inlet configurations. However, since the design discharge for outlet control results in full flow, the entrance loss for a culvert flowing under outlet control is usually a small fraction of the headwater requirements. Extensive inlet configurations, which considerably increase the cost of culverts, are therefore unnecessary for culvert's under outlet control conditions. With culverts flowing under inlet control, however, the culverts' hydraulic capacity depends only on the inlet configuration and headwater depth. A suitable inlet configuration can therefore result in full or nearly full flow of a culvert under inlet control. This significantly increases the capacity of the culvert.

The design charts provided in *Hydraulic Design of Highway Culverts* include charts for improved inlet configurations, such as bevel-edged, side-tapered, and slope-tapered, that help to increase the culvert capacity.[17]

Figure 15.26   Beveled Edges

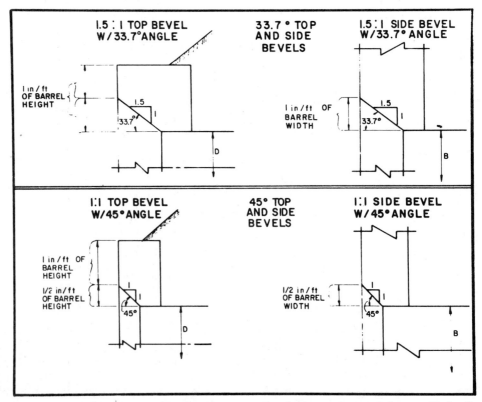

Source: Reproduced from J. M. Normann, R. J. Houghtalen, and W. J. Johnston, *Hydraulic Design of Highway Culverts*, Report No. FHWA–IP–85–15, U.S. Department of Transportation, Office of Implementation, McLean, Va., September 1985.

Bevel-Edged Inlets. Figure 15.26 shows different beveled-edged configurations. The bevel edge is similar to a chamfer, except that a chamfer is usually much smaller. It has been estimated that the addition of bevels to a culvert having a square edge inlet will increase the culvert capacity by 15 percent to 20 percent. As a minimum, therefore, all culverts operating under inlet control should be fitted with bevels. Use of bevels on culverts under outlet control is also recommended, as the entrance loss coefficient may be reduced by up to 40 percent.

Tapered Inlets. Tapered inlets increase the capacity of culverts under inlet control, mainly by reducing the contraction at the inlet control section. Tapered inlets are more effective than bevel-edged inlets on culverts flowing under inlet control, but have similar results as bevels when used on culverts under outlet control. Since tapered inlets are more expensive, they are not recommended for culverts under outlet control. Design charts are available for rectangular box and circular pipes for two types of tapered inlets: side-tapered and slope-tapered.

Figure 15.27 shows different designs of a *side-tapered inlet*. It consists of an enlarged face section that uniformly reduces to the culvert barrel size by tapering the side walls. The inlet flow of the side taper is formed by extending the flow of culvert barrel outward, with the face section being approximately the same height as the barrel height. The throat section is the intersection of the culvert barrel and the tapered side walls. Either the throat or face section may act as the inlet control section depending on the inlet design. When the throat acts on the inlet control section, the headwater depth is measured from the throat section invert HW_t, and when the face acts as the inlet control section, the headwater depth is measured from the face section invert HW_f. See Figure 15.27(a). It is advantageous for the throat section to be the primary control section since the throat is usually lower than the face, resulting in a higher head on the throat for a specified headwater elevation. Figure 15.27(b) and (c) show two ways of increasing the effectiveness of the side-tapered inlet. In Figure 15.27(b), the throat section is depressed below the stream bed and a depression constructed between the two wing walls. When this type of construction is used, it is recommended that the culvert barrel floor be extended upstream a minimum distance of $D/2$, before the steep upward slope begins. In Figure 15.27(c) a sump is constructed upstream from the face section, with the dimensional requirements given. When the side-tapered inlet is constructed as in Figure 15.27(b) or (c), a crest is formed upstream at the intersection of the stream bed and the depression slope. This crest may act as a weir if its length is too short. It should therefore be ascertained that the crest does not control the flow at the design flow and headwater.

The *slope-tapered inlet* is similar to the side-tapered inlet, in that it also has an enlarged face section, which is gradually reduced to the barrel size at the throat section by sloping the sidewalls. The slope-tapered inlet, however, also has a uniform vertical drop (fall) between the face and throat sections. (See Figure 15.28.) A third section known as the bend is also placed at the intersection of the inlet slope and barrel slope as shown in Figure 15.28.

Any one of three sections may act as the primary control section of the slope-tapered section. These are the face, the bend, and the throat. Design procedures, which are beyond the scope of this book, for the dimensions of the throat

Figure 15.27   Side-Tapered Inlets

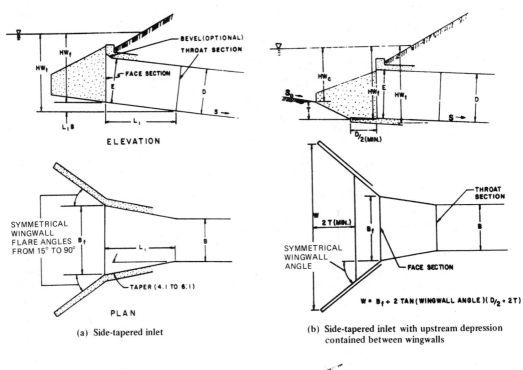

(a) Side-tapered inlet

(b) Side-tapered inlet with upstream depression
contained between wingwalls

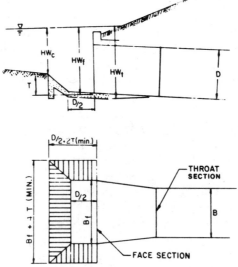

(c) Side-tapered inlet with upstream sump

Source: Reproduced from J. M. Normann, R. J. Houghtalen, and W. J. Johnston, *Hydraulic Design of Highway Culverts*, Report No. FHWA–IP–85–15, U.S. Department of Transportation, Office of Implementation, McLean, Va., September 1985.

Figure 15.28 Slope-Tapered Inlet with Vertical Face

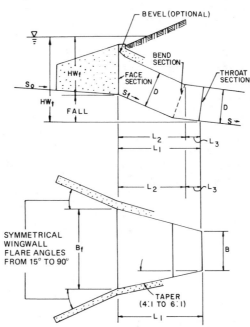

Source: Reproduced from J. M. Normann, R. J. Houghtalen, and W. J. Johnston, *Hydraulic Design of Highway Culverts*, Report No. FHWA–IP–85–15, U.S. Department of Transportation, Office of Implementation, McLean, Va., September 1985.

and face sections are given in *Hydraulic Design of Highway Culverts*.[18] The only criterion given for the size of the bend section is that it should be located a minimum distance from the throat. Again, the slope-tapered inlet is most efficient when the throat acts as the primary control section.

SUBSURFACE DRAINAGE

Subsurface drainage systems are provided within the pavement structure to drain water in one or more of the following forms:

- Water that has permeated through cracks and joints in the pavement to the underlying strata
- Water that has moved upward through the underlying soil strata as a result of capillary action
- Water that exists in the natural ground below the water table, usually referred to as ground water

The subsurface drainage system must be an intergral part of the total drainage system, since the subsurface drains must operate in consonant with the surface drainage system to obtain an efficient overall drainage system.

The design of subsurface drainage should be carried out as an integral part of the complete design of the highway, since inadequate subsurface drainage may also have detrimental effects on the stability of slopes and pavement performance. However, certain design elements of the highway such as geometry and material properties are required for the design of the subdrainage system. Thus, the procedure usually adopted for subdrainage design is first to determine the geometric and structural requirements of the highway based on standard design practice and then to subject these to a subsurface drainage analysis to determine the subdrainage requirements. In some cases, the subdrainage requirements determined from this analysis will require some changes in the original design.

It is extremely difficult, if not impossible, to develop standard solutions for solving subdrainage problems because of the many different situations that engineers come across in practice. Therefore, basic methods of analysis are given that can be used as tools to identify solutions for subdrainage problems. The experience gained from field and laboratory observations for a particular location, coupled with good engineering judgment, should always be used in conjunction with the design tools provided. Before presenting the design tools, discussions of the effects on the highway of an inadequate subdrainage system and the different subdrainage systems are first presented.

Effect of Inadequate Subdrainage

Inadequate subdrainage on a highway will result in the accumulation of uncontrolled subsurface water within the pavement structure and/or right of way, which can result in poor performance of the highway or outright failure of sections of the highway. The effects of inadequate subdrainage fall into two classes: poor pavement performance and instability of slopes.

Pavement Performance

If the pavement structure and subgrade are saturated with underground water, the pavement's ability to resist traffic load is considerably reduced, resulting in one or more of several problems, which can lead to a premature destruction of the pavement if remedial actions are not taken in time. In Portland cement concrete pavement, for example, inadequate subdrainage can result in excessive repeated deflections of the pavement (see Pumping of Rigid Pavements in Chapter 19), which will eventually lead to cracking.

When asphaltic concrete pavements are subjected to excessive uncontrolled subsurface water due to inadequate subdrainage, very high pore pressures are developed within the untreated base and subbase layers (see Chapter 18 for base and subbase definitions), resulting in a reduction of the pavement strength and thereby its ability to resist traffic load.

Another common effect of poor pavement performance due to inadequate subdrainage is frost action. As described later, this phenomenon requires that the base and/or subbase material be a frost-susceptible soil and that an adequate amount of subsurface water is present in the pavement structure. Under these conditions, during the active freezing period, subsurface water will move upward by capillary action toward the freezing zone and subsequently freeze to form lenses of ice. Continuous growth of the ice lenses due to the capillary action of the subsurface water can result in considerable heaving of the overlying pavement. This eventually leads to serious pavement damage, particularly if differential frost heaving occurs. Frost action also has a detrimental effect on pavement performance during the spring thaw period. During this period, the ice lenses formed during the active freeze period gradually thaw from the top down, resulting in the saturation of the subgrade soil, which results in a substantial reduction of pavement strength.

Slope Stability

The presence of subsurface water in an embankment or cut can cause an increase of the stress to be resisted and a reduction of the shear strength of the soil forming the embankment or cut. This can lead to the condition where the stress to be resisted is greater than the strength of the soil, resulting in sections of the slope crumbling down or a complete failure of the slope.

Highway Subdrainage Systems

Subsurface drainage systems are usually classified into five general categories:

- Longitudinal drains
- Transverse drains
- Horizontal drains
- Drainage blankets
- Well systems

Longitudinal Drains

Subsurface longitudinal drains usually consist of pipes laid in trenches, within the pavement structure and parallel to the center line of the highway. These drains can be used to lower the water table below the pavement structure, as shown in Figure 15.29, or to remove any water that is seeping into the pavement structure, as shown in Figure 15.30. In some cases, when the water table is very high and the highway is very wide, it may be necessary to use more than two rows of longitudinal drains to achieve the required reduction of the water table below the pavement structure (see Figure 15.31).

Transverse Drains

Transverse drains are placed transversely below the pavement, usually in a direction perpendicular to the center line, although they may be skewed to form the herringbone configuration. An example of the use of transverse drains is shown in Figure 15.32, where they are used to drain ground water that has infiltrated through the joints of the pavement. One disadvantage of transverse drains is that

Figure 15.29 Symmetrical Longitudinal Drains Used to Lower Water Table

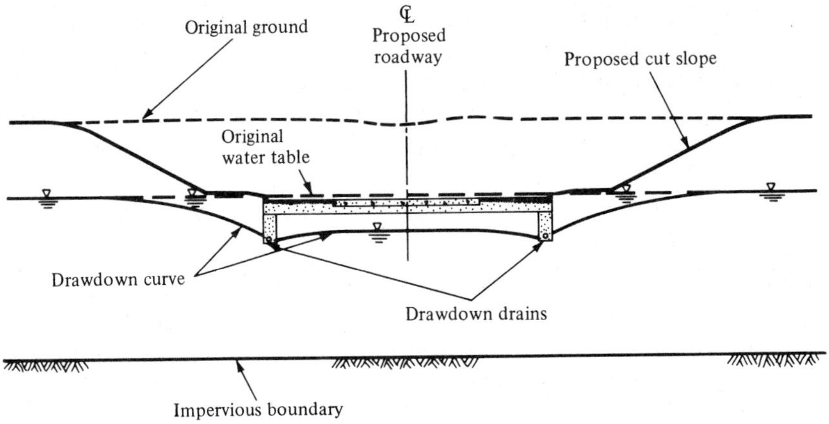

Original ground

℄
Proposed
roadway

Proposed cut slope

Original
water table

Drawdown curve

Drawdown drains

Impervious boundary

Source: Redrawn from *Highway Subdrainage Design*, Report No. FHWA–TS–80–224, U.S. Department of Transportation, Washington, D.C., August 1980.

they can cause unevenness of the pavement when used in areas susceptible to frost action, where general frost heaving occurs. The unevenness is due to the general heaving of the whole pavement, except at the transverse drains.

Horizontal Drains

Horizontal drains are used to relieve pore pressures at slopes of cuts and embankments on the highway. They usually consist of small diameter, perforated pipes inserted into the slopes of the cut or fill. The subsurface water is collected by the pipes and is then discharged at the face of the slope through paved spillways to longitudinal ditches.

Figure 15.30 Longitudinal Collector Drain Used to Remove Water Seeping into Pavement Structural Section

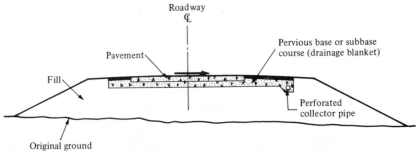

Roadway
℄

Pervious base or subbase
course (drainage blanket)

Pavement

Fill

Perforated
collector pipe

Original ground

Source: Redrawn from *Highway Subdrainage Design*, Report No. FHWA–TS–80–224, U.S. Department of Transportation, Washington, D.C., August 1980.

Figure 15.31   Multiple Longitudinal Drawdown Drain Installation

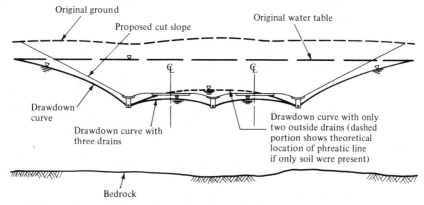

Source: Redrawn from *Highway Subdrainage Design*, Report No. FHWA–TS–80–224, U.S. Department of Transportation, Washington, D.C., August 1980.

Figure 15.32   Transverse Drains on Superelevated Curves

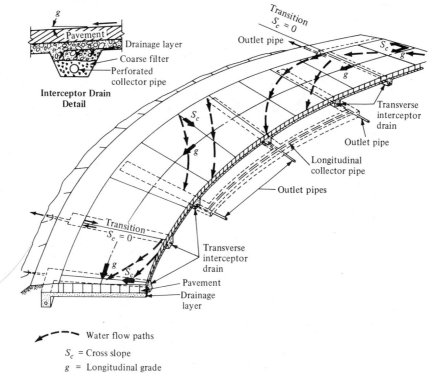

S_c = Cross slope
g = Longitudinal grade

Source: Redrawn from *Highway Subdrainage Design*, Report No. FHWA–TS–80–224, U.S. Department of Transportation, Washington, D.C., August 1980.

Drainage Blankets

A drainage blanket is a layer of material that has a very high coefficient of permeability, usually greater than 30 ft/day, and is laid beneath or within the pavement structure, such that its width and length in the flow direction are much greater than its thickness. The coefficient of permeability is the constant of proportionality of the relationship between the flow velocity and the hydraulic gradient between two points in the material (see Chapter 16). Drainage blankets can be used to facilitate the flow of subsurface water away from the pavement, as well as to facilitate the flow of ground water that has seeped through cracks into the pavement structure or subsurface water from artesian sources. A drainage blanket can also be used in conjunction with longitudinal drains to improve the stability of cut slopes by controlling the flow of water on the slopes, thereby

Figure 15.33   Applications of Horizontal Drainage Blankets

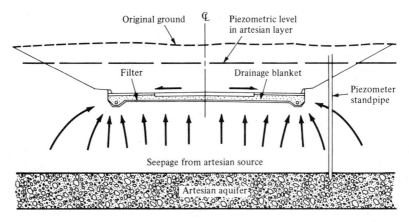

(a)

(b)

Source: Redrawn from *Highway Subdrainage Design*, Report No. FHWA–TS–80–224, U.S. Department of Transportation, Washington, D.C., August 1980.

preventing the formation of a slip surface. However, drainage blankets must be properly designed to be effective. Figure 15.33 shows two drainage blanket systems.

Well Systems

A well system consists of a series of vertical wells, drilled into the ground, into which ground water flows, thereby reducing the water table and releasing the pore pressure. When used as a temporary measure for construction, the water collected in the wells is continuously pumped out or it may be left to overflow. A more common construction, however, usually includes a drainage layer either at the top or bottom of the wells to facilitate the flow of the water collected.

Design of Subsurface Drainage

The design procedure for subsurface drainage involves the following.

1. Summarize the available data.
2. Determine the quantity of water for which the subdrainage system is being designed.
3. Determine the drainage system required.
4. Determine the capacity and spacing of longitudinal and transverse drains and select filter material, if necessary.
5. Evaluate the design with respect to economic feasibility and long-term performance.

Summarize Available Data

The data that should be identified and summarized can be divided into the four following classes:

- The flow geometry
- The materials' properties
- The hydrologic and climatic characteristics
- Miscellaneous information

The flow geometry is given by the existing subsurface characteristics of the area in which the highway is located and the geometric characteristics of the highway. These are used to determine whether any special subdrainage problems exist and what conditions must be considered in developing solutions for these problems.

The main material property required is the permeability, since this is the property that indicates the extent to which water will flow through the material.

Hydrologic and climatic characteristics will indicate precipitation rates, the sources of subsurface water, and the possibility of frost action occurring.

Miscellaneous information includes all other information that will aid in the design of an effective and economic subdrainage system, including information of any impact the subdrainage system may have on future construction and any difficulties identified that may preclude the construction of a subdrainage system.

Determination of Discharge Quantity

The net amount of water to be discharged consists of the following components:

- Water due to infiltration
- Ground water
- Water due to thawing of ice lenses
- Water flowing vertically from the pavement structure

Water Due to Infiltration, q_i. This is the amount of surface water that infiltrates into the pavement structure through cracks in the pavement surface. It is extremely difficult to calculate this amount of water exactly, since the rate of infiltration depends on the intensity of the design storm, the frequency and size of the cracks and/or joints in the pavement, the moisture conditions of the atmosphere, and the permeability characteristics of the materials below the pavement surface.

FHWA recommends the use of the following empirical relationship to estimate the infiltration rate:[19]

$$q_i = I_c\left(\frac{N_c}{W} + \frac{W_c}{WC_s}\right) + K_p \tag{15.22}$$

where

q_i = design infiltration rate, ft³/day/ft² of drainage layer
I_c = crack infiltration rate, ft³/day/ft of crack
   = 2.4 ft³/day/ft (recommended for most design)
N_c = number of contributing longitudinal cracks or joints
W_c = length of contributing transverse cracks (ft)
W = width of granular base or subbase subjected to infiltration (ft)
C_s = spacing of the transverse cracks or joints (ft)
K_p = rate of infiltration
   = coefficient of permeability through the uncracked pavement surface, ft³/day/ft²

The suggested value of 2.4 for I_c should normally be used, but local experience should also be relied on to increase or decrease this value, if deemed necessary.

The value of N_c is usually taken as $N + 1$ for new pavements, where N is the number of traffic lanes. Local experience should be used to determine a value for C_s, although a value of 40 has been suggested for new bituminous concrete pavements. The rate of infiltration for Portland cement concrete and well-compacted, dense, graded asphaltic concrete pavements is usually very low and can therefore be taken as zero. However, when there is evidence of high infiltration rates, these should be determined from laboratory tests.

Example 15–8 Computing Infiltration Rate of a Flexible Pavement

Determine the infiltration rate for a new two-lane flexible pavement with the following characteristics.

> Lane width $= 11$ ft
> Shoulder width $= 8$ ft
> Number of contributing longitudinal cracks $(N_c) = (N + 1) = 3$
> Length of contributing transverse cracks $(W_c) = 20$ ft
> K_p (from laboratory tests) $= 0.03$
> $W = 38$ ft
> Spacing of transverse cracks $(C_s) = 35$ ft

Assuming $I_c = 2.4$ ft^3/day/ft^2, then from Eq. 15.22

$$q_i = 2.4 \left[\frac{3}{38} + \frac{20}{38(35)} \right] + 0.03$$
$$= 0.23 + 0.03$$
$$= 0.26 \text{ ft}^3/\text{day}/\text{ft}^2$$

Ground water. When it is not possible to intercept the flow of ground water or lower the water table sufficiently before the water reaches the pavement, it is necessary to determine the amount of ground water seepage that will occur. Figures 15.29 and 15.33(a) demonstrate the two possible sources of ground water of interest in this case. Figure 15.29 shows a case of gravity drainage, whereas Figure 15.33(a) shows a case of artesian flow. A simple procedure to estimate the ground water flow rate due to gravity drainage is to use the chart shown in Figure 14.34. In this case the *radius of influence* L_i is first determined as

$$L_i = 3.8(H - H_o) \tag{15.23}$$

where

> $H_o =$ thickness of subgrade below the drainage pipe (ft)
> $H =$ thickness of subgrade below the natural water table (ft)
> $H - H_o =$ amount of draw down (ft)

The chart shown in Figure 15.34 is then used to determine the total quantity of upward flow (q_2) from which the average inflow rate is determined as

$$q_g = \frac{q_2}{0.5W} \tag{15.24}$$

Figure 15.34   Chart for Determining Flow Rate in Horizontal Drainage Blanket

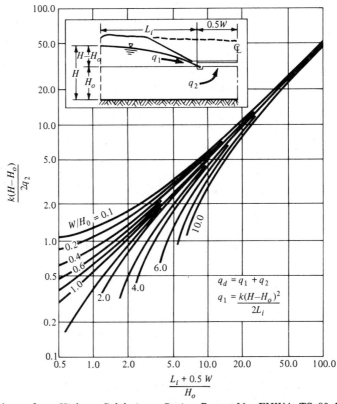

Source: Redrawn from *Highway Subdrainage Design*, Report No. FHWA–TS–80–224, U.S. Department of Transportation, Washington, D.C., August 1980.

where

   q_g = design inflow rate for gravity drainage ft³/day/ft² of drainage layer
   q_2 = total upward flow into one-half of the drainage blanket, ft³/day/linear ft
       of roadway
   W = width of drainage layer

For the case of the artesian flow, the average inflow rate is simply estimated using Darcy's law as

$$q_a = K \frac{\Delta H}{H_o}$$ **(15.25)**

where

   q_a = design inflow rate from artesian flow (ft³/day/ft² of drainage area)
   ΔH = excess hydraulic head (ft)
   H_o = thickness of the subgrade soil between the drainage layer and the
       artesian aquifer
   K = coefficient of permeability

Example 15–9 Computing Average Inflow Rate Due to Gravity Drainage

Using the chart shown in Figure 15.34, determine the average inflow rate (q_g) due to gravity drainage, as shown in Figure 15.35, for the following data.

Thickness of subgrade below drainage pipe $(H_o) = 15$ ft
Coefficient of permeability of native soil $(K) = 0.4$ ft/day
Width of drainage area $= 40$ ft
Drawdown $(H - H_o) = 8$ ft
Radius of influence $(L_i) = 3.8 \times 8$ (from Eq. 15.23)
$$= 30.4 \text{ ft}$$

$$\frac{L_i + 0.5W}{H_o} = \frac{30.4 + 0.5 \times 40}{15} = 3.36$$

$$\frac{W}{H_o} = \frac{40}{15} = 2.67$$

Entering the chart at

$$\frac{L_i + 0.5W}{H_o} = 3.36$$

and

$$W/H_o = 2.67$$

Figure 15.35 Rigid Pavement Section in Cut Dimensions and Details for Example 15–9

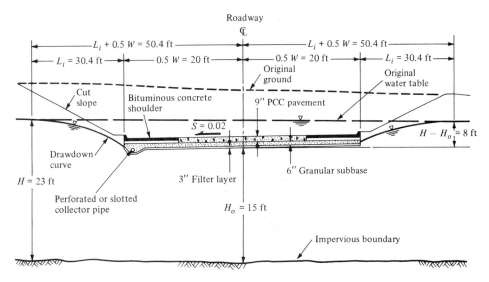

we obtain

$$\frac{K(H - H_o)}{2q_2} \approx 1.3$$

$$q_2 = \frac{0.4 \times 8}{2 \times 1.3} = 1.23 \text{ ft}^3/\text{day}/\text{ft}^2$$

$$q_g = \frac{1.23}{0.5 \times 40} = 0.062 \text{ ft}^3/\text{day}/\text{ft}^2$$

Example 15–10   Computing Average Inflow Rate Due to Artesian Flow

Determine the average flow rate of ground water into a pavement drainage layer due to artesian flow constructed on a subgrade soil having a coefficient of permeability of 0.05 ft/day if a piezometer installed at the site indicates an excess hydraulic head of 10 ft. The thickness of the subgrade soil between the drainage layer and the artesian aquifer is 20 ft (see Figure 15.36).

In this case,

$$q_a = \frac{0.05(10)}{20} = 0.025 \text{ ft}^3/\text{day}/\text{ft}^2$$

Note that flow net analysis may be used to estimate the flow rates due to both gravity flow and artesian flow. Flow net analysis is beyond the scope of this book,

Figure 15.36   Artesian Flow of Groundwater into a Pavement Drainage Layer—Dimensions and Details for Example 15–10

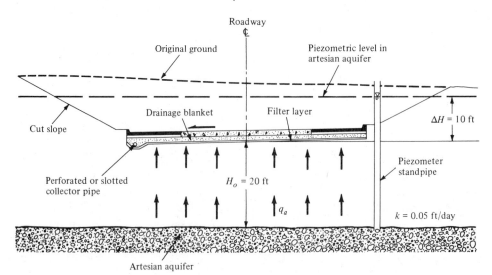

but good estimates of these flows can be obtained by the methods presented. The amount of ground water flow due to any one source is usually small, but ground water flow should not be automatically neglected, as the sum of ground water flow from all sources may be significant. It is therefore essential that estimates be made of the inflow from all sources of ground water that influence the pavement structure.

Water from Ice Lenses.   As explained earlier, frost action results in ice lenses forming within the pavement subgrade. The extent to which this occurs during the active freeze period is highly dependent on the frost susceptibility of the subgrade material. These ice lenses thaw during the spring thaw period, and it is necessary to properly drain this water from the pavement environment. However, the rate of seepage of this water through the soil depends on several factors, which include the permeability of the soil, the thawing rate, and the stresses imposed on the soil. It is very difficult to determine the extent to which each of these factors affects the flow rate, which makes it extremely difficult to develop an exact method for determining the flow rate. However, an empirical method for determining the

Figure 15.37   Chart for Estimating Design Inflow Rate of Melt Water from Ice Lenses

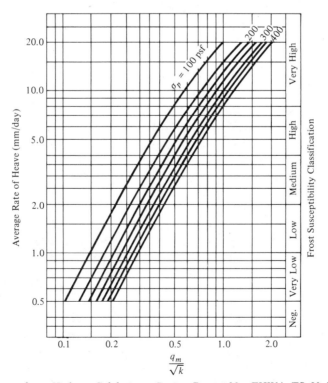

Source: Redrawn from *Highway Subdrainage Design*, Report No. FHWA–TS–80–224, U.S. Department of Transportation, Washington, D.C., August 1980.

design flow rate q_m has been developed, using the chart shown in Figure 15.37. This requires the determination from laboratory tests of the average rate of heave of the soil due to frost action, or the classification of the soil with respect to its susceptibility to frost action and the stress σ_p imposed on the subgrade soil. The stress imposed on the subgrade in pounds per square foot due to the weight of a 1 ft square column of the pavement structure is usually taken as the value of σ_p. The design flow rate q_m obtained from the chart is the average flow during the first full day of thawing. This rate is higher than those for subsequent days, as the rate of seepage decreases with time. Although the use of such a high value of flow is rather conservative, such a value may also cause the soil to be saturated for a period of up to 6 hr after thawing. In cases where saturation for this period of time is unacceptable, measures to increase the rate of drainage of the thawed water from the soil must be used. For example, a thick enough drainage layer that consists of a material with a suitable permeability that will never allow saturation to occur can be provided.

Example 15–11 Computing Flow Rate Due to Thawing of Ice Lenses

Determine the flow rate due to water thawing from ice lenses below a pavement structure that consists of a 6 in. concrete surface and a 9 in. granular base. The subgrade material is silty sand with a high frost susceptibility classification and a coefficient of permeability of 0.075 ft/day.

Assuming that concrete weighs 150 lb/ft³ and the granular material weighs 130 lb/ft³, then

$$\sigma_p = 0.5 \times 150 + 0.75 \times 130 = 172.5 \text{ lb/ft}^3$$

Entering the chart in Figure 15.37 at the midlevel of the high range of frost susceptibility and projecting to a stress of 172.5 on the σ_p charts and reading off the value of $q_m/\sqrt{k}$, we obtain

$$q_m/\sqrt{k} \approx 0.55$$
$$q_m = (\sqrt{0.075})0.55 = 0.15 \text{ ft}^3/\text{day}$$

Note that if the average rate of heave of the subgrade soil is determined from either laboratory tests or local experience based on observations of frost action, then this value may be used instead of the frost susceptibility classification.

Vertical Outflow, q_v. In some cases, the total amount of water accumulated within the pavement structure can be reduced because of the vertical seepage of some of the accumulated water through the subgrade. When this occurs, it is necessary to estimate the amount of this outflow in order to determine the net inflow for which the subdrainage system is to be provided. The procedure for estimating this flow involves the use of flow net diagrams, which is beyond the scope of this book. This procedure is discussed in detail in *Highway Subdrainage Design*.[20] As will be seen later, however, the use of the vertical outflow to reduce the inflow is applicable only when there is neither ground water inflow nor frost action.

Net Inflow. The net inflow is the sum of inflow rates from all sources less any amount attributed to vertical outflow through the underlying soil. Note, however, that all of the different flows discussed earlier do not necessarily occur at the same time. For example, it is unlikely that flows from thawed water and ground water will occur at the same time, since soils susceptible to frost action will have very low permeability when frozen. Similarly, downward vertical outflows will never occur at the same time as upward inflow from any other source. Downward vertical outflow will therefore only occur when there is no inflow due to ground water. A set of relationships for estimating the net inflow rate (q_n) have been developed, taking into consideration the different flows that occur concurrently and are given in Eqs. 15.26 through 15.30.[21]

$$q_n = q_i \tag{15.26}$$

$$q_n = q_i + q_g \tag{15.27}$$

$$q_n = q_i + q_a \tag{15.28}$$

$$q_n = q_i + q_m \tag{15.29}$$

$$q_n = q_i - q_v \tag{15.30}$$

Guidelines for using these equations are given in Table 15.14.

Table 15.14 Guidelines for Using Eqs. 15.26 Through 15.30 to Compute Net Inflow, q_n, for Design of Pavement Drainage

Highway Cross Section	Ground Water Inflow	Frost Action	Net Inflow Rate, q_n, Recommended for Design
Cut	Gravity	Yes	Max. of Eqs. 15.27 and 15.29
		No	Eq. 15.27
Cut	Artesian	Yes	Max. of Eqs. 15.28 and 15.29
		No	Eq. 15.28
Cut	None	Yes	Eq. 15.29
		No	Eq. 15.26
Cut	None	Yes	Eq. 15.29
		No	Eq. 15.30
Fill	None	Yes	Eq. 15.29
		No	Eq. 15.30

Source: Adapted from *Highway Subdrainage Design*, Report No. FHWA–TS–80–224, U.S. Department of Transportation, Washington, D.C., August 1980.

Design of Drainage Layer

The design of the drainage layer involves either the determination of the maximum depth of flow H_m when the permeability of the material k_d is known, or the determination of the required permeability of the drainage material when the maximum flow depth is stipulated. In each case, however, both the slope S of the drainage layer along the flow path and the length L of the flow path must be known. The flow through a drainage layer at full depth is directly related to the *coefficient of transmissibility*, which is the product of k_d and the depth H_d of the drainage layer. This relationship may be used to determine the characteristics of the drainage layer required, or alternatively, the graphical solution presented in Figure 15.38 may be used. The chart shown in Figure 15.38 is based on steady inflow, uniformly distributed across the surface of the pavement section. This condition does not normally occur in practice, but a conservative result is usually obtained when the chart is used in combination with the procedure presented herein for determining the net inflow rate q_n.

Figure 15.38   Chart for Estimating Maximum Depth of Flow Caused by Steady Inflow

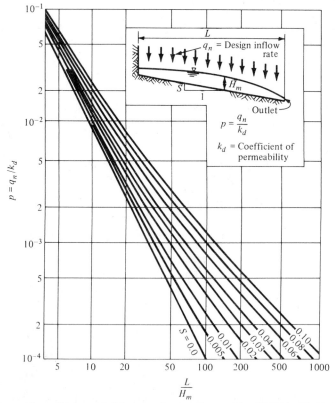

Source: Redrawn from *Highway Subdrainage Design*, Report No. FHWA–TS–80–224, U.S. Department of Transportation, Washington, D.C., August 1980.

Example 15–12 Computing Required Depth for a Drainage Layer

Determine the depth required for a drainage layer to carry a net inflow of 0.50 ft³/day/ft² if the permeability of the drainage material is 2000 ft/day. The drainage layer will be laid at a slope of 2 percent and the length of the flow path is 40 ft.

$$p = q_n/k_d \qquad \text{(from Figure 15.38)}$$

$$= \frac{0.50}{2000} = 2.50 \times 10^{-4}$$

Entering the chart at $p = 2.50 \times 10^{-4}$ and projecting horizontally to the slope of 0.02, we determine L/H_m as 130.

$$H_m = \frac{40}{130} \text{ (required depth of drainage layer)}$$

$$\approx 0.31 \text{ ft}$$

$$= 3.7 \text{ in.} \qquad \text{(say, 4 in.—i.e., } H_d)$$

Note that $H_d > H_m$.

Filter Requirements. The provision of a drainage layer consisting of coarse material allows for the flow of water from the fine-grained material of the subgrade soil to the coarse drainage layer. This may result in the fine-grained soil particles being transmitted to the coarse soil and eventually clogging the voids of the coarse-grained soil. When this occurs, the permeability of the coarse-grained soil is significantly reduced, thereby making the drainage layer less effective. This intrusion of fine particles into the voids of the coarse material can be minimized if the coarse material has certain filter criteria. In cases where these criteria are not satisfied by the drainage material, a protective filter must be provided between the subgrade and the drainage layer to prevent clogging of the drainage layer.

The following criteria have been developed for soil materials used as filters.

$$(D_{15})_{\text{filter}} \leq 5(D_{85})_{\text{protected soil}}$$
$$(D_{15})_{\text{filter}} \geq 5(D_{15})_{\text{protected soil}}$$
$$(D_{50})_{\text{filter}} \leq 25(D_{50})_{\text{protected soil}}$$
$$(D_{5})_{\text{filter}} \geq 0.074 \text{ mm}$$

where D_i is the grain diameter that is larger than the ith percent of the soil grains—that is, the ith percent size on the grain-size distribution curve. (See Chapter 16.)

Design of Longitudinal Collectors

Circular pipes are generally used for longitudinal collectors and are usually constructed of either porous concrete, perforated corrugated metal, or vitrified clay.

The pipes are laid in trenches located at depths that will allow the drainage of the subsurface water from the pavement structure. The trenches are then backfilled with porous granular material to facilitate free flow of the subsurface water into the drains.

Design of the longitudinal collectors involves the determination of the pipe location, the pipe diameter, and the identification of a suitable backfill material.

Pipe Location. Shallow trenches within the subbase layer may be used in locations where the depth of frost penetration is insignificant and where the drawdown of the water table is low, as shown in Figure 15.39. In cases where the depth of frost penetration is high or the water table is high, thereby requiring a high drawdown, it is necessary to locate the pipe in a deeper trench below the subbase layer, as shown in Figure 15.40. Note, however, that the deeper the trench, the higher is the construction cost of the system. The lateral location of the pipe depends on whether the shoulder is also to be drained. If the shoulder is to be drained, the pipe should be located close to the edge of the shoulder as shown in Figures 15.39(b) and 15.40(b), but if shoulder drainage is not required, the pipe is located just outside the pavement surface as shown in Figures 15.39(a) and 15.40(a).

Pipe Diameter. The diameter D_p of the collector pipe depends on the gradient g, the amount of water per running foot q_d that should be transmitted through the pipe, Manning's roughness coefficent of the pipe material, and the distance between the outlets L_o. The chart shown in Figure 15.41 can be used either to determine the minimum pipe diameter when the flow depth, distance between the

Figure 15.39 Typical Location of Shallow Longitudinal Collector Pipes

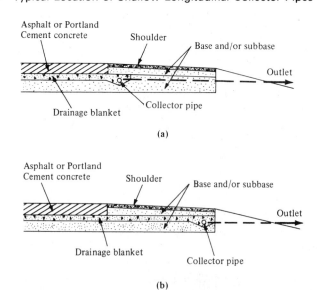

Source: Redrawn from *Highway Subdrainage Design*, Report No. FHWA–TS–80–224, U.S. Department of Transportation, Washington, D.C., August 1980.

Figure 15.40   Typical Location of Deep Longitudinal Collector Pipes

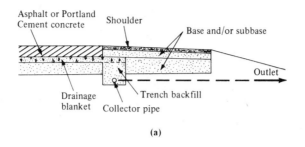

(a)

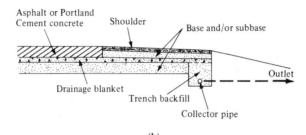

(b)

Source: Redrawn from *Highway Subdrainage Design*, Report No. FHWA–TS–80–224, U.S. Department of Transportation, Washington, D.C., August 1980.

outlets, and the gradient are specified or to determine the maximum spacing between outlets for different combinations of gradient and pipe diameters.

In using the chart, it is first necessary to determine the amount of flow q_d from q_n as

$$q_d = (q_n)/(L) \qquad (15.31)$$

where

q_n = net inflow as determined earlier
L = the length of the flow path

Note that some variation of L may occur along the highway. The average of all values of L associated with a given pipe may therefore be used for that pipe. The use of the chart is demonstrated in the example given in Figure 15.41.

Backfill Material.   The material selected to backfill the pipe trench should be coarse enough to permit the flow of water into the pipe and also fine enough to prevent the infiltration of the drainage aggregates into the pipe. The following criteria can be used to select suitable filter material.

For slotted pipes $(D_{85})_{\text{filter}} > 1/2$ slot width
For circular holes $(D_{85})_{\text{filter}} >$ hole diameter

Figure 15.41   Nomogram Relating Collector Pipe Size with Flow Rate, Outlet Spacing, and Pipe Gradient

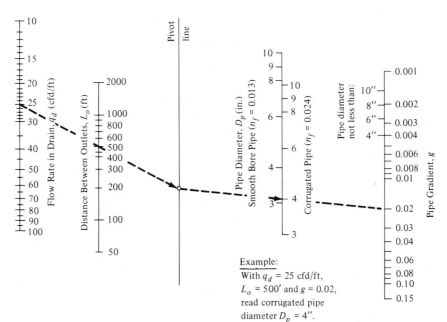

Source: Redrawn from *Highway Subdrainage System*, Report No. FHWA–TS–80–224, U.S. Department of Transportation, Washington, D.C., August 1980.

ECONOMIC ANALYSIS

The economic analysis normally carried out is similar to those carried out for open channel and culvert designs. In this case, however, the cost of the subdrainage system is highly dependent on the cost and availability of suitable drainage materials and the cost of the pipes used as longitudinal and transverse drains.

SUMMARY

The provision of adequate drainage facilities on a highway is fundamental and essential to the effective performance of the highway. Unfortunately, the operation of any drainage system consists of very complex hydraulic phenomena, which make it very difficult to develop exact mathematical equations for design or analysis. The analysis and design of drainage facilities are therefore mainly based on empirical relations that have been developed from extensive test results. In addition, the stochastic nature of rainfall occurrence makes it infeasible for drainage facilities to be designed for the worst-case situation. The material presented in this chapter, however, gives the reader the basic principles of analysis and design currently in use. However, the use of any methods presented should go hand in hand with experience that has been gained from local conditions.

The use of these procedures coupled with sound judgment will result in drainage facilities that effectively serve the highway.

PROBLEMS

15–1 Briefly describe the main differences between surface drainage and subsurface drainage.

15–2 What is the difference between supercritical and critical flow? Under what conditions will either of these occur?

15–3 A 170 acre rural drainage area consists of four different watershed areas as follows:

Steep grass covered area = 40 percent
Cultivated area = 25 percent
Forested area = 30 percent
Turf meadows = 5 percent

Using the rational formula, determine the run-off rate for a storm of 100 yr frequency. Assume that the rainfall intensity curves in Figure 15.2 are applicable to this drainage area and the following land characteristics apply.

Overland flow length = 0.5 mi
Average slope of overland area = 3 percent

15–4 Using the TR–55 method, determine the depth of run-off for a 24 hr, 100 yr precipitation of 9 in. if the soil can be classified as group B and the watershed is contoured pasture with good hydrologic condition and an antecedent soil condition III.

15–5 Determine the depth of run-off for a 24 hr, 100 yr precipitation of 9 in. for an antecedent moisture condition III, if the following land uses and soil conditions exist.

Area Fraction	Land/Use Condition	Soil Group
0.30	Wooded/fair condition	D
0.25	Small grain/straight row/good condition	D
0.20	Pasture/contoured/fair condition	D
0.15	Meadow/good condition	D
0.10	Farmstead	D

15–6 Determine the peak discharge that will occur for the conditions indicated in problem 15–5 if the drainage area is 0.5 mi^2, and the time of concentration is 1.6 hr.

15–7 A trapezoidal channel of 2 : 1 side slope and 5 ft bottom width, discharges a flow of 275 ft^3/sec. If the channel slope is 2.5 percent and the Manning coefficient is 0.03, determine (a) flow velocity, (b) flow depth, and (c) type of flow.

15–8 A 6 ft wide rectangular channel lined with rubble masonry is required to carry a flow of 300 ft^3/sec. If the slope of the channel is 2 percent and $n = 0.015$, determine (a) flow depth, (b) flow velocity, and (c) type of flow.

15–9 Determine a suitable rectangular flexible lined channel to resist erosion for a maximum flow of 20 ft^3/sec if the channel slope is 2 percent.

15–10 A trapezoidal channel of 2 : 1 side slope and 5 ft bottom width is to be used to discharge a flow of 200 ft^3/sec. If the channel slope is 2 percent and the Manning coefficient is 0.015, determine the minimum depth required for the channel. Is the flow supercritical or subcritical?

15–11 Determine whether a 5 ft × 5 ft reinforced concrete box culvert with 45° flared wingwalls and beveled edge at top of inlet carrying a 50 yr flow rate of 200 ft^3/sec will operate under inlet or outlet control for the following conditions. Assume $k_e = 0.5$.

> Design headwater elevation $(EL_{hd}) = 105$ ft
> Elevation of stream bed at face of invert = 99.55 ft
> Tailwater depth = 4.75 ft
> Approximate length of culvert = 200 ft
> Slope of stream = 1.5 percent
> $n = 0.012$

15–12 Repeat problem 15–11 using a 6 ft, 6 in. diameter circular pipe culvert with $k_e = 0.05$.

15–13 Determine the ground water infiltration rate for a new two-lane pavement with the following characteristics

> Lane width = 12 ft
> Shoulder width = 10 ft
> Length of contributing transverse cracks $(W_c) = 20$ ft
> Rate of infiltration $(K_p) = 0.05$ ft^3/day/ft^2
> Spacing of transverse cracks = 30 ft

15–14 In addition to the infiltration determined in problem 15–13, ground water seepage due to gravity also occurs. Determine the thickness of a suitable drainage layer required to transmit the net inflow to a suitable outlet.

> Thickness of subgrade below drainage pipe = 12 ft
> Coefficient of permeability of native soil = 0.35 ft/day
> Height of water table above impervious layer = 21 ft
> Slope of drainage layer = 2 percent
> Permeability of drainage area = 2000 ft/day
> Length of flow path = 44 ft

REFERENCES

1. *Urban Hydrology for Small Watersheds*, Technical Release No. 55, U.S. Department of Agriculture, Soil Conservation Service, Engineering Division, Washington, D.C., January 1975.

2. *Computer Program for Project Formulation—Hydrology*, Technical Release No. 20 (SCS–TR–20), U.S. Department of Agriculture, Soil Conservation Service, Engineering Division, Washington, D.C., January 1975.

3. Ibid.

4. *Design Charts for Open Channel Flow*, Hydraulic Design Series No. 3, U.S. Department of Transportation, Federal Highway Administration, Washington, D.C., December 1980.

5. *Design of Stable Channels with Flexible Linings*, Hydraulic Engineering Circular No. 15, U.S. Department of Transportation, Federal Highway Administration, Washington, D.C., 1975.

6. Ibid.

7. Ibid

8. *Design Charts for Open Channel Flows.*

9. *Hydraulic Design of Highway Culverts*, U.S. Department of Transportation, Federal Highway Administration, Washington, D.C., September 1985.

10. Ibid.

11. Ibid.

12. Ibid.

13. *Design of Stable Channels with Flexible Linings.*

14. *Hydraulic Design of Highway Culverts.*

15. P. D. Wlashchin, M. M. Chatfield, A. H. Lowe, and R. G. Magalong, *Hydraulic Design of Improved Inlets for Culverts Using Programmable Calculators*, Calculator Design Series 1, 2 & 3 for the Compucorp-325 Scientist, the HP-65 and the TI-59 Programmable Calculators, U.S. Department of Transportation, Federal Highway Administration, Hydraulics Branch, Bridge Division, Office of Engineering, Washington, D.C., October 1980.

16. *Hydraulic Analysis by Pipe Arch and Elliptical Shape Culverts Using Programmable Calculators*, Calculator Design Series 4, U.S. Department of Transportation, Federal Highway Administration, Hydraulics Branch, Bridge Division, Office of Engineering, Washington, D.C., March 1982.

17. *Hydraulic Design of Highway Culverts.*

18. Ibid.

19. *Highway Subdrainage Design*, Report No. FHWA–TS–80–224, U.S. Department of Transportation, Federal Highway Administration, Washington, D.C., August 1980.

20. Ibid.

21. Ibid.

PART 5

Materials and Pavement Design

Highway pavements are constructed of either asphalt or concrete and ultimately rest on native soil. The engineer must be familiar with the properties and structural characteristics of materials that will be used in constructing or rehabilitating a roadway segment. The engineer must also be familiar with methods and theories for the design of heavy-duty asphaltic and concrete pavements, as well as various treatment strategies for low-volume roads.

CHAPTER 16

Soil Engineering for Highway Design

Highway engineers are interested in the basic engineering properties of soils because soils are used extensively in highway construction. Soil properties are of significant importance when a highway is to carry high traffic volumes with a large percentage of trucks. They are also of importance when high embankments are to be constructed and when the soil is to be strengthened and used as intermediate support for the highway pavement. Thus, several transportation agencies have developed detailed procedures for investigating soil materials used in highway construction.

This chapter presents a summary of current knowledge of the characteristics and engineering properties of soils that are important to highway engineers, including the origin and formation of soils, soil identification, and soil testing methods.

Procedures for improving the engineering properties of soils will be discussed in Chapter 18, Design of Flexible Pavements.

SOIL CHARACTERISTICS

The basic characteristics of a soil may be described in terms of its origin and formation and its grain size and shape. It will be seen later in this chapter that the principal engineering properties of any soil are mainly related to the basic characteristics of that soil.

Origin and Formation of Soils

Soil can be defined from the civil engineering point of view as the loose mass of mineral and organic materials that cover the solid crust of granitic and basaltic rocks of the earth. Soil is mainly formed by weathering and other geologic processes that occur on the surface of the solid rock at or near the surface of the earth. Weathering is the result of physical and chemical actions, mainly due to atmospheric factors that change the structure and composition of the rocks.

Weathering occurs through either physical or chemical means. Physical weathering, sometimes referred to as mechanical weathering, causes the disintegration of the rocks into smaller particle sizes by the action of forces exerted on the rock. These forces may be due to running water, wind, freezing and thawing, and the activity of plants and animals. Chemical weathering occurs as a result of oxidation, carbonation, and other chemical actions that decompose the minerals of the rocks.

Soils may be described as residual or transported. Residual soils are weathered in place and are located directly above the original material from which they were formed. Transported soils are those that have been moved by water, wind, glaciers, and so forth, and are located away from their parent materials.

The geological history of any soil deposit has a significant effect on the engineering properties of the soils. For example, sedimentary soils, which are formed by the action of water, are usually particles that have settled from suspension in a lake, river, or ocean. These soils range from beach or river sands to marine clays. Soils that are formed by the action of wind are known as aeolian soils and are typically loess. Their voids are usually partially filled with water, and when submerged in water, the soil structure collapses.

Soils may also be described as organic when the particles are mainly composed of organic matter, or as inorganic, when the particles are mainly composed of mineral materials.

Surface Texture

The texture of a soil can be described in terms of its appearance, which depends mainly on the shapes and sizes of the soil particles and their distribution in the soil mass. For example, soils consisting mainly of silts and clays with very small particle sizes are known as fine-textured soils, whereas soils consisting mainly of sands and gravel with much larger particles are known as coarse-textured soils. The individual particles of fine-textured soils are usually invisible to the naked eye, whereas those of coarse-textured soils are visible to the naked eye.

It will be seen later in this chapter that the engineering properties of a soil are related to its texture. For example, the presence of water in fine-textured soils results in significant reduction in their strength, whereas this does not happen with coarse-textured soils. Soils can therefore be divided into two main categories based on their texture. Coarse-grained soils are sometimes defined as those with particle sizes greater than 0.05 mm, such as sands and gravel, and fine-grained soils are those with particle sizes less than 0.05 mm, such as silts and clays. The dividing line of 0.05 mm (0.075 mm has also been used) is selected because that is normally the smallest grain size that can be seen by the naked eye. Since there is a wide range of particle sizes in soils, both the coarse-grained soils and fine-grained soils may be further subdivided as will be shown later under soil classification.

The particle size distribution of soils can be determined by conducting a sieve analysis (sometimes known as mechanical analysis) on a soil sample if the particles are sufficiently large. This is done by shaking a sample of air-dried soil through a set of sieves with progressively smaller openings. The smallest practical opening of these sieves is 0.075 mm; this sieve is designated No. 200. Other sieves include No. 140 (0.106 mm), No. 100 (0.15 mm), No. 60 (0.25 mm), No. 40 (0.425 mm), No. 20 (0.85 mm), No. 10 (2.0 mm), No. 4 (4.75 mm), and several others with openings increasing up to 125 mm or 5 in.

For soils containing particle sizes smaller than the lower limit, the hydrometer analysis is used. A representative sample of the air-dried soil is sieved through the No. 10 sieve and a sieve analysis carried out on the portion of soil retained. This will give a distribution of the coarse material. A portion of the material that passes through the No. 10 sieve is suspended in water, usually in the presence of a deflocculating agent, and is then left standing until the particles gradually settle to the bottom. A hydrometer is used to determine the specific gravity of the suspension at different times. The specific gravity of the suspension after any time (t) from the start of the test is used to determine the maximum particle sizes in the suspension as

$$D = \sqrt{\frac{18\eta}{\gamma_s - \gamma_w}\left(\frac{y}{t}\right)} \qquad (16.1)$$

where

D = maximum diameter of particles in suspension at depth (y)—that is, all particles in suspension at depth y have diameters less than D

η = coefficient of viscosity of the suspending medium (in this case water) in poises

γ_s = unit weight of soil particles

γ_w = unit weight of water

The above expression is based on Stoke's law.

At the completion of this test, which lasts up to 24 hours, the sample of soil used for the hydrometer test is then washed over a No. 200 sieve. The portion retained in the No. 200 sieve is then oven-dried and sieved through Nos. 20, 40, 60, and 140 sieves. The combination of the results of the sieve analysis and the hydrometer test is then used to obtain the particle size distribution of the soil. This is usually plotted as the cumulative percentage by weight of the total sample less than a given sieve size or computed grain diameter, versus the logarithm of the sieve size or grain diameter. Figure 16.1 shows examples of particle size distributions of three different soil samples taken from different locations.

The natural shape of a soil particle is either round, angular, or flat. This natural shape is usually an indication of the strength of the soil, particularly for larger soil particles. Round particles are found in deposits of streams and rivers and have been subjected to extensive wear and therefore are generally strong. Flat and flaky particles have not been subjected to similar action and are usually weak. Fine-grained soils generally have flat and flaky-shaped particles, whereas coarse-grained soils generally have round or angular-shaped particles. Soils with

Figure 16.1   Particle Size Distribution of Different Soils

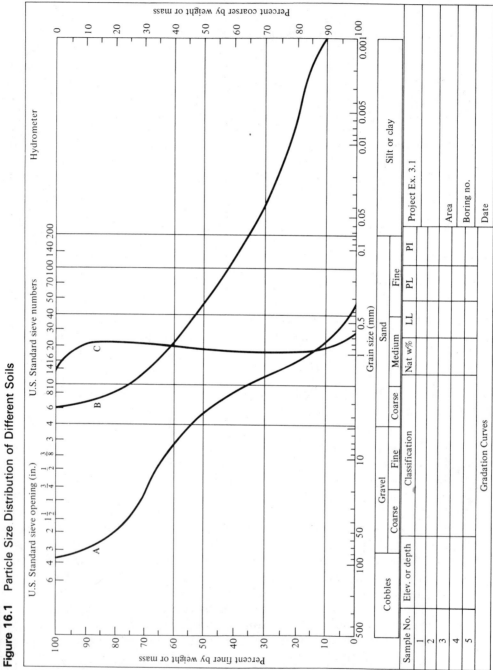

angular-shaped particles have more resistance to deformation than those with round particles, since the individual angular-shaped particles tend to lock together, whereas the rounded particles tend to roll over each other.

BASIC ENGINEERING PROPERTIES OF SOILS

Highway engineers must be familiar with those basic engineering properties of soils that influence their behavior when subjected to external loads. The determination of how a specific soil deposit will behave when subjected to an external load is rather complicated because soil deposits may have heterogeneous properties. Highway engineers must always keep in mind that the behavior of any soil depends on the conditions of that soil at the time it is being tested.

Phase Relations

A soil mass generally consists of solid particles of different minerals with spaces between them. The spaces can be filled with air and/or water. Soils are therefore considered as three-phase systems that consist of air, water, and solids. Figure 16.2 schematically illustrates the three components of a soil mass of total volume V. The volumes of air, water, and solids are V_a, V_w, and V_s, respectively, and their weights are W_a, W_w, and W_s, respectively. The volume V_v is the total volume of the space occupied by air and water, generally referred to as void.

Porosity

The relative amount of voids in any soil is an important quantity that influences some aspects of soil behavior. This amount can be measured in terms of the porosity of the soil, which is defined as the ratio of the volume of voids to the total volume of the soil and is designated as n as shown in Eq. 16.2.

$$n = \frac{V_v}{V}$$

(16.2)

Figure 16.2 Schematic of the Three Phases of a Soil Mass

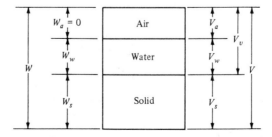

Void Ratio

The amount of voids can also be measured in terms of the void ratio, which is defined as the ratio of the volume of voids to the volume of solids and is designated as e as shown in Eq. 16.3.

$$e = \frac{V_v}{V_s} \qquad (16.3)$$

Combining Eqs. 16.2 and 16.3 we obtain

$$e = \frac{nV}{V_s} = \frac{n}{V_s}(V_s + V_v) = n(1 + e)$$

and

$$n = \frac{e}{1 + e} \qquad (16.4)$$

Similarly,

$$e = \frac{n}{1 - n} \qquad (16.5)$$

Moisture Content

The quantity of water in a soil mass is expressed in terms of the moisture content, which is defined as the ratio of the weight of water W_w in the soil mass to the oven-dried weight of solids W_s expressed in percentage. It is given as

$$w = \frac{W_w}{W_s} \, 100 \qquad (16.6)$$

where w is the moisture content.

The moisture content of a soil mass can be determined in the laboratory by obtaining the weight of water in the soil and the weight of the dry solids, after completely drying the soil in an oven at a temperature between 100° and 110°C. The weight of the water W_w is obtained by first obtaining the weight W_1 of the soil and container. The soil and container are then placed in the oven. The weight of the dry soil and container is again obtained after repeated drying until there is no further reduction in weight. This weight W_2 is that of the dry soil and container. The weight of water W_w is therefore given as

$$W_w = W_1 - W_2$$

The weight of the empty container W_c is then obtained and the weight of the dry soil is given as

$$W_s = W_2 - W_c$$

and the moisture content is obtained as

$$w\ (\%) = \frac{W_w}{W_s}\ 100 = \frac{W_1 - W_2}{W_2 - W_c}\ 100 \tag{16.7}$$

Degree of Saturation

The degree of saturation is the percentage of void space occupied by water and is given as

$$S = \frac{V_w}{V_v}\ 100 \tag{16.8}$$

where S is the degree of saturation.

The soil is saturated when the void is fully occupied with water, that is, $S = 100$ percent, and partially saturated when the voids are only partially occupied with water.

Density of Soil

A very useful soil property for highway engineers is the density of the soil. The density is the ratio that relates the mass side of the phase diagram to the volumetric side. Three densities are commonly used in soil engineering. These are total or bulk density γ, dry density γ_d, and submerged or buoyant density γ'.

Total Density.   The total (or bulk) density is the ratio of the weight of a given sample of soil to the volume or

$$\gamma = \frac{W}{V} = \frac{W_s + W_w}{V_s + V_w + V_a} \qquad \text{(weight of air is negligible)} \tag{16.9}$$

The total density for saturated soils is the saturated density and is given as

$$\gamma_{\text{sat}} = \frac{W}{V} = \frac{W_s + W_w}{V_s + V_w} \tag{16.10}$$

Dry Density.   The dry density is the density of the soil with the water removed. It is given as

$$\gamma_d = \frac{W_s}{V} = \frac{W_s}{V_s + V_w + V_a} = \frac{\gamma}{1 + w} \tag{16.11}$$

The dry density is often used to evaluate how well earth embankments have been compacted and is therefore an important quantity in highway engineering.

Submerged Density.   The submerged density is the density of the soil when submerged in water and it is the difference between the saturated density and the density of water or

$$\gamma' = \gamma_{\text{sat}} - \gamma_w \tag{16.12}$$

where γ_w is the density of water.

Specific Gravity of Soil Particles

The specific gravity of soil particles is the ratio of density of the soil particles to the density of distilled water.

Other Useful Relationships

The basic definitions presented above can be used to derive other useful relationships. For example, the bulk density can be given as

$$\gamma = \frac{G_s + Se}{1 + e}\,\gamma_w \qquad\qquad (16.13)$$

where

γ = total or bulk density
S = degree of saturation
γ_w = density of water
G_s = specific gravity of the soil particles
e = void ratio

Example 16–1   Determining Soil Characteristics Using the Three-Phase Principle

The wet weight of a specimen of soil is 340 g and the dried weight is 230 g. The volume of the soil before drying is 210 cc. If the specific gravity of the soil particles is 2.75, determine the void ratio, porosity, degree of saturation, and dry density. Figure 16.3 is a schematic of the soil mass.

Since the weight of the air can be taken as zero, the weight of the water is

$$W_w = 340 - 230 = 110 \text{ g}$$

Therefore, the volume of water is

$$V_w = (110/1.00) = 110 \text{ cc} \qquad \text{(since density of water} = 1 \text{ gm/cc)}$$

Figure 16.3   Schematic of Soil Mass for Example 16–1

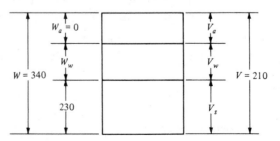

For the volume of solids,

$$V_s = \frac{W_s}{\gamma_s} = \frac{230}{2.75} \qquad (\gamma_s = \text{specific gravity} \times \text{density of water} = 2.75 \times 1)$$

$$= 83.64 \text{ cc}$$

For the volume of air,

$$V_a = 210 - 110 - 83.64 = 16.36 \text{ cc}$$

For the void ratio,

$$e = \frac{V_w + V_a}{V_s} = \frac{110 + 16.36}{83.64} = 1.51$$

For the porosity,

$$n = \frac{V_v}{V} = \frac{110 + 16.36}{210} = 0.60$$

For the degree of saturation,

$$S = \frac{V_w}{V_v} = \frac{110}{110 + 16.36} = 0.87 \qquad \text{or } 87\%$$

For the dry density,

$$\gamma_d = \frac{W_s}{V} = \frac{230}{210} = 1.095 \text{ g/cc}$$

Example 16–2   Determining Soil Characteristics Using the
Three-Phase Principle

The moisture content of a specimen of soil is 28 percent, and the bulk density is 115 lb/ft³. If the specific gravity of the soil particles is 2.75, determine the void ratio and the degree of saturation.

The weight of 1 ft³ of the soil is 115 lb—that is,

$$W = 115 \text{ lb} = W_s + W_w = W_s + wW_s = W_s(1 + 0.28)$$

Therefore,

$$W_s = \frac{115}{1.28} = 89.8 \text{ lb}$$

$$W_w = 0.28 \times 89.8 = 25.2 \text{ lb}$$

$$V_s = \frac{W_s}{\gamma_s} \quad (\gamma_s = \text{specific gravity} \times \text{density of water} = 2.75 \times 62.4)$$

$$= \frac{89.8}{2.75 \times 62.4} = 0.52 \text{ ft}^3$$

$$V_w = \frac{W_w}{\gamma_w} = \frac{25.2}{62.4} = 0.4 \text{ ft}^3$$

$$V_a = V - V_s - V_w = 1 - 0.52 - 0.4 = 0.08 \text{ ft}^3$$

Therefore, the void ratio is

$$e = \frac{V_w + V_a}{V_s} = \frac{0.40 + 0.08}{0.52} = 0.92$$

and the degree of saturation is

$$S = \frac{V_w}{V_v} 100 = \frac{0.4}{0.4 + 0.08} 100 = 83.3\%$$

Atterberg Limits

Clay soils with very low moisture content will be in the form of solids. As the water content increases, however, the solid soil gradually becomes plastic—that is, the soil can easily be molded into different shapes without breaking up. Continuous increase of the water content will eventually bring the soil to a state where it can flow as a viscous liquid. The stiffness or consistency of the soil at any time therefore depends on the state at which the soil is, which in turn depends on the amount of water present in the soil. The water content levels at which the soil changes from one state to the other are the Atterberg limits. They are the shrinkage limit (SL), plastic limit (PL), and liquid limit (LL), as illustrated in Figure 16.4. These are important limits of engineering behavior because they facilitate the comparison of the water content of the soil with those at which the soil changes from one state to another. They are used in the classification of fine-grained soils and are extremely useful, since they correlate with the engineering behaviors of such soils.

Shrinkage Limit (SL)

When a saturated soil is slowly dried, the volume shrinks but the soil continues to contain moisture. Continuous drying of the soil, however, will lead to a moisture content at which further drying will not result in additional shrinkage. The volume of the soil will stay constant, and further drying will be accompanied by air entering the voids. The moisture content at which this occurs is the SL of the soil.

Figure 16.4 Consistency Limits

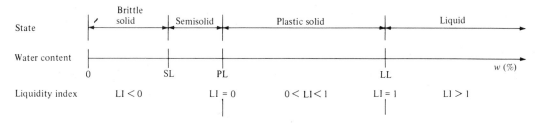

$$\text{LI} = \text{liquidity index} = \frac{w - \text{plastic limit (PL)}}{\text{plasticity index (PI)}}$$

Plastic Limit (PL)

The PL is defined as the moisture content at which the soil crumbles when it is rolled down to a diameter of 1/8 in. The moisture content is higher than the PL if the soil can be rolled down to diameters less than 1/8 in., and the moisture content is lower than the PL if the soil crumbles before it can be rolled to 1/8 in. diameter.

Liquid Limit (LL)

The LL is defined as the moisture content at which the soil will flow and close a groove of 1/2 in. within it, after the standard LL equipment has been dropped 25 times. The equipment used for LL determination is shown in Figure 16.5. This device was developed by Casagrande, who worked to standardize the Atterberg limit tests. It is difficult in practice to obtain the exact moisture content at which the groove will close at exactly 25 blows. The test is therefore conducted for

Figure 16.5 Schematic of the Casagrande Liquid Limit Apparatus

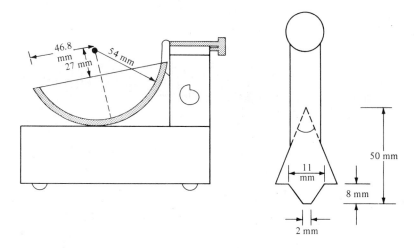

different moisture contents and the number of blows required to close the groove for each moisture content recorded. A graph of moisture content versus the logarithm of the number of blows (usually a straight line known as the flow curve) is then drawn. The moisture content at which the flow curve crosses 25 blows is the LL.

The range of moisture content over which the soil is in the plastic state is the difference between the LL and the PL and is known as the plasticity index (PI).

$$PI = LL - PL \qquad (16.14)$$

where

PI = plasticity index
LL = liquid limit
PL = plastic limit

Liquidity Index (LI)

Since both the PL and LL can only be determined on remolded soils, it is quite possible that the limits determined may not apply to the undisturbed soil, since the structure of the soil particles in the undisturbed state may be different from that in the disturbed state. It is therefore possible that an undisturbed soil will not be in the liquid state if its moisture content is 35 percent even though the LL was found to be 35 percent. The LI is used to reflect the properties of the natural soil and is defined as

$$LI = \frac{w_n - PL}{PI} \qquad (16.15)$$

where

LI = liquidity index
w_n = natural moisture content of the soil

A soil with an LI less than zero will have a brittle fracture when sheared, and a soil with an LI between zero and 1 will be in a plastic state. When LI is greater than 1, the soil will be in a state of viscous liquid if sheared.

Soils with LIs greater than 1 are known as quick clays. They stay relatively strong if undisturbed but become very unstable and can even flow like liquid if they are sheared.

Permeability

The permeability of a soil is the property that describes how water flows through the soil. It is usually given in terms of the coefficient of permeability (K), which is the constant of proportionality of the relationship between the flow velocity and the hydraulic gradient between two points in the soil. This relationship was first determined by a French engineer named D'Arcy, and is given as

$$u = Ki \qquad (16.16)$$

where

u = velocity of water in the soil

i = hydraulic gradient

$= \dfrac{h}{l}$ (head loss h per unit length l)

K = coefficient of permeability

The coefficient of permeability of a soil can be determined in the laboratory by conducting either a constant head or falling head test, or in the field by pumping tests.[1,2]

Clays and fine-grained soils have very low permeability; thus hardly any flow of water occurs in these soils. Coarse-grained soils, such as gravel and sands, have high permeability, which allows for water to flow easily in them. Soils with high permeability are therefore generally stable, both in the dry and saturated states. Thus, coarse-grained soils make excellent subgrade materials. Note, however, that capillary action may occur in some permeable soils such as "dirty" gravel, which may cause serious stability problems. Capillary action is the movement of free moisture by capillary forces through small diameter openings in the soil mass into pores that are not full of water. Although the moisture can move in any direction, the upward movement usually causes the most serious problems, since this may cause weakness or lead to frost heave. This is discussed further under frost action in soils.

Shear Strength

The shear strength of soils is of particular importance to the highway engineer because soil masses will usually fail in shear under highway loads. The shear strength of a soil depends on the cohesion and the angle of internal friction and is expressed as

$$S = C + \sigma \tan \phi \qquad (16.17)$$

where

S = shear strength

C = cohesion (expressed in pounds per square foot)

ϕ = angle of internal fracture

σ = normal stress on the shear plane

The degree of importance of either the cohesion or the angle of internal friction depends on the type of soil. In fine-grained soils such as clays, the cohesion component is the major contributor to the shear strength. In fact, it is usually assumed that the angle of internal friction of saturated clays is zero, which makes the shearing resistance on any plane of these soils equal to the cohesion C. Factors that affect the shear strength of cohesive soils include the geologic deposit, moisture content, drainage conditions, and density.

In coarse-grained soils such as sands, the shear strength is achieved mainly through the internal resistance to sliding as the particles roll over each other. The angle of internal friction is therefore important. The value of the angle of internal friction depends on the density of the soil mass, the shape of the individual soil particles, and the surface texture. In general, the angle of internal friction is high when the density is high. Similarly, soils with rough particles such as angular sand grains will have a high angle of internal friction.

The shearing strength of a soil deposit may be obtained in the laboratory by conducting either the triaxial test, the unconfined compression test, or the direct shear test. These tests may be conducted either on the undisturbed soil or on remolding soils. Note, however, that in using remolded samples, the remolding should represent conditions similar to those in the field. Details of each of these tests can be found in *Soil Testing for Engineers* and *Soil Mechanics Laboratory Manual*.[3,4]

The in-situ shearing strengths of soils can also be obtained directly by conducting either the plate bearing test or a cone penetration test. These tests are described in *Foundation Engineering Handbook*.[5]

CLASSIFICATION OF SOILS FOR HIGHWAY USE

Soil classification is a method by which soils are systematically categorized according to their probable engineering characteristics. It therefore serves as a means of identifying suitable subbase materials and predicting the probable behavior of a soil when used as subgrade material. (See definitions of subgrade and subbase in Chapter 18.) The classification of a given soil is determined by conducting relatively simple tests on disturbed samples of the soil; the results are then correlated with field experience.

Note, however, that although the engineering properties of a given soil to be used in highway construction can be predicted reliably from its classification, this should not be regarded as a substitute for the detailed investigation of the soil properties. Classifying the soil should be considered as a means of obtaining a general idea of how the soil will behave if used as a subgrade or subbase material.

The most commonly used classification system for highway purposes is the American Association of State Highway and Transportation Officials (AASHTO) classification system. The Unified Soil Classification System (USCS) is also used to a lesser extent in this country. A slightly modified version of the USCS is used fairly extensively in the United Kingdom.

AASHTO Soil Classification System

The AASHTO Classification System is based on the Public Roads Classification System that was developed from the results of extensive research conducted by the Bureau of Public Roads, now known as the Federal Highway Administration. Several revisions have been made to the system since it was first published. The system has been described by AASHTO as a means for determining the relative quality of soils for use in embankments, subgrades, subbases, and bases.[6]

In the current publication, soils are classified into seven groups, A-1 through A-7, with several subgroups, as shown in Table 16.1. The classification of a given soil is based on its particle size distribution, LL, and PI. Soils are evaluated within each group by using an empirical formula to determine the group index (GI) of the soils, given as

$$GI = (F - 35)[0.2 + 0.005(LL - 40)] + 0.01(F - 15)(PI - 10) \quad \textbf{(16.18)}$$

where

GI = group index

F = % of soil particles passing 0.075 mm (No. 200) sieve in whole number based on material passing 75 mm (3 in.) sieve

LL = liquid limit expressed in whole number

PI = plasticity index expressed in whole number

Table 16.1 AASHTO Soil Classification System

General classification	Granular Materials (35% or Less Passing No. 200)							Silt-Clay Materials (More than 35% Passing No. 200)			
	A-1			A-2							A-7
Group classification	A-1-a	A-1-b	A-3	A-2-4	A-2-5	A-2-6	A-2-7	A-4	A-5	A-6	A-7-5, A-7-6
Sieve analysis, percent passing:											
No. 10	50 max	–	–	–	–	–	–	–	–	–	–
No. 40	30 max.	50 max.	51 min.	–	--	–	–	–	–	–	–
No. 200	15 max.	25 max.	10 max.	35 max.	35 max.	35 max.	35 max.	36 min.	36 min.	36 min.	36 min.
Characteristics of fraction passing No. 40:											
Liquid limit		–	–	40 max.	41 min.	40 max.	41 min.	40 max.	41 min.	40 max.	41 min.
Plasticity index	6 max.		N.P.	10 max.	10 max.	11 min.	11 min.	10 max.	10 max.	11 min.	11 min.*
Usual types of significant constituent materials	Stone fragments, gravel and sand		Fine sand	Silty or clayey gravel and sand				Silty soils		Clayey soils	
General rating as subgrade	Excellent to good							Fair to poor			

*Plasticity index of A-7-5 subgroup is equal to or less than L.L. minus 30. Plasticity index of A-7-6 subgroup is greater than L.L. minus 30.

Source: Adapted from *Standard Specifications for Transportation Materials and Methods of Sampling and Testing*, 14th ed., Washington, D.C.: The American Association of State Highways and Transportation Officials, copyright 1986. Used by permission.

The GI is determined to the nearest whole number. A value of zero should be recorded when a negative value is obtained for the GI. Also in determining the GI for A-2-6 and A-2-7 subgroups, the LL part of the Eq. 16.18 is not used, that is, only the second term of the equation is used.

Under the AASHTO system, granular soils fall into classes A-1 to A-3. A-1 soils consist of well-graded granular materials, whereas A-2 soils contain significant amounts of silts and clays, and A-3 soils are clean, but poorly graded sands.

Classifying soils under the AASHTO system will consist of first determining the particle size distribution and Atterberg limits of the soil and then reading Table 16.1 from left to right to find the correct group. The correct group is the first one from the left that fits the particle size distribution and Atterberg limits and should be expressed in terms of group designation and the GI. Examples are A-2-6(4) and A-6(10).

In general, the suitability of a soil deposit for use in highway construction can be summarized as follows.

1. Soils classified as A-1-a, A-1-b, A-2-4, A-2-5, and A-3 can be used satisfactorily as subgrade or subbase material if properly drained. (See definitions of subgrade and subbase in Chapter 18.) In addition, such soils must be properly compacted and covered with an adequate thickness of pavement (base and/or surface cover) for the surface load to be carried.

2. Materials classified as A-2-6, A-2-7, A-4, A-5, A-6, A-7-5, and A-7-6 will require a layer of subbase material if used as subgrade. If these are to be used as embankment materials, special attention must be given to the design of the embankment.

3. When soils are properly drained and compacted, their value as subgrade material decreases as the GI increases. For example, a soil with a GI of 0 (an indication of a good subgrade material) will be better as a subgrade material than one with GI of 20 (an indication of a poor subgrade material).

Example 16-3   Classifying a Soil Sample Using the AASHTO Method
The following data were obtained for a soil sample.

Mechanical Analysis

Sieve No.	Percent Finer	Plasticity Tests:
4	97	LL = 48%
10	93	PL = 26%
40	88	
100	78	
200	70	

Using the AASHTO method for classifying soils, determine the classification of the soil and state whether this material is suitable in its natural state for use as a subbase material.

- Since more than 35 percent of the material passes the No. 200 sieve, the soil is either A-4, A-5, A-6, or A-7.
- The LL is greater than 40 percent, therefore the soil cannot be in group A-4 or A-6. Thus, it is either A-5 or A-7.
- The PI is 22 percent (48 − 26), which is greater than 10 percent, therefore eliminating group A-5.
- Thus, the soil is A-7-5 or A-7-6.
- (LL − 30) = 18 < PI (22%). Therefore the soil is A-7-6, since the plasticity index of A-7-5 soil subgroup is less than (LL − 30). The GI is given as

$$(70 - 35)[0.2 + 0.005(48 - 40)] + 0.01(70 - 15)(22 - 10) = 8.4 + 6.6 = 15$$

The soil is A-7-6 (15) and therefore unsuitable as a subbase material in its natural state.

Unified Soil Classification System (USCS)

The original USCS system was developed during World War II for use in airfield construction. That system has been modified several times to obtain the current version, which can also be applied to other types of construction such as dams and foundations. The fundamental premise used in the USCS system is that the engineering properties of any coarse-grained soil depend on its particle size distribution, whereas those for a fine-grained soil depend on its plasticity. Thus, the system classifies coarse-grained soils on the basis of grain size characteristics and fine-grained soils according to plasticity characteristics.

Table 16.2 lists the USCS definitions for the four major groups of materials, consisting of coarse-grained, fine-grained, organic soils, and peat. Material that is retained in the 75 mm (3 in.) sieve is recorded, but only that which passes is used for the classification of the sample. Soils with more than 50 percent of their particles being retained on the No. 200 sieve are coarse-grained, and those with less than 50 percent of their particles retained are fine-grained soils (see Table 16.3). The coarse-grained soils are subdivided into gravels (G) and sands (S). Soils having more than 50 percent of their particles larger than 75 mm—that is, retained on No. 4 sieve—are gravels and those with more than 50 percent of their particles smaller than 75 mm—that is, pass through No. 4 sieve—are sands. The gravels and sands are further divided into four subgroups, each based on grain size distribution and the nature of the fine particles in them. They can therefore be classified as either well graded (W), poorly graded (P), silty (M), or clayey (C). Gravels can be described as either well-graded gravel (GW), poorly graded gravel (GP), silty gravel (GM), or clayey gravels (GC), and sands can be described as well-graded sand (SW), poorly graded sand (SP), silty sand (SM), or clayey sand (SC). A gravel or sandy soil is described as well graded or poorly graded, depending on the values of two shape parameters known as the coefficient of uniformity, C_u, and the coefficient of curvature, C_c, given as

$$C_u = \frac{D_{60}}{D_{10}} \qquad (16.19)$$

and

$$C_c = \frac{(D_{30})^2}{D_{10} \times D_{60}} \qquad (16.20)$$

Table 16.2   USCS Definition of Particle Sizes

Soil Fraction or Component	Symbol	Size Range
1. Coarse-grained soils		
Gravel	G	75 mm to No. 4 sieve (4.75 mm)
Coarse		75 mm to 19 mm
Fine		19 mm to No. 4 sieve (4.75 mm)
Sand	S	No. 4 (4.75 mm) to No. 200 (0.075 mm)
Coarse		No. 4 (4.75 mm) to No. 10 (2.0 mm)
Medium		No. 10 (2.0 mm) to No. 40 (0.425 mm)
Fine		No. 40 (0.425 mm) to No. 200 (0.075 mm)
2. Fine-grained soils		
Fine		Less than No. 200 sieve (0.075 mm)
Silt	M	(No specific grain size—use Atterberg limits)
Clay	C	(No specific grain size—use Atterberg limits)
3. Organic soils	O	(No specific grain size)
4. Peat	Pt	(No specific grain size)

Gradation Symbols	Liquid Limit Symbols
Well graded, W	High LL, H
Poorly graded, P	Low LL, L

Source: Adapted from *The Unified Soil Classification System*, Technical Memorandum No. 3-357, U.S. Army Engineers Waterways Experiment Station, Vicksburg, Miss., 1960.

Table 16.3   Unified Soil Classification System

Major Divisions			Group Symbols	Typical Names		Laboratory Classification Criteria
Coarse-grained soils (More than half of material is larger than No. 200 sieve size)	Gravels (More than half of coarse fraction is larger than No. 4 sieve size)	Clean gravels (Little or no fines)	GW	Well-graded gravels, gravel-sand mixtures, little or no fines	Determine percentages of sand and gravel from grain-size curve. Depending on percentage of fines (fraction smaller than No. 200 sieve size), coarse-grained soils are classified as follows: Less than 5 per cent — GW, GP, SW, SP; More than 12 per cent — GM, GC, SM, SC; 5 to 12 per cent — Borderline cases requiring dual symbols[b]	$C_u = \dfrac{D_{60}}{D_{10}}$ greater than 4; $C_c = \dfrac{(D_{30})^2}{D_{10} \times D_{60}}$ between 1 and 3
			GP	Poorly graded gravels, gravel-sand mixtures, little or no fines		Not meeting all gradation requirements for GW
		Gravels with fines (Appreciable amount of fines)	GM[a] d / u	Silty gravels, gravel-sand-silt mixtures		Atterberg limits below "A" line or P.I. less than 4 → Above "A" line with P.I. between 4 and 7 are borderline cases requiring use of dual symbols
			GC	Clayey gravels, gravel-sand-clay mixtures		Atterberg limits below "A" line with P.I. greater than 7
	Sands (More than half of coarse fraction is smaller than No. 4 sieve size)	Clean sands (Little or no fines)	SW	Well-graded sands, gravelly sands, little or no fines		$C_u = \dfrac{D_{60}}{D_{10}}$ greater than 6; $C_c = \dfrac{(D_{30})^2}{D_{10} \times D_{60}}$ between 1 and 3
			SP	Poorly graded sands, gravelly sands, little or no fines		Not meeting all gradation requirements for SW
		Sands with fines (Appreciable amount of fines)	SM[a] d / u	Silty sands, sand-silt mixtures		Atterberg limits above "A" line or P.I. less than 4 → Limits plotting in hatched zone with P.I. between 4 and 7 are borderline cases requiring use of dual symbols
			SC	Clayey sands, sand-clay mixtures		Atterberg limits above "A" line with P.I. greater than 7
Fine-grained soils (More than half of material is smaller than No. 200 sieve)	Silts and clays (Liquid limit less than 50)		ML	Inorganic silts and very fine sands, rock flour, silty or clayey fine sands, or clayey silts with slight plasticity		
			CL	Inorganic clays of low to medium plasticity, gravelly clays, sandy clays, silty clays, lean clays		
			OL	Organic silts and organic silty clays of low plasticity		
	Silts and clays (Liquid limit greater than 50)		MH	Inorganic silts, micaceous or diatomaceous fine sandy or silty soils, elastic silts		
			CH	Inorganic clays of high plasticity, fat clays		
			OH	Organic clays of medium to high plasticity, organic silts		
	Highly organic soils		Pt	Peat and other highly organic soils		

Plasticity Chart

Plasticity index vs. Liquid limit. "A" line. Regions: CH, OH and MH, CL, CL-ML, ML and OL. Liquid limit axis 0–100; Plasticity index axis 0–60.

[a]Division of GM and SM groups into subdivisions of d and u are for roads and airfields only. Subdivision is based on Atterberg limits; suffix d used when L.L. is 28 or less and the P.I. is 6 or less; the suffix u used when L.L. is greater than 28.

[b]Borderline classifications, used for soils possessing characteristics of two groups, are designated by combinations of group symbols. For example: GW-GC, well-graded gravel-sand mixture with clay binder.

Source: Adapted from Hans F. Winterkorn and Hsai-Yang Fang, ed., *Foundation Engineering Handbook*, Van Nostrand Reinhold Company, New York, 1975. Copyright © 1975 by Litton Educational Publishing, Inc.

Table 16.4   Comparable Soil Groups in the AASHTO and USCS Systems

Soil Group in Unified System	Comparable Soil Groups in AASHTO System		
	Most Probable	Possible	Possible but Improbable
GW	A-1-a	— A-2-6, A-2-7	A-2-4, A-2-5,
GP	A-1-a	A-1-b	A-3, A-2-4 A-2-5, A-2-6, A-2-7
GM	A-1-b, A-2-4, A-2-5, A-2-7	A-2-6	A-4, A-5 A-6, A-7-5, A-7-6, A-1-a
GC	A-2-6, A-2-7	A-2-4, A-6	A-4, A-7-6, A-7-5
SW	A-1-b	A-1-a	A-3, A-2-4, A-2-5, A-2-6, A-2-7
SP	A-3, A-1-b	A-1-a	A-2-4, A-2-5, A-2-6, A-2-7
SM	A-1-b, A-2-4, A-2-5, A-2-7	A-2-6, A-4, A-5	A-6, A-7-5, A-7-6, A-1-a
SC	A-2-6, A-2-7	A-2-4, A-6 A-4, A-7-6	A-7-5
ML	A-4, A-5	A-6, A-7-5	—
CL	A-6, A-7-6	A-4	—
OL	A-4, A-5	A-6, A-7-5, A-7-6	—
MH	A-7-5, A-5	—	A-7-6
CH	A-7-6	A-7-5	—
OH	A-7-5, A-5	—	A-7-6
Pt	—	—	—

Source: Adapted from T. K. Liu, *A Review of Engineering Soil Classification Systems—Special Procedures for Testing Soil and Rock for Engineering Purposes*, 5th ed., ASTM Special Technical Publication 479, American Society for Testing and Materials, Easton, Md., 1970.

where

D_{60} = grain diameter at 60% passing

D_{30} = grain diameter at 30% passing

D_{10} = grain diameter at 10% passing

Gravels are described as well graded if C_u is greater than 4 and C_c is between 1 and 3. Sands are described as well graded if C_u is greater than 6 and C_c is between 1 and 3.

The fine-grained soils, which are defined as those having more than 50 percent of their particles passing the No. 200 sieve, are subdivided into clays (C) or silt (M), depending on the PI and LL of the soil. A plasticity chart, shown in Table 16.3, is used to determine whether a soil is silty or clayey. The chart is a plot of PI versus LL, from which a dividing line known as the "A" line, which generally separates the more clayey materials from the silty materials, was developed. Soils with plots of LLs and PIs below the "A" line are silty soils, whereas those with plots above the "A" line are clayey soils. Organic clays are an exception to this general rule since they plot below the "A" line. Organic clays, however, generally behave similarly to soils of lower plasticity.

Classification of coarse-grained soils as silty or clayey also depends on their LL plots. Only coarse-grained soils with more than 12 percent fines (that is, passes No. 200 sieve) are so classified (see Table 16.3). Those soils with plots below the "A" line or with a PI less than 4 are silty gravel (CM) or silty sand (SM), and those with plots above the "A" line with a PI greater than 7 are classified as clayey gravels (GC) or clayey sands (SC).

The organic, silty, and clayey soils are further divided into two groups, one having a relatively low LL (L) and the other having a relatively high LL (H). The dividing line between high LL soils and low LL soils was arbitrarily set at 50 percent.

Fine-grained soils can be classified as either silt with low plasticity (ML), silt with high plasticity (MH), clays with high plasticity (CH), clays with low plasticity (CL), or organic silt with high plasticity (OH).

Table 16.3 gives the complete layout of the USCS and Table 16.4 shows an approximate correlation between the AASHTO system and USCS.

Example 16–4 Classifying a Soil Sample Using the Unified Soil Classification System

The results obtained from a mechanical analysis and a plasticity test on a soil sample are shown below. Classify the soil using the USCS and state whether or not it can be used in the natural state as a subbase material. To solve this problem, the grain size distribution curve is first plotted as shown in Figure 16.6. Since more than 50 percent (62 percent) of the soil passes No. 200 sieve, the soil is fine grained. The plot of the limits on the plasticity chart is below the "A" line (PI = 40 − 30 = 10); therefore, it is either silt or organic clay. The LL is however

Figure 16.6   Grain Size Distribution Curve for Examples 16–5 and 16–6

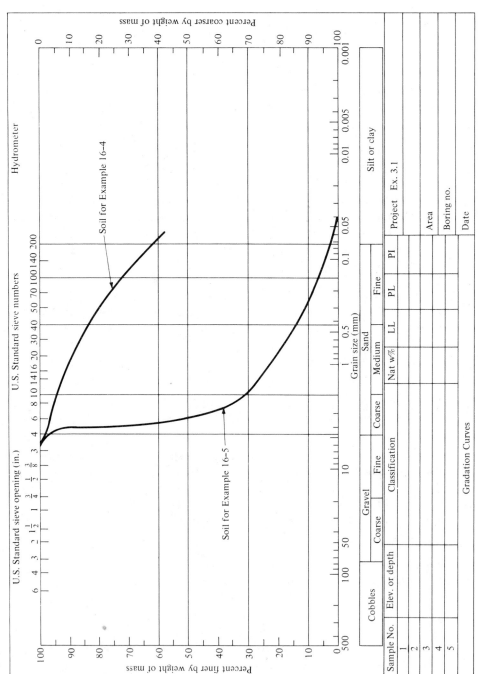

less than 50 percent (40 percent); therefore, it is low LL. This soil can be classified as ML or OL, which is probably equivalent to an A-4 or A-5 soil in the AASHTO classification system. It is therefore not useful as a subbase material.

Mechanical Analysis

Sieve No.	Percent Passing (by weight)	Plasticity Tests:
4	98	LL = 40%
10	93	PL = 30%
40	85	
100	73	
200	62	

Example 16–5   Classifying a Soil Sample Using the Unified Soil Classification System

Repeat example 16–4 for the data shown below.

Mechanical Analysis

Sieve No.	Percent Passing (by weight)	Plasticity Tests:
4	95	LL = nonplastic
10	30	PL = nonplastic
40	15	
100	8	
200	3	

The grain size distribution curve is also plotted in Figure 16.6. Since only 3 percent of the particles pass through No. 200 sieve, the soil is coarse grained. Since more than 50 percent pass through the No. 4 sieve, the soil is classified as sand. Because the soil is nonplastic, it is necessary to determine its coefficient of uniformity C_u and coefficient of curvature C_c. From the particle size distribution curve,

$$C_u = \frac{D_{60}}{D_{10}} = \frac{3.8}{0.25} = 15.2 > 6$$

$$C_c = \frac{(D_{30})^2}{D_{10} \times D_{60}} = \frac{2^2}{0.25 \times 3.8} = 4.2 > 3$$

This sand is not well graded and is classified as SP and can therefore be used as a subbase material, if properly drained and compacted.

SOIL SURVEYS FOR HIGHWAY CONSTRUCTION

Soil surveys for highway construction entail the investigation of the soil characteristics on the highway route and the identification of suitable soils for use as subbase and fill materials. Soil surveys are therefore normally an integral part of preliminary location surveys, since the soil conditions may significantly affect the location of the highway. A detailed soil survey is always carried out on the final highway location.

The first step in any soil survey is in the collection of existing information on the soil characteristics of the area in which the highway is to be located. Such information can be obtained from geological and argicultural soil maps, existing aerial photographs, and an examination of excavations and existing roadway cuts. It is also usually helpful to review the design and construction of other roads in the area. The information obtained from these sources can be used to develop a general understanding of the soil conditions in the area and to identify any unique problems that may exist. The extent of additional investigation usually depends on the amount of existing information that can be obtained.

The next step is to obtain and investigate enough soil samples along the highway route to identify the boundaries of the different types of soils so that a soil profile can be drawn. Samples of each type of soil along the route location are obtained by auger boring or from test pits for laboratory testing. Samples are usually taken at different depths down to about 5 ft. In cases where rock locations are required, depths may be increased. The engineering properties of the samples are then determined and used to classify the soils. It is important that the characteristics of the soils in each hole be systematically recorded, including the depth, location, thickness, texture, and so forth. It is also important that the location of the water level be noted. These data are then used to plot a detailed soil profile along the highway (see Figure 16.7).

Figure 16.7   Soil Profile Along a Section of Highway

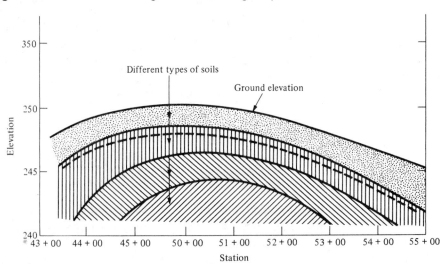

Geophysical Methods of Soil Exploration

Soil profiles can also be obtained from one of two geophysical methods of soil exploration, known as the resistivity and seismic methods.

Resistivity Method

The resistivity method is based on the difference in electrical conductivity or resistivity of different types of soils. An electrical field is produced in the ground by means of two current electrodes, as shown in Figure 16.8, and the potential drop between the two intermediate or potential electrodes is then recorded. The apparent resistivity of the soil to a depth approximately equal to the spacing "*A*" is then computed. The resistivity equipment used is usually designed such that the apparent resistivity can be directly read on the potentiometer. Data for the soil profile are obtained by moving the electrode along the center line of the proposed highway without changing the spacing. The apparent resistivity is then determined along the highway within a depth equal to the spacing "*A*." The resistivities obtained are then compared with known values of different soils by calibrating the instrument using locally exposed materials.

Seismic Method

The seismic method is used to identify the location of rock profiles or dense strata underlying softer materials (Figure 16.9 shows the layout for the seismic method). It is conducted by inducing impact or shock waves into the soil by either striking a plate located on the surface with a hammer or exploding small charges in the soil. Listening devices known as geophones then pick up the shock waves. The time lapse of the wave traveling to the geophone is then used to calculate the velocity of the wave in the surface soil. Some of the shock waves can be made to

Figure 16.8   Electrical Resistivity Method of Soil Exploration

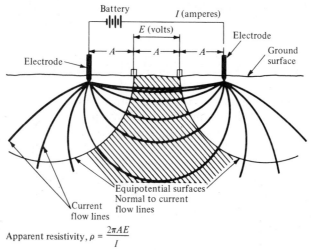

Apparent resistivity, $\rho = \dfrac{2\pi AE}{I}$

Source: Redrawn from Hans F. Winterkorn and Hsai-Yang Fang, ed., *Foundation Engineering Handbook*, Van Nostrand Reinhold Company, New York, 1975. Copyright © 1975 by Litton Educational Publishing, Inc.

Figure 16.9 Soil Exploration by the Seismic Method

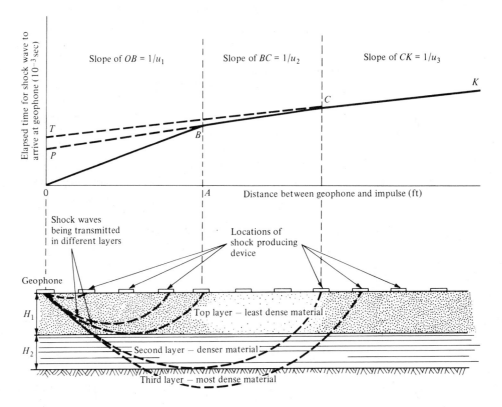

pass from the surface stratum into underlying layers and then back into the surface stratum by moving the shock point away from the geophone. This permits the computation of the wave velocity in the underlying material.

The seismic test is conducted by moving the shock-producing device along the proposed center line of the highway and the shock is produced at known distances from the geophone. A graph of the time it takes the shock waves to arrive at the geophone versus the distance of the geophone from the shock point is then drawn. These points will be on a straight line as long as the shock wave travels through the same soil material. When the distance of shock point from the geophone becomes large enough, the waves travel through the denser material and a break is observed as shown in Figure 16.9. The inverse of the slope of each straight line will give the velocity of the wave within each layer. It can be shown that for three layers, as shown in Figure 16.9, the depths of the first and second soil layers, H_1 and H_2, are

$$H_1 = \frac{\overline{OP}u_1}{2 \cos \alpha} \qquad \textbf{(16.21)}$$

and

$$H_2 = \frac{\overline{PT}u_2}{2 \cos \beta} \qquad\qquad (16.22)$$

where

$\overline{OP}$ = time obtained from plot (see Figure 16.9)

u_1 = velocity of wave in upper stratum

    = 1/slope of first straight line

u_2 = velocity of wave in underlying stratum

    = 1/slope of second straight line

u_3 = velocity of wave in third stratum

    = 1/slope of the third straight line

$\overline{PT}$ = time obtained from plot (see Figure 16.9)

α = first refraction angle

$\sin \alpha = u_1/u_2$

β = second refraction angle

$\sin \beta = u_2/u_3$

The type of material within each stratum can be identified by comparing the wave velocity within each stratum, with known values of wave velocity for different types of soils. Representative values of wave velocities for different types of soils are given in Table 16.5. Note that the seismic method can be used only for cases where the underlying soil is denser than the overlying soil—that is, when u_2 is greater than u_1.

Table 16.5   Representative Values of Wave Velocities for Different Types of Soils

Material	Velocity (m/sec)
Sand	200–2,000
Loess	300– 600
Alluvium	500–2,000
Loam	800–1,800
Clay	1,000–2,800
Marl	1,800–3,800
Sandstone	1,400–4,300
Limestone	1,700–6,400
Slate and shale	2,300–4,600
Granite	4,000–5,700
Quartzite	6,100

Source: Adapted from L. D. Leet, *Earth Waves*, Harvard University Press, Cambridge, Mass. and John Wiley & Sons, New York, 1950. Reprinted by permission.

Example 16–6   Estimating Depth and Soil Type of Each Soil Stratum Using the Seismic Method

The seismic method of exploration was used to establish the soil profile along the proposed center line of a highway. The table below shows part of the results obtained. Estimate the depth of each stratum of soil at this section and suggest the type of soil in each.

Distance of Impulse to Geophone (ft)	Time for Wave Arrival (10^{-3} sec)
20	32
40	60
60	88
80	94
100	100
120	106
140	112
160	116
180	117
200	118.5
220	120
250	122

The plot of the data is shown in Figure 16.10. From Figure 16.10,

$$u_1 = \frac{OA}{AB} = \frac{60}{88 \times 10^{-3}} = 681.8 \text{ ft/sec} \approx 207.6 \text{ m/sec}$$

From Table 16.5, the soil in the first stratum is possibly sand.

$$u_2 = \frac{BD}{CD} = \frac{140 - 60}{(112 - 88) \times 10^{-3}} = 3333 \text{ ft/sec} \approx 1016.2 \text{ m/sec}$$

Figure 16.10   Solution for Example 16–6

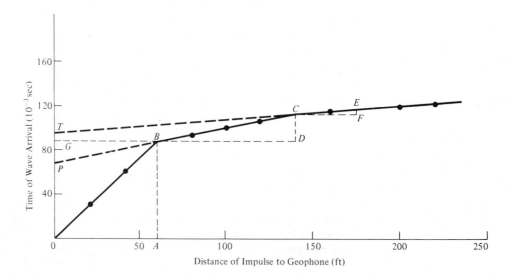

From Table 16.5, the soil in the second stratum can be either alluvium, loam, or clay.

$$u_3 = \frac{CF}{EF} = \frac{175 - 140}{(116 - 112) \times 10^{-3}} = 8750 \text{ ft/sec} \approx 2667.7 \text{ m/sec}$$

The wave velocity in this material is very high, which indicates some type of rock, such as sandstone, limestone, or slate and shale.

$$H_1 = \frac{\overline{OP}u_1}{2 \cos \alpha}$$

$$\sin \alpha = \frac{u_1}{u_2} = \frac{681.8}{3333}$$

$$= 0.205, \alpha = 11.8°$$

$$\cos \alpha = 0.979$$

$$\overline{OP} = 68 \times 10^{-3} \text{ sec} \qquad \text{(from Figure 16.10)}$$

$$H_1 = \frac{(68 \times 10^{-3})681.8}{2 \times 0.979} = 23.68 \text{ ft}$$

$$H_2 = \frac{\overline{PT}u_2}{2 \cos \beta}$$

$$\overline{PT} = 26 \times 10^{-3} \text{ sec} \qquad \text{(from Figure 16.10)}$$

$$\sin \beta = \frac{u_2}{u_3} = \frac{3333}{8750} = 0.381$$

$$\beta = 22.39°$$

$$\cos \beta = 0.925$$

$$H_2 = \frac{26 \times 10^{-3} \times 3333}{2 \times 0.925} = 46.84 \text{ ft}$$

SOIL COMPACTION

When soil is to be used as embankment or subbase material in highway construction, it is essential that the material be placed in uniform layers and compacted to a high density. Proper compaction of the soil will reduce to a minimum subsequent settlement and volume change, thereby enhancing the strength of the embankment or subbase. Compaction is achieved in the field by using hand-operated tampers, sheepsfoot rollers, rubber-tired rollers, or other types of rollers.

The strength of the compacted soil is directly related to the maximum dry density achieved through compaction. The relationship between dry density and

Figure 16.11   Typical Moisture Density Relationship for Soils

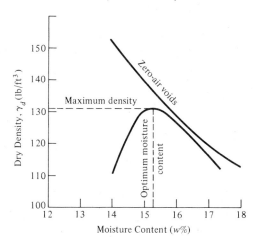

moisture content for practically all soils takes the form shown in Figure 16.11. It can be seen from this relationship that for a given compactive effort, the dry density attained is low at low moisture contents. The dry density increases with increase in moisture content to a maximum value, when an optimum moisture content is reached. Further increase in moisture content results in a decrease in the dry density attained. This phenomenon is due to the effect of moisture on the soil particles. At low moisture content, the soil particles are not lubricated and friction between adjacent particles prevents the densification of the particles. As the moisture content is increased, larger films of water develop on the particles, making the soil more plastic and easier for the particles to be moved and densified. When the optimum moisture content is reached, however, the maximum practical degree of saturation (where $S < 100\%$) is attained. The degree of saturation at the optimum moisture content cannot be increased by further compaction because of the presence of entrapped air in the void spaces and around the particles. Further addition of moisture therefore results in the voids being overfilled with water, with no accompanying reduction in the air. The soil particles are separated, resulting in a reduction in the dry density. The zero-air void curve shown in Figure 16.11 is the theoretical moisture density curve for a saturated soil and zero air voids, where the degree of saturation is 100 percent. This curve is usually not attained in the field, since zero air void cannot be attained as explained earlier. Points on the curve may be calculated from Eq. 16.23 as

$$\gamma_d = \frac{\gamma_w G_s}{1 + w G_s} \qquad (16.23)$$

where

γ_w = density of water

G_s = specific gravity of soil particles

w = moisture content of soil

γ_d = dry density of soil

Although this curve is theoretical, the distance between it and the test moisture–density curve is of importance, since this distance is an indication of the amount of air voids remaining in the soil at different moisture contents. The farther away a point on the moisture density curve is from the zero-air void curve, the more air voids remain in the soil and the higher is the likelihood of expansion or swelling if the soil is subjected to flooding. It is therefore better to compact at the higher moisture content—that is, wet side of optimum moisture content—if a given dry density other than the optimum is required.

Optimum Moisture Content

The determination of the optimum moisture content of any soil to be used as embankment or subgrade material is necessary before any field work is commenced. Most highway agencies now use dynamic or impact tests to determine the optimum moisture content and maximum dry density. In each of these tests, samples of the soil to be tested are compacted in layers to fill a specified size mold. Compacting effort is obtained by dropping a hammer of known weight and dimensions from a specified height a specified number of times for each layer. The moisture content of the compacted material is then obtained and the dry density determined from the measured weight of the compacted soil and the known volume of the mold. The soil is then broken down or another sample of the same soil is obtained. The moisture content is then increased and the test repeated. The process is repeated until a reduction in the density is observed. Usually a minimum of four or five individual compaction tests are required. A plot of dry density versus moisture content is then drawn from which the optimum moisture content is obtained. The two types of tests commonly used are the standard AASHTO or the modified AASHTO.

Table 16.6 shows details for the standard AASHTO, designated T99, and the modified AASHTO, designated T180. Most transportation agencies use the standard AASHTO test.

Table 16.6 Details of the Standard AASHTO and Modified AASHTO Tests

Test Details	Standard AASHTO (T99)	Modified AASHTO (T180)
Diameter of mold (in.)	4 or 6	4 or 6
Height of sample (in.)	5 cut to 4.58	5 cut to 4.58
Number of lifts	3	5
Blows per lift	25 or 56	25 or 56
Weight of hammer	5.5	10
Diameter of compacting surface	2	2
Free fall distance (in.)	12	18
Net volume (ft^3)	1/30 or 1/13.33	1/30 or 1/13.33

Figure 16.12 Effect of Compactive Effort in Dry Density

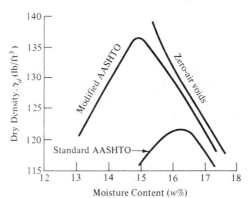

Effect of Compacting Effort

Compacting effort is a measure of the mechanical energy imposed on the soil mass during compaction. In the laboratory it is given in units of ft·lb/in.^{3} or ft·lb/ft^3, whereas in the field it is given in terms of the number of passes of a roller of known weight and type. The compactive effort in the standard AASHTO test, for example, is approximately calculated as

$$\frac{(5.5)\text{lb}(1\text{ ft})(3)(25)}{1/30\text{ ft}^3} = 12,375\text{ ft·lb/ft}^3 \qquad \text{or } 7.16\text{ ft·lb/in.}$$

Note that the optimum moisture content and maximum dry density attained depend on the compactive effort used. Figure 16.12 shows that as the compactive effort increases so does the maximum dry density. Also the compactive effort required to obtain a given density increases as the moisture content of the soil decreases.

Example 16–7   Determining Maximum Dry Density and Optimum Moisture Content

The table below shows results obtained from a standard AASHTO compaction test on six samples, 4 in. in diameter of a soil to be used as fill for a highway. Determine the maximum dry density and the optimum moisture content of the soil.

Sample No.	Weight Compacted Soil, W (lb)	Moisture Content, w (%)
1	4.16	4.0
2	4.39	6.1
3	4.60	7.8
4	4.68	10.1
5	4.57	12.1
6	4.47	14.0

Figure 16.13   Moisture–Density Relationship for Example 16–7

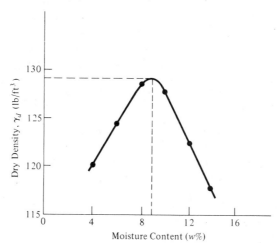

Since we are using the standard AASHTO test, 4 in. in diameter, the volume of each sample is $1/30$ ft^3. The calculated dry densities are calculated as shown below.

Sample No.	Bulk density, γ 30W (lb/ft^3)	Moisture Content, w (%)	Dry Density, γ_d lb/ft^3 $\left(\dfrac{\gamma}{1+w}\right)$
1	124.80	4.0	120.0
2	131.70	6.1	124.1
3	138.00	7.8	128.0
4	140.40	10.1	127.5
5	137.10	12.0	122.4
6	134.10	14.0	117.6

Figure 16.13 shows the plot of dry density versus moisture content, from which it is determined that maximum dry density is 129 lb/ft^3 and the optimum moisture content is 9 percent.

Field Compaction Procedures and Equipment

A brief description of the field compaction procedures and the equipment used for embankment compaction is useful at this point as it enables the reader to see how the theory of compaction is applied in the field.

Field Compaction Procedures

The first step in the construction of a highway embankment is the identification and selection of a suitable material. This is done by obtaining samples from economically feasible borrow pits or borrow areas and testing them in the laboratory to determine the group of each. It has been shown earlier that, based on the AASHTO system of classification, materials classified as A-1, A-2-4, A-2-5, and A-3 are usually suitable embankment materials. In cases where it is necessary to use materials in other groups, special consideration should be given to the design and construction. For example, soils in groups A-4 and A-6 can be used for embankment construction if the embankment height is low, the field compaction process is carefully controlled, and the embankment is located where the moisture content is not expected to exceed that at which the construction was undertaken. A factor that also significantly influences the selection of any material is whether that material can be economically transported to the construction site. Having identified suitable materials, their optimum moisture contents and maximum dry densities are determined.

Embankment Formation.   Highway embankments are formed by spreading thin layers of uniform thickness of the material and compacting each layer at or near the optimum moisture content. End dumping of the material from trucks is not recommended. The process of constructing one layer at a time facilitates obtaining uniform strength and moisture content in the embankment. End dumping or compaction of thick layers, on the other hand, may result in variable strengths within the embankment, which could lead to differential settlement between adjacent areas.

Most states stipulate a thickness of 6 to 12 in. for each layer, although the thickness may be increased to 24 in. when the lower portion of an embankment consists mainly of large boulders.

All transportation agencies have their own requirements for the minimum density in the field. Some of these are based on the AASHTO specifications for transportation materials.[7] Table 16.7 gives AASHTO recommendations in terms of the relative density, which is given as a percentage of the maximum dry density obtained from the standard AASHTO (T99) test. Some agencies base their requirements on the maximum dry density obtained in the laboratory. For example, when the maximum dry density obtained in the laboratory is less than $100 \, \text{lb/ft}^3$, the required field density is 100 percent of the laboratory density. When the maximum dry density obtained in the laboratory is $100 \, \text{lb/ft}^3$ or greater but less than $120 \, \text{lb/ft}^3$, 95 percent is required in the field, and so forth. The former practice of specifying the number of passes for different types of compacting equipment is not widely used at the present time.

Some transportation agencies also have specifications for the moisture content to be used during compaction. These specifications are usually given in general terms, although limits above and below the optimum moisture content have been given.

Control of Embankment Construction

The construction control of an embankment entails frequent and regular checks of the dry density and the moisture content of materials being compacted. The bulk density is obtained directly from measurements obtained in the field and the

Table 16.7 AASHTO-Recommended Minimum Requirements for Compaction of Embankments and Subgrades

	Minimum Relative Density		
	Embankments		
AASHTO Class of Soil	Height Less Than 50 ft	Height Greater Than 50 ft	Subgrade
A-1, A-3	≥ 95	≥ 95	100
A-2-4, A-2-5	≥ 95	≥ 95	100
A-2-6, A-2-7	> 95	—[a]	≥ 95[b]
A-4, A-5, A-6, A-7	> 95	—[a]	≥ 95[b]

[a]Use of these materials requires special attention to design and construction.

[b]Compaction at 95 percent of T99 moisture content.

Source: Adapted from *Standard Specifications for Transportation Materials and Methods of Sampling and Testing*, 14th ed., Washington, D.C.: The American Association of State Highway and Transportation Officials, copyright 1986. Used by permission.

dry density is then calculated from the bulk density and the moisture content. The laboratory moisture–density curve is then used to determine whether the dry density obtained in the field is in accordance with the laboratory results for the compactive effort used. These tests are conducted by using either a destructive method or a nondestructive method.

Destructive Methods. In determining the bulk density by the destructive method, a cylindrical hole of about 4 in. in diameter and a depth equal to that of the layer is excavated. The material obtained from the hole is immediately sealed in a container. Care should be taken not to lose any of the excavated material. The total weight of the excavated material is obtained, usually in the field laboratory, and the moisture content determined. The compacted volume of the excavated material is then measured by determining the volume of the excavated hole.

The moisture content is determined by either rapidly drying the soil in a field oven or by facilitating evaporation of the moisture by adding some volatile solvent material such as alcohol and igniting it. The volume of the excavated hole may be obtained by one of three methods: sand replacement, oil, or balloon. In the sand replacement method, the excavated hole is carefully filled with standard sand from a jar originally filled with the standard sand. The jar can be opened and closed by a valve. When the hole is completely filled with sand, the valve is closed and the weight of the remaining sand in the jar is determined. The weight of the quantity of sand used to fill the excavated hole is then obtained by subtracting the weight of the sand remaining in the jar from the weight of the sand required to fill the jar. The volume of the quantity of standard sand used to fill the excavated hole (that is, volume of hole) is then obtained from a previously established relationship between the weight and volume of the standard sand.

In the oil method, the volume of the hole is obtained by filling the excavated hole with a heavy oil of known specific gravity.

In the balloon method, a balloon is placed in the excavated hole and then filled with water. The volume of water required to fill the hole is the volume of the excavated hole.

The destructive methods are all subject to errors. For example, in the soil replacement method, adjacent vibration will increase the density of the sand in the excavated hole, thereby indicating a larger volume hole. Large errors in the volume of the hole will be obtained if the balloon method is used in holes having uneven walls, and large errors may be obtained if the heavy oil method is used in coarse sand or gravel material.

Nondestructive Method. The nondestructive method involves the direct measurement of the in-situ density and moisture content of the compacted soil, using nuclear equipment. The density is obtained by measuring the scatter of gamma radiation by the soil particles since the amount of scatter is proportional to the bulk density of the soil. A calibration curve for the particular equipment is then used to determine the bulk density of the soil. A plot of the amount of scatter in materials of known density measured by the equipment versus the density of the materials is generally used in the calibration curve. The moisture content is also obtained by measuring the scatter of neutrons emitted in the soil. This scatter is due mainly to the presence of hydrogen atoms. The assumption is made that most of the hydrogen is in the form of water, which allows the amount of scatter to be related to the amount of water in the soil. The moisture content is then obtained directly from a calibrated gauge.

One advantage of the nondestructive method is that results are obtained speedily, which is essential if corrective actions are necessary. Another advantage is that more tests can be carried out, which facilitates the use of statistical methods in the control process. The main disadvantages are that a relatively high capital expenditure is required to obtain the equipment, and field personnel are exposed to dangerous radioactive material, making it imperative that strict safety standards be enforced when nuclear equipment is used.

Field Compaction Equipment

Compaction equipment used in the field can be divided into two main categories. The first category includes the equipment used for speading the material to the desired layer or lift thickness, and the second category includes the equipment used to compact each layer of material.

Spreading Equipment. Spreading of the material to the required thickness is done by bulldozers and motor graders. Several types and sizes of graders and dozers are now available on the market. The equipment used for any specific project will depend on the size of the project. A typical motor grader is shown in Figure 16.14.

Compacting Equipment. Rollers are used for field compaction and apply either a vibrating force or an impact force on the soil. The type of roller used for any particular job depends on the type of soil to be compacted.

Figure 16.14 Typical Motor Grader

Source: Photo courtesy of Machinery Distribution, Inc., Mitsubishi Construction Equipment, Houston, Tex.

A smooth wheel or drum roller applies contact pressure of up to 55 lb/in.^{2} over 100 percent of the soil area in contact with the wheel. This type of roller is generally used for finish rolling of subgrade material and can be used for all types of soil material except rocky soils. Figure 16.15(a) shows a typical smooth wheel roller. The rubber-tired roller is another type of contact roller, consisting of a heavily loaded wagon with rows of four to six tires placed close to each other. The pressure in the tires may be up to 100 lb/in.^{2} They are used for both granular and cohesive materials. Figure 16.15(b) shows a typical rubber-tired roller.

One of the most frequently used rollers is the sheepsfoot. This roller has a drum wheel that can be filled with water. The drum wheel has several protrusions, which may be round or rectangular in shape, ranging from 5 to 12 in.^{2} in area. The protrusions penetrate the loose soil and compact from the bottom to the top of each layer of soil, as the number of passes increases. Contact pressures ranging from 200 to 1000 lb/in.^{2} can be obtained from sheepsfoot rollers, depending on the size of the drum and whether or not it is filled with water. The sheepsfoot roller is used mainly for cohesive soils. Figure 16.16 shows a typical sheepsfoot roller.

Tamping foot rollers are similar to sheepsfoot rollers in that they also have protrusions that are used to obtain high contact pressures, ranging from 200 to 1200 lb/in.^{2}. The feet of the tamping foot rollers are specially hinged to obtain a kneading action while compacting the soil. As with sheepsfoot rollers, tamping foot rollers compact from the bottom of the soil layer. Tamping foot rollers are mainly used for compacting fine-grained soils.

Figure 16.15   Typical Smooth Wheel and Rubber Tired Rollers

(a) Smooth wheel roller

Source: Photo courtesy of BOMAG-USA, Springfield, Ohio.

(b) Rubber-tired roller

Source: Photo courtesy of Caterpillar, Inc., Peoria, Ill.

Figure 16.16 Typical Sheepsfoot Roller

Source: Photo courtesy of Caterpillar, Inc., Peoria, Ill.

The smooth wheel and tamping foot rollers can be altered to vibrating rollers by attaching a vertical vibrator to the drum, producing a vibrating effect that makes the smooth wheel and tamping foot rollers more effective on granular soils. Vibrating plates and hammers are also available for use in areas where the larger drum rollers cannot be used.

SPECIAL SOIL TESTS FOR PAVEMENT DESIGN

Apart from the tests discussed so far, there are a few special soil tests that are sometimes undertaken to determine the strength or supporting value of a given soil if used as a subgrade or subbase material. The results obtained from these tests are used individually in the design of some pavements, depending on the pavement design method used (see Chapter 18). The two most commonly used tests under this category are the California Bearing Ratio Test and Hveem Stabilometer Test.

California Bearing Ratio (CBR) Test

This test is commonly known as the CBR test and involves the determination of the load-deformation curve of the soil in the laboratory using the standard CBR testing equipment shown in Figure 16.17. It was originally developed by the California Division of Highways prior to World War II and used in the design of some highway pavements. The test has now been modified and is standardized under the AASHTO designation of T193. The test is conducted as follows.

Step 1. Disturbed samples of the selected soil material are compacted at different moisture contents in molds of 6 in. diameter and about 6 in. high using the

Figure 16.17   CBR Testing Equipment

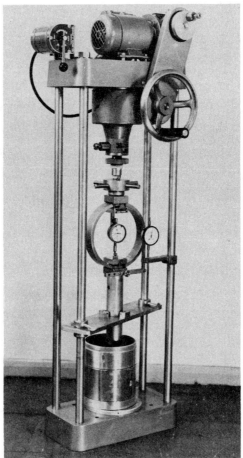

Source: Wykeham Farrancè, Inc., Raleigh, N.C., 1983.

standard AASHTO compacting method. The curve of dry density versus moisture content is then plotted. The sample having the highest dry density is selected for the CBR test.

Step 2.   The selected compacted sample, still in the mold, is immersed in water for four days to obtain a saturation condition similar to what may occur in the field. During this period, the sample is loaded with a surcharge, usually 10 lb or greater, that simulates the estimated weight of pavement material the soil will support. Any expansion of the soil sample due to soaking is measured.

Step 3.   The sample is then removed from the water and allowed to drain for about 15 min.

Step 4. The sample still carrying the surcharge weight is then subjected to penetration by the piston of the standard CBR equipment. The loads that cause different penetrations are recorded (in lb/in.²), and a load-penetration curve drawn. The CBR is then determined as

$$\text{CBR} = \frac{\text{(unit load for 0.1 piston penetration in test specimen) (lb/in.}^2\text{)}}{\text{(unit load for 0.1 piston penetration in standard crushed rock)}} \times 100 \quad \textbf{(16.24)}$$

The unit load for 0.1 piston in standard crushed rock is usually taken as 1000 lb/in.², which gives the CBR as

$$\text{CBR} = \frac{\text{(unit load for 0.1 piston penetration in test sample}}{1000} \times 100 \quad \textbf{(16.25)}$$

Eqs. 16.24 and 16.25 show that the CBR gives the relative strength of a soil with respect to crushed rock, which is considered as an excellent coarse base material.

The main criticism of the CBR test is that it does not correctly simulate the shearing forces imposed on subbase and subgrade materials as they support highway pavements. For example, it is possible to obtain a relatively high CBR value for a soil containing rough or angular coarse material and some amount of troublesome clay if the coarse material resists penetration of the piston by keeping together in the mold. When such a material is used in highway construction, however, the performance of the soil may be poor, due to the lubrication of the soil mass by the clay, which reduces the shearing strength of the soil mass.

Hveem Stabilometer Test

This test is used to determine the resistance value R of the soil to the horizontal pressure obtained by imposing a vertical stress of 160 lb/in. on a sample of the soil. The value of R may then be used to determine the pavement thickness above the soil to carry the estimated traffic load. (See Chapter 18.) The test was first conceived in the 1930s and was used to obtain the stability of laboratory and field samples of bituminous pavements. (See Chapter 17.) It has now been modified and made suitable for subgrade materials and is designated as T190 by AASHTO. The current test procedure includes the determination of the swelling potential of subgrade materials due to the absorption of water. The test procedure described here is that used by the California Department of Transportation, and is referred to as Test Method 301 in their materials manual. The procedure consists of three phases: determination of the exudation pressure, determination of the expansion pressure, and determination of the resistance value R (stabilometer test). In order to carry out the test, four cylindrical specimens of 4 in. diameter and 2.5 in. high are prepared at different moisture contents by compacting samples of the soils in steel molds. The compaction is achieved by tamping or kneading the soil.

Exudation Pressure. This is the compressive stress that will exude water from the compacted specimen. Each specimen is pressed in the steel mold by applying a vertical load until water exudes from the soil. The base of the exudation equipment has several electrical circuits wired into it in parallel, and the exuded water completes these circuits. The pressure that exudes enough water to activate five or six of these circuits is the exudation pressure. Several tests in California have indicated that soils supporting highway pavements will exude moisture under pressure of about 300 lb/in.^{2}. The moisture content of the stabilometer test (R value) samples is therefore set to that moisture content that exudes water at a pressure of 300 lb/in.^{2}.

Expansion Pressure. At the completion of the exudation test, a perforated brass plate is placed on each sample in the steel mold and covered with water. The samples are then left to stand for a period of 16–20 hr, during which time any expansion of the soil is prevented by applying a load. The pressure that prevents any expansion is measured and recorded as the expansion pressure.

This pressure indicates the load and therefore the thickness of material required above this soil to prevent any swelling if the soil is inundated with water when used as a subgrade material.

Resistance Value, *R*. A schematic of the Hveem stabilometer used in this phase of the test is shown in Figure 16.18. At the completion of the expansion test, the specimen is put into a flexible sleeve and placed in the stabilometer as shown in the figure. Vertical pressure is applied gradually on the specimen at a speed of 0.05 in./min until a pressure of 160 lb/in.^{2} is attained. The corresponding horizontal pressure is immediately recorded. To correct for any distortion of the results due to the surface roughness of the sample, the penetration of the flexible diaphragm into the sample is measured. This is done by reducing the vertical load on the specimen by half and also reducing the horizontal pressure to 5 lb/in.^{2},

Figure 16.18 Hveem Stabilometer

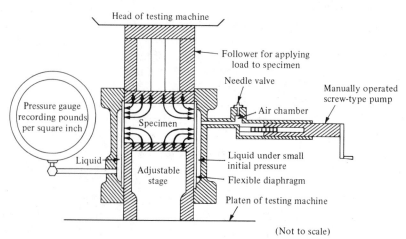

(Not to scale)

using the screw-type pump. The number of turns of the pump required to increase the horizontal pressure to 100 lb/in.^{2} is then recorded. The soil's resistance value is given as

$$R = 100 - \frac{100}{\dfrac{2.5}{D}\left(\dfrac{P_v}{P_h} - 1\right) + 1} \tag{16.26}$$

where

R = resistance value

P_v = vertical pressure (160 lb/in.^{2})

P_h = horizontal pressure at P_v of 160 lb/in.^{2} (lb/in.^{2})

D = number of turns of displacement pump

FROST ACTION IN SOILS

When the ambient temperature falls below freezing for several days, it is quite likely that the water in soil pores will freeze. Since the volume of water increases by about 10 percent when it freezes, the first problem is the increase in volume of the soil. The second problem is that the freezing can cause ice crystals and lenses that can develop to several centimeters thick to form in the soil. These two problems can result in heaving of the subgrade (frost heave), which may result in significant structural damage to the pavement.

In addition, the ice lenses melt during the spring (spring thaw), resulting in a considerable increase in the water content of the soil. This increase in water significantly reduces the strength of the soil, causing structural damage of the highway pavement known as "spring break up."

In general three conditions must exist for severe frost action to occur:

1. Ambient temperature must be lower than freezing for several days.
2. The shallow water table that provides capillary water to the frost line must be available.
3. The soil must be susceptible to frost action.

The first condition is a natural phenomenon and cannot be controlled by humans. Frost action will therefore be more common in cold areas than in warm areas if all other conditions are the same. The second condition requires that the ground water table be within the height of the capillary rise, so that water will be continuously fed to the growing ice lenses.

The third condition requires that the soil material be of such quality that relatively high capillary pressures can be developed, but at the same time the flow of water through its pores is restricted. Granular soils are therefore not susceptible to frost action because they have a relatively high coefficient of permeability. Clay soils are also not highly susceptible to frost action because they have very low permeability so that not enough water can flow during a freezing period to allow the formulation of ice lenses. Sandy or silty clays or cracked clay soils near the surface may, however, be susceptible to frost action. Silty soils are most susceptible to frost action. It has been determined that 0.02 mm is the critical grain size for frost susceptibility. For example, gravels with 5 percent of 0.02 mm particles are in general susceptible to frost action, whereas well-graded soils with only 3 percent by weight of their material finer than 0.02 mm are susceptible, and fairly uniform soils must contain at least 10 percent of 0.02 mm particles to be frost susceptible. Soils with less than 1 percent of their material finer than the critical size are rarely affected by frost action.

Current measures taken to prevent frost action, as discussed in Chapter 15, include removal of frost susceptible soils to the depth of the frost line and replacing them with gravel material, lowering the water table by installing adequate drainage facilities, use of impervious membranes or chemical additives, and restriction of truck traffic on some roads during the spring thaw.

SUMMARY

Selection of suitable soils to be used as the foundation for the highway pavement surface is of primary importance in the design and construction of any highway. Use of unsuitable material will often result in premature failure of the pavement surface and reduction of the ability of the pavement to carry the design traffic load. Chapters 18 and 19 will show that the types of material used for the base, subbase, and/or subgrade of a highway significantly influence the depth of these materials used and also the thickness of the pavement·surface.

It is therefore important that transportation engineers involved in the design and/or maintenance of highway pavements be familiar with the engineering properties of soils and the procedures through which the suitability of any soil for highway construction can be determined. A summary of some of the current procedures and techniques used for soil testing and identification is presented in this chapter. The techniques presented are those currently used in highway pavement design.

One of the most important tasks in highway pavement construction is the control of embankment compaction, and discussion of the different methods used in the control of embankment compaction is presented. A brief discussion of the different types of equipment used in the compaction of embankments is also presented to cover all aspects of embankment compaction.

PROBLEMS

16–1 The weight of a sample of saturated soil before drying is 3 lb and after drying is 2.2 lb. If the specific gravity of the soil particles is 2.7, determine: (a) moisture content, (b) void ratio, (c) porosity, (d) bulk density, and (e) dry density.

16–2 Detemine the void ratio of a soil if its bulk density is 120 lb/ft³ and has a moisture content of 25 percent. The specific gravity of the soil particles is 2.7. Also determine its dry density and degree of saturation.

16–3 Given a soil that has a bulk density of 135 lb/ft³ and a dry density of 120 lb/ft³, and the specific gravity of the soil particles is 2.75, determine: **(a)** moisture content, **(b)** degree of saturation, **(c)** void ratio, and **(d)** porosity.

16–4 Derive from first principles an expression in terms of γ, γ_s, γ_w, and w for **(a)** dry density, **(b)** void ratio, and **(c)** degree of saturation.

16–5 A liquid limit test conducted in the laboratory on a sample of soil gave the following results listed below. Determine the liquid limit of this soil from a plot of the flow curve.

Number of Blows	Moisture Content (%)
40	42.0
35	42.8
30	43.2
28	43.6
20	45.0

16–6 The results obtained from the mechanical analysis of six different soils is shown in the table below. Using a suitable semilogarithmic paper, plot the grain size distribution curve and classify each soil by the AASHTO method. Which of these soils are suitable for subbase material? State the reasons for your conclusion in each case.

U.S. Standard No. or Particle Size	Percentage Passing					
	Soil A	Soil B	Soil C	Soil D	Soil E	Soil F
75 mm (3 in.)	100	100	100			
38 mm (1 ½ in.)	75	98	100			
19 mm (¾ in.)	52	95	98			
No. 4	35	90	90	100		
No. 10	30	85	83	99		
No. 20	18	78	75	97	100	
No. 40	15	75	65	80	99	
No. 50	12	73	45	60	80	
No. 100	8	70	40	10	70	100
No. 140	6	64	36	5	65	100
No. 200	5	40	30	< 1	40	100
0.040 mm	4	28	24	—	20	98
0.020 mm	3	20	18	—	10	95
0.010 mm	2	14	15	—	8	
0.005 mm	1	12	12	—	6	80
0.002 mm	< 1	8	11	—	2	50
0.001 mm	—	2	10	—	< 1	40
Liquid limit	10	42	35	—	38	43
Plasticity index	4	9	12	N.P.	8	12

16–7 Repeat problem 16–6 using the USCS.

16–8 The results listed below are from a compaction test on samples of a soil that is to be used for embankment construction on a highway. Draw the dry density versus moisture content curve for this soil to determine its optimum moisture content. Draw the zero-air voids curve and comment on the distance between this curve and the dry density versus moisture content curve. If the specifications on this project call for compaction to a relative density of 95 percent, what is the probable moisture content if compaction is carried out in the field? Give reasons for your selection.

Sample No.	Moisture Content (% by wt. of dry soil)	Bulk Density (lb/ft²)
1	4.8	135.1
2	7.5	145.0
3	7.8	146.8
4	8.91	146.4
5	9.72	145.3

16–9 Describe the steps you will take to identify and select suitable subbase material for an urban expressway in your area.

16–10 The results obtained from a seismic study along a section of the center line of a highway are shown below. Estimate the depths of the different strata of soil and suggest the type of soil in each stratum.

Distance of Impulse to Geophone (ft)	Time for Wave Arrival (10^{-3} sec)
25	20
50	40
75	60
100	68
125	74
150	82
175	84
200	86
225	88
250	90

REFERENCES

1. Hans F. Winterkorn and Hsai-Yang Fang, *Foundation Engineering Handbook*, Van Nostrand Reinhold, New York, 1975.

2. K. Terzaghi and R. B. Peck, *Soil Mechanics in Engineering Practice*, John Wiley & Sons, New York, 1967.

3. William T. Lambe, *Soil Testing for Engineers*, John Wiley & Sons, New York, 1957.

4. Das Braja, *Soil Mechanics Laboratory Manual*, Engineering Press, 1982.

5. Winterkorn and Fang, *Foundation Engineering Handbook*.

6. *Standard Specifications for Transportation Materials and Methods of Sampling Testing*, 14th ed., American Association of State Highways and Transportation Officials, Washington, D.C., 1986.

7. *Standard Specifications for Transportation Materials*.

Bituminous Materials

Bituminous materials are widely used all over the world in highway construction. These hydrocarbons are found in natural deposits or are obtained as a product of the distillation of crude petroleum. The bituminous materials used in highway construction are either asphalts or tars.

All bituminous materials consist primarily of bitumen, have strong adhesive properties, and have colors ranging from dark brown to black. They vary in consistency from liquid to solid; thus, they are divided into liquids, semisolids, and solids. The solid form is usually hard and brittle at normal temperatures but will flow when subjected to long, continuous loading.

The liquid form is obtained from the semisolid or solid forms by heating, dissolving in solvents, or breaking the material into minute particles and dispersing them in water with an emulsifier to form asphalt emulsion.

This chapter presents a description of the different types of bituminous materials used in highway construction, the processes by which they are obtained, and the tests required to determine those properties that are pertinent to highway engineering. The chapter also includes a description of a commonly used method of mix design to obtain a paving material known as asphaltic concrete.

SOURCES OF ASPHALTIC MATERIALS

Asphaltic materials are obtained from seeps or pools of natural deposits in different parts of the world or as a product of the distillation of crude petroleum.

Natural Deposits

Natural deposits of asphaltic materials occur as native asphalts or as rock asphalts. The largest deposit of native asphalt is known to have existed in Iraq several thousand years ago. Native asphalts have also been found on the island of Trinidad, Bermudez, and the La Brea asphalt pits in Los Angeles, California.

Native asphalts, after being softened with petroleum fluxes, were at one time used extensively as binders in highway construction. The properties of native asphalts vary from one deposit to another, particularly with respect to the amount of insoluble material the asphalt contains. The Trinidad deposit, for example, contains about 40 percent insoluble organic and inorganic materials, whereas the Bermudez material contains about 6 percent of such material.

Rock asphalts are natural deposits of sandstone or limestone rocks filled with asphalt. Deposits have been found in California, Texas, Oklahoma, and Alabama. The amount of asphaltic material varies from one deposit to another and can be as low as 4.5 percent and as high as 18 percent. Rock asphalt can be used to surface roads, after the mined or quarried material has been suitably processed. This process includes adding suitable mineral aggregates, asphaltic binder, and oil, which facilitates the flowing of the material. Rock asphalt is not widely used because of its high transportation costs.

Petroleum Asphaltic Materials

The asphaltic materials obtained from the distillation of petroleum are in the form of different types of asphalts. These include asphalt cements, slow-curing liquid asphalts, medium-curing liquid asphalts, rapid-curing liquid asphalts, and asphalt emulsions. The quantity of asphalt obtained from crude petroleum is dependent on the American Petroleum Institute (API) gravity of the petroleum. In general, large quantitites of asphalt are obtained from crude petroleums with low API gravity. Before discussing the properties and uses of the different types of petroleum asphalts, we first describe the refining processes used to obtain them.

Refining Processes

The refining processes used to obtain petroleum asphalts can be divided into two main groups, namely, fractional distillation and destructive distillation (cracking). The fractional distillation processes involve the separation of the different materials in the crude petroleum without significant changes in the chemical composition of each material. The destructive distillation processes involve the application of high temperature and pressure, resulting in chemical changes.

Fractional Distillation. The fractional distillation process removes the different volatile materials in the crude oil at successively higher temperatures until the petroleum asphalt is obtained as residue. Steam or vacuum is used to gradually increase the temperature.

Steam distillation is a continuous flow process in which the crude petroleum is pumped through tube stills or stored in batches, and the temperature is increased gradually to facilitate the evaporation of the different materials at different temperatures. Tube stills are more efficient than batches and are therefore preferred in modern refineries.

Figure 17.1 is a flow chart that shows the interrelationships of the different materials that can be obtained from the fractional distillation of crude petroleum. Immediately after increasing the temperature of the crude in the tube still, it is injected into a bubble tower, which consists of a vertical cylinder into which are built several trays or platforms one above the other. The first separation of

Figure 17.1 Simplified Flow Chart of Recovery and Refining of Petroleum Asphalt

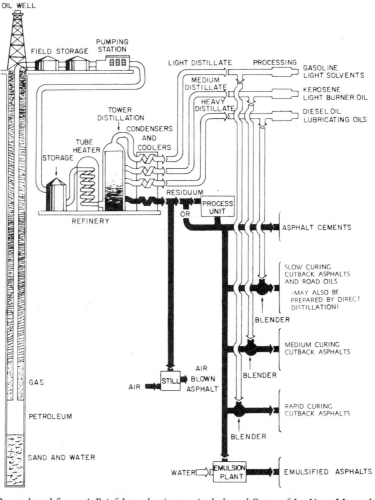

Source: Reproduced from *A Brief Introduction to Asphalt and Some of Its Uses*, Manual Series No. 5, The Asphalt Institute, College Park, Md., 1977.

materials occurs in this tower. The lighter fractions of the evaporated materials collect on the top tray and the heavier fractions collect in successive trays, with the heaviest residue containing asphalt remaining at the bottom of the distillation tower. The products obtained during this first phase of separation are gasoline, kerosene distillate, diesel fuel, lubricating oils, and the heavy residual material that contains the asphalt (see Figure 17.1). The various fractions collected are stored and further refined into specific grades of petroleum products. Note that a desired consistency of the residue can be obtained by continuing the distillation process. Attainment of the desired consistency is checked by measuring the temperature of the residue or observing the character of the distillate. The residue becomes harder the longer the distillation process is continued.

It can be seen from Figure 17.1 that further processing of the heavy residue obtained after the first separation will give asphalt cement of different penetration grades—slow-curing, and rapid-curing asphalts—depending on the additional processing carried out. Emulsified asphalts can also be obtained. A description of each of these different types of asphalt materials will be given later.

Cracking processes are used when larger amounts of the light fractions of the materials such as motor fuels are required. Intense heat and high pressures are applied to produce chemical changes in the material. Although several specific methods of cracking exist, the process generally involves the application of temperatures as high as 1100°F and pressure higher than 735 lb/in.^{2} to obtain the desired effect. The asphaltic material obtained from cracking is not widely used in paving because it is more susceptible to weather changes than is that produced from fractional distillation.

DESCRIPTION AND USES OF BITUMINOUS BINDERS

It is necessary at this point to describe the different bituminous binders and identify the type of construction for which each is used. The commonly used bituminous binders are asphalt cement, slow-curing asphalts, medium-curing cutback asphalts, rapid-curing cutback asphalts, and asphalt emulsion. Road tars are also a type of bituminous material but are not now widely used in highway construction.

Asphalt Cements

Asphalt cements are obtained after separation of the lubricating oils as shown in Figure 17.1. They are semisolid hydrocarbons with certain physiochemical characteristics that make them good cementing agents. They are also very viscous, and when used as a binder for aggregates in pavement construction, it is necessary to heat both the aggregates and the asphalt cement prior to mixing the two materials. For several decades the particular grade of asphalt cement has been designated by its penetration and viscosity, both of which give an indication of the consistency of the material at a given temperature. The penetration is the distance in 1/10 of mm that a standard needle will penetrate a given sample, under specific conditions of loading, time, and temperature. The softest grade used for highway pavement construction has a penetration value of 200–300 and the hardest has a penetration value of 60–70. Recently, however, viscosity has been used more often than penetration to grade asphalt cements.

Asphalt cements are used for a variety of purposes, as shown in Table 17.1. The use of a given sample depends on its grade. The procedures for determining the grades of asphalt cements through standard penetration and viscosity tests will be described in subsequent sections of this chapter.

Slow-Curing Asphalts

Slow-curing (SC) asphalts can be obtained directly as *slow-curing straight run asphalts* through the distillation of crude petroleum, as shown in Figure 17.1, or as *slow-curing cutback asphalts* by *cutting back* asphalt cement with a heavy distillate such as diesel oil. They have lower viscosities than asphalt cement and are very slow to harden. Slow-curing asphalts are usually designated as SC-70, SC-250, SC-800, or SC-3000, where the numbers are related to the approximate kinematic viscosity in centistokes at 60°C (140°F). Their uses are also shown in Table 17.1. Specifications for the use of these asphalts are not now included in American Association of State Highway and Transportation Official's (AASHTO) *Standard Specification for Transportation Materials.*[1]

Medium-Curing Cutback Asphalts

Medium-curing (MC) asphalts are produced by *fluxing* or *cutting back*, the residual asphalt (usually 120–150 penetration) with light fuel oil or kerosene. The term *medium* refers to the medium volatility of the kerosene type dilutent used. Medium-curing cutback asphalts harden faster than slow-curing liquid asphalts, although the consistencies of the different grades are similar to those of the slow-curing asphalts. However, the MC-30 is a unique grade in this series as it is very fluid and has no counterpart in the SC and RC series.

The fluidity of medium-curing asphalts depends on the amount of solvent in the material. MC-3000, for example, may have only 20 percent of the solvent by volume, whereas MC-70 may have up to 45 percent. These medium-curing asphalts can be used for the construction of pavement bases, surfaces, and surface treatments, as shown in Table 17.1.

Rapid-Curing Cutback Asphalts

Rapid-curing (RC) cutback asphalts are produced by blending asphalt cement with a petroleum distillate that will easily evaporate, thereby facilitating a quick change from the liquid form at time of application to the consistency of the original asphalt cement. Gasoline or naphtha is generally used as the solvent for this series of asphalts.

The grade of rapid-curing asphalt required dictates the amount of solvent to be added to the residual asphalt cement. For example, RC-3000 requires about 15 percent of distillate, whereas RC-70 requires about 40 percent. These grades of asphalt can be used for jobs similar to those for which the MC series is used. Specifications for the use of these asphalts are given in AASHTO's *Standard Specifications for Transportation Materials.*[2]

Blown Asphalts

Blown asphalt is obtained by blowing air through the semisolid residue obtained during the latter stages of the distillation process. The process involves stopping

Table 17.1 Typical Uses of Asphalt

TYPE OF CONSTRUCTION	ASPHALT CEMENTS														
	VISCOSITY GRADED -ORIGINAL					VISCOSITY GRADED -RESIDUE					PENETRATION GRADED				
	AC-40	AC-20	AC-10	AC-5	AC-2.5	AR-16000	AR-8000	AR-4000	AR-2000	AR-1000	40-50	60-70	85-100	120-150	200-300
ASPHALT-AGGREGATE MIXTURES															
ASPHALT CONCRETE AND HOT LAID PLANT MIX															
PAVEMENT BASE AND SURFACES															
HIGHWAYS		X	X	X	X¹	X	X	X	X^A	X¹		X	X	X	X¹
AIRPORTS		X	X	X			X	X				X	X		
PARKING AREAS		X	X			X	X	X				X	X		
DRIVEWAYS		X	X				X	X				X	X		
CURBS	X	X	X			X	X				X	X	X		
INDUSTRIAL FLOORS	X	X				X	X				X	X			
BLOCKS	X										X				
GROINS	X	X									X	X			
DAM FACINGS	X	X									X	X			
CANAL AND RESERVOIR LININGS	X	X									X	X			
COLD-LAID PLANT MIX															
PAVEMENT BASE AND SURFACES															
OPEN-GRADED AGGREGATE*															
WELL-GRADED AGGREGATE*															
PATCHING, IMMEDIATE USE															
PATCHING, STOCKPILE															
MIXED-IN-PLACE (ROAD MIX)															
PAVEMENT BASE AND SURFACES															
OPEN-GRADED AGGREGATE*															
WELL-GRADED AGGREGATE*															
SAND				X	X									X	X
SANDY SOIL				X	X									X	X
PATCHING, IMMEDIATE USE															
PATCHING, STOCKPILE															
ASPHALT-AGGREGATE APPLICATIONS															
SURFACE TREATMENTS															
SINGLE SURFACE TREATMENT				X	X									X	X
MULTIPLE SURFACE TREATMENT				X	X									X	X
AGGREGATE SEAL				X	X				X					X	X
SAND SEAL															
SLURRY SEAL															
PENETRATION MACADAM															
PAVEMENT BASES															
LARGE VOIDS			X										X		
SMALL VOIDS				X										X	
ASPHALT APPLICATIONS															
SURFACE TREATMENT															
FOG SEAL															
PRIME COAT, OPEN SURFACES															
PRIME COAT, TIGHT SURFACES															
TACK COAT															
DUST LAYING															
MULCH															
MEMBRANE															
CANAL AND RESERVOIR LININGS	X										X				
EMBANKMENT ENVELOPES	X	X									X	X			
CRACK FILLING															
ASPHALT PAVEMENTS															
PORTLAND CEMENT CONCRETE PAVEMENTS	X⁴										X⁴				

* Evaluation of emulsified asphalt-aggregate system required
to determine the proper grade of emulsified asphalt to use.

Source: Reproduced from *A Brief Introduction to Asphalt and Some of Its Uses*, Manual Series No. 5, The Asphalt Institute, College Park, Md., 1977.

Table 17.1—*Continued*

Column groups: **CUTBACK ASPHALTS** — RAPID CURING (RC), MEDIUM CURING (MC), SLOW CURING (SC); **EMULSIFIED ASPHALTS#** — ANIONIC, CATIONIC.

RC 70	RC 250	RC 800	RC 3000	MC 30	MC 70	MC 250	MC 800	MC 3000	SC 70	SC 250	SC 800	SC 3000	RS-1	RS-2	MS-1	MS-2	MS-2h	SS-1	SS-1h	CRS-1	CRS-2	CMS-2	CMS-2h	CSS-1	CSS-1h
						X							X	X	X					X	X				
	X				X	X	X		X	X	X							X	X		X			X	X
	X	X			X	X	X			X	X							X	X					X	X
					X	X			X	X															
	X	X	X				X	X	X	X			X	X	X					X	X				
	X						X	X	X	X								X	X		X			X	X
	X	X			X	X	X											X	X					X	X
X	X	X					X											X	X					X	X
	X	X			X	X	X			X	X							X	X					X	X
					X	X			X	X								X	X					X	X
	X	X	X				X	X					X	X						X	X				
	X	X	X					X					X	X						X	X				
	X	X	X				X	X					X	X						X	X				
	X					X	X						X	X						X	X				
																X	X						X	X	
		X	X										X	X						X	X				
	X												X	X						X	X				
																		X[2]	X[2]					X[2]	X[2]
X	X				X	X																			
X					X													X							
X													X					X[2]	X[2]		X			X[2]	X[2]
						X												X[2]	X[2]					X[2]	X[2]
																		X[2]	X[2]						
																		X[3]	X[3]					X[3]	X[3]

1   FOR USE IN COLD CLIMATES
   A   FOR USE IN BASES ONLY IN COLD CLIMATES
2   DILUTED WITH WATER
3   SLURRY MIX
4   RUBBER ASPHALT COMPOUNDS
\#   Emulsified asphalts shown are AASHTO and ASTM grades and may not include all grades produced in all geographical areas.

the regular distillation while the residue is in the liquid form, and then transferring it into a tank known as a converter. The material is then maintained at a high temperature while air is blown through it. This is continued until the required properties are achieved. Blown asphalts are relatively stiff when compared with other types of asphalts and can maintain a firm consistency at the maximum temperature normally experienced when exposed to the environment.

Blown asphalt is not generally used as a paving material. However, it is very useful as a roofing material, for automobile undercoating, and as a joint filler for concrete pavements. If a catalyst is added during the air-blowing process, the material obtained will usually maintain its plastic characteristics even at temperatures much lower than that at which ordinary asphalt cement will become brittle. The elasticity of catalytically blown asphalt is similar to that of rubber, and it is used for canal lining.

Asphalt Emulsion

Emulsified asphalts are produced by breaking asphalt cement, usually of 100–250 penetration range, into minute particles and dispersing them in water with an emulsifier. These minute particles have like electrical charges and therefore do not coalesce. They remain in suspension in the liquid phase as long as the water does not evaporate or the emulsifier does not break. Asphalt emulsions therefore consist of asphalt, which makes up about 55 percent to 70 percent by weight, water, and an emulsifying agent, which in some cases may contain a stabilizer.

Asphalt emulsions are generally classified as anionic, cationic, or nonionic. The first two types have electrical charges surrounding the particles, whereas the third type is neutral. Classification as anionic or cationic is based on the electrical charges that surround the asphalt particles. Emulsions containing negatively charged particles of asphalt are classified as anionic and those having positively charged particles of asphalts are classified as cationic. The anionic and cationic asphalts are generally used in highway maintenance and construction, although it is likely that the nonionics may be used in the future as emulsion technology advances.

Each of the above categories is further divided into three subgroups, based on how rapidly the asphalt emulsion will return to the state of the original asphalt cement. These subgroups are rapid setting (RS), medium setting (MS), and slow setting (SS). A cationic emulsion is identified by placing the letter "C" in front of the emulsion type; no letter is placed in front of anionic and nonionic emulsions. For example, CRS-2 denotes a cationic emulsion, and RS-2 denotes either anionic or nonionic emulsion.

Asphalt emulsions are used for several purposes, as shown in Table 17.1. Note, however, that since anionic emulsions contain negative charges, they are more effective in treating aggregates containing electropositive charges such as limestone, whereas cationic emulsions are more effective with electronegative aggregates such as those containing a high percentage of siliceous material. Also note that ordinary emulsions must be protected during very cold spells as they will break down if frozen. Three grades of high-float, medium-setting anionic emulsions designated as HFMS have been developed and are used mainly in cold and hot plant mixes, road mixes, and coarse aggregate seal coats. These high-float emulsions have one significant property, that is, they can be laid at relatively thicker films, without a high probability of run off.

Specifications for the use of emulsified asphalts are given in AASHTO M140[3] and ASTM D977.[4]

Road Tars

Tars are obtained from the destructive distillation of organic materials such as coal. Their properties are significantly different from petroleum asphalts. In general, they are more susceptible to weather conditions when compared with similar grades of asphalts, and they set more quickly when exposed to the atmosphere. Because tars are rarely used now for highway pavements, only a brief discussion of the subject is included.

The American Society for Testing Materials (ASTM) has classified road tars into three general categories based on the method of production.

1. Gashouse coal tars are produced as a by-product in gashouse retorts in the manufacture of illuminating gas from bituminous coals.
2. Coke-oven tars are produced as a by-product in coke ovens in the manufacture of coke from bituminous coal.
3. Water-gas tars are produced by cracking oil vapors at high temperatures in the manufacture of carburated water gas.

Road tars have also been classified by AASHTO into 14 grades: RT-1 through RT-12, RTCB-5 and RTCB-6. RT-1 has the lightest consistency and can be used effectively at normal temperatures for prime or tack coat (described later in this chapter). The viscosity of each grade increases as the number designation increases to RT-12, which is the most viscous. RTCB-5 and RTCB-6 are suitable for application during cold weather as they are produced by cutting back the specific grade of tar with easily evaporating solvent. Detailed specifications for the use of tars are given by AASHTO Designation M52-78.[5]

PROPERTIES OF ASPHALTIC MATERIALS

The properties of asphaltic materials pertinent to pavement construction can be classified into four main categories:

- Consistency
- Durability
- Rate of curing
- Resistance to water action

Consistency

The consistency properties of an asphaltic material are usually considered under two conditions: (1) variation of consistency with temperature, and (2) consistency at a specified temperature.

Variation of Consistency with Temperature

The consistency of any asphaltic material changes as the temperature changes. The change in consistency of different asphaltic materials may differ considerably even for the same amount of temperature change. For example, if a sample of blown semisolid asphalt and a sample of semisolid regular paving-grade asphalt with the same consistency at a given temperature are heated to a high enough temperature, the consistencies of the two materials will be different at the high temperatures, with the regular paving-grade asphalt being much softer than the blown asphalt. Further increase in temperature will eventually result in the paving asphalt becoming liquid at a temperature much lower than that at which the blown asphalt becomes liquid. If these two asphalts are then gradually cooled down to about the freezing temperature of water, the blown asphalt will be much softer than the paving-grade asphalt. Thus, the consistency of the blown asphalt is less affected by temperature changes than the consistency of regular paving-grade asphalt. This property of asphaltic materials is known as temperature susceptibility. The temperature susceptibility of a given asphalt depends on the crude oil from which the asphalt is obtained, although the variation in temperature susceptibility of paving-grade asphalts from different crudes is not as high as that between regular paving-grade asphalt and blown asphalt.

Consistency at a Specified Temperature

As stated earlier, the consistency of an asphaltic material will vary from solid to liquid depending on the temperature of the material. It is therefore essential that when the consistency of an asphalt material is given, the associated temperature also be given. Different ways for measuring consistency are presented later in this chapter.

Durability

When asphaltic materials are exposed to environmental elements, natural deterioration gradually takes place and eventually the materials lose their plasticity and become brittle. This change is caused primarily by chemical and physical reactions that take place in the material. This natural deterioration of the asphaltic material is known as weathering. For paving asphalt to act successfully as a binder, the weathering must be minimized as much as possible. The ability of an asphaltic material to resist weathering is described as the durability of the material. Some of the factors that influence weathering are oxidation, volatilization, temperature, exposed surface area, and age hardening. These factors are discussed briefly in the following sections.

Oxidation

Oxidation is the chemical reaction that takes place when the asphaltic material is attacked by the oxygen in the air. This chemical reaction causes gradual hardening and eventually permanent hardening and considerable loss of the plastic characteristics of the material.

Volatilization

Volatilization is the evaporation of the lighter hydrocarbons from the asphaltic material. The loss of these lighter hydrocarbons also causes loss of the plastic characteristics of the asphaltic material.

Temperature

It has been shown that temperature has a significant effect on the rate of oxidation and volatilization. The higher the temperature, the higher is the rate of oxidation and volatilization. The relationship between temperature increase and increase in rate of oxidation and volatilization is, however, not linear; the percentage increase in rate of oxidation and volatilization is usually much greater than the percentage increase in temperature that causes the increase in oxidation and volatilization. It has been postulated that the rate of organic and physical reactions in the asphaltic material approximately doubles for each $10°C$ ($50°F$) increase in temperature.

Surface Area

The exposed surface of the material also influences its rate of oxidation and volatilization. There is a direct relationship between surface area and rate of oxygen absorption and loss due to evaporation in grams per cubic centimeter per minute. An inverse relationship, however, exists between volume and rate of oxidation and volatilization. This means that the rate of hardening is directly proportional to the ratio of the surface area to the volume.

This fact is taken into consideration when asphalt concrete mixes are designed for pavement construction in that the air voids are kept to the practicable minimum required for stability to reduce the area exposed to oxidation.

Age Hardening

If a sample of asphalt is heated and then allowed to cool, its molecules will be rearranged to form a gel-like structure. This will cause continuous hardening of the asphalt over a period of time even though it is protected from other factors such as oxidation and volatilization that cause hardening. This process of hardening with time is known as age hardening. The rate at which age hardening occurs is relatively high during the first few hours after cooling but gradually decreases and becomes almost negligible after about a year. Age hardening does not seem to have a significant effect on pavements, except when they are laid as a thin mat.

Rate of Curing

Curing is defined as the process through which an asphaltic material increases its consistency as it loses solvent by evaporation.

Rate of Curing of Cutbacks

As discussed earlier, the rate of curing of any cutback asphaltic material depends on the distillate used in the cutting-back process. This is an important characteristic of cutback materials, as the rate of curing indicates the time that should elapse before a cutback will attain a consistency that is thick enough for the binder to perform satisfactorily. The rate of curing is affected by both inherent and external factors. The important inherent factors are

- Volatility of the solvent
- Quantity of solvent in the cutback and
- Consistency of the base material

The more volatile the solvent is, the faster it can evaporate from the asphaltic material, therefore, the higher the curing rate of the material. This is why gasoline or naphtha is generally used for rapid-curing cutbacks, whereas light fuel oil or kerosene is used for medium-curing cutbacks.

For any given type of solvent, the smaller the quantity used, the less time is required for it to evaporate and therefore the faster the asphalt material will cure. Also, the higher the penetration of the base asphalt, the longer it takes for the asphalt cutback to cure.

The important external factors that affect curing rate are

- Temperature
- Ratio of surface area to volume
- Wind velocity across exposed surface

These three external factors are directly related to the rate of curing in that the higher any of these factors is, the higher is the rate of curing. Unfortunately these factors cannot be controlled or predicted in the field, which makes it extremely difficult to predict the expected curing time. The curing rates of different asphaltic materials are therefore usually compared with the assumption that the external factors are held constant.

Rate of Curing for Asphalt Emulsions

The curing and adhesion characteristics of emulsions used for pavement construction (anionic and cationic) depend on the rate at which the water evaporates from the mixture. When weather conditions are favorable, the water is relatively rapidly displaced and curing progresses rapidly. When weather conditions include high humidity, low temperature, or rainfall immediately following the application of the emulsion, its ability to properly cure is adversely affected. Although the effect of surface and weather conditions on proper curing is more critical for anionic emulsions, favorable weather conditions are also required to obtain optimum results for cationic emulsions. A major advantage of cationic emulsions is that they release their water more readily.

Resistance to Water

When asphaltic materials are used in pavement construction, it is important that the asphalt continues to adhere to the aggregates even with the presence of water. If this bond between the asphalt and the aggregates is lost, the asphalt will strip from the aggregates, resulting in the deterioration of the pavement. The asphalt must therefore sustain its ability to adhere to the aggregates even in the presence of water. In hot-mix, hot-laid asphaltic concrete, where the aggregates are thoroughly dried before mixing, stripping does not normally occur and no preventive action is usually taken. However, when water is added to a hot-mix, cold-laid asphaltic concrete, commercial antistrip additives are usually added to improve the asphalt's ability to adhere to the aggregates.

Temperature Effect on Volume of Asphaltic Materials

The volume of asphalt is significantly affected by changes in temperature. The volume increases with increase in temperature and decreases with decrease in temperature. The rate of change in volume is given as the coefficient of expansion, which is the volume change in a unit volume of the material for a unit change in temperature. Because of this variation of volume with temperature, the volumes of asphaltic materials are usually given for a temperature of 60°F (15.6°C). Volumes measured at other temperatures are converted to the equivalent volumes at 60°F by using appropriate multiplication factors published by ASTM in Petroleum Measurement tables (ASTM D-1250).

TESTS FOR ASPHALTIC MATERIALS

Several tests are conducted on asphaltic materials to determine their consistency and quality to ascertain whether materials used in highway construction meet the prescribed specifications. Some of these specifications given by AASHTO and ASTM have been referred to earlier and some are listed in Tables 17.2 and 17.3. Standard specifications have also been published by the Asphalt Institute for the types of asphalts used in pavement construction.[6] Procedures for selecting representative samples of asphalt for testing have also been standardized and are given in MS-18 by the Asphalt Institute[7] and in D140 by ASTM.[8]

Some of the tests used to identify the quality of a given sample of asphalt on the basis of the properties discussed will now be described, followed by the description of some general tests.

Table 17.2 Specifications for Rapid Curing of Cutback Asphalts

	RC-70		RC-250		RC-800		RC-3000	
	Min.	Max.	Min.	Max.	Min.	Max.	Min.	Max.
Kinematic Viscosity at 60 C (140 F) (See Note 1) centistokes	70	140	250	500	800	1600	3000	6000
Flash point (Tag, open-cup), degrees C (F)	. . .	. . .	27 (80)	. . .	27 (80)	. . .	27 (80)	. . .
Water, percent	. . .	0.2	. . .	0.2	. . .	0.2	. . .	0.2
Distillation test:								
Distillate, percentage by volume of total distillate to 360 C (680 F)								
to 190 C (374 F)	10	. . .	. . .	. . .	. . .	. . .	. . .	. . .
to 225 C (437 F)	50	. . .	35	. . .	15	. . .	. . .	. . .
to 260 C (500 F)	70	. . .	60	. . .	45	. . .	25	. . .
to 315 C (600 F)	85	. . .	80	. . .	75	. . .	70	. . .
Residue from distillation to 360 C (680 F) volume percentage of sample by difference	55	. . .	65	. . .	75	. . .	80	. . .
Tests on residue from distillation:								
Absolute viscosity at 60 C (140 F) (See Note 3) poises	600	2400	600	2400	600	2400	600	2400
Ductility, 5 cm./min. at 25 C (77 F) cm	100	. . .	100	. . .	100	. . .	100	. . .
Solubility in Trichloroethylene, percent	99.0	. . .	99.0	. . .	99.0	. . .	99.0	. . .
Spot test (See Note 2) with:								
Standard naphtha	Negative for all grades							
Naphtha - xylene solvent, - percent xylene	Negative for all grades							
Heptane - xylene solvent, - percent xylene	Negative for all grades							

NOTE 1. As an alternate, Saybolt-Furol viscosities may be specified as follows:
Grade RC-70—Furol viscosity at 50 C (122 F)—60 to 120 sec.
Grade RC-250—Furol viscosity at 60 C (140 F)—125 to 250 sec.
Grade RC-800—Furol viscosity at 82.2 C (180 F)—100 to 200 sec.
Grade RC-3000—Furol viscosity at 82.2 C (180 F)—300 to 600 sec.

NOTE 2. The use of the spot test is optional. When specified, the Engineer shall indicate whether the standard naphtha solvent, the naphtha xylene solvent or the heptane xylene solvent will be used in determining compliance with the requirement, and also, in the case of the xylene solvents, the percentage of xylene to be used.

NOTE 3. In lieu of viscosity of the residue, the specifying agency, at its option, can specify penetration at 100 g; 5s at 25C (77 F) of 80-120 for Grades RC-70, RC-250, RC-800, and RC-3000. However, in no case will both be required.

Source: Reproduced from *Standard Specifications for Transportation Materials and Methods of Sampling and Testing*, 14th ed., Washington, D.C.: The American Association of State Highway and Transportation Officials, copyright 1986. Used by permission.

Consistency Tests

The consistency of asphaltic materials is important in pavement construction because the consistency at a specified temperature will indicate the grade of the material. It is important that the temperature at which the consistency is determined be specified, since temperature significantly affects the consistency of asphaltic materials. As stated earlier, asphaltic materials can exist either in liquid, semisolid, or solid states. This wide range dictates the necessity for more than one method of determining consistency of asphaltic materials. The property generally used to describe the consistency of asphaltic materials in the liquid state is the

Table 17.3 Specifications for Medium Curing of Cutback Asphalts

	MC-30		MC-70		MC-250		MC-800		MC-3000	
	Min.	Max.	Min.	Max.	Min.	Max.	Min.	Max.	Min.	Max.
Kinematic Viscosity at 60 C (140 F) centistokes (See Note 1)	30	60	70	140	250	500	800	1600	3000	6000
Flash point (Tag. open-cup), degrees C (F)	38 (100)	...	38 (100)	...	66 (150)	...	66 (150)	...	66 (150)	...
Water percent	...	0.2	...	0.2	...	0.2	...	0.2	...	0.2
Distillation test: Distillate percentage by volume of total distillate to 360 C (680 F)										
to 225 C (437 F)	...	25	0	20	0	10	...	...	...	...
to 260 C (500 F)	40	70	20	60	15	55	0	35	0	15
to 315 C (600 F)	75	93	65	90	60	87	45	80	15	75
Residue from distillation to 360 C (680 F) Volume percentage of sample by difference	50	...	55	...	67	...	75	...	80	...
Tests on residue from distillation: Absolute viscosity at 60 C (140 F) (See Note 4) poises	300	1200	300	1200	300	1200	300	1200	300	1200
Ductility, 5 cm/cm., cm. (See Note 2)	100	...	100	...	100	...	100	...	100	...
Solubility in Trichloroethylene, percent	99.0	...	99.0	...	99.0	...	99.0	...	99.0	...
Spot test (See Note 3) with:										
Standard naphtha	Negative for all grades									
Naphtha - xylene solvent, - percent xylene	Negative for all grades									
Heptane - xylene solvent, - percent xylene	Negative for all grades									

NOTE 1. As an alternate, Saybolt Furol viscosities may be specified as follows:
Grade MC-70—Furol viscosity at 50 C (122 F)—60 to 120 sec.
Grade MC-30—Furol viscosity at 25 C (77 F)—75 to 150 sec.
Grade MC-250—Furol viscosity at 60 C (140 F)—125 to 250 sec.
Grade MC-800—Furol viscosity at 82.2 C (180 F)—100 to 200 sec.
Grade MC-3000—Furol viscosity at 82.2 C (180 F)—300 to 600 sec.

NOTE 2. If the ductility at 25 C (77 F) is less than 100, the material will be acceptable if its ductility at 15.5 C (60 F) is more than 100.

NOTE 3. The use of the spot test is optional. When specified, the Engineer shall indicate whether the standard naphtha solvent, the naphtha xylene solvent, or the heptane xylene solvent will be used in determining compliance with the requirement, and also, in the case of the xylene solvents, the percentage of xylene to be used.

NOTE 4. In lieu of viscosity of the residue, the specifying agency, at its option, can specify penetration 100 g; 5s at 25 C (77 F) of 120 to 250 for Grades MC-30, MC-70, MC-250, MC-800, and MC-3000. However, in no case will both be required.

Source: Reproduced from *Standard Specifications for Transportation Materials and Methods of Sampling and Testing*, 14th ed., Washington, D.C.: The American Association of State Highway and Transportation Officials, copyright 1986. Used by permission.

viscosity, which can be determined by conducting either the Saybolt Furol viscosity test, or the kinematic viscosity test. Tests used for asphaltic materials in the semisolid and solid states include the penetration test and the float test. The ring-and-ball softening point test, which is not often used in highway specifications, may also be used for blown asphalt.

Saybolt Furol Viscosity Test

Figure 17.2 shows the Saybolt Furol Viscometer. The principal part of the equipment is the standard viscometer tube, which is 5 in. long and about 1 in. in

Figure 17.2   Saybolt Furol Viscometer

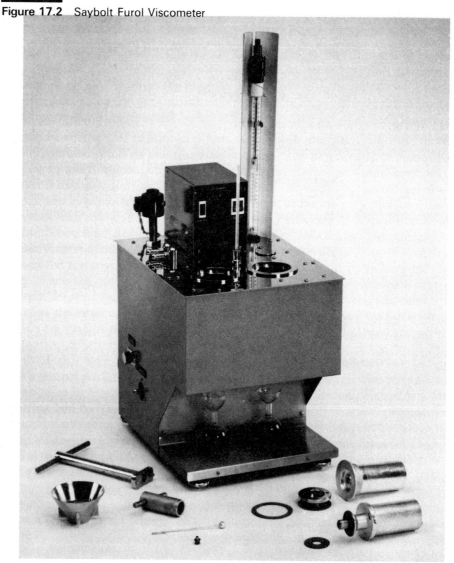

Source: Photo courtesy of STANHOPE-SETA, Ltd., Surrey, England.

diameter. An orifice of specified shape and dimensions is provided at the bottom of the tube. The orifice is closed with a stopper and the tube filled with a quantity of the material to be tested. The standard tube is then placed in a larger oil or water bath, fitted with an electric heater and a stirring device. The material in the tube is then brought to the specified temperature by heating the oil or water bath. Immediately upon reaching the prescribed temperature, the stopper is removed, and the time in seconds for exactly 60 milliliters of the asphaltic material to flow through the orifice is recorded. This time is the Saybolt Furol viscosity in units of seconds at the specified temperature. Temperatures at which asphaltic materials for highway construction are tested include 25°C (77°F), 50°C (122°F), and 60°C (140°F). It is apparent that the higher the viscosity of the material the longer it takes for a given quantity to flow through the orifice. Details of the equipment and procedures for conducting the Saybolt Furol test are given in AASHTO T72–83.[9]

Kinematic Viscosity Test

Figure 17.3 shows the equipment used to determine the kinematic viscosity, which is defined as the absolute viscosity divided by the density. The test uses a capillary viscometer tube to measure the time it takes the asphalt sample to flow at a specified temperature between timing marks on the tube. Three types of viscometer tubes are shown in Figure 17.4. Flow between the timing marks in the Zeitfuch's cross-arm viscometer is induced by gravitational forces, whereas flow in the Asphalt Institute vacuum viscometer and Cannon-Manning vacuum viscometer is induced by creating a partial vacuum.

When the cross-arm viscometer is used, the test is started by placing the viscometer tube in a thermostatically controlled constant temperature bath, as

Figure 17.3 Kinematic Viscosity Apparatus

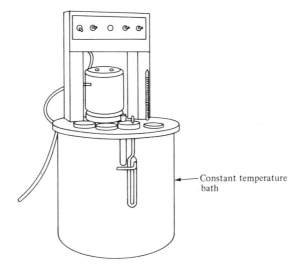

Constant temperature bath

Figure 17.4   Different Types of Kinematic Viscosity Tubes

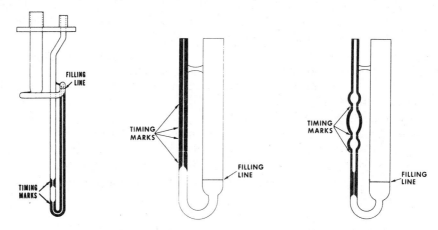

(a) Zeitfuch's cross-arm viscometer   (b) Asphalt Institute vacuum viscometer   (c) Canmen-Manning vacuum viscometer

Source: Reproduced from *A Brief Introduction to Asphalt and Some of Its Uses*, Manual Series No. 5, The Asphalt Institute, College Park, Md., 1977.

shown in Figure 17.3. A sample of the material to be tested is then preheated and poured into the large side of the viscometer tube (Figure 17.4a) until the filling line level is reached. The temperature of the bath is then brought to 135°C (275°F) and some time is allowed for the viscometer and the asphalt to reach a temperature of 135°C (275°F). Flow is then induced by applying a slight pressure to the large opening or a partial vacuum to the efflux (small) opening of the viscometer tube. This causes an initial flow of the asphalt over the siphon section just above the filling line. Continuous flow is then induced by the action of gravitational forces. The time it takes for the material to flow between two timing marks is recorded. The kinematic viscosity of the material in units of centistokes is obtained by multiplying the time in seconds by a calibration factor for the viscometer used. The calibration of each viscometer is carried out by using standard calibrating oils with known viscosity characteristics. The factor for each viscometer is usually furnished by the manufacturer. This test is described in detail in AASHTO Designation T201.[10]

The test may also be conducted at a temperature of 60°C (140°F) as described in AAHSTO Designation T202, using either the Asphalt Institute vacuum viscometer (Figure 17.4b) or the Cannon-Manning vacuum viscometer (Figure 17.4c). In this case, flow is induced by applying a prescribed vacuum through a vacuum control device attached to a vacuum pump. The product of the time interval and the calibration factor in this test gives the absolute viscosity of the material in poises.

Penetration Test

The penetration test gives an empirical measurement of the consistency of a material in terms of the distance a standard needle sinks into that material under a prescribed loading and time. Although more fundamental tests are now being substituted for this test, it may still be included in specifications for viscosity of asphalt cements to ensure the exclusion of materials with very low penetration values at 25°C (77°F).

Figure 17.5 shows a typical penetrometer and a schematic of the standard penetration test. A sample of the asphalt cement to be tested is placed in a container, which in turn is placed in a temperature-controlled water bath. The sample is then brought to the prescribed temperature of 25°C (77°F) and the standard needle, loaded to a total weight of 100 gm, is left to penetrate the sample of asphalt for the prescribed time of exactly 5 sec. The penetration is given as the distance in units of 0.1 mm that the needle penetrates the sample. For example, if the needle penetrates a distance of exactly 20 mm, the penetration is 200.

The penetration test can also be conducted at 0°C (32°F) or at 4°C (39.2°F) with the needle loaded to a total weight of 200 g and penetrations allowed for 60 sec. Details of the penetration test are given in AASHTO Designated T49[11] and ASTM Test D5.[12]

Float Test

The float test is used to determine the consistency of semisolid asphalt materials that are more viscous than grade 3000 or have penetration higher than 300, since these materials cannot be tested conveniently using either the Saybolt Furol viscosity test or the penetration test.

The float test is conducted with the apparatus schematically shown in Figure 17.6. It consists of an aluminum saucer (float), a brass collar that is open at both ends, and a water bath. The brass collar is filled with a sample of the material to be tested and then is attached to the bottom of the float and chilled to a temperature of 5°C (41°F) by immersing it in ice water. The temperature of the water bath is brought to 50°C (122°F) and the collar (still attached to the float) is placed in the water bath, which is kept at 50°C (122°F). The heat gradually softens the sample of asphaltic material in the collar, until eventually the water forces its way through the plug into the aluminum float. The time in seconds that expires between the instant the collar is placed in the water bath and that at which the water forces its way through the bituminous plug is the *float test value*, and it is a measure of the consistency. It is readily apparent that the higher the float-test value, the stiffer the material. Details of the float test are given in ASTM D139.[13]

Ring-and-Ball Softening Point Test

The ring-and-ball softening point test is used to measure the susceptibility of blown asphalt to temperature changes by determining the temperature at which the material will be adequately softened to allow a standard ball to sink through it. Figure 17.7 shows the apparatus commonly used for this test. It consists principally of a small brass ring of $\frac{5}{8}$ in. inside diameter and $\frac{1}{4}$ in. high, a steel ball $\frac{3}{8}$ in. in diameter, and a water or glycerin bath. The test is conducted by first placing a sample of the material to be tested in the brass ring, which is cooled and immersed

Figure 17.5 Standard Penetration Test and Equipment

(a) Penetrometer

Source: Photo courtesy of Wykeham Farrancé, Inc., Raleigh, N.C., 1983.

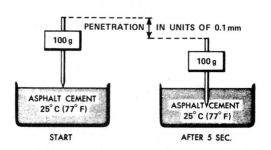

(b) Needle penetration

Source: Reproduced from *A Brief Introduction to Asphalt and Some of Its Uses*, Manual Series No. 5, The Asphalt Institute, College Park, Md., 1977.

Figure 17.6 Float Test

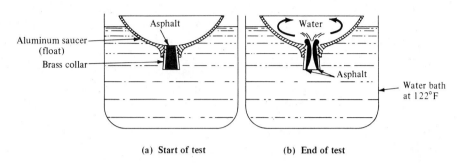

Asphalt

Aluminum saucer (float)

Brass collar

Water

Asphalt

Water bath at 122°F

(a) Start of test   (b) End of test

in the water or glycerin bath maintained at a temperature of 5°C (41°F). The ring is immersed to a depth such that its bottom is exactly 1 in. above the bottom of the bath. The temperature of the bath is then gradually increased causing the asphalt to soften, eventually permitting the ball to sink to the bottom of the bath. The temperature in degrees Fahrenheit at which the asphaltic material touches the bottom of the bath is recorded as the softening point. The resistance characteristic to temperature shear can also be evaluated using this test.[14]

Figure 17.7 Ring-and-Ball Softening Point Test

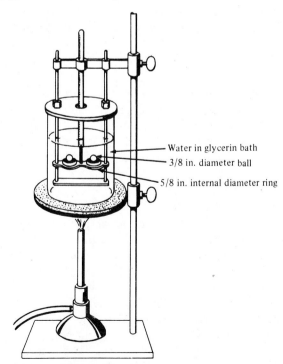

Water in glycerin bath
3/8 in. diameter ball
5/8 in. internal diameter ring

Source: Reproduced from *A Brief Introduction to Asphalt and Some of Its Uses*, Manual Series No. 5, The Asphalt Institute, College Park, Md., 1977.

Durability Tests

When asphaltic materials are used in the construction of roadway pavements, they are subjected to changes in temperature (freezing and thawing) and other weather conditions over a period of time. These changes cause natural weathering of the material, which may lead to loss of plasticity, cracking, abnormal surface abrasion, and eventual failure of the pavement. One test used to evaluate the susceptibility characteristics of asphaltic materials to changes in temperature and other atmospheric factors is the thin-film oven test.

Thin-Film Oven Test (TFO)

This is actually not a test but a procedure that provides for a measure of the changes that take place in an asphalt during the hot-mix process by subjecting the asphaltic material to hardening conditions similar to those in a normal hot-mix plant operation. The consistency of the material is determined before and after the TFO procedure, using either the penetration or a viscosity test, to estimate the amount of hardening that will take place in the material when used to produce plant hot-mix.

The procedure is performed by pouring 50 cc of the material into a cylindrical flat-bottom pan, 5.5 in. (14 cm) inside diameter and 3/8 in. (1 cm) high. The pan containing the sample is then placed on a rotating shelf in an oven and rotated for five hours while the temperature is kept at 163°C (325°F). The amount of penetration after the TFO test is then expressed as a percentage of that before the test to determine percent of penetration retained. The minimum allowable percent of penetration retained is usually specified for different grades of asphalt cement.

Rate of Curing

Tests for curing rates of cutbacks are based on inherent factors, which can be controlled. These tests compare different asphaltic materials on the assumption that the external factors are held constant. Volatility and quantity of solvent are commonly used to indicate the rate of curing. The volatility and quantity of solvent may be determined from the distillation test; tests for consistency were described earlier.

Distillation Test for Cutbacks

Figure 17.8 is a schematic of the apparatus used in the distillation test. The apparatus consists principally of a flask in which the material is heated, a condenser, and a graduated cylinder for collecting the condensed material. A sample of 200 cc of the material to be tested is measured and poured into the flask and the apparatus is set up as shown in Figure 17.8. The material is then brought to boiling point by heating it with the burner. The evaporated solvent is condensed and collected in the graduated cylinder. The temperature in the flask is continuously monitored and the amount of solvent collected in the graduated cylinder recorded when the temperature in the flask reaches 190°C (374°F), 225°C (437°F), 260°C (500°F), and 316°C (600°F). The amount of condensate collected

Figure 17.8   Distillation Test for Cutbacks

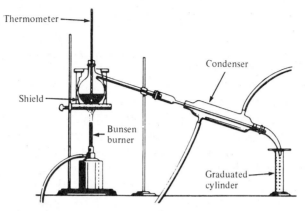

Thermometer

Condenser

Shield

Bunsen
burner

Graduated
cylinder

Source: Reproduced from *A Brief Introduction to Asphalt and Some of Its Uses*, Manual Series No. 5, The Asphalt Institute, College Park, Md., 1977.

at the different specified temperatures gives an indication of the volatility characteristics of the solvent. The residual in the flask is the base asphalt used in preparing the cutback. Details of this test are given in AASHTO Designation T78.[15]

Distillation Test for Emulsions

The distillation test for emulsions is similar to that described for cutbacks. A major difference, however, is that the glass flask and Bunsen burner are replaced with an aluminum alloy still and a ring burner. Any potential problem that may arise from the foaming of the emulsified asphalt as it is being heated to a maximum of 260°C (500°F) is prevented by using this equipment.

Note, however, that the results obtained from the use of this method to recover the asphaltic residue and to determine the properties of the asphalt base stock used in the emulsion may not always be accurate as significant changes can occur in the properties of the asphalt. These changes are due to

- The inorganic salts from an aqueous phase concentrating in the asphaltic residue
- The emulsifying agent and stabilizer concentrating in the asphaltic residue

These changes, which are mainly due to the increase in temperature, do not occur in field application of the emulsion since the temperature in the field is usually much less than that used in the distillation test. The emulsion in the field, therefore, breaks either electrochemically or by evaporation of the water. An alternative method to determine the properties of the asphalt after it is cured on the pavement surface is to evaporate the water at subatmospheric pressure and lower temperatures. Such a test is designated AASHTO T59[16] or ASTM D244.[17]

Other General Tests

Several other tests are routinely conducted on asphaltic materials used for pavement construction either to obtain specific characteristics for design purposes (for example, specific gravity) or to obtain additional information that aids in determining the quality of the material. Some of the more common routine tests are now described briefly.

Specific Gravity Test

The specific gravity of asphaltic materials is used mainly to determine the weight of a given volume of material or vice versa, to determine the amount of voids in compacted mixes and to correct volumes measured at high temperatures. Specific gravity is defined as the ratio of the weight of a given volume of the material to the weight of the same volume of water. The specific gravity of bituminous materials, however, changes with temperature, which dictates that the temperature at which the test is conducted should be indicated. For example, if the test is conducted at 20°C (68°F) and the specific gravity determined is 1.41, this should be recorded as 1.41/20°C. Note that both the asphaltic material and the water should be at the same temperature. The usual temperature at which the specific gravities of asphaltic materials are determined is 25°C (77°F).

Figure 17.9   Pycnometers for Determining Specific Gravity of Asphaltic Materials

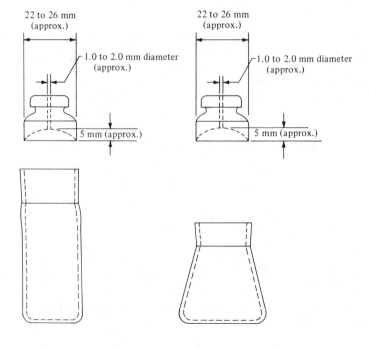

The test is normally conducted with a pycnometer, examples of which are shown in Figure 17.9. The dry weight (W_1) of the pycnometer and stopper is first obtained and then filled with distilled water at the prescribed temperature. The weight (W_2) of the water and pycnometer is determined. If the material to be tested can easily flow into the pycnometer, the pycnometer is completely filled with the material at the specified temperature after pouring out the water. The weight W_3 is then obtained. The specific gravity of the asphaltic material is then given as

$$G_b = \frac{W_3 - W_1}{W_2 - W_1} \qquad (17.1)$$

where G_b is the specific gravity of the asphaltic material and W_1, W_2, and W_3 are in grams.

If the asphaltic material cannot easily flow, a small sample of the material is heated gradually to facilitate flow and then poured into the pycnometer and left to cool to the specified temperature. The weight W_4 of pycnometer and material is then obtained. Water is then poured into the pycnometer to completely fill the remaining space not occupied by the material. The weight W_5 of the filled pycnometer is obtained. The specific gravity is then given as

$$G_b = \frac{W_4 - W_1}{(W_2 - W_1) - (W_5 - W_4)} \qquad (17.2)$$

Ductility Test

Ductility is the distance in centimeters a standard sample of asphaltic material will stretch before breaking when tested on standard ductility test equipment 25°C (77°F). The result of this test indicates the extent to which the material can be deformed without breaking. This is an important characteristic for asphaltic materials, although the exact value of ductility is not as important as the existence or nonexistence of the property in the material.

Figure 17.10 shows the ductility test apparatus. The test is used mainly for semisolid or solid materials, which are first gently heated to facilitate flow and then are poured into a standard mold to form a briquette of at least 1 cm² in cross section. The material is then allowed to cool to 25°C (77°F) in a water bath. The prepared sample is then placed in the ductility machine shown in Figure 17.10 and extended at a specified rate of speed until the thread of material joining the two ends breaks. The distance (in centimeters) moved by the machine is the ductility of the material. The test is fully described in AASHTO Designation T51[18] and ASTM D113.[19]

Solubility Test

The solubility test is used to measure the amount of impurities in the asphaltic material. Since asphalt is nearly 100 percent soluble in certain solvents, the portion of any asphaltic material that will be effective in cementing aggregates

Figure 17.10 Apparatus for Ductility Test

Source: Photo courtesy of Wykeham Farrancé, Inc., Raleigh, N.C., 1983.

together can be determined from the solubility test. Insoluble materials include free carbon, salts, and other inorganic impurities.

The test is conducted by dissolving a known quantity of the material in a solvent such as trichloroethylene and then filtering it through a Gooch Crucible. The material retained in the filter is dried and weighed. The test results are given in terms of the percent of the asphaltic material that dissolved in the solvent.

Flash-Point Test

The flash point of an asphaltic material is the temperature at which its vapors will ignite instantaneously in the presence of an open flame. Note that the flash point is normally lower than the temperature at which the material will burn.

The test can be conducted by using either the Tagliabue open-cup apparatus or the Cleveland open-cup apparatus, which is shown in Figure 17.11. The Cleveland open-cup test is more suitable for materials with higher flash points, whereas the Tagliabue open-cup is more suitable for materials with relatively low flash points, such as cutback asphalts. The test is conducted by partly filling the cup with the asphaltic material and gradually increasing its temperature at a specified rate. A small open flame is passed over the surface of the sample at regular intervals as the temperature increases. The increase in temperature will cause evaporation of volatile materials from the material being tested, until a sufficient quantity of volatile materials is present to cause an instantaneous flash when the open flame is passed over the surface. The minimum temperature at which this

Figure 17.11   Apparatus for Cleveland Open-Cup Test

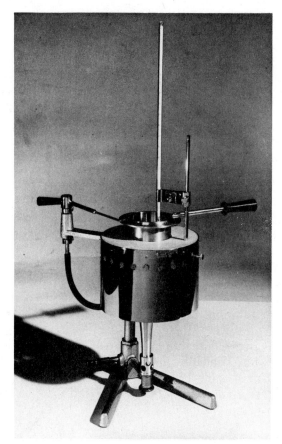

Source: Photo courtesy of Wykeham Farrancè, Inc., Raleigh, N.C., 1983.

occurs is the flash point. It can be seen that this temperature gives an indication of the temperature limit at which extreme care should be taken, particularly when heating is done over open flames in open containers.

Loss-on-Heating Test

The loss-on-heating test is used to determine the amount of material that evaporates from a sample of asphalt under a specified temperature and time. The result indicates whether an asphaltic material has been contaminated with lighter materials. The test is conducted by pouring 50 g of the material to be tested into a standard cylindrical tin and leaving it in an oven for 5 hr at a temperature of 163°C (325°F). The weight of the material remaining in the tin is then determined and the loss in weight expressed as a percentage of the original weight. The penetration of the sample may also be determined before and after the test to determine the loss of penetration due to the evaporation of the volatile material.

Figure 17.12   Particle Charge Test for Emulsions

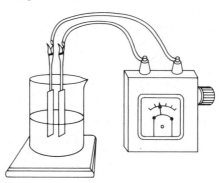

This loss in penetration may be used as an indication of the weathering characteristics of the asphalt. Details of the test procedures are given in AASHTO Designation T47.[20]

Water Content Test

The presence of large amounts of water in asphaltic materials used in pavement construction is undesirable, and to ensure that only a limited quantity of water is present, specifications for these materials usually include the maximum percentage of water by volume that is allowable. A quantity of the sample to be tested is mixed with an equal quantity of a suitable distillate in a distillation flask that is connected with a condenser and a trap for collecting the water. The sample is then gradually heated in the flask, eventually causing all the water to evaporate and be collected. The quantity of water in the sample is then expressed as a percentage of the total sample volume.

Demulsibility Test for Emulsion

The demulsibility test is used to indicate the relative susceptibility of asphalt emulsions to breaking down (coalescing) when in contact with aggregates. Asphalt emulsions are expected to break immediately when they come in contact with the aggregate so as to prevent the material from being washed away by rain that may occur soon after application. A high degree of demulsibility is therefore required for emulsions used for surface treatments such as RS1 and RS2. A relatively low degree of demulsibility is required for emulsions used for mixing coarse aggregates to avoid the materials peeling off before placing, and a very low degree is required for materials produced for mixing fine aggregates. Since calcium chloride will coagulate minute particles of asphalt, it is used in the test for anionic rapid-setting (RS) emulsions, but dioctyl sodium sulfosuccinate is used instead of calcium chloride for testing cationic rapid-setting (CRS) emulsions. The test is conducted by thoroughly mixing the required standard solution with the asphalt emulsion and the mixture passed through a No. 14 wire cloth sieve that will retain the asphalt particles that have coalesced. The quantity of asphalt retained in the sieve is a measure of the amount of breakdown that has occurred.

Demulsibility is expressed in percentage as $(A/B)100$, where A is the average weight of demulsibility residue from three tests of each sample of emulsified asphalt and B is the weight of residue by distillation in 100 gm of the emulsified asphalt.

The strength of the test solution used and the minimum value of demulsibility required are prescribed in the relevant specifications. Details of this test are described in AASHTO T59[21] and ASTM D244.[22]

Sieve Test for Emulsions

The sieve test is conducted on asphalt emulsions to determine to what extent the material has emulsified and the suitability of the material for application through pressure distributors. The test is conducted by passing a sample of the material through a No. 20 sieve and determining what percentage by weight of the material is retained in the sieve. A maximum value of 0.10 percent or lower is usually specified.

Particle Charge Test for Emulsions

The particle charge test is used to identify CRS and CMS grades of emulsions. The test is conducted by immersing an anode electrode and a cathode electrode in a sample of the material to be tested and then passing an electric current through the system as shown in Figure 17.12. The electrodes are then examined after some time to identify which one contains an asphalt deposit. If a deposit occurs on the cathode electrode, the emulsion is cationic.

ASPHALTIC CONCRETE

Asphaltic concrete is a uniformly mixed combination of asphalt cement, coarse aggregate, fine aggregate, and other materials, depending on the type of asphalt concrete. The different types of asphaltic concretes commonly used in pavement construction are hot-mix, hot-laid and cold-mix, cold-laid. Asphaltic concrete is the most popular paving material used in the United States. When used in the construction of highway pavements, it must be able to resist deformation from imposed traffic loads, be skid resistant even when the pavement is wet, and should not be easily affected by weathering forces. The degree to which an asphaltic concrete achieves these characteristics is mainly dependent on the design of the mix used in producing the concrete. Three categories of asphaltic concrete will now be described, together with an appropriate mix design procedure.

Hot-Mix, Hot-Laid Asphaltic Concrete

Hot-mix, hot-laid asphaltic concrete is produced by properly blending asphalt cement, coarse aggregate, fine aggregate, and filler (dust) at temperatures ranging from about 175°F to 325°F, depending on the type of asphalt cement used. Suitable types of asphaltic materials include AC-20, AC-10, and AR-8000 with penetration grades of 60–70, 85–100, 120–150, and 200–300, as shown in Table 17.1. Hot-mix, hot-laid asphaltic concrete is normally used for high-type pavement construction and can be described as open-graded, coarse-graded, dense-graded, or fine-graded mixtures. When produced for high-type surfacing, maximum sizes of aggregates normally range from $\frac{3}{8}$ in. to $\frac{3}{4}$ in. for open-graded mixtures, $\frac{1}{2}$ in. to $\frac{3}{4}$ in. for coarse-graded mixtures, $\frac{1}{2}$ in. to 1 in. for dense-graded mixtures, and $\frac{1}{2}$ in. to $\frac{3}{4}$ in. for fine-graded mixtures. When used as base, maximum sizes of aggregates are usually $\frac{3}{4}$ in. to $1\frac{1}{2}$ in. for open- and coarse-graded, 1 in. to $1\frac{1}{2}$ in. for dense-grades, and $\frac{3}{4}$ in. for fine-graded mixtures. As stated earlier, the extent to which an asphalt concrete meets the desired characteristics for highway pavement construction is mainly dependent on the mix design, which involves the selection and proportioning of the different material components. Note, however, that in designing hot-mix asphaltic concrete, a favorable balance must be struck between a highly stable product and a durable one. This is necessary because achieving a very highly stable product may result in a product of lower durability, and vice versa. The overall objective of the mix design is therefore to determine an optimum blend of the different components that will satisfy the requirements of the given specifications.

Aggregate Gradation

Aggregates are usually categorized as crushed rock, sand, and filler. The rock material is predominantly coarse aggregate retained in a No. 8 sieve, sand is predominantly fine aggregate passing the No. 8 sieve, and filler is predominantly mineral dust that passes the No. 200 sieve. It is customary for gradations of the combined aggregate and the individual fractions to be specified. Table 17.4 gives suggested grading requirements of aggregate material based on ASTM Designation 3515. The first phase in any mix design is the selection and combination of aggregates to obtain a gradation within the limits prescribed. This is sometimes referred to as mechanical stabilization.

The procedure used to select and combine aggregates will be illustrated in the following example.

Example 17–1 Determining Proportions of Different Aggregates to Obtain a Required Gradation

Table 17.5 gives the specifications for the aggregates and mix composition for highway pavement asphaltic concrete, and Table 17.6 shows the results of a sieve analysis of samples from the materials available. We must determine the proportions of the separate aggregates that will give a gradation within the specified limits.

It can be seen that the amount of the different sizes selected should not only give a mix that meets the prescribed limits, but should be such that allowance is made for some variation during actual production of the mix. It can also be seen from Table 17.5 that to obtain the required specified gradation, some combination of all three materials is required, as the coarse and fine aggregates do not

Table 17.4 Examples of Composition of Asphalt Paving Mixtures

Sieve Size	Asphalt Concrete					Sand Asphalt	Sheet Asphalt
	Mix Designation and Nominal Maximum Size of Aggregate						
	1½ in. (2A) (37.5 mm)	1 in. (3A) (25.0 mm)	¾ in. (4A) (19.0 mm)	½ in. (5A) (12.5 mm)	⅜ in. (6A) (9.5 mm)	No. 4 (7A) (4.75 mm)	No. 16 (8A) (1.18 mm)
	Grading of Total Aggregate (Coarse Plus Fine, Plus Filler if Required) Amounts Finer Than Each Laboratory Sieve (Square Opening), weight percent						
2½ in. (63 mm)	...	...	...	...	...	...	...
2 in. (50 mm)	100	...	...	...	...	...	...
1½ in. (37.5 mm)	90 to 100	100	...	...	...	...	...
1 in. (25.0 mm)	...	90 to 100	100	...	...	...	...
¾ in. (19.0 mm)	60 to 80	...	90 to 100	100	...	...	...
½ in. (12.5 mm)	...	60 to 80	...	90 to 100	100	...	...
⅜ in. (9.5 mm)	...	...	60 to 80	...	90 to 100	100	...
No. 4 (4.75 mm)	20 to 55	25 to 60	35 to 65	45 to 70	60 to 80	80 to 100	100
No. 8[a] (2.36 mm)	10 to 40	15 to 45	20 to 50	25 to 55	35 to 65	65 to 100	95 to 100
No. 16 (1.18 mm)	...	...	...	...	...	40 to 80	85 to 100
No. 30 (600 μm)	...	...	...	...	...	20 to 65	70 to 95
No. 50 (300 μm)	2 to 16	3 to 18	3 to 20	5 to 20	6 to 25	7 to 40	45 to 75
No. 100 (150 μm)	...	...	...	...	...	3 to 20	20 to 40
No. 200[b] (75 μm)	0 to 5	1 to 7	2 to 8	2 to 9	2 to 10	2 to 10	9 to 20
Asphalt Cement, weight percent of Total Mixture[c]							
	3½ to 8	4 to 8½	4 to 9	4½ to 9½	5 to 10	7 to 12	8½ to 12
Suggested Coarse Aggregate Sizes							
	4 and 67	5 and 7 or 57	67 or 68 or 6 and 8	7 or 78	8		

[a]In considering the total grading characteristics of an asphalt paving mixture the amount passing the No. 8 (2.36 mm) sieve is a significant and convenient field control point between fine and coarse aggregate. Gradings approaching the maximum amount permitted to pass the No. 8 (2.36-mm) sieve will result in pavement surfaces having comparatively fine texture, while gradings approaching the minimum amount passing the No. 8 (2.36-mm) sieve will result in surfaces with comparatively coarse texture.

[b]The material passing the No. 200 (75-μm) sieve may consist of fine particles of the aggregates or mineral filler, or both. It shall be free from organic matter and clay particles and have a plasticity index not greater than 4 when tested in accordance with Method D423 and Method D424.

[c]The quantity of asphalt cement is given in terms of weight percent of the total mixture. The wide difference in the specific gravity of various aggregates, as well as a considerable difference in absorption, results in a comparatively wide range in the limiting amount of asphalt cement specified. The amount of asphalt required for a given mixture should be determined by appropriate laboratory testing or on the basis of past experience with similar mixtures, or by a combination of both.

Reprinted by permission from ASTM Designation 3515, Standard Specifications for Hot-Mixed, Hot-Laid Bituminous Paving Mixtures.

Source: Reproduced from *1985 Annual Book of ASTM Standards*, Section 4, Vol. 04.03, American Society for Testing and Materials, Philadelphia, Pa., 1985.

Table 17.5   Required Limits for Mineral Aggregates Gradation and Mix Composition for an Asphaltic Concrete for Example 17–1

Passing Sieve Designation	Retained on Sieve Designation	Percent by Weight
$\frac{3}{4}$ in. (19.0 mm)	$\frac{1}{2}$ in.	0–5
$\frac{1}{2}$ in. (12.5) mm)	$\frac{3}{8}$ in.	8–42
$\frac{3}{8}$ in. (9.5 mm)	No. 4	8–48
No. 4 (4.75 mm)	No. 10	6–28
Total coarse aggregates	No. 10	48–65
No. 10 (2.0 mm)	No. 40	5–20
No. 40 (0.425 mm)	No. 80	9–30
No. 80 (0.180 mm)	No. 200	5–8
No. 200 (0.075 mm)	—	2–6
Total fine aggregate and filler	Passing No. 10	35–50
Total mineral aggregate in asphalt concrete		90–95
Asphalt cement in asphalt concrete		5–7
Total mix		100

together meet the requirement of 5 percent to 8 percent by weight of filler material. A trial mix is therefore selected arbitrarily within the prescribed limits. Let this mix be

Coarse aggregates = 55%   (48–65% specified)
Fine aggregates = 39%   (35–50% specified)
Filler = 6%   (5–8% specified)

Table 17.6   Sieve Analysis of Available Materials for Example 17–1

Passing Sieve Designation	Retained on Sieve Designation	Percent by Weight		
		Coarse Aggregate	Fine Aggregate	Mineral Filler
$\frac{3}{4}$ in. (19.0 mm)	$\frac{1}{2}$ in.	5	—	—
$\frac{1}{2}$ in. (12.5 mm)	$\frac{3}{4}$ in.	35	—	—
$\frac{3}{8}$ in. (9.5 mm)	No. 4	38	—	—
No. 4 (4.75 mm)	No. 10	17	8	—
No. 10 (2.00 mm)	No. 40	5	30	—
No. 40 (0.425 mm)	No. 80	—	35	5
No. 80 (0.180 mm)	No. 200	—	26	35
No. 200 (0.075 mm)	—	—	1	60
Total		100	100	100

Table 17.7 Computation of Percentages of Different Aggregate Sizes for Example 17–1

Passing Sieve Size	Retained on Sieve Size	Percent by Weight			
		Coarse Aggregate	Fine Aggregate	Mineral Filler	Total Aggregate
3/4 in.	1/2 in.	0.55 × 5 = 2.75	—	—	2.75
1/2 in.	3/8 in.	0.55 × 35 = 19.25	—	—	19.25
3/8 in.	No. 4	0.55 × 38 = 20.90	—	—	20.90
No. 4	No. 10	0.55 × 17 = 9.35	0.39 × 8 = 3.12	—	12.47
No. 10	No. 40	0.55 × 5 = 2.75	0.39 × 30 = 11.70	—	14.45
No. 40	No. 80		0.39 × 35 = 13.65	0.06 × 5 = 0.3	13.95
No. 80	No. 200		0.39 × 26 = 10.14	0.06 × 35 = 2.10	12.24
No. 200	—		0.39 × 1 = 0.39	0.06 × 60 = 3.60	3.99
Total		55.0	39.0	6.0	100.00

The selected proportions are then used to determine the combination of the different sizes, as shown in Table 17.7. The calculation is based on the fundamental equation for the percentage of material P passing a given sieve for the aggregates 1, 2, 3 is given as

$$P = aA_1 + bA_2 + cA_3 + \cdots \qquad (17.3)$$

where A_1, A_2, A_3 is equal to the percentage of material passing a given sieve for aggregates 1, 2, 3, and a, b, c is equal to the proportions of aggregates 1, 2, 3 used in the combination, and where $a + b + c \ldots = 100$. Note that this is true for any number of aggregates combined.

It can be seen that the combination obtained, as shown in the last column of Table 17.7, meets the specified limits as shown in the last column of Table 17.5. The trial combination is therefore acceptable. Note, however, that the first trial may not always meet the specified limits. In such cases, other combinations must be tried until a satisfactory one is obtained.

Several graphical methods have been developed for obtaining a suitable mixture of different aggregates to obtain a desired gradation.[23] These methods tend to be rather complicated when the number of batches of aggregates is high. They generally can be of advantage over the trial-and-error method described here when the number of aggregates is not more than two. When the number of aggregates is three or higher, the trial-and-error method is preferable, although the graphical methods can also be used.

Asphalt Content

Having determined a suitable mix of aggregates, the next step is to determine the optimum percentage of asphalt that should be used in the asphalt concrete mixture. This percentage should, of course, be within the prescribed limits. The gradation of the aggregates determined earlier and the optimum amount of asphalt cement determined combine to give the proportions of the different materials to be used in producing the hot-mix, hot-laid concrete for the project under consideration. These determined proportions are usually referred to as the *job-mix formula*.

Two commonly used methods to determine the optimum asphalt content are the Marshall method and the Hveem method. The Marshall method is described here since it is more widely used.

Marshall Method Procedure. The original concepts of this method were developed by Bruce Marshall, who was then a bituminous engineer with the Mississippi State Highway Department. The original features have been improved by the U.S. Corps of Engineers, and the test is now standardized and described in detail in ASTM Designation D1559.[24] Test specimens of 4 in. in diameter and $2\frac{1}{2}$ in. in height are used in this method. They are prepared by a specified procedure of heating, mixing, and compacting the mixture of asphalt and aggregates, which is then subjected to a stability-flow test and a density-voids analysis. The stability is defined as the maximum load resistance, N, in pounds that the specimen will achieve at 140°F under specified conditions. The flow is the total movement of the specimen, in units of 1/100 in. during the stability test, as the load is increased from zero to the maximum.

Test specimens for the Marshall method are prepared for a range of asphalt contents within the prescribed limits. Usually the asphalt content is measured by 0.5 percent increments from the minimum prescribed, ensuring that at least two are below the optimum and two above the optimum so that the curves obtained from the result will indicate a well-defined optimum. For example, for a specified amount of 5–7 percent, mixtures of 5, 5.5, 6, 6.5, and 7 are prepared. At least three specimens are provided for each asphalt content to facilitate the provision of adequate data. For this example of five different asphalt contents, a total minimum number of 15 specimens is therefore required. The amount of aggregates required for each specimen is about 1.2 kg.

A quantity of the aggregates having the designed gradation is dried at a temperature between 105°C (221°F) and 110°C (230°F) until a constant weight is obtained. The mixing temperature for this procedure is set as the temperature that will produce a kinematic viscosity of 170 ± 20 centistokes or a Saybolt Furol viscosity of 85 ± 10 sec in the asphalt. The compacting temperature is that which will produce a kinematic viscosity of 280 ± 30 centistokes or Saybolt Furol viscosity of 160 ± 15 sec. These temperatures are determined and recorded.

The specimens containing the appropriate amounts of aggregates and asphalt are then prepared by thoroughly mixing and compacting each mixture. The compactive effort used is either 35, 50, or 75 blows of the hammer falling a distance of 18 in., depending on the design traffic category. After the application on one face, the sample mold is reversed and the same number of blows is applied to the other face of the sample. The specimen is then allowed to cool and tested for stability and flow after determining its bulk density.

The bulk density of the sample is usually determined by weighing the sample in air and in water. It may be necessary to coat samples made from open-graded mixtures with paraffin before determining the density. The bulk specific gravity G_{bcm} of the sample—that is, compacted mixture—is given as

$$G_{bcm} = \frac{W_a}{W_a - W_w} \tag{17.4}$$

where

$\quad W_a$ = weight of sample in air (g)

$\quad W_w$ = weight of sample in water (g)

Stability Test

In conducting the stability test, the specimen is immersed in a bath of water at a temperature of $60°C \pm 1°C$ ($140°F \pm 1.8°F$) for a period of 30 to 40 min. It is then

Figure 17.13 Marshall Stability Equipment

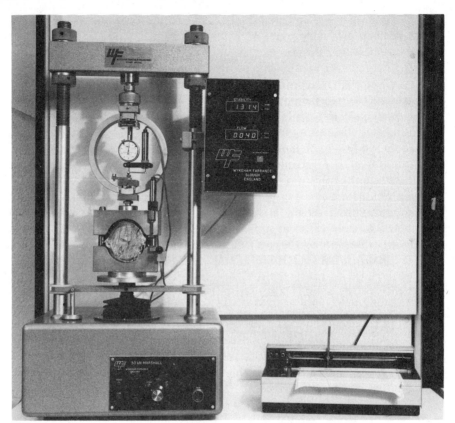

Source: Photo courtesy of Wykeham Farrancé, Inc., Raleigh, N.C., 1983.

placed in the Marshall stability testing machine, as shown in Figure 17.13, and loaded at a constant rate of deformation of 2 in. (5 mm) per minute until failure occurs. The total load, N, in pounds that causes failure of the specimen at 60°C (140°F) is noted as the Marshall stability value of the specimen. The total amount of deformation in units of 1/100 in. that occurs up to the point the load starts decreasing is also recorded as the flow value. The total time between removing the specimen from the bath and completion of the test should not exceed 30 sec.

Analysis of Results from Marshall Test. The first step in the analysis of the results is the determination of the average bulk specific gravity for all test specimens having the same asphalt content. The average unit weight of each mixture is then obtained by multiplying its average specific gravity by the density of water γ_w. A smooth curve that represents the best fit of plots of unit weight versus percentage of asphalt is then drawn as shown in Figure 17.14(a). This curve is used to obtain the bulk specific gravity values that are used in further computations as shown in Example 17–2.

Figure 17.14 Marshall Test Property Curves for Example 17–2

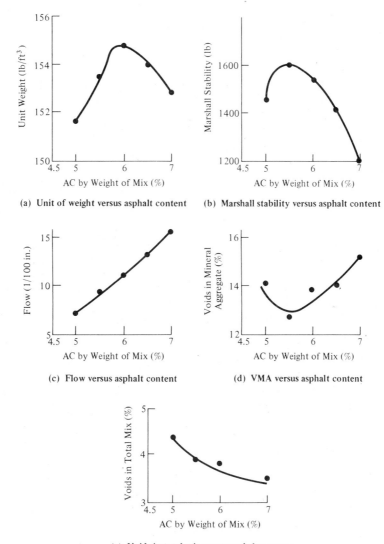

(a) Unit of weight versus asphalt content

(b) Marshall stability versus asphalt content

(c) Flow versus asphalt content

(d) VMA versus asphalt content

(e) Voids in total mix versus asphalt content

The percent air voids, percent voids in the mineral aggregate, and the absorbed asphalt in pounds of the dry aggregate are then calculated as shown below.

In order to compute the percent air voids, the percent voids in the mineral aggregate, and the absorbed asphalt, it is first necessary to compute the bulk specific gravity of the aggregate mixture, the apparent specific gravity of the aggregate mixture, the effective specific gravity of the aggregate mixture, and the maximum specific gravity of the paving mixtures for different asphalt contents.

Figure 17.15 Schematic Showing Voids in Mineral Aggregate, Air Voids, and Effective Asphalt Content in Compacted Asphalt Paving Mixture

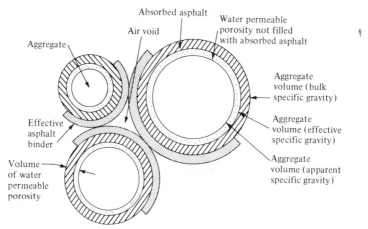

Source: Redrawn from *Mix Design Methods for Asphalt Concrete and Other Hot-Mix Types*, Manual Series No. 2, The Asphalt Institute, College Park, Md., May 1984.

These different measures of the specific gravity of the aggregates take into consideration the variation with which mineral aggregates can absorb water and asphalt. See Figure 17.15.

Bulk Specific Gravity of Aggregate.   The bulk specific gravity is defined as the weight in air of a unit volume (including all normal voids) of a permeable material at a selected temperature, divided by the weight in air of the same density of gas-free distilled water at the same selected temperature.

Since the aggregate mixture consists of different fractions of coarse aggregate, fine aggregate, and mineral fillers with different specific gravities, the bulk specific gravity of the total aggregate in the paving mixture is given as

$$G_{bam} = \frac{P_{ca} + P_{fa} + P_{mf}}{\dfrac{P_{ca}}{G_{bca}} + \dfrac{P_{fa}}{G_{bfa}} + \dfrac{P_{mf}}{G_{bmf}}} \tag{17.5}$$

where

G_{bam} = bulk specific gravity of aggregates in the paving mixture

P_{ca}, P_{fa}, P_{mf} = percent by weight of coarse aggregate, fine aggregate, and mineral filler, respectively, in the paving mixture. (Note that $P_{ca}, P_{fa},$ and P_{mf} could be found as a percentage of the paving mixture or as a percentage of the total aggregates only. The same results will be obtained for G_{bam}.)

$G_{bca}, G_{bfa}, G_{bmf}$ = bulk specific gravities of coarse aggregate, fine aggregate, and mineral filler, respectively

It is not easy to determine accurately the bulk specific gravity of the mineral filler. The apparent specific gravity may therefore be used with very little error resulting.

Apparent Specific Gravity of Aggregates. The apparent specific gravity is defined as the ratio of the weight in air of an impermeable material to the weight of an equal volume of distilled water at a specified temperature. The apparent specific gravity of the aggregate mix is therefore obtained as

$$G_{ama} = \frac{P_{ca} + P_{fa} + P_{mf}}{\dfrac{P_{ca}}{G_{aca}} + \dfrac{P_{fa}}{G_{afa}} + \dfrac{P_{mf}}{G_{amf}}} \tag{17.6}$$

where

G_{ama} = apparent specific gravity of the aggregate mixture

P_{ca}, P_{fa}, P_{mf} = percent by weight of coarse aggregate, fine aggregate, and mineral filler, respectively, in the mixture

$G_{aca}, G_{afa}, G_{amf}$ = apparent specific gravities of coarse aggregate, fine aggregate, and mineral filler, respectively

Effective Specific Gravity of Aggregate. The effective specific gravity of the aggregates is normally based on the maximum specific gravity of the paving mixture. It is therefore the specific gravity of the aggregates when all void spaces in the aggregate particles are included, with the exception of those that are filled with asphalt. (See Figure 17.15.) It is given as

$$G_{ea} = \frac{100 - P_{ac}}{(100/G_{mp}) - (P_{ac}/G_{ac})} \tag{17.7}$$

where

G_{ea} = effective specific gravity of the aggregates

100 = total percent by weight of base mixture

G_{mp} = maximum specific gravity of paving mixture (no air voids)

P_{ac} = asphalt percent by total weight of paving mixture

G_{ac} = specific gravity of the asphalt

Maximum Specific Gravity of the Paving Mixture. The maximum specific gravity of the paving mixture G_{mp} assumes that there are no air voids in the asphalt concrete. Although the G_{mp} can be determined in the laboratory by conducting the standard test (ASTM Designation D2041) the best accuracy is attained at mixtures near the optimum asphalt content. Since it is necessary to determine the G_{mp} for all samples, some of which contain much lower or much higher quantities than the optimum asphalt content, the following procedure can be used to determine the G_{mp} for each sample.

The ASTM Designation D2041 test is conducted on all specimens containing a selected asphalt cement content, and the mean of these determined. This value is then used to determine the effective specific gravity of the aggregates using Eq. 17.7. The effective specific gravity of the aggregates can be considered as being constant, since varying the asphalt content in the paving mixture does not significantly vary the asphalt absorption. The effective specific gravity obtained is then used to determine the maximum specific gravity of the paving mixtures with different asphalt cement contents using Eq. 17.8.

$$G_{mp} = \frac{100}{(P_{ta}/G_{ea}) + (P_{ac}/G_{ac})} \tag{17.8}$$

where

G_{mp} = maximum specific gravity of paving mixture (asphalt concrete)

100 = total percent by weight of loose mixture

P_{ta} = percent by weight of aggregates in paving mixture (asphalt concrete)

P_{ac} = percent by weight of asphalt in paving mixture (asphalt concrete)

G_{ea} = effective specific gravity of the aggregates (assumed to be constant for different asphalt cement contents)

G_{ac} = specific gravity of asphalt

Once these different specific gravities have been determined, the asphalt absorption, the effective asphalt content, the percent voids in mineral aggregates (VMA), and the percent air voids in the compacted mixture can all be determined.

Asphalt absorption is the percent by weight of the asphalt that is absorbed by the aggregates based on the total weight of the aggregates. This is given as

$$P_{aa} = 100 \frac{G_{ea} - G_{bam}}{G_{bam}G_{ea}} G_{ac} \tag{17.9}$$

where

P_{aa} = amount of asphalt absorbed as a percentage of the total weight of aggregates

G_{ea} = effective specific gravity of the aggregates

G_{bam} = bulk specific gravity of the aggregates

G_{ac} = specific gravity of the asphalt

Effective Asphalt Content. The effective asphalt concrete is the difference between the total amount of asphalt in the mixture and that absorbed into the aggregate particles. The effective asphalt content is therefore that which coats the outside of the aggregate particles and influences the pavement performance. It is given as

$$P_{cac} = P_{ac} - \frac{P_{aa}}{100} P_{ta} \tag{17.10}$$

where

P_{cac} = effective asphalt content in paving mixture (percent by weight)

P_{ac} = percent by weight of asphalt in paving mixture

P_{ta} = aggregate percent by weight of paving mixture

P_{aa} = amount of asphalt absorbed as a percentage of the total weight of aggregates

Percent Voids in Compacted Mineral Aggregates. The percent voids in compacted mineral aggregates VMA is the percentag of void spaces between the granular particles in the compacted paving mixture, including the air voids and the volume occupied by the effective asphalt content. It is usually calculated as a percentage of the bulk volume of the compacted mixture, based on the bulk specific gravity of the aggregates. It is given as

$$VMA = 100 - \frac{G_{bcm}P_{ta}}{G_{bam}} \qquad (17.11)$$

where

VMA = percent voids in compacted mineral aggregates

G_{bcm} = bulk specific gravity of compacted mixture (asphalt concrete)

G_{bam} = bulk specific gravity of aggregate

P_{ta} = aggregate percent by weight of total paving mixture (asphalt concrete)

Percent Air Voids in Compacted Mixture. This is the ratio of the volume of the small air voids between the coated particles to the total volume of the mixture expressed as a percentage. It can be obtained from

$$P_{av} = 100 \frac{G_{mp} - G_{bcm}}{G_{mp}} \qquad (17.12)$$

where

P_{av} = percent air voids in compacted paving mixture

G_{mp} = maximum specific gravity of the compacted paving mixture

G_{bcm} = bulk specific gravity of the compacted paving mixture

Four additional separate smooth curves are drawn from plots of percent voids in total mix versus percent of asphalt, percent voids in mineral aggregate versus percent of asphalt, Marshall stability versus percent of asphalt, and flow versus percent of asphalt. These graphs are then used to select the asphalt contents for maximum stability, maximum unit weight, and percent voids in total mix within the limits specified (usually the median of the limits). The average of the asphalt contents is the optimum asphalt content. The stability and flow for this optimum content can then be obtained from the appropriate graphs to determine whether the required criteria are met. Suggested criteria for these test limits are given in Table 17.8. The above analysis is illustrated in Example 17–2.

Table 17.8 Suggested Criteria for Test Limits

(a) Maximum and Minimum Values

Design Method	Heavy Traffic*		Medium Traffic*		Light Traffic*	
	Min.	Max.	Min.	Max.	Min.	Max.
Marshall						
Number of compaction blows, each end of specimen	75		50		35	
Stability, lb	1,500	—	750	—	500	—
(N)	(6,672)		(2,224)		(3,336)	
Flow, units of 0.01 in.						
(0.25 mm)	8	16	8	18	8	20
Percent air voids						
Surfacing or leveling	3	5	3	5	3	5
Base	3	8	3	8	3	8
Percent voids in mineral aggregate			(See Table below)			
Hveem						
Stabilometer value	37	—	35	—	30	—
Swell, in.	—	.030	—	.030	—	.030
(mm)	—	(0.762)	—	(0.762)	—	(0.762)
Percent voids	4	—	4	—	4	—

(b) Minimum Percent Voids in Mineral Aggregate, *VMA*

U.S. Standard Sieve Designation**	Nominal Maximum Particle Size		Minimum Voids in Mineral Aggregate (%)
	(in.)**	(mm)**	
63 mm ($2\frac{1}{2}$ in.)	2.5	63.0	10.0
50 mm (2 in.)	2.0	50.0	11.0
38.1 mm ($1\frac{1}{2}$ in.)	1.5	38.1	11.5
25.0 mm (1 in.)	1.0	25.0	12.5
19.0 mm ($\frac{3}{4}$ in.)	0.750	19.0	13.0
12.5 mm ($\frac{1}{2}$ in.)	0.500	12.5	14.5
9.5 mm ($\frac{3}{8}$ in.)	0.375	9.5	15.5
4.75 mm (No. 4)	0.187	4.75	17.5
2.36 mm (No. 8)	0.093	2.36	20.0
1.18 mm (No. 16)	0.0469	1.18	23.0

Note: Criteria applicable only when testing is done in conformance with methods outlined in the Asphalt Institute Publication. *Mix Design Methods for Asphalt Concrete and Other Hot-Mix Types* (MS-2). All criteria, not stability value alone, must be considered in designing an asphalt paving mix.

*Traffic Classification (See Chapter 18 for determination of ESAL.)

   Light: Traffic conditions resulting in a Design Traffic Number (ESAL) less than 10^4.
   Medium: Traffic conditions resulting in a Design Traffic Number (ESAL) between 10^4 and 10^6.
   Heavy: Traffic conditions resulting in a Design Traffic Number (ESAL) above 10^6.

**Standard Specification for Wire Cloth Sieves for Testing Purposes, ASTM Designation E11 (AASHTO Designation M92).

Source: Adapted from *Mix Design Methods for Asphalt Concrete and Other Hot-Mix Types*, Manual Series No. 2, The Asphalt Institute, College Park, Md., May 1984.

Evaluation and Adjustment of Mix Design

As stated earlier, the overall objective of the mix design is to determine an optimum blend of different components that will satisfy the requirements of the given specifications, as shown in Table 17.8. This mixture should have

- An adequate amount of asphalt to ensure a durable pavement
- An adequate mix stability to prevent unacceptable distortion and displacement when traffic load is applied
- Adequate voids in the total compacted mixture to permit a small amount of compaction when traffic load is applied without loss of stability, blushing, and bleeding, but at the same time low enough to prevent harmful penetration of air and moisture into the compacted mixture
- Adequate workability to facilitate placement of the mix without segregation

When the mix design for the optimum asphalt content does not satisfy all the requirements given in Table 17.8, it is necessary to adjust the original blend of aggregates. Trial mixes can be adjusted by using the following general guidelines.

Low Voids and Low Stability.   In this situation, the voids in the mineral aggregates can be increased by adding more coarse aggregates. Alternatively, the asphalt content can be reduced, but this is done only if the asphalt content is higher than that normally used and the excess is not required as a replacement for the amount absorbed by the aggregates. Care should be taken when the asphalt content is to be reduced as this can also lead to a decrease in durability and an increase in permeability of the pavement.

Low Voids and Satisfactory Stability.   This mix can cause reorientation of particles and additional compaction of the pavement with time as traffic load is imposed on the pavement. This in turn may lead to instability or flushing of the pavement. Mixes with low voids should therefore be altered by adding more aggregates.

High Voids and Satisfactory Stability.   When voids are high, it is likely that the permeability of the pavement will also be high, which will allow water and air to circulate through the pavement, resulting in premature hardening of the asphalt. High voids should therefore be reduced to acceptable limits even though the stability is satisfactory. This can be achieved by increasing the amount of mineral dust filler in the mix.

Satisfactory Voids and Low Stability.   This condition suggests low quality aggregates; the quality should be improved.

High Voids and Low Stability.   It may be necessary to carry out two steps in this case. The first step is to adjust the voids as discussed earlier. When this adjustment does not also improve the stability, the second step is to consider the improvement of the aggregate quality.

Example 17–2   Designing an Asphalt Concrete Mixture

In designing an asphalt concrete mixture for a highway pavement to support very heavy traffic, data in Table 17.9 showing the aggregate characteristics and Table 17.10 showing data obtained using the Marshall method were used. Determine the optimum asphalt content for this mix for the specified limits given in Table 17.8.

The bulk specific gravity of the mix for each asphalt cement content is determined by calculating the average value for the specimens with the same asphalt cement content using Eq. 17.4.

For 5% asphalt content, the average bulk specific gravity is given as

$$G_{bcm} = \frac{1}{3}\left(\frac{1325.6}{1325.6 - 780.1} + \frac{1325.4}{1325.4 - 780.3} + \frac{1325.0}{1325.0 - 779.8}\right)$$

$$= \frac{1}{3}(2.43 + 2.43 + 2.43)$$

$$= 2.43$$

Therefore, bulk density = $2.43 \times 62.4 = 151.6$ lb/ft³.

For 5.5% asphalt content,

$$G_{bcm} = \frac{1}{3}\left(\frac{1331.3}{1331.3 - 789.6} + \frac{1330.9}{1330.9 - 789.3} + \frac{1331.8}{1331.8 - 790.0}\right)$$

$$= \frac{1}{3}(2.46 + 2.46 + 2.46)$$

$$= 2.46$$

Therefore, bulk density = 153.5 lb/ft³.

For 6.0% asphalt content,

$$G_{bcm} = \frac{1}{3}\left(\frac{1338.2}{1338.2 - 798.6} + \frac{1338.5}{1338.5 - 798.3} + \frac{1338.1}{1338.1 - 797.3}\right)$$

$$= \frac{1}{3}(2.48 + 2.48 + 2.47)$$

$$\cong 2.48$$

Therefore, bulk density = 154.8 lb/ft³.

Table 17.9   Aggregate Characteristics for Example 17–2

Aggregate Type	Percent by Weight of Total Paving Mixture	Bulk Specific Gravity
Coarse	52.3	2.65
Fine	39.6	2.75
Filler	8.1	2.70

Note: The nominal maximum particle size in the aggregate mixture is 1 in.

Table 17.10 Marshall Test Data for Example 17-2

Asphalt % by Weight of Total Mix	Weight of Specimen (g)						Stability (lb)			Flow (1/100 in.)		
	in Air			in Water								
	1	2	3	1	2	3	1	2	3	1	2	3
5.0	1325.6	1325.4	1325.0	780.1	780.3	779.8	1460	1450	1465	7	7.5	7
5.5	1331.3	1330.9	1331.8	789.6	789.3	790.0	1600	1610	1595	10	9	9.5
6.0	1338.2	1338.5	1338.1	798.6	798.3	797.3	1560	1540	1550	11	11.5	11
6.5	1343.8	1344.0	1343.9	799.8	797.3	799.9	1400	1420	1415	13	13	13.5
7.0	1349.0	1349.3	1349.8	798.4	799.0	800.1	1200	1190	1210	16	15	16

Maximum Specific Gravity of Paving Mixture

Asphalt (%)	
5	2.54
5.5	2.56
6.0	2.58
6.5	2.56
7.0	2.54

For 6.5% asphalt content,

$$G_{bcm} = \frac{1}{3} \left(\frac{1343.8}{1343.8 - 799.8} + \frac{1344.0}{1344.0 - 797.3} + \frac{1343.9}{1343.9 - 799.9} \right)$$

$$= \frac{1}{3}(2.47 + 2.46 + 2.47)$$

$$\cong 2.47$$

Therefore, bulk density = 154.1 lb/ft³.
For 7.0% asphalt content,

$$G_{bcm} = \frac{1}{3} \left(\frac{1349.0}{1349.0 - 798.4} + \frac{1349.3}{1349.3 - 799} + \frac{1349.8}{1349.8 - 800.1} \right)$$

$$= \frac{1}{3}(2.45 + 2.45 + 2.46)$$

$$\cong 2.45$$

Therefore, bulk density = 152.9 lb/ft³.
Average bulk density is then plotted against asphalt content as shown in Figure 17.14(a). Similarly, the average stability and flow for each asphalt cement content are as follows.

%	Stability	Flow
5.0	1458	7.2
5.5	1602	9.5
6.0	1550	11.2
6.5	1412	13.2
7.0	1200	15.7

These values are plotted in Figure 17.14(b) and (c).
We now have to compute percent voids in the mineral aggregate VMA and the percent voids in the compacted mixture for each asphalt cement mixture. From Eq. 17.11,

$$VMA = 100 - \frac{G_{bcm}P_{ta}}{G_{bam}}$$

where

G_{bcm} = bulk specific gravity of the compacted paving mixture
P_{ta} = percent of aggregate by total weight of mixture
G_{bam} = bulk specific gravity of aggregates in the mixture

For 5% asphalt content,

$G_{bcm} = 2.43$
$P_{ta} = 95.0$

From Eq. 17.5,

$$G_{bam} = \frac{P_{ca} + P_{fa} + P_{mf}}{(P_{ca}/G_{bca}) + (P_{fa}/G_{bfa}) + (P_{mf}/G_{bmf})}$$

Determining P_{ca}, P_{fa}, and P_{mf} in terms of total aggregates,

$P_{ca} = 0.523 \times 95.0 = 49.7$
$P_{fa} = 0.396 \times 95.0 = 37.6$
$P_{mf} = 0.081 \times 95.0 = 7.7$

Therefore,

$$G_{bam} = \frac{49.7 + 37.6 + 7.7}{(49.7/2.65) + (37.6/2.75) + (7.7/2.70)} = 2.69$$

$$P_{ta} = (100 - 5) = 95$$

Therefore,

$$VMA = 100 - \frac{2.43 \times 95}{2.69} = 14.18$$

For 5.5% asphalt content,

$P_{ca} = 0.523 \times 94.5 = 49.4$
$P_{fa} = 0.396 \times 94.5 = 37.4$
$P_{mf} = 0.081 \times 94.5 = 7.7$

Therefore,

$$G_{bam} = \frac{49.4 + 37.4 + 7.7}{(49.4/2.65) + (37.4/2.75) + (7.7/2.70)} = 2.69$$

Therefore,

$$VMA = 100 - \frac{2.46 \times 94.5}{2.69} = 13.58$$

For 6% asphalt cement,

$P_{ca} = 0.523 \times 94 = 49.2$
$P_{fa} = 0.396 \times 94 = 37.2$
$P_{mf} = 0.081 \times 94 = 7.6$

Therefore,

$$G_{bam} = \frac{49.2 + 37.2 + 7.7}{(49.2/2.65) + (37.2/2.75) + (7.6/2.70)} = 2.69$$

Therefore,

$$VMA = 100 - \frac{2.48 \times 94}{2.69} = 13.34$$

For 6.5% asphalt content,

$P_{ca} = 0.523 \times 93.5 = 48.9$
$P_{fa} = 0.396 \times 93.5 = 37.0$
$P_{mf} = 0.081 \times 93.5 = 7.6$

Therefore,

$$G_{bam} = \frac{48.9 + 37.0 + 7.6}{\dfrac{48.9}{2.65} + \dfrac{37.0}{2.75} + \dfrac{7.6}{2.7}} = 2.69$$

Therefore,

$$VMA = 100 - \frac{2.47 \times 93.5}{2.69} = 14.15$$

For 7.0% asphalt content,

$P_{ca} = 0.523 \times 93.0 = 48.6$
$P_{fa} = 0.396 \times 93.0 = 36.8$
$P_{mf} = 0.081 \times 93.0 = 7.5$

Therefore,

$$G_{bam} = \frac{48.6 + 36.8 + 7.5}{\dfrac{48.6}{2.65} + \dfrac{36.8}{2.75} + \dfrac{7.5}{2.7}} = 2.69$$

Therefore,

$$VMA = 100 - \frac{2.45 \times 93}{2.69} = 15.30$$

A plot of VMA versus asphalt content is shown in Figure 17.14(d).

We now have to determine the percentage of air voids in each of the paving mixtures. From Eq. 17.12,

$$P_{av} = 100 \frac{G_{mp} - G_{bcm}}{G_{mp}}$$

where

P_{av} = percent air voids in compacted mixture
G_{mp} = maximum specific gravity of the paving mixture
G_{bcm} = bulk specific gravity of the compacted mixture

For 5% asphalt content,

$$P_{av} = 100 \frac{2.54 - 2.43}{2.54} = 4.33$$

For 5.5% asphalt content,

$$P_{av} = 100 \frac{2.56 - 2.46}{2.56} = 3.91$$

For 6.0% asphalt content,

$$P_{av} = 100 \frac{2.58 - 2.48}{2.58} = 3.88$$

For 6.5% asphalt content,

$$P_{av} = 100 \frac{2.56 - 2.47}{2.57} = 3.50$$

For 7.0% asphalt content,

$$P_{av} = 100 \frac{2.54 - 2.45}{2.54} = 3.54$$

A plot of P_{av} versus asphalt content is shown in Figure 17.14(e).

The asphalt content that meets the design requirements for unit weight, stability, and percent air voids is then selected from the appropriate plot in Figure 17.14. The asphalt content having the maximum value of unit weight and stability is selected from each of the respective plots.

1. Maximum unit weight = 6.0% [Figure 17.14(a)]
2. Maximum stability = 5.5% [Figure 17.14(b)]
3. Percent air voids in compacted mixture using mean of limits—that is, $(3 + 5)/2 = 4$, 5.4% [Figure 17.14(e)]. (Note the limits of 3 percent and 5 percent given in Table 17.8.)

The optimum asphalt content is determined as the average.

Therefore, the optimum asphalt cement content is

$$\frac{6.0 + 5.5 + 5.4}{3} = 5.6\%$$

The properties of the paving mixture containing the optimum asphalt content should now be determined from Figure 17.14 and compared with the suggested criteria given in Table 17.8. The values for this mixture are

Unit weight $= 153.8$ lb/ft^3
Stability $= 1600$ lb
Flow $= 9.5$ units of 0.01 in.
Percent void total mix $= 3.9$
Percent voids in mineral aggregates $= 13$

This mixture meets all the requirements given in Table 17.8 for stability, flow, percent voids in total mix, and percent voids in mineral aggregates.

Example 17–3 Computing the Percent of Asphalt Absorbed

Using the information given in Example 17–2, determine the asphalt absorbed for the optimum mix. The maximum specific gravity for this mixture is 2.57, and the specific gravity of the asphalt cement is 1.02. From Eq. 17.9 the absorbed asphalt is given as

$$P_{aa} = 100 \, \frac{G_{ea} - G_{bam}}{G_{bam} \times G_{ea}} \, G_{ac}$$

where

P_{aa} = absorbed asphalt percent by weight of aggregate
G_{ea} = effective specific gravity of the aggregates
G_{bam} = bulk specific gravity of the aggregates
G_{ac} = specific gravity of the asphalts

It is first necessary to determine the effective specific gravity of aggregates using Eq. 17.7—that is,

$$G_{ea} = \frac{100 - P_{ac}}{(100/G_{mp}) - (P_{ac}/G_{ac})}$$

where

G_{ea} = effective specific gravity of the paving mixture
G_{mp} = maximum specific gravity of the paving mixture
P_{ac} = asphalt percent by weight of total weight of paving mixture
G_{ac} = specific gravity of the asphalt

$$G_{ea} = \frac{100 - 5.6}{(100/2.57) - (5.6/1.02)} = 2.82$$

The bulk specific gravity of the aggregates in this mixture is

$$G_{bam} = \frac{0.523 \times 94.4 + 0.396 \times 94.4 + 0.081 \times 94.4}{\dfrac{0.523 \times 94.4}{2.65} + \dfrac{0.396 \times 94.4}{2.75} + \dfrac{0.081 \times 94.4}{2.7}} = 2.69$$

The asphalt absorbed is

$$P_{aa} = 100 \frac{2.82 - 2.69}{2.69 \times 2.82} 1.02 \cong 1.75\%$$

Hot-Mix, Cold-Laid Asphaltic Concrete

Asphaltic concretes in this category are manufactured hot and then shipped and immediately laid, or they can be stockpiled for use at a future date. Thus, they are suitable for small jobs for which it may be uneconomical to set up a plant. They are also a suitable material for patching high-type pavements. The Marshall method of mix design can be used for this type of asphalt concrete, but high-penetration asphalt is normally used. The most suitable asphalt cements have been found to have penetrations within the lower limits of the 200–300 penetration grade.

Hot-mix, cold-laid asphaltic concretes are produced by first thoroughly drying the different aggregates in a central hot-mix plant and then separating them into several bins containing different specified sizes. One important factor in the production of this type of asphalt concrete mix is that the manufactured product should be discharged at a temperature of 170°F ± 10°. To achieve this, the aggregates are cooled to approximately 180°F after they are dried and before they are placed into the mixer. Based on the job-mix formula, the exact amount of the aggregates from each bin is then weighed and placed in the mixer. The different sizes of the aggregates are then thoroughly mixed together and dried. About 0.75 percent by weight of a medium-curing cutback asphalt (MC-30), to which a wetting agent has been added, is then mixed with the aggregates for another 10 sec. The high-penetration asphalt cement and water are then simultaneously added to the mixture. The addition of the water is necessary to ensure that the material remains workable after it has cooled down to normal temperatures. The amount of asphalt cement added is the optimum amount obtained from the mix design, but the amount of water depends on whether the material is to be used within a few days or whether it is to be stockpiled for periods up to several months. When the material is to be used within a few days, 2 percent of water by weight is used, but if it is to be stockpiled for a long period, 3 percent of water is used. The mixture is then thoroughly mixed for about 45 sec to produce a uniform mix.

Cold-Mix, Cold-Laid Asphaltic Concrete

Emulsified asphalts and low viscosity cutback asphalts are used to produce cold mix asphaltic concretes. They also can be used immediately after production or stockpiled for use at a later date. The production process is similar to that of the hot mix, except that the mixing is done at normal temperatures, and it is not always necessary to dry the aggregates. However, saturated aggregates and aggregates with surface moisture should be dried before mixing. The type and grade of asphaltic material used depends on whether the material is to be stockpiled for a long time, the use of the material, and the gradation of the aggregates.

Table 17.1 shows a suitable type of asphaltic material for different types of cold mixes.

Seal Coats

Seal coats are usually single applications of asphaltic material that may or may not contain aggregates. The three types of seal coats commonly used in pavement maintenance are fog seal, slurry seal, and aggregate seals.

Fog Seal

Fog seal is a thin application of emulsified asphalt, usually with no aggregates added. Slow-setting emulsions, such as SS-1, SS-1H, CSS-1, and CSS-1H are normally used for fog seals. The emulsion is sprayed at a rate of 0.1 to 0.2 gal/yd^2 after it has been diluted with clean water. Fog seals are mainly used to

- Reduce the infiltration of air and water into the pavement
- Prevent the progressive separation of aggregate particles from the surface downward or from the edges inward (raveling) in a pavement. (Raveling is mainly caused by insufficient compaction during construction, which was carried out in wet or cold weather conditions.)
- Bring the surface of the pavement to its original state

Slurry Seal

Slurry seal is a uniformly mixed combination of a slow-setting asphalt emulsion (usually SS-1), fine aggregate, mineral filler, and water. The mixing can be carried out in a conventional plastic mixer or in a wheelbarrow, if the quantity required is small. It is usually applied with an average thickness of $\frac{1}{16}$ to $\frac{1}{8}$ in.

Slurry seal is used as a low-cost maintenance material for pavements carrying light traffic. Note, however, that although the application of a properly manufactured slurry seal coat will fill cracks of about $\frac{1}{4}$ in. or more and provide a fine-textured surface, the existing cracks will appear through the slurry seal in a short time.

Aggregate Seals

Aggregate seals are obtained by spraying asphalt, immediately covering it with aggregates, and then rolling the aggregates into the asphalt. Asphalts used for aggregate seals are usually the softer grades of paving asphalt and the heavier grades of liquid asphalts. Aggregate seals can be used to restore the surface of old pavements.

Prime Coats

Prime coats are obtained by spraying asphaltic binder materials onto nonasphalt base courses. Prime coats are mainly used to

- Provide a waterproof surface on the base
- Fill capillary voids in the base
- Facilitate the bonding of loose mineral particles
- Facilitate the adhesion of the surface treatment to the base

Medium-curing cutbacks are normally used for prime coating, with MC-30 recommended for priming a dense flexible base and MC-70 for more granular-type base materials. The rate of spray is usually between 0.2 and 0.35 gal/yd^2 for the MC-30 and between 0.3 and 0.6 gal/yd^2 for the MC-70. The amount of asphaltic binder used, however, should be the maximum that can be completely absorbed by the base within 24 hr of application under favorable weather conditions. The base course must contain a nominal amount of water to facilitate the penetration of the asphaltic material into the base. It is therefore necessary to lightly spray the surface of the base course with water just before the application of the prime coat if its surface has become dry and dusty.

Tack Coats

A tack coat is a thin layer of asphaltic material sprayed over an old pavement to facilitate the bonding of the old pavement and a new course, which is to be placed over the old pavement. In this case, the rate of application of the asphaltic material should be limited, since none of this material is expected to penetrate the old pavement. Asphalt emulsions such as SS-1, SS-1H, CSS-1, and CSS-1H are normally used for tack coats after they have been thinned with an equal amount of water. Rate of application varies from 0.05 to 0.15 gal/yd^2 of the thinned material. Rapid-curing cutback asphalts such as RC-70 may also be used as tack coats.

Sufficient time must elapse between the application of the tack coat and the application of the new course to allow for adequate curing of the material through the evaporation of most of the dilutent in the asphaltic emulsion. This curing process usually takes several hours in hot weather but can take more than 24 hr in cooler weather. When the material is satisfactorily cured, it becomes a highly viscous, tacky film.

Surface Treatments

Asphalt surface treatments are obtained by applying a quantity of asphaltic material and suitable aggregates on a properly constructed flexible base course to provide a suitable wearing surface for traffic. Surface treatments are used to protect the base course and to eliminate the problem of dust on the wearing surface. They can be applied as a single course with thicknesses varying from $\frac{1}{2}$ to $\frac{3}{4}$ in. or a multiple course with thicknesses varying from $\frac{7}{8}$ to 2 in.

A single-course asphalt treatment is obtained by applying a single course of asphaltic material and a single course of aggregates. The rate of application of the asphaltic material for a single course varies from 0.13 to 0.42 gal/yd^2, depending

Table 17.11   Quantities of Materials for Bituminous Surface Treatments

NOTE—The values are typical design or target values and are not necessarily obtainable to the precision indicated.

Surface Treatment		Aggregate			Bituminous Material[A]
Type	Application	Size No.[B]	Nominal Size (Square Openings)	Typical Rate of Application, ft³/yd²	Typical Rate of Application, gal/yd²
Single	initial	5	1 in. to ½ in.	0.50	0.42
		6	¾ in. to ⅜ in.	0.36	0.37
		7	½ in. to No. 4	0.23	0.23
		8	⅜ in. to No. 8	0.17	0.19
		9	No. 4 to No. 16	0.11	0.13
Double	initial	5	1 in. to ½ in.	0.50	0.42
	second	7	½ in. to No. 4	0.25	0.26
Double	initial	6	¾ in. to ⅜ in.	0.36	0.37
	second	8	⅜ in. to No. 8	0.18	0.20
Triple	initial	5	1 in. to ½ in.	0.50	0.42
	second	7	½ in. to No. 4	0.25	0.26
	third	9	No. 4 to No. 16	0.13	0.14
Triple	initial	6	¾ in. to ⅜ in.	0.36	0.37
	second	8	⅜ in. to No. 8	0.18	0.20
	third	9	No. 4 to No. 16	0.13	0.14

[A] Experience has shown that these quantities should be increased slightly (5 to 10 %) when the bituminous material to be used was manufactured for application with little or no heating.
[B] According to Specification D 448.

Source: Reproduced from *1985 Annual Book of ASTM Standards*, Section 4, Vol. 04.03, American Society for Testing and Materials, Philadelphia, Pa., 1985.

on the gradation of the aggregates used; the rate of application of the aggregates varies from 0.11 ft³/yd² to 0.50 ft³/yd². Multiple-course asphalt surface treatments can be obtained either as a double asphalt surface treatment consisting of two courses of asphaltic material and aggregates or as a triple asphalt treatment consisting of three layers. The multiple-course surface treatments are constructed by first placing a uniform layer of coarse aggregates over an initial application of the bituminous materials and then applying one or more layers of bituminous materials and smaller aggregates, with each layer having a thickness that is approximately equal to the nominal maximum size of the aggregates used for that layer. The maximum aggregate size of each layer subsequent to the initial layer is usually taken as one-half that of the aggregates used in the preceding layer. The recommended rates of application of the asphaltic material and the aggregates are shown in Table 17.11.

SUMMARY

Asphaltic concrete pavement is becoming the most popular type of pavement used in highway construction. It is envisaged that the use of asphalt in highway construction will continue to increase, particularly with the additional knowledge that will be obtained from research conducted over the next few years. The engineering properties of different asphaltic materials is therefore of significant importance in highway engineering.

This chapter has presented information on the different types of asphaltic materials, their physical characteristics, and some of the tests usually conducted on these materials when used in the maintenance and/or construction of highway pavements. A mix design method for asphalt concrete (Marshall method) has also been provided so that the reader will understand the principles involved in determining the optimum mix for asphalt concrete. The chapter contains sufficient material on the subject to enable the reader to become familiar with the fundamental engineering properties and characteristics of those asphaltic materials used in pavement engineering.

PROBLEMS

17–1 Briefly describe the process of distillation by which asphalt cement is produced from crude petroleum. Also describe in detail how you would obtain asphaltic binders that can be used to coat highly silaceous aggregates.

17–2 Results obtained from laboratory tests on a sample of rapid-curing asphalt cement (RC-250) are shown below. Determine whether the properties of this material meet the Asphalt Institute specifications for this type of material. If not, where do the differences lie?

> Kinematic viscosity at 140°F (60°C) = 260 (centistokes)
> Flash point (Tag. open-cup) °F = 75
> Distillation Test
>   Distillate percent by volume of total distillate to 680°F (360°C)
>   to 437°F (225°C) = 35
>   to 500°F (200°C) = 54
>   to 600°F (316°C) = 75
> Residue from distillation to 680°F (360°C) by volume = 64
> Tests on residue from distillation
>   Ductility at 77°F (25°C) = 95 (cm)
>   Absolute viscosity at 140°F (60°C) = 750 poises
>   Solubility (%) = 95

17–3 The table below shows the particle size distributions of two aggregates A and B, which are to be blended to produce an acceptable aggregate for use in manufacturing an asphalt concrete for highway pavement construction. If the required limits of particle size distribution for the mix are as shown in the table, determine a suitable ratio for blending aggregates A and B to obtain an acceptable combined aggregate.

| | Percent Passing by Weight | | |
Sieve Size	A	B	Required Mix
$\frac{3}{4}$ in. (19 mm)	100	98	96–100
$\frac{3}{8}$ in. (9.5 mm)	80	76	65–80
No. 4 (4.25 mm)	50	45	40–55
No. 10 (2.00 mm)	43	33	35–40
No. 40 (0.425 mm)	20	30	15–35
No. 200 (0.075 mm)	4	8	5–8

17–4 The table below shows four different types of aggregates that will be used to produce a blended aggregate for use in the manufacture of asphaltic concrete. Determine the bulk specific gravity of the aggregate mix.

Material	Percent by Weight	Bulk Specific Gravity
A	35	2.58
B	40	2.65
C	15	2.60
D	10	2.55

17–5 If the specific gravity of the asphalt cement used in a sample of asphalt concrete mix is 1.01, the maximum specific gravity of the mix is 2.50, and the mix contains 6.5 percent by weight of asphalt cement, determine the effective specific gravity of the mixture.

17–6 The table below lists data used in obtaining a mix design for an asphaltic concrete paving mixture. If the maximum specific gravity of the mixture is 2.41 and the bulk specific gravity is 2.35, determine: (a) the bulk specific gravity of aggregates in the paving mixture, (b) the asphalt absorbed, (c) the effective asphalt content of the paving mixture, and (d) the percent voids in the mineral aggregate VMA.

Material	Specific Gravity	Mix Composition by Weight of Total Mix
Asphalt cement	1.02	6.40
Coarse aggregate	2.51	52.35
Fine aggregate	2.74	33.45
Mineral filler	2.69	7.80

17–7 The aggregate mix used for the design of an asphaltic concrete mixture consists of 42 percent coarse aggregates, 51 percent fine aggregates, and 7 percent mineral fillers. If the respective bulk specific gravities of these materials are 2.60, 2.71, and 2.69 and the effective specific gravity of the aggregates is 2.82, determine the optimum asphalt content as a percentage of the total mix if results obtained using the Marshall method are shown in the following table. The specific gravity of the asphalt is 1.02. (Use Table 17.8 for required specifications.)

Percent Asphalt	Weight of Specimen (g) in Air	in Water	Stability (lb)	Flow (1/100 in.)
5.5	1325.3	785.6	1796	13
6.0	1330.1	793.3	1836	14
6.5	1336.2	800.8	1861	16
7.0	1342.0	804.5	1818	20
7.5	1347.5	805.1	1701	25

17–8 Determine the asphalt absorption of the optimum mix of problem 7–7.

REFERENCES

1. *Standard Specifications for Transportation Materials and Methods of Sampling and Testing*, 14th ed., American Association of State Highway and Transportation Officials, Washington, D.C., 1986.

2. Ibid.

3. Ibid.

4. *Annual Book of Standards, Section 4, Construction*, Vol. 4, No. 3, *Road and Paving Materials*, American Society for Testing and Materials, Philadelphia, Pa., 1985.

5. *Standard Specifications for Transportation Materials.*

6. *Specifications for Paving and Industrial Asphalts (SS2)*, Asphalt Institute, College Park, Md., 1983.

7. *Sampling Asphalt Products for Specifications Compliance MS-18*, Asphalt Institute, College Park, Md., 1981.

8. *Annual Book of Standards.*

9. *Standard Specifications for Transportation Materials.*

10. Ibid.

11. Ibid.

12. *Annual Book of Standards.*

13. Ibid.

14. *Mix Design Methods for Asphalt Concrete and Other Hot-Mix Types*, Asphalt Institute Manual Series No. 2 (MS-2), Asphalt Institute, College Park, Md., May 1984.

15. *Standard Specifications for Transportation Materials.*

16. Ibid.

17. *Annual Book of Standards.*

18. *Standard Specifications for Transportation Materials.*

19. *Annual Book of Standards.*

20. *Standard Specifications for Transportation Materials.*

21. Ibid.

22. *Mix Design Methods.*

23. Ibid.

24. *Annual Book of Standards.*

Design of Flexible Pavements

Highway pavements are divided into two main categories: rigid and flexible. The wearing surface of a rigid pavement is usually constructed of Portland cement concrete such that it acts like a beam over any irregularities in the underlying supporting material. The wearing surface of flexible pavements, on the other hand, is usually constructed of bituminous materials such that it remains in contact with the underlying material even when minor irregularities occur. Flexible pavements usually consist of a bituminous surface underlaid with a layer of granular material and a layer of a suitable mixture of coarse and fine materials. Traffic loads are transferred by the wearing surface to the underlying supporting materials through the interlocking of aggregates, the frictional effect of the granular materials, and the cohesion of the fine materials.

Flexible pavements are further divided into three subgroups: high type, intermediate type, and low type. High-type pavements have wearing surfaces that adequately support the expected traffic load without visible distress due to fatigue and are not susceptible to weather conditions. Intermediate-type pavements have wearing surfaces that range from surface treated to those with qualities just below that of high-type pavements. Low-type pavements are used mainly for low-cost roads and have wearing surfaces that range from untreated to loose natural materials to surface-treated earth.

This chapter deals with the design of high-type pavements, although the methodologies presented can also be used for some intermediate-type pavements.

STRUCTURAL COMPONENTS OF A FLEXIBLE PAVEMENT

Figure 18.1 shows the components of a flexible pavement consisting of the subgrade or prepared roadbed, the subbase, the base, and the wearing surface. The performance of the pavement depends on the satisfactory performance of each component, which requires proper evaluation of the properties of each component separately.

Subgrade (Prepared Road Bed)

The subgrade is usually the natural material located along the horizontal alignment of the pavement and serves as the foundation of the pavement structure. The subgrade may also consist of a layer of selected borrow materials, well compacted to prescribed specifications as discussed in Chapter 16. It may also be necessary to treat the subgrade material to achieve certain strength properties required for the type of pavement being constructed. This will be discussed later.

Subbase Course

The subbase component is located immediately above the subgrade and consists of material of a superior quality to that which is generally used for subgrade construction. The requirements for subbase materials are usually given in terms of the gradation, plastic characteristics, and strength as discussed in Chapter 16. When the quality of the subgrade material meets the requirements of the subbase material, the subbase component may be omitted. In cases where suitable subbase material is not readily available, the available material can be treated with other materials to achieve the necessary properties. This process of treating soils to improve their engineering properties is known as stabilization. Some of the different methods of soil stabilization will be discussed later in this chapter.

Base Course

The base course lies immediately above the subbase. It is placed immediately above the subgrade if a subbase course is not used. This course usually consists of granular materials such as crushed stone, crushed or uncrushed slag, crushed or uncrushed gravel, and sand. The specifications for base course materials usually include stricter requirements than those for subbase materials, particularly with respect to their plasticity, gradation, and strength. Materials that do not have the required properties can be used as base materials if they are properly stabilized with Portland cement, asphalt, or lime. In some cases, high quality base course materials may also be treated with asphalt

Figure 18.1   Schematic of a Flexible Pavement

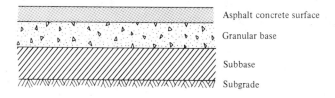

or Portland cement to improve the stiffness characteristics of heavy-duty pavements.

Surface Course

The surface course is the upper course of the road pavement and is constructed immediately above the base course. The surface course in flexible pavements usually consists of a mixture of mineral aggregates and asphaltic materials. It should be capable of withstanding high tire pressures, resisting the abrasive forces due to traffic, providing a skid-resistant driving surface, and preventing the penetration of surface water into the underlying layers. The thickness of the wearing surface can vary from 3 in. to more than 6 in., depending on the expected traffic on the pavement. It was shown in Chapter 17 that the quality of the surface course of a flexible pavement depends on the mix design of the asphalt concrete used.

SOIL STABILIZATION

Soil stabilization is the treatment of natural soil to improve its engineering properties. Soil stabilization methods can be divided into two categories, namely, mechanical and chemical. Mechanical stabilization is the blending of different grades of soils to obtain a required grade. This type of stabilization was discussed in Chapter 17. Chemical stabilization is the blending of the natural soil with chemical agents. Several blending agents have been used to obtain different effects, as shown in Table 18.1. The most commonly used agents are Portland cement, asphalt binders, and lime. It is necessary at this point to define some of the terms commonly used in the field of soil stabilization—particularly when Portland cement, lime, or asphalt are used—to help the reader fully understand Table 18.1.

Cement-stabilized soil is a mixture of water, soil, and measured amounts of Portland cement, thoroughly mixed and compacted to a high density and then allowed to cure for a specific period, during which it is protected from loss of moisture.

Soil cement is a hardened material obtained by mechanically compacting a mixture of finely crushed soil, water, and a quantity of Portland cement that will make the mixture meet certain durability requirements.

Cement-modified soil is a semihardened or unhardened mixture of water, Portland cement, and finely crushed soil. This mixture has less cement than the soil-cement mixture.

Plastic soil cement is a hardened material obtained by mixing finely crushed soil, Portland cement, and a quantity of water, such that at the time of mixing and placing, a consistency similar to that of mortar is obtained.

Table 18.1 Soil Types and Stabilization Methods Best Suited for Specific Applications

Purpose	Soil Type	Recommended Stabilization Methods
1. Subgrade Stabilization		
A. Improved load-carrying	Coarse granular	SA, SC, MB, C
and stress-distributing	Fine granular	SA, SC, MB, C
characteristics	Clays of low PI	C, SC, CMS, LMS, SL
	Clays of high PI	SL, LMS
B. Reduce frost	Fine granular	CMS, SA, SC, LF
susceptibility	Clays of low PI	CMS, SC, SL, CW, LMS
C. Waterproof and		
improve run-off	Clays of low PI	CMS, SA, CW, LMS, SL
D. Control shrinkage	Clays of low PI	CMS, SC, CW, C, LMS, SL
and swell	Clays of high PI	SL
E. Reduce resiliency	Clays of high PI	SL, LMS
	Elastic silts and clays	SC, CMS
2. Base Course Stabilization		
A. Improve substandard	Fine granular	SC, SA, LF, MB
materials	Clays of low PI	SC, SL
B. Improve load-carrying	Coarse granular	SA, SC, MB, LF
and stress-distributing	Fine granular	SC, SA, LF, MB
characteristics		
C. Reduce pumping	Fine granular	SC, SA, LF, MB membranes
3. Shoulders (unsurfaced)		
A. Improve load-carrying		See section 1A above;
ability	All soils	also MB
B. Improve durability	All soils	See section 1A above
C. Waterproof and		
improve run-off	Plastic soils	CMS, SL, CW, LMS
D. Control shrinkage and swell	Plastic soils	See section 1E above
4. Dust Palliative	Fine granular	CMS, CL, SA, oil, or bituminous surface spray
	Plastic soils	CL, CMS, SL, LMS
5. Ditch Lining	Fine granular	PSC, CS, SA
	Plastic soils	PSC, CS
6. Patching and Reconstruction	Granular soils	SC, SA, LF, MB

C = Compaction	LMS = Lime modified soil
CMS = Cement modified soil	MB = Mechanical blending
CL = Chlorides	PSC = Plastic soil cement
CS = Chemical solidifiers	SA = Soil asphalt
CW = Chemical waterproofers	SC = Soil cement
LF = Lime fly ash	SL = Soil lime

Source: Adapted from *Materials Testing Manual*, AFM 88–51, Department of the U.S. Air Force, Washington, D.C., February 1966.

Soil-lime is a mixture of lime, water, and fined-grained soil. If the soil contains silica and alumina, pozzolanic reaction occurs, resulting in the formation of a cementing-type material. Clay minerals, quartz, and feldspars are all possible sources of silica and alumina in typical fine-grained soils.

The stabilization process of each of the commonly used materials is briefly discussed. An in-depth discussion of soil stabilization is given in *Soil Stabilization in Pavement Structures.*[1]

Cement Stabilization

Cement stabilization of soils usually involves the addition of 5 percent to 14 percent Portland cement by volume of the compacted mixture to the soil being stabilized. This type of stabilization is used mainly to obtain the required engineering properties of soils that are to be used as base course materials. Although the best results have been obtained when well-graded granular materials were stabilized with cement, the Portland Cement Association has indicated that nearly all types of soils can be stabilized with cement.[2]

The procedure for stabilizing soils with cement involves

- Pulverizing the soil
- Mixing the required quantity of cement with the pulverized soil
- Compacting the soil cement mixture
- Curing the compacted layer

Pulverizing the Soil

The soil to be stabilized should first be thoroughly pulverized to facilitate the mixing of the cement and soil. When the existing material on the roadway is to be used, the roadway is scarified to the required depth, using a scarifier attached to a grader. When the material to be stabilized is imported to the site, the soil is evenly spread to the required depth above the subgrade and then pulverized by using rotary mixers. Sieve analysis is conducted on the soil during pulverization, and pulverization is continued (except for gravel) until all material can pass through a 1 in. sieve and not more than 20 percent is retained on the No. 4 sieve. A suitable moisture content must be maintained during pulverization, which may require air drying of wet soils or the addition of water to dry soils.

Mixing of Soil and Cement

The first step in the mixing process is to determine the amount of cement to be added to the soil by conducting laboratory experiments to determine the minimum quantity of cement required. The required tests are described by the Portland Cement Association.[3] These tests generally include the classification of the soils to be stabilized, using the American Association of State Highway and Transportation Official's (AASHTO) classification system, from which the range of cement content can be deduced based on past experience. Samples of soil–cement mixtures containing different quantities of cement within the range deduced are then prepared, and the American Society for Testing and Materials moisture–density relationship test (ASTM Designation D558) is conducted on these samples. The results are used to determine the amount of cement to be used for preparing test specimens that are then used to conduct durability tests. The durability tests are described in "Standard Methods of Wetting-and-Drying Tests of Compacted Soil–Cement Mixtures," ASTM Designation D559, and "Standard Methods of Freezing-and-Thawing Tests of Compacted Soil–Cement Mixtures," ASTM Designation D560.[4] The quantity of cement required to achieve satisfactory stabilization is obtained from the results of these tests. Table 18.2 gives suggested cement contents for different types of soils.

Having determined the quantity of cement required, the mixing can be carried out either on the road, *road mixing*, or in a central plant, *plant mixing*. Plant mixing is primarily used when the material to be stabilized is borrowed from another site.

With road mixing the cement is usually delivered in bulk in dump or hopper trucks and spread by a spreader box or some other type of equipment that will provide a uniform amount over the pulverized soil. Enough water is then added to achieve a moisture content that is 1 percent or 2 percent higher than the optimum required for compaction, and the soil, cement, and water are properly blended to obtain a uniform mixture of soil cement. Blending at a moisture content slightly higher than the compaction optimum moisture content allows for loss of water by evaporation during the mixing process.

Table 18.2 Cement Requirements for Various Soils

AASHTO Soil Classification	Unified Soil Classification[a]	Usual Range in Cement Requirement[b] (%/vol.)	Usual Range in Cement Requirement[b] (%/wt.)	Estimated Cement Content and That Used in Moisture-Density Test (%/wt.)	Cement Contents for Wet-Dry and Freeze-Thaw Tests (%/wt.)
A-1-a	GW, GP, GM, SW, SP, SM	5–7	3–5	5	3–5–7
A-1-b	GM, GP, SM, SP	7–9	5–8	6	4–6–8
A-2	GM, GC, SM, SC	7–10	5–9	7	5–7–9
A-3	SP	8–12	7–11	9	7–9–11
A-4	CL, ML	8–12	7–12	10	8–10–12
A-5	ML, MH, CH	8–12	8–13	10	8–10–12
A-6	CL, CH,	10–14	9–15	12	10–12–14
A-7	OH, MH, CH	10–14	10–16	13	11–13–15

[a]Based on correlation presented by the Air Force.

[b]For most A horizon soils, the cement should be increased 4 percentage points if the soil is dark grey to grey and 6 percentage points if the soil is black.

Source: Adapted from *Soil Cement Inspector's Manual*, Portland Cement Association, Skokie, Ill., 1963.

In plant mixing, the borrowed soil is properly pulverized and mixed with the cement and water in either a continuous or batch mixer. It is not necessary to add more than the quantity of water that will bring the moisture content to the optimum for compaction, since moisture loss in plant mixing is relatively small. The soil–cement mixture is then delivered to the site in trucks and uniformly spread. The main advantage of plant mixing over road mixing is that the proportioning of cement, water, and soil can be easily controlled.

Compacting the Soil–Cement Mixture

It is essential that compaction of the mixture be carried out before the mixture begins to set. Specifications given by the Federal Highway Administration (FHWA) stipulate that the length of time between the addition of water and the compaction of the mix at the site should not be greater than 2 hours for plant mixing and not more than 1 hour for road mixing. The soil–cement mixture is initially compacted with sheepsfoot rollers, or with pneumatic-tired rollers when the soil (for example, very sandy soil) cannot be effectively compacted with a sheepsfoot roller. The uppermost layer of 1 to 2 in. in depth is usually compacted with a pneumatic-tired roller, and the final surface compaction is carried out with a smooth-wheel roller.

Curing the Compacted Layer

Moisture loss in the compacted layer must be prevented before setting is completed because the moisture is required for the hydration process. This is achieved by applying a thin layer of bituminous material such as RC-250 or MC-250. Tars and emulsions can also be used.

Bituminous Stabilization

Bituminous stabilization is carried out to achieve one or both of the following:

- Waterproofing of natural materials
- Binding of natural materials

Waterproofing the natural material through bituminous stabilization aids in maintaining the water content at a required level by providing a membrane that impedes the penetration of water, thereby reducing the effect of any surface water that may enter the soil when it is used as a base course. In addition, surface water is prevented from seeping into the subgrade, which protects the subgrade from failing due to increase in moisture content.

Binding improves the durability characteristics of the natural soil by providing an adhesive characteristic, whereby the soil particles adhere to each other, increasing cohesion.

Several types of soils can be stabilized with bituminous materials, although it is generally required that less than 25 percent of the material passes the No. 200

sieve. This is necessary because the smaller soil particles tend to have extremely large surface areas per unit volume and require a large amount of bituminous material for the soil surfaces to be adequately coated. It is also necessary to use soils that have a plastic index (PI) of less than 10 because difficulty may be encountered in mixing soils with a high PI, which may result in the plastic fines swelling on contact with water and thereby losing strength.

The mixing of the soil and bituminous materials also can be done in a central or moveable plant (plant mixing) or at the roadside (road mixing). In plant mixing, the desired amounts of water and bituminous material are automatically fed into the mixing hoppers, whereas in road mixing, the water and bituminous material are measured and applied separately using a pressure distributor. The materials are then thoroughly mixed in the plant when plant mixing is used or by rotary speed mixers or suitable alternative equipment when road mixing is used.

The material is then evenly spread in layers of uniform thickness, usually not greater than 6 in. and not less than 2 in. Each layer is properly compacted until the required density is obtained using a sheepsfoot roller or a pneumatic-tired roller. The mixture must be completely aerated before compaction to ensure the removal of all volatile materials.

Lime Stabilization

Lime stabilization is one of the oldest processes of improving the engineering properties of soils and can be used for stabilizing both base and subbase materials. In general, the oxides and hydroxides of calcium and magnesium are considered as lime, but the materials most commonly used for lime stabilization are calcium hydroxide $C_a(OH)_2$, dolomite $C_a(OH)_2 + MgO$. The dolomite, however, should not have more than 36 percent by weight of magnesium oxide (MgO) to be acceptable as a stabilizing agent.

Clayey materials are most suitable for lime stabilization, but these materials should also have PI values less than 10 for the lime stabilization to be most effective. When lime is added to fine-grained soil, cation exchange takes place, with the calcium and magnesium in the lime replacing the sodium and potassium in the soil. The tendency to swell as a result of increase in moisture content is therefore immediately reduced. The PI value of the soil is also reduced. Pozzolanic reaction may also occur in some clays, resulting in the formation of cementing agents that increase the strength of the soil. When silica or alumina is present in the soil, a significant increase in strength may be observed over a long period of time. An additional effect is that lime causes flocculation of the fine particles, thereby increasing the effective grain size of the soil.

The percentage of lime used for any project depends on the type of soil being stabilized. The determination of the quantity of lime is usually based on an analysis of the effect that different lime percentages have on the reduction of plasticity and the increase in strength on the soil. The PI is most commonly used for testing the effect on plasticity, whereas the unconfined compression test, the

Hveem Stabilometer test, or the California bearing-ratio (CBR) test can be used to test for the effect on strength. Most fine-grained soil can, however, be effectively stabilized with 3 percent to 10 percent of lime, based on the dry weight of the soil.

GENERAL PRINCIPLES OF FLEXIBLE PAVEMENT DESIGN

In the design of flexible pavements, the pavement structure is usually considered as a multilayered elastic system, with the material in each layer characterized by certain physical properties that may include the modulus of elasticity, the resilient modulus, and the Poisson ratio. It is usually assumed that the subgrade layer is infinite in both the horizontal and vertical directions, whereas the other layers are finite in the vertical direction and infinite in the horizontal direction. The application of a wheel load causes a stress distribution, which can be represented as shown in Figure 18.2. The maximum vertical stresses are compressive and occur directly under the wheel load. These decrease with increase in depth from the surface. The maximum horizontal stresses also occur directly under the wheel load but can be either tensile or compressive as shown in Figure 18.2(c). When the load and pavement thickness are within certain ranges, horizontal compressive stresses will occur above the neutral axis, whereas horizontal tensile stresses will occur below the neutral axis. The temperature distribution within the pavement structure, as shown in Figure 18.2(d), will also have an effect on the magnitude of the stresses. The design of the pavement is therefore generally

Figure 18.2 Typical Stress and Temperature Distributions in a Flexible Pavement Under a Wheel Load

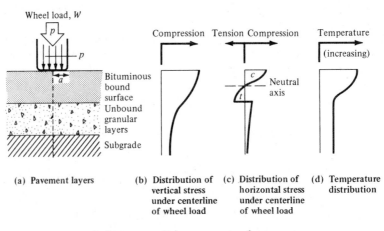

p = wheel pressure applied on pavement surface

a = radius of circular area over which wheel load is spread

c = compressive horizontal stress

t = tensile horizontal stress

Figure 18.3 Spread of Wheel Load Pressure Through Pavement Structure

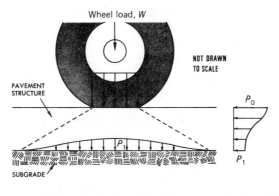

(a) Stress distribution within different
  components of the pavement

(b) General form
  of stress
  reduction

Source: Reproduced from *Thickness Design—Asphalt Pavements for Highways and Streets*, Manual Series No. 1, The Asphalt Institute, College Park, Md., September 1981.

based on strain criteria that limit both the horizontal and vertical strains below those that will cause excessive cracking and excessive permanent deformation. These criteria are considered in terms of repeated load applications since it is known that the accumulated repetitions of the traffic loads are of significant importance to the development of cracks and permanent deformation of the pavement.

Figure 18.4 Schematic of Tensile and Compressive Stresses in Pavement Structure

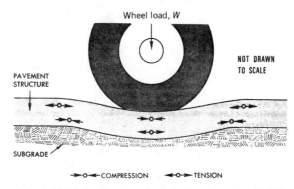

Source: Reproduced from *Thickness Design—Asphalt Pavements for Highways and Streets*, Manual Series No. 1, The Asphalt Institute, College Park, Md., September 1981.

The availability of highly sophisticated computerized solutions for multi-layered systems, coupled with recent advances in materials evaluation, has led to the development of several design methods that are based wholly or partly on theoretical analysis.[5-11] The more commonly used design methods are the Asphalt Institute method, the American Association of State Highway and Transportation Officials (AASHTO) method, and the California method. These are presented in the following sections.

Asphalt Institute Design Method

In the Asphalt Institute design method, the pavement is represented as a multi-layered elastic system. The wheel load W is assumed to be applied through the tire as a uniform vertical pressure p_0, which is then spread by the different components of the pavement structure and eventually applied on the subgrade as a much lower stress p_1. This is shown in Figure 18.3. Experience, established theory, and test data are then used to evaluate two specific stress–strain conditions. The first, shown in Figure 18.3(b), is the general way in which the stress p_0 is reduced to p_1 within the depth of the pavement structure and the second, shown in Figure 18.4, is the tensile and compressive stresses and strains imposed on the asphalt due to the deflection caused by the wheel load. Thickness design charts were developed, based on criteria for maximum tensile strains at the bottom of the asphalt layer and maximum vertical compressive strains at the top of the subgrade layer. The procedure originally involved the use of a computer program, but the charts presented here can be used without computers. A detailed description of the method is given in *Thickness Design—Asphalt Pavements for Highways and Streets*.[12]

Design Procedure

The principle adopted in the design procedure is to determine the minimum thickness of the asphalt layer that will adequately withstand the stresses that develop for the two strain criteria discussed earlier—that is, the vertical compressive strain at the surface of the subgrade and the horizontal tensile strain at the bottom of the asphalt layer. Design charts have been prepared for a range of traffic loads. This range is usually adequate for normal traffic volumes encountered in practice. However, when this range is exceeded, the computer version should be used.

The procedure consists of five main steps.

1. Select or determine input data.
2. Select surface and base materials.
3. Determine minimum thickness required for input data.
4. Evaluate feasibility of staged construction and prepare stage construction plan, if necessary.
5. Carry out economic analyses of alternative designs and select the best design.

Step 1: Design Inputs. The design inputs in this method are traffic characteristics, subgrade engineering properties, and subbase and base engineering properties.

Traffic Characteristics. The traffic characteristics are determined in terms of the number of repetitions of an 18,000 lb (80 kilonewtons (kN)) single-axle load applied to the pavement on two sets of dual tires. This is usually referred to as the *equivalent single-axle load* (ESAL). The dual tires are represented as two circular plates each 4.52 in. in diameter, spaced 13.57 in. apart. This representation corresponds to a contact pressure of 70 lb/in.^{2}. The use of an 18,000 lb axle load is based on the results of experiments that have shown that the effect of any load on the performance of a pavement can be represented in terms of the number of single applications of an 18,000 lb single axle. A series of equivalency factors used in this method for axle loads are given in Table 18.3.

To determine the ESAL, the number of different types of vehicles such as cars, buses, single-unit trucks, and multiple-unit trucks expected to use the facility during its lifetime must be known. The distribution of the different types of vehicles expected to use the proposed highway can be obtained from results of classification counts that are taken by state highway agencies at regular intervals. In cases where these data are not available, estimates can be made from Table 18.4, which gives representative values for the United States. When the axle load of each vehicle type is known, these can then be converted to equivalent 18,000 lb loads using the equivalency factors given in Table 18.3. The equivalent 18,000 lb load can also be determined from the vehicle type, if the axle load is unknown, by using a truck factor for that vehicle type. The truck factor is defined as the number of 18,000 lb single-load applications caused by a single passage of a vehicle. These have been determined for each class of vehicle from the expression

$$\text{truck factor} = \frac{\sum \left(\text{number of axles} \times \text{load equivalency factor} \right)}{\text{number of vehicles}}$$

Table 18.5 gives values of truck factors for different classes of vehicles; however, these factors are based on data collected prior to the deregulation of the trucking industry. Recent analysis of limited data has shown that truck factors can be as high as 5.0, even though the maximum value given in Table 18.5 is only 1.99. Therefore, it is advisable, when truck factors are to be used, to collect data on axle loads for the different types of vehicles expected to use the proposed highway and to determine realistic values of truck factors from that data.

The total ESAL applied on the highway during its design period can be determined only after the design period and traffic growth factors are known. The design period is the number of years the pavement will effectively continue to carry the traffic load without requiring an overlay. Flexible highway pavements are usually designed for a 20-year period. Since traffic volume does not remain constant over the design period of the pavement, it is essential that the rate of growth be determined and applied when calculating the total ESAL. Annual growth rates can be obtained from regional planning agencies or from state

Table 18.3 Load Equivalency Factors

Gross Axle Load		Load Equivalency Factors		Gross Axle Load		Load Equivalency Factors	
kN	lb	Single Axles	Tandem Axles	kN	lb	Single Axles	Tandem Axles
4.45	1,000	0.00002		182.5	41,000	23.27	2.29
8.9	2,000	0.00018		187.0	42,000	25.64	2.51
13.35	3,000	0.00072		191.3	43,000	28.22	2.75
17.8	4,000	0.00209		195.7	44,000	31.00	3.00
22.25	5,000	0.00500		200.0	45,000	34.00	3.27
26.7	6,000	0.01043		204.5	46,000	37.24	3.55
31.15	7,000	0.0196		209.0	47,000	40.74	3.85
35.6	8,000	0.0343		213.5	48,000	44.50	4.17
40.0	9,000	0.0562		218.0	49,000	48.54	4.51
44.5	10,000	0.0877	0.00688	222.4	50,000	52.88	4.86
48.9	11,000	0.1311	0.01008	226.8	51,000		5.23
53.4	12,000	0.189	0.0144	231.3	52,000		5.63
57.8	13,000	0.264	0.0199	235.7	53,000		6.04
62.3	14,000	0.360	0.0270	240.2	54,000		6.47
66.7	15,000	0.478	0.0360	244.6	55,000		6.93
71.2	16,000	0.623	0.0472	249.0	56,000		7.41
75.6	17,000	0.796	0.0608	253.5	57,000		7.92
80.0	18,000	1.000	0.0773	258.0	58,000		8.45
84.5	19,000	1.24	0.0971	262.5	59,000		9.01
89.0	20,000	1.51	0.1206	267.0	60,000		9.59
93.4	21,000	1.83	0.148	271.3	61,000		10.20
97.8	22,000	2.18	0.180	275.8	62,000		10.84
102.3	23,000	2.58	0.217	280.2	63,000		11.52
106.8	24,000	3.03	0.260	284.5	64,000		12.22
111.2	25,000	3.53	0.308	289.0	65,000		12.96
115.6	26,000	4.09	0.364	293.5	66,000		13.73
120.0	27,000	4.71	0.426	298.0	67,000		14)54
124.5	28,000	5.39	0.495	302.5	68,000		15.38
129.0	29,000	6.14	0.572	307.0	69,000		16.26
133.5	30,000	6.97	0.658	311.5	70,000		17.19
138.0	31,000	7.88	0.753	316.0	71,000		18.15
142.3	32,000	8.88	0.857	320.0	72,000		19.16
146.8	33,000	9.98	0.971	325.0	73,000		20.22
151.2	34,000	11.18	1.095	329.0	74,000		21.32
155.7	35,000	12.50	1.23	333.5	75,000		22.47
160.0	36,000	13.93	1.38	338.0	76,000		23.66
164.5	37,000	15.50	1.53	342.5	77,000		24.91
169.0	38,000	17.20	1.70	347.0	78,000		26.22
173.5	39,000	19.06	1.89	351.5	79,000		27.58
178.0	40,000	21.08	2.08	356.0	80,000		28.99

Note: kN converted to lb are within 0.1 percent of lb shown.

Source: Reproduced from *Thickness Design—Asphalt Pavements for Highways and Streets*, Manual Series No. 1, The Asphalt Institute, College Park, Md., September 1981.

Table 18.4 Distribution of Trucks on Different Classes of U.S. Highways

Truck Class	Percent Trucks									
	Interstate Rural		Other Rural		All Rural		All Urban		All Systems	
	Average	Range	Average	Range	Average	Range	Average	Range	Average	Range
Single-unit trucks										
2-axle, 4-tire	39	17-64	58	40-80	47	23-66	61	33-84	49	26-67
2-axle, 6-tire	10	5-15	11	4-18	10	4-16	13	4-26	11	5-20
3-axle or more	2	1-4	4	1-6	2	1-4	3	1-7	3	1-5
All single-units	51	30-71	73	50-88	59	36-77	77	55-94	63	36-81
Multiple-unit trucks										
3-axle	1	<1-2	1	<1-3	1	1-3	1	<1-4	1	<1-2
4-axle	5	1-10	3	<1-8	4	1-10	4	1-13	4	1-10
5-axle or more*	43	24-59	23	8-40	36	16-57	18	5-37	32	15-56
All multiple-units	49	31-71	27	13-50	41	23-66	23	6-44	37	20-67
All trucks	100		100		100		100		100	

Note: Compiled from data supplied by the Highway Statistics Division, Federal Highway Administration.

*Including full-trailer combinations in some states.

Source: Reproduced from *Thickness Design—Asphalt Pavements for Highways and Streets,* Manual Series No. 1, The Asphalt Institute, College Park, Md., September 1981.

highway departments. These are usually based on traffic volume counts over several years. It is also advisable to determine annual growth rates separately for trucks and passenger vehicles since these may be significantly different in some cases. The overall growth rate in the United States is between 3 percent and 5 percent per year, although growth rates of up to 10 percent per year have been suggested for some interstate highways. Table 18.6 shows growth factors for different growth rates and design periods, which can be used to determine the total ESAL over the design period.

The portion of the total ESAL acting on the design lane is used in the determination of pavement thickness. Either lane of a 2-lane highway can be considered as the design lane, whereas for multilane highways, the outside lane is considered. The identification of the design lane is important because in some cases more trucks will travel in one direction than in the other or trucks may travel heavily loaded in one direction and empty in the other direction. Thus, it is necessary to determine the relevant proportion of trucks on the design lane.

Table 18.5 Distribution of Truck Factors on Different Classes of U.S. Highways

Vehicle Type	Truck Factors									
	Rural Systems						Urban Systems		All Systems	
	Interstate Rural		Other Rural		All Rural		All Urban			
	Average	Range	Average	Range	Average	Range	Average	Range	Average	Range
Single-unit trucks										
2-axle, 4-tire	0.02	0.01–0.06	0.02	0.01–0.09	0.03**	0.02–0.08	0.03**	0.01–0.05	0.02	0.01–0.07
2-axle, 6-tire	0.19	0.13–0.30	0.21	0.14–0.34	0.20	0.14–0.31	0.26	0.18–0.42	0.21	0.15–0.32
3-axle or more	0.56	0.09–1.55	0.73	0.31–1.57	0.67	0.23–1.53	1.03	0.52–1.99	0.73	0.29–1.59
All single-units	0.07	0.02–0.16	0.07	0.02–0.17	0.07	0.03–0.16	0.09	0.04–0.21	0.07	0.02–0.17
Tractor semi-trailers										
3-axle	0.51	0.30–0.86	0.47	0.29–0.82	0.48	0.31–0.80	0.47	0.24–1.02	0.48	0.33–0.78
4-axle	0.62	0.40–1.07	0.83	0.44–1.55	0.70	0.37–1.34	0.89	0.60–1.64	0.73	0.43–1.32
5-axle or more*	0.94	0.67–1.15	0.98	0.58–1.70	0.95	0.58–1.64	1.02	0.69–1.69	0.95	0.63–1.53
All multiple units	0.93	0.67–1.38	0.97	0.67–1.50	0.94	0.66–1.43	1.00	0.72–1.58	0.95	0.71–1.39
All trucks	0.49	0.34–0.77	0.31	0.20–0.52	0.42	0.29–0.67	0.30	0.15–0.59	0.40	0.27–0.63

Note: Compiled from data supplied by the Highway Statistics Division, Federal Highway Administration.

*Including full-trailer combinations in some states.

**For values to be used when the number of heavy trucks is low, see original source.

Source: Reproduced from *Thickness Design—Asphalt Pavements for Highways and Streets*, Manual Series No. 1, The Asphalt Institute, College Park, Md., September 1981.

When data are not available to make this determination, percentages given in Table 18.7 can be used. The procedure for determining the design ESAL is demonstrated in Examples 18–1 and 18–2.

Example 18–1   Computing Accumulated Equivalent Single-Axle Load for a Proposed 8-Lane Highway Using Load Equivalency Factors

An 8-lane divided highway is to be constructed on a new alignment. Traffic volume forecasts indicate that the average annual daily traffic (AADT) in both directions during the first year of operation will be 12,000, with the following vehicle mix and axle loads.

Passenger cars (1000 lb/axle) = 50 percent
2-axle single-unit trucks (5000 lb/axle) = 33 percent
3-axle single-unit trucks (7000 lb/axle) = 17 percent

Table 18.6 Growth Factors

Design Period, Years (n)	Annual Growth Rate, Percent (r)							
	No Growth	2	4	5	6	7	8	10
1	1.0	1.0	1.0	1.0	1.0	1.0	1.0	1.0
2	2.0	2.02	2.04	2.05	2.06	2.07	2.08	2.10
3	3.0	3.06	3.12	3.15	3.18	3.21	3.25	3.31
4	4.0	4.12	4.25	4.31	4.37	4.44	4.51	4.64
5	5.0	5.20	5.42	5.53	5.64	5.75	5.87	6.11
6	6.0	6.31	6.63	6.80	6.98	7.15	7.34	7.72
7	7.0	7.43	7.90	8.14	8.39	8.65	8.92	9.49
8	8.0	8.58	9.21	9.55	9.90	10.26	10.64	11.44
9	9.0	9.75	10.58	11.03	11.49	11.98	12.49	13.58
10	10.0	10.95	12.01	12.58	13.18	13.82	14.49	15.94
11	11.0	12.17	13.49	14.21	14.97	15.78	16.65	18.53
12	12.0	13.41	15.03	15.92	16.87	17.89	18.98	21.38
13	13.0	14.68	16.63	17.71	18.88	20.14	21.50	24.52
14	14.0	15.97	18.29	19.16	21.01	22.55	24.21	27.97
15	15.0	17.29	20.02	21.58	23.28	25.13	27.15	31.77
16	16.0	18.64	21.82	23.66	25.67	27.89	30.32	35.95
17	17.0	20.01	23.70	25.84	28.21	30.84	33.75	40.55
18	18.0	21.41	25.65	28.13	30.91	34.00	37.45	45.60
19	19.0	22.84	27.67	30.54	33.76	37.38	41.45	51.16
20	20.0	24.30	29.78	33.06	36.79	41.00	45.76	57.28
25	25.0	32.03	41.65	47.73	54.86	63.25	73.11	98.35
30	30.0	40.57	56.08	66.44	79.06	94.46	113.28	164.49
35	35.0	49.99	73.65	90.32	111.43	138.24	172.32	271.02

Note: Factor $= [(1 + r)^n - 1]/r$, where $r = \dfrac{\text{rate}}{100}$ and is not zero. If annual growth is zero, growth factor = design period.

Source: Reproduced from *Thickness Design—Asphalt Pavements for Highways and Streets*, Manual Series No. 1, The Asphalt Institute, College Park, Md., September 1981.

The vehicle mix is expected to remain the same throughout the design life of the pavement. If the expected annual traffic growth rate is 4 percent for all vehicles, determine the design ESAL, given a design period of 20 years.

A general equation for the accumulated ESAL for each category of axle load is obtained as

$$\text{ESAL}_i = f_d \times G_{jt} \times \text{AADT}_i \times 365 \times N_i \times F_{Ei}$$

where

$\text{ESAL}_i =$ equivalent accumulated 18,000 lb (80 kN) single-axle load for the axle category i

$f_d =$ design lane factor

$G_{jt} =$ growth factor for a given growth rate j and design period t

$\text{AADT}_i =$ first year annual average daily traffic for axle category i

$N_i =$ number of axles on each vehicle in category i

$F_{Ei} =$ load equivalency factor for axle category i

Table 18.7 Percentage of Total Truck Traffic on Design Lane

Number of Traffic Lanes (Two Directions)	Percentage of Trucks in Design Lane
2	50
4	45 (35–48)*
6 or more	40 (25–48)*

*Probable range.

Source: Adapted from *Thickness Design—Asphalt Pavements for Highways and Streets*, Manual Series No. 1, The Asphalt Institute, College Park, Md., September 1981.

The following data apply.

Growth factor = 29.78 (from Table 18.6)
Percent truck volume on design lane = 45 (assumed, from Table 18.7)
Load equivalency factors (from Table 18.3)
 Passenger cars (1000 lb/axle) = 0.00002
 2-axle single-unit trucks (5000 lb/axle) = 0.00500
 3-axle single-unit trucks (7000 lb/axle) = 0.01960
Number of equivalent accumulated axle loads in the design lane
 Passenger cars = $0.45 \times 29.78 \times 12{,}000 \times 0.5 \times 365 \times 2 \times 0.00002$
  $= 0.001 \times 10^6$
 2-axle single-unit trucks = $0.45 \times 29.78 \times 12{,}000 \times 0.33 \times 365 \times 2$
  $\times 0.00500 = 0.1937 \times 10^6$
 3-axle single-unit trucks = $0.45 \times 29.78 \times 12{,}000 \times 0.17 \times 365 \times 3$
  $\times 0.0196 = 0.5867 \times 10^6$

Thus,

$$\text{total ESAL} = 0.781 \times 10^6$$

It can be seen that the contribution of passenger cars to the ESAL is negligible. Passenger cars are therefore omitted when computing ESAL values. This example illustrates the conversion of axle loads to ESAL, using axle load equivalency factors.

Example 18–2 Computing Accumulated Equivalent Single-Axle Load for a Proposed 2-Lane Highway Using Truck Factors

This projected vehicle mix for a proposed 2-lane rural highway during its first year of operation is given below. If the first year AADT will be 3000, the annual growth rate is 5 percent, and the design period is 20 yr, determine the accumulated ESAL, using the truck factors given in Table 18.5.

> Passenger cars = 66 percent
> Single-unit trucks
>> 2-axle, 4-tire = 18 percent
>> 2-axle, 6-tire = 8 percent
>> 3-axle or more = 4 percent
> Tractor semitrailers and combination
>> 3 axle = 3 percent
>> 4 axle = 1 percent

Since the axle loads are not given, the truck factor is used to compute the respective ESALs.

A general equation for accumulated ESAL for each category of truck is obtained as

$$\text{ESAL}_i = \text{AADT}_i \times 365 \times f_i \times G_{jt} \times f_d$$

and the accumulated ESAL for all categories of axle loads is

$$\text{ESAL} = \sum_{i=1}^{n} [\text{ESAL}_i]$$

where

ESAL_i = equivalent accumulated 18,000 lb axle load for truck category i

AADT_i = first year annual average daily traffic for vehicles in truck category i

f_i = truck factor for vehicles in truck category i

G_{jt} = growth factor for a given growth rate j and design period t

f_d = design lane factor

ESAL = equivalent accumulated 18,000 lb axle loads for all vehicles

n = number of truck categories

The solution of this problem is given in Table 18.8, which also demonstrates a tabular format for determining ESAL. Passenger vehicles are not considered in the calculations since their contribution to the ESAL is negligible.

The use of load equivalency factors for facilities such as residential streets, parkways, and parking lots, where the traffic is primarily automobiles with only occasional trucks, will result in low ESALs and therefore thin pavements that may not be capable of withstanding the occasional heavy traffic or environmental

Table 18.8 Tabular Solution of Example 18–2

Vehicle Type	Number of Vehicles During 1st year 1	Truck Factor (from Table 18.5) 2	Growth Factor for 5% Annual Growth Rate (from Table 18.6) 3	ESAL (1 × 2 × 3) 4
Single Unit Trucks				
2 axle, 4 tire	98,550	0.02	33.06	65,161
2 axle, 6 tire	43,800	0.21	33.06	304,086
3 axle or more	21,900	0.73	33.06	528,530
Tractor Semitrailers and Combinations				
3 axle	16,425	0.47	33.06	255,215
4 axle	5,475	0.83	33.06	150,233
Total				1,303,225

Note: Passenger vehicles are not considered here because their effect is negligible. Values under column 1 are obtained from the first year AADT and the design lane factor of 0.5. For example, for 2 axle, 6 tire single-unit trucks, the number of vehicles during first year = 3000 × 365 × 0.08 × 0.5 = 43,800.

effects. Thus, a reasonably accurate estimate of the occasional truck traffic for residential streets and parking lots should be made and included in the determination of the design ESAL. In cases where this is not possible, the minimum pavement thickness recommended by the Asphalt Institute should be used.[13] When it can be ascertained that the heavy truck representation on parkways restricted to automobiles and buses is less than 2 percent of the total traffic, Asphalt Institute suggests that the design ESAL may be obtained by multiplying the total traffic by the truck factor of 0.06.[14]

Subgrade Engineering Properties. As stated earlier, the subgrade consists of either the natural soil or soil that has been imported to form an embankment. This borrowed soil may be in its natural state or may be improved by stabilization. Stabilization of the natural soil existing on the alignment is not usually done, except when it is found that the material in its natural state cannot support heavy construction equipment. Stabilization of in situ subgrade material is therefore usually carried out to provide an adequate support platform for heavy construction equipment and is not considered in the design of the pavement.

The main engineering property required for the subgrade is its resilient modulus, which gives the resilient characteristic of the soil when it is repeatedly loaded with an axial load. It is determined in the laboratory by loading specially prepared samples of the soil with a deviator stress of fixed magnitude, frequency, and load duration while the specimen is triaxially loaded in a triaxial chamber. The method of conducting this test is described in detail in the Asphalt Institute's *Soils*

Manual for the Design of Asphalt Pavement Structures.[15] To facilitate the use of the more direct CBR and Hveem Stabilometer tests described in Chapter 15, the Asphalt Institute has determined conversion factors that can be used to convert the CBR and R values to the resilient modulus values. These are given as

$$M_r(MP_a) = 10.342 \text{ CBR}$$

$$M_r(\text{lb/in.}^2) = 1500 \text{ CBR}$$

or

$$M_r(MP_a) = 7.963 + 3.826 \,(R \text{ value})$$

$$M_r(\text{lb/in.}^2) = 1155 + 555 \,(R \text{ value})$$

where M_r is the equivalent resilient modulus.

The above conversion factors should be used only for materials that can be classified under the unified classification system as CL, CH, ML, SC, SM, and SP or when the resilient modulus is less than 30,000 lb/in.^{2}. For materials with higher values, direct measurement is recommended.[16]

Subbase and Base Engineering Properties. The material used for subbase or base courses must meet certain requirements, which are given in terms of the CBR, liquid limit, PI, particle size distribution (maximum percentage passing the No. 200 sieve), and minimum sand equivalent. The sand equivalent test is a rapid field test that shows the relative proportions of fine dust or claylike material in soils or graded aggregates. A detailed description of this test is given in *Soils Manual for the Design of Asphalt Pavement Structures.*[17] Table 18.9 gives the requirements for those engineering properties for soils that can be used as base or subbase in this method.

Table 18.9 Untreated Aggregate Base and Subbase Quality Requirements

Test	Test Requirements	
	Subbase	Base
CBR, minimum	20	80
or		
R value, minimum	55	78
Liquid limit, maximum	25	25
Plasticity index, maximum, or	6	NP
Sand equivalent, minimum	25	35
Passing No. 200 sieve, maximum	12	7

Source: Adapted from *Thickness Design—Asphalt Pavements for Highways and Streets*, Manual Series No. 1, The Asphalt Institute, College Park, Md., September 1981.

Step 2: Surface and Base Materials. The designer is free to select either an asphalt concrete surface or an emulsified asphalt surface, along with an asphalt concrete base, an emulsified asphalt base, or an untreated aggregate base and subbase for the underlying layers. This, of course, will depend on the material that is economically available.

However, the Asphalt Institute recommends certain grades of asphalt cement that should be used for different temperature conditions, as shown in Table 18.10. The grade of asphalt used should be selected primarily on the basis of its ability to satisfactorily coat the aggregates at the given temperatures.

Step 3: Minimum Thickness Requirements. The minimum thickness required for the design ESAL and the type of surface, base, and subbase selected is obtained by using the computer program DAMA or by entering the appropriate table or chart with the design ESAL and M_r of the subgrade and selecting the required minimum thickness, as demonstrated for different types of pavements in the following sections.

Full-Depth Asphalt Concrete. Pavements of this type use asphalt mixtures for all courses above the subgrade. The minimum thickness is determined from Figure 18.5 by extracting the pavement thickness for the design ESAL and subgrade M_r, both of which were determined in step 1. The procedure for determining the minimum depth required for this type of pavement is illustrated in Example 18–3.

Table 18.10 Recommended Asphalt Grades for Different Temperature Conditions

Temperature Condition	Asphalt Grades*	
Cold, mean annual air temperature $\leq 7°C$ (45°F)	AC-5, AR-2000, 120/150 pen.	AC-10 AR-4000 85/100 pen.
Warm, mean annual air temperature between 7°C (45°F) and 24°C (75°F)	AC-10, AR-4000, 85/100 pen.	AC-20 AR-8000 60/70 pen.
Hot, mean annual air temperature $\geq 24°C$ (75°F)	AC-20, AR-8000, 60/70 pen.	AC-40 AR-16000 40/50 pen.

*Both medium setting (MS) and slow setting (SS) emulsified asphalts are used in emulsified asphalt base mixes. They can be either of two types: cationic (ASTM D2397 or AASHTO M208) or anionic (ASTM D977 or AASHTO M140).

The grade of emulsified asphalt is selected primarily on the basis of its ability to satisfactorily coat the aggregate. This is determined by coating and stability tests (ASTM D244, AASHTO T59). Other factors important in the selection are the water availability at the job site, anticipated weather at the time of construction, the mixing process to be used, and the curing rate.

Source: Adapted from *Thickness Design—Asphalt Pavements for Highways and Streets*, Manual Series No. 1, The Asphalt Institute, College Park, Md., September 1981.

Example 18–3 Computing Thickness for a Full-Depth Asphalt Pavement

A full-depth asphalt pavement is to be constructed to carry an ESAL of 2,172,042. If the CBR for the subgrade is 10, determine the depth required for the asphalt layer.

$$M_r = 10 \times 1500 \text{ lb/in.}^2 = 15,000$$
$$\text{ESAL} = 2,172,042 = 2.172 \times 10^6$$

Using Figure 18.5, the depth required for a full-depth asphalt layer $= 9$ in.

Asphalt Concrete Surface and Emulsified Asphalt Base. These pavements have asphalt concrete as surface material and emulsified asphalt as base material. Three mix types of emulsified asphalt are used in this design, and they are defined by the Asphalt Institute as:

Type I: Emulsified asphalt mixes made with processed, dense-graded aggregates

Type II: Emulsified asphalt mixes made with semiprocessed, crusher-run, pit-run, or bank-run aggregate

Type III: Emulsified asphalt mixes made with sands or silty sands

Table 18.11 shows the recommended minimum thickness of asphalt concrete over types II and III emulsified asphalt bases. For pavements constructed with type I emulsified asphalt base, a surface treatment will be adequate. The depth of the emulsified asphalt base is determined as the difference between the total thickness (asphalt concrete surface and emulsified asphalt base) as obtained from the design charts and the minimum required thickness of the asphalt concrete, as

---

Table 18.11 Minimum Thickness of Asphalt Concrete over Emulsified Asphalt Bases

Traffic Level ESAL	Type II and Type III*	
	(mm)	*(in.)*
10^4	50	2
10^5	50	2
10^6	75	3
10^7	100	4
$> 10^7$	130	5

*Asphalt concrete, or Type I emulsified asphalt mix with a surface treatment, may be used over Type II or Type III emulsified asphalt base courses.

Source: Adapted from *Thickness Design—Asphalt Pavements for Highways and Streets*, Manual Series No. 1, The Asphalt Institute, College Park, Md., September 1981.

Figure 18.5   Design Chart for Full-Depth Asphalt

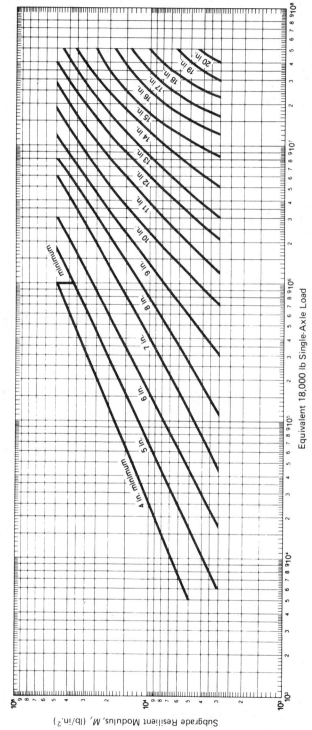

Source: Reproduced from *Thickness Design—Asphalt Pavements for Highways and Streets*, Manual Series No. 1, The Asphalt Institute, College Park, Md., September 1981.

Figure 18.6 Design Chart for Emulsified Asphalt Mix Type I

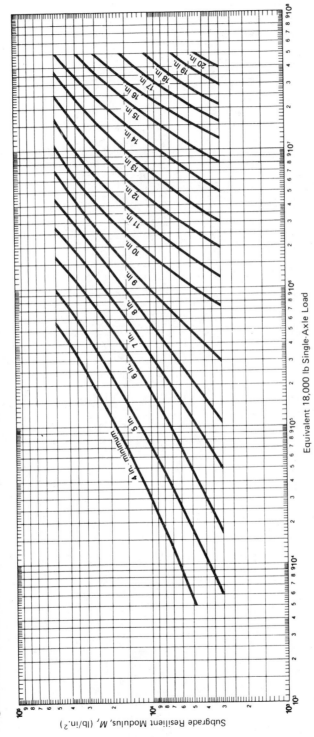

Equivalent 18,000 lb Single-Axle Load

Subgrade Resilient Modulus, M_r (lb/in.²)

Source: Reproduced from *Thickness Design—Asphalt Pavements for Highways and Streets*, Manual Series No. 1, The Asphalt Institute, College Park, Md., September 1981.

Figure 18.7 Design Chart for Emulsified Asphalt Mix Type II

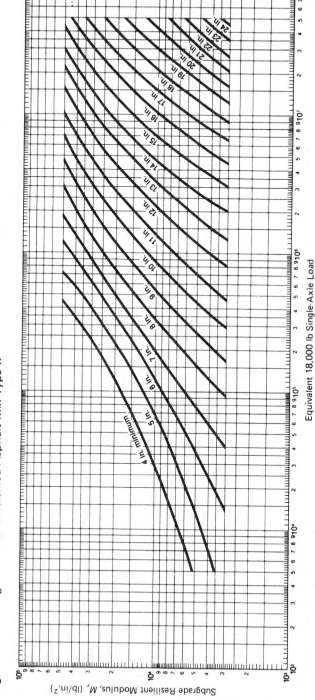

Equivalent 18,000 lb Single-Axle Load

Subgrade Resilient Modulus, M_r (lb/in.²)

Source: Reproduced from *Thickness Design—Asphalt Pavements for Highways and Streets*, Manual Series No. 1, The Asphalt Institute, College Park, Md., September 1981.

Figure 18.8 Design Chart for Emulsified Asphalt Mix Type III

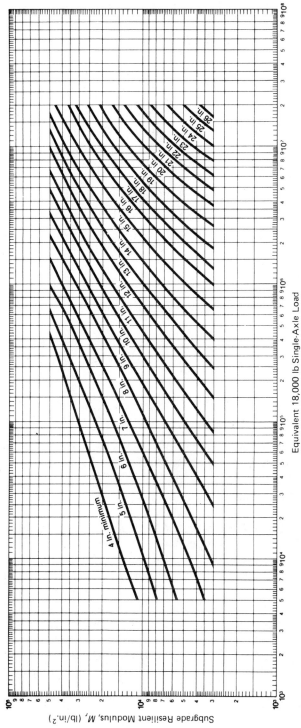

Equivalent 18,000 lb Single-Axle Load

Subgrade Resilient Modulus, M_r (lb/in.²)

Source: Reproduced from *Thickness Design—Asphalt Pavements for Highways and Streets*, Manual Series No. 1, The Asphalt Institute, College Park, Md., September 1981.

obtained from Table 18.11. This depth should not be less than the recommended minimum. Figures 18.6 through 18.8 give design charts for minimum depths of types I, II, and III emulsified asphalt bases. These charts were developed based on a 6 month curing period. The procedure for the design of this type of pavement is illustrated in Example 18–4.

Example 18–4   Designing a Pavement Consisting of Asphalt Concrete Surface and Emulsified Asphalt Base

Design a suitable pavement to carry a design ESAL of 1,303,225 on a subgrade having a resilient modulus of 15,000 lb/in.^{2}.

a.   Determine the thicknesses for the surface and base courses for a pavement consisting of an asphaltic concrete surface and type II emulsified asphalt base.

b.   Repeat (**a**) for type III emulsified asphalt base.

Solution for **a**,

> ESAL $= 1,303,225$
> $M_r = 15,000$ lb/in.^{2}
> Minimum asphalt concrete surface depth $= 3$ in. (from Table 18.11)
> Total thickness (surface and base) $= 10$ in. (from Figure 18.7)
> Base thickness $= (10 - 3)$ in. $= 7$ in.

Solution for **b**,

> Minimum asphalt concrete surface depth $= 3$ in. (from Table 18.11)
> Total pavement thickness $= 13$ in. (from Figure 18.8)
> Base thickness $= 10$ in.

Asphalt Concrete Surface and Untreated Aggregate Base.   These pavements consist of a layer of asphalt concrete over untreated aggregate base and subbase courses. As stated earlier, it is not always necessary to use a subbase course. Figures 18.9 through 18.14 give design charts for determining the thicknesses of asphalt concrete surfaces for different pavements constructed of asphalt concrete and untreated aggregates. These charts are given for different base thicknesses and are based on the quality requirements for base and subbase materials given in Table 18.9. The Asphalt Institute also recommends that the base course be not less than 6 in. Table 18.12 also gives the minimum recommended thicknesses for the asphalt concrete surface over the untreated aggregate base. These values depend on the design ESAL. In using the design charts, minimum thicknesses should not be extrapolated into higher traffic regions. The procedure for carrying out this design is illustrated in Example 18–5.

Figure 18.9 Design Chart for Pavements with Asphalt Concrete Surface and Untreated Aggregate Base 4 in. Thick

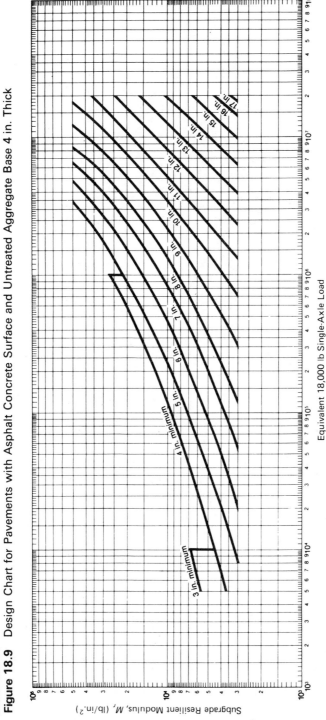

Equivalent 18,000 lb Single-Axle Load

Source: Reproduced from *Thickness Design—Asphalt Pavements for Highways and Streets*, Manual Series No. 1, The Asphalt Institute, College Park, Md., September 1981.

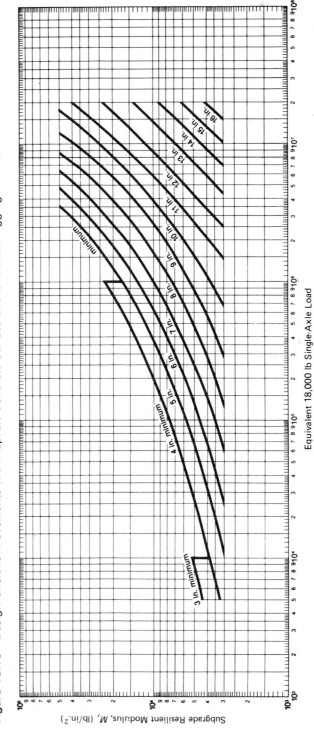

Figure 18.10 Design Chart for Pavements with Asphalt Concrete Surface and Untreated Aggregate Base 6 in. Thick

Source: Reproduced from *Thickness Design—Asphalt Pavements for Highways and Streets*, Manual Series No. 1, The Asphalt Institute, College Park, Md., September 1981.

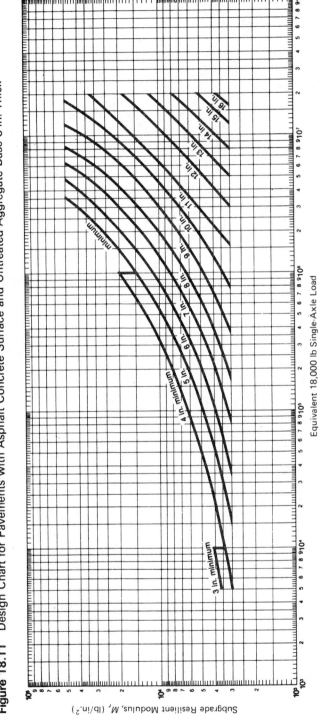

Figure 18.11   Design Chart for Pavements with Asphalt Concrete Surface and Untreated Aggregate Base 8 in. Thick

Source: Reproduced from *Thickness Design—Asphalt Pavements for Highways and Streets*, Manual Series No. 1, The Asphalt Institute, College Park, Md., September 1981.

Figure 18.12   Design Chart for Pavements with Asphalt Concrete Surface and Untreated Aggregate Base 10 in. Thick

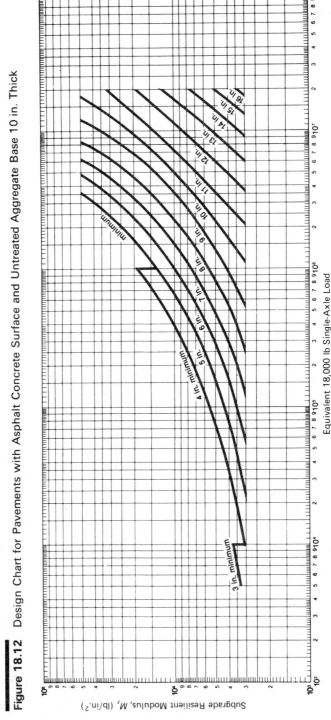

Source: Reproduced from *Thickness Design—Asphalt Pavements for Highways and Streets*, Manual Series No. 1, The Asphalt Institute, College Park, Md., September 1981.

Figure 18.13   Design Chart for Pavements with Asphalt Concrete Surface and Untreated Aggregate Base 12 in. Thick

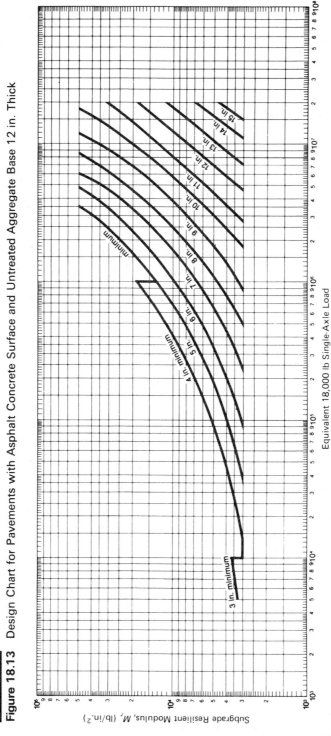

Equivalent 18,000 lb Single-Axle Load

Source: Reproduced from *Thickness Design—Asphalt Pavements for Highways and Streets*, Manual Series No. 1, The Asphalt Institute, College Park, Md., September 1981.

Figure 18.14   Design Chart for Pavements with Asphalt Concrete Surface and Untreated Aggregate Base 18 in. Thick

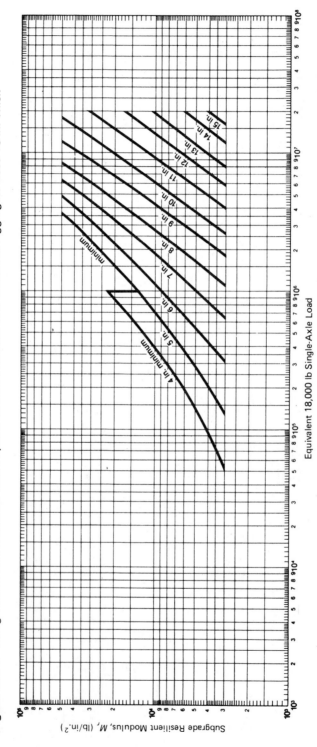

Source: Reproduced from *Thickness Design—Asphalt Pavements for Highways and Streets*, Manual Series No. 1, The Asphalt Institute, College Park, Md., September 1981.

Table 18.12 Minimum Thickness of Asphalt Concrete over Untreated Aggregate Base

Traffic ESAL	Traffic Condition	Minimum Thickness of Asphalt Concrete
10^4	Light traffic parking lots, driveways, and light traffic rural roads	75 mm (3.0 in.)*
10^4 but less than 10^6	Medium truck traffic	100 mm (4.0 in.)
10^6	Medium to heavy truck traffic	125 mm (5.0 in.) or greater

*For full-depth asphalt concrete or emulsified asphalt pavements, a minimum thickness of 100 mm (4 in.) applies in this traffic region, as shown on the design charts.

Source: Adapted from *Thickness Design—Asphalt Pavements for Highways and Streets*, Manual Series No. 1, The Asphalt Institute, College Park, Md., September 1981.

Example 18–5   Designing a Pavement Consisting of Asphalt Concrete Surface and Untreated Aggregate Base

Design a suitable pavement to carry the design ESAL of Example 18–4 on a subgrade having a resilient modulus of 15,000 lb/in.², with the pavement constructed of an asphalt concrete surface and a 6 in. untreated granular base course.

The minimum depth of the asphalt concrete surface is obtained from Figure 18.10 as 6 in. The design will therefore consist of 6 in. of asphalt concrete surface plus 6 in. of untreated granular base course.

An alternative design can be made with a 12 in. base. The depth of the asphalt concrete surface required for this is 5 in. as obtained from Figure 18.13. Since the minimum requirement for the depth of the base material is 6 in., the pavement may consist of 5 in. of asphalt concrete surface, 6 in. of untreated aggregate base course, and 6 in. of untreated aggregate subbase course.

Step 4: Feasibility of Planning Staged Construction.   Staged construction of a pavement structure involves construction in steps at specific times to meet the increase in traffic. When used as a design tool, it is presumed that a subsequent stage of the pavement will be constructed before the preceding stage shows any serious signs of distress. This method of construction usually is used when adequate funds are not available to construct the pavement to the required depth for the full design period of, say, 20 years. In such a case, the pavement may be constructed in two stages, with the first stage designed on an ESAL based on the first 10 years, and the first section overlayed sometime later to increase the depth to that required for the original design ESAL of 20 years. To ascertain that the planned overlay is placed before any serious distress occurs on the first stage, it is assumed that the cumulative damage in the first layer before the overlay will not be higher than 60 percent of its life. This ensures that at least 40 percent of the life of the first layer is still remaining.

The feasibility of staged construction should be evaluated when capital expenditure funds for the full depth of pavement are not available and when traffic growth rates are expected to be higher in the future.

The procedure for the design of a pavement using staged construction is demonstrated as follows. Let n_1 = design ESAL for the first stage and n_2 = design ESAL for the second stage. Then adjust n_1 and n_2 to take into account the hypothesis that not more than 60 percent of the cumulative damage of the first layer occurs before the second layer is applied. Thus, N_1 = adjusted design ESAL for stage 1 and N_2 = adjusted design ESAL for stage 2. It is assumed that the response of the pavement is elastic and that the damage at any stage is proportional to the ratio of the actual ESAL for that stage to the allowable ESAL for the thickness selected for that stage.

If D_1 = proportion of the life of the pavement expended (damage) at end of stage 1, then

$$D_1 = \frac{n_1}{N_1}$$

where

n_1 = actual accumulated ESALs for stage 1

N_1 = allowable number of ESALs for the initial thickness (h_1) selected for stage 1

At the end of stage 1, the proportion of the life of the pavement remaining is $(1 - D_1)$.

$$\frac{n_2}{N_2} = (1 - D_1)$$

where

n_2 = design ESAL for stage 2

N_2 = adjusted design ESAL for stage 2

If $D_1 = 0.6$, then

$$N_1 = 1.67n_1$$

$$N_2 = 2.50n_2$$

If h_1 and h_2 = design thicknesses for stage 1 and stage 2, respectively, then h_s = thickness to be added at stage 2—that is, $h_2 - h_1$. Thus, the respective design thicknesses are determined through the following steps.

1. Determine n_1 and n_2.
2. Determine N_1 and N_2.
3. Using appropriate charts, determine h_1 and h_2.
4. Determine h_s.

Example 18-6 Designing a Full-Depth Asphalt Concrete Pavement for a Two-Stage Construction

A full-depth asphalt concrete pavement is to be constructed in two stages. The design period is 20 years, and the second stage will be constructed 10 years after the first stage. If the ESAL on the design lane during the first year is 60,000 and the growth rate for all vehicles is 5 percent, determine the asphalt thicknesses for the first and second stages of construction if the subgrade resilient modulus is 15,000 $lb/in.^2$.

First we must determine the ESAL for the first 10 years and the ESAL for the 20 year design period.

> Growth rate $= 5$ percent
> Growth factor for 10 yr $= 12.58$ (from Table 18.6)
> Growth factor for 20 yr $= 33.06$ (from Table 18.6)
> $n_1 = 60,000 \times 12.58 = 754,800$
> $n_2 = 60,000(33.06 - 12.58) = 1,228,800$ (Note that the ESAL is given, not the AADT)
> $N_1 = 754,800 \times 1.67 = 1,260,516$
> $N_2 = 1,228,800 \times 2.5 = 3,072,000$

From Figure 18.5, we obtain the following.

> Required depth for first stage $(h_1) = 7.5$ in.
> Required depth for first and second stages $(h_2) = 9.5$ in.
> Depth of overlay $h_s = 2$ in.

Step 5: Economic Analysis and Design Selection. Examples 18-3 through 18-6 have demonstrated that several alternative designs can be obtained for the same design ESAL and subgrade resilient modulus, based on the type of pavement and whether or not planned stage construction is used. When alternative pavements are designed, it is necessary to carry out an economic evaluation of these alternatives, as presented in Chapter 11, to determine the best alternative.

AASHTO Design Method

The AASHTO method for design of highway pavements is based primarily on the results of the AASHTO road test that was conducted in Ottawa, Illinois. It was a cooperative effort carried out under the auspices of 49 states, the District of Columbia, Puerto Rico, the Bureau of Public Roads, and several industry groups. Tests were conducted on short-span bridges and test sections of flexible and rigid pavements constructed on A-6 subgrade material. The pavement test sections consisted of two small loops and four larger ones, each being a 4-lane divided highway. The tangent sections consisted of a successive set of pavement lengths of different designs, each length being at least 100 ft. The principal flexible pavement sections were constructed of asphaltic concrete surface, a well-graded crushed limestone base, and a uniformly graded sand-gravel subbase. Three levels of surface thicknesses ranging from 1 to 6 in. were used in combination with three levels of base thicknesses ranging from 0 to 9 in. Each of these nine combinations were then combined with three levels of subbase thicknesses ranging from 0 to 6 in. In addition to the crushed limestone bases, some special sections had bases constructed with either a well-graded uncrushed gravel, a bituminous plant mixture, or cement-treated aggregate.

Test traffic consisting of both single-axle and tandem-axle vehicles were then driven over the test sections until several thousand load repetitions had been made. Ten different axle-arrangement and axle-load combinations were used, with single-axle loads ranging from 2,000 to 30,000 lb, and tandem-axle loads ranging from 24,000 to 48,000 lb. Data were then collected on the pavement condition with respect to extent of cracking and amount of patching required to maintain the section in service. The longitudinal and transverse profiles were also obtained to determine the extent of rutting, surface deflection caused by loaded vehicles moving at very slow speeds, pavement curvature at different vehicle speeds, stresses imposed on the subgrade surface, and temperature distribution in the pavement layers. These data were then thoroughly analyzed, the results of which formed the basis for the AASHTO method of pavement design.

AASHTO initially published an interim guide for the design of pavement structures in 1961, which was revised in 1972.[18] A further revision was published in 1986, incorporating new developments and specifically addressing pavement rehabilitation.[19] Each edition of the guide specifically mentions that the design procedure presented cannot possibly include all conditions that relate to any one specific site. It is therefore recommended that, in using the guide, local experience be used to augment the procedures given in the guide.

Design Considerations

The factors considered in the AASHTO procedure for the design of flexible pavement as presented in the 1986 guide are

- Pavement performance
- Traffic
- Roadbed soils (subgrade material)
- Materials of construction
- Environment
- Drainage
- Reliability

Pavement Performance.   The primary factors considered under pavement performance are the structural and functional performance of the pavement. Structural performance is related to the physical condition of the pavement with respect to factors that have a negative impact on the capability of the pavement to carry the traffic load. These factors include cracking, faulting, raveling, and so forth. Functional performance is an indication of how effectively the pavement serves the user. The main factor considered under functional performance is riding comfort.

To quantify pavement performance, a concept known as the *serviceability performance* was developed.[20,21] Under this concept, a procedure was developed to determine the present serviceability index (PSI) of the pavement, based on its roughness and distress, which were measured in terms of extent of cracking, patching, and rut depth for flexible pavements. The original expression developed gave the PSI as a function of the extent and type of cracking and patching and the slope variance in the two wheel paths, which is a measure of the variations in the longitudinal profile. The mean of the ratings of individual engineers with wide experience in all facets of highway engineering was used to relate the PSI with the factors considered. The scale ranges for 0 to 5, where 0 is the lowest PSI and 5 is the highest. A detailed discussion of PSI is given in Chapter 20.

The serviceability indices are used in the design procedure: the initial serviceability index (p_i), which is the serviceability index immediately after the construction of the pavement, and the terminal serviceability index (p_t), which is the minimum acceptable value before resurfacing or reconstruction is necessary. In the AASHTO road test, a value of 4.2 was used for p_i for flexible pavements. AASHTO, however, recommends that each agency determine more reliable levels for p_i based on existing conditions. Recommended values for the terminal serviceability index are 2.5 or 3.0 for major highways and 2.0 for highways with a lower classification. In cases where economic constraints restrict capital expenditures for construction, the p_t can be taken as 1.5 or the performance period may be reduced. This low value should, however, be used only in special cases on selected classes of highways.[22]

Traffic.   The treatment of traffic load in the AASHTO design method is similar to that presented for the Asphalt Institute method, in that the traffic load application is given in terms of the number of 18,000 lb single-axle loads (ESALs). The procedure presented earlier is used to determine the design ESAL. The equivalence factors used in this case, however, are based on the terminal serviceability index to be used in the design and the structural number (SN) (see definition of SN under Structural Design). Table 18.13 gives traffic equivalence factors of 2.5 for p_t.

Roadbed Soils (Subgrade Material).   The 1986 AASHTO guide also uses the resilient modulus (M_r) of the soil to define its property. However, the method allows for the conversion of the CBR or R value of the soil to an equivalent M_r value using the following conversion factors.

M_r (lb/in.^{2}) = 1500 CBR   (for fine-grain soils with soaked CBR of 10 or less)
M_r lb/in.^{2} = 1000 + 555 × R value   (for R ≤ 20)

Table 18.13 Traffic Equivalence Factors for Flexible Pavements ($p_t = 2.5$)

Axle Load		Structural Number (SN)					
Kip	kN	1	2	3	4	5	6
(a) Single Axles							
2	8.9	0.0004	0.0004	0.0003	0.0002	0.0002	0.0002
4	17.8	0.003	0.004	0.004	0.003	0.002	0.002
6	26.7	0.01	0.02	0.02	0.01	0.01	0.01
8	35.6	0.03	0.05	0.05	0.04	0.03	0.03
10	44.5	0.08	0.10	0.12	0.10	0.09	0.08
12	53.4	0.17	0.20	0.23	0.21	0.19	0.18
14	62.3	0.33	0.36	0.40	0.39	0.36	0.34
16	71.2	0.59	0.61	0.65	0.65	0.62	0.61
18	80.1	1.00	1.00	1.00	1.00	1.00	1.00
20	89.0	1.61	1.57	1.49	1.47	1.51	1.55
22	97.9	2.48	2.38	2.17	2.09	2.18	2.30
24	106.8	3.69	3.49	3.09	2.89	3.03	3.27
26	115.7	5.33	4.99	4.31	3.91	4.09	4.48
28	124.6	7.49	6.98	5.90	5.21	5.39	5.98
30	133.4	10.31	9.50	7.94	6.83	6.97	7.79
32	142.3	13.90	12.82	10.52	8.85	8.88	9.95
34	151.2	18.41	16.94	13.74	11.34	11.18	12.51
36	160.1	24.02	22.04	17.73	14.38	13.93	15.50
38	169.0	30.90	28.30	22.61	18.06	17.20	18.98
40	177.9	39.26	35.89	28.51	22.50	21.08	23.04
(b) Tandem Axles							
10	44.5	0.01	0.01	0.01	0.01	0.01	0.01
12	53.4	0.02	0.02	0.02	0.02	0.01	0.01
14	62.3	0.03	0.04	0.04	0.03	0.03	0.02
16	71.2	0.04	0.07	0.07	0.06	0.05	0.04
18	80.1	0.07	0.10	0.11	0.09	0.08	0.07
20	89.0	0.11	0.14	0.16	0.14	0.12	0.11
22	97.9	0.16	0.20	0.23	0.21	0.18	0.17
24	106.8	0.23	0.27	0.31	0.29	0.26	0.24
26	115.7	0.33	0.37	0.42	0.40	0.36	0.34
28	124.6	0.45	0.49	0.55	0.53	0.50	0.47
30	133.4	0.61	0.65	0.70	0.70	0.66	0.63
32	142.3	0.81	0.84	0.89	0.89	0.86	0.83
34	151.2	1.06	1.08	1.11	1.11	1.09	1.08
36	160.1	1.38	1.38	1.38	1.38	1.38	1.38
38	169.0	1.75	1.73	1.69	1.68	1.70	1.73
40	177.9	2.21	2.16	2.06	2.03	2.08	2.14
42	186.8	2.76	2.67	2.49	2.43	2.51	2.61
44	195.7	3.41	3.27	2.99	2.88	3.00	3.16
46	204.6	4.18	3.98	3.58	3.40	3.55	3.79
48	213.5	5.08	4.80	4.25	3.98	4.17	4.49

Source: Adapted from *AASHTO Guide for Design of Pavement Structures*, Washington, D.C.: The American Association of State Highway and Transportation Officials, copyright 1986. Used by permission.

Note that the first term in the expression for converting R value to M_r is 1000 and not 1155 as used by the Asphalt Institute. Actually the Asphalt Institute developed the general relationship

$$M_r\,(\text{lb/in.}^2) = A + B \times (R \text{ value})$$

where A varies from 772 to 1155 and B varies from 369 to 555. The Asphalt Institute uses the maximum values of A and B, whereas AASHTO suggests using 1000 and 555, respectively.

Materials of Construction. The materials used for construction can be classified under three general groups: those used for subbase construction, those used for base construction, and those used for surface construction.

Subbase Construction Materials. The quality of the material used is determined in terms of the layer coefficient, a_3, which is used to convert the actual thickness of the subbase to an equivalent SN. The sandy gravel subbase course material used in the AASHTO road test was assigned a value of 0.11. Table 18.14 gives a summary of the results of a survey of values of layer coefficients assigned to different materials in different states. Layer coefficients are usually assigned, based on the description of the material used. Note, however, that due to the widely different environmental, traffic, and construction conditions, it is essential that each design agency develop layer coefficients appropriate to the conditions that exist in its own environment.

Charts correlating the layer coefficients with different soil engineering properties have been developed.[23] Figure 18.15 shows one such chart for granular subbase materials.

Base Course Construction Materials. Materials selected should satisfy the general requirements for base course materials given earlier in this chapter and in Chapter 16. A structural layer coefficient, a_2, for the material used should also be determined. This can be done using Figure 18.16.

Surface Course Construction Materials. The most commonly used material is a hot plant mix of asphalt cement and dense-graded aggregates with a maximum size of 1 in. The procedure discussed in Chapter 17 for the design of asphalt mix can be used. The structural layer coefficient (a_1) for the surface course can be extracted from Table 18.14 or Figure 18.17, which relates the structural layer coefficient of a dense-graded asphalt concrete surface course with its resilient modulus at 68°F.

Environment. Temperature and rainfall are the two main environmental factors used in evaluating pavement performance in the AASHTO method. The effects of temperature on asphalt pavements include stresses induced by thermal action, changes in the creep properties, and the effect of freezing and thawing of the subgrade soil as discussed in Chapters 15 and 16. The effect of rainfall is mainly due to the penetration of the surface water into the underlying material. If penetration occurs, the properties of the underlying materials may be significantly

Table 18.14 Summary of Structural Coefficients Used for Different Pavement Components

Component	Alabama	Arizona	Delaware	Massachusetts	Minnesota	Montana	Nevada	New Hampshire
Surface Courses								
Plant mix (high stab.)	0.44	.35–.44	.35–.40	0.44	0.315	.30–.40	.30–.35	0.38
Road mix (low stab.)	0.20	.25–.38				0.20	.17–.25	0.20
Sand asphalt	0.40	.25			Plant mix (low stab.) 0.28			0.20
Base courses Untreated	limestone 0.14, slag 0.14, sandstone 0.13, granite 0.12	sand and gravel, well graded 0.14, cinders .12–.14, sandy gravel, mostly sand .11–.13	waterbound macadam 0.20, crusher run .14, quarry waste 0.11, select borrow 0.08	crushed stone .14	crushed rock (Cl. 5 and 6 gravel) 0.14, sandy gravel 0.07	select surf 0.10, crushed gravel 0.12–0.14	crushed gravel .10–.12, crushed rock .13–.16	crushed gravel 0.10, bank run gravel .07, crushed stone 0.14
Cement treated 650 psi or more	0.23	500 psi (3.5 MPa) .25–.30	soil-cement			400 psi (2.8 MPa) or more 0.20		gravel
400 to 650 psi	0.20	300–500 psi (2.1–3.5 MPa) .18–.25	0.20					0.17
400 psi or less	0.15	less than 300 psi (2.1 MPa) 0.15				.15		

Lime treated							
Bituminous treated	coarse graded 0.30 / sand 0.25 / sand-gravel .25–.34 / sand .20	asph. stab. .10	black base 0.34 / penetrated crushed stone .29	0.175–0.21	.15–.20 / plant mix 0.30 / bit. stab. 0.20	plant mix .25–.34	bit. conc. 0.34 / gravel 0.24
Subbase	sand and sandy clay 0.11 / chert low P.I. 0.10 / top soil 0.09 / float gravel vel. 0.09 / sand and silty clay 0.05 / sand-gravel, well graded 0.14 / cr. stone or cinders 0.12 / sand and silty clay .05–.10	select borrow 0.08	gravel 0.11 / select material 0.08	sandy gravel (Cl. 3 and 4 gravel) 0.105 / selected granular (12% minus 0.075 mm (#200)) 0.07	sand 0.05 / sp. borrow 0.07	gravel type 1 .09–.11 / select material .05–.09	sand-gravel 0.05

Notes: 1. Indiana, Iowa, Montana, New Jersey, Tennessee, and Puerto Rico—conform to AASHTO Guides.

2. North Carolina—conforms to AASHTO Guides, except 0.30 for bituminous-treated base.

3. North Dakota—conforms to AASHTO Guides, except 0.30 for bituminous-aggregate base.

4. Maine—conforms to AASHTO Guides with some modification. No further information.

5. Maryland—substitution values for materials to replace design thickness of asphalt hot-mix are the AASHTO structural coefficients expressed in equivalent values, in inches.

Source: Adapted from *AASHTO Interim Guide for Design of Pavement Structures, 1972*, Chapter III Revised, 1981, Washington, D.C.: The American Association of State Highway and Transportation Officials, copyright 1972. Used by permission.

Figure 18.15 Variation in Granular Subbase Layer Coefficient, a_3, with Various Subbase Strength Parameters

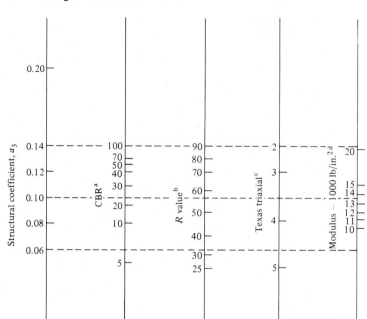

[a] Scale derived from correlations from Illinois.
[b] Scale derived from correlations obtained from The Asphalt Institute, California, New Mexico, and Wyoming.
[c] Scale derived from correlations obtained from Texas.
[d] Scale derived on NCHRP project 128, 1972.

Source: Redrawn from *AASHTO Guide for Design of Pavement Structures*, Washington, D.C.: The American Association of State Highway and Transportation Officials, copyright 1986. Used by permission.

altered. In Chapter 15, different ways of preventing water penetration were discussed. This effect is, however, taken into consideration in the design procedure, and the methodology used is presented later under drainage.

The effect of temperature, particularly with regard to the weakening of the underlying material during the thaw period, is considered a major factor in determining the strength of the underlying materials used in the design. Test results have shown that the normal modulus (that is, modulus during summer and fall seasons) of materials susceptible to frost action can reduce by 50 percent to 80 percent during the thaw period. Also resilient modulus of a subgrade material may also vary during the year, even when there is no specific thaw period. This occurs in areas subject to very heavy rains during specific periods of the year. It is likely that the strength of the material will be affected during the periods of heavy rains.

The procedure used to take into consideration the variation during the year of the resilient modulus of the roadbed soil is to determine an effective annual

Figure 18.16   Variation in Granular Base Layer Coefficient, a_2, with Various Subbase Strength Parameters

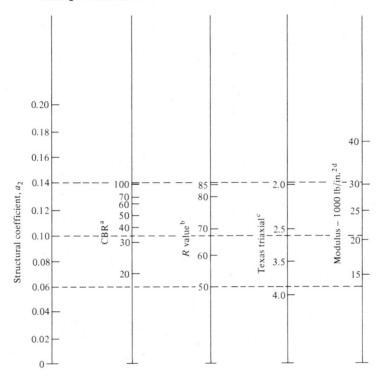

[a] Scale derived by averaging correlations obtained from Illinois.
[b] Scale derived by averaging correlations obtained from California,
New Mexico, and Wyoming.
[c] Scale derived by averaging correlations obtained from Texas.
[d] Scale derived on NCHRP project 128, 1972.

Source: Redrawn from *AASHTO Guide for Design of Pavement Structures*, Washington, D.C.: The American Association of State Highway and Transportation Officials, copyright 1986. Used by permission.

roadbed soil resilient modulus. The change in the PSI of the pavement during a full 12 month period will then be the same if the effective resilient modulus is used for the full period or if the appropriate resilient modulus for each season is used. This means that the effective resilient modulus is equivalent to the combined effect of the different seasonal moduli during the year.

The AASHTO guide suggests two methods for determining the effective resilient modulus. Only the first method is described here. In this method, a relationship between resilient modulus of the soil material and moisture content is developed using laboratory test results. This relationship is then used to determine the resilient modulus for each season based on the estimated in situ moisture content during the season being considered. The whole year is then divided into the different time intervals that correspond with the different seasonal resilient modulus. The AASHTO guide suggests that it is not necessary to use a time

Figure 18.17   Chart for Estimating Structural Layer Coefficient of Dense-Graded Asphalt Concrete Based on the Elastic (Resilient) Modulus

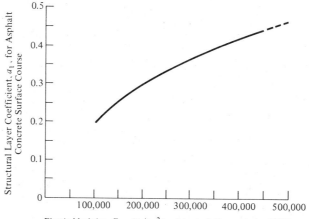

Elastic Modulus, E_{AC} (lb/in.^{2}), of Asphalt Concrete (at 68°F)

Source: Redrawn from *AASHTO Guide for Design of Pavement Structures*, Washington, D.C.: The American Association of State Highway and Transportation Officials, copyright 1986. Used by permission.

interval less than one-half month. The relative damage u_f for each time period is then determined from the chart in Figure 18.18, using the vertical scale, or the equation given in the chart. The mean relative damage $\bar{u}_f$ is then computed, and the effective subgrade resilient modulus is determined using the chart and the value of $\bar{u}_f$.

Example 18–7   Computing Effective Resilient Modulus

Figure 18.18 shows roadbed soil resilient modulus M_r for each month estimated from laboratory results correlating M_r with moisture content. Determine the effective resilient modulus of the subgrade.

Note that in this case, the moisture content does not vary within any one month. The solution of the problem is given in Figure 18.18. The value of u_f for each M_r is obtained directly from the chart. The mean relative damage $\bar{u}_f$ is 0.133, which in turn gives an effective resilient modulus of 7250 lb/in.^{2}.

Drainage.   The effect of drainage on the performance of flexible pavements is considered in the 1986 guide with respect to the effect water has on the strength of the base material and roadbed soil. The approach used is to provide for the rapid drainage of the free water (noncapillary) from the pavement structure by providing a suitable drainage layer, as shown in Figure 18.19, and by modifying the structural layer coefficient. The modification is carried out by incorporating a factor m_i for the base and subbase layer coefficients (a_2 and a_3). The m_i factors are based on the percentage of time during which the pavement structure will be nearly saturated, and the quality of drainage, which is dependent on the time it

Figure 18.18 Chart for Estimating Effective Roadbed Soil Resilient Modulus for Flexible Pavements Designed Using the Serviceability Criteria

Month	Roadbed Soil Modulus M_r (lb/in.²)	Relative Damage u_f
Jan.	22000	0.01
Feb.	22000	0.01
Mar.	5500	0.25
Apr.	5000	0.30
May	5000	0.30
June	8000	0.11
July	8000	0.11
Aug.	8000	0.11
Sept.	8500	0.09
Oct.	8500	0.09
Nov.	6000	0.20
Dec.	22000	0.01
Summation: $\Sigma u_f =$		1.59

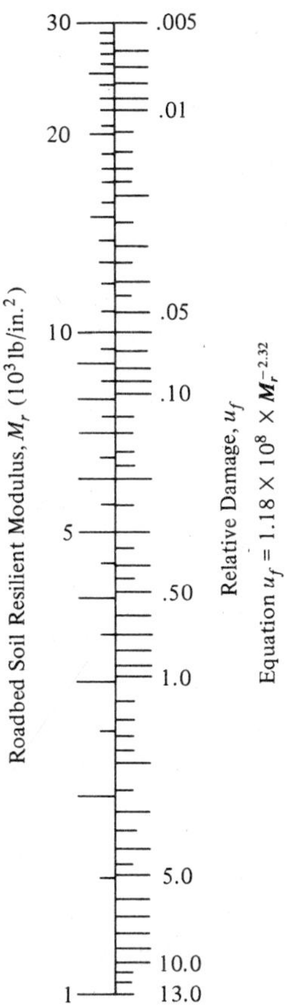

Average $\bar{u}_f = \dfrac{\Sigma u_f}{n} = \dfrac{1.59}{12} = 0.133$

Effective Roadbed Soil Resilient Modulus, M_r (lb/in.²) = 7250 (corresponds to $\bar{u}_f$)

Source: Redrawn from *AASHTO Guide for Design of Pavement Structures*, Washington, D.C.: The American Association of State Highway and Transportation Officials, copyright 1986. Used by permission.

takes to drain the base layer to 50 percent of saturation. Table 18.15 gives the general definitions of the different levels of drainage quality, and Table 18.16 gives recommended m_i values for different levels of drainage quality.

Reliability. It has been pointed out that the cumulative ESAL is an important input to any pavement design method. However, the determination of this input is usually based on assumed growth rates, which may not be accurate. Most design methods do not consider this uncertainty, but the 1986 AASHTO guide proposes the use of a reliability factor that considers the possible uncertainties in traffic prediction and performance prediction. A detailed discussion of the

Figure 18.19   Example of Drainage Layer in Pavement Structure

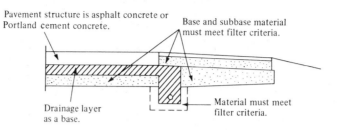

Pavement structure is asphalt concrete or
Portland cement concrete.

Base and subbase material
must meet filter criteria.

Drainage layer
as a base.

Material must meet
filter criteria.

(a) **Base is used as the drainage layer.**

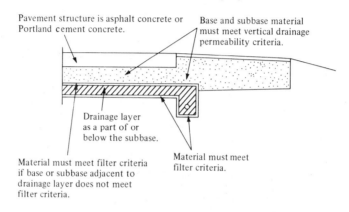

Pavement structure is asphalt concrete or
Portland cement concrete.

Base and subbase material
must meet vertical drainage
permeability criteria.

Drainage layer
as a part of or
below the subbase.

Material must meet filter criteria
if base or subbase adjacent to
drainage layer does not meet
filter criteria.

Material must meet
filter criteria.

(b) **Drainage layer is part of or below the subbase.**

Note: Filter fabrics may be used in lieu of filter
        material, soil, or aggregate, depending on
        economic considerations.

Source: Redrawn from *AASHTO Guide for Design of Pavement Structures*, Washington, D.C.: The
American Association of State Highway and Transportation Officials, copyright 1986. Used by
permission.

Table 18.15   Definition of Drainage Quality

Quality of Drainage	Water Removed Within*
Excellent	2 hours
Good	1 day
Fair	1 week
Poor	1 month
Very poor	(water will not drain)

*Time required to drain the base layer to 50 percent saturation.

Source: Adapted from *AASHTO Guide for Design of Pavement Structures*, Washington, D.C.: The
American Association of State Highway and Transportation Officials, copyright 1986. Used by
permission.

Table 18.16 Recommended m_i Values

Quality of Drainage	Percent of Time Pavement Structure Is Exposed to Moisture Levels Approaching Saturation			
	Less Than 1 Percent	1–5 Percent	5–25 Percent	Greater Than 25 Percent
Excellent	1.40–1.35	1.35–1.30	1.30–1.20	1.20
Good	1.35–1.25	1.25–1.15	1.15–1.00	1.00
Fair	1.25–1.15	1.15–1.05	1.00–0.80	0.80
Poor	1.15–1.05	1.05–0.80	0.80–0.60	0.60
Very poor	1.05–0.95	0.95–0.75	0.75–0.40	0.40

Source: Adapted from *AASHTO Guide for Design of Pavement Structures*, Washington, D.C.: The American Association of State Highway and Transportation Officials, copyright 1986. Used by permission.

development of the approach used is beyond the scope of this book, however, a general description of the methodology is presented to allow the incorporation of reliability in the design process, if so desired by a designer. Reliability design levels ($R\%$), which determine assurance levels that the pavement section designed using the procedure will survive for its design period, have been developed for different types of highways. For example, a 50 percent reliability design level implies a 50 percent chance for successful pavement performance—that is, the probability of design performance success is 50 percent. Table 18.17 shows suggested reliability levels, based on a survey of the AASHTO pavement design task force. Reliability factors, $F_R \geq 1$, based on the reliability level selected and the overall variation, S_o^2, have also been developed. S_o^2 accounts for the chance variation in the traffic forecast and the chance variation in actual pavement performance for a given design period traffic, W_{18}.

Table 18.17 Suggested Levels of Reliability for Various Functional Classifications

Functional Classification	Recommended Level of Reliability	
	Urban	Rural
Interstate and other freeways	85–99.9	80–99.9
Other principal arterials	80–99	75–95
Collectors	80–95	75–95
Local	50–80	50–80

Note: Results based on a survey of the AASHTO Pavement Design Task Force.

Source: Adapted from *AASHTO Guide for Design of Pavement Structures*, Washington, D.C.: The American Association of State Highway and Transportation Officials, copyright 1986. Used by permission.

The reliability factor F_R is given as

$$\log_{10} F_R = -Z_R S_o \qquad (18.1)$$

where

Z_R = standard normal variate for a given reliability ($R\%$)
S_o = estimated overall standard deviation

Table 18.18 gives values of Z_R for different reliability levels R.

Overall standard deviation ranges have been identified for flexible and rigid pavements as

	Standard Deviation, S_o
Flexible pavements	0.40–0.5
Rigid pavements	0.30–0.40

These values were derived through a detailed analysis of existing data. However, very little data presently exist for certain design components, such as drainage. A methodology for improving these estimates is presented in the 1986 AASHTO guide, which may be used when additional data are available.

Table 18.18   Standard Normal Deviation (Z_R) Values Corresponding to Selected Levels of Reliability

Reliability (R%)	Standard Normal Deviation, Z_R
50	−0.000
60	−0.253
70	−0.524
75	−0.674
80	−0.841
85	−1.037
90	−1.282
91	−1.340
92	−1.405
93	−1.476
94	−1.555
95	−1.645
96	−1.751
97	−1.881
98	−2.054
99	−2.327
99.9	−3.090
99.99	−3.750

Source: Adapted from *AASHTO Guide for Design of Pavement Structures*, Washington, D.C.: The American Association of State Highway and Transportation Officials, copyright 1986. Used by permission.

Structural Design

The objective of the design using the AASHTO method is to determine a flexible pavement SN adequate to carry the projected design ESAL. It is left to the designer to select the type of surface used, which can be either asphalt concrete, a single surface treatment, or a double surface treatment. This design procedure is used for ESALs greater than 50,000 for the performance period. The design for ESALs less than this is usually considered under low-volume roads.

The 1986 AASHTO guide gives the expression for SN as

$$SN = a_1D_1 + a_2D_2m_2 + a_3D_3m_3 \tag{18.2}$$

where

m_i = drainage coefficient for layer i

a_1, a_2, a_3 = layer coefficients representative of surface, base, and subbase course, respectively

D_1, D_2, D_3 = actual thickness in inches of surface, base, and subbase courses, respectively

The basic design equation given in the 1986 guide is

$$\log_{10} W_{18} = Z_R S_o + 9.36 \log_{10}(SN + 1) - 0.20 + \frac{\log_{10}[\Delta PSI/(4.2 - 1.5)]}{0.40 + [1094/(SN + 1)^{5.19}]}$$

$$+ 2.32 \log_{10} M_r - 8.07 \tag{18.3}$$

where

W_{18} = predicted number of 18,000 lb (80 kN) single-axle load applications

Z_R = standard normal deviation for a given reliability

S_o = overall standard deviation

SN = structural number indicative of the total pavement thickness

$\Delta PSI = p_i - p_t$

p_i = initial serviceability index

p_t = terminal serviceability index

M_r = resilient modulus in lb/in.^{2}

Eq. 18.3 can be solved for SN using a computer program or the chart in Figure 18.20. The use of the chart is demonstrated by the example solved on the chart and in the solution of Example 18–8.

Example 18–8 Designing a Flexible Pavement Using the AASHTO Method

A flexible pavement for an urban interstate highway is to be designed using the 1986 AASHTO guide procedure to carry a design ESAL of 2×10^6. It is estimated that it takes about a week for water to be drained from within the pavement and

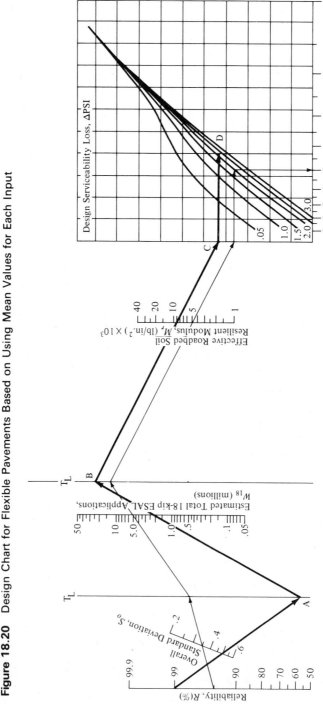

Figure 18.20   Design Chart for Flexible Pavements Based on Using Mean Values for Each Input

Source: Redrawn from *A Policy on Geometric Design of Highways and Streets*, Washington, D.C.: The American Association of State Highway and Transportation Officials, copyright 1984. Used by permission.

the pavement structure will be exposed to moisture levels approaching saturation for 30 percent of the time. The following additional information is available.

Resilient modulus of asphalt concrete at $68°F = 450,000$ lb/in.^{2}
CBR value of base course material $= 100$, $M_r = 31,000$ lb/in.^{2}
CBR value of subbase course material $= 22$, $M_r = 13,500$ lb/in.^{2}
CBR value of subgrade material $= 6$

Determine a suitable pavement structure, M_r of subgrade $= 6 \times 1500$ lb/in.$^2 = 9000$ lb/in.^{2}.

Since the pavement is to be designed for an interstate highway, the following assumptions are made.

Reliability level $(R) = 99$ percent (range is 80–99.9 from Table 18.17)
Standard deviation $(S_o) = 0.49$ (range is 0.4–0.5)
Initial serviceability index $p_i = 4.5$
Terminal serviceability index $p_t = 2.5$

The nomograph in Figure 18.20 is used to determine the design SN through the following steps.

Step 1. Draw a line joining the reliability level of 99 percent and the overall variance S_o of 0.49 and extend this line to intersect the first T_L line at point A.

Step 2. Draw a line joining point A to the ESAL of 2×10^6 and extend this line to intersect the second T_L line at point B.

Step 3. Draw a line joining point B and resilient modulus (M_r) of the roadbed soil and extend this line to intersect the design serviceability loss chart at point C.

Step 4. Draw a horizontal line from point C to intersect the design serviceability loss (ΔPSI) curve at point D. In this problem $\Delta PSI = 4.5 - 2.5 = 2$.

Step 5. Draw a vertical line to intersect the design SN and read off this value. SN $= 4.4$.

Step 6. Determine the appropriate structure layer coefficient for each construction material.

a. Resilient value of asphalt cement $= 450,000$ lb/in.^{2}. From Figure 18.17, $a_1 = 0.44$.

b. CBR of base course material $= 100$. From Figure 18.16, $a_2 = 0.14$.

c. CBR of subbase course material $= 22$. From Figure 18.15, $a_3 = 0.10$.

Step 7. Determine appropriate drainage coefficient m_i. Since only one set of conditions is given for both the base and subbase layers, the same value will be used for m_1 and m_2. The time required for water to drain from within pavement $= 1$ week, and from Table 18.15, drainage quality is fair. The percentage of time pavement structure will be exposed to moisture levels approaching saturation = 30, and from Table 18.16, $m_i = 0.80$.

Step 8.   Determine appropriate layer thicknesses from Eq. 18.3

$$SN = a_1D_1 + a_2D_2m_2 + a_3D_3m_3$$

It can be seen that several values of D_1, D_2, and D_3 can be obtained to satisfy the SN value of 4.40. Layer thicknesses, however, are usually rounded up to the nearest 0.5 in.

The selection of the different layer thicknesses should also be based on constraints associated with maintenance and construction practices, so that a practical design is obtained. For example, it is normally impractical and uneconomical to construct any layer with a thickness less than some minimum value. Table 18.19 lists minimum thicknesses suggested by AASHTO.

Taking into consideration that a flexible pavement structure is a layered system, the determination of the different thicknesses should be carried out as indicated in Figure 18.21. The required SN above the subgrade is first determined, then the required SNs above the base and subbase layers are determined using the appropriate strength of each layer. The minimum allowable thickness of each layer can then be determined using the differences of the computed SNs as shown in Figure 18.21.

Using the appropriate values for M_r in Figure 18.20, we obtain $SN_3 = 4.4$ and $SN_2 = 3.8$. Note that when SN is assumed to compute ESAL, the assumed and computed SN values must be approximately equal. If these are significantly different, the computation must be repeated with a new assumed SN.

We know

$$M_r \text{ for base course} = 31,000 \text{ lb/in.}^2$$

Using this value in Figure 18.20, we obtain

$$SN_1 = 2.6$$

Table 18.19   AASHTO Recommended Minimum Thicknesses of Highway Layers

Traffic, ESALs	Minimum Thickness (in.)	
	Asphalt Concrete	Aggregate Base
Less than 50,000	1.0 (or surface treatment)	4
50,001–150,000	2.0	4
150,001–500,000	2.5	4
500,001–2,000,000	3.0	6
2,000,001–7,000,000	3.5	6
Greater than 7,000,000	4.0	6

Source: Adapted from *AASHTO Guide for Design of Pavement Structures*, Washington, D.C.: The American Association of State Highway and Transportation Officials, copyright 1986. Used by permission.

Figure 18.21   Procedure for Determining Thicknesses of Layers Using a Layered Analysis Approach

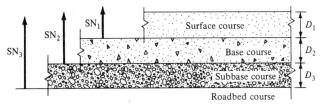

Roadbed course

Source: Redrawn from *AASHTO Guide for Design of Pavement Structures*, Washington, D.C.: The American Association of State Highway and Transportation Officials, copyright 1986. Used by permission.

giving

$$D_1 = \frac{2.6}{0.44} = 5.9 \text{ in.}$$

Using 6 in. for the thickness of the surface course,

$D^*_1 = 6$ in.

$SN^*_1 = a_1 D^*_1 = 0.44 \times 6 = 2.64$

$D^*_2 \geq \dfrac{SN_2 - SN^*_1}{a_2 m_2} \geq \dfrac{3.8 - 2.64}{0.14 \times 0.8} \geq 10.36 \text{ in.}$ (use 12 in.)

$SN^*_2 = 0.14 \times 0.8 \times 12 + 2.64 = 1.34 + 2.64$

$D^*_3 = \dfrac{SN_3 - SN^*_2}{a_3 m_3} = 4.4 - \dfrac{(2.64 + 1.34)}{0.1 \times 0.8} = 5.25 \text{ in.}$ (use 6 in.)

$SN^*_3 = 2.64 + 1.34 + 6 \times 0.8 \times 0.1 = 4.46$

The pavement will therefore consist of 6 in. asphalt concrete surface, 12 in. granular base, and 6 in. subbase.

California (Hveem) Design Method

This method was originally developed in the 1940s, based on a combination of information obtained from some test roads, theory, and experience.[24] It is widely used in the western states. It has been modified several times to take into consideration changes in traffic characteristics. The objective of the original design method was to avoid plastic deformation and the distortion of the pavement

*An asterisk with *D* on SN indicates that it represents the value actually used, which must be equal to or greater than the required value.

surface, but a later modification includes the reduction to a minimum of early fatigue cracking due to traffic load.

The factors considered are

- Traffic load
- Strength of subgrade material
- Strength of construction materials

The traffic load is initially calculated as the ESAL—that is, the total number of 18,000 lb axle loads in one direction—as discussed earlier, and then converted to a traffic index (TI), where

$$TI = 9.0\left(\frac{ESAL}{10^6}\right)^{0.119} \tag{18.4}$$

In determining the ESAL, passenger vehicles and pickups are not considered.

The strength of the subgrade material is given in terms of the resistance value R of the subgrade soil, obtained from the Hveem Stabilometer test described in Chapter 16.

The strength characteristics of each construction material (asphalt concrete, base, and subbase materials) are given in terms of a gravel equivalent factor (G_f), which has been determined for different types of materials. The factors given for asphalt concrete depend on the TI, as shown in Table 18.20.

Structural Design

The objective of the design is to determine the total thickness of material required above the subgrade to carry the projected traffic load. This thickness is determined in terms of gravel equivalent (GE) in feet, which is given as

$$GE = 0.0032(TI)(100 - R) \tag{18.5}$$

where

GE = thickness of material required above a given layer in terms of gravel equivalent in feet

TI = traffic index

R = resistance value of the supporting layer material, normally determined at an exudation pressure of 300 lb/in.²

The actual depth of each layer can then be determined by dividing GE for that layer by the GE factor G_f for the material used in the layer. A check is usually

Table 18.20 Gravel Equivalent Factors for Different Types of Materials

Material	G_f
Cement-treated base	
Class A	1.7
B	1.2
Lime-treated base	1.2
Untreated aggregate base	1.1
Aggregate subbase	1.0
Asphalt concrete for TI of	
$\leqslant 5.0$	2.50
5.5–6.0	2.32
6.5–7.0	2.14
7.5–8.0	2.01
8.5–9.0	1.89
9.5–10.0	1.79
10.5–11.0	1.71
13.5–14.0	1.52

Source: Adapted from *California Department of Transportation Design Manual*, California Division of Highways, Sacramento, Calif., 1975.

made to ascertain that the thickness obtained is adequate for the expansion pressure requirements.

Example 18–9 Designing a Flexible Pavement Using the Hveem Method

A flexible pavement is to be designed to carry a 1.5×10^6 ESAL during its design period. Using the California (Hveem) method, determine appropriate thicknesses for an asphalt concrete surface and a granular base course over a subgrade soil, for which stabilometer test data are shown in Table 18.21. The R value for the untreated granular base material is 70.

The design is carried out through the following steps.

Step 1. Determine TI using Eq. 18.4

$$TI = 9.0\left(\frac{1.5 \times 10^6}{10^6}\right)^{0.119} = 9.44 \quad \text{(say 9.5)}$$

Step 2. Determine the asphalt concrete surface thickness over the granular base. Use Eq. 18.5 to determine GE, with $R = 70$ and TI $= 9.5$.

$$GE = 0.0032(9.5)(100 - 70) = 0.91 \text{ ft}$$

Table 18.21   Stabilometer Test Data for Subgrade Soil of Example 18–9

Moisture Content (%)	R Value	Exudation Pressure (lb/in.)	Expansion Pressure (lb/in.)	GE (ft) (from Eq. 8.5)	Expansion Pressure Thickness (ft)
25.2	48	550	1.03	1.6	1.13
25.8	37	412	0.30	1.9	0.33
29.3	15	80	0.00	2.6	0.00

The G_f factor from Table 18.20 is 1.79.

$$\text{actual depth required} = 0.91/1.79 = 0.51 \text{ ft}$$
$$= 6.1 \text{ in.} \quad (\text{say, } 6.0 \text{ in.})$$

Step 3. Determine base thickness. The total thickness required above the subgrade should first be determined for an exudation pressure of 300 lb/in.² This is obtained by plotting a graph of GE (given in Table 18.21) for all layers above the subgrade versus the exudation pressure as shown in Figure 18.22(a). The GE is obtained as 2.05 ft and G_f for untreated aggregate base = 1.1. The base thickness is determined from the difference in total GE required over subgrade material and the GE provided by the asphalt base.

$$\text{GE of base layer} = 2.05 - \frac{6.0}{12} \times 1.79$$

Therefore,

$$\text{base thickness} = \left(2.05 - \frac{6.0}{12} \times 1.79\right)/1.1 \text{ ft}$$
$$= 12.6 \text{ in.} \quad (\text{say, } 12\tfrac{1}{2} \text{ in.})$$

Step 4. Check to see whether strength design satisfies expansion pressure requirements. The moisture content that corresponds with the strength design thickness (that is, at exudation pressure of 300 lb/in.²) is obtained by plotting the GE values for all layers above the subgrade versus the moisture content (given in Table 18.21) as shown in curve B in Figure 18.22(b). This moisture content is determined as 27 percent. The thickness required to resist expansion at 27 percent moisture content is then determined from curve C, which is a plot of expansion pressure thickness against moisture content. Thickness required to prevent expansion at 27 percent moisture content is 0.2 ft, which shows that strength design satisfies expansion pressure requirements, since the total thickness of the materials above the subgrade is much higher than 0.2 ft.

Figure 18.22   Plot of Test Data for Example 18–9

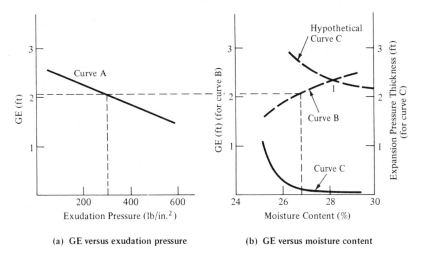

(a) **GE versus exudation pressure**   (b) **GE versus moisture content**

When the thickness based on strength design does not satisfy the expansion pressure requirements, a balanced design should be adopted, where the design point is the intersection of the curves of GE (for strength) versus moisture content and GE (for expansion pressure requirement) versus moisture content. A hypothetical case is shown in Figure 18.22, where the balanced point is point 1, indicating that the total thickness GE must be increased to about 2.4 ft and compaction of the embankment must take place at a moisture content of 28 percent rather than at 27 percent.

SUMMARY

The design of flexible pavements basically involves the determination of the strength characteristics of the materials of the pavement surface and underlying materials, and then determining the respective thicknesses of the subbase (if any), base course, and pavement surface that should be placed over the native soil. The pavement is therefore usually considered as a multilayered elastic system. The thicknesses provided should be adequate to prevent excessive cracking and permanent deformation beyond certain limits. These limits are considered in terms of required load characteristics, which can be determined as the number of repetitions of 18,000 lb single-axle loads the pavement is expected to carry during its design life.

The availability of computerized solutions for multilayered systems has led to the development of several design methods. Those commonly used in the United States are the Asphalt Institute method, the AASHTO method, and the California (Hveem) method, all of which are presented in this chapter.

PROBLEMS

18–1 An axle weight study on a section of a highway gave the following data on axle-load distribution:

Axle Load Group: Single (1000 lb)	No. of Axles per 1000 Vehicles	Axle Load Group: Tandem (1000 lb)	No. of Axles per 1000 Vehicles
<3	678	<6	18
3–7	600	6–12	236
7–8	175	12–18	170
8–12	500	18–24	120
12–16	150	24–30	152
16–18	60	30–32	66
18–20	40	32–34	30
20–22	7	34–36	12
22–24	4	36–38	4
24–26	3	38–40	1

Determine the truck factor for this section of the highway.

18–2 A 6-lane divided highway is to be designed to replace an existing highway. The present AADT (both directions) of 6000 vehicles is expected to grow at 5 percent per annum. Determine the design ESAL if the design life is 20 years from now and the vehicle mix is

Passenger cars (1000 lb/axle) = 60 percent
2-axle single-unit trucks (5000 lb/axle) = 30 percent
3-axle single-unit trucks (7000 lb/axle) = 10 percent

18–3 Conduct a survey of the district engineers in your state to determine the extent to which stabilized bases have been and are being used. Identify the predominant stabilization methods used and write a report on the reason for stabilization and the method of construction.

18–4 A section of a 2-lane rural highway is to be realigned and replaced by a 4-lane highway with a full-depth asphalt pavement. The AADT (both ways) on the existing section can be represented by 500 ESAL. It is expected that construction will be completed 5 years from now. If the traffic growth rate is 5 percent and the CBR of the subgrade on the new alignment is 10, determine the depth of the asphalt pavement, using the Asphalt Institute method. Take the design life of the pavement as 20 years.

18–5 The predicted traffic mix of a proposed 4-lane urban expressway is

Passenger cars = 69 percent
Single-unit trucks
    2-axle, 4-tire = 20 percent
    2-axle, 6-tire = 5 percent
    3-axle or more = 4 percent
Tractor semitrailers and combinations
    3-axle = 2 percent

The projected AADT during the first year of operation is 3500 (both directions). If the traffic growth rate is estimated at 4 percent and the CBR of the subgrade is 9, determine the depth of a full-asphalt pavement using the Asphalt Institute method, and $n = 20$ years.

18–6 Using the information given in problem 18–2, design a suitable pavement consisting of an asphalt concrete surface and a type III emulsified asphalt base on a subgrade having a CBR of 10.

18–7 Repeat problem 18–6 for a pavement consisting of an asphalt concrete surface and 6 in. untreated aggregate base.

18–8 Repeat problem 18–4 using four different depths of untreated aggregate bases ranging from the minimum 6 in. to a maximum of 12 in. Contact highway contractors in your area and obtain the rates for providing and properly laying asphalt concrete surface and untreated granular base. With these rates determine the cost for constructing the different pavement designs if the highway section is 5 mi long and lane width is 12 ft. Which design will you select for construction?

18–9 The design life of a proposed 4-lane, urban expressway is 20 yr. The estimated ESAL (both directions) during the first year of operation is 150,000 with a growth rate of 5 percent. Design a full-depth asphalt concrete (using the Asphalt Institute method) to be constructed in two stages, with the second stage to be constructed 10 yr after the first stage. The resilient modulus of the subgrade is 15,000 lb/in.^{2}.

18–10 The traffic on the design lane of a proposed 4-lane rural interstate highway consists of 40 percent trucks. If classification studies have shown that the truck factor can be taken as 0.45, design a suitable flexible pavement using the 1986 AASHTO procedure if the AADT on the design lane during first year of operation is 1000, and $p_i = 4.2$.

> Growth rate = 4 percent
> Design life = 20 yr
> Reliability level = 95 percent
> Standard deviation = 0.45

The pavement structure will be exposed to moisture levels approaching saturation 20 percent of the time, and it will take about 1 week for drainage of water. CBR of the subgrade material is 7. CBR of the base and subbase are 70 and 22, respectively, and M_r for the asphalt, concrete, base, and subbase are 450,000, 28,000, and 13,500 lb/in.^{2}, respectively.

18–11 Repeat problem 18–10, with the subgrade M_r values for each month from January through December being 20,000, 20,000, 6,000, 6,000, 6,000, 9,000, 9,000, 9,000, 9,500, 9,500, 8,000, and 20,000 lb/in.^{2}, respectively. The pavement structure will be exposed to moisture levels approaching saturation for 20 percent of the time, and it will take about 4 weeks for drainage of water from the pavement. Use untreated sandy gravel for subbase and untreated granular material for the base course.

18–12 An existing 2-lane rural highway is to be replaced by a 4-lane divided highway on a new alignment. Construction of the new highway will commence 2 years from now and is expected to take 3 years to complete. The design life of the pavement is 20 years. The present AADT (two way) on the 2-lane highway is 3600, consisting of 60 percent passenger cars, 20 percent 2-axle, 6-tire single-unit trucks, 15 percent 3-axle, and 5 percent 4-axle tractor semitrailer combinations. Design a flexible pavement consisting of asphalt concrete surface and lime-treated base, using the California (Hveem) method. The results of a stabilometer test on the subgrade soil are

Moisture Content (%)	R Value	Exudation Pressure (lb/in.²)	Expansion Pressure Thickness (ft)
19.8	55	575	1.00
22.1	45	435	0.15
24.9	16	165	0.10

Assume the R value for the lime-treated base is 80 and a traffic growth rate of 5 percent annually.

REFERENCES

1. *Soil Stabilization in Pavement Structures, A Users Manual*, Vols. I and II, U.S. Department of Transportation, Federal Highway Administration, Washington, D.C., October 1979.

2. *Soil Cement Laboratory Handbook*, Portland Cement Association, Skokie, Ill., 1971.

3. Ibid.

4. *Annual Book of ASTM Standards*, Section 4, American Society for Testing and Materials, Philadelphia, Pa., 1985.

5. C. M. Gerrard and L. I. Wardle, "Rational Design of Surface Pavement Layers," *Australia Road Research* 10(1980).

6. H. F. Southgate, R. C. Deen, and J. G. Moyer, *Strain Energy Analysis of Pavement Designs for Heavy Trucks*, Transportation Research Record 949, National Research Council, Transportation Research Board, Washington, D.C., 1983.

7. G. M. Dorman and C. T. Metcalf, "Design Curves for Flexible Pavements Based on Layered Systems Theory," *Highway Research Record* 71(1964).

8. *Thickness Design—Asphalt Pavements for Highways and Streets*, MS-1, Asphalt Institute, College Park, Md., September 1981.

9. *Soils Manual for the Design of Asphalt Pavement Structures*, MS-10, Asphalt Institute, College Park, Md., March 1978.

10. *AASHTO Interim Guide for Design of Pavement Structures*, American Association of State Highway and Transportation Officials, Washington, D.C., 1972, Chapter III, revised 1981.

11. *AASHTO Guide for Design of Pavement Structures*, American Association of State Highway and Transportation Officials, Washington, D.C., 1986.

12. *Thickness Design*.

13. Ibid.

14. Ibid.

15. *Soils Manual for Design of Asphalt Pavement Structures.*

16. Ibid.

17. Ibid.

18. *AASHTO Interim Guide for Design of Pavement Structures.*

19. *AASHTO Guide for Design of Pavement Structures.*

20. W. Carey and P. Irick, *The Pavement Serviceability–Performance Concept,* Highway Research Board Special Report 61E, AASHTO Road Test, Highway Research Board, Washington, D.C., 1962.

21. W. Carey and P. Irick, "The Pavement Serviceability–Performance Concept," *Highway Research Record* 250(1960).

22. *AASHTO Guide for Design of Pavement Structures.*

23. C. J. Van Til, B. F. McCullough, B. A. Vallerga, and R. G. Hicks, *Evaluation of AASHTO Interim Guides for Design of Pavement Structures,* NCHRP Report 128, National Research Council, Transportation Research Board, Washington, D.C., 1972.

24. F. W. Hveem and R. Carmany, "The Factors Underlying the Rational Design of Pavements," *Highway Research Board Proceedings* 28(1948).

CHAPTER 19

Design of Rigid Pavements

Rigid highway pavements are normally constructed of Portland cement concrete and may or may not have a base course between the subgrade and the concrete surface. When a base course is used in rigid pavement construction, it is usually referred to as a subbase course. It is, however, common for only the concrete surface to be referred to as the rigid pavement even where there is a base course. In this text, the terms rigid pavement and concrete pavement are synonymous. Rigid pavements have some flexural strength that permits them to sustain a beamlike action across minor irregularities in the underlying material. Thus, the minor irregularities may not be reflected in the concrete pavement. Properly designed and constructed rigid pavements have long service lives and are usually less expensive to maintain than the flexible pavements.

Thickness of highway concrete pavements normally range from 6 in. to 13 in. and can be plain jointed, simply reinforced, continuously reinforced, or prestressed. Different types of rigid pavements are described later in this chapter. These pavement types are usually constructed to carry heavy traffic loads, although they have been used for residential and local roads.

It is beyond the scope of this book to cover all aspects of rigid pavement design, but the topics discussed cover the major aspects sufficiently for an understanding of the design principles of the basic types: plain, simply reinforced, and continuous. The topics covered include a description of the basic types of rigid pavements and the materials used, a discussion of the stresses imposed by traffic wheel loads and temperature differentials on the concrete pavement, and a description of two design methods.

MATERIALS USED IN RIGID PAVEMENTS

The Portland cement concrete commonly used for rigid pavements consists of Portland cement, coarse aggregate, fine aggregate, and water. Steel reinforcing rods may or may not be used, depending on the type of pavement being constructed. A description of the quality requirements for each of the basic materials is presented in the following sections.

Portland Cement

Portland cement is manufactured by crushing and pulverizing a carefully prepared mix of limestone, marl, and clay or shale and burning the mixture at a high temperature of about 2800°F to form a clinker. The clinker is then allowed to cool, a small quantity of gypsum is added, and the mixture ground until more than 90 percent of the material passes the No. 200 sieve. The main chemical constituents of the material are tricalcium silicate (C_3S), dicalcium silicate (C_2S), and tetracalcium alumino ferrite (C_4AF).

The material is usually transported in 1 ft^3 bags, each weighing 94 lb, although it can also be transported in bulk for large projects.

Most highway agencies use either the American Society for Testing Materials (ASTM) specifications (ASTM Designation C150)[1] or the American Association of State Highway and Transportation Officials (AASHTO) specifications (AASHTO Designation M85)[2] for specifying Portland cement quality requirements used in their projects. The AASHTO specifications list five main types of Portland cement.

Type I is suitable for general concrete construction, where no special properties are required. A manufacturer will supply this type of cement when no specific type is requested.

Type II is suitable for use in general concrete construction, where the concrete will be exposed to moderate action of sulphate or where moderate heat of hydration is required.

Type III is suitable for concrete construction that requires a high concrete strength in a relatively short time. It is sometimes referred to as *high early strength cement*.

Types IA, IIA, and IIIA are similar to types I, II, and III, respectively, but contain a small amount (4 percent to 8 percent of total mix) of entrapped air. This is achieved during production by thoroughly mixing the cement with air-entraining agents and grinding the mixture. In addition to the properties listed for types I, II, and III, types IA, IIA, and IIIA are more resistant to calcium chloride and de-icing salts and are therefore more durable.

Type IV is suitable for projects where low heat of hydration is necessary, and type V is used in concrete construction projects where the concrete will be exposed to high sulphate action. Table 19.1 shows the proportions of the different chemical constituents for each of the five types of cement.

Coarse Aggregates

The coarse aggregates used in Portland cement concrete are inert materials that do not react with cement and are usually comprised of crushed gravel, stone, or blast furnace slag. The coarse aggregates may be any one of the three materials or a combination of any two or all three. One of the major requirements for coarse aggregates used in Portland cement concrete is the gradation of the material. The material is well graded, with the maximum size specified. Material retained in a No. 4 sieve is considered coarse aggregate. Table 19.2 shows three gradation requirements for different maximum sizes as stipulated by ASTM.

Table 19.1   Proportions of Chemical Constituents and Strength Characteristics for the Different Types of Portland Cement

	Cement Type				
	I and IA	II and IIA	III and IIIA	IV	V
Silicon dioxide (SiO$_2$) Min. percent	—	20.0	—	—	—
Aluminum oxide (Al$_2$O$_3$) Max. percent	—	6.0	—	—	—
Ferric oxide (Fe$_2$O$_3$) Max. percent	—	6.0	—	6.5	—
Magnesium oxide (MgO) Max. percent	6.0	6.0	6.0	5.0	6.0
Sulfur trioxide (SO$_3$) Max. percent					
When 3 CAO·Al$_2$O is 8 percent or less	3.0	3.0	3.5	2.3	2.3
When 3CAO·Al$_2$O$_3$ is more than 8 percent	3.5	—	4.5	—	—
Loss on ignition, Max. percent	3.0	3.0	3.0	2.5	3.0
Insoluble residue, Max. percent	0.75	0.75	0.75	0.75	0.75
Tricalcium silicate (3CaO·SiO$_2$) Max. percent	—	55	—	35	—
Dicalcium silicate (2CaO·SiO$_2$) Min. percent	—	—	—	40	—
Tricalcium aluminate (3CaO·Al$_2$O$_3$) Max. percent	—	8	15	7	5
Tetracalcium aluminoferrite plus twice the tricalcium aluminate (4CaO· Al$_2$O$_3$·Fe$_2$O$_3$) + 2(3CaO·Al$_2$O$_3$), or solid solution (4CaO·Al$_2$O$_3$·Fe$_2$O$_3$ + 2CaO· Fe$_2$O$_3$), as applicable, Max. percent	—	—	—	—	20.0

Source: Adapted from *Standard Specifications for Transportation Materials and Methods of Sampling and Testing*, 14th ed., Washington, D.C.: The American Association of State Highway and Transportation Officials, copyright 1986. Used by permission.

Coarse aggregates must be clean. This is achieved by specifying the maximum percentage of deleterious substances allowed in the material. Other quality requirements include the ability of the aggregates to resist abrasion and the soundness of the aggregates.

A special test, known as the Los Angeles Rattler Test (AASHTO Designation T96), is used to determine the abrasive quality of the aggregates.[3] In this test, a sample of the coarse aggregate retained in the No. 8 sieve and a specified number of standard steel spheres are placed in a hollow steel cylinder 28 in. in internal diameter and 20 in. long that is closed at both ends. The cylinder, which also contains a steel shelf that projects 3 1/2 in. radially inward, is mounted with its axis in the horizontal position. The cylinder is then rotated 500 times at a specified speed. The sample of coarse aggregate is then removed and sieved on a No. 12 sieve. The portion of the material retained on the No. 12 sieve is weighed, and the difference between the original weight is the loss in weight. This loss in weight is expressed as a percentage of the original weight. Maximum permissible

Table 19.2 Gradation Requirements for Aggregates in Portland Cement Concrete (ASTM Designation C33)

Sieve Designation	Percent Passing by Weight		
	Aggregate Designation		
	2 in. to No. 4 (357)	$1\frac{1}{2}$ in. to No. 4 (467)	1 in. to No. 4 (57)
$2\frac{1}{2}$ in. (63 mm)	100	—	—
2 in. (50 mm)	95–100	100	—
$1\frac{1}{2}$ in. (37.5 mm)	—	95–100	100
1 in. (25.0 mm)	35–70	—	95–100
$\frac{3}{4}$ in. (19.0 mm)	—	35–70	—
$\frac{1}{2}$ in. (12.5 mm)	10–30	—	25–60
$\frac{3}{8}$ in. (9.5 mm)	—	10–30	—
No. 4 (4.75 mm)	0–5	0–5	0–10
No. 8 (2.36 mm)	—	—	0–5

Source: Adapted from *ASTM Standards, Concrete and Mineral Aggregates*, Vol. 04.02, American Society for Testing and Materials, Philadelphia, Pa., October 1985.

loss in weight is from 30 percent to 60 percent, depending on the specifications used. A maximum of 40 percent to 50 percent has, however, proved to be generally acceptable.

Soundness is defined as the ability of the aggregate to resist breaking up, due to freezing and thawing. This property can be determined in the laboratory by first sieving a sample of the coarse aggregate through a No. 4 sieve, then immersing in water the portion retained in the sieve. The sample is then frozen in the water for 2 hours and thawed for $\frac{1}{2}$ hr. This alternate freezing and thawing is repeated, between 20 and 50 times. The sample is then dried and sieved again to determine the change in particle size. Sodium or magnesium sulphate may be used instead of water.[4]

Fine Aggregates

Sand is mainly used as the fine aggregate in Portland cement concrete. Specifications for this material usually include grading requirements, soundness, and cleanliness. Standard specifications for the fine aggregates for Portland cement concrete (AASHTO Designation M6) give grading requirements normally adopted by state highway agencies (see Table 19.3).[5]

The soundness requirement is usually given in terms of the maximum permitted loss in the material after 5 alternate cycles of wetting and drying in the soundness test. A maximum of 10 percent weight loss is usually specified.

Cleanliness is usually specified in terms of the maximum amounts of different types of deleterious materials contained in the fine aggregates. For example, maximum amount of silt (material passing No. 200 sieve) is usually specified

Table 19.3   AASHTO Recommended Particle Size Distribution for Fine Aggregates Used in Portland Cement Concrete

Sieve Designation	Percent Passing by Weight
$\frac{3}{8}$ in. (9.5 mm)	100
No. 4 (4.75 mm)	95–100
No. 16 (1.18 mm)	45–80
No. 50 (0.30 mm)	10–30
No. 100 (0.15 mm)	2–10

Source: Adapted from *Standard Specifications for Transportation Materials and Methods of Sampling and Testing*, 14th ed., Washington, D.C.: The American Association of State Highway and Transportation Officials, copyright 1986. Used by permission.

within a range of 2 percent to 5 percent of the total fine aggregates. Since the presence of large amounts of organic material in the fine aggregates may reduce the hardening properties of the cement, a standard test (AASHTO Designation T21) is also usually specified as part of the cleanliness requirements. In this test, a sample of the fine aggregates is mixed with sodium hydroxide solution. If organic material is present, the sodium hydroxide will change to a dark color. When this occurs, the sand can be used only if the strength developed by 2 in. cubes made with this sand is at least 95 percent of that developed by similar cubes made with the same sand, after washing it in a 3 percent hydroxide solution.

Water

The main water requirement stipulated is that the water used should also be suitable for drinking. This requires that the quantity of organic matter, oil, acids, and alkalies should not be greater than the allowable amount in drinking water.

Reinforcing Steel

Steel reinforcing may be used in concrete pavements to reduce the amount of cracking that occurs, as a load transfer mechanism at joints, or as a means of tying two slabs together. Steel reinforcement used to control cracking is usually referred to as temperature steel, whereas steel rods used as load transfer mechanisms are known as dowel bars, and those used to connect two slabs together are known as tie bars.

Temperature Steel

Temperature steel is provided in the form of a bar mat or wire mesh consisting of longitudinal and transverse steel wires welded at regular intervals. The mesh is usually placed about 3 in. below the slab surface. The cross-sectional area of the steel provided per foot width of the slab depends on the size and spacing of the steel wires forming the mesh. The amount of steel required depends on the length of the pavement between expansion joints, the maximum stress desired in the concrete pavement, the thickness of the pavement, and the moduli of elasticity of the concrete and steel. Eq. 19.16 (developed later in this chapter) can be used to determine the area of steel required if the length of the slab is fixed. Steel areas obtained by this equation for concrete slabs less than 45 ft may, however, be inadequate and therefore should be compared with the following general guidelines for the minimum cross-sectional area of the temperature steel.

1. Cross-sectional area of longitudinal steel should be at least equal to 0.1 percent of the cross-sectional area of the slab.
2. Longitudinal wires should not be less than No. 2 gauge, spaced at a maximum distance of 6 in.
3. Transverse wires should not be less than No. 4 gauge, spaced at a maximum distance of 12 in.

Temperature steel does not prevent cracking of the slab but does control the crack widths because the steel acts as a tie holding the edges of the cracks together. This helps to maintain the shearing resistance of the pavement, thereby maintaining its capacity to carry traffic load, even though the flexural strength is not improved.

Dowel Bars

Dowel bars are used mainly as load-transfer mechanisms across joints and provide flexural, shearing, and bearing resistance. The dowel bars must be of a much larger diameter than the wires used in temperature steel. Size selection is mainly based on experience. Diameters of 1 to $1\frac{1}{2}$ in. and lengths of 2 to 3 ft have been used, with the bars usually spaced at 1 ft centers across the width of the slab. At least one end of the bar should be smooth and lubricated to facilitate free expansion.

Tie Bars

Tie bars are used to tie two sections of the pavement together and therefore should be either deformed bars or should contain hooks to facilitate the bonding of the two sections of the concrete pavement with the bar. These bars are usually much smaller in diameter than the dowel bars and are spaced at larger centers. Typical diameter and spacing for these bars is $\frac{3}{4}$ in. and 3 ft, respectively.

JOINTS IN CONCRETE PAVEMENTS

Different types of joints are placed in concrete pavements to limit the stresses induced by temperature changes and to facilitate proper bonding of two adjacent sections of pavement when there is a time lapse between their construction (for example, at the end of one day's work to the beginning of the next day's work). These joints can be divided into four basic categories.

- Expansion joints
- Contraction joints
- Hinge joints
- Construction joints

Expansion Joints

When concrete pavement is subjected to an increase in temperature, it will expand, resulting in an increase in length of the slab. When the temperature is sufficiently high, the slab may buckle or "blow-up" if it is sufficiently long and no provision is made to accommodate the increased length. Therefore, expansion joints are usually placed transversely, at regular intervals, to provide adequate space for the slab to expand. These joints are placed across the full width of the slab and are $\frac{3}{4}$ to 1 in. wide in the longitudinal direction. They must create a distinct break throughout the depth of the slab. The joint space is filled with a compressible filler material that permits the slab to expand. Filler materials can be cork, rubber, bituminous materials, or bituminous fabrics.

A means of transferring the load across the joint space must be provided since there are no aggregates that will develop an interlocking mechanism. The load-transfer mechanism is usually a smooth dowel bar that is lubricated on one side. An expansion cap is also usually installed, as shown in Figure 19.1, to provide a space for the dowel to occupy during expansion.

Some states no longer use expansion joints because of the inability of the load-transfer mechanism to adequately transfer the load. Other states continue to use expansion joints and may even use them in place of construction joints.

Contraction Joints

When concrete pavement is subjected to a decrease in temperature, the slab will contract if it is free to move. Prevention of this contraction movement will induce tensile stresses in the concrete pavement. Contracting joints are therefore placed transversely at regular intervals across the width of the pavement to release some of the tensile stresses that are so induced. A typical contraction joint is shown in Figure 19.2. It may be necessary in some cases to install a load-transfer mechanism in the form of a dowel bar when there is doubt about the ability of the interlocking grains to transfer the load.

Hinge Joints

Hinge joints are used mainly to reduce cracking along the center line of highway pavements. Figure 19.3 shows a typical hinge joint (keyed joint) suitable for single-lane-at-a-time construction.

Figure 19.1   Typical Expansion Joint

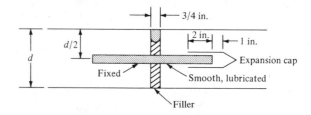

Figure 19.2 Typical Contraction Joint

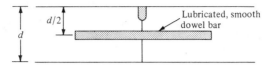

Figure 19.3 Typical Hinge Joint (Keyed Joint)

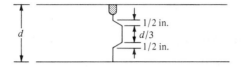

Figure 19.4 Typical Butt Joint

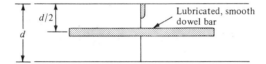

Construction Joints

Construction joints are placed transversely across the pavement width to provide suitable transition between concrete laid at different times. For example, a construction joint is usually placed at the end of a day's pour to provide suitable bonding with the start of the next day's pour. A typical butt construction joint is shown in Figure 19.4. In some cases, as shown in Figure 19.3, a keyed construction joint may also be used in the longitudinal direction when only a single lane is constructed at a time. In this case, alternative lanes of the pavement are cast, and the key is formed by using metal formwork that has been cast with the shape of the groove or by attaching a piece of metal or wood to a wooden formwork. An expansion joint can be used in lieu of a transverse construction joint in cases where the construction joint falls at or near the same position as the expansion joint.

TYPES OF RIGID HIGHWAY PAVEMENTS

Rigid highway pavements can be divided into three general types: plain concrete pavements, simply reinforced concrete pavements, and continuously reinforced concrete pavements. The definition of each pavement type is related to the amount of reinforcement used.

Plain Concrete Pavement

Plain concrete pavement has no temperature steel or dowels for load transfer. However, steel tie bars are often used to provide a hinge effect at longitudinal joints and to prevent the opening of these joints. Plain concrete pavements are used mainly on low volume highways or when cement stabilized soils are used as subbase. Joints are placed at relatively shorter distances (10 to 20 ft) than with the other types of concrete pavements to reduce the amount of cracking. In some cases, the transverse joints of plain concrete pavements are skewed about 4 to 5 ft in plan, such that only one wheel of a vehicle passes through the joint at a time. This helps to provide a smoother ride.

Simply Reinforced Concrete Pavement

Simply reinforced concrete pavements have dowels for the transfer of traffic loads across joints, with these joints spaced at larger distances, ranging from 30 to 100 ft. Temperature steel is used throughout the slab, with the amount dependent on the length of the slab. Tie bars are also commonly used at longitudinal joints.

Continuously Reinforced Concrete Pavement

Continuously reinforced concrete pavements have no transverse joints, except construction joints or expansion joints when they are necessary at specific positions, such as at bridges. These pavements have a relatively high percentage of steel, with the minimum usually at 0.6 percent of the cross section of the slab. They also contain tie bars across the longitudinal joints.

PUMPING OF RIGID PAVEMENTS

Pumping is an important phenomenon associated with rigid pavements. Pumping is the discharge of water and subgrade (or subbase) material through joints, cracks, and along the pavement edges. It is primarily caused by the repeated deflection of the pavement slab in the presence of accumulated water beneath it. The mechanics of pumping can be explained best by considering the sequence of events that lead to it.

The first event is the formation of void space beneath the pavement. This void forms from either the combination of the plastic deformation of the soil, due to imposed loads and the elastic rebound of the pavement after it has been deflected by the imposed load, or warping of the pavement, which occurs as a result of temperature gradient within the slab. Water then accumulates in the space after many repetitions of traffic load. The water may be infiltrated from the surface through joints and the pavement edge or, to a lesser extent, ground water may settle in the void. If the subgrade or base material is granular, the water will freely drain through the soil. If the material is fine-grained, however, the water is not easily discharged, and additional load repetitions will result in the soil going into suspension with the water to form a slurry. Further load repetitions and deflections of the slab will result in the slurry being ejected to the surface (pumping). Pumping action will then continue, with the result that a relatively large void space is formed underneath the concrete slab. This results in faulting of the joints and eventually the formation of transverse cracks or the breaking of the corners of the slab. Joint faulting and cracking is therefore progressive, since formation of a crack facilitates the pumping action.

Visual manifestations of pumping include:

- Discharge of water from cracks and joints
- Spalling near the center line of the pavement and a transverse crack or joint
- Mud boils at the edge of the pavement
- Pavement surface discoloration (caused by the subgrade soil)
- Breaking of pavement at the corners

Design Considerations for Preventing Pumping

A major design consideration for preventing pumping is the reduction or elimination of expansion joints since pumping is usually associated with these joints. This is the main reason why current design practices limit the number of expansion joints to a minimum. Since pumping is also associated with fine-grained soils, another design consideration is either to replace soils that are susceptible to pumping with a nominal thickness of granular or sandy soils or to improve them by stabilization. Current design practices therefore usually include the use of 3 to 6 in. layers of granular subbase material at areas along the pavement alignment where the subgrade material is susceptible to pumping or stabilizing the susceptible soil with asphaltic material or Portland cement. The Portland Cement Association method of rigid pavement design indirectly considers this phenomenon in the erosion analysis.

STRESSES IN RIGID PAVEMENTS

Stresses are developed in rigid pavements as a result of several factors, including the action of traffic wheel loads, the expansion and contraction of the concrete due to temperature changes, yielding of the subbase or subgrade supporting the concrete pavement, and volumetric changes. For example, traffic wheel loads will induce flexural stresses that are dependent on the location of the vehicle wheels relative to the edge of the pavement, whereas expansion and contraction may induce tensile and compressive stresses, which are dependent on the range of temperature changes in the concrete pavement. These different factors that can induce stress in concrete pavement have made the theoretical determination of stresses rather complex, requiring the following simplifying assumptions.

1. Concrete pavement slabs are considered as unreinforced concrete beams. Any contribution made to the flexural strength by the inclusion of reinforcing steel is neglected.

2. The combination of flexural and direct tensile stresses will inevitably result in transverse and longitudinal cracks. The provision of suitable crack control in the form of joints, however, controls the occurrence of these cracks, thereby maintaining the beam action of large sections of the pavement.

3. The supporting subbase and/or subgrade layer acts as an elastic material in that it deflects at the application of the traffic load and recovers at the removal of the load.

Stresses Induced by Bending

The ability of rigid pavement to sustain a beamlike action across irregularities in the underlying materials suggests that the theory of bending is fundamental to the analysis of stresses in such pavements. The theory of a beam supported on an elastic foundation can therefore be used to analyze the stresses in the pavement when it is externally loaded. Figure 19.5 shows the deformation sustained by a beam on an elastic foundation when it is loaded externally. The stresses developed in the beam may be analyzed by assuming that a reactive pressure (p), which is proportional to the deflection, is developed as a result of the applied load. This pressure is given as

$$p = ky \qquad (19.1)$$

where

p = reactive pressure at any point beneath the beam (lb/in.²)
y = deflection at the point (in.)
k = modulus of subgrade reaction (lb/in.³)

The modulus of subgrade reaction is the stress (in lb/in.²) that will cause an inch deflection of the underlying soil. Eq. 19.1 assumes that k (lb/in.³) is constant, which implies that the subgrade is elastic. This assumption is, however, valid for only a limited range of different factors. Research has shown that the value of k depends on certain soil characteristics such as density, moisture, soil texture, and other factors that influence the strength of the soils. The k value of a particular soil will also vary with the size of the loaded area and the amount of deflection.

Figure 19.5 Deformation of a Beam on Elastic Foundation

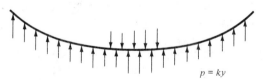

$p = ky$

The modulus of subgrade reaction is directly proportional to the loaded area and inversely proportional to the deflection. In pavement design, however, minor changes in k do not have significant impact on design results, and an average value is usually assumed. The plate-bearing test is commonly used for determining the value of k in the field.

A general relationship between the moment and the radius of curvature of a beam is given as

$$\frac{1}{R} = \frac{M}{EI} \tag{19.2}$$

where

$R =$ radius of curvature
$M =$ moment in beam
$E =$ modulus of elasticity
$I =$ moment of intertia

The general differential equation relating the moment at any section of a beam with the deflection at that section is given as

$$M = EI\frac{d^2y}{dx^2} \tag{19.3}$$

whereas the basic differential equation for the deflection on an elastic foundation is given as

$$EI\frac{d^4y}{dx^4} = -ky = q \tag{19.4}$$

and the basic differential equation for a slab is given as

$$M_x = \left[\frac{Eh^3}{12(1-\mu^2)}\right]\frac{d^2w}{dx^2}$$

or

$$M_y = \left[\frac{Eh^3}{12(1-\mu^2)}\right]\frac{d^2w}{dy^2} \tag{19.5}$$

where

$h =$ thickness of slab
$\mu =$ Poisson ratio of concrete
$w =$ deflection of the slab at a given point
$M_x =$ bending moment at a point about the x axis
$M_y =$ bending moment at a point about the y axis

The *EI* term in Eq. 19.4 is called the stiffness of the beam, whereas the stiffness of the slab is given by the term within the square brackets of Eq. 19.5. This term is usually denoted as *D*, where

$$D = \frac{E_c h^3}{12(1 - \mu^2)} \tag{19.6}$$

In developing expressions for the stresses in a concrete pavement, Westergaard made use of the radius of relative stiffness, which depends on the stiffness of the slab and the modulus of subgrade reaction of the soil.[6] It is given as

$$\ell = \sqrt[4]{\frac{E_c h^3}{12(1 - \mu^2)k}} \tag{19.7}$$

where

ℓ = radius of relative stiffness (in.)
E_c = modulus of elasticity of the concrete pavement (lb/in.^{2})
h = thickness of pavement (in.)
μ = Poisson ratio of the concrete pavement
k = modulus of subgrade reaction (lb/in.^{3})

It will be seen later that the radius of relative stiffness is an important parameter in the equations used to determine various stresses in the concrete pavement.

Stresses Due to Traffic Wheel Loads

The basic equations for determining flexural stresses in concrete pavements due to traffic wheel loads were first developed by Westergaard.[7] Although several theoretical developments have been made since then, the Westergaard equations are still considered a fundamental tool for evaluating the stresses on concrete

Figure 19.6   Critical Locations of Wheel Loads on Concrete Pavements

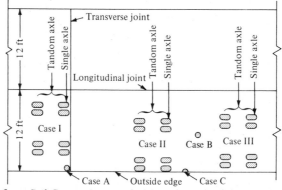

Source: Redrawn from *Soil Cement Inspector's Manual*, Portland Cement Association, Skokie, Ill., 1963.

pavements. Westergaard considered three critical locations of the wheel load on the concrete pavement in developing the equations. These locations are shown in Figure 19.6 and are described as follows.

Case A. Load is applied at the corner of a rectangular slab. This provides for the cases when the wheel load is applied at the intersection of the pavement edge and a transverse joint. However, this condition is not common anymore because pavements are now much wider. No equation is therefore presented for this case.

Case B. Load is applied at the interior of the slab at a considerable distance from its edges.

Case C. Load is applied at the edge of the slab at considerable distance away from any corner.

The locations shown as cases I, II, and III in Figure 19.6 are the critical locations presently used for the relatively wide pavements now being constructed.

The equations for determining these stresses were developed taking into consideration the different day and night temperature conditions that may exist. During the day, the temperature is higher at the surface of the slab than at the bottom. This temperature gradient through the depth of the slab will create a tendency for the slab edges to curl downward. During the night, however, the temperature at the bottom of the slab is higher than at the surface, thereby reversing the temperature gradient, which results in the slab edges tending to curl upward. The equations for stresses due to traffic load reflect this phenomenon of concrete pavements.

The original equations developed by Westergaard were modified, using the results of full-scale tests conducted by the Bureau of Public Roads.[8] These modified equations for the different loading conditions are as follows.

1. Edge loading when the edges of the slab are warped upward at night

$$\sigma_e = \frac{0.572P}{h^2}\left[4\log_{10}\left(\frac{\ell}{b}\right) + \log_{10} b\right] \qquad (19.8)$$

2. Edge loading when the slab is unwarped or when the edge of the slab is curled downward in daytime

$$\sigma_e = \frac{0.572P}{h^2}\left[4\log_{10}\left(\frac{\ell}{b}\right) + 0.359\right] \qquad (19.9)$$

3. Interior loading

$$\sigma_i = \frac{0.316P}{h^2}\left[4\log_{10}\left(\frac{\ell}{b}\right) + 1.069\right] \qquad (19.10)$$

where

σ_e = maximum stress (lb/in.²) induced in the bottom of the slab, directly
under the load P and applied at the edge and in a direction parallel to the
edge

σ_i = maximum tensile stress (lb/in.²) induced at the bottom of the slab
directly under the load P applied at the interior of the slab

P = applied load in pounds, including allowance for impact

h = thickness of slab (in.)

ℓ = radius of relative stiffness

$= \sqrt[4]{E_c h^3 / [12(1 - \mu^2)k]}$

E_c = modulus of elasticity of concrete (lb/in.²)

μ = Poisson ratio for concrete = 0.15

k = subgrade modulus (lb/in.³)

b = radius of equivalent distribution of pressure (in.)

$= \sqrt{1.6a^2 + h^2} - 0.675h$ (for $a < 1.724h$)

$= a$ (for $a > 1.724h$)

a = radius of contact area of load (in.) (Contact area usually assumed
as a circle for interior and corner loadings and semicircle for edge
loading.)

Example 19–1 Computing Tensile Stress Resulting from a Wheel Load on a
Rigid Pavement

Determine the tensile stress imposed by a wheel load of 900 lb imposed during the
day and located at the edge of a concrete pavement with the following dimensions
and properties.

Pavement thickness = 6 in.
$\mu = 0.15$
$E = 5 \times 10^6$ lb/in.²
$k = 130$ lb/in.³
radius of loaded area = 3 in.

From Eq. 19.9

$$\sigma_e = \frac{0.572P}{h^2} \left[4 \log_{10} \left(\frac{\ell}{b} \right) + 0.359 \right]$$

$a = 3$ in. $< 1.724h$

where

$h = 6$ in.

$b = \sqrt{1.6a^2 + h^2} - 0.675h = \sqrt{1.6 \times 3^2 + 6^2} - 0.675 \times 6 = 7.1 - 4.05$

$= 3.05$

$\ell = \sqrt[4]{(5 \times 10^6 \times 6^3)/[12(1 - 0.15^2)130]} = 29.0$

$\sigma_e = [(0.572 \times 900)/6^2][4 \log_{10}(29/3.05) + 0.359] = 14.3(4 \times 0.978 + 0.359)$

$= 61.07$ lb/in.²

Stresses Due to Temperature Effects

The tendency of the slab edges to curl downward during the day and upward during the night as a result of temperature gradients is resisted by the weight of the slab itself. This resistance tends to keep the slab in its original position, resulting in stresses being induced in the pavement. Compressive and tensile stresses are therefore induced at the top and bottom of the slab, respectively, during the day, whereas tensile stresses are induced at the top and compressive stresses at the bottom during the night.

Under certain conditions these curling stresses may have values high enough to cause cracking of the pavement. They may also reduce the subgrade support beneath some sections of the pavement, which can result in a considerable increase of the stresses due to traffic loads over those for pavements with uniform pavement support. Studies have shown that curling stresses can be higher than 200 lb/in.² for 10 ft slabs and much higher for wider slabs. One of the main purposes of longitudinal joints is to limit the slab width by dividing the concrete pavement into individual slabs 11 or 12 ft wide.

Tests have shown that maximum temperature differences between the top and bottom of the slab depend on the thickness of the slab, and that these differences are about 2.5° to 3°F per inch thickness for 6 to 9 in. thick slabs. The temperature differential also depends on the season, with maximum differentials occurring during the day in the spring and summer months. Another factor that affects the temperature differential is the latitude of the location of the slab. The surface temperature of the pavement tends to be high if the angle of incidence of the sun's rays is high, as in areas near the equator.

These curling stresses can be determined from Eqs. 19.11 and 19.12. Note, however, that curling stresses are not normally taken into consideration in pavement thickness design, as joints and steel reinforcement are normally used to reduce the effect of such stresses.

$$\sigma_{xe} = \frac{C_x E_c et}{2} \tag{19.11}$$

and

$$\sigma_{xi} = \frac{E_c et}{2}\left(\frac{C_x + \mu C_y}{1 - \mu^2}\right) \tag{19.12}$$

where

σ_{xe} = maximum curling stress in lb/in.² at the edge of the slab in the direction of the slab length due to temperature difference between the top and bottom of the slab

σ_{xi} = maximum curling stress in lb/in.² at the interior of the slab in the direction of the slab length due to temperature difference between the top and bottom of the slab

E_c = modulus of elasticity for concrete (lb/in.²)

μ = Poisson ratio for concrete

C_x, C_y = coefficients that are dependent on the radius of relative stiffness of the concrete and can be obtained from Figure 19.7.

e = thermal coefficient of expansion and contraction of concrete per degree fahrenheit

t = temperature difference between the top and bottom of the slab in degrees fahrenheit

Temperature changes in the slab will also result in expansion (for increased temperature) and contraction (for reduced temperature). The provision of suitable expansion/contraction joints in the slab reduces the magnitude of these stresses but does not entirely eliminate them since considerable resistance to the free horizontal movement of the pavement will still be offered by the subgrade, due to friction action between the bottom of the slab and the top of the subgrade. The magnitude of these stresses depends on the length of the slab, the type of concrete pavement, the magnitude of the temperature changes, and the coefficient of friction between the pavement and the subgrade. Tensile stresses greater than 100 lb/in.² have been reported for an average temperature change of 40°F in a slab of 100 ft long.

Figure 19.7   Values of C_x and C_y for Use in Formulas for Curling Stresses

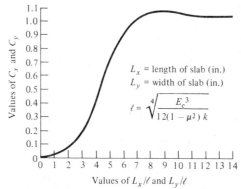

Values of L_x/ℓ and L_y/ℓ

L_x = length of slab (in.)
L_y = width of slab (in.)

$$\ell = \sqrt[4]{\frac{E_c{}^3}{12(1 - \mu^2)\,k}}$$

Source: Redrawn from Royall D. Bradbury, *Reinforced Concrete Pavements,* Wire Reinforcement Institute, Washington, D.C., 1938.

If only the effect of uniform temperature drop is being considered, the maximum spacing of contraction joints required to ensure that the maximum stress does not exceed a desired value p_c can be determined by equating the force in the slab and the force developed due to friction, as follows.

Considering a unit width of the pavement, the frictional force (F) developed due to a uniform drop in temperature is

$$F = f \frac{h}{12} (1) \frac{L}{2} \gamma_c$$

where

f = coefficient of friction between the bottom of concrete pavement and subgrade

h = thickness of concrete pavement (in.)

L = length of pavement between contraction joints

γ_c = density of concrete (lb/ft^3)

The force (P) developed in the concrete due to stress p_c is given as

$$P = p_c(12)(1)h$$

where p_c is the maximum desired stress in the concrete in lb/in.^{2}.

Equating these two forces gives

$$f \frac{h}{12} (1) \frac{L}{2} \gamma_c = p_c(12)(1)(h) \tag{19.13}$$

and

$$L = \frac{288 p_c}{f \gamma_c} \tag{19.14}$$

The effects of temperature changes can also be reduced by including reinforcing steel in the concrete pavement. The additional force developed by the steel may also be taken into consideration in determining L. If A_s is the total cross-sectional area of steel per foot width of slab, Eq. 19.13 becomes

$$f \frac{h}{12} (1) \frac{L}{2} \gamma_c = p_c(12h + nA_s) \tag{19.15}$$

or

$$L = \frac{24 p_c}{h \gamma_c f} (12h + nA_s) \tag{19.16}$$

where n = modular ratio = E_s/E_c.

THICKNESS DESIGN OF RIGID PAVEMENTS

The main objective in rigid pavement design is to determine the thickness of the concrete slab that will be adequate to carry the projected traffic load for the design period. Several design methods have been developed over the years, some of which are based on the results of full-scale road tests, others on theoretical development of stresses on layered systems, and others on the combination of the results of tests and theoretical development. However, two methods are used extensively: the AASHTO method[9] and the Portland Cement Association (PCA) method.[10]

AASHTO Design Method

The AASHTO method for rigid pavement design is based mainly on the results obtained from the AASHO road test (Chapter 18). The design procedure was initially published in the early 1960s but was revised in the 1970s and early 1980s. A further revision has been carried out since then, incorporating new developments.[11] The design procedure provides for the determination of the pavement thickness, the amount of steel reinforcement when used, and the design of joints. It is suitable for plain concrete, simply reinforced concrete, and continuously reinforced concrete pavements. The design procedure for the longitudinal reinforcing steel in continuously reinforced concrete pavements is, however, beyond the scope of this book and is therefore not presented. Interested readers should refer to the AASHTO guide for this procedure.[12]

Design Considerations

The factors considered in the AASHTO procedure for the design of rigid pavements as presented in the 1986 guide are

- Pavement performance
- Subgrade strength
- Subbase strength
- Traffic
- Concrete properties
- Drainage
- Reliability

Pavement Performance.   Pavement performance is considered in the same way as for flexible pavement as presented in Chapter 18. The initial serviceability index (P_i) may be taken as 4.5 and the terminal serviceability may also be selected by the designer.

Subbase Strength.   The guide allows the use of either graded granular materials or suitably stabilized materials for the subbase layer. Table 19.4 gives recommended specifications for six types of subbase materials. AASHTO suggests that

Table 19.4 Recommended Particle Size Distributions for Different Types of Subbase Materials

Sieve Designation	Type A	Type B	Type C (Cement Treated)	Type D (Lime Treated)	Type E (Bituminous Treated)	Type F (Granular)
Sieve analysis percent passing						
2 in.	100	100	—	—	—	—
1 in.	—	75–95	100	100	100	100
3/8 in.	30–65	40–75	50–85	60–100	—	70–100
No. 4	25–55	30–60	35–65	50–85	55–100	55–100
No. 10	15–40	20–45	25–50	40–70	40–100	30–70
No. 40	8–20	15–30	15–30	25–45	20–50	8–25
No. 200	2–8	5–20	5–15	5–20	6–20	
(The minus No. 200 material should be held to a practical minimum)						
Compressive strength lb/in.^{2} at 28 days			400–750	100		
Stability						
Hveem stabilometer					20 min.	
Hubbard field					1000 min.	
Marshall stability					500 min.	
Marshall flow					20 max.	
Soil constants						
Liquid limit	25 max.	25 max.				25 max.
Plasticity index[a]	N.P.	6 max.	10 max.[b]		6 max.[b]	6 max.

[a] As performed on samples prepared in accordance with AASHTO Designation T87.
[b] These values apply to the mineral aggregate prior to mixing with the stabilizing agent.
Source: Adapted from *Standard Specifications for Transportation Materials and Methods of Sampling and Testing*, 14th ed., Washington, D.C.: The American Association of State Highway and Transportation Officials, copyright 1986. Used by permission.

the first five types—A through E—can be used within the upper 4 in. layer of the subbase, whereas type F can be used below the uppermost 4 in. layer. Special precautions should be taken when certain conditions exist. For example, when A, B, and F materials are used in areas where the pavement may be subjected to frost action, the percentage of fines should be reduced to a minimum. Subbase thickness is usually not less than 6 in. and should be extended 1 to 3 ft outside the edge of the pavement structure.

Subgrade Strength. The strength of the subgrade is given in terms of the Westergaard modulus of subgrade reaction k, which is defined as the load in pounds per square inch on a loaded area, divided by the deformation in inches. Values of k can be obtained by conducting a plate-bearing test in accordance with the AASHTO test Designation T222 using a 30 in. diameter plate. Estimates of k values can also be made either from experience or by correlating with other tests. Figure 19.8 shows an approximate interrelationship of bearing values obtained from different types of tests.

The guide also provides for the determination of an effective modulus of subgrade reaction, which depends on (1) the seasonal effect on the resilient modulus of the subgrade, (2) the type and thickness of the subbase material used, (3) the effect of potential erosion of the subbase, and (4) whether bedrock lies within 10 ft of the subgrade surface. The seasonal effect on the resilient modulus of the subgrade was discussed in Chapter 18, and a procedure similar to that used in flexible pavement design is used here to take into consideration the variation of the resilient modulus during a 12 month period.

Since different types of subbase materials have different strengths, the type of material used is an important input in the determination of the effective modulus of subgrade reaction. In estimating the composite modulus of subgrade reaction, the subbase material is defined in terms of its elastic modulus E_{SB}. Also, it is necessary to consider the combination of material types and the required thicknesses because this serves as a basis for determining the cost-effectiveness of the pavement. The chart in Figure 19.9 is used to estimate the composite modulus of subgrade reaction for the type of subbase material, based on its elastic modulus, its resilient modulus, and the thickness of the subbase.

The effective k value also depends on the potential erosion of the subbase material. This effect is included by the use of a factor (see Table 19.5) for the loss of support (LS) in determining the effective k value. This factor is used to reduce the effective modulus of subgrade reaction, as shown in Figure 19.10.

The presence of bedrock, within a depth of 10 ft of the subgrade surface and extending over a significant length along the highway alignment, may result in an increase of the overall modulus of subgrade reaction. This effect is taken into consideration by adjusting the effective modulus subgrade using the chart in Figure 19.11. The procedure is demonstrated in the solution of Example 19–2.

Figure 19.8 Approximate Interrelationship of Soil Classification and Bearing Values

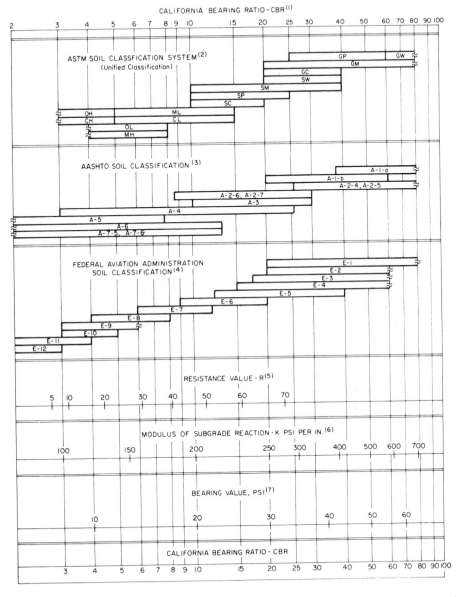

(1) For the basic idea, see O. J. Porter, "Foundations for Flexible Pavements," Highway Research Board *Proceedings of the Twenty-second Annual Meeting*, 1942, Vol. 22, pages 100-136.

(2) ASTM Designation D2487.

(3) "Classification of Highway Subgrade Materials," Highway Research Board *Proceedings of the Twenty-fifth Annual Meeting*, 1945, Vol. 25, pages 376-392.

(4) *Airport Paving*, U.S. Department of Commerce, Federal Aviation Agency, May 1948, pages 11-16. Estimated using values given in FAA *Design Manual for Airport Pavements.*(Formerly used FAA Classification; Unified Classification now used.)

(5) C. E.Warnes, "Correlation Between *R* Value and *k* Value," unpublished report, Portland Cement Association, Rocky Mountain-Northwest Region, October 1971 (best-fit correlation with correction for saturation).

(6) See T. A. Middlebrooks and G. E. Bertram, "Soil Tests for Design of Runway Pavements," Highway Research Board *Proceedings of the Twenty-second Annual Meeting*, 1942, Vol. 22, page 152.

Source: Reproduced from Robert G. Packard, *Thickness Design for Concrete Highway and Street Pavements*, Portland Cement Association, Skokie, Ill., 1984.

Figure 19.9   Chart for Estimating Composite Modulus of Subgrade Reaction, k_∞, Assuming a Semi-Infinite Subgrade Depth (For practical purposes, a semi-infinite depth is considered to be greater than 10 ft below the surface of the subgrade.)

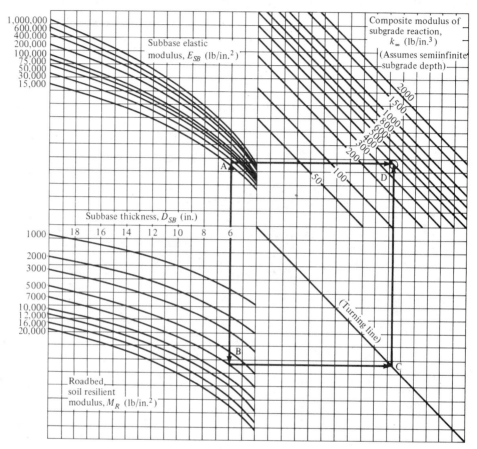

Example:

D_{SB} = 6 in.

E_{SB} = 20,000 lb/in.^{2}

M_R = 7,000 lb/in.^{2}

Solution: k_∞ = 400 lb/in.^{3}

Source: Redrawn from *AASHTO Guide for Design of Pavement Structures*, Washington, D.C.: The American Association of State Highway and Transportation Officials, copyright 1986. Used by permission.

Figure 19.10 Correction of Effective Modulus of Subgrade Reaction for Potential Loss of Subbase Support

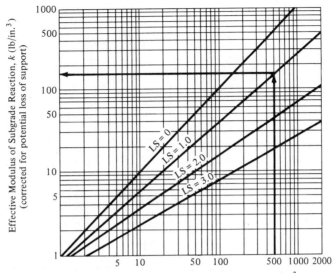

Effective Modulus of Subgrade Reaction, k (lb/in.^{3})

Source: Redrawn from *AASHTO Guide for Design of Pavement Structures*, Washington, D.C.: The American Association of State Highway and Transportation Officials, copyright 1986. Used by permission.

Table 19.5 Typical Ranges of Loss of Support Factors for Various Types of Materials

Type of Material	Loss of Support (LS)
Cement-treated granular base ($E = 1,000,000$ to $2,000,000$ lb/in.^{2})	0.0 to 1.0
Cement aggregate mixtures ($E = 500,000$ to $1,000,000$ lb/in.^{2})	0.0 to 1.0
Asphalt-treated base ($E = 350,000$ to $1,000,000$ lb/in.^{2})	0.0 to 1.0
Bituminous stabilized mixtures ($E = 40,000$ to $300,000$ lb/in.^{2})	0.0 to 1.0
Lime stabilized ($E = 20,000$ to $70,000$ lb/in.^{2})	1.0 to 3.0
Unbound granular materials ($E = 15,000$ to $45,000$ lb/in.^{2})	1.0 to 3.0
Fine-grained or natural subgrade materials ($E = 3,000$ to $40,000$ lb/in.^{2})	2.0 to 3.0

Note: E in this table refers to the general symbol for elastic or resilient modulus of the material.

Source: Adapted from B. F. McCullough and Gary E. Elkins, *CRC Pavement Design Manual*, Austin Research Engineers, Inc., Austin, Tex., October 1979.

Figure 19.11 Chart to Modify Modulus of Subgrade Reaction to Consider Effects of Rigid Foundation Near Surface (Within 10 ft)

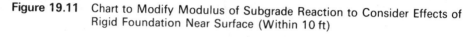

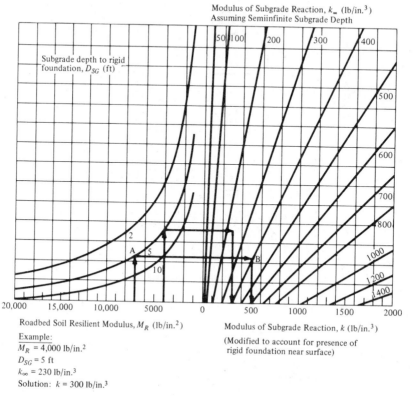

Modulus of Subgrade Reaction, k_∞ (lb/in.^{3})
Assuming Semiinfinite Subgrade Depth

Subgrade depth to rigid foundation, D_{SG} (ft)

Roadbed Soil Resilient Modulus, M_R (lb/in.^{2})

Example:
$\overline{M_R}$ = 4,000 lb/in.^{2}
D_{SG} = 5 ft
k_∞ = 230 lb/in.^{3}
Solution: k = 300 lb/in.^{3}

Modulus of Subgrade Reaction, k (lb/in.^{3})
(Modified to account for presence of rigid foundation near surface)

Source: Redrawn from *AASHTO Guide for Design of Pavement Structures*, Washington, D.C.: The American Association of State Highway and Transportation Officials, copyright 1986. Used by permission.

Example 19–2 Computing Effective Modulus of Subgrade Reaction for a Rigid Pavement Using AASHTO Method

A 6 in. layer of cement-treated granular material is to be used as subbase for a rigid pavement. The monthly values for the roadbed soil resilient modulus and the subbase elastic (resilient) modulus are given in columns 2 and 3 of the Table 19.6. If the rock depth is located 5 ft below the subgrade surface and the projected slab thickness is 9 in., estimate the effective modulus of subgrade reaction, using the AASHTO method.

Note that this is the example given in the 1986 AASHTO guide. Also note that the values for the modulus of the roadbed and subbase materials should be determined as discussed in Chapter 18, and the corresponding values shown in columns 2 and 3 of Table 19.6 should be for the same seasonal period.

Table 19.6 Data for and Solution to Example 19–2
Total Subbase
   Type: Granular
   Thickness (in.): 6
   Loss of Support, LS: 1.0
Depth to Rigid Foundation (ft): 5
Projected Slab Thickness (in.): 9

(1)	(2)	(3)	(4)	(5)	(6)
			Composite	k Value (E_{SB})	Relative
	Roadbed	Subbase	k Value	on Rigid	Damage,
	Modulus,	Modulus,	($lb/in.^2$)	Foundation	u_r
Month	M_R ($lb/in.^2$)	E_{SB} ($lb/in.^2$)	(Fig. 19.9)	(Fig. 19.11)	(Fig. 19.12)
Jan.	20,000	50,000	1100	1350	0.35
Feb.	20.000	50,000	1100	1350	0.35
Mar.	2,500	15,000	160	230	0.86
April	4,000	15,000	230	300	0.78
May	4,000	15,000	230	300	0.78
June	7,000	20,000	400	500	0.60
July	7,000	20,000	400	500	0.60
Aug.	7,000	20,000	400	500	0.60
Sept.	7,000	20,000	400	500	0.60
Oct.	7,000	20,000	400	500	0.60
Nov.	4,000	15,000	230	300	0.78
Dec.	20,000	50,000	1100	1350	0.35
					$\sum u_r = 7.25$

Average: $\bar{u}_r = \dfrac{\sum u_r}{n} = \dfrac{7.25}{12} = 0.60$

Effective modulus of subgrade reaction, k ($lb/in.^3$) = 500
Corrected for loss of support: k ($lb/in.^3$) = 170

Having determined the different moduli as given in columns 2 and 3 of Table 19.6, the next step is to estimate the composite subgrade modulus for each of the seasonal periods considered. At this stage the effect of the existence of bedrock within 10 ft from the subgrade surface is not considered, and it is assumed that the subgrade is of infinite depth. The composite subgrade k is then determined for each seasonal period using the chart in Figure 19.9. The procedure involves the following steps.

Step 1. Estimate k_∞. For example, in September roadbed modulus $M_R = 7,000$ $lb/in.^2$ and subbase modulus $E_{SB} = 20,000$ $lb/in.^2$.

1. Enter the chart in Figure 19.9 at subbase thickness of 6 in. and draw a vertical line to intersect the subbase elastic modulus graph of 20,000 $lb/in.^2$ at A and the roadbed modulus graph of 7,000 $lb/in.^2$ at B.
2. Draw a horizontal line from B to intersect the turning line at C.

3. Draw a vertical line upward from C to intersect the horizontal line drawn from A, as shown. This point of intersection D is the composite modulus of subgrade reaction.

$$k_\infty = 400 \ \text{lb/in.}^3$$

Step 2. Adjust k_∞ for presence of rockbed within 10 ft of subgrade surface. This involves the use of the chart in Figure 19.11. This chart takes into account the depth below the subgrade surface at which the rockbed is located, the resilient modulus of the subgrade soil, and the composite modulus of subgrade reaction (k_∞) determined earlier. The adjustment is made as follows.

1. Enter Figure 19.11 at roadbed soil resilient modulus M_R of 7000 lb/in.². Draw a vertical line to intersect the graph corresponding to the depth of the rockbed below the subgrade surface. In this case it is 5 ft. The intersection point is A.

Figure 19.12 Chart for Estimating Relative Damage to Rigid Pavements Based on Slab Thickness and Underlying Support

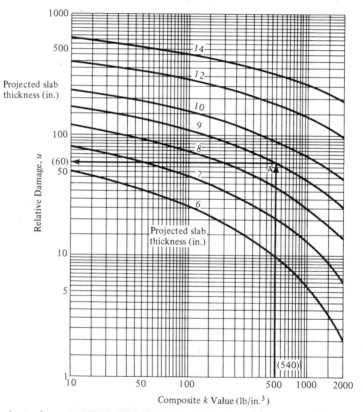

Source: Redrawn from *AASHTO Guide for Design of Pavement Structures*, Washington, D.C.: The American Association of State Highway and Transportation Officials, copyright 1986. Used by permission.

2. From A, draw a horizontal line to intersect the appropriate k_∞ line. In this case the intersection point is B.

3. From B, draw a vertical line downward to determine the adjusted modulus of subgrade reaction k. This is obtained as 500 lb/in.^{2}. Enter the k values in column 5.

Step 3. Determine the effective modulus of subgrade reaction by determining the average damage $(\bar{u}_r)$. The steps involved are similar to those for flexible pavements, but in this case Figure 19.12 is used.

1. Enter Figure 19.12 at the k value obtained in step 2 (that is, 500 lb/in.^{2}) and project a vertical line to intersect the graph representing the projected slab thickness of 9 in. at point A.

2. From A, project a horizontal line to determine the appropriate u_r. In this case $u_r = 60$ percent or 0.60. These values are recorded in column 6 of Table 19.6.

3. Determine the mean u_r as shown in Table 19.6, that is, $\bar{u}_r = 0.6$.

4. Using the mean u_r of 0.6, obtain the effective modulus of subgrade reaction from Figure 19.12 as 500 lb/in.^{2}.

Step 4. Adjust the effective modulus of subgrade reaction determined in step 3 to account for the potential loss of subbase support due to erosion, using Figure 19.10. The LS factor given in Table 19.5 is used. Since the subbase consists of a cement-treated granular material, LS $= 1$ and $k = 500$ lb/in^2. The corrected modulus of subgrade reaction is 170 lb/in^3.

Traffic. The treatment of traffic load is similar to that presented for flexible pavements, in that the traffic load application is given in terms of the number of 18,000 lb equivalent single-axle loads (ESALs). ESAL factors depend on the slab thickness and the terminal serviceability index of the pavement. Tables 19.7 and 19.8 give ESAL factors for rigid pavements with a terminal serviceability index of 2.5. Since the ESAL factor depends on the thickness of the slab, it is therefore necessary to assume the thickness of the slab at the start of the computation. This assumed value is used to compute the number of accumulated ESALs, which in turn is used to compute the required thickness. If the computed thickness is significantly different from the assumed thickness, the accumulated ESAL should be recomputed. This procedure should be repeated until the assumed and computed thicknesses are approximately the same.

Concrete Properties. The concrete property is given in terms of its flexural strength (modulus of rupture) at 28 days. The flexural strength at 28 days of the concrete to be used in construction should be determined by conducting a three-point loading test as specified in AASHTO Designation T97.

Drainage. The drainage quality of the pavement is considered by introducing a factor (C_d) into the performance equation. This factor depends on the quality of

Table 19.7 ESAL Factors for Rigid Pavements, Single Axles, and P_t of 2.5

Axle Load (kip)	Slab Thickness, D (in.)								
	6	7	8	9	10	11	12	13	14
2	.0002	.0002	.0002	.0002	.0002	.0002	.0002	.0002	.0002
4	.003	.002	.002	.002	.002	.002	.002	.002	.002
6	.012	.011	.010	.010	.010	.010	.010	.010	.010
8	.039	.035	.033	.032	.032	.032	.032	.032	.032
10	.097	.089	.084	.082	.081	.080	.080	.080	.080
12	.203	.189	.181	.176	.175	.174	.174	.173	.173
14	.376	.360	.347	.341	.338	.337	.336	.336	.336
16	.634	.623	.610	.604	.601	.599	.599	.599	.598
18	1.00	1.00	1.00	1.00	1.00	1.00	1.00	1.00	1.00
20	1.51	1.52	1.55	1.57	1.58	1.58	1.59	1.59	1.59
22	2.21	2.20	2.28	2.34	2.38	2.40	2.41	2.41	2.41
24	3.16	3.10	3.22	3.36	3.45	3.50	3.53	3.54	3.55
26	4.41	4.26	4.42	4.67	4.85	4.95	5.01	5.04	5.05
28	6.05	5.76	5.92	6.29	6.61	6.81	6.92	6.98	7.01
30	8.16	7.67	7.79	8.28	8.79	9.14	9.35	9.46	9.52
32	10.8	10.1	10.1	10.7	11.4	12.0	12.3	12.6	12.7
34	14.1	13.0	12.9	13.6	14.6	15.4	16.0	16.4	16.5
36	18.2	16.7	16.4	17.1	18.3	19.5	20.4	21.0	21.3
38	23.1	21.1	20.6	21.3	22.7	24.3	25.6	26.4	27.0
40	29.1	26.5	25.7	26.3	27.9	29.9	31.6	32.9	33.7
42	36.2	32.9	31.7	32.2	34.0	36.3	38.7	40.4	41.6
44	44.6	40.4	38.8	39.2	41.0	43.8	46.7	49.1	50.8
46	54.5	49.3	47.1	47.3	49.2	52.3	55.9	59.0	61.4
48	66.1	59.7	56.9	56.8	58.7	62.1	66.3	70.3	73.4
50	79.4	71.7	68.2	67.8	69.6	73.3	78.1	83.0	87.1

Source: Adapted from *AASHTO Guide for Design of Pavement Structures*, Washington, D.C.: The American Association of State Highway and Transportation Officials, copyright 1986. Used by permission.

the drainage as described in Chapter 18 (see Table 18.14) and the percent of time the pavement structure is exposed to moisture levels approaching saturation. Table 19.9 gives AASHTO recommended values for C_d.

Reliability. Reliability considerations for rigid pavement are similar to those for flexible pavement as presented in Chapter 18. Reliability levels, $R\%$, and the overall standard deviation, S_o, are incorporated directly in the design charts.

Design Procedure

The objective of the design is to determine the thickness of the concrete pavement that is adequate to carry the projected design ESAL. The basic equation developed in the 1986 AASHTO design guide[13] for the pavement thickness is given as:

$$
\log_{10} W_{18} = Z_R S_o + 7.35 \log_{10} (D + 1) - 0.06 + \frac{\log_{10} [\Delta PSI/(4.5 - 1.5)]}{1 + [(1.624 \times 10^7)/(D + 1)^{8.46}]}
$$

$$
+ (4.22 - 0.32 P_t) \log_{10} \left\{ \frac{S'_c C_d}{215.63 J} \left(\frac{D^{.75} - 1.132}{D^{.75} - [18.42/(E_c/k)^{.25}]} \right) \right\}
$$

$$
(19.17)
$$

where

Z_R = standard normal variant corresponding to the selected level of reliability

S_o = overall standard deviation (see Chapter 18)

W_{18} = predicted number of 18 kip ESAL applications that can be carried by the pavement structure after construction

D = thickness of concrete pavement to the nearest half-inch

ΔPSI = design serviceability loss = $p_i - p_t$

p_i = initial serviceability index

p_t = terminal serviceability index

E_c = elastic modulus of the concrete to be used in the construction

S'_c = modulus of rupture of the concrete to be used in construction

J = load transfer coefficient = 3.2 (assumed)

C_d = drainage coefficient

Eq. 19.17 can be solved for the thickness D of the pavement in inches by using either a computer program or the two charts in Figures 19.13 and 19.14. The use of a computer program facilitates the iteration necessary, since D has to be assumed to determine the effective modulus of subgrade reaction and the ESAL factors used in the design.

Table 19.8 ESAL Factors for Rigid Pavements, Tandem Axles, and p_t of 2.5

Axle Load (kip)	Slab Thickness, D (in.)								
	6	7	8	9	10	11	12	13	14
2	.0001	.0001	.0001	.0001	.0001	.0001	.0001	.0001	.0001
4	.0006	.0006	.0005	.0005	.0005	.0005	.0005	.0005	.0005
6	.002	.002	.002	.002	.002	.002	.002	.002	.002
8	.007	.006	.006	.005	.005	.005	.005	.005	.005
10	.015	.014	.013	.013	.012	.012	.012	.012	.012
12	.031	.028	.026	.026	.025	.025	.025	.025	.025
14	.057	.052	.049	.048	.047	.047	.047	.047	.047
16	.097	.089	.084	.082	.081	.081	.080	.080	.080
18	.155	.143	.136	.133	.132	.131	.131	.131	.131
20	.234	.220	.211	.206	.204	.203	.203	.203	.203
22	.340	.325	.313	.308	.305	.304	.303	.303	.303
24	.475	.462	.450	.444	.441	.440	.439	.439	.439
26	.644	.637	.627	.622	.620	.619	.618	.618	.618
28	.855	.854	.852	.850	.850	.850	.849	.849	.849
30	1.11	1.12	1.13	1.14	1.14	1.14	1.14	1.14	1.14
32	1.43	1.44	1.47	1.49	1.50	1.51	1.51	1.51	1.51
34	1.82	1.82	1.87	1.92	1.95	1.96	1.97	1.97	1.97
36	2.29	2.27	2.35	2.43	2.48	2.51	2.52	2.52	2.53
38	2.85	2.80	2.91	3.03	3.12	3.16	3.18	3.20	3.20
40	3.52	3.42	3.55	3.74	3.87	3.94	3.98	4.00	4.01
42	4.32	4.16	4.30	4.55	4.74	4.86	4.91	4.95	4.96

44	5.26	5.01	5.16	5.48	5.75	5.92	6.01	6.06	6.09
46	6.36	6.01	6.14	6.53	6.90	7.14	7.28	7.36	7.40
48	7.64	7.16	7.27	7.73	8.21	8.55	8.75	8.86	8.92
50	9.11	8.50	8.55	9.07	9.68	10.14	10.42	10.58	10.66
52	10.8	10.0	10.0	10.6	11.3	11.9	12.3	12.5	12.7
54	12.8	11.8	11.7	12.3	13.2	13.9	14.5	14.8	14.9
56	15.0	13.8	13.6	14.2	15.2	16.2	16.8	17.3	17.5
58	17.5	16.0	15.7	16.3	17.5	18.6	19.5	20.1	20.4
60	20.3	18.5	18.1	18.7	20.0	21.4	22.5	23.2	23.6
62	23.5	21.4	20.8	21.4	22.8	24.4	25.7	26.7	27.3
64	27.0	24.6	23.8	24.4	25.8	27.7	29.3	30.5	31.3
66	31.0	28.1	27.1	27.6	29.2	31.3	33.2	34.7	35.7
68	35.4	32.1	30.9	31.3	32.9	35.2	37.5	39.3	40.5
70	40.3	36.5	35.0	35.3	37.0	39.5	42.1	44.3	45.9
72	45.7	41.4	39.6	39.8	41.5	44.2	47.2	49.8	51.7
74	51.7	46.7	44.6	44.7	46.4	49.3	52.7	55.7	58.0
76	58.3	52.6	50.2	50.1	51.8	54.9	58.6	62.1	64.8
78	65.5	59.1	56.3	56.1	57.7	60.9	65.0	69.0	72.3
80	73.4	66.2	62.9	62.5	64.2	67.5	71.9	76.4	80.2
82	82.0	73.9	70.2	69.6	71.2	74.7	79.4	84.4	88.8
84	91.4	82.4	78.1	77.3	78.9	82.4	87.4	93.0	98.1
86	102.	92.	87.	86.	87.	91.	96.	102.	108.
88	113.	102.	96.	95.	96.	100.	105.	112.	119.
90	125.	112.	106.	105.	106.	110.	115.	123.	130.

Source: Adapted from *AASHTO Guide for Design of Pavement Structures*, Washington, D.C.: The American Association of State Highway and Transportation Officials, copyright 1986. Used by permission.

Table 19.9 Recommended Values for Drainage Coefficient, C_d, for Rigid Pavements

	Percent of Time Pavement Structure is Exposed to Moisture Levels Approaching Saturation			
Quality of Drainage	Less Than 1 Percent	1–5 Percent	5–25 Percent	Greater Than 25 Percent
Excellent	1.25–1.20	1.20–1.15	1.15–1.10	1.10
Good	1.20–1.15	1.15–1.10	1.10–1.00	1.00
Fair	1.15–1.10	1.10–1.00	1.00–0.90	0.90
Poor	1.10–1.00	1.00–0.90	0.90–0.80	0.80
Very poor	1.00–0.90	0.90–0.80	0.80–0.70	0.70

Source: Adapted from *AASHTO Guide for Design of Pavement Structures*, Washington, D.C.: The American Association of State Highway and Transportation Officials, copyright 1986. Used by permission.

Example 19–3 Designing a Rigid Pavement Using the AASHTO Method

The use of the charts is demonstrated with the example given in Figure 19.13. In this case, input values for segment 1 of the chart (Figure 19.13) are

- Effective modulus of subgrade reaction, $k = 72$ lb/in.^{3}
- Mean concrete modulus of rupture, $S'_c = 650$ lb/in.^{2}

Figure 19.13 Design Chart for Rigid Pavements Based on Using Mean Values for Each Input Variable (Segment 1)

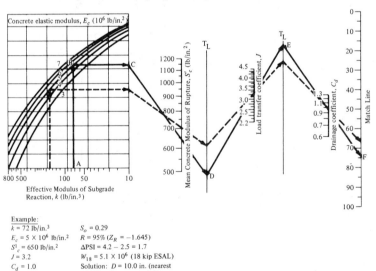

Example:
$\overline{k} = 72$ lb/in.^{3} $S_o = 0.29$
$E_c = 5 \times 10^6$ lb/in.^{2} $R = 95\%$ ($Z_R = -1.645$)
$S'_c = 650$ lb/in.^{2} ΔPSI $= 4.2 - 2.5 = 1.7$
$J = 3.2$ $W_{18} = 5.1 \times 10^6$ (18 kip ESAL)
$C_d = 1.0$ Solution: $D = 10.0$ in. (nearest half-in., from segment 2)

Source: Redrawn from *AASHTO Guide for Design of Pavement Structures*, Washington, D.C.: The American Association of State Highway and Transportation Officials, copyright 1986. Used by permission.

Figure 19.14   Design Chart for Rigid Pavements Based on Using Mean Values for Each Input Variable (Segment 2)

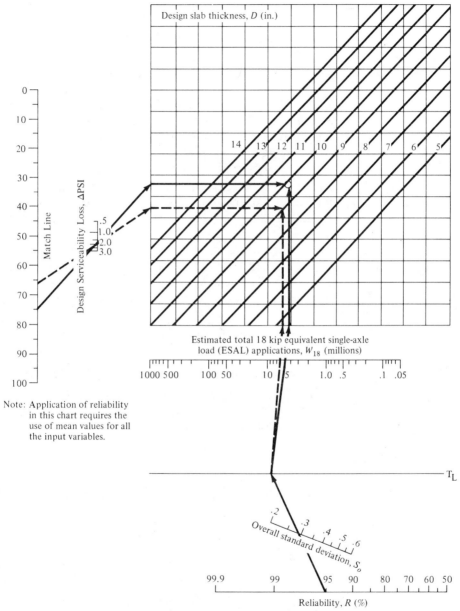

Design slab thickness, D (in.)

Match Line

Design Serviceability Loss, ΔPSI

.5
1.0
2.0
3.0

14 13 12 11 10 9 8 7 6 5

Estimated total 18 kip equivalent single-axle load (ESAL) applications, W_{18} (millions)

1000 500   100 50   10 5   1.0 .5   .1 .05

Note: Application of reliability in this chart requires the use of mean values for all the input variables.

T_L

.2   .3   .4   .5   .6

Overall standard deviation, S_o

99.9   99   95   90   80   70   60 50

Reliability, R (%)

Source: Redrawn from *AASHTO Guide for Design of Pavement Structures*, Washington, D.C.: The American Association of State Highway and Transportation Officials, copyright 1986. Used by permission.

- Load transfer coefficient, $J = 3.2$
- Drainage coefficient, $C_d = 1.0$

These values are used to determine a value on the match line (solid line ABCDEF), as shown in Figure 19.13. Input parameters for segment 2 (Figure 19.14) of the chart are

- Match line value determined in segment 1 (74)
- Design serviceability loss $\Delta PSI = 4.5 - 2.5 = 2.0$
- Reliability, $R\% = 95$ percent ($Z_R = 1.645$)
- Overall standard deviation, $S_o = 0.29$
- Cumulative 18 kip ESAL (5×10^6)

The required thickness of the concrete slab is then obtained, as shown in Figure 19.14, as 10 in. (nearest half-inch).

Note that when the thickness obtained from solving Eq. 19.7 analytically or by use of Figures 19.13 and 19.14 is significantly different from that originally assumed to determine the effective subgrade modulus and to select the ESAL factors, the whole procedure has to be repeated until the assumed and designed values are approximately the same, emphasizing the importance of using a computer program to facilitate the necessary iteration.

Example 19–4 Designing a Rigid Pavement Using the AASHTO Method

Using the data and effective subgrade modulus obtained in Example 19–2, determine whether the pavement will be adequate on a rural expressway for a 20 yr analysis period and the following design criteria.

$P_i = 4.5$
$P_t = 2.5$
ESAL on design lane during 1st yr of operation $= 0.2 \times 10^6$
Traffic growth rate $= 4$ percent
Concrete elastic modulus, $E_c = 5 \times 10^6$ lb/in.^{2}
Mean concrete modulus of rupture $= 700$ lb/in.^{2}
Drainage conditions are such that $C_d = 1.0$
$R = 0.95$ ($Z_R = 1.645$)
$S_o = 0.30$ (for rigid pavements $S_o = 0.3 - 0.4$)
Growth factor $= 29.78$ (from Table 18.6)
$k = 170$ (from Example 19–2)
Assume $D = 9$ in.   (from Example 19–2)
ESAL over design period $= 0.2 \times 10^6 \times 29.78 = 6 \times 10^6$

The depth of concrete required is obtained from Figures 19.13 and 19.14. The dashed lines represent the solution and a depth of 9 in. is obtained. The pavement is therefore adequate.

PCA Design Method

The PCA method for concrete pavement design is based on a combination of theoretical studies, results of model and full-scale tests, and experience gained from the performance of concrete pavements normally constructed and carrying normal traffic loads. Tayabji and Colley have reported on some of these studies and tests.[14,15] The design procedure was initially published in 1961 but was revised in 1984.[16] The procedure provides for the determination of the pavement thickness for plain concrete, simply reinforced concrete, and continuously reinforced concrete pavements.

Design Considerations

The basic factors considered in the PCA design method are

- Flexural strength of the concrete
- Subgrade and subbase support
- Traffic load

Flexural Strength of Concrete. The flexural strength of the concrete used in this procedure is given in terms of the modulus of rupture obtained by the third-point method (ASTM Designation C78).[17] The average of the 28 day test results is used as input by the designer. The design charts and tables, however, incorporate the variation of the concrete strength from one point to another in the concrete slab and the gain in strength with age.

Subgrade and Subbase Support. The Westergaard modulus of subgrade reaction (k) is used to define the subgrade and subbase support. This can be determined by performing a plate-bearing test or correlating with other test results, using the chart in Figure 19.8. No specific correction is made for the reduced value of k during the spring thaw period, but it is suggested that normal summer or fall k values should be used. The modulus of subgrade reaction can be increased by adding a layer of untreated granular material over the subgrade. An approximate value of the increased k can be obtained from Table 19.10(a).

Cement-stabilized soils can also be used as subbase material when the pavement is expected to carry very heavy traffic or when the subgrade material has a low value of modulus of subgrade reaction. Suitable soil materials for cement stabilization are those classified under the AASHTO Soil Classification System as A-1, A-2-4, A-2-5, and A-3. The amount of cement used should be based on standard ASTM laboratory freeze–thaw and wet–dry tests, ASTM Designation D560[18] and D559,[19] respectively, and weight loss criteria. It is permissible, however, to use other methods that will produce an equivalent quality material. Suggested k values for this type of stabilized material are given in Table 19.10(b).

Table 19.10   Design k Values for Untreated and Cement-Treated Subbases

(a) Untreated Granular Subbases

Subgrade k Value (lb/in.^{3})	Subbase k Value (lb/in.^{3})			
	4 in.	6 in.	9 in.	12 in.
50	65	75	85	110
100	130	140	160	190
200	220	230	270	320
300	320	330	370	430

(b) Cement-Treated Subbases

Subgrade k Value (lb/in.^{3})	Subbase k Value (lb/in.^{3})			
	4 in.	6 in.	9 in.	12 in.
50	170	230	310	390
100	280	400	520	640
200	470	640	830	—

Source: Adapted from Robert G. Packard, *Thickness Design for Concrete Highway and Street Pavements*, Portland Cement Association, Skokie, Ill., 1984.

Traffic Load.   The traffic load is computed in terms of the cumulated number of single and tandem axles of different loads projected for the design period of the pavement. The information required to determine cumulated numbers are the average daily traffic (ADT), the average daily truck traffic (ADTT) (in both directions), and the axle load distribution of truck traffic. Only trucks with six or more tires are included in this design. It can be assumed that truck volume is the same in each direction of travel. When there is reason to believe that truck volume varies in each direction, an adjustment factor can be used as discussed in Chapter 18.

The design also incorporates a load safety factor (LSF), which is used to multiply each axle load. The recommended LSF values are:

- 1.2 for interstate and multilane projects with uninterrupted traffic flow and high truck volumes
- 1.1 for highways and arterials with moderate truck volume
- 1.0 for roads and residential streets with very low truck volume

The LSF can be increased to 1.3 if the objective is to maintain a higher-than-normal pavement serviceability level throughout the design life of the pavement. The design procedure also provides for a factor of safety of 1.1 or 1.2 over and above the LSF to allow for unexpected truck traffic.

Design Procedure

The procedure is based on a detailed finite-element computer analysis of stresses and deflections of the pavement at edges, joints, and corners.[20,21] Factors such as finite slab dimensions, position of axle load, load transfer at transverse joints or cracks, and load transfer at pavement and concrete shoulder joints are considered. Load transfer characteristics were modeled using the diameter and modulus of elasticity of the dowels at dowel joints.

A spring stiffness value is used for keyway joints, aggregate interlocks, and cracks in continuously reinforced concrete pavements to represent the load deflection characteristics at such joints. Field and laboratory test results were used to determine the spring stiffness value.

Analysis of different axle load positions on the pavement revealed that the case II position of Figure 19.6 is critical for flexural stresses, whereas case I

Figure 19.15 Allowable Load Repetitions for Fatigue Analysis Based on Stress Ratio

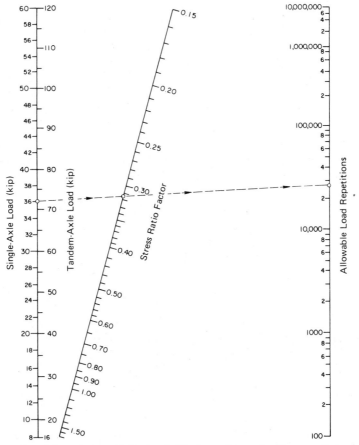

Source: Reproduced from Robert G. Packard, *Thickness Design for Concrete Highway and Street Pavements*, Portland Cement Association, Skokie, Ill., 1984.

Table 19.11   Equivalent Stress Values for Single Axles and Tandem Axles (Without Concrete Shoulder)

Slab Thickness (in.)	k of Subgrade–Subbase (lb/in.³) (Single Axle/Tandem Axle)						
	50	100	150	200	300	500	700
4	825/679	726/585	671/542	634/516	584/486	523/457	484/443
4.5	699/586	616/500	571/460	540/435	498/406	448/378	417/363
5	602/516	531/436	493/399	467/376	432/349	390/321	363/307
5.5	526/461	464/387	431/353	409/331	379/305	343/278	320/264
6	465/416	411/348	382/316	362/296	336/271	304/246	285/232
6.5	417/380	367/317	341/286	324/267	300/244	273/220	256/207
7	375/349	331/290	307/262	292/244	271/222	246/199	231/186
7.5	340/323	300/268	279/241	265/224	246/203	224/181	210/169
8	311/300	274/249	255/223	242/208	225/188	205/167	192/155
8.5	285/281	252/232	234/208	222/193	206/174	188/154	177/143
9	264/264	232/218	216/195	205/181	190/163	174/144	163/133
9.5	245/248	215/205	200/183	190/170	176/153	161/134	151/124
10	228/235	200/193	186/173	177/160	164/144	150/126	141/117
10.5	213/222	187/183	174/164	165/151	153/136	140/119	132/110
11	200/211	175/174	163/155	154/143	144/129	131/113	123/104
11.5	188/201	165/165	153/148	145/136	135/122	123/107	116/98
12	177/192	155/158	144/141	137/130	127/116	116/102	109/93
12.5	168/183	147/151	136/135	129/124	120/111	109/97	103/89
13	159/176	139/144	129/129	122/119	113/106	103/93	97/85
13.5	152/168	132/138	122/123	116/114	107/102	98/89	92/81
14	144/162	125/133	116/118	110/109	102/98	93/85	88/78

Source: Reproduced from Robert G. Packard, *Thickness Design for Concrete Highway and Street Pavements*, Portland Cement Association, Skokie, Ill., 1984.

position is critical for deflection, but with one set of wheels at or near the corner of the free edge and the transverse joint.

The design procedure consists of two parts: fatigue analysis and erosion analysis. The objective of fatigue analysis is to determine the minimum thickness of the concrete required to control fatigue cracking. This is done by comparing the expected axle repetitions with the allowable repetitions for each axle load and ensuring that the cumulative repetitions are less than the allowable. Allowable axle repetitions depend on the stress ratio factor, which is the ratio of the equivalent stress of the pavement to the modulus of rupture of the concrete. The equivalent stress of the pavement depends on the thickness of the slab and the subbase–subgrade k. The chart in Figure 19.15 can be used to determine the allowable load repetitions based on the stress ratio factor. Tables 19.11 and 19.12 give equivalent stress values for pavements without concrete shoulders and with concrete shoulders, respectively.

The objective of the erosion analysis is to determine the minimum thickness of the pavement required to control foundation and shoulder erosion, pumping, and faulting. These pavement distresses are more closely related to deflection, as

Table 19.12   Equivalent Stress Values for Single Axles and Tandem Axles (With Concrete Shoulder)

Slab Thickness (in.)	k of Subgrade–Subbase (lb/in.³) (Single Axle/Tandem Axle)						
	50	100	150	200	300	500	700
4	640/534	559/468	517/439	489/422	452/403	409/388	383/384
4.5	547/461	479/400	444/372	421/356	390/338	355/322	333/316
5	475/404	417/349	387/323	367/308	341/290	311/274	294/267
5.5	418/360	368/309	342/285	324/271	302/254	276/238	261/231
6	372/325	327/277	304/255	289/241	270/225	247/210	234/203
6.5	334/295	294/251	274/230	260/218	243/203	223/188	212/180
7	302/270	266/230	248/210	236/198	220/184	203/170	192/162
7.5	275/250	243/211	226/193	215/182	201/168	185/155	176/148
8	252/232	222/196	207/179	197/168	185/155	170/142	162/135
8.5	232/216	205/182	191/166	182/156	170/144	157/131	150/125
9	215/202	190/171	177/155	169/146	158/134	146/122	139/116
9.5	200/190	176/160	164/146	157/137	147/126	136/114	129/108
10	186/179	164/151	153/137	146/129	137/118	127/107	121/101
10.5	174/170	154/143	144/130	137/121	128/111	119/101	113/95
11	164/161	144/135	135/123	129/115	120/105	112/95	106/90
11.5	154/153	136/128	127/117	121/109	113/100	105/90	100/85
12	145/146	128/122	120/111	114/104	107/95	99/86	95/81
12.5	137/139	121/117	113/106	108/99	101/91	94/82	90/77
13	130/133	115/112	107/101	102/95	96/86	89/78	85/73
13.5	124/127	109/107	102/97	97/91	91/83	85/74	81/70
14	118/122	104/103	97/93	93/87	87/79	81/71	77/67

Source: Reproduced from Robert G. Packard, *Thickness Design for Concrete Highway and Street Pavements*, Portland Cement Association, Skokie, Ill., 1984.

will be seen later. The erosion criterion is mainly based on the rate of work expended by an axle load in deflecting a slab, as it was determined that a useful correlation existed between pavement performance and the product of the corner deflection and the pressure at the slab–subgrade interface. It is suggested that the erosion criterion be used mainly as a guideline and be modified, using experience based on local conditions of climate and drainage.

The erosion analysis is similar to that of fatigue analysis, except that an erosion factor is used instead of the stress factor. The erosion factor is also dependent on the thickness of the slab and the subgrade–subbase k. Tables 19.13 through 19.16 give erosion factors for different types of pavement construction. Figures 19.16 and 19.17 are charts that can be used to determine the allowable load repetitions based on erosion.

The minimum thickness that satisfies both analyses is the design thickness. Design thicknesses for pavements carrying light traffic and pavements with doweled joints carrying medium traffic will usually be based on fatigue analysis, whereas design thicknesses for pavements with undoweled joints carrying medium or heavy traffic and pavements with doweled joints carrying heavy traffic will normally be based on erosion analysis.

Table 19.13   Erosion Factors for Single Axles and Tandem Axles (Doweled Joints, (Without Concrete Shoulder)

Slab Thickness (in.)	k of Subgrade–Subbase (lb/in.^{3}) (Single Axle/Tandem Axle)					
	50	100	200	300	500	700
4	3.74/3.83	3.73/3.79	3.72/3.75	3.71/3.73	3.70/3.70	3.68/3.67
4.5	3.59/3.70	3.57/3.65	3.56/3.61	3.55/3.58	3.54/3.55	3.52/3.53
5	3.45/3.58	3.43/3.52	3.42/3.48	3.41/3.45	3.40/3.42	3.38/3.40
5.5	3.33/3.47	3.31/3.41	3.29/3.36	3.28/3.33	3.27/3.30	3.26/3.28
6	3.22/3.38	3.19/3.31	3.18/3.26	3.17/3.23	3.15/3.20	3.14/3.17
6.5	3.11/3.29	3.09/3.22	3.07/3.16	3.06/3.13	3.05/3.10	3.03/3.07
7	3.02/3.21	2.99/3.14	2.97/3.08	2.96/3.05	2.95/3.01	2.94/2.98
7.5	2.93/3.14	2.91/3.06	2.88/3.00	2.87/2.97	2.86/2.93	2.84/2.90
8	2.85/3.07	2.82/2.99	2.80/2.93	2.79/2.89	2.77/2.85	2.76/2.82
8.5	2.77/3.01	2.74/2.93	2.72/2.86	2.71/2.82	2.69/2.78	2.68/2.75
9	2.70/2.96	2.67/2.87	2.65/2.80	2.63/2.76	2.62/2.71	2.61/2.68
9.5	2.63/2.90	2.60/2.81	2.58/2.74	2.56/2.70	2.55/2.65	2.54/2.62
10	2.56/2.85	2.54/2.76	2.51/2.68	2.50/2.64	2.48/2.59	2.47/2.56
10.5	2.50/2.81	2.47/2.71	2.45/2.63	2.44/2.59	2.42/2.54	2.41/2.51
11	2.44/2.76	2.42/2.67	2.39/2.58	2.38/2.54	2.36/2.49	2.35/2.45
11.5	2.38/2.72	2.36/2.62	2.33/2.54	2.32/2.49	2.30/2.44	2.29/2.40
12	2.33/2.68	2.30/2.58	2.28/2.49	2.26/2.44	2.25/2.39	2.23/2.36
12.5	2.28/2.64	2.25/2.54	2.23/2.45	2.21/2.40	2.19/2.35	2.18/2.31
13	2.23/2.61	2.20/2.50	2.18/2.41	2.16/2.36	2.14/2.30	2.13/2.27
13.5	2.18/2.57	2.15/2.47	2.13/2.37	2.11/2.32	2.09/2.26	2.08/2.23
14	2.13/2.54	2.11/2.43	2.08/2.34	2.07/2.29	' 2.05/2.23	2.03/2.19

Source: Reproduced from Robert G. Packard, *Thickness Design for Concrete Highway and Street Pavements*, Portland Cement Association, Skokie, Ill., 1984.

Example 19–5   Designing a Rigid Pavement Using the PCA Method

The procedure is demonstrated by the following example, which is one of the many solutions provided by PCA. The following project and traffic data are available.

4-lane interstate highway
Rolling terrain in rural location
Design period = 20 yr
Axle loads and expected repetitions are shown in Figure 19.18
Subbase–subgrade $k = 130$ lb/in.^{3}
Concrete modulus of rupture = 650

Determine minimum thickness of a pavement with doweled joints and without concrete shoulders.

Table 19.14 Erosion Factors for Single Axles and Tandem Axles (Aggregate Interlock Joints, Without Concrete Shoulder)

Slab Thickness (in.)	k of Subgrade–Subbase (lb/in.³) (Single Axle/Tandem Axle)					
	50	100	200	300	500	700
4	3.94/4.03	3.91/3.95	3.88/3.89	3.86/3.86	3.82/3.83	3.77/3.80
4.5	3.79/3.91	3.76/3.82	3.73/3.75	3.71/3.72	3.68/3.68	3.64/3.65
5	3.66/3.81	3.63/3.72	3.60/3.64	3.58/3.60	3.55/3.55	3.52/3.52
5.5	3.54/3.72	3.51/3.62	3.48/3.53	3.46/3.49	3.43/3.44	3.41/3.40
6	3.44/3.64	3.40/3.53	3.37/3.44	3.35/3.40	3.32/3.34	3.30/3.30
6.5	3.34/3.56	3.30/3.46	3.26/3.36	3.25/3.31	3.22/3.25	3.20/3.21
7	3.26/3.49	3.21/3.39	3.17/3.29	3.15/3.24	3.13/3.17	3.11/3.13
7.5	3.18/3.43	3.13/3.32	3.09/3.22	3.07/3.17	3.04/3.10	3.02/3.06
8	3.11/3.37	3.05/3.26	3.01/3.16	2.99/3.10	2.96/3.03	2.94/2.99
8.5	3.04/3.32	2.98/3.21	2.93/3.10	2.91/3.04	2.88/2.97	2.87/2.93
9	2.98/3.27	2.91/3.16	2.86/3.05	2.84/2.99	2.81/2.92	2.79/2.87
9.5	2.92/3.22	2.85/3.11	2.80/3.00	2.77/2.94	2.75/2.86	2.73/2.81
10	2.86/3.18	2.79/3.06	2.74/2.95	2.71/2.89	2.68/2.81	2.66/2.76
10.5	2.81/3.14	2.74/3.02	2.68/2.91	2.65/2.84	2.62/2.76	2.60/2.72
11	2.77/3.10	2.69/2.98	2.63/2.86	2.60/2.80	2.57/2.72	2.54/2.67
11.5	2.72/3.06	2.64/2.94	2.58/2.82	2.55/2.76	2.51/2.68	2.49/2.63
12	2.68/3.03	2.60/2.90	2.53/2.78	2.50/2.72	2.46/2.64	2.44/2.59
12.5	2.64/2.99	2.55/2.87	2.48/2.75	2.45/2.68	2.41/2.60	2.39/2.55
13	2.60/2.96	2.51/2.83	2.44/2.71	2.40/2.65	2.36/2.56	2.34/2.51
13.5	2.56/2.93	2.47/2.80	2.40/2.68	2.36/2.61	2.32/2.53	2.30/2.48
14	2.53/2.90	2.44/2.77	2.36/2.65	2.32/2.58	2.28/2.50	2.25/2.44

Source: Reproduced from Robert G. Packard, *Thickness Design for Concrete Highway and Street Pavements*, Portland Cement Association, Skokie, Ill., 1984.

Figure 19.18 is used in working out the example through the following steps.

Step 1: Fatigue Analysis.

1. Select a trial thickness (9.5 in.).
2. Complete all the information at the top of the form as shown.
3. Determine projected number of single-axle and tandem-axle repetitions in the different weight groups for the design period. This is done by determining the ADT, the percentage of truck traffic, the axle load distribution of the truck traffic, and the proportion of trucks on the design lane. The actual cumulative expected repetitions for each range of axle load (without converting to 18,000 lb equivalent axle load) is then computed, using the procedure presented in Chapter 18.
4. Complete columns 1, 2, and 3. Note that column 2 is column 1 multiplied by the LSF, which, in this case is 1.2, as the design is for a 4-lane interstate highway.

Table 19.15   Erosion Factors for Single Axles and Tandem Axles (Doweled Joints, Concrete Shoulder)

Slab Thickness (in.)	k of Subgrade–Subbase (lb/in.³) (Single Axle/Tandem Axle)					
	50	100	200	300	500	700
4	3.28/3.30	3.24/3.20	3.21/3.13	3.19/3.10	3.15/3.09	3.12/3.08
4.5	3.13/3.19	3.09/3.08	3.06/3.00	3.04/2.96	3.01/2.93	2.98/2.91
5	3.01/3.09	2.97/2.98	2.93/2.89	2.90/2.84	2.87/2.79	2.85/2.77
5.5	2.90/3.01	2.85/2.89	2.81/2.79	2.79/2.74	2.76/2.68	2.73/2.65
6	2.79/2.93	2.75/2.82	2.70/2.71	2.68/2.65	2.65/2.58	2.62/2.54
6.5	2.70/2.86	2.65/2.75	2.61/2.63	2.58/2.57	2.55/2.50	2.52/2.45
7	2.61/2.79	2.56/2.68	2.52/2.56	2.49/2.50	2.46/2.42	2.43/2.38
7.5	2.53/2.73	2.48/2.62	2.44/2.50	2.41/2.44	2.38/2.36	2.35/2.31
8	2.46/2.68	2.41/2.56	2.36/2.44	2.33/2.38	2.30/2.30	2.27/2.24
8.5	2.39/2.62	2.34/2.51	2.29/2.39	2.26/2.32	2.22/2.24	2.20/2.18
9	2.32/2.57	2.27/2.46	2.22/2.34	2.19/2.27	2.16/2.19	2.13/2.13
9.5	2.26/2.52	2.21/2.41	2.16/2.29	2.13/2.22	2.09/2.14	2.07/2.08
10	2.20/2.47	2.15/2.36	2.10/2.25	2.07/2.18	2.03/2.09	2.01/2.03
10.5	2.15/2.43	2.09/2.32	2.04/2.20	2.01/2.14	1.97/2.05	1.95/1.99
11	2.10/2.39	2.04/2.28	1.99/2.16	1.95/2.09	1.92/2.01	1.89/1.95
11.5	2.05/2.35	1.99/2.24	1.93/2.12	1.90/2.05	1.87/1.97	1.84/1.91
12	2.00/2.31	1.94/2.20	1.88/2.09	1.85/2.02	1.82/1.93	1.79/1.87
12.5	1.95/2.27	1.89/2.16	1.84/2.05	1.81/1.98	1.77/1.89	1.74/1.84
13	1.91/2.23	1.85/2.13	1.79/2.01	1.76/1.95	1.72/1.86	1.70/1.80
13.5	1.86/2.20	1.81/2.09	1.75/1.98	1.72/1.91	1.68/1.83	1.65/1.77
14	1.82/2.17	1.76/2.06	1.71/1.95	1.67/1.88	1.64/1.80	1.61/1.74

Source: Reproduced from Robert G. Packard, *Thickness Design for Concrete Highway and Street Pavements*, Portland Cement Association, Skokie, Ill., 1984.

5. Determine the equivalent stresses for single axle and tandem axle. Table 19.11 is used in this case since there is no concrete shoulder. Interpolating for $k = 130$, for single axles and 9.5 in. thick slab

$$\text{equivalent stress} = 215 - \frac{215 - 200}{50} \times 30$$

$$= 206$$

Similarly, for the tandem axles,

$$\text{equivalent stress} = 205 - \frac{205 - 183}{50} \times 30$$

$$= 192$$

Record these values in spaces provided at 8 and 11, respectively.

6. Determine the stress ratio, which is the equivalent stress divided by the modulus of rupture.

a. For single axles,

$$\text{stress ratio} = \frac{206}{650} = 0.317$$

Table 19.16   Erosion Factors for Single Axles and Tandem Axles (Aggregate Interlock
Joints, Concrete Shoulder)

Slab Thickness (in.)	k of Subgrade–Subbase (lb/in.³) (Single Axle/Tandem Axle)					
	50	100	200	300	500	700
4	3.46/3.49	3.42/3.39	3.38/3.32	3.36/3.29	3.32/3.26	3.28/3.24
4.5	3.32/3.39	3.28/3.28	3.24/3.19	3.22/3.16	3.19/3.12	3.15/3.09
5	3.20/3.30	3.16/3.18	3.12/3.09	3.10/3.05	3.07/3.00	3.04/2.97
5.5	3.10/3.22	3.05/3.10	3.01/3.00	2.99/2.95	2.96/2.90	2.93/2.86
6	3.00/3.15	2.95/3.02	2.90/2.92	2.88/2.87	2.86/2.81	2.83/2.77
6.5	2.91/3.08	2.86/2.96	2.81/2.85	2.79/2.79	2.76/2.73	2.74/2.68
7	2.83/3.02	2.77/2.90	2.73/2.78	2.70/2.72	2.68/2.66	2.65/2.61
7.5	2.76/2.97	2.70/2.84	2.65/2.72	2.62/2.66	2.60/2.59	2.57/2.54
8	2.69/2.92	2.63/2.79	2.57/2.67	2.55/2.61 ·	2.52/2.53	2.50/2.48
8.5	2.63/2.88	2.56/2.74	2.51/2.62	2.48/2.55	2.45/2.48	2.43/2.43
9	2.57/2.83	2.50/2.70	2.44/2.57	2.42/2.51	2.39/2.43	2.36/2.38
9.5	2.51/2.79	2.44/2.65	2.38/2.53	2.36/2.46	2.33/2.38	2.30/2.33
10	2.46/2.75	2.39/2.61	2.33/2.49	2.30/2.42	2.27/2.34	2.24/2.28
10.5	2.41/2.72	2.33/2.58	2.27/2.45	2.24/2.38	2.21/2.30	2.19/2.24
11	2.36/2.68	2.28/2.54	2.22/2.41	2.19/2.34	2.16/2.26	2.14/2.20
11.5	2.32/2.65	2.24/2.51	2.17/2.38	2.14/2.31	2.11/2.22	2.09/2.16
12	2.28/2.62	2.19/2.48	2.13/2.34	2.10/2.27	2.06/2.19	2.04/2.13
12.5	2.24/2.59	2.15/2.45	2.09/2.31	2.05/2.24	2.02/2.15	1.99/2.10
13	2.20/2.56	2.11/2.42	2.04/2.28	2.01/2.21	1.98/2.12	1.95/2.06
13.5	2.16/2.53	2.08/2.39	2.00/2.25	1.97/2.18	1.93/2.09	1.91/2.03
14	2.13/2.51	2.04/2.36	1.97/2.23	1.93/2.15	1.89/2.06	1.87/2.00

Source: Reproduced from Robert G. Packard, *Thickness Design for Concrete Highway and Street Pavements*, Portland Cement Association, Skokie, Ill., 1984.

b.   For tandem axles,

$$\text{stress ratio} = \frac{192}{650} = 0.295$$

Record these values in spaces provided at 9 and 12, respectively.

7.   Using Figure 19.15, determine the allowable load repetitions for each axle load based on fatigue analysis. Note that the axle loads used are those in column 2 (that is, after multiplying by the LSF). For example, for the first row of single axles, the axle load to be used is 36. Draw a line joining 36 kip on the single-axle load axis (left side) and 0.317 on the stress ratio factor line. Extend that line to the allowable load repetitions line and read allowable repetitions as 27,000. Repeat this for all axle loads, noting that for tandem axles, the scale is on the right side of the axle load axis. Complete column 4.

Figure 19.16 Allowable Load Repetitions for Erosion Analysis Based on Erosion
Factors (Without Concrete Shoulder)

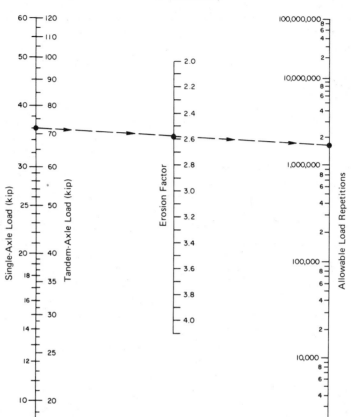

Source: Reproduced from Robert G. Packard, *Thickness Design for Concrete Highway and
Street Pavements*, Portland Cement Association, Skokie, Ill., 1984.

8. Determine the fatigue percentage for each axle load, which is an indication
   of the resistance consumed by the expected number of axle load repetitions.

$$\text{fatigue percentage} = \frac{\text{column 3}}{\text{column 4}} \times 100$$

Complete column 5.

9. Determine total fatigue resistance consumed by summing up column 5
   (single and tandem axles). If this total does not exceed 100 percent, the
   assumed thickness is adequate for fatigue resistance for the design period.
   The total in this example is 62.8 percent, which shows that 9.5 in. is adequate for fatigue resistance.

Figure 19.17   Allowable Load Repetitions for Erosion Analysis Based on Erosion
Factor (With Concrete Shoulder)

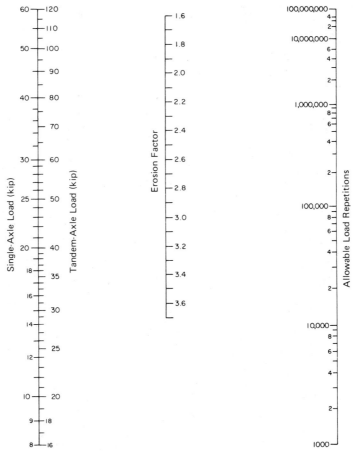

Source: Reproduced from Robert G. Packard, *Thickness Design for Concrete Highway and Street Pavements*, Portland Cement Association, Skokie, Ill., 1984.

Step 2: Erosion Analysis.   Items 1 through 4 in the fatigue analysis are the same for the erosion analysis, therefore, they will not be repeated. The following additional procedures are necessary.

1.   Determine the erosion factor for the single-axle loads and tandem-axle loads. Note that the erosion factors depend on the type of pavement construction—that is, doweled or undoweled joints, with or without concrete shoulders. The construction type in this project is doweled joints without shoulders. Table 19.13 is therefore used. For single axles, erosion factor = 2.59. For tandem axles, erosion factor = 2.79. Record these values as items 10 and 13, respectively.

Figure 19.18   Pavement Thickness Calculation

Project _Design 1A, four-lane Interstate, Rural_

Trial thickness _9.5_ in.

Subbase-subgrade k _130_ pci

Modulus of rupture, MR _650_ psi

Load safety factor, LSF _1.2_

Doweled joints: yes _✓_ no _____

Concrete shoulder: yes _____ no _✓_

Design period _20_ years

4 in. untreated subbase

Axle load, kips	Multiplied by LSF 1.2	Expected repetitions	Fatigue analysis		Erosion analysis	
			Allowable repetitions	Fatigue, percent	Allowable repetitions	Damage, percent
1	2	3	4	5	6	7

8. Equivalent stress _206_   10. Erosion factor _2.59_

9. Stress ratio factor _0.317_

Single Axles

30	36.0	6,310	27,000	23.3	1,500,000	0.4
28	33.6	14,690	77,000	19.1	2,200,000	0.7
26	31.2	30,140	230,000	13.1	3,500,000	0.9
24	28.8	64,410	1,200,000	5.4	5,900,000	1.1
22	26.4	106,900	Unlimited	0	11,000,000	1.0
20	24.0	235,800	//	0	23,000,000	1.0
18	21.6	307,200	//	0	64,000,000	0.5
16	19.2	422,500			Unlimited	0
14	16.8	586,900			//	0
12	14.4	1,837,000			//	0

11. Equivalent stress _192_   13. Erosion factor _2.79_

12. Stress ratio factor _0.295_

Tandem Axles

52	62.4	21,320	1,100,000	1.9	920,000	2.3
48	57.6	42,870	Unlimited	0	1,500,000	2.9
44	52.8	124,900	//	0	2,500,000	5.0
40	48.0	372,900	//	0	4,600,000	8.1
36	43.2	885,800			9,500,000	9.3
32	38.4	930,700			24,000,000	3.9
28	33.6	1,656,000			92,000,000	1.8
24	28.8	984,900			Unlimited	0
20	24.0	1,227,000			//	0
16	19.2	1,356,000				
				Total 62.8		Total 38.9

Source: Redrawn from Robert G. Packard, _Thickness Design for Concrete Highway and Street Pavements_, Portland Cement Association, Skokie, Ill., 1984.

2. Determine the allowable axle repetitions for each axle load based on erosion analysis using either Figure 19.16 or Figure 19.17. In this problem, Figure 19.16 will be used as the pavement has no concrete shoulder. Enter these values under column 6.

3. Determine erosion damage percent for each axle load, that is, divide column 3 by column 6. Enter these values in column 7.

4. Determine the total erosion damage by summing column 7 (single and tandem axles). In this problem total drainage is 38.9 percent.

The results indicate that 9.5 in. is adequate for both fatigue and erosion analysis. Since the total damage for each analysis is much lower than 100 percent, the question may arise whether a 9 in. thick pavement would not be adequate. If the calculation is repeated for a 9 in. thick pavement, it will show that the fatigue condition will not be satisfied since total fatigue consumption will be about 245 percent. In order to achieve the most economic section for the design period, trial runs should be made until the minimum pavement thickness that satisfies both analyses is obtained.

SUMMARY

This chapter has presented the basic design principles for rigid highway pavements and the application of these principles in the AASHTO and PCA methods. Rigid pavements have some flexural strength and can therefore sustain a beam-like action across minor irregularities in the underlying material. The flexural strength of the concrete used is therefore an important factor in the design of the pavement thickness. Although reinforcing steel may sometimes be used to reduce cracks in rigid pavement, it is not considered as contributing to the flexural strength of the pavement.

Both the AASHTO and PCA methods of design are iterative, indicating the importance of computers in carrying out the design. Design charts are included in this chapter, which can be used if computer facilities are not readily available.

A phenomenon that may have significant effect on the life of a concrete pavement is pumping. The PCA method indirectly considers this phenomenon in the erosion consideration. When rigid pavements are to be constructed over fine-grain soils that are susceptible to pumping, effective means to reduce pumping, such as limiting expansion joints or stabilization of the subgrade, should be considered.

PROBLEMS

19-1 Describe the basic types of highway concrete pavements giving the conditions under which each type will be constructed.

19–2 Determine the tensile stress imposed at night by a wheel load of 750 lb located at the edge of a concrete pavement with the following dimensions and properties.

Pavement thickness $= 8\frac{1}{2}$ in.
$\mu = 0.15$
$E_c = 4.2 \times 10^6$ lb/in.^{2}
$k = 200$ lb/in.^{3}
Radius of loaded area $= 3$ in.

19–3 Repeat problem 19–2 for the load located at the interior of the slab.

19–4 Determine the maximum distance of contraction joints for a plain concrete pavement if the maximum allowable tensile stress in the concrete is 50 lb/in.^{2} and the coefficient of friction between the slab and the subgrade is 1.7. Assume uniform drop in temperature. Weight of concrete is 144 lb/ft^3.

19–5 Repeat problem 19–4, with the slab containing temperature steel in the form of welded wire mesh consisting of 0.125 in.^{2} steel/ft width. The modulus of steel E_s is 30×10^6 lb/in.^{2}, $E_c = 5 \times 10^6$ lb/in.^{2}, and $h = 6$ in.

19–6 A concrete pavement is to be constructed for a 4-lane urban expressway on a subgrade with an effective modulus of subgrade reaction k of 100 lb/in.^{3}. The accumulated equivalent axle load for the design period is 3.25×10^6. The initial and terminal serviceability are 4.5 and 2.5, respectively. Using the AASHTO design method determine a suitable thickness of the concrete pavement, if the working stress of the concrete is 600 lb/in.^{2} and the modulus of elasticity is 5×10^6 lb/in.^{2}. Take the overall standard deviation, S_o, as 0.30, the load transfer coefficient J as 3.2, the drainage coefficient as 0.9, and $R = 95\%$.

19–7 An existing rural 4-lane highway is to be replaced by a 6-lane divided expressway (3 lanes in each direction). Traffic volume data on the highway indicate that the AADT (both directions) during the first year of operation is 24,000 with the following vehicle mix and axle loads.

Passenger cars $= 50$ percent
2-axle single-unit trucks (12,000 lb/axle) $= 40$ percent
3-axle single-unit trucks (16,000 lb/axle) $= 10$ percent

The vehicle mix is expected to remain the same throughout the design life of 20 years, although traffic is expected to grow at a rate of 3.5 percent annually. Using the AASHTO design procedure, determine the minimum depth of concrete pavement required for the design period of 20 years.

$P_i = 4.5$	$J = 3.2$
$P_t = 2.5$	$C_d = 1.0$
$S'_c = 650$ lb/in.^{2}	$S_o = 0.3$
$E_c = 5 \times 10^6$ lb/in.^{2}	$R = 95$ percent
$k = 130$	

19–8 Repeat Example 19–7 in the text for a pavement containing doweled joints, 6 in. untreated subbase, and concrete shoulders, using the PCA design method.

19–9 Repeat problem 19–7 using the PCA design method, for a pavement containing doweled joints and concrete shoulders. The modulus of rupture of the concrete used is 600 lb/in.^{2}.

19–10 Repeat problem 19–7, using the PCA design method if the subgrade k value is 50 and a 6 in. stabilized subbase is used. The modulus of rupture of the concrete is 600 lb/in.^{2} and the pavement has aggregate interlock joints (that is, no dowels) and a concrete shoulder.

19–11 Repeat problem 19–10, using PCA design method, assuming the pavement has doweled joints and no concrete shoulders.

REFERENCES

1. *Annual Book of ASTM Standards, Concrete and Mineral Aggregates*, Vol. 04.02, American Society for Testing and Materials, Philadelphia, Pa., October 1985.

2. *Standard Specifications for Transportation Materials and Methods of Sampling and Testing*, 14th ed., American Association of State Highway and Transportation Officials, Washington, D.C., 1986.

3. Ibid.

4. Ibid.

5. Ibid.

6. H. M. Westergaard, *Theory of Concrete Pavement Design*, Proceedings, Highway Research Board, Washington, D.C., 1927.

7. Ibid.

8. L. W. Teller and E. C. Sutherland, "The Structural Design of Concrete Pavements," *Public Roads*, Vol. 17, Bureau of Public Roads, Washington, D.C., 1936.

9. *AASHTO Guide for Design of Pavement Structures*, American Association of State Highway and Transportation Officials, Washington, D.C., 1986.

10. *Thickness Design for Concrete Highway and Street Pavements*, Portland Cement Association, Skokie, Ill., 1984.

11. *AASHTO Guide for Design of Pavement Structures*.

12. Ibid.

13. Ibid.

14. S. D. Tayabji and B. E. Colley, "Analysis of Jointed Concrete Pavements," Report prepared by the Construction Technology Laboratories of the Portland Cement Association for the Federal Highway Administration, October 1981.

15. S. D. Tayabji and B. E. Colley, "Improved Rigid Pavement Joints," Paper presented at the annual meeting of the Transportation Research Board, January 1983.

16. *Thickness Design for Concrete Highway and Street Pavements*.

17. *Annual Book of ASTM Standards*.

18. Ibid.

19. Ibid.

20. Tayabji and Colley, "Analysis of Jointed Concrete Pavements."

21. Tayabji and Colley, "Improved Rigid Pavement Joints."

chapter 20

Pavement Rehabilitation Management

The nation's highway system is largely in place, and little in the way of major new highway construction is anticipated during the coming decades. Accordingly, rehabilitation of the existing system has become a major activity for highway and transportation agencies. In this chapter we describe the various approaches for pavement rehabilitation management and the data required to evaluate pavement condition.

PROBLEMS OF HIGHWAY REHABILITATION

A major problem that faces highway and transportation agencies is that the amount of funds they receive is usually insufficient to adequately repair and rehabilitate every roadway section that deteriorates. The problem is further complicated in that roads may be in poor condition but still usable, making it easy to defer repair projects until conditions become unacceptable. Roadway deterioration usually is not the result of poor design and construction practices but is caused by the inevitable wear and tear that occurs over a period of years. The gradual deterioration of a pavement occurs because of variations in climate and increasing truck and automobile traffic. Just as a piece of cloth eventually tears asunder if a small hole is not immediately repaired, so will a roadway unravel if its surface is allowed to deteriorate. Lack of funds or materials often limit timely repair and rehabilitation of transportation facilities, causing a greater problem with more serious pavement defects and higher costs.

Because funds and personnel are often inadequate to accommodate needs, the dilemma faced by many transportation agencies is to balance its programs between preventive maintenance activities and projects requiring immediate corrective action. If preventive maintenance is neglected, then projects will be selected based on the extent of deterioration as seen by the user. The traveling public is unwilling to tolerate the condition of a pavement when the ride is extremely rough, vibrations cause damage to their vehicles, accidents occur, and user costs increase significantly. Usually preventive maintenance, carried out in an orderly and systematic way, will be the least expensive approach in the long run. However, when funds are extremely limited, agencies often respond to the most pressing and severe problems or the ones that generate the most vocal complaints.

Approaches to Pavement Rehabilitation

The term *pavement rehabilitation management* is used to describe the various strategies that can be used to decide on a pavement restoration and rehabilitation policy. At one extreme there is the "squeaking wheel" approach, wherein projects are selected that have created the greatest attention. At the other extreme is a system wherein all roads are repaired on a regular schedule, for which money is no object. In realistic terms, pavement rehabilitation or repair strategies are plans that establish minimum standards for pavement condition and seek to establish the type of treatment required and the time frame for project completion. Rehabilitation management strategies include consideration of items such as pavement life, first cost, annual maintenance costs, benefits, safety, physical, environmental, and economic constraints, and life cycle owner and user costs.

The general topic of pavement management includes design, construction, maintenance, and rehabilitation. In essence, pavement management includes all activities involved in furnishing the pavement portion of a transportation system, such as planning, design, construction, maintenance, and rehabilitation. A total framework for pavement management is shown in Figure 20.1. Note that this process includes elements of network and project management and requires technical analysis in areas of planning, economics, design, and construction—topics that were described in earlier chapters of this textbook.

In this chapter we confine ourselves to the subject of rehabilitation management, which is defined as any improvement made to an existing pavement after initial construction, excluding improvements to shoulders (that is, widening or surfacing) or bridges (which is a topic more appropriate to structural design). Pavement rehabilitation can be both preventive and corrective.

Importance of Pavement Condition Data

The first step in the process of pavement rehabilitation management is to secure data about the condition of each pavement section in the system. Originally, condition data were obtained by visual inspections that established the type, extent, and severity of pavement condition. These inspections were subjective and relied heavily on judgment and experience for determining pavement condition and program priorities. Although such an approach can be appropriate under certain circumstances, possibilities exist for variations among inspectors, and experience is not easily transferable. In more recent years, visual ratings have been supplemented with standardized testing equipment to measure road roughness, pavement deflection, and skid resistance.

Figure 20.1   Framework for Total Pavement Management System

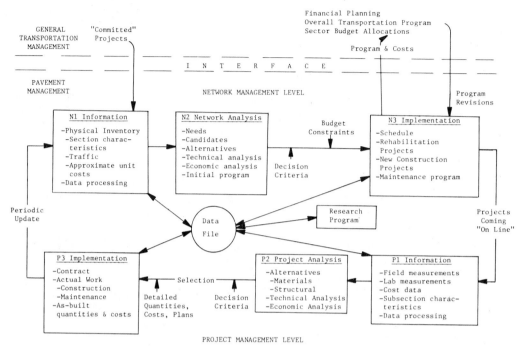

Source: Reproduced from W. R. Hudson, R. Haas, and R. D. Pedigo, *Pavement Management System Development*, NCHRP Report 215, Transportation Research Board, Washington, D.C., November 1979.

Pavement condition data are used for the following purposes.

1. *Establishing Project Priorities:* The data on pavement condition are used in various ways to establish the relative condition of each pavement and to establish project priorities. There is no single method used and each state selects that combination of measures it considers most appropriate.

2. *Establishing Options:* Pavement condition data can also be used to develop a rehabilitation program on an annual or 5-year basis. Data about pavement condition, in terms of type, extent, and severity, can be used to determine which of the available rehabilitation options (ranging from minor repairs to complete pavement recycling) should be selected.

3. *Forecasting Performance:* By use of correlations between pavement performance indicators and variables such as traffic loadings, it is possible to predict the likely future condition of any given pavement section. This information is useful for preparing long-range budget estimates of the cost to maintain the highway system at a minimum standard of performance or to determine future consequences of various funding levels.

METHODS FOR DESCRIBING ROADWAY CONDITION

Four characteristics of pavement condition that are used in evaluating rehabilitation needs are: (1) pavement roughness (rideability), (2) pavement distress (surface condition), (3) pavement deflection (structural failure), and (4) skid resistance (safety).

Measuring Pavement Roughness

Pavement roughness refers to irregularities in the pavement surface that affect the smoothness of a ride. The serviceability of a roadway was initially defined in the AASHO road test, a national pavement research project. Two terms were defined: (1) present serviceability rating (PSR), and (2) present serviceability index (PSI).

PSR is a number grade given to a pavement section based on the ability of that pavement to serve its intended traffic. The PSR rating is established by observation and requires judgment on the part of the individual doing the rating. In the original AASHO road test, a panel of raters drove on each test section in a vehicle of their choice and rated the performance of each section on the basis of how well the road section would serve if the rater were to drive his car over a similar road all day long. The ratings range between 0 and 5, with 5 being very good and 0 being impassable. Figure 20.2 illustrates the PSR as used in the AASHO road test. Serviceability ratings are based on the user's perception of pavement performance and are determined from the average rating of a panel of road users.

PSI is a value for pavement condition determined as a surrogate for PSR and is based on physical measurements. PSI is not based on panel ratings. The primary measure of PSI is pavement roughness. The PSI is an objective means of estimating the PSR, which is subjective.

Figure 20.2   Inidvidual Present Serviceability Rating

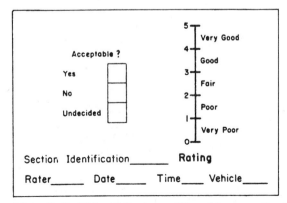

Source: Reproduced from R. G. Hicks, *Collection and Use of Pavement Condition Data,* NCHRP Synthesis 76, Transportation Research Board, Washington, D.C., July 1981.

Figure 20.3   Typical Pavement Performance Curve

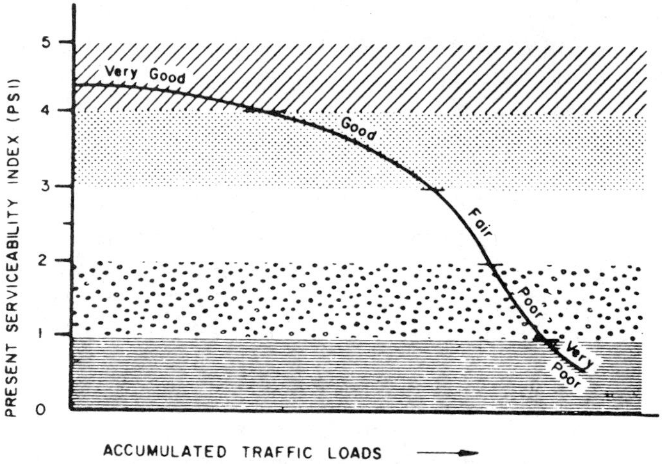

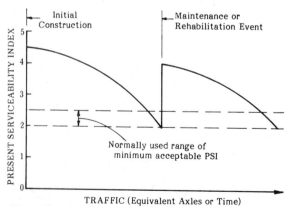

Source: Reproduced from Dale E. Peterson, *Evaluation of Pavement Maintenance Strategies*, NCHRP Synthesis 77, Transportation Research Board, Washington, D.C., September 1981.

The performance of a pavement can be described in terms of its PSI and traffic loading over time, as is illustrated in Figure 20.3. For example, when a pavement is originally constructed it is in very good condition with a PSI value of 4.5 (see Figure 20.4). Then as the number of traffic loadings increases, the PSI declines to a value of 2, which is normally the minimum acceptable range. After the pavement section is maintained or rehabilitated, the PSI value increases to 4, and as traffic loads increase, the PSI declines again until it reaches 2 and rehabilitation is again required.

Figure 20.4   Performance History for Pavement Using PSI

Source: Reproduced from R. G. Hicks, *Collection and Use of Pavement Condition Data*, NCHRP Synthesis 76, Transportation Research Board, Washington, D.C., July 1981.

Subsequent to the introduction of the PSI concept and the AASHTO road test, considerable development of equipment for measuring road roughness took place. Various designs were adopted such as the U.S. Bureau of Roads roughometer, the CHLOE profilometer, and precise leveling. Other devices include car ride meters (the most common are the Mays, PCA, and Cox), which measure vehicle response to pavement roughness, and still others, such as the surface dynamics profilometer and the U.S. Air Force laser profilometer. Essentially these devices measure the extent of deviation of the road profile, and from this, its roughness or PSI can be inferred.

A survey of rehabilitation programming in eight states found that four used the Mays ride meter and four used the Cox road meter.[1] For example, the Arizona Department of Transportation (ADOT) uses the Mays ride meter, which reports results in inches per mile (with a greater number of inches per mile indicating a rougher road). These readings are correlated with panel ratings on a scale from 0 to 5. Using the panel's ratings, levels of service of desirable and undesirable have been established where desirable is less than 165 in./mi and undesirable is greater than 256 in./mi. These levels of performance play an important role in data analysis and in optimization models used to establish maintenance priorities. Regression equations that correlate PSI with road roughness data have also been used.

Most agencies use some type of car-mounted meter for measuring ride quality. The advantages are relatively low cost, simplicity and ease of operation, capability for acquisition of large amounts of data, adequate repeatability, and output correlated with PSI. Disadvantages are the equipment must be calibrated at frequent intervals to insure repeatability, ride meters cannot measure pavement profiles, and results depend on physical characteristics of the vehicle on which the meter is mounted. Meters towed in trailers can eliminate some of these problems.

Measuring Pavement Distress

The term *pavement distress* refers to the condition of a pavement surface in terms of its general appearance. A perfect pavement is level, and has a continuous and unbroken surface. In contrast a distressed pavement may be fractured, distorted, or disintegrated. These three basic categories of distress can be further subdivided. For example, fractures can be seen as cracks or as spalling (chipping of the pavement surface). Cracks can be further described as generalized, transverse, longitudinal, alligator, and block. A pavement distortion may be evidenced by ruts or corrugation of the surface. Pavement disintegration can be observed as raveling (loosening of pavement structure), stripping of the pavement from the subbase, and surface polishing. The types of distress data collected for flexible and rigid pavements vary from one state to another. Figure 20.5 lists the three pavement distress groups, the measure of distress, and the probable cause.

Most agencies use some measure of cracking in evaluating the condition of flexible pavements. The most common measures are transverse, longitudinal, and alligator cracks. Distortion is usually measured by determining the extent of rutting, and disintegration is measured by the amount of raveling. Each state or

Figure 20.5   Pavement Distress Groups and Their Causes

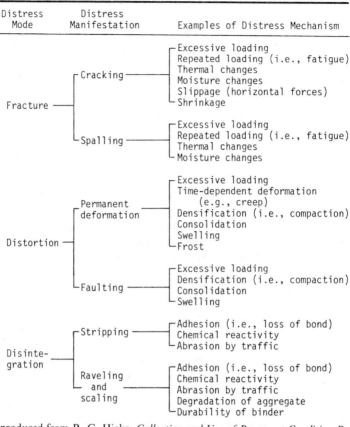

Distress Mode	Distress Manifestation	Examples of Distress Mechanism
Fracture	Cracking	Excessive loading Repeated loading (i.e., fatigue) Thermal changes Moisture changes Slippage (horizontal forces) Shrinkage
	Spalling	Excessive loading Repeated loading (i.e., fatigue) Thermal changes Moisture changes
Distortion	Permanent deformation	Excessive loading Time-dependent deformation (e.g., creep) Densification (i.e., compaction) Consolidation Swelling Frost
	Faulting	Excessive loading Densification (i.e., compaction) Consolidation Swelling
Disintegration	Stripping	Adhesion (i.e., loss of bond) Chemical reactivity Abrasion by traffic
	Raveling and scaling	Adhesion (i.e., loss of bond) Chemical reactivity Abrasion by traffic Degradation of aggregate Durability of binder

Source: Reproduced from R. G. Hicks, *Collection and Use of Pavement Condition Data,* NCHRP Synthesis 76, Transportation Research Board, Washington, D.C., July 1981.

federal agency has its own procedures for measuring pavement distress, consequently, there is a wide range of methods used to conduct distress surveys. Typically, each agency has a procedural manual that defines each element of distress to be observed, with instructions as to how these are to be rated on a given point scale. As an example, the pavement condition data form for the Washington Department of Transportation is illustrated in Figure 20.6. Note that the form distinguishes between bituminous and Portland cement concrete pavements. For bituminous pavements, the items observed are corrugations, alligator cracking, raveling, longitudinal cracking, transverse cracking, and patching. For Portland cement concrete, the measures are cracking, raveling, joint spalling, faulting, and patching.

Typically, distress data are obtained by trained observers who make subjective judgments about pavement condition based on predetermined factors. Often photographs are used for making judgments. Figure 20.7 shows the guide used to assist raters in estimating cracking. Although some agencies use full sampling, pavement sections of about 300 ft often are randomly selected to represent each mile of road. Measurements are usually made on a regular schedule about every 1–3 yr. After the data are recorded, the results are condensed into a single number, called a distress (or defect) rating (DR). A perfect pavement usually is given a score of 100 and if distress is observed, points are subtracted. The general equation is

$$DR = 100 - \sum_{i=1}^{n} d_{t_i}$$

where

d_{t_i} = the number of points assigned to distress type t_i

n = number of distress types used in rating method

One of the major problems with condition or distress surveys is the variability in results due to the subjective procedures used. Other causes of error are variations in the condition of the highway segment observed, changes in the evaluation procedure, and changes in observed location from year to year. These variations can be minimized by using the same pavement location each year, observing pavement condition at regular intervals of about 1 mi, using teams of raters to reach consensus on numerical values, keeping procedures simple and easy to understand and use, and scheduling regular training sessions.

Example 20–1 Computing Distress Rating of a Pavement Section

A pavement rating method for a certain state uses the following elements in its evaluation process: longitudinal or alligator cracking, rutting, pushing, raveling, and patching. Determine the DR for a given section of pavement if observed values are 8, 2, 7, 4, and 5.

$$DR = 100 - \sum_{i=1}^{n} d_{t_i}$$
$$= 100 - (8 + 2 + 7 + 4 + 5)$$
$$= 100 - 26$$
$$= 74$$

Measuring Pavement Structural Condition

The structural adequacy of a pavement is measured by either nondestructive means, which measure deflection under static or dynamic loadings, or destructive tests, which involve removing sections of the pavement and testing these in the

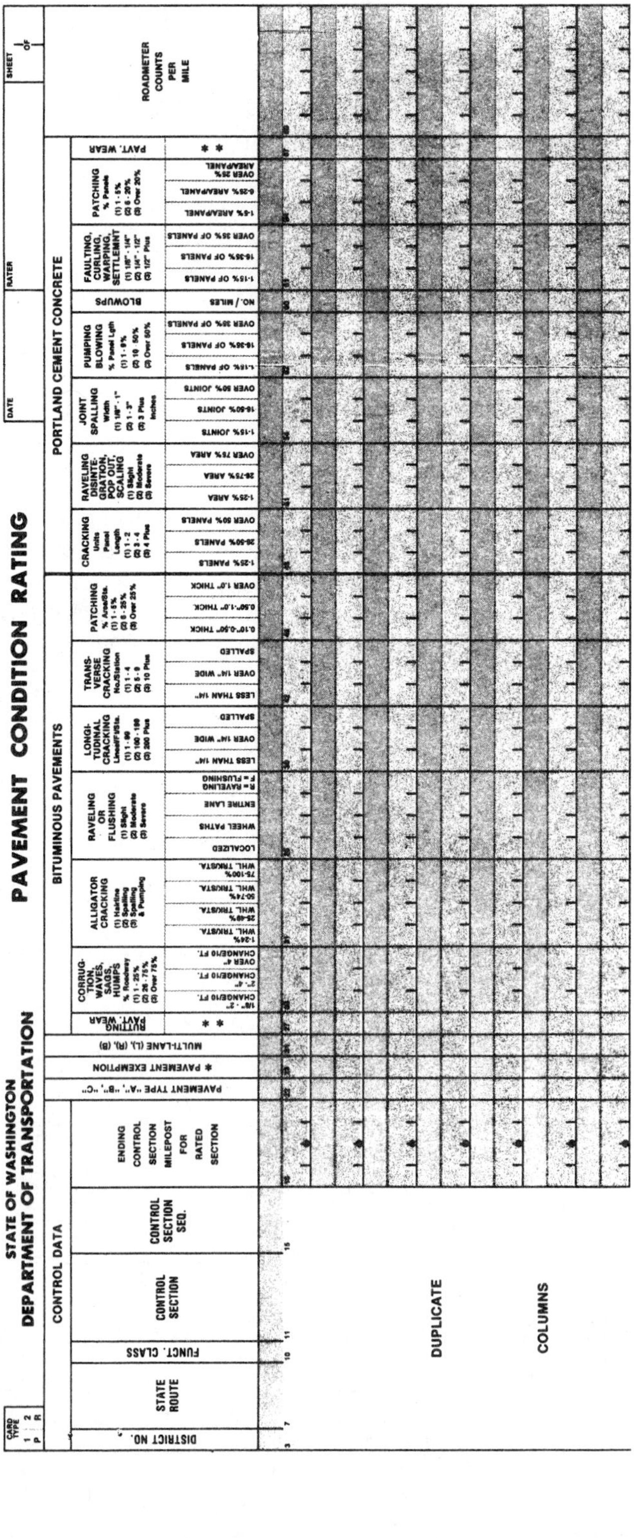

Figure 20.6 Pavement Condition Rating Form for State of Washington

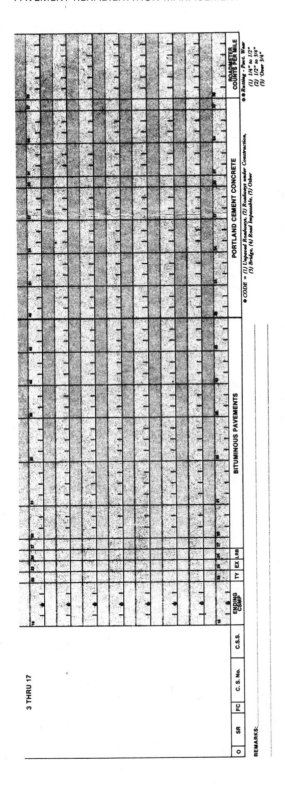

Figure 20.7   Guide Used to Estimate Pavement Cracking in Arizona

Source: Reproduced from R. G. Hicks, *Collection and Use of Pavement Condition Data*, NCHRP Synthesis 76, Transportation Research Board, Washington, D.C., July 1981.

laboratory. Structural condition evaluations are rarely used by state agencies for monitoring network pavement condition, due to the expense involved. However, nondestructive evaluations, which gather deflection data, are used by some agencies on a project basis for pavement design purposes and to develop rehabilitation strategies.

Nondestructive structural evaluation is based on the premise that measurements can be made on the surface of the pavement and from these measurements in situ characteristics can be inferred about the structural adequacy of the

Figure 20.8 Benkelman Beam

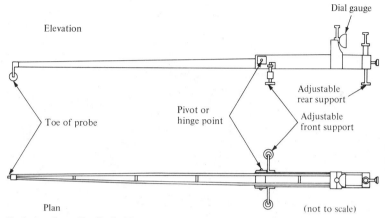

Elevation

Dial gauge

Toe of probe

Pivot or
hinge point

Adjustable
rear support

Adjustable
front support

Plan

(not to scale)

Source: Redrawn from R. C. G. Haas, *Pavement Management Guide*, Roads and Transportation Association of Canada, Ottawa, 1977.

pavement. The three basic nondestructive test methods are: (1) measurement of static deflections, (2) measurement of deflections due to dynamic or repeated loads, and (3) measurement of density of pavement layers by nuclear radiation. (The last method is used primarily to evaluate individual pavement layers during construction and not for pavement evaluation purposes.)

The most common method for measuring static deflections is the Benkelman Beam, Figure 20.8, which is a simple hand-operated device designed to measure deflection responses of a flexible pavement to a standard wheel load. A probe point is placed between two dual tires and the motion of the beam is observed on a dial, which records the maximum deflection. Other static devices that are used include the traveling deflectometer, the plate-bearing test, and the lacroix deflectograph. Most of these devices are based on the Benkelman Beam principle, in which pavement deflections due to a static or slowly moving load are measured manually or by automatic recording devices.

The most common dynamic loading method for measuring pavement deflections is the Dynaflect. This device, Figure 20.9, basically consists of a dynamic cyclical force generator mounted on a two-wheeled trailer, a control unit, a sensor assembly, and a sensor calibration unit. The system provides rapid and precise measurements of roadway deflections, which in this test are caused by forces generated by unbalanced flywheels rotating in opposite directions. A vertical force of 1000 lb is produced at the loading wheels, and deflections are measured at five points on the pavement surface, located 1 ft apart.

Deflection data from Dynaflect tests are used to correlate with Benkelman Beam measurements and for the calculation of curvature indices used in pavement structural analysis. Table 20.1 lists examples of agency practice for structural evaluation. As noted in this table, the Dynaflect and Benkelman Beam methods are most common, but they are primarily used for design purposes and not for pavement rehabilitation management.

Figure 20.9   Dynaflect During Measurement Operations (Upper) and with Trailer Body Removed (Lower)

Source: Reproduced from N. M. Moore, D. I. Hanson, and J. W. Hall, "An Introduction to Nondestructive Structural Evaluation of Pavements," *Transportation Research Circular*, No. 189, Transportation Research Board, Washington, D.C., January 1978.

Measuring Skid Resistance

Safety characteristics of a pavement are another measure of its condition, and highway agencies continually monitor this aspect to insure that roadway sections are operating at the highest possible level of safety. The principal measure of pavement safety is its skid resistance. Other elements contributing to the extent in which pavements perform safely are rutting (which causes water to collect that creates hydroplaning) and adequacy of visibility of pavement markings.

Skid-resistance data are collected to monitor and evaluate the effectiveness of a pavement in preventing or reducing skid-related accidents. Skid data are used by highway agencies to identify pavement sections with low skid resistance, to develop priorities for rehabilitation, and to evaluate the effectiveness of various pavement mixtures and surface types.

The coefficient of sliding function between a tire and pavement depends on factors such as weather conditions, pavement texture, tire condition, and speed.

Table 20.1 Current Practices for Structural Evaluation

Agency	Comments
Arizona	Annual Dynaflect deflections at three locations per mile as routine measure up to 1980 Now only used for design purposes Recently purchased a falling-weight deflectometer
California	Dynaflect deflections used in design but not in monitoring system
Florida	Dynaflect reflections for design and some monitoring purposes Recently used a falling-weight deflectometer in a research study
New York	Deflection data obtained for research purposes only
Ontario	Data collected on selective basis only Both the Dynaflect and Benkelman Beam in use
Pennsylvania	Road Rater deflections used to evaluate selected sections that have reached terminal serviceability (flexible pavements only)
U.S. Air Force (CERL)	No single device used Structural evaluation is presently based on measurement of field CBR and k values, and various other material properties
Utah	Dynaflect deflection measurements used to predict remaining life based on projected 18 kip loads One test per mile with temperature corrections (candidate projects are tested more extensively for overlay design)
Washington	Benkelman Beam deflections used for selected locations but not for routine monitoring

Source: Adapted from R. G. Hicks, *Collection and Use of Pavement Condition Data*, NCHRP Synthesis 76, Transportation Research Board, Washington, D.C., July 1981.

Since skidding characteristics are not solely dependent on the pavement condition, it is necessary to standardize testing procedures and in this way eliminate all factors but the pavement. The basic formula for friction factor f is

$$f = \frac{L}{N}$$

where

L = lateral or frictional force required to cause two surfaces to move tangentially to each other.

N = force perpendicular to the two surfaces

When skid tests are performed, they must conform to specified standards set by the American Society for Testing and Materials (ASTM). The test results produce a skid number (SK*) where

$$SK = 100f$$

The SK is usually obtained by measuring the forces obtained with a towed trailer riding on a wet pavement, equipped with standarized tires. The principal methods of testing are (1) locked-wheel trailers, (2) Yaw mode trailers, and (3) the British Portable Tester. Table 20.2 lists the types of skid tests used by a representative group of agencies.

Table 20.2   Current Practices for Measuring Skid Resistance

Agency	Comments
Arizona	Mu-Meter for 500 ft in each mile of entire system on annual basis
California	Measured periodically with locked-wheel trailer manufactured by K. J. Law, Inc.
Florida	Skid resistance measured with locked-wheel trailer Approximately 35 to 40 percent of interstate and primary network evaluated each year
New York	Skid trailer covers entire system about every 3 years Test conducted every 0.1 or 0.2 mi Data used separately, mostly in connection with accident surveillance and analysis
Ontario	Skid resistance measurements made on selective basis
Pennsylvania	Skid resistance measured with locked-wheel trailer Measurements made on every other 250 ft segment or approximately 10 tests per mi Data evaluated separately from other condition information
U.S. Air Force (CERL)	Mu-Meter used approximately every 5 years
Utah	Mu-Meter used on wet pavement 0.1 mi sections measured at every milepost (closer intervals where low SN measured)
Washington	Skid trailer measurements made with locked-wheel trailer manufactured by K. J. Law, Inc. High acccident locations checked; 1 mi sections routinely measured every other year Data considered separately

Source: Adapted from R. G. Hicks, *Collection and Use of Pavement Condition Data*, NCHRP Synthesis 76, Transportation Research Board, Washington, D.C., July 1981.

---

*Skid number is also referred to as SN. However, SK is used in this text to avoid confusion with the term *structural number* (SN) used in pavement design.

Locked-wheel trailers are the most widely used skid measuring devices. The test involves wetting the pavement surface and pulling a two-wheel trailer whose wheels have been locked in place. The test is conducted at 40 mph, with standard tires each with seven grooves. The locking force is measured and from this an SK is obtained.

The Yaw mode test is done with the wheels turned at a specified angle, to simulate the effects of cornering. The most common device for this test is a Mu-Meter, which uses two wheels turned at 7.5°. The trailer is pulled in a straight line on a wetted surface with both wheels locked. Since both wheels cannot be in the wheel paths, friction values may be higher than obtained using a locked-wheel trailer.

The British Portable Tester consists of a pendulum with a spring-loaded rubber shoe. This portable device, which can be used in the field or laboratory, operates by having the pendulum drop and the shoe slide over the pavement area. A friction number, called the British Pendulum Number (BPN) is determined.

Figure 20.10 Skid Data for Various Pavement Surface Types

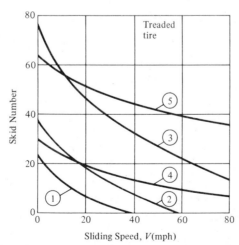

Source: Redrawn from H. W. Kummer and W. E. Meyer, *Tentative Skid Resistance Requirements for Main Rural Highways*, NCHRP Report 37, Transportation Research Board, Washington, D.C., 1967.

This test is not widely used in the United States and results have not correlated well with locked-wheel trailers.

The most common method for determining skid resistance is the locked-wheel trailer, and most state highway agencies own one or more of these devices. Several states use the Mu-Meter—a Yaw-mode device. Figure 20.10 illustrates typical skid results for various pavement conditions. Skid resistance data are not typically used in developing rehabilitation programs. Rather they are used to monitor the safety of the highway system and to assist in reducing potential accident locations.

Example 20–2   Measuring Skid Resistance

A 10,000 lb load is placed on two tires of a locked-wheel trailer. At a speed of 30 mph a force of 5000 lb is required to move the device. Determine the SK and the surface type, assuming that treaded tires were used.

$$SK = 100f = (100)\frac{L}{N}$$

$$= 100 \times \frac{5000}{10,000} = 50$$

From Figure 20.10, at 30 mph and SK = 60, the surface type is coarse textured and gritty.

PAVEMENT REHABILITATION

A variety of methods can be used to rehabilitate pavements or to correct deficiencies in a given pavement section, including using overlays, sealing cracks, using seal coats, and repairing potholes.

Classification of Techniques

Rehabilitation techniques are classified as (1) corrective, which involves the permanent or temporary repair of deficiencies on an as needed basis, or (2) preventive, which involves surface applications of either structural or nonstructural improvements intended to keep the quality of the pavement above a predetermined level. Corrective maintenance is analogous to repairing a small hole in a cloth, whereas preventive maintenance can be thought of as sewing a large patch or replacing the lining in a suit. Just as with the sewing analogy, corrective measures can serve as prevention measures as well. To illustrate, sealing a crack is done to correct an existing problem, but it also prevents further deterioration that would occur if the crack were not repaired. Similarly, a "chip seal coat," which is a layer of gravel placed on a thin coating of asphalt, often is used to correct a skid problem but also helps prevent further pavement deterioration.

Rehabilitation Strategies

Pavement rehabilitation strategies can be categorized in a variety of ways. One approach is in terms of the problem being solved, such as skid resistance, surface drainage, unevenness, roughness, or cracking. Another approach is in terms of the type of treatment used, such as surface treatment, overlay, or recycle. A third approach is in terms of the type of surface that will result from the process, such as asphalt overlay, rock seal coat, or liquid seal coat. The latter system is the most commonly used because it enables the designer to consider each maintenance alternative in terms of a final product and then select the most appropriate one in terms of results desired and cost. Table 20.3 is a partial list of some of the pavement repair strategies used in California for both flexible and rigid pavements. Also listed are proper and improper uses of these strategies, their service life, and cost per lane mile.

For example, the table shows that asphalt concrete overlays (repair strategy 3) are used to restore the structural adequacy of a pavement, restore its surface texture, and improve ride quality. It is used where cracking is due to loads, ride score > 45, rut depth $> \frac{3}{4}$ in., and coarse aggregate ravels. (Ride score is defined in the pavement rehabilitation programming example later in this chapter.) This treatment is not used on steep grades, it will last 10 yr and cost $12,500/lane mi/0.1 ft. In contrast, a rock seal coat (repair strategy 11) is used to waterproof pavements and decrease crack spalling. Its proper use is to seal dried-out pavements and correct skids but should not be used if the road is curved, where high volumes of traffic occur, to heal cracks, or where ride score > 45. This treatment lasts only 1–3 yr and costs $3000/lane mi.

To illustrate how this information can be used, consider a situation in which a surface treatment is desired that will serve to waterproof an existing pavement. Two repair strategies are considered: (1) a rock seal coat and (2) a liquid seal coat. (These are strategies 11 and 15 in Table 20.3.) As is noted in the table, the service life for both strategies is 1–3 yr. However, strategy 11 costs $3000/lane mi per year, whereas strategy 15 costs only $300/lane mi per year. However, the rock seal coat will also serve to improve the skid resistance characteristics, whereas the liquid seal coat should not be used if the skid number is low. Neither treatment should be used if the ride score is > 45. In this instance, if the road is in generally good condition and simply requires rejuvenation, then the liquid seal coat is appropriate. As noted in the final column of Table 20.3, both treatments are extensively used.

Thus, the repair strategy information provides a means by which appropriate strategies for the problem at hand can be identified and the most economical one (based on annual cost and life expectancy) can be selected.

Table 20.3   Pavement Repair Strategies for Flexible and Rigid Pavements

Flexible:

REPAIR STRATEGY	FUNCTION (OBJECTIVE)	PROPER USE	IMPROPER USE	SERVICE LIFE	*1976/77 COST PER LANE MILE	CALIFORNIA'S EXPERIENCE
1. LANE RECONSTRUCTION	RESTORE STRUCTURAL ADEQUACY	A. WHERE MORE COST EFFECTIVE THAN ALTERNATES B. RIDE SCORE >45 C. VERTICAL GRADE CONSTRAINTS		20 YR	$90,000 ①	EXTENSIVE
2. PCC OVERLAYS	RESTORE STRUCTURAL ADEQUACY	WHERE MORE COST EFFECTIVE THAN ALTERNATES (0.55' MINIMUM THICKNESS)		10 YR	$65,000 ②	LIMITED
3. AC OVERLAYS	A. RESTORE STRUCTURAL ADEQUACY B. REPAIR CRACKED PAVEMENT C. RESTORE SURFACE TEXTURE D. IMPROVE RIDE QUALITY	A. LOAD ASSOCIATED CRACKING B. RIDE SCORE >45 C. RUT DEPTH >3/4" D. COARSE AGGREGATE RAVEL	A. VERTICAL GRADE CONSTRAINTS	10 YR	$12,500/0.10'	EXTENSIVE
4. INVERTED OVERLAYS	A. RESTORE STRUCTURAL ADEQUACY B. REPAIR CRACKED PAVEMENT C. RESTORE SURFACE TEXTURE D. IMPROVE RIDE QUALITY	A. WHERE MORE COST EFFECTIVE THAN CONVENTIONAL OVERLAY B. PROVIDE DRAINAGE BLANKET	A. FREEZE-THAW AREAS	10 YR TARGET	$35,000 ③	LIMITED EXPERIMENTAL INSTALLATIONS
5. PAVEMENT REINFORCING FABRIC & OVERLAY	A. RESTORE STRUCTURAL ADEQUACY B. REPAIR CRACKED PAVEMENT C. RESTORE SURFACE TEXTURE D. IMPROVE RIDE QUALITY E. WATER RESISTANT MEMBRANE	A. WHERE MORE COST EFFECTIVE THAN CONVENTIONAL OVERLAY B. VERTICAL GRADE CONSTRAINTS		10 YR TARGET	$35,000 ④	LIMITED EXPERIMENTAL INSTALLATIONS
6. RUBBERIZED ASPHALT INTERLAYER & OVERLAY	A. RESTORE STRUCTURAL ADEQUACY B. REPAIR CRACKED PAVEMENT C. RESTORE SURFACE TEXTURE D. IMPROVE RIDE QUALITY E. WATER RESISTANT MEMBRANE	A. WHERE MORE COST EFFECTIVE THAN CONVENTIONAL OVERLAYS B. VERTICAL GRADE CONSTRAINTS		10 YR TARGET	$35,000 ⑤	LIMITED EXPERIMENTAL INSTALLATIONS
7. HOT RECYCLING	A. RESTORE STRUCTURAL ADEQUACY B. REPAIR CRACKED PAVEMENT C. RESTORE SURFACE TEXTURE D. CONSERVE NATURAL RESOURCES	A. WHERE MORE COST EFFECTIVE THAN ALTERNATES B. VERTICAL GRADE CONSTRAINTS	AIR QUALITY CONSTRAINT AT PLANT	10 YR	$24,000/0.10	NONE TO DATE
8A. HEATER REMIX 8B. CUTLER PROCESS	A. RESTORE STRUCTURAL ADEQUACY B. REPAIR CRACKED PAVEMENT C. RESTORE SURFACE TEXTURE	A. WHERE MORE COST EFFECTIVE THAN ALTERNATES B. VERTICAL GRADE CONSTRAINTS	AIR QUALITY CONSTRAINT AT SITE	5-10 YR	$25,000 ⑥	HEATER REMIX-EXTENSIVE CUTLER PROCESS-NONE TO DATE
9. COLD PLANING	A. CONFORM TO ELEVATION CONTROL B. REMOVE DETERIORATED AND/OR CONTAMINATED LAYER	A. WHERE MORE COST EFFECTIVE THAN ALTERNATIVES B. PREPARE FOR OVERLAY C. AIR QUALITY CONTROLS PRECLUDE HOT PLANING D. VERTICAL GRADE CONTROLS		NOT APPLICABLE	$15,000 ⑦	MODERATE

#	Item	Function	Use	Limitations	Service Life	Cost	Extent of Use
10.	RUBBERIZED ASPHALT CHIP SEAL COAT	A. WATERPROOF PAVEMENT B. DECREASE CRACK SPALLING	A. SEAL DRIED OUT PAVEMENT B. STAGE CONSTRUCTION PRECEDING A PLANNED OVERLAY C. FINE AGGREGATE RAVEL	A. HIGH DEGREE OF ROAD CURVATURE B. HIGH VOLUME TURNING MOVES C. >HAIRLINE CRACKS D. HEAL CRACKS UNLESS FILLED E. RIDE SCORE >45	UNKNOWN ESTIMATE 2-5YR	$10,000	LIMITED EXPERIMENTAL INSTALLATIONS
11.	ROCK SEAL COAT	A. WATERPROOF PAVEMENT B. DECREASE CRACK SPALLING C. TEXTURE SURFACE	A. SEAL DRIED OUT PAVEMENT B. FINE AGGREGATE RAVEL C. SKID RESISTANCE CORRECTION	A. HIGH DEGREE OF ROAD CURVATURE B. HIGH VOLUME TURNING MOVES C. >HAIRLINE CRACKS D. HEAL CRACKS UNLESS FILLED E. RIDE SCORE >45	1-3 YR	$3,000	EXTENSIVE
12.	OPEN GRADED SEAL COAT	A. DECREASE CRACK SPALLING B. TEXTURE SURFACE C. CORRECT BLEEDING	A. SEAL DRIED OUT PAVEMENT B. COARSE AGGREGATE RAVEL C. SKID RESISTANCE CORRECTION D. CORRECT BLEEDING	A. HEAL CRACKS B. >HAIRLINE CRACKS UNLESS FILLED C. RIDE SCORE >45 D. FREQUENT TIRE CHAIN USE REQUIRED	5 YR	$5,000	EXTENSIVE
13.	SLURRY SEALS	A. STOP RAVEL B. WATERPROOF PAVEMENT C. DECREASE CRACK SPALLING D. TEXTURE SURFACE	A. SEAL DRIED OUT PAVEMENT B. FINE OR COARSE AGGREGATE RAVEL	A. HEAL CRACKS B. RIDE SCORE >45 C. >HAIRLINE CRACKS UNLESS FILLED D. FREQUENT TIRE CHAIN USE REQUIRED	4 YR	$4,000	LIMITED
14.	SEAL COAT WITH (SAND) COVER	A. WATERPROOF PAVEMENT B. DECREASE CRACK SPALLING C. STOP RAVEL D. RESTORE BINDER FLEXIBILITY	A. SEAL DRIED OUT PAVEMENT B. FINE AGGREGATE RAVEL	A. HEAL CRACKS B. COARSE RAVEL C. RIDE SCORE >45 D. LOW TO MODERATE SKID NUMBER E. RUTTING F. HIGH. IMPER. PVMT.	1-3 YR	$1,500	EXTENSIVE
15.	LIQUID SEAL COAT	A. WATERPROOF PAVEMENT B. DECREASE CRACK SPALLING C. STOP RAVEL D. RESTORE BINDER FLEXIBILITY	A. SEAL DRIED OUT PAVEMENT B. FINE AGGREGATE RAVEL	A. HEAL CRACKS B. COARSE RAVEL C. RIDE SCORE >45 D. LOW TO MODERATE SKID NUMBER E. RUTTING F. HIGH. IMPER. PVMT.	1-3 YR	$300	EXTENSIVE
16.	BINDER MODIFIERS (REJUVENATING AGENT)	A. √ WATERPROOF PAVEMENT B. DECREASE CRACK SPALLING C. STOP RAVEL D. RESTORE BINDER FLEXIBILITY	A. SEAL DRIED OUT PAVEMENT B. FINE AGGREGATE RAVEL	A. HEAL CRACKS B. COARSE RAVEL C. RIDE SCORE >45 D. LOW TO MODERATE SKID NUMBER E. RUTTING F. HIGHLY IMPERMEABLE PAVEMENT	1-3 YR	$500	EXTENSIVE

Assumptions for Flexible Pavement Cost Estimates: ① 0.35' AC over 0.70' Class A CTB ② 0.55' PCC ③ 0.08' O.G. with 0.20' AC ④ Reinforcing Fabric with 0.20' AC ⑤ Rubberized Chip Seal with 0.20' AC ⑥ Scarify, Add Rejuvenator and 0.08' AC ⑦ 0.10' Depth

Continued

Table 20.3—*Continued*

Rigid:

REPAIR STRATEGY	FUNCTION (OBJECTIVE)	PROPER USE	IMPROPER USE	SERVICE LIFE	*1976-66 COST PER LANE MILE	CALIFORNIA'S EXPERIENCE
1. LANE RECONSTRUCTION	RESTORE STRUCTURAL ADEQUACY	A. >10% THIRD STAGE CRACKING B. WHERE MORE COST EFFECTIVE THAN ALTERNATES. C. VERTICAL GRADE CONSTRAINTS D. RIDE SCORE >45		20 YR.	$100,000	LIMITED
2. PCC OVERLAYS	RESTORE STRUCTURAL ADEQUACY	A. >10% THIRD STAGE CRACKING B. WHERE MORE COST EFFECTIVE THAN ALTERNATES. C. NO VERTICAL GRADE CONSTRAINTS D. WHEN TRAFFIC HANDLING PERMITS E. RIDE SCORE >45		20 YR.	$65,000①	LIMITED
3. AC OVERLAYS	A. RESTORE STRUCTURAL ADEQUACY B. REPAIR CRACKED PAVEMENT C. RESTORE SURFACE TEXTURE D. IMPROVE RIDE QUALITY	A. >10% THIRD STAGE CRACKING B. WHERE MORE COST EFFECTIVE THAN ALTERNATES. C. FAULTING IF SLABS STABILIZED D. NO VERTICAL GRADE CONSTRAINTS E. RIDE SCORE >45	ROCKING SLABS	10 YR.	$12,500/0.10'	MODERATE
4. INVERTED OVERLAYS	A. RESTORE STRUCTURAL ADEQUACY B. REPAIR CRACKED PAVEMENT C. RESTORE SURFACE TEXTURE D. IMPROVE RIDE QUALITY	A. >10% THIRD STAGE CRACKING B. WHERE MORE COST EFFECTIVE THAN ALTERNATES. C. FAULTING IF SLABS STABILIZED D. NO VERTICAL GRADE CONSTRAINTS E. RIDE SCORE >45 F. PROVIDE DRAINAGE BLANKET	FREEZE-THAW AREAS	10 YR.	$40,000②	LIMITED
5. PAVEMENT REINFORCING FABRIC & OVERLAY	A. REPAIR BADLY CRACKED PAVEMENT B. WATER RESISTANT MEMBRANE	A. WHERE MORE COST EFFECTIVE THAN ALTERNATES. B. VERTICAL GRADE CONSTRAINTS		10 YR.	$35,000③	ONE EXPERIMENTAL INSTALLATION
6. SLAB REPLACEMENT	REPLACE RANDOM SLABS WHICH ARE SEVERELY DISTRESSED	A. LESS THAN 35 SLABS PER MILE B. WHERE MORE COST EFFECTIVE THAN OVERLAY OR RECONSTRUCTION C. SEVERE CRACK SPALLING D. RIDE SCORE >45		5 YRS.	$15,000④	MODERATE

Treatment	Purpose	Criteria	Condition	Life Expectancy	Cost	Experience
7. MUDJACKING	A. FILL CAVITIES UNDER PAVEMENT B. RESTORE PAVEMENT GRADELINE	A. IMPROVE RIDE SCORE B. FAULTED OR VERTICALLY DISPLACED SLABS C. WHERE MORE COST EFFECTIVE THAN ALTERNATES	BADLY CRACKED	5-10 YR.	$ 55,000 ⑤	EXTENSIVE
8. SUBSEALING	FILL CAVITIES UNDER PAVEMENT	FAULTED PAVEMENT	BADLY CRACKED SLABS	5-10 YR.	$55,000	NO RECENT EXPERIENCE
9. GRINDING	A. RELIEVE FAULTING B. IMPROVE RIDE QUALITY	A. FAULTING > 1/4" B. RIDE SCORE > 45	> 10% THIRD STAGE CRACKING	MORE THAN 5 YR.	$20,000	EXTENSIVE
10. PAVEMENT SUBDRAINAGE	DEWATER STRUCTURAL SECTION	A. AT LOWER PAVEMENT EDGE B. IN WET CLIMATE C. INDICATIONS OF FAULTING AND/OR PUMPING		UNKNOWN EST 10-15 YR.	$20,000/MILE	LIMITED EXPERIMENTAL INSTALLATIONS
11. CRACK FILLING	WATERPROOF PAVEMENT	A. CLEAN CRACKS ≥ 1/4" WIDE B. APPROPRIATE SEALANT	A. DIRTY CRACKS B. < 1/4" CRACKS	1-2 YR.	$ 200	EXTENSIVE
12. GROOVING	A. REDUCE HYDROPLANING B. IMPROVE VEHICLE TRACKING	A. ABNORMAL RATE OF WET PAVEMENT ACCIDENTS DUE TO HYDROPLANING	BADLY CRACKED PAVEMENT	10-15 YR.	$ 5,000	EXTENSIVE

* COSTS DO NOT INCLUDE TRAFFIC HANDLING

Assumptions for Rigid Pavement Repair Cost Estimates: ① 0.60' PCC ② 0.08' O.G. plus 0.25' AC ③ Reinforcing Fabric with 0.20' AC ④ 30 Slabs/Mile @ $500 each (12" Depth PCC) ⑤ $8 Per Square Yard

Source: Reproduced from Dale E. Peterson, *Evaluation of Pavement Maintenance Strategies,* NCHRP Synthesis 77, Transportation Research Board, Washington, D.C., September 1981.

Alternatives for Repair and Rehabilitation

Figure 20.11 illustrates a variety of pavement repair and rehabilitation alternatives and differentiates between preventive and corrective approaches. For example, preventive strategies for pavement surfaces include fog-seal asphalt, rejuvenators, joint sealing, seal coats (with aggregate) and a thin blanket. This figure is helpful in understanding the purpose for which a given treatment is intended.

Table 20.4 lists a variety of pavement rehabilitation techniques for both flexible and rigid pavements and indicates if the technique is commonly used in corrective or preventive situations and its degree of effectiveness. For example, patching is always considered to be corrective and can be effective if properly done. Many patching materials are available. At the other extreme, overlays are both corrective and preventive and considered to be an effective technique. Surface treatments can be either preventive or corrective and are considered an effective means of maintaining roads on a regular basis.

Another method of describing pavement rehabilitation alternatives is to tabulate the possible types of deficiencies and show the most appropriate treatment. Table 20.5 is a summary of repair methods for flexible and rigid pavements. The distress types are shown in terms of severity (moderate or severe) and whether or not the repair is considered temporary or permanent. For example, for flexible pavements alligator cracking is repaired by removal and replacement of the surface course, by permanent patching or scarifying, and mixing the surface materials with asphalt.

Figure 20.11 Pavement Maintenance and Rehabilitation Alternatives

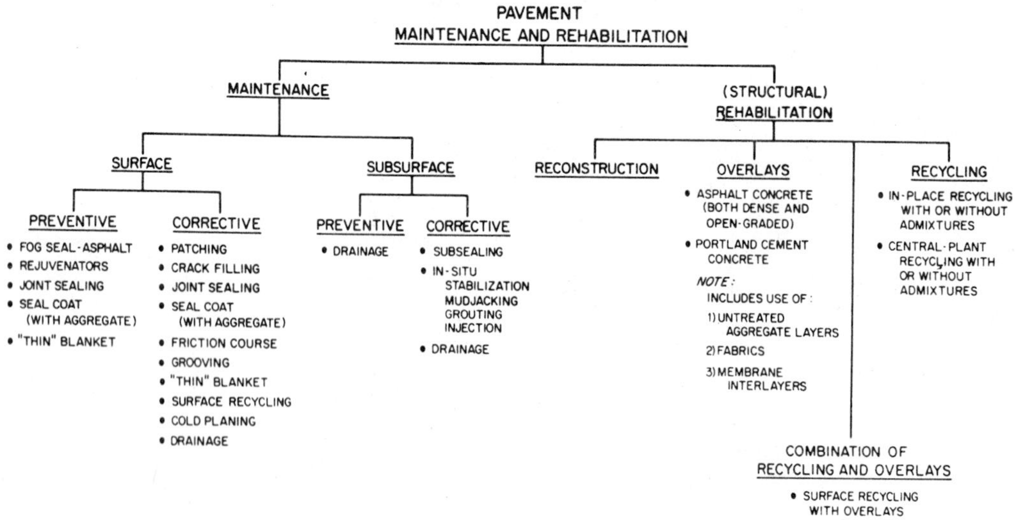

Source: Reproduced from C. L. Monismith, "Pavement Evaluation and Overlay Design Summary of Methods," *Transportation Research Record 700,* Transportation Research Board, Washington, D.C., 1979.

Table 20.4 Flexible and Rigid Pavement Maintenance Techniques

Treatment Type	Techniques	Corrective or Preventive	Effectiveness	Remarks
Patching	Temporary Permanent Spot seal (spray) Cold mix Hot mix Level	All patching is considered corrective. Some agencies said patching could be preventive in some cases when preventing more serious deterioration.	When properly done, patching is considered to be effective and to serve purpose. Temporary patching is considered moderately effective to serve short period until permanent repairs can be made.	There is a wide range of materials that can be used for patching. See NCHRP Synthesis 64 (56).
Surface Treatments	Seal coating with cover aggregate, chips, etc. Sand seal coat Flush coating (black seal, liquid asphalt, etc.) Dilute sealing (fog seal) Slurry seal Rejuvenators (binder modifiers, Reclamite, etc.) Hot-weather sanding Open-graded seal coat	Approximately 70% of replies indicated that surface treatments were preventive. Others indicated that surface treatments were corrective for skid resistance, etc. Some said surface treatments serve dual function.	Most of those providing ratings said surface treatments are effective for purpose intended. Several agencies have a schedule of every so many years for placing a surface treatment.	There is a considerable variation in materials and techniques used by various agencies for surface treatments. Local conditions vary extensively (climate, traffic, etc.) and must be considered for each treatment used.
Crack Maintenance	Crack cleaning Crack sealing Crack filling	Crack sealing is used as a preventive technique about 2/3 of the time. It is used 1/3 of the time as a corrective technique. Some consider it to serve a dual function.	Crack sealing is considered to be relatively effective. It has a fairly short life (1 to 2 yr) and so must be repeated often.	There are several different materials that can be used for crack sealing. Rubber asphalt appears to be the most cost-effective if properly used.
Surface Planing	Burn/plane Cold plane	Corrective	Generally effective in correcting corrugations and in reducing effect of high asphalt content or soft mix. May excessively harden asphalt.	There are different types of equipment that can be used. See NCHRP Synthesis 54 (60).
Other Localized Repairs	Blankets Base repair Remove and replace	Corrective	No evaluation given.	None
Recycling	Plant recycling In-place recycling (cold) In-place recycling (hot)	Corrective	Very few agencies identified this technique. Those that did considered it effective.	A relatively new technique that uses different softening agents, different equipment, etc.
Overlays	Thick overlays Thin overlays Pavement reinforcing Fabric and overlay Rubberized asphalt interlayer and overlayer Inverted overlays	Most agencies (69%) consider overlays to be a corrective technique. Thirty-one percent considered them to be preventive. Some believe they serve a dual function.	All agencies rating overlays said they were effective.	The timing of when an overlay is placed determines whether it is preventive or corrective. If the overlay is placed after failure, then it becomes corrective. Some agencies consider overlays to be rehabilitation rather than maintenance.

Continued

Table 20.4—*Continued*

Treatment Type	Techniques	Corrective or Preventive	Effectiveness	Remarks
Patching	Temporary patching Permanent patching PCC patching	Corrective	Temporary patching considered fair to poor. Permanent patching rated good.	There are numerous patching materials available. NCHRP Synthesis 45 (57) gives data on some.
Crack Maintenance	Crack cleaning Crack sealing	Classed as preventive and/or corrective	Rated as generally effective.	There was limited input on crack maintenance in rigid pavements.
Joints	Pressure relief joints	Corrective and/or preventive	Rated as effective in helping maintain rigid pavement and in protecting pavement and bridges.	Limited input.
	Repair joints	Corrective	Effective	Limited input.
	Cleaning joints Sealing joints	Mostly classed corrective; some considered preventive	Generally effective depending on materials and workmanship	There are a number of different materials that can be used.
Faulting Repair	Grinding (planing)	Corrective	Effective	The most common repair method identified was mudjacking.
	Mudjacking (pressure grouting)	Mostly corrective, some preventive	Moderately effective	
	Subsealing	Preventive	Not rated	
Blowup Repair	Temporary	Corrective	Temporary	There was minimal input in this area.
	Remove and replace	Corrective	Effective	
Surface Planing	Grinding	Corrective	Effective for correcting ride but may not solve basic problem creating roughness.	Limited input from agencies.
	Grooving	Corrective	Effective for correcting skid resistance.	
Other Localized Repairs	Drainage Correction Underdrain installation	Preventive	Rated effective	It was felt by some agencies that water was serious problem and that proper drainage of system would extend performance.
	Remove and replace Surface/base replacement	Corrective	Effective	Limited input.
Overlays	Thick bituminous overlay Thin bituminous overlay Thick PCC overlay Thin PCC overlay Pavement reinforcing fabric and overlap Rubberized asphalt interlayer and overlay	Corrective	Generally effective; thicker overlays considered more effective.	There was limited input on overlays over portland cement concrete pavement.

Source: Reproduced from Dale E. Peterson, *Evaluation of Pavement Maintenance Strategies*, NCHRP Synthesis 77, Transportation Research Board, Washington, D.C., September 1981.

Table 20.5 Summary of Repair Methods for Flexible and Rigid Pavements

Flexible Pavement

Distress	Seal Coat	Spray Patching	Spread Blotting Material	Remove with Blade or Heater-Planer	Removal and Replacement of Surface Course	Asphaltic Concrete Leveling and Surfacing	Temporary Hole Filling	Permanent Patching	Scarifying and Mixing	Crack Filling	Crack Filling with Slurry	Asphalt Emulsion Slurry Seal
	1	1	1	1	2	3	4	4	5	6	7	8
Abrasion	MT					BP		BP				
Bleeding			MB		SP							
Char		MB			SP							
Indentation				MB		SP						
Loss of Cover Agg.	BB	MT	SP			SP						
Polished Aggregates	BT			BB		BP						
Pothole		MT					BT	BP				
Raveling	BB			SP								
Streaking				BB	SP	SP						
Weathering	BB											MT
Corrugation				MB		MB			BP			
Alligator Cracking					MT			SP	SP			
Contraction Cracking					SP				SP	BT	BT	BT
Edge Cracking						SP				MT	BT	
Edge Joint Cracking						SP		SP		MT		
Lane Joint Cracking					SP				SP	MT	MT	MT
Reflection Cracking										BT		
Shrinkage Cracking	MT				SP						SB	
Slippage Cracking		MT			SP			SP				
Depression						SP						
Butting					SP	MB	MT	SP				
Shoving						MT		SP				
Upheaval					MT			SP				
Utility Cut Depression						MB		SP				

Rigid Pavement

Distress	Asphalt Emulsion Slurry Seal	Joint Filling	Thin Asphaltic Concrete Overlay	Concrete Patching Shallow	Concrete Patching Deep	Slab Jacking
	8	9	10	11	11	12
Crazing	MT			SP		
Joint Filler Extrusion/Stripping		BB				
Scaling	MT		BP	SP	SP	
Buckling Blow-Up					BP	
Shattering Blow-Up					BP	
Corner Cracking		MT			BP	MP
Diagonal Cracking		MT			BP	BP
Longitudinal Cracking		MT			SP	BP
Transverse Cracking		MT			SP	BP
2nd Stage Cracking		MT			SP	BP
Progressive Cracking		MT			SP	SP
Random Cracking	MT	ST			SP	
"D" Cracking	MT		BP	MP		
Faulting		BT				BP
Joint Failure		BT		BP	BP	
Pumping					BP	
Spalling			MT	MP	BP	

Severity of Distress: (M) moderate, (S) severe, (B) both
Permanency of Repair: (T) temporary, (P) permanent, (B) both.

Source: Adapted from Dale E. Peterson, *Evaluation of Pavement Maintenance Strategies*, NCHRP Synthesis 77, Transportation Research Board, Washington, D.C., September 1981.

PAVEMENT REHABILITATION PROGRAMMING

In the previous sections, we described how pavement condition is measured and the alternatives and strategies available to repair and rehabilitate these surfaces. In this section we describe the process used to decide upon a specific program for rehabilitation. The program requires decisions about the type of repair or restoration technique that should be used for a given pavement section and the timing (or programming) of the project. These decisions consider the design life of the pavement, its cost, and other physical and environmental conditions.

Methods Used to Develop Pavement Improvement Programs

Transportation agencies use various methods for selecting a program of pavement rehabilitation. Among these are (1) economic cost analysis using present worth of net benefits, (2) cost-effectiveness analysis, (3) traditional allocation process using average life span with funds allocated annually, (4) sufficiency ratings, (5) visual inspections and judgment, and (6) as part of an overall pavement management system. Table 20.6 lists the strategies used by 35 states, 3 Canadian provinces, and 8 toll road authorities.

Table 20.6 Bases Used for Selecting Pavement Management Strategies

Bases for Pavement Maintenance Strategies	States[a]	Canadian Provinces[b]	Toll Roads[c]
Rational engineering concepts	21	2	6
Cost-effectiveness	20	2	4
Traditional allocation	13	0	0
Fund availability	10	0	0
Maintenance management system	4	1	0
Visual observations, periodic inspections	5	1	1
Optimal use of existing maintenance personnel and equipment	1	1	1
Experience	1	1	1
Previous budgets, history	2	0	0

Note: Responses to the questionnaire were received from 35 states, 3 Canadian provinces, and 8 toll roads.

[a]Other bases identified by states: priority sequence, safety, rideability, cost necessity, protection of investment, sufficiency ratings, management principles, pavement management system, and planning data.

[b]By Canadian provinces: potential energy shortfall, serviceability index.

[c]By toll roads: maintain in good condition at all times, pride in work.

Source: Adapted from Dale E. Peterson, *Evaluation of Pavement Maintenance Strategies*, NCHRP Synthesis 77, Transportation Research Board, Washington, D.C., September 1981.

Economic evaluation of alternative pavement strategies is the appropriate method to use when selecting the best or optimum maintenance program. Other factors should also be considered and a decision made based on judgment and previous experience. Questions that should be asked when evaluating alternative pavement strategies are: How suitable is the method for immediate needs? How will the pavement perform in the future? When will additional treatments be required? What are the costs, the effects on traffic during repair, its appearance and effects on road users?

Figure 20.12 Decision-Making Process for Pavement Improvements

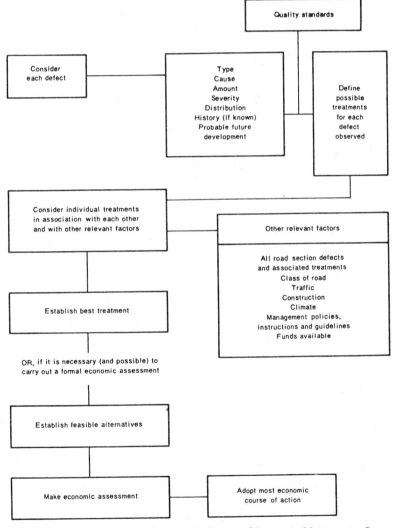

Source: Reproduced from Dale E. Peterson, *Evaluation of Pavement Maintenance Strategies*, NCHRP Synthesis 77, Transportation Research Board, Washington, D.C., 1981.

A flow chart that describes the decision-making process is shown in Figure 20.12. The process begins by examining each defect and describing the type, cause, and extent of the problem. Then possible treatments are selected, using data as described in the previous section. Each treatment is considered in terms of the conditions at the site and other available guidelines and policies to establish the most appropriate treatment under the circumstances. Once the feasible alternatives are established, it is possible to conduct an economic evaluation and then to select a course of action that meets the criteria established and the availability of funds. Establishing priorities of project improvements is most often based on subjective considerations, including the road type and category, traffic volume, and present pavement condition.

Example of Pavement Rehabilitation Programming

To illustrate the process of pavement rehabilitation programming, the procedures followed by the California Department of Transportation (CALTRANS) are described. This approach is comprehensive and typifies methods used by those states most active in this area.

Two types of data are collected for use in establishing a pavement maintenance program: pavement distress survey and a ride survey. In addition, average daily traffic (ADT) volumes are measured and used to establish priorities. (Pavement deflection data are used only for design.) Distress data are obtained by field observation, using highly trained observers. Ride survey data are obtained, using a custom manufactured Cox ultrasound road meter to determine ride quality. A ride score is determined as the sum of $\frac{1}{8}$ in. vertical movements accumulated over a measured distance divided by 50 times length in miles. It has been established by observations that if the ride score is 45 or greater, the pavement section should be considered for improvement because of the poor ride. Thus a ride score of 45 is a trigger value that delineates a good ride from a bad one and is useful in establishing priorities. Ride surveys and distress surveys are conducted at the same time on a 2 yr rotating basis.

Analysis of the data provides guidance concerning the need for repair. For example, if longitudinal and transverse cracks are wider than $\frac{1}{4}$ in., repairs are deemed necessary. Also ride score values exceeding 45 establish the need for repair. The use of trigger values establishes that a pavement section should be repaired. Further analysis is now required to determine the appropriate type of repair and the order of priorities.

Determination of the type of repair required is based on a decision tree analysis. The tree path is determined by trigger values and these lead to an appropriate repair strategy. Figure 20.13 illustrates a decision tree for a flexible pavement section in which alligator/block cracking has been observed. For example, if a given pavement section has class B cracking (alligator cracking in wheel paths), with less than 10 percent cracked and more than 10 percent of the area patched, then the decision tree path indicates that the type of repair should be a thin asphalt cement overlay with local dig-outs.

Figure 20.13 Decision Tree for Alligator Cracking

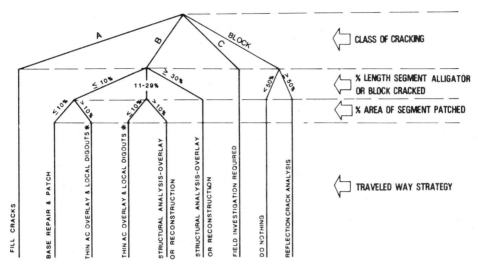

LEGEND

A = LONGITUDINAL CRACKING IN WHEEL PATH(S)
B = ALLIGATOR CRACKING IN WHEEL PATH(S)
C = SPECIAL OR UNUSUAL ALLIGATOR CRACKING
BLOCK = BLOCK CRACKING IN MAJORITY OF LANE WIDTH

* THIN AC OVERLAY = < 0.10' DENSE GRADED OR OPEN GRADED MIX

Source: Reproduced from R. G. Hicks, *Collection and Use of Pavement Condition Data,* NCHRP Synthesis 76, Transportation Research Board, Washington, D.C., July 1981.

A similar analysis is carried out for sections along the traveled way and in adjoining lanes and shoulders. This may result in different pavement rehabilitation strategies for several of the sections. Figure 20.14 illustrates the results for various sections along a single traveled way. As can be seen, alligator cracking may require strategy A, longitudinal cracking, strategy B, and so forth. These results are compared and a *dominant* strategy is selected, one that will correct all the defects in that lane. A similar analysis is conducted for the other lanes of the roadway and a dominant strategy is produced for each. These dominant strategies for each lane are then compared and a single pavement repair strategy is chosen (called a *compatible* strategy) that will correct as many defects as possible for the entire roadway segment (see Figure 20.15). The rehabilitation strategy selected is then assigned a cost and life, which is used in the next step of the procedure. The process described for selecting pavement strategies is computerized and reports are produced that furnish data about each lane, including dominant strategy, location, ADT, road type, and cost.

The final step in the process is to establish priorities for needed rehabilitation work that has been previously identified. This step is necessary since sufficient funds are not always available to address each pavement need. California's

Figure 20.14   Results of Flexible Pavement Evaluation Along One Lane of a Highway

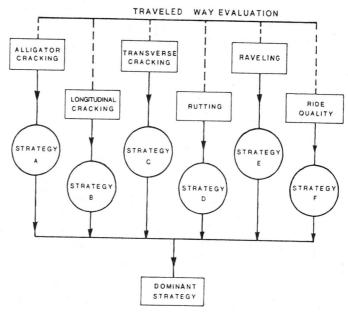

Source: Reproduced from R. G. Hicks, *Collection and Use of Pavement Condition Data*, NCHRP Synthesis 76, Transportation Research Board, Washington, D.C., July 1981.

Figure 20.15   Results of Flexible Pavement Evaluation for All Lanes of a Highway

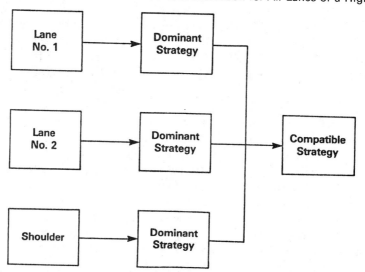

Source: Reproduced from *Pavement Management Rehabilitation Programming: Eight States' Experiences*, U.S. Department of Transportation, Washington, D.C., August 1983.

Figure 20.16 Priority Matrix for Pavement Maintenance

Pavement Problem Type		Candidate Category		
		ADT Range		
		>5,000	1,000 to 5,000	<1,000
Ride ≥ 45	Major Structural Problem and Unacceptable Ride Flex: Allig.B = 11-29% & Patch>10% or Allig.B≥30% Rigid: 3rd Stg. Crk.≥10%	1	2	11
	Minor Structural Problem and Unacceptable Ride Flex: Allig.B = 11-29% & Patching<10% Allig.B<10% & Patching>10%	3	4	12
	Unacceptable Ride Only	5	6	
Ride < 45	Major Structural Problem Only Flex: Allig.B = 11-29% & Patch>10% or Allig.B≥30% Rigid: 3rd Stg. Crk.≥10%	7	8	13
	Minor Structural Problem Only Flex: Allig.B = 11-29% & Patching<10% Allig.B<10% & Patching>10%	9	10	14

Source: Reproduced from *Pavement Management Rehabilitation Programming: Eight States' Experiences*, U.S. Department of Transportation, Washington, D.C., August 1983.

prioritization method is based on three variables: ride score, distress rating, and ADT. These figures are combined and a priority matrix is used, as illustrated in Figure 20.16. Priority rankings increase with traffic volume and severity of the problem. For example, if ADT on a roadway section (with no surface problems) exceeds 5000 and its ride quality exceeds 45, the section will be assigned a priority ranking of 5. However, if traffic volumes are below 1000, it is not considered. There are 14 possible rankings. If ADT exceeds 1000, a high ride score will assure a ranking of at least 6, which reflects CALTRANS' commitment to serve as many users as possible. Also a high ADT will increase the priority for a similar category of pavement distress.

When pavement sections are given the same priority ranking a "tie-breaker" value is computed, which is then added to the priority numbers of each of the tied sections. This value is cost per mile divided by ADT. The factor gives higher priority to projects with a lower per-mile cost and with a higher ADT. The priority ranking process produces a computer-generated report that shows the priority ranking for each candidate segment within the state.

The rehabilitation management process that has been developed in California is used for (1) making cost-effectiveness comparisons among repair strategies, (2) determining repair costs by district, and (3) determining the appropriate statewide funding level for all categories of pavement repair programs. The results of this process are used at the district level to establish specific programs.

SUMMARY

This chapter has described the procedures used to develop a pavement rehabilitation program. The specifics may vary from one state to another, but there is agreement on the need for a rational and objective process to insure that funds are efficiently used for pavement improvements. Some agencies have recently been adapting pavement rehabilitation strategies because of funding limitations and the need to use formalized procedures. The benefits to the agency are (1) improved performance monitoring, (2) a rational basis for legislative support, (3) determination of various funding level consequences, (4) improved administrative credibility, and (5) engineering input in policy decisions.

PROBLEMS

20–1 What is meant by the term *pavement rehabilitation management*? Describe three strategies used by public agencies to develop restoration and rehabilitation programs.

20–2 What are the three principal uses for pavement condition data?

20–3 What is the difference between PSI and PSR?

20–4. Draw a sketch showing the relationship between pavement condition (expressed as PSI) and time for a service life of 20 yr, if the PSI values range from 4.5 to 2.5 in a 6 year period, and then the pavement is resurfaced such that the PSI is increased to 4.0. After another 6 years, the PSI has reached 2.0, but with a rehabilitated pavement in place, PSI is increased to 4.5. At the end of its service life the PSI value is 3.4.

20–5 Describe the four characteristics of pavement condition used to evaluate whether a pavement should be rehabilitated, and if so, the appropriate treatment required.

20–6 A given pavement rating method uses five distress types to establish the DR. These are corrugation, alligator cracking, raveling, longitudinal cracking, and patching. For a stretch of highway, the number of points assigned to each category were: 6, 4, 2, 4, 3, and 3. Determine the DR for the section.

20–7 Describe the methods used to determine static and dynamic deflection of pavements. To what extent are these tests used in pavement rehabilitation management?

20–8 A 5000 lb load is placed on two tires, which are then locked in place. A force of 2400 lb is necessary to cause the trailer to move at a speed of 20 mph. Determine the value of the skid number. If treaded tires were used, how would you characterize the pavement type?

20–9 Differentiate between corrective and preventive rehabilitation techniques. Cite three examples of surface treatments in each category. What is the best preventive technique for subsurface maintenance?

20–10 Describe the techniques used to repair flexible and rigid pavements and their effectiveness for the following treatment types: **(a)** patching, **(b)** crack maintenance, and **(c)** overlays.

20–11 A pavement section has lost its seal, due to drying out, and the fine aggregates have raveled. The ride score is 27. The section is part of a road where tire chains are used and there are considerable vehicle turning movements. The skid number is 83. Which repair strategy would you recommend? Explain your answer.

20–12 Describe six methods that can be used to select a program of pavement rehabilitation. Under what circumstances would each method be used?

20–13 A pavement section has been observed to exhibit alligator cracking in the wheel paths (Class B) for approximately 20 percent of its length. About 15 percent of the section has been previously patched. The ride score is 32. Use a decision tree analysis to determine the appropriate treatment required.

20–14 If the ADT on the roadway section described in problem 20–13 is 3500 and cost per mile is $5000, what priority ranking would the road receive?

20–15 Another road segment has been examined and is given the same priority ranking as determined in problem 14. If the ADT is 4000 and the cost per mile is 10 percent greater than the alternative, what is the priority ranking of each road segment?

REFERENCE

1. *Pavement Management Rehabilitation Programming: Eight States' Experiences*, U.S. Department of Transportation, Federal Highway Administration, Washington, D.C., August 1983.

ADDITIONAL READINGS

Haas, R. C. G., *Pavement Management Guide*, Roads and Transportation Association of Canada, Ottawa, 1977.

Haas, R., and W. R. Hudson, *Pavement Management Systems*, McGraw-Hill Book Company, New York, 1978.

Hicks, R. G., *Collection and Use of Pavement Condition Data*, NCHRP Synthesis 76, National Research Council, Transportation Research Board, Washington, D.C., July 1981.

Hudson, W. R., R. Haas, and R. D. Pedigo, *Pavement Management System Development*, NCHRP Report 215, National Research Council, Transportation Research Board, Washington, D.C., November 1979.

Kummer, H. W., and W. E. Meyer, *Tentative Skid Resistance Requirements for Main Rural Highways*, NCHRP Report 37, National Research Council, Transportation Research Board, Washington, D.C., 1967.

McGhee, K. H., *Development of a Pavement Management System for Virginia*, Virginia Highway and Transportation Research Council, Charlottesville, Va., January 1984.

Monismith, C. L., *Pavement Evaluation and Overlay Design Summary of Methods*, Transportation Research Record 700, National Research Council, Transportation Research Board, Washington, D.C., 1979.

Moore, N. M., D. I. Hanson, and J. W. Hall, *An Introduction to Nondestructive Structural Evaluation of Pavements*, Transportation Research Circular 189,

National Research Council, Transportation Research Board, Washington, D.C., January 1978.

Peterson, Dale E., *Evaluation of Pavement Maintenance Strategies*, NCHRP Synthesis 77, National Research Council, Transportation Research Board, Washington, D.C., September 1981.

Peterson, E., *Life Cycle Cost Analysis of Pavements*, NCHRP Synthesis 122, National Research Council, Transportation Research Board, Washington, D.C., December 1985.

Appendixes

Critical Values for the Student's *t* Distribution

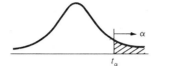 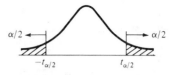

Degrees of Freedom	Level of Significance for One-Tailed Test							
	.250	.100	.050	.025	.010	.005	.0025	.0005
	Level of Significance for a Two-Tailed Test							
	.500	.200	.100	.050	.020	.010	.005	.001
1.	1.000	3.078	6.314	12.706	31.821	63.657	27.321	536.627
2.	.816	1.886	2.920	4.303	6.965	9.925	14.089	31.599
3.	.765	1.638	2.353	3.182	4.541	5.841	7.453	12.924
4.	.741	1.533	2.132	2.776	3.747	4.604	5.598	8.610
5.	.727	1.476	2.015	2.571	3.365	4.032	4.773	6.869
6.	.718	1.440	1.943	2.447	3.143	3.707	4.317	5.959
7.	.711	1.415	1.895	2.365	2.998	3.499	4.029	5.408
8.	.706	1.397	1.860	2.306	2.896	3.355	3.833	5.041
9.	.703	1.383	1.833	2.262	2.821	3.250	3.690	4.781
10.	.700	1.372	1.812	2.228	2.764	3.169	3.581	4.587
11.	.697	1.363	1.796	2.201	2.718	3.106	3.497	4.437
12.	.695	1.356	1.782	2.179	2.681	3.055	3.428	4.318
13.	.694	1.350	1.771	2.160	2.650	3.012	3.372	4.221
14.	.692	1.345	1.761	2.145	2.624	2.977	3.326	4.140
15.	.691	1.341	1.753	2.131	2.602	2.947	3.286	4.073
16.	.690	1.337	1.746	2.120	2.583	2.921	3.252	4.015
17.	.689	1.333	1.740	2.110	2.567	2.898	3.222	3.965
18.	.688	1.330	1.734	2.101	2.552	2.878	3.197	3.922
19.	.688	1.328	1.729	2.093	2.539	2.861	3.174	3.883
20.	.687	1.325	1.725	2.086	2.528	2.845	3.153	3.850
21.	.686	1.323	1.721	2.080	2.518	2.831	3.135	3.819
22.	.686	1.321	1.717	2.074	2.508	2.819	3.119	3.792
23.	.685	1.319	1.714	2.069	2.500	2.807	3.104	3.768
24.	.685	1.318	1.711	2.064	2.492	2.797	3.091	3.745
25.	.684	1.316	1.708	2.060	2.485	2.787	3.078	3.725
26.	.684	1.315	1.706	2.056	2.479	2.779	3.067	3.707
27.	.684	1.314	1.703	2.052	2.473	2.771	3.057	3.690
28.	.683	1.313	1.701	2.048	2.467	2.763	3.047	3.674
29.	.683	1.311	1.699	2.045	2.462	2.756	3.038	3.659
30.	.683	1.310	1.697	2.042	2.457	2.750	3.030	3.646
35.	.682	1.306	1.690	2.030	2.438	2.724	2.996	3.591
40.	.681	1.303	1.684	2.021	2.423	2.704	2.971	3.551
45.	.680	1.301	1.679	2.014	2.412	2.690	2.952	3.520
50.	.679	1.299	1.676	2.009	2.403	2.678	2.937	3.496
55.	.679	1.297	1.673	2.004	2.396	2.668	2.925	3.476
60.	.679	1.296	1.671	2.000	2.390	2.660	2.915	3.460
65.	.678	1.295	1.669	1.997	2.385	2.654	2.906	3.447
70.	.678	1.294	1.667	1.994	2.381	2.648	2.899	3.435
80.	.678	1.292	1.664	1.990	2.374	2.639	2.887	3.416
90.	.677	1.291	1.662	1.987	2.368	2.632	2.878	3.402
100.	.677	1.290	1.660	1.984	2.364	2.626	2.871	3.390
125.	.676	1.288	1.657	1.979	2.357	2.616	2.858	3.370
150.	.676	1.287	1.655	1.976	2.351	2.609	2.849	3.357
200.	.676	1.286	1.653	1.972	2.345	2.601	2.839	3.340
∞	.6745	1.2816	1.6448	1.9600	2.3267	2.5758	2.8070	3.2905

Source: Reproduced from Richard H. McCuen, *Statistical Methods for Engineers*, copyright © 1985. Reprinted by permission of Prentice-Hall, Inc., Englewood Cliffs, N.J.

Equations for Computing Regression Coefficients

Let a dependent variable Y and an independent variable x be related by an estimated regression function

$$Y = a + bx \qquad \text{B.1}$$

Let Y_i be an estimate and y_i be an observed value of Y for a corresponding value x_i for x. Estimates of a and b can be obtained by minimizing the sum of the squares of the differences (R) between Y_i and y_i for a set of observed values where

$$R = \sum_{i=1}^{n} (y_i - Y_i)^2 \qquad \text{B.2}$$

Substituting $(a + bx_i)$ for Y_i in Eq. B.2 we obtain

$$R = \sum_{i=1}^{n} (y_i - a - bx_i)^2 \qquad \text{B.3}$$

Differentiating R partially with respect to a, and then with respect to b, and equating each to zero, we obtain

$$\frac{\partial R}{\partial a} = -2 \sum_{i=1}^{n} (y_i - a - bx_i) = 0 \qquad \text{B.4}$$

$$\frac{\partial R}{\partial b} = -2 \sum_{i=1}^{n} x_i(y_i - a - bx_i) = 0 \qquad \text{B.5}$$

From Eq. B.4 we obtain

$$\sum_{i=1}^{n} (y_i) = na + b \sum_{i=1}^{n} (x_i) \qquad \text{B.6}$$

giving

$$a = \frac{1}{n} \sum_{i=1}^{n} y_i - \frac{b}{n} \sum_{i=1}^{n} (x_i) \qquad \text{B.7}$$

From Eq. B.5 we obtain

$$\sum_{i=1}^{n} x_i y_i = a \sum_{i=1}^{n} x_i + b \sum_{i=1}^{n} x_i^2 \qquad \text{B.8}$$

Substituting for a, we obtain

$$b = \frac{\sum_{i=1}^{n} x_i y_i - \frac{1}{n} \left(\sum_{i=1}^{n} x_i \right) \left(\sum_{i=1}^{n} y_i \right)}{\sum_{i=1}^{n} x_i^2 - \frac{1}{n} \left(\sum_{i=1}^{n} x_i \right)^2} \qquad \text{B.9}$$

where

n = number of sets of observation
x_i = ith observation for x
y_i = ith observation for y

Eqs. B.8 and B.9 may be used to obtain estimated values for a and b in Eq. B.1. To test the suitability of the regression function obtained, the coefficient of determination, R^2, which indicates to what extent values of Y_i obtained from the regression function agree with observed values y_i, is determined from the expression

$$R^2 = \frac{\sum_{i=1}^{n} (Y_i - \bar{y})^2}{\sum_{i=1}^{n} (y_i - \bar{y})^2} \qquad \text{B.12}$$

which is also written as

$$R^2 = \frac{\left(\sum_{i=1}^{n} x_i y_i - n\bar{x}\bar{y} \right)^2}{\left(\sum_{i=1}^{n} x_i^2 - n\bar{x}^2 \right) \left(\sum_{i=1}^{n} y_i^2 - n\bar{y}^2 \right)} \qquad \text{B.13}$$

The closer to 1 is the value of R^2, the more suitable is the estimated regression function for the data.

Index

The letter *t* following a page number indicates a table; the letter *f* indicates a figure.